# Developmental Biology

A COMPREHENSIVE SYNTHESIS

Volume 1

# Oogenesis

# Developmental Biology
## A COMPREHENSIVE SYNTHESIS



---

Volume 1 OOGENESIS
Edited by Leon W. Browder

*Forthcoming*

Volume 2 THE CELLULAR BASIS OF MORPHOGENESIS
Edited by Leon W. Browder

Volume 3 THE CELL SURFACE IN DEVELOPMENT AND CANCER
Edited by Malcolm S. Steinberg

Volume 4 MANIPULATION OF MAMMALIAN DEVELOPMENT
Edited by Ralph B. L. Gwatkin

# Developmental Biology

## A COMPREHENSIVE SYNTHESIS

### Volume 1

# Oogenesis

Edited by
LEON W. BROWDER
*University of Calgary*
*Calgary, Alberta, Canada*

PLENUM PRESS • NEW YORK AND LONDON

Library of Congress Cataloging in Publication Data

Main entry under title:

Developmental biology.

Includes bibliographies and index.
Contents: v. 1. Oogenesis.
1. Developmental biology—Collected works. I. Browder, Leon W.
QH491.D426 1985 574.3 85-3406
ISBN 0-306-41866-5 (v. 1)

*Cover illustration:* Acridine Orange fluorescence micrograph of insect ovarian follicle. Micrograph prepared by Lois Nichols. Courtesy of Spencer J. Berry, Department of Biology, Wesleyan University.


A Division of Plenum Publishing Corporation
233 Spring Street, New York, N.Y. 10013



Printed in the United States of America

# Contributors

**Rosemary Bachvarova** Department of Cell Biology and Anatomy, Cornell University Medical College, New York, N.Y. 10021

**Spencer J. Berry** Department of Biology, Hall-Atwater and Shanklin Laboratories, Wesleyan University, Middletown, Connecticut 06457

**Bruce P. Brandhorst** Department of Biology, McGill University, Montreal, Quebec H3A 1B1 Canada

**Anna Ruth Brummett** Biology Department, Oberlin College, Oberlin, Ohio 44074

**Marie-Christine Dabauvalle** Institute of Cell and Tumor Biology, German Cancer Research Center, D-6900 Heidelberg 1, Federal Republic of Germany

**James N. Dumont** Biology Division, Oak Ridge National Laboratory, Oak Ridge, Tennessee 37830

**Lee Engstrom** Developmental Biology Center, Department of Developmental Genetics and Anatomy, Case Western Reserve University, Cleveland, Ohio 44106

**John J. Eppig** The Jackson Laboratory, Bar Harbor, Maine 04609

**Franco Giorgi** Istituto di Istología e Embriología, 56100 Pisa, Italy

**Richard G. Kessel** Department of Biology, University of Iowa, Iowa City, Iowa 52242

**Kenneth D. Konrad** Developmental Biology Center, Department of Developmental Genetics and Anatomy, Case Western Reserve University, Cleveland, Ohio 44106

**Angela Krämer** Division of Molecular Biology and Biochemistry, Institute of Cell and Tumor Biology, German Cancer Research Center, D-6900 Heidelberg 1, Federal Republic of Germany

**Anthony P. Mahowald** Developmental Biology Center, Department of Developmental Genetics and Anatomy, Case Western Reserve University, Cleveland, Ohio 44106

**James L. Maller** Department of Pharmacology, University of Colorado School of Medicine, Denver, Colorado 80262

**Norbert Perrimon** Developmental Biology Center, Department of Developmental Genetics and Anatomy, Case Western Reserve University, Cleveland, Ohio 44106

**John H. Postlethwait** Department of Biology, University of Oregon, Eugene, Oregon 97403

**Ulrich Scheer** Institute of Cell and Tumor Biology, German Cancer Research Center, D-6900 Heidelberg 1, Federal Republic of Germany

**Allen W. Schuetz** Department of Population Dynamics, Johns Hopkins School of Hygiene and Public Health, Baltimore, Maryland 21205

**Robin A. Wallace** C. V. Whitney Laboratory, St. Augustine, Florida 32086

# Preface

This book launches a new continuing series in developmental biology. Any new publishing venture in science can only be justified if it satisfies a need and fulfills that need in an exemplary way. I believe that there is an acute need for comprehensive treatises on specialized topics in developmental biology—especially because this is a very diverse field that receives contributions from a wide variety of disciplines. It is difficult to remain abreast of progress in multifaceted topics and to understand how the various aspects of a topic interrelate. The volumes that will comprise this series are designed to facilitate those processes.

Each volume of this series will attempt to convey the current status of research and thought on a specific topic in development, place current research within historical perspective, and lay the intellectual groundwork for future research. These volumes are intended to provide an in-depth background on these topics for advanced undergraduates, graduate students, and senior scientists. An investigator considering research in one of these areas will get an overview of the field as well as discover the status of recent research in the area.

The extent to which these volumes achieve their goal will be a measure of their usefulness to the scientific community. Contributors are selected who are authorities in their fields and who can effectively communicate their knowledge, understanding, and ideas. Every effort has been made to ensure that the writing is clear and logical. Where useful, high-quality illustrations are used to clarify the text. The references accompanying each chapter are selected to permit access to the primary literature in that area. One key word in the title of this series is "Comprehensive." An attempt has been made to cover the most significant aspects of each topic. However, most topics are open ended, and value judgments must be made about the relative significance of possible subjects to include in a book of moderate length; some areas must therefore be omitted.

The choice of oogenesis as a topic for the first volume of this series was made for a number of reasons. Clearly, differentiation of the oocyte lays the groundwork for development of the zygote. RNA and protein are synthesized, stored, and organized in the oocyte such that development can be initiated at fertilization of the egg without immediate nuclear involvement. Energy sources may be packaged in the oocyte to reduce or eliminate the need of the embryo to acquire the means to obtain food from exogenous sources. The oocyte acquires a number of specializations that enable it to function in fertilization and to respond to the sperm by initiating development. Thus, the oocyte becomes a highly specialized cell through a complex differentiation process. However, the ramifications of oocyte differentiation are extensive, since the program for development of the zygote begins in the oocyte.

Oocyte differentiation is very precisely regulated and is dependent upon growth of the organism and is correlated with sexual behavior, likelihood of fertilization, and suitability of the environment for development of the zygote. Regulation of oogenesis is mediated by hormones, which have numerous profound effects on the oocyte, oocyte accessory cells, and other reproductive tissues.

Because of its many aspects, the study of oocyte differentiation is one of the most integrated areas of investigation in developmental biology. Molecular biologists, cell biologists, morphologists, and physiologists are actively involved in probing the events that occur and the modes of their control. It is intended that this volume will provide the reader with a comprehensive understanding of the research conducted on oogenesis from a variety of perspectives. It is hoped that the following chapters will convey the excitement of contemporary research on oogenesis and stimulate investigators to further our understanding of this vital process.

The concept for this series was developed in consultation with Mr. Kirk Jensen, Life Sciences Editor of Plenum Publishing Corporation. Mr. Jensen has remained enthusiastic about this project through its long gestation and has been immeasurably helpful at all stages in the development of this book. His commitment to excellence in academic publishing is reflected in Plenum's biology books. It is intended that this series will further that tradition. I am fortunate to have been assisted in developing this series by an editorial board of outstanding developmental biologists. Their contribution is greatly appreciated.

I have also been greatly assisted in working on this series and in organizing this volume by my secretary, Miss Margaret Hunik, and by my wife, Sandy, whose knowledge of publishing and patience with me have greatly eased my burden.

Leon Browder

# Contents

## I. Physiological and Morphological Aspects

Chapter 9 · Functional Organization of the Amphibian Oocyte Nucleus

Ulrich Scheer and Marie-Christine Dabauvalle



Chapter 10 · 5 S Ribosomal Gene Transcription During *Xenopus* Oogenesis

Angela Krämer



Chapter 11 · Gene Expression During Oogenesis and Oocyte Development in Mammals

Rosemary Bachvarova

Chapter 12 · Informational Content of the Echinoderm Egg

Bruce P. Brandhorst



Chapter 13 · Genetic Analysis of Oogenesis and the Role of Maternal Gene Expression in Early Development

Kenneth D. Konrad, Lee Engstrom, Norbert Perrimon, and Anthony P. Mahowald

# I

# Physiological and Morphological Aspects

Chapter 1

# Local Control Mechanisms During Oogenesis and Folliculogenesis

ALLEN W. SCHUETZ

## 1. Introduction

In most animal species a single cell, the female gamete, is the primary or only cellular link between the present and successive generations; the female gamete constitutes an irreplaceable element of life. Thus, the origin, differentiation, qualities, and functions of this essential cell are of fundamental interest and importance to developmental biology and are related directly or indirectly to multiple human activities—social, personal, scientific, medical, and agricultural.

Differentiation of a precisely organized fully mature and functioning egg is the culmination of a long, complex series of specific and unique cellular and molecular events that occur within the cytoplasm and nucleus as well as extracellularly after segregation of the germ cell line and meiotic commitment (Monroy *et al.* 1983). These events are programmed to take place at distinct times over the course of oogenesis during the life cycle of an individual animal (Figs. 1 and 2). Typically, a major component of this differentiation process takes place while oocytes reside within the ovary and, in most animals, occurs in close association with somatic tissues of varying complexity. These somatic units comprise a morphological entity called the **ovarian follicle** that undergoes simultaneous structural and functional differentiation.

Oocytes at the time of formation in the gonad and for a considerable period thereafter have few, if any, of the properties of fertilizable gametes. They undergo limited or no development even if fertilized. Thus, the gonads and their specialized follicular subcompartments serve as the primary repository and vehicles for the formation and growth of germ cells. During their residence in these tissues, oocytes are endowed with many or

ALLEN W. SCHUETZ • Department of Population Dynamics, Johns Hopkins School of Hygiene and Public Health, Baltimore, Maryland 21205.

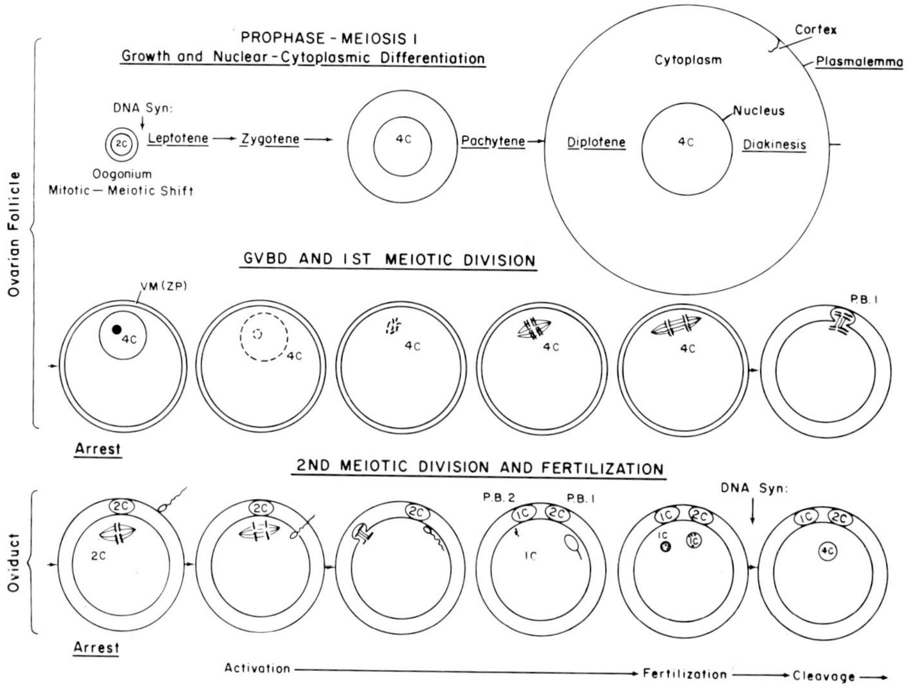


**Figure 1.** Diagrammatic representation of the sequence of major morphological transformations that occur in the cytoplasm and nucleus of oocytes from the time of their origin and through growth, meiotic maturation, fertilization, and activation. Typically, oocyte differentiation begins in the ovary and is completed external to the ovary (e.g., oviduct, seawater) at the time of fertilization. In most species, oocytes characteristically remain arrested at prophase I of meiosis (germinal vesicle stage) in the ovary until gonadotropic hormone(s) or some other stimulus triggers meiosis reinitiation. (Modified from Schuetz, 1974.)

all of the properties that will permit normal development to proceed after fertilization or parthenogenetic activation. Intraovarian development of the oocyte is terminated at the time of ovulation, but oocyte maturation normally continues and proceeds, depending upon the species, in various extraovarian environments, such as the oviducts. Significantly, the consummation of meiosis in oocytes from most species does not occur until the fertilizing sperm interacts with the oocyte to reactivate and complete the process. Thus, it is necessary to consider events before and after sperm penetration as integral, rather than as separate, parts of the oocyte maturation process. In attempting to understand how a fully mature oocyte is produced, one needs to elucidate how, over the course of oogenesis, germ cell differentation is initiated, mediated, maintained, and regulated during the female gamete's sojourn in two (intra- and extraovarian) very different environments (Fig. 3).

Considerable research effort concentrated in previous years on the identification and characterization of the hierarchy of extragonadal molecules (gonadotropin hormones and hypothalamic releasing hormones) that stimulate and control gonad and follicular development. Likewise, the endocrine secretory products of the gonads (particularly steroid hormones) have been studied extensively, as have their functions and modes of action in peripheral target tissues and, to a more limited extent, within the gonad (Fig. 4). In addition, the process of ovulation has been examined, although with insufficient regard for the role of the oocyte in the process. Thus, hormonal secretions, gametogenic functions, and the ovulatory process have been investigated essentially as independent events.

With the discovery that gonadal products in starfish and amphibians (e.g., 1-methyladenine and steroids) directly stimulate reinitiation of meiosis in fully grown prophase I-arrested oocytes in the germinal vesicle stage, a direct molecular link between somatic endocrine secretions and morphogenesis and function of the gametic cell within the gonads was established for the first time (see Chapter 6). Early studies of Detlaff *et al.* (1964) and Kanatani (1964) provide historical benchmarks concerning our

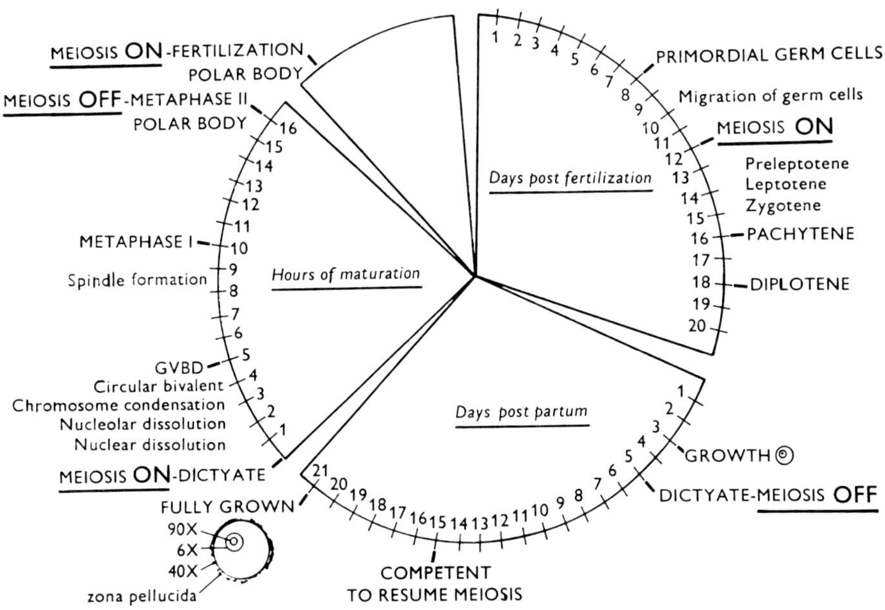


**Figure 2.** Diagrammatic representation of certain aspects of oogenesis during the life cycle of the mouse. (From Schultz and Wasserman, 1977.)

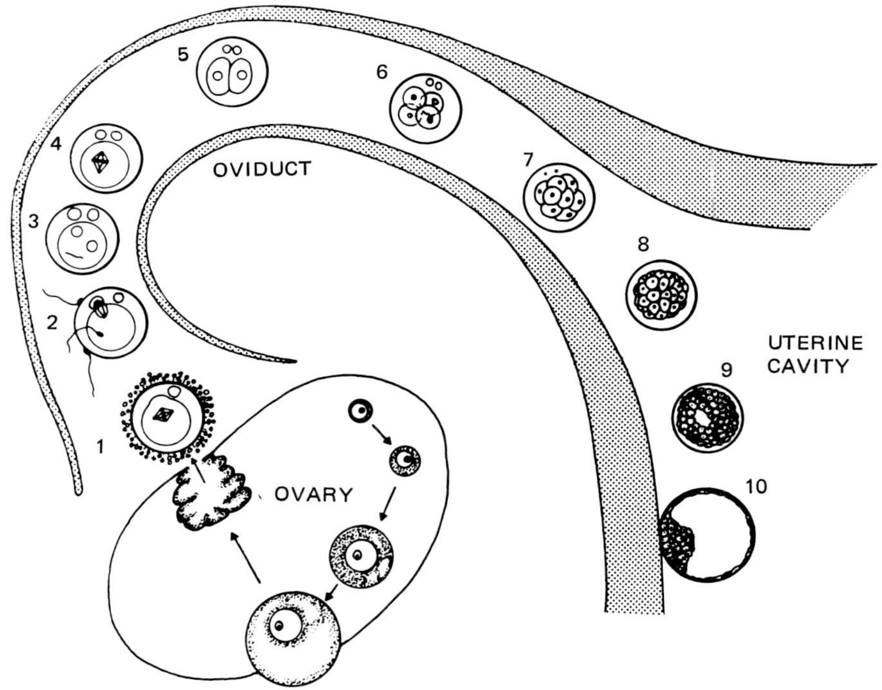


KEY:

1: egg released from ovary with first polar body and second metaphase spindle

2: sperm entry into egg, second polar body forming

3: male and female pronucleus formation. Sperm tail in egg cytoplasm

4: first cleavage metaphase spindle

5: 2-cell stage

6: 4-cell stage

7: 8-cell stage

8: morula

9: early blastocyst, blastocoele cavity forming

10: blastocyst starting to implant

**Figure 3.** Diagrammatic representation of follicular growth, ovulation, fertilization, and preimplantation development. (From Whittingham, 1979.)

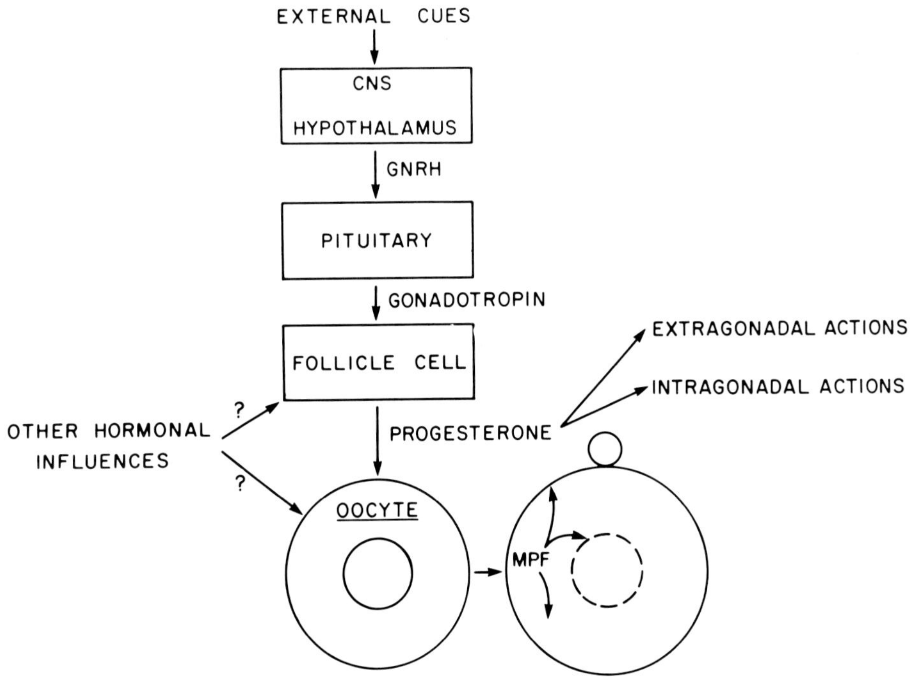


**Figure 4.** Schematic outline of the hierarchy of major endocrine and cellular interactions involved in regulating ovarian function and the oocyte maturation process. Primarily based on data obtained from amphibians but applicable to varying degrees to most vertebrate species. CNS, central nervous system; GNRH, gonadotropin releasing hormone; MPF, maturation promoting factor.

knowledge about the mode of action of gonadotropic hormones in reinitiating meiosis as well as a base with which to judge the extent of our subsequent progress. As a result of this new knowledge and the use of defined hormones, it became possible to (1) examine and characterize the role of specific regulatory molecules at different stages of oocyte and follicle differentiation; (2) define what aspects of cellular differentation can be attributed to a single molecular trigger or inducer; (3) elucidate the manner in which the regulatory substances are produced and controlled as well as act and interact at the cellular and molecular level; and (4) determine how the somatic and gametic compartments of the gonads interact.

Considerable effort has been directed toward elucidating how germ cell differentiation proceeds and is controlled at the local (intra- and extraoocytic) level within the gonad and specifically within the ovarian follicle. Such basic information is required if we are to understand normal as well as abnormal ovarian and gametic functions. Thus, reanalysis of the structural and functional changes and interactions associated with the coordinated differentiation of oocytes and follicles has been necessitated and is under way.

Ideas and concepts of intraovarian and intraoocyte regulatory mechanisms have expanded greatly and are being modified continually in the light of new experimental data. However, in attempting to understand the physiological mechanism controlling oocyte development at the local level, we are continually confronted with several key questions:

1. To what extent do alterations in somatic tissue structure and function account for or result from the changes (growth, nutrition, meiosis, death) that occur in oocytes?
2. To what extent are changes in somatic tissue structure and function related to ovarian processes that may or may not directly involve differentiation of the oocyte, follicle growth, ovulation, or peripheral action of hormones?
3. To what extent are changes in the oocyte mediated by intracellular versus extracellular mechanisms?
4. How are internal and external signals transduced and coordinated to bring about differentiation of follicles with similar properties?
5. To what extent and by what mechanisms do follicles within the gonad interact with each other?
6. How can cellular differentiation and regulatory factors be experimentally analyzed and quantitated at the local level?

Significantly, much of our understanding of these events has resulted from the use of *in vitro* culture procedures that permit selection and maintenance of oocytes, follicles, or follicular components individually or as a group at discrete stages of differentiation. Such methods, however, also present problems that need to be considered in interpreting the physiological significance of any changes observed (see Section 7). Consequently, it is no longer sufficient to know that hormones are present or are produced within the gonads: We must also ask whether and how they function at the oocyte, follicle, ovarian, and/or extraovarian levels. Alternatively, a description of the morphological and biochemical anatomy of the oocyte and follicle does not define the physiological processes that bring about and control differentiation.

Oocytes grow, develop, and mature in accordance with genetic, physiological, biochemical, and environmental rules and cues characteristic of each species. Typically, oocyte growth and differentiation occur coordinately with the unique process of meiosis, which is distinguished by universally defined and accepted changes in the structure and composition of the chromosomes (Riley *et al.*, 1977) (see Fig. 1). Particularly dramatic and important are the changes in the oocyte nucleus **(germinal vesicle)** over the course of the process. Usually, oocyte development proceeds only to a certain point in the ovary, and much attention has been directed at elucidating how meiotic arrest and subsequent reinitiation are controlled. In many cases, **germinal vesicle breakdown (GVBD)** and associated chromosomal changes that mark the completion of prophase I have been considered or interpreted to indicate a complete or fully mature oocyte. It must be stated emphatically, however, that these important events are only portions of the sequence of events that lead to complete oocyte development. Local regulating mechanisms that precede and succeed GVBD play major functional roles in establishing the full developmental potential of the oocyte. This chapter attempts to integrate and interpret our present understanding relating to the nature, regulation, and functional roles of local cellular and molecular factors that operate to bring about complete maturation of the female germ cells (oocytes). Rather than provide a detailed analysis of the literature, an attempt is made to provide some broader perspective to the continuum of local biochemical, structural, and functional processes (ovarian and extraovarian) and problems controlling the differentiation of the female germ cell. Further details can be obtained from a number of excellent sources (Wilson, 1925; Biggers and Schuetz, 1972; Porter and Whelan, 1983; Jones, 1978; Zuckerman and Weir, 1977; Davidson, 1968; Slack, 1983; Jagiello and Vogel, 1981; Rolland *et al.*, 1982)

In order to study oocyte differentiation, it is necessary to define the stage of follicle and oocyte development being analyzed. To do so presents a considerable problem, since only a relatively few roughly defined stages have been identified cytologically; these obviously overlay the molecular and biochemical processes occurring continuously at all levels of organization and over a considerable period of time. The extent to which results obtained by different investigators are directly comparable is sometimes questionable because of our inability to define stages precisely.

Increased interest in, and use of, techniques for *in vitro* fertilization, gamete storage, embryo transfer, and genetic engineering necessitates a more detailed and precise understanding of the composition and vital functions of gametes as well as the processes involved in their gradual evolution to a functional (normal or abnormal) state (Whittingham 1979). Oocyte development—like embryological development—is a dynamic process involving the dimensions of time, space, and form. Whereas oogenesis is concerned with the sequestering and packaging of developmental information in an organized way, embryogenesis constitutes an unfolding and utilization of the developmental potential stored in the oocyte. Oocyte (cytoplasmic and nuclear) maturation can thus be envisioned as the mechanism whereby the transition from oogenesis to embryogenesis is mediated.

## 2. Choice of Animals

Compelling scientific, economic, and social reasons can be given for analyzing the local mechanism(s) that control oogenesis in a wide range of vertebrate and invertebrate organisms. This will be apparent from the diverse range of species described in this and subsequent chapters of this book.

In order to provide some focus, this chapter concentrates on information obtained by the author in investigating oogenesis in animals from several phyla and classes, using echinoderms, amphibians, and mammals, as well as limited data derived from birds and fish. This seems appropriate since these animals (1) have provided much basic information concerning local control mechanisms; (2) exhibit many common as well as unique features of oogenesis; and (3) cover a broad span of the evolutionary scale and thus provide insights concerning the structural, functional, and biochemical evolution of the increasingly complex process of folliculogenesis. Significantly, oocytes of each of these and all other species are confronted with similar problems in growing and maturing to the point at which fertilization and development can proceed.

The unique properties, amenability to experimental analysis, availability, and cost of these different species have influenced their use as experimental animals. For example, in starfish, the entire gonad or its component oocyte or follicle populations remain both viable in seawater and responsive for extended periods to starfish gonadotropic polypeptide, i.e., radial nerve factor (RNF) or maturation-inducing hormone, 1-methyladenine (1-MA). Thus, the entire gamut of events relating to ovarian (ovulation, hormone induction, spawning) and extraovarian (fertilization, cleavage, aging, parthenogenesis) processes can be induced and studied at appropriate times within the course of several hours. These events, all of which can be observed in a single dish of seawater, occur in a highly synchronized population of oocytes. Such oocytes are relatively large and clear, permitting detailed cytological analysis of meiotic and developmental stages with simple, inexpensive microscopic observations. Sea urchins also provide excellent material for the study of fertilization mechanisms but are not as suitable for the investigation of ovarian events because of a heterogeneous population of oocytes, their small size, and the fact that oocytes complete meiosis before spawning. In contrast to the process in starfish, however, sea urchin gametogenesis continues after spawning in animals that are properly maintained and fed.

In fish, amphibians, and birds, the ovary is organized such that individual follicles develop essentially as individual units that can be identified readily and separated from each other for experimental purposes. Furthermore, because of their relatively large size, such units are amenable to microsurgery and micromanipulation (e.g., enucleation, cytoplasmic, and nuclear transfer and injection studies). The presence of large amounts of yolk in such oocytes makes detailed analysis of the meiotic stages difficult; such studies typically require fixation and histological examination. Yolk-filled oocytes, however, make these animals very useful for the study of vitellogenesis and follicular and oocyte growth processes. In birds, which typically release a single oocyte per ovulation, the follicles and oocytes are arranged in a hierarchy under controlled environment conditions. Thus, one can readily predict the sequence of preovulatory development, this feature being particularly useful in correlating the progressive events of follicular differentiation with specific cellular and biochemical changes such as steroidogenesis.

When considering oogenesis in the most studied amphibians (*Xenopus laevis*, *Rana pipiens*), one finds that each species has many common as well as unique features that are particularly useful for analyzing different aspects of this process. In *Rana*, ovulation typically occurs only once per year, during the spring season, and the follicles and oocytes grow and differentiate as a single cohort during the summer months. Thus, by the time the animals enter natural hibernation, a large population (several thousand) of uniformly sized, hormone-responsive follicles exist in arrested meiosis within the ovary. These animals can be maintained in the cold for months with a minimum of care. By contrast, *Xenopus* exhibits no hibernation period, and oogenesis proceeds continuously throughout the year. Significantly, when properly fed, *Xenopus* ovaries contain, at any one time, a wide range of follicle and oocyte sizes. Oocyte growth processes and vitellogenesis can thus be readily studied in different-size follicles at a single time. That local intrafollicular factors in each follicle control the process of oocyte and follicle growth is dramatically illustrated by the fact that a multitude of growing follicles in amphibians exist simultaneously in the ovary in the presence of a constant level of circulating yolk platelet precursor **(vitellogenin)**. Excessive oocyte and follicle growth has been associated with a parasitic infection of amphibian oocytes and further attests to the fact that local growth control mechanisms in individual follicles are amenable to modification (Schuetz *et al.*, 1978).

In mammals, oocyte processes have been studied most extensively in common laboratory animals, including rats, mice, rabbits, hamsters, and—increasingly—in lower primates and humans. Rats and hamsters have well-defined estrous cycles that can be monitored readily via the changes in cells removed from the vagina. Furthermore, the timing and release of gonadotropic hormones from the pitutary gland in rats are well synchronized and can be predictably manipulated. By contrast, mice have very fragile estrous cycles. Significantly, follicular growth, pituitary hormone secretions, ovulation, and even mating can be synchronized in such animals by appropriate injections of exogenous gonadotropic hormones, i.e., pregnant mare serum gonadotropin (PMSG) and human chorionic gonadotropin (hCG) (Strauss and Meyer, 1962). Furthermore, when immature prepuberal animals are so treated, follicle development can be studied in the absence of preexisting corpora lutea. In addition, superovulation and superfecundation can be achieved when appropriate doses of gonadotropins are used—as long as endocrine inbalances that interfere with postfertilization development and implantation are not produced (Schuetz, 1971b; Miller and Armstrong, 1982).

Alternatively, polyovular animals such as the pig have a multitude of large ovarian follicles that contain considerable amounts of follicular fluid and thus provide a useful model for studies of follicular fluid formation, composition, and associated biological activities. Such tissues can be obtained often through local packing plants.

Mammalian ovarian follicles are typically embedded in a fibrous, tough connective tissue that presents considerable difficulty for collecting and culturing. To circumvent such problems, many studies have been performed using oocytes obtained by random puncturing of the follicles. Under these conditions, the material released includes germinal vesicle-stage oocytes completely denuded of cumulus cells as well gametes surrounded by cumulus cells in various stages of differentiation. The extent to which these morphological variations reflect differences in follicular and oocyte stages is of importance in attempting to delineate the process of oocyte and follicle growth and maturation.

## 3. Folliculogenesis

**Folliculogenesis,** i.e., differentiation of the follicle, is an integral and fundamental feature of oogenesis that serves several structural and functional purposes:

1. Physically separates oocytes from each other
2. Individualizes the development of each oocyte
3. Establishes a barrier to exclude germ cells from direct contact with the nervous, blood vascular, and lymphatic systems
4. Provides a vehicle for expelling (ovulating) each oocyte from the ovary
5. Makes possible a multitude of local microenvironments and different control mechanisms
6. Provides a mechanism by which multiple stages of development can be established and selected

The development and complexity of somatic components of the ovarian follicle vary between species as well as in characteristic ways over the entire period of oogenesis in each species. Thus, in starfish, a single layer of follicle cells makes up the entire follicle, whereas in vertebrates such as fish, amphibians, birds, and mammals at least three tissue layers of varying complexity are generally identified within the follicle wall (Fig. 5). These include an inner layer of **granulosa** or **follicle cells,** an intermediate **theca layer**(s), and finally an outer layer of **interstitial** (stromal) or epithelial cells.

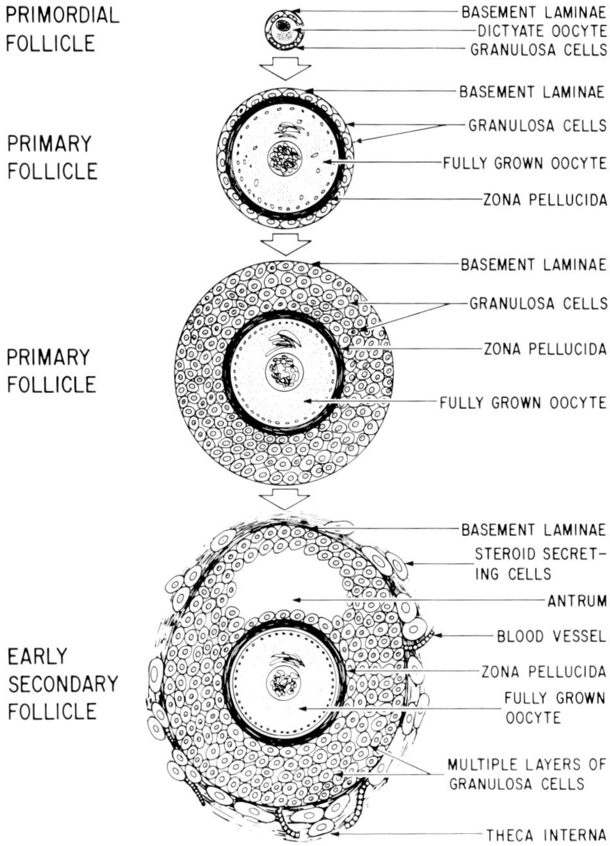


**Figure 5.** Morphological changes that occur during primordial, primary, and early secondary follicle growth. When a primordial follicle begins to grow, the dictyate oocyte initiates a series of ultrastructural and biochemical changes that culminate in the formation of a fully grown and physiologically differentiated ovum surrounded by a glycoprotein coat called the zona pellucida and a single layer of cuboidal granulosa cells. This follicular unit measures ~150 μm in diameter and is called a primary follicle. A steady increase in granulosa cell mitosis causes the primary follicle to grow to a diameter of 200 μm. During this period, the granulosa cells develop follicle-stimulating hormone (FSH), estrogen, and testosterone receptors and become physiologically coupled by gap junctions. The fully differentiated primary follicle migrates into the medulla, where it acquires a theca tissue. In response to FSH, it is triggered to start developing into a secondary follicle. (From Erickson, 1978.)

In some follicles, two thecal layers are recognized—**theca interna** and **theca externa**—and a basal lamina typically is present between the granulosa and thecal tissues in many species. Mammalian ovarian follicles undergo a unique type of further differentiation with accumulation of large amounts of follicular fluid within the antrum, which is formed between the cells of the granulosa cell layer. Follicular fluid constituents are derived essentially from two sources: (1) blood serum as a transudate, and (2) locally produced hormones and other secretory products (Edwards, 1974; McNatty, 1978).

Although in mammals, it is generally recognized that follicular growth and differentiation continue beyond completion of oocyte growth, there is little agreement as to how many stages of follicular development exist over the course of folliculogenesis. Using as criteria the number and arrangement of the granulosa cells in the largest cross section of each follicle, Pederson and Peters (1968) defined some 10 stages of follicular development in the mouse.

Initial stages of oogenesis and gonadal differentiation typically occur in the absence of any organized follicular structure or attached follicle cells, and all oocytes do not proceed simultaneously through development. Ovaries at later stages of gonadal differentation contain a mixture of oocyte and follicle populations of variable numbers with the same or similar identifiable features, e.g., structural, physiological, biochemical (Byersing, 1982). By definition, such follicles, having common features, comprise a cohort, and their number and characteristics vary with each species. The structure of the gonad at any particular time can be characterized by the total number of oocytes and follicles present either individually or as cohorts. In hibernating *Rana*, two distinct classes or cohorts of follicles are identifiable to the naked eye: large, fully grown pigmented oocytes and small, previtellogenic, nonpigmented oocytes. However, in most animals, including other amphibians *(Xenopus laevis)*, many more follicular cohorts are simultaneously present.

Differentiating oocytes and follicles exhibit considerable heterogeneity in structure, composition, and organelle distribution over the course of oocyte growth and maturation; these differences have considerable physiological and developmental significance (Dumont, 1972). In amphibians, ovulation occurs at a predetermined structurally specialized region of the follicle. In mammals, smooth muscle cells localized at the base of the ovarian follicles have been implicated in the ovulatory process (Martin and Talbot, 1981a). How these regional specializations arise remains largely unexplained. Furthermore, in amphibian oocytes, the animal and vegetal hemispheres are distinguished visibly by the presence of melanin granules in the animal pole cytoplasm. Functionally, the animal pole is the normal site of fertilization, is preferentially sensitive to maturation-inducing hormones (Cloud and Schuetz, 1977), and contains a wide range of proteins not present in the vegetal pole (Moen and Namenwirth, 1977). The plane of bilateral symmetry in embryos also appears to be fixed by the site of sperm penetration in the animal pole (Clavert, 1962; Manes and Barbieri, 1976). In addition, the animal and vegetal cytoplasms confer quite different roles on cells, into which they are partitioned during cleavage. Less obvious, but possibly equally as important, regional differences in the composition and structure of more homogeneous-appearing and smaller-size mammalian oocytes have been described (Nicosia *et al.*, 1977).

How and when gamete polarity arises and is established during gametogenesis is of considerable importance; evidence from a number of systems indicates that electrical currents may be involved (Løvtrup *et al.*, 1978). Using an extracellular vibrating electrode, Robinson (1979) found consistent differences in the current pattern around the surface of denuded immature *Xenopus* oocytes (Fig. 6). It was observed that current enters the oocyte in the animal pole and exits from the vegetal pole. Within minutes of treatment with progesterone, the current decreases dramatically; other maturation-inducing agents (gonadotropins, lanthanum) produce similar changes. Furthermore, ion substitution and drug studies suggest that chlo-

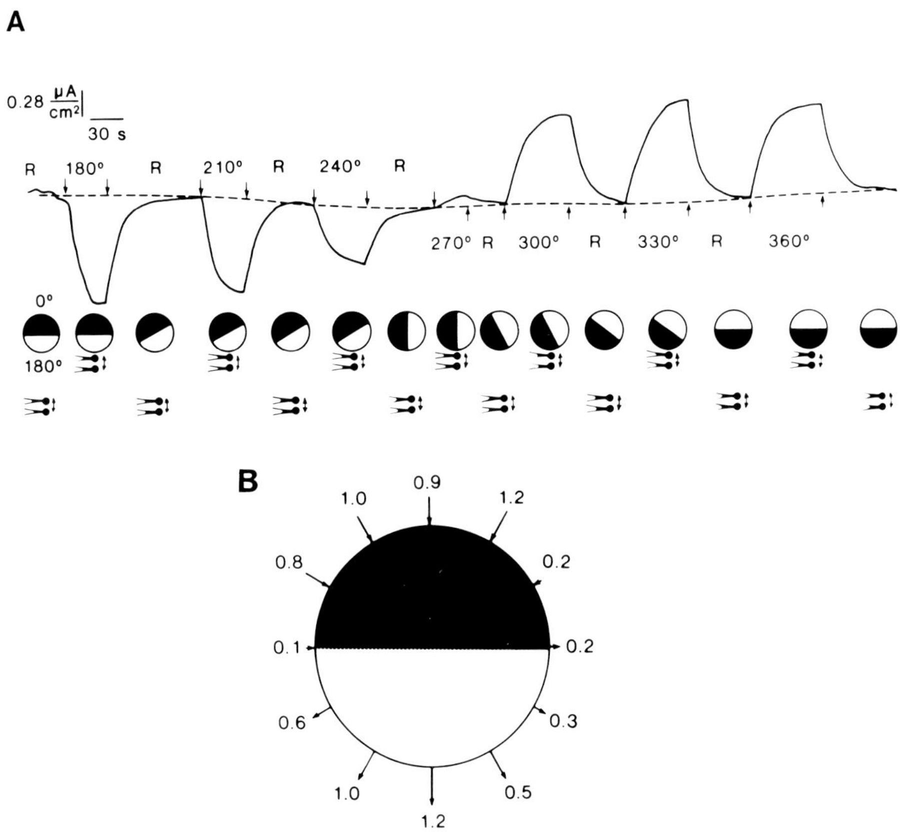


**Figure 6.** Demonstration of the current pattern around the *Xenopus* oocyte surface. (A) Mapping of current around a fully grown oocyte with a vibrating probe. The graph is a tracing of a portion of a chart recording. A downward deflection of the trace from the reference level represents current out of the oocyte; an upward deflection is current into the oocyte. The animal pole is taken as 0; R indicates times at which the probe was moved to a reference position far from the oocyte. (B) Summary of the mapping shown in A. The lengths of the arrows are proportional to the magnitude of the current (μA). (From Robinson, 1979.)

ride is the major current-carrying ion and that membrane permeability to these ions is controlled by calcium. Experimental evidence suggests that establishment of polarity in amphibian oocytes is partially due to endogenous factors. For example, Wallace and Misulovin (1978) demonstrated that during *in vitro* growth, amphibian oocytes acquire a normal (i.e., polarized) pigment pattern in the absence of the follicular structure and are capable of undergoing hormone-induced maturation. These results clearly indicate that intraoocyte mechanisms direct much of the process of oocyte differentiation.

Cellular and molecular processes that transform the amorphous stromal cells into differentiated somatic cell layers of the follicle over the course of oogenesis remain largely unknown. The manner in which the ovary (follicles and oocytes) changes over time must be a reflection of how long each follicle stage persists, the duration of the transition period between stages, and any premature loss or malfunction of follicles that may occur, regardless of cause. Aberrant folliculogenesis has been linked directly to luteal phase dysfunction, infertility, and parthenogenetic activation of follicular oocytes in mammals (Eppig, 1978; DiZerega and Hodgen, 1981a). Furthermore, since in most mammals there is a finite number of oocytes and follicles, the relative proportion of each cohort changes with time until they all essentially disappear (menopause in humans). In other species, oogenesis continues throughout adult life. In mammals, mathematical models have been used in an attempt to estimate, from histological sections of whole gonads, the dynamics of follicle population change and to delineate the factors regulating population shifts. Faddy *et al.* (1976) used a death/migration or compartmental type of mathematical mode to analyze changes in different follicle populations (i.e., of different sizes) over time in intact and hypophysectomized mice. Information collected using five size classes of follicles was then used to estimate the mean inflow and outflow of follicles from a particular class over the life-span. On the basis of such data, it was concluded that medium-size follicles with two layers of granulosa cells may be the initial site of gonadotropic hormonal action.

The constantly changing pattern of follicle and oocyte development thus presents a continual problem from an experimental and analytical point of view. The investigator finds each follicle is in a unique and unstable state. Likewise, after its extrusion from the ovary, the oocyte is also in an unstable state, as exemplified by its limited period of fertilizability, susceptibility to aging, parthenogenetic activation, polyspermic fertilization, and atresia. How to cope with these problems and to sort out which cellular and molecular events are involved in initiating cellular change rather than merely responding to such change presents a considerable technical and analytical challenge of fundamental importance.

In view of the diversity of stages and functions, analysis of changes that occur in well-defined cohorts or clusters of follicles and oocytes (individually or collectively) has provided most of the important information concerning local mechanisms that regulate intra- and extra-follicular as well as intra- and extraoocyte differentiation and maturation. Attempts to delineate local control mechanisms using whole gonads or mixed populations of follicular or oocyte cohorts leads to uninterpretable data. These studies emphasize the need for sensitive micromethods that permit analysis of physiological and biochemical processes in individual follicles and oocytes.

An interesting recent development has been the demonstration that amphibian ovarian follicles and oocytes respond to a number of biologically active substances: inducers of oocyte maturation, ovulation inhibitors, and oocyte maturation inhibitor (OMI) present in follicular fluid collected from mammalian, including human, ovarian follicles (Schuetz and Rock, 1982; Cameron *et al.*, 1983). Extensive experiments have also demonstrated that homologous and foreign nuclei, mRNAs and DNAs, retain functional activity after their microinjection into amphibian oocytes (Gurdon *et al.*, 1976, 1979; Gurdon, 1981). Thus, amphibian oocytes and follicles can be used as living test tubes to study processes of translation, transcription, nuclear reprogramming, and gene regulation in development, as well as the mechanism of hormonal action (Korn and Gurdon, 1981; DiBerardino, 1980). Amphibian follicles and oocytes therefore provide a readily available, cheap, sensitive, and uniform assay system for characterizing biologically active substances in follicular fluid, for assessing various intracellular regulatory factors and for analyzing oocyte and follicular differentiation.

The previously mentioned data provide striking evidence that information obtained from one species can be used to study similar processes in another species. Utilization of the hamster egg to analyze human sperm function (diagnose male infertility) is yet another example of an application of this principle (Yanagimachi *et al.*, 1976).

## 4. Follicle (Granulosa) Cells and Oogenesis

Extensive evidence, both direct and indirect, demonstrates that somatic follicle cells, particularly those enveloping and directly attached to the oocyte, execute multiple and variable functions throughout life and at distinct stages of gamete and follicle development in most animals. A single layer of follicle cells exists in many species, including starfish and amphibians. As a consequence of formation of the fluid-filled antrum during folliculogenesis in mammals, two types of granulosa cells can be identified: (1) **mural granulosa cells,** which form the outer border of the antral cavity, remain within the follicle after ovulation, and contribute to corpus luteum formation; and (2) **cumulus granulosa cells,** which remain associated with the oocyte during the ovulatory process and surround the oocyte within the oviduct for variable periods of time. Significantly, the cells appear to mediate regulatory functions as well as to contribute directly to the growth and differentiation of specific oocyte structures or to prevent them from undergoing degeneration.

During the early fetal period of gonadal development in mammals, the lack of follicle cells in the ovarian cortex has been correlated with extensive oocyte death and atresia (Ohno and Smith, 1964). By contrast, migration of oocytes to the medullary region of the gonad and the presence of a single layer of closely applied follicle cells have been correlated with oocyte viability. It is unclear, however, as to whether and by what mechanisms the follicle cells act to prevent oocyte death. It is of particular interest that survival of follicles and oocytes in the differentiating gonad is markedly affected by alterations in the nutritional and physiological status of the individual organism (Sakakura *et al.*, 1979; Chen *et al.*, 1981; Mattison and Ross, 1983). Oocytes of adult mammals are also affected by the follicle cells. In adult mice of the LT/SV strain, Stevens and Varnum (1974) noted that a high incidence of spontaneous ovarian teratomas derived from oocytes that complete the first meiotic division is associated with a deficiency in the number of granulosa cell layers (see Chapter 7). The capacity of oocytes derived from such follicles to undergo embryonic development after *in vitro* culture is considerably improved when the oocytes are cultured in the presence of cumulus cells (Eppig, 1978, 1982a). The ability of germinal vesicle-stage mouse oocytes and early embryonic stages to be maintained in culture has been linked to the capacity of cumulus cells to convert glucose to usable smaller high-energy intermediates such as pyruvate (Biggers *et al.*, 1967). Selective uptake of amino acids, cyclic nucleotides, and other compounds into oocytes from cumulus cells has also been reported by a number of investigators (Heller and Schultz, 1980; Colonna and Mangia, 1983).

Considerable evidence has demonstrated that follicle cells are necessary intermediates for gonadotropic hormonal induction of oocyte maturation in a number of vertebrate and invertebrate species. Quite simply, control of the oocyte maturation process, at the ovarian level in most species, can be envisioned as a two-step, two-hormone stimulation–secretion sequence involving the somatic follicle cells (Fig. 7). In support of this hypothesis are several important observations:

1. Gonadotropins induce oocyte maturation in the presence but not the absence of follicle cells
2. Gonadal intermediates (1-MA, progesterone) replicate gonadotropic hormone effects in the follice and oocyte, and produce mature, fertilizable oocytes that undergo embryonic development
3. Interruption of somatic cell secretions interferes with gonadotropic but not gonadal hormone induction of maturation.

In mammals, the question of whether oocyte maturation is controlled by some positive or negative signal emanating from follicle cells remains

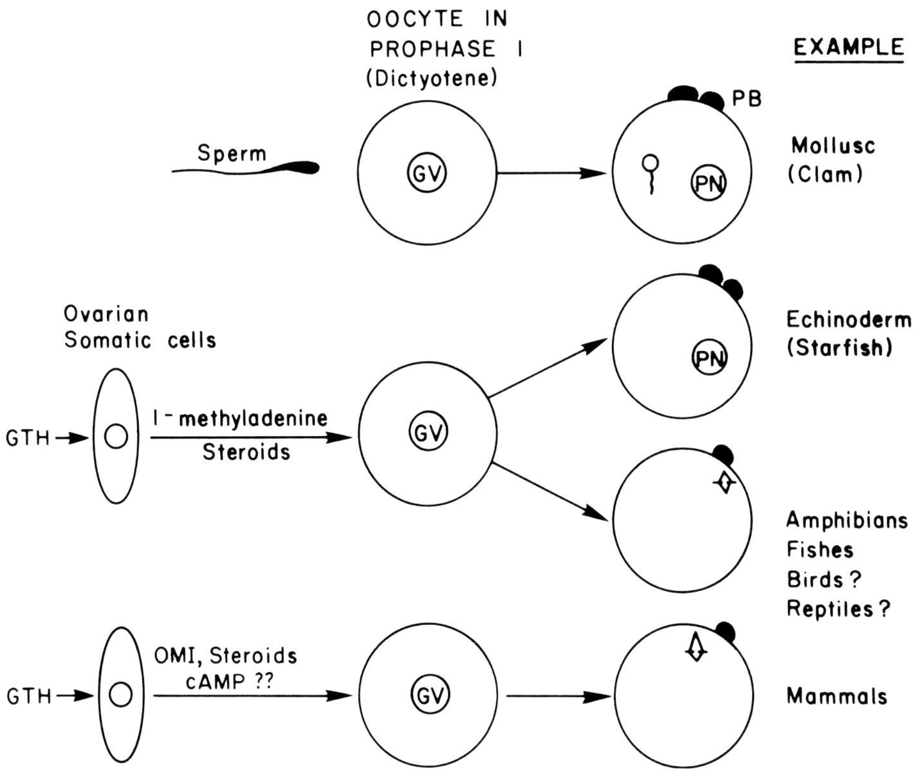


**Figure 7.** Schematic illustration of different stimuli thought to induce meiosis reinitiation (GVBD) in different animal species. (From Schuetz, 1979).

controversial and unresolved (Schuetz, 1979b; Hillensjo *et al.*, 1980; Tsafriri *et al.*, 1982). Only recently, studies in rats demonstrated the necessity for cumulus cells in mediation of gonadotropin induction of oocyte germinal vesicle breakdown in cumulus–oocyte complexes prevented from maturing spontaneously (N. Dekel and Beers, 1978, 1980). In sheep, the presence of cumulus cells markedly increases the developmental capacity of oocytes after gonadotropin, i.e., luteinizing hormone (LH), stimulation (Crosby *et al.*, 1981). Similar effects have been observed in rats (Shalgi *et al.*, 1979).

After the LH surge in mid-cycle, the cumulus follicular cells in mammals undergo dramatic changes in morphology (mucification, expansion, interruption of gap junctions, Gilula *et al.*, 1978; Eppig, 1982b; Dekel *et al.*, 1978) and biochemistry (steroids, prostaglandins, and hyaluronic acid secretions). The secretory activities of these cells furthermore vary, depending on the stage of follicle differentiation at the time of collection and on the *in vitro* culture conditions. Interestingly, the hormone (steroids, prostaglandins) secretory function of cumulus cells extends into the postovulatory period (Schuetz and Dubin, 1981). Whether such secretions play a role in providing for local control of events in the oviduct during the period of fertilization and early development is not known. Such steroids may extend the intrafollicular conditions and further ensure the full developmental potential of oocytes before the corpus luteum becomes fully functional and peripheral levels of progesterone become elevated.

It has generally been assumed that because of their common embryological origin, cumulus and mural granulosa cells in the mammalian follicle have the same functions. One could therefore study cumulus cell function by studying the mural granulosa cells. This supposition appears to be inaccurate, in view of recent evidence establishing that cumulus and granulosa cells have contrasting functions. Studies on LH binding to these different cells demonstrate that LH receptors are much less prevalent on cumulus than on mural granulosa cells and that oocytes contain few, if any, such receptors. In addition to LH receptor differences, granulosa cells secrete plasminogen activator, whereas cumulus cells do not (Gilula *et al.*, 1978). Granulosa cells also secrete glycosaminoglycans, whereas cumulus cells secrete hyaluronic acids (Yanagishita, *et al.*, 1979) while mucifying. Furthermore, the life-spans of these two cell types are very different. In mated pregnant animals, granulosa cells are transformed into greatly enlarged luteinized cells that form and persist as corpora luteal cells for the duration of pregnancy. By contrast, cumulus cells have a short-lived existence and proceed to exhibit degenerative cytological changes with or without fertilization. A variety of evidence suggests that cumulus cells also consist of heterogeneous (structural, functional) cell populations.

Several particularly important questions are raised by these studies:

1. When do local differences in the structure and function of the graulosa cells arise?
2. To what extent are the different functions of the cell types attributable to the presence of gap junctions and products (metabolic secretions) of the oocyte?
3. To what extent do the oocyte and somatic cells control the local environment independently or cooperatively?
4. Since the follicle cells appear to undergo progressive changes with time, how is functional differentiation programmed, and where is this control exerted (cytoplasm, genome)?
5. Are the cumulus follicular cells and their secretory products all that are required to obtain fully mature oocytes? Sheep oocytes matured within the ovarian follicle undergo better development than do those matured outside the follicle (Crosby *et al.*, 1981). Thus, the possibility exists that mural granulosa cells or other components of the ovarian follicle are required alone or along with the cumulus cells to achieve complete oocyte maturation.

## 5. Role of Steroids and Hormone Interactions in Follicular and Oocyte Differentiation

Direct effects of steroids on oocyte maturation and ovulation in amphibians and fish are typically evaluated by using cultured ovarian follicles or oocytes. Such *in vitro* bioassays have been used extensively to examine hormonal structure–biological activity (GVBD) relationships and to establish typical hormone-dose–response curves. Results of such studies show that a wide range of steroids may be biologically active at some dose, whereas others are without biological activity on the same parameter of oocyte differentiation or function. For example, estrogens are ineffective in initiating GVBD, whereas progesterone and other closely related steroids actively induce the process. Although such studies establish that exogenously applied or locally produced steroid hormones can regulate directly the further differentiation of fully-grown germinal vesicle stage oocytes, this does not mean that steroids act alone or that their involvement is limited to a single follicle cohort or oocyte stage. In fact, examination of the literature indicates that steroid secretions of one type or another are involved, implicated, or cannot be ruled out as physiological regulators at any stage of germ cell differentiation—embryonic, prepubertal, or adult (see Naftolin and Butz, 1981).

The fact that complete (sex-reversed) or incomplete (intersex) alteration of the normal pattern of sexual differentiation can be achieved in a number of species (fish, amphibians) dramatically illustrates the pervasive role of steroids in redirecting cellular differentiation as well as the lability of such cells to the action of the hormones. The degree of sex reversal observed depends largely on the amount and type of steroid used to treat embryonic or larval stages (for review, see Yamamoto, 1969; Harris and Edwards, 1970; Gallien, 1974; Donaldson and Hunter, 1982). Thus, treatment of genetic male fish with estrogens leads to a total reorganization of the gonad such that follicles and oocytes rather than sperm are produced (Yamamoto and Kajishima, 1968; Hunter *et al.*, 1982). Reversal in the opposite direction can also be achieved by treating females with typical male hormones. In some species, oocytes or sperm are alternately produced at particular times of the life cycle as the result of spontaneous sex inversion (protandry or protogyny) (Reinboth, 1980). Steroids and steroid metabolic enzymes and various pathways thus appear to be involved in normal as well as abnormal gonadal and germ cell differentiation (Fig. 8). How these processes and pathways are regulated is poorly understood.

Striking evidence that gonadal hormones, particularly estrogens, play a key and largely indirect role in oocyte growth has come from studies of vitellogenesis and oocyte growth in birds and amphibians. Secreted estrogens stimulate vitellogenin synthesis in the liver, and the phosphoprotein is then transported via the blood to the ovary. Here it is incorporated by micropinocytosis into the oocyte cytoplasm, where it undergoes further

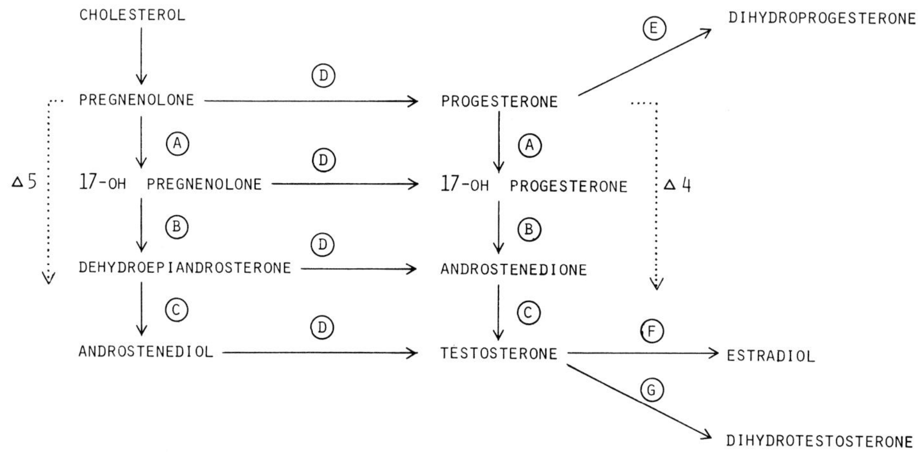


**Figure 8.** Schematic presentation of some of the precursors, products, enzymes, and metabolic pathways involved in ovarian steroid biosynthesis and metabolism subsequent to cholesterol formation. (A) 17-Hydroxylase: (B) Desmolase: (C) 17$\beta$-Dehydrogenase: (D) 3$\beta$-Dehydrogenase and $\Delta$5–3 ketosteroid isomerase: (E) 20 ($\alpha$ or $\beta$) Dihydroxy-dehydrogenase: (F) Aromatase: (G) 5$\alpha$-Reductase. (For a detailed description of these pathways, see Feder, 1981.)

biochemical and structural alterations in association with yolk platelet formation (see Chapter 3). Likewise, progesterone secretion by the ovary has been linked to oviductal secretions of jelly that envelop ovulated maturing oocytes and prepare them for fetilization. Progesterone, in contrast to estrogens, also appears to play a major role in regulating oocyte growth by terminating vitellogenin incorporation across the cell membrane and into the cytoplasm, presumably via the production of a cytoplasmic factor (Schuetz *et al.*, 1974, 1977).

What role steroid hormones play in mammalian oocyte maturation has been and remains a particularly controversial topic (Tsafriri *et al.*, 1982; Schuetz, 1979b) in spite of the fact that the LH stimulus that triggers oocyte maturation *in vivo* or *in vitro* also stimulates a rapid and massive synthesis of wide variety of steroids (estrogens, androgens, progestins) and other factors that accumulate and remain in follicular fluid for variable periods. These factors include certain plasma proteins including steroid-binding proteins, enzymes (plasminogen, proteases), mucopolysaccharides (hyaluronic acid, chondroitin sulfate), pituitary hormones, nonsteroidal factors or activities (OMI, inhibin, luteinization stimulator and inhibitor, follicle-stimulating hormone inhibitor), prostaglandins, gonadotropin releasing hormone (GNRH), and relaxin (Franchimont and Channing, 1981; Channing *et al.*, 1982). To what extent any of the steroids or other components of follicular fluid are products and/or mediators of follicular and oocyte differentiation remains largely unresolved.

Thus, a functional relationship between steroid secretions or these other follicular products and the oocyte maturation process has been difficult to establish. Evidence for lack of involvement of steroids in this phase of oocyte development in mammals is very tenuous and is based largely on several negative findings:

1. Failure of steroids added to the culture medium to induce meiosis reinitiation in follicle-enclosed oocytes
2. Failure of steroid synthesis blockers added to culture medium to prevent LH-induced meiosis reinitiation in follicle-enclosed oocytes
3. Failure of steroids to overcome the inhibiting effects of cyclic adenosine monophosphate (cAMP) or other inhibitors of spontaneous oocyte maturation (GVBD), which occurs when immature oocytes are released from their follicles and cultured *in vitro*
4. The suggestion that meiosis reinitation may be under negative control, i.e., that it occurs as a result of the loss of inhibitory molecules (cAMP, OMI) originating in the somatic cells of the ovarian follicle somatic cells and acting on or in the oocyte

Alternative possibilities should be considered. In particular, it may be that steroids and metabolic blockers are not accessible to oocytes within the ovarian follicle in sufficient amounts and rapidly enough to be effective (Ekholm and Magnusson, 1979). Maturation-inducing activity of selected steroids and doses is typically tested using isolated oocytes prevented from undergoing spontaneous oocyte maturation by a variety of inhibitors. Extensively used in this regard are cAMP derivatives and various other inhibitors that act to maintain or increase intraoocyte levels of cAMP or to prevent the drop in intracellular cAMP. It is particularly relevant that these same metabolic inhibitors or treatments that inhibit spontaneous oocyte maturation also prevent steroid-induced oocyte maturation in amphibian oocytes. One must conclude that even though sufficient steroid is present in an oocyte that can mature, maturation does not necessarily result. Oocyte sensitivity to steroids is therefore a critical factor requiring more thorough investigation. An additional factor contributing to the controversy is that meiosis reinitiation (GVBD) in many cases is used as the only indicator of fully mature oocytes. This event must be considered as one of many essential steps in the oocyte maturation process.

In contrast to the previously cited data, a number of studies strongly implicate steroids in the completion of oocyte maturation in at least certain mammals. Moor and Trounson (1977) observed that oocytes undergoing spontaneous nuclear maturation *in vitro* did not undergo normal development after transfer to oviducts of inseminated ewes. Addition of FSH and LH to cultured follicle enclosed oocytes was found to improve the incidence of blastocyst formation; however, best development occurred when estradiol-17B was added to the culture medium. Interestingly, increasing the amount of FSH and LH utilized drastically reduced blastocyst formation and increased the fragmentation rate. The results clearly suggest that optimum levels of hormones are required to obtain normal development.

McGaughey (1977) observed that the lack of steroids during the later stages of maturation increased the incidence of chromosomal abnormalities at metaphase I and II. Moor *et al.* (1980) altered follicular steroidogenesis during *in vitro* culture of gonadotropin treated ovine follicles and subsequently assessed the developmental capacity of the isolated oocytes after their transfer to oviducts and insemination of host animals. Exposure of follicles to aminoglutethimide (inhibitor of cholesterol conversion to pregnenolone) markedly reduced steroid secretion as well as the number of oocytes with morphologically normal pronuclei. Furthermore, a large number of the oocytes failed to complete meiosis and remained at metaphase I or II despite sperm penetration. Abnormalities of male pronuclear formation were also observed. Addition of steroids along with the inhibitor at the time of culture initiation increased the number of oocytes with normal pronuclei, whereas delaying addition of the metabolic inhibitor until the latter stages of follicle culture markedly reduced the incidence of abnormalities. The results suggest that essential steroid synthesis occurs during the early period of follicle culture in response to the gonadotropic hormones and subsequent to GVBD. Interestingly, addition of steroids along with aminoglutethemide restored female, but not male, pronuclear development. Similar studies were also performed using an enzyme inhibitor of 17 $\alpha$-hydroxylase that suppresses estrogen and testosterone, but not progesterone, secretion. Such oocytes likewise exhibited a low incidence of fertilization, which could be reversed by addition of steroids or delayed addition of inhibitor. It was concluded that oocytes require a specific intrafollicular steroid environment for complete maturation and that alterations in this profile during the maturation period produce gross abnormalities at fertilization. This evidence clearly suggests that more than one steroid may be required for complete maturation of the oocytes. Abnormalities of the nuclear development were interpreted to be the result of the lack of cytoplasmic factors.

Considerable indirect evidence linking steroids to oocyte maturation also comes from studies of natural or induced follicular atresia in a number of model systems (reviewed by Ingram, 1962; Byskov, 1978). In mammals, the vast majority of ovarian oocytes and follicles degenerate before ever being ovulated. Atretic events presumably occur in most species, but they are poorly documented. In all cases, however, the mechanisms controlling these processes are poorly understood in large part because there are no generally accepted criteria of atresia.

In particular, it is unclear whether atresia is initiated in the oocyte, the surrounding follicular tissues, or both somatic and germinal cells. Thus, treatment (e.g., hypophysectomy, anesthesia, steroid injections) of animals that prevents or interrupts the normal course of follicular differentiation and ovulation has been shown to lead to marked changes in the steroid secretory activity of the gonads and individual follicles (Bigl and Greenwald, 1981; Hillier and Ross, 1979). Carson *et al.*, (1981) found differences in the steroids (estradiol, testosterone, androstenedione) present in the follicular fluids of atretic and nonatretic sheep follicles and concluded that changes in the pattern of steroidogenesis were some of the earliest follicular changes to be correlated with atresia. Bagnell *et al.* (1982) analyzed the effects of androgen and estrogen on follicular atresia in hypophysectomized rats injected with gonadotropins and found that 5$\alpha$-dihydrotestosterone (DHT) induced follicular atresia, whereas simultaneous injection of estradiol prevented such effects. These results suggest that DHT induces atresia as a result of interfering with the local production and/or action of estradiol within the ovary.

Significantly, it remains to be determined whether steroid effects on atresia are partially or wholly mediated through effects on the oocyte directly or represent more general action on the entire follicle to alter its functional ability. Further studies of local effects of steroids within the ovarian follicle will be needed in order to address these questions. Elucidation of the local causes of atresia however has considerable theoretical importance. Some investigators believe that atresia represents an abnormal process within the follicle or oocyte and serves to remove defective gametes before fertilization. Others hold that atresia results simply because the oocyte maturation and the ovulatory processes become asynchronous (Schuetz,1972a). In this case, normal maturation of oocytes would occur within the ovarian follicle, but degenerative changes similar to those seen in unfertilized oviductal oocytes would occur in the absence of ovulation. Interestingly, oocytes recovered from follicles exhibiting morphological evidence of advanced stages of atresia retain apparently normal developmental capacity (Hay *et al.*, 1976; Moor and Trounson, 1977). Atretic mechanisms may also be relevant to the process of intrafollicular aging of oocytes. Further research on these problems is clearly needed.

Steroid hormone involvement in germ cell differentiation is not limited to vertebrate species. Reviewing data obtained using invertebrates, Sandor (1980) concluded that steroids are universal biomolecules. Progesterone, estradiol and estrone, as well as the steroidogenic enzyme 3 beta-hydroxy steroid dehydrogenase (3$\beta$-HSD) have been detected and measured in starfish gonads, their levels changing in relationship to the reproductive cycle of animals (Schoenmakers,1980). Limited evidence indicates that estrogens *in vitro* and *in vivo* may stimulate vitellogenesis and oocyte growth. Other workers have demonstrated the existence of numerous other steroids in the starfish, including compounds called asterosaponins (Goad *et al*, 1972; Ikegami, 1974). Asterosaponin A, which consists of a sulfated steroid ring conjugated to a sugar moiety, has been isolated from the jelly layer of starfish oocytes, where it may play some role in the sperm agglutination reaction (Uno and Hoshi, 1978). The specific functional roles of gonadal steroids in invertebrates thus remains largely unknown. However, it is of interest from a comparative and evolutionary perspective that, although the same steroids are present in starfish as in higher organisms, these hormones do not act at the same stages of germ cell differentiation (GVBD). An entirely different type of hormone serves this funtion, i.e., 1-MA (consult Barrington, 1979).

On the basis of these studies, it appears that steroids contribute directly and indirectly to the proper development of the oocytes. Several crucial questions are raised by these data:

1. Do different steroids and hormones act at the same or different times and affect different cellular components involved in oocyte maturation?
2. Do combinations of hormones affect a single parameter of oocyte maturation?
3. Does a single hormone affect many aspects of oocyte maturation, and are the hormone doses the same for some or all such cellular parameters?
4. Do the hormonal requirements for different stages of oocyte maturation vary over time?

The answers to these questions will provide greater understanding of the oocyte maturation process and will resolve some of the apparently contradictory interpretations of experimental results in different systems.

In particular, the role of hormonal interactions in controlling induction of oocyte maturation requires experimental analysis. Because individual steroidal hormones can induce oocyte maturation *in vitro*, it is generally thought that a single steroid probably controls meiosis reinitiation under normal physiological conditions. Thus, little attention has been given to questions of whether simultaneous and/or sequential interactions between hormones are relevant to induction of meiosis. Recent studies, however, demonstrate that synergistic as well as antagonist interactions occur between steroids as well as nonsteroidal hormones such as insulin, which was shown to stimulate oocyte growth and maturation of *Xenopus* oocytes (Wallace and Misulovin, 1978; El-Etr *et al*, 1979; Hanocq-Quertier and Baltus, 1982; Thibier *et al.*, 1982). Effects of insulin on the oocyte are of particular interest, because they suggest for the first time that polypeptides can act on the oocyte independent of somatic cells. Subsequent studies on *Rana*, however, demonstrated that the actions of insulin on oocyte maturation were probably mediated in part by low-level stimulation of progesterone secretion by the follicle wall (Lessman and Schuetz, 1981, 1982). Significantly, synergistic effects of progesterone and insulin were consistently observed in oocytes denuded of follicle cells. Denuded oocytes obtained from certain animals were furthermore found to be insensitive to massive insulin doses, although they remained responsive (matured) to progesterone alone or in combination with insulin. Similar results have been reported in *Xenopus* oocytes denuded of somatic cells by removal of the vitelline envelope (Hirai *et al.*, 1983). The combined results strongly suggest that insulin acts at a number of sites within the ovarian follicle to stimulate oocyte maturation. Significantly, it appears that insulin can act directly on the oocyte; however, its action may require the presence of critical levels of endogenous steroidal hormone with which it can synergize.

Whether such hormonal interactions occur is particularly relevant, since it is becoming increasingly evident that numerous and variable amounts of steroids and other factors are produced at different stages of differentiation by ovarian follicles and/or their cellular components. Amphibian ovarian tissues simultaneously secrete estradiol, testosterone, and progesterone, and it appears that these secretions are follicular stage specific (Fortune and Tsang, 1981; Fortune, 1983). Similarly, in mammals, cumulus–oocyte complexes (preovulatory and postovulatory) as well as whole follicles are capable of secreting a range of steroids and prostaglandins during *in vitro* culture (McNatty *et al.*, 1980).

The relationship between estrogen and progesterone secretions and their interactions during folliculogenesis is particularly interesting and complex. In nonmammalian species, follicular estrogens are necessary for vitellogenesis and a major portion of oocyte growth, whereas progesterone is required for maturation. Furthermore, estrogens are reported to inhibit the action of pituitary hormone and pregnenolone in inducing oocyte maturation in *Rana pipiens* (Schuetz, 1972c; Snyder and Schuetz, 1973; Spiegel et al., 1978). Thus, it has been generally assumed that estrogen interference with pituitary-induced maturation occurs only at the level of the follicle wall somatic cells.

More recently, Lin and Schuetz (1983) reexamined this question and concluded that estrogens can modulate oocyte responsiveness to low concentrations of progesterone (Fig. 9). Pretreatment of follicles with estrogen increases the efficiency of inhibition; however, the inhibitory effects of estrogen are readily overcome by increasing the dose of progesterone. By contrast, increased amounts of frog pituitary gland extracts did not overcome the inhibitory effects of the estrogen.

Other studies have established that estrogens inhibit frog pituitary homogenate (FPH)-induced production of endogenous levels of progesterone (Lin and Schuetz, unpublished). Thus, estrogens appear to modulate follicular responsiveness to gonadotropins at two sites—either by regulating the production of maturation-inducing steroid in the follicle wall or by modifying the response of the oocyte to previously formed steroid (Fig. 10).

Obviously, under *in vivo* conditions, changing estrogen levels presumably reflect alterations in enzyme substrates and the activity of specific steroidogenic enzymes, particularly aromatase, which converts testosterone to estrogens. A decrease in aromatase activity in salmon ovarian follicles has also been correlated with induction of oocyte maturation (Young *et al.*,

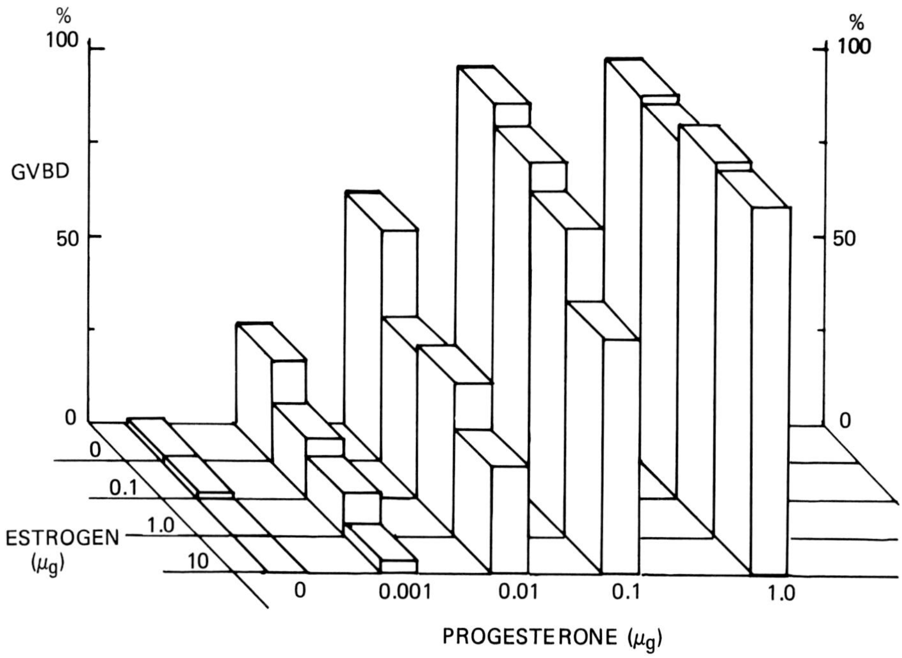


**Figure 9.** Three-dimensional representation of the effects of various combinations of 17β-estradiol and progesterone on induction of oocyte germinal vesicle breakdown during *in vitro* culture of amphibian ovarian follicles. Twenty follicles from an individual animal were tested with each hormone combination. Each block represent the mean response of results obtained from 10–18 animals. Follicles were exposed to different doses of estradiol for an 8-h period before the addition of progesterone. Results demonstrate that estradiol inhibits, in a dose-dependent fashion, the oocyte maturation-inducing activity at lower doses of progesterone tested. At higher concentrations of progesterone, inhibitory effects of estrogen are no longer evident. (From data of Lin and Schuetz, 1983.)

1983). Fortune and Vincent (1983) observed that progesterone inhibits aromatase enzyme activity in cultured rat granulosa cells and suggested that progesterone secretions from ovulatory follicles might act to suppress nearby developing follicles. Possibly, progesterone also acts locally within individual follicles to bring about rapid interruption of estrogen secretion. In mammalian oocytes, attainment of maturational competence (GVBD) also has been linked to previous estrogen exposure and during folliculogenesis where estrogen secretion typically precedes progesterone secretion. It seems necessary in all these species that in order to understand follicular differentiation and the local role of steroids and other hormones in follicular and oocyte growth and maturation, we will need to consider what steroids are produced, in what amounts, in what sequence, and by what tissues. As studies *in vitro* have shown, different isolated tissue components of the follicle wall clearly have variable steroidogenic activity at different stages of follicle differentiation. This is illustrated in Fig. 11 using data obtained from hens. Furthermore, considerable evidence from a number of species demonstrates that complex cellular interactions are required for the synthesis of particular steroidal hormones. Such studies have provided evidence for the two-cell theory of estrogen secretion involving theca and granulosa cells (Fig. 12). Equally important, procedures for evaluating the relative biological importance of hormonal mixtures and fluctuations in hormonal secretions need to be developed. Delineation of the intracellular molecular mechanisms controlling the shift between estrogenic and progestational phases of follicular differentiation should provide particularly relevant information.

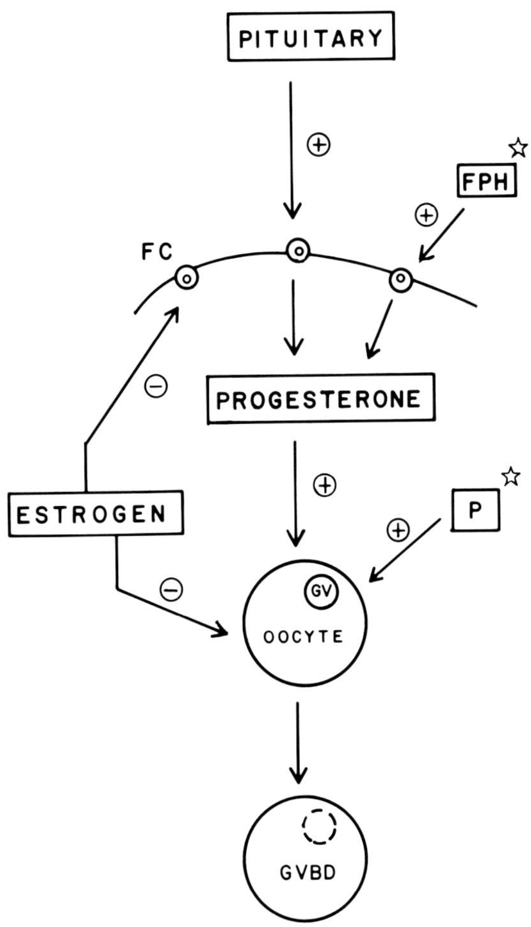


**Figure 10.** Proposed intrafollicular sites of estrogen action in regulating oocyte maturation in *Rana pipiens*. Frog pituitary homogenate (FPH) stimulates (+) progesterone (P) secretion by the follicle cells (FC), in turn causing the oocyte germinal vesicle (GV) breakdown (GVBD). Addition of estrogen to follicles *in vitro* inhibits (−) progesterone production by the follicle cells as well as progesterone induction of oocyte GVBD.

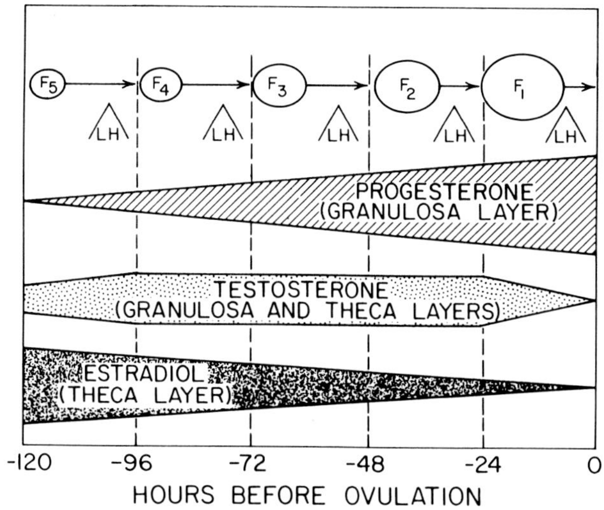


**Figure 11.** Schematic representation of changes in steroid concentration of granulosa and theca layers during follicular maturation. The luteinizing hormone represents the LH surge, which occurs at 4–6 hr during the ovulatory cycle in chickens. (From Bahr *et al.*, 1983.)

## 6. Intercellular Transfer of Hormonal Signals of Differentiation

From the previous discussion, it is clear that the follicle cells attached to the oocyte are in a unique position to influence the germ cell and are the source of a variety of hormones and other factors that may or do affect growth and maturation of the gamete. Little is known, however, concerning the mechanisms and routes of intercellular transfer of such regulatory molecules between somatic cells and oocytes. In recent years, considerable attention has focused on the possible physiological role of the specialized structural relationship that exists between these two cell types. In the species being considered here, long cytoplasmic processes (macrovilli) of the somatic cells extend across the vitelline envelope and impinge upon or penetrate the surface of growing immature oocytes in a characteristic way.

In starfish, desmosomelike junctions consisting of a prejunctional (follicle cell) and postjunctional component (oocyte) have been described (Schroeder, 1981). These villar processes lack normal cytoplasmic organelles and inclusions and occasionally contain microfilamentous elements. It is estimated that some 6500 follicle cell processes impinge on an equal number of postjunctional specializations on the surface of each oocyte (Fig. 13). It has been proposed that such villi are the conducts through which 1-MA is transferred to postulated hormonal receptors on the oocyte cell surface, thereby triggering oocyte maturation (Fig. 14). Interestingly, rupture

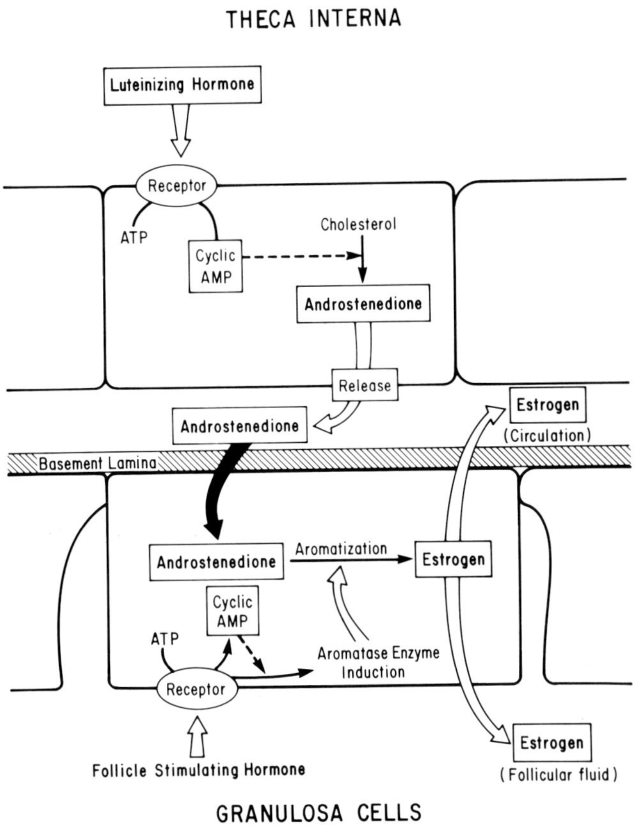


**Figure 12.** Current hypothesis proposed to explain the two-cell theory of follicle estrogen production. In response to luteinizing hormone (LH) stimulation, the steroid-secreting cells in the theca interna secrete androstenedione. The androstenedione diffuses across the basement membrane of the follicle and enters the granulosa cells, where it is aromatized into estrogen by the aromatase enzymes induced in the granulosa cells by follicle-stimulating hormone (FSH). The newly synthesized estrogen is released into the follicular fluid and peripheral circulation. (From Erickson, 1978.)

of these intercellular attachments is produced by exposure to 1-MA or to calcium-free seawater and is highly correlated with the incidence of spontaneous oocyte maturation (Schuetz and Cloud, 1978). Connection between the follicle cell and oocyte, however, does not appear to be required for 1-MA release, since isolated follicle cells release oocyte maturation-inducing substance (presumably 1-MA) after exposure to calcium ions (Cloud and Schuetz, 1973).

In amphibians and mammals, ovarian follicle cell gap junctional complexes exist along macrovilli that penetrate the oocyte cytoplasm and tightly fuse the two cells together (Gilula *et al.*, 1978; H. Dekel *et al.*, 1978;

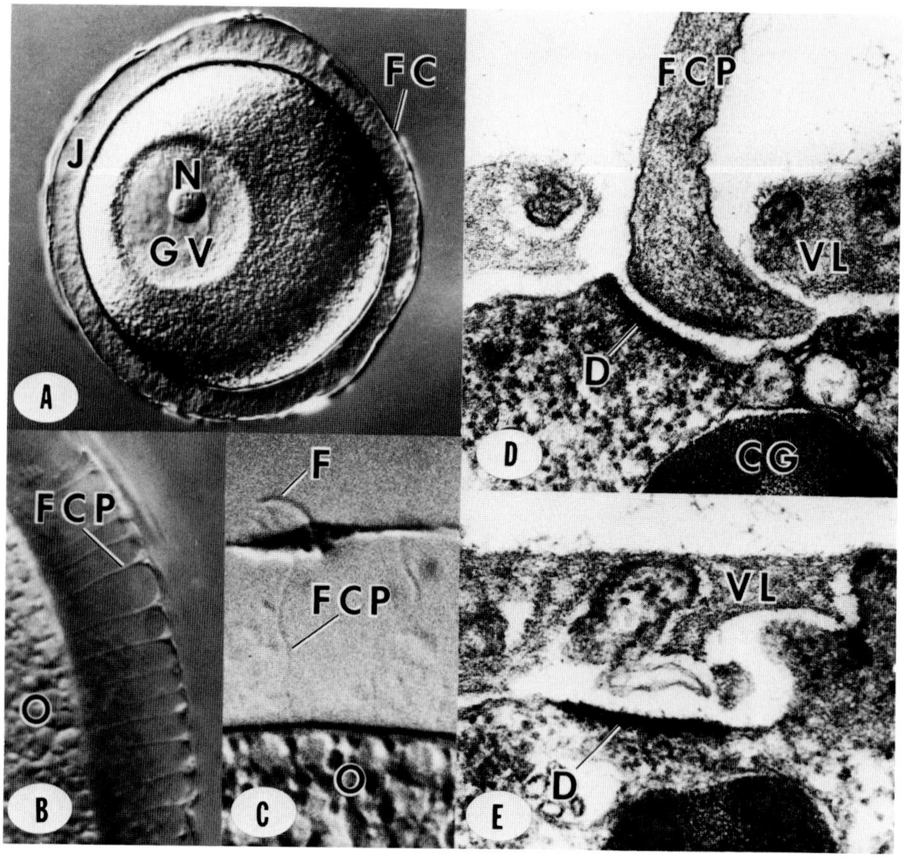


**Figure 13.** Oocyte–follicle cell association in starfish. (A) Light micrograph of a living starfish oocyte and its surrounding layer of very thin follicle cells. Follicle cells are present as a complete sheath early in the season and tend to be lost later in the season, when oocytes are removed from an ovary for experiments. Slightly compressed. Nomarski optics, ×185. (B) At higher magnification, numerous thin, 15-μm-long processes from the follicle cells are seen to pass through the jelly coat and vitelline layer and to impinge upon the surface of the oocyte. Nomarski optics, ×1,050. (C) Additional details of a follicle cell, including its single flagellum and a follicle cell process. Unstained 3-μm-thick Epon section. Nomarski optics, ×2100. (D) Transmission electron micrograph of the follicle cell–oocyte junction of a fully grown immature oocyte. The follicle cell process penetrates the jelly coat and vitelline layer, terminating in a slight swelling. A 20- to 35-nm-wide gap separates the follicle cell and the oocyte. A subplasma membrane density characterizes each side of the junction; the postjunctional density belonging to the oocyte is more pronounced. ×56,000. (E) The postjunctional density of an immature oocyte persists after follicle cells are removed by washing in calcium-free seawater. ×56,000. CG, cortical granule; D, density; F, flagellum; FC, follicle cell; FCP, follicle cell process; GV, germinal vesicles; J, jelly coat; N, nucleolus, O, oocytes; VL, vitelline layer. (From Schroeder, 1981. Reproduced from *The Journal of Cell Biology*, by copyright permission of The Rockefeller University Press.)

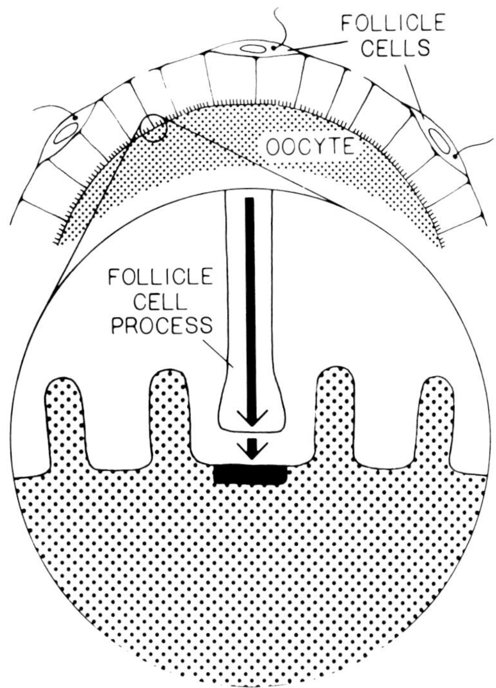


**Figure 14.** Diagrammatic representation of the hypothetical pathway (arrows) and receptor structure (dense bar) of the maturation-inducing "hormone" of starfish, 1-MA. 1-Methyladenine is produced in the body of the follicle cell, transported centripetally within the follicle cell process, and transmitted across the narrow junctional space separating follicle cell and oocyte, where it is finally recognized by receptor molecules associated with the postjunctional complex on the oocyte surface. This model may be applied to the communication between follicle cells and oocytes in all vertebrates, including humans. (From Schroeder, 1981. Reproduced from *The Journal of Cell Biology*, by copyright permission of The Rockefeller University Press.)

Browne *et al.*, 1979). A striking feature of such gap junctions is that they provide a direct opening for the two-way transfer of small-molecular-weight ($\sim$1000 $M_r$) substances between the cytoplasms of the two cell types. In amphibians, the follicle cells typically become dissociated from the oocytes during gonadotropin- or steroid- induced maturation (Schuetz, 1972b). This event appears to be under the control of a secretory product—presumably progesterone of the follicle cells. Follicle cell detachment from the oocyte also results after treatment of follicles or defolliculated oocytes with calcium-free medium, cytoplasmic injections of maturation-promoting factor, or simply long-term *in vitro* culture. Thus, it appears that the extracellular environment as well as intracellular factors derived from the follicle cells or oocyte have some controlling influence on these intercellular junctions.

In mammals, it has been suggested that gap junctions exert some control on the oocyte maturation process by providing direct access to the oocyte cytoplasm of molecules that cannot readily cross the cell membrane. Such intercytoplasmic channels have been proposed as conduits for transfer of the postulated oocyte maturation inhibitor (OMI) or cAMP from either the cumulus cells or the surrounding follicular fluid environment to the oocyte cytoplasm, where they could act to prevent meiosis reinitiation.

In such a model, interrupting gap junctions could lead to reinitiation of meiosis because of the lack of inhibitory molecules. However, in mammals, the experimental evidence indicates that rupture of the connections occurs some hours after germinal vesicle breakdown and is correlated with mucification and expansion of the cumulus cells. Thus, it does not appear likely that interruption of these connections leads to maturation induction as a result of interrupting transfer of OMI to the oocyte (Tsafriri *et al.*, 1982; Eppig, 1982b; Schultz *et al.*, 1983).

Gap junctions may not be the only mechanism involved in the transfer of the hormone signal to the oocyte from the somatic cells. Considerable evidence indicates that steroids are released from steroidogenic cells by some active process that involves steroid binding to proteins (Willcox and Alison, 1982). Calcium ions and calmodulin also appear to play a role in controlling progesterone production and secretion by granulosa cells (Carnegie and Tsang, 1983; Tsang and Carnegie, 1983).

Alternatively, other data suggest that steroid–protein complexes are packaged intracellulary within membrane-enclosed organelles and then released from the cells by exocytosis (Gemmell and Stacy, 1979; Parry *et al.*, 1980). Recent data have demonstrated the presence of a single progesterone-binding protein in granulosa cells obtained from preovulatory rat ovarian follicles (Schreiber and Hsueh, 1979). Two progesterone-binding proteins of considerable specificity have also been described in cytosol preparations of bovine corpora lutea (Willcox, 1983). Definitive evidence is lacking as to whether these binding proteins are involved in the secretory function of the cells. The possibility exists that they may act as receptors to mediate the effects of progesterone within the granulosa cell itself or contribute to the pool of steroid-binding proteins of considerable specificity localized in the follicular fluid (Mahajan and Little, 1978; Fleming and McGaughey, 1982) (Fig. 15).

The extent to which any, all, or none of the processes discussed in this section are involved in the transfer of a steroidal or other type of signal to mammalian or amphibian oocytes remains to be elucidated. The answer to this question, however, has considerable implications concerning the mode of action of the steroidal hormones on oocytes. Our present understanding of the mechanisms involved in hormonal–oocyte interactions has been primarily derived from studies involving pure hormones. If hormones are transferred through gap junctions at the ends of macrovilli, direct access to preferred sites of action or metabolism within the cytoplasm may be provided. Such localized targeting could greatly reduce the amount of hormone required to initiate physiological changes in the oocyte. Furthermore, in this case, the steroid "receptors" could be at sites distant from the oocyte surface. Simple diffusion gradients may also be involved in steroidal hormone transfer with lipids, which are incorporated in considerable amounts by most oocytes. Amphibian oocytes are capable of incorporating massive amounts of unconjugated steroids. If cell transfer greatly reduces

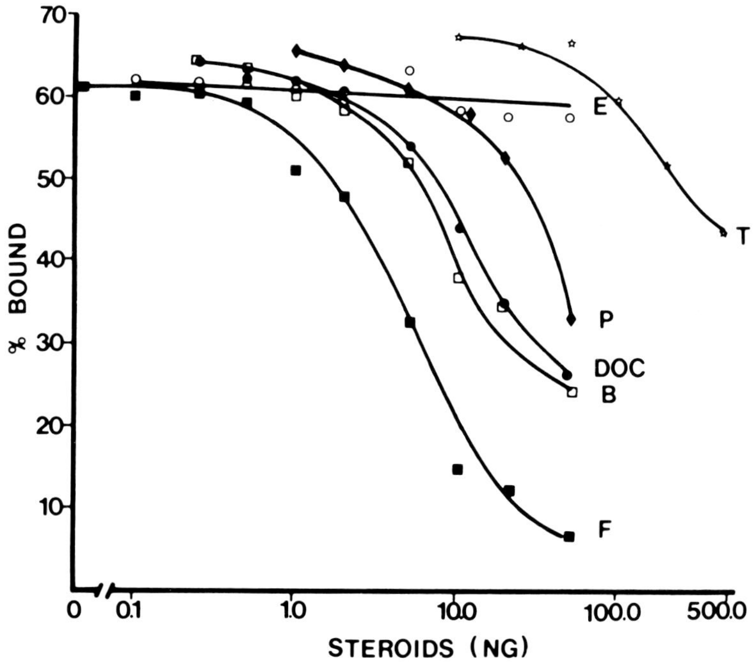


**Figure 15.** Specificity of steroid-binding protein in follicular fluid. Dose–response competition curves of the binding of tritiated cortisol and various amounts of different steroids using 83.3 $\times 10^3$ dpm of [$^3$H] cortisol and 1: 100 dilution of follicular fluid. Letters on curves are as follows: F, cortisol; B, corticosterone; DOC, 11-deoxycorticosterone; P, progesterone; E, 17$\beta$-estradiol; T, testosterone. (From Mahajan and Little, 1978.)

the hormonal requirement for inducing physiological change, the question arises as to how many and which follicle cells would be required to secrete sufficient hormone to induce maturation.

Although gap junctions provide a mechanism for bidirectional transfer of small molecular-weight molecules, little attention has been given to the possible role of gap junctions in controlling cumulus (follicle) cell function via regulatory molecules originating in the oocyte cytoplasm. It is apparent that much more critical information about structure, formation, and degradation functions of gap junctions is needed before we can understand the processes of oocyte and follicle growth and differentiation and the intercellular transfer of informational molecules (Larsen *et al.*, 1982).

## 7. Spontaneous Cellular Differentiation: The Dilemma of Regulation

A central and continuing dilemma regarding the local mechanisms that regulate oocyte and follicular maturation revolves around the phenomenon of "spontaneous" *in vitro* cellular differentiation, which occurs subsequent to that of whole follicles or their various components (oocytes, granulosa cells, theca, cumulus cells, or ovulated gem cells) being placed into culture. Typically during *in vitro* culture, some or all cells differentiate or transform such that they change their structure in a way that—although it resembles what happens *in vivo*—occurs without apparent involvement of the normal physiological factors believed to initiate a control process *in vivo*. The time course of *in vitro* cellular differentiation may or may not closely reflect what occurs *in vivo* but in many cases is considerably delayed. Well-documented examples of "spontaneous" cellular differentiation include (1) germinal vesicle breakdown and meiotic progression, (2) ovulation, (3) pseudogastrulation, (4) parthenogenetic activation (metaphase II), (5) luteinization and progesterone secretion by granulosa cells, (6) mucification and expansion of cumulus cells and secretion of hyaluronic acid, and (7) oocyte growth. Thus, cell division and chromosomal events as well as hormone, protein, and glycosaminoglycan secretions all occur under appropriate circumstances.

The best-characterized example of spontaneous differentiation is reinitiation of meiosis in fully grown germinal vesicle-stage oocytes. Interruption of meiotic arrest occurs when oocytes are released into culture medium before endogenous release of gonadotropic hormones. The extent to which *in vitro* differentiation continues varies in relation to many factors including culture conditions, follicular sources, previous hormonal treatment, and time of recovery from the donor animals. In mammals, lack of oxygen has been shown to inhibit spontaneous GVBD, and it has been suggested that increased concentrations of oxygen in the intrafollicular environment could trigger meiotic maturation (Zeilmaker and Verhamme, 1974). Such an idea is consistent with the observation that the gonadotropin stimulus for oocyte maturation leads to hyperemia and increased vascular permeability around the ovarian follicles. Oocytes undergoing GVBD may proceed to metaphase I or II before their development arrests again, or they may even undergo parthenogenetic activation.

Significantly, spontaneous cellular differentiation occurs over a wide range of species (vertebrate and invertebrate), including echinoderms, amphibians, and mammals. In amphibians, starfish, and some mammals, the mechanisms controlling spontaneous *in vitro* oocyte maturation appear to be intimately linked to the presence of closely attached somatic cells and of cell–cell interactions initiated during culture (Schuetz and Cloud, 1978). Tearing of the ovarian wall of *Asterias forbesii*—a starfish species that exhibits extensive spontaneous maturation—releases ovarian follicles into seawater and typically results in meiosis reinitiation and follicle disintegration. Detachment and removal of the somatic cells from the oocyte by prewashing in calcium-free seawater before their release into seawater prevents spontaneous maturation without obliterating oocyte responsiveness to 1-MA. Furthermore, isolated follicle cells collected in calcium-free seawater release maturation-inducing activity into the medium after exposure to calcium ions (Cloud and Schuetz, 1973). Calcium levels in seawater are normally inadequate to trigger meiosis reinitiation in such oocytes. It is unclear whether the calcium ions cause release of preformed homone or stimulate rapid synthesis and release of 1-MA. Many follicles in this species, however, do not spontaneously mature but can be induced to do so in the presence of 1-MA. Thus, the critical factors that distinguish spontaneous from nonspontaneous maturation may be the level of the hormone in the follicle cells, their sensitivity to calcium, and the level of sensitivity of oocytes (stage of differentiation) to released hormones. Variations in any one or various combinations of these factors could possibly explain why spontaneous maturation does or does not occur in a particular follicle or oocyte. Gradation in such factors rather than the existence of uniquely different mechanisms appears to be the most likely explanation for the relative incidence of spontaneous maturation in different animal species or between animals.

Spontaneous maturation also occurs in amphibian ovarian follicles in varying degrees, depending on the species being studied. Vilain *et al.*, (1980) observed considerable spontaneous maturation in *Xenopus laevis*, *Pleuordeles waltlii*, and *Ambystoma mexicanum* oocytes after enzymatic, mechanical, or manual defolliculation. Spontaneous maturation occurs to a minor degree in *Rana* oocytes, but it is inhibited by complete removal of the follicular epithelium or by treatment with EDTA or estradiol. The latter interferes with progesterone production and activity (Lin and Schuetz, unpublished data). It it also important to consider the possibility that the *in vitro* culture system accomplishes what normally occurs *in vivo*. In amphibians, hypophysectomy experiments have established that oocyte growth is dependent on gonadotropic hormone secretions. However, it has been shown that extensive growth and differentiation of small oocytes can be accomplished when defolliculated oocytes are cultured in the presence of vitellogenin and insulin (Wallace and Misulovin, 1978) in the absence of gonadotropic hormones. Clearly, this represents spontaneous growth but does not negate the role of gonadotropins in *in vivo* oocyte growth. Such data strongly support the contention that gonadotropic effects on oocyte growth are mediated indirectly through follicular somatic cells rather than directly on the oocyte.

In mammals, it is generally assumed that spontaneous maturation occurs without gonadotropic stimuli, but that such stimuli are required for steroidogenesis. The possibility, does exist, however, that the cumulus cells produce or are induced to produce *in vitro* low levels of maturation-inducing hormones in amounts sufficient to induce oocyte differentiation. This idea is supported by the fact that granulosa cells collected from long-term hypophysectomized animals have the capacity to secrete low levels of steroid hormones during *in vitro* culture (Taya and Greenwald, 1980). Likewise, preovulatory rat ovarian follicles cultured in serum from

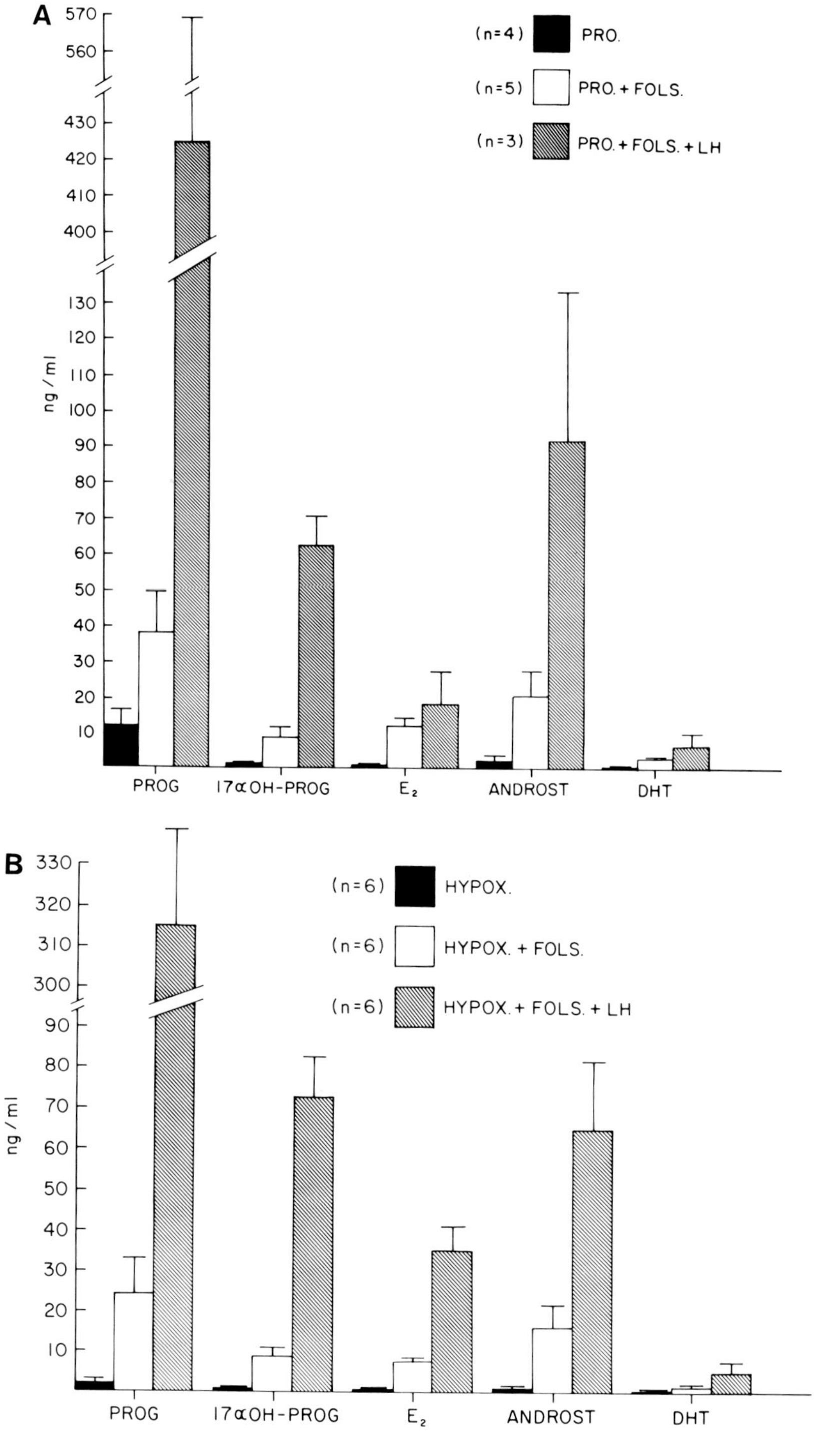
A
(n=4) PRO.
(n=5) PRO. + FOLS.
(n=3) PRO. + FOLS. + LH
ng / ml
PROG
17αOH-PROG
E2
ANDROST
DHT
B
(n=6) HYPOX.
(n=6) HYPOX. + FOLS.
(n=6) HYPOX. + FOLS. + LH
ng / ml
PROG
17αOH-PROG
E2
ANDROST
DHT

hypophysectomized animals secrete considerable amounts of steroid during *in vitro* culture (Fig. 16). In cultured cumulus–oocyte complexes, it was further observed that the steroid secretory activity of cumulus–oocyte complexes is dependent on the amount of serum present in the culture medium, the gaseous atmosphere, and the cumulus cell component (Fig. 17). Thus, the composition of the culture medium and the conditions of culture have a profound effect on cumulus cell secretory processes. Furthermore, cumulus–oocyte complexes collected after ovulation secrete much greater quantities of hormones than those collected from preovulatory follicles and not exposed to LH (Schuetz and Dubin, unpublished). The state of differentiation (previous gonadotropin exposure) of the somatic cells at the time they are collected thus appears to be of primary importance in determining their hormone secretory capacity *in vitro*. This interpretation is consistent with the wide variability seen in amphibian ovarian follicles and oocytes of different animals relative to (1) steroid induction of oocyte maturation, (2) progesterone production, (3) FPH and insulin sensitivity, and (4) timing of GVBD.

## 8. Germinal Vesicle Breakdown and Meiotic Progression

The germinal vesicle envelope serves to segregate the chromosomes and nucleoplasm from the cytoplasm and appears to restrict physical and metabolic interaction between these two components. Disintegration or breakdown of the nucleus (GBVD) is a landmark event that occurs during maturation (nuclear and cytoplasmic) of the oocyte and is triggered by a number of mechanisms in different species (Fig. 7). As a consequence of GVBD, the nucleoplasm mixes with the cytoplasm, and the physiological and physical properties of the cytoplasm are extensively altered.

Despite the importance of GVBD to the maturation process, the mechanisms mediating this event remain poorly understood. Our ideas concerning the GVBD control mechanisms are probably best understood as a result of the analysis of steroid- (progesterone) induced GVBD in amphibian oocytes (*Rana pipiens, Xenopus laevis*). Progesterone does not bring about nuclear disintegration directly but rather acts through the formation of cytoplasmic maturation promoting factor(s) (MPF). Such factors result from a complex cascade of biochemical reactions occurring at or near the oocyte cell membrane and involve calcium ions, calmodulin, the adenyl cyclase

**Figure 16.** Production of steroids (ng/ml medium) by rat graafian follicles after an 8-hr incubation in serum obtained from proestrus (A) or hypophysectomized (B) rats. Various culture vessels contain control medium (■); medium plus follicles, (□); or medium, follicles plus 10 μg LH (▨). Data show that follicles secrete a range of steroids in the presence or absence of gonadotropins during *in vitro* culture. However, LH markedly stimulates the steroid-secretory activity of the follicles. (From Readhead *et al.*, 1979.)

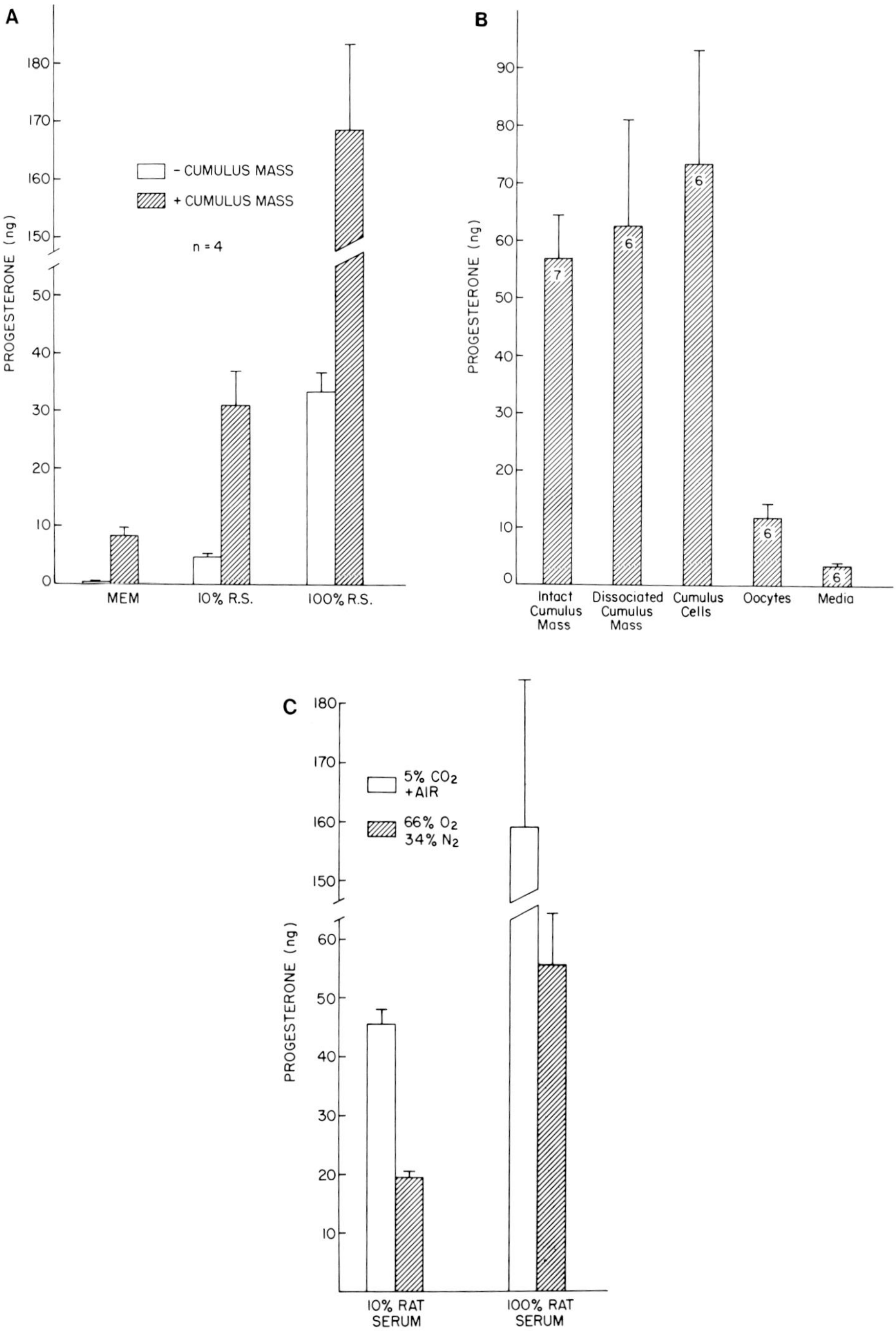


**Figure 17.** Steroid-secretory activity of cultured cumulus–oocyte complexes. (A) Progesterone production *in vitro* by cumulus masses incubated in Eagle's MEM or in MEM supple-

system, cAMP, and protein phosphorylation (Masui and Clarke, 1979; Maller, 1983; Baulieu *et al.*, 1978) (see Chapter 6). For practical purposes, since GVBD is triggered, we can define four stages to this process: (1) an initial hormone or stimulus-dependent period (time needed to acquire sufficient stimulation to induce GVBD); (2) a poststimulus, hormone-independent, pre-GVBD period; (3) nuclear disintegration period; and (4) a post-GVBD stage of oocyte differentiation.

On the basis of direct microscopic examination and/or electron microscopic evaluation of fixed and unfixed samples, seven distinct maturation stages have been described in *Xenopus* oocytes (Huchon *et al.*, 1981). Nuclear breakdown begins at the basal region (toward the oocyte center) of the germinal vesicle and is associated with the formation of a fibrillar network composed of microtubular bundles and a microtubule organizing center similar to that seen in mammalian oocytes. Marked undulations in the nuclear envelope as well as disruption of the nucleolus occur in starfish oocytes and appear to be characteristic features of GVBD in most organisms (Longo *et al.*, 1982).

In *Xenopus* oocytes undergoing steroid-induced GVBD, a white maturation spot typically appears in the animal pole, the result of movement of pigment granules, and a concentrated ring of pigment surrounded by another depigmented area is formed. Although this white spot has been used extensively as an indicator of maturation (i.e., GVBD), it has been demonstrated (Bellé *et al.*, 1982) that elevated $CO_2$ levels inhibit GVBD but not formation of the white spot. Thus, direct visual inspection of dissected oocytes or histological analysis is required in order to assess unequivocally whether the germinal vesicle has disintegrated. The $CO_2$ appears to inhibit MPF formation in the cytoplasm. Nuclear migration without maturation induction also results after prolonged treatment of immature oocytes in calcium-free medium and EDTA. These results suggest that migration of the germinal vesicle and white spot formation occur independent of MPF production.

In fish as well as amphibian oocytes, migration of the germinal vesicle to the periphery of the animal pole is an important common feature of the

---

mented with 10% rat (diestrus) serum or 100% rat (diestrus) serum. The progesterone levels in media not containing cumulus masses represent the endogenous level of hormone in the media after 4.5 hr of culture. (From Schuetz and Dubin, 1981). (B) Progesterone production by intact and dissociated cellular components of cumulus–oocyte complexes. Cumulus cells and oocytes were physically separated after their dissociation by hyaluronidase. The group labeled dissociated cumulus mass represents masses treated with hyaluronidase but not separated from each other before incubation. Incubation was for 4.5 hr in Eagle's MEM supplemented with 10% rat serum. Data represent the mean (nanograms), ±SEM; *N* = number of flasks or animals. (From Schuetz and Dubin, 1981.) (C) Effect of different gaseous atmospheres on progesterone secretion by cumulus–oocyte complexes during *in vitro* culture.

hormone-induced maturation process (Iwamatsu and Ohta, 1977). On the basis of cytological changes in the annulate lamellae, mitochondria, the vitelline surface, and germinal vesicle, three stages of maturation have been described in fish oocytes.

Stimulation of protein synthesis is an integral and necessary precedent for GVBD induced by steroids and other substances (lanthanum, mianserine, organomercurials, insulin) that mimic effects of progesterone. The protein synthesis requirement is readily demonstrated by addition of protein synthesis inhibitors (cycloheximide, puromycin) to hormone-treated oocytes. Typically, addition of such inhibitors for some period after hormonal stimulation prevents GVBD. In order to exert its effects, the inhibitor need not be added at the time of progesterone, suggesting that protein synthesis is delayed for some time after hormone exposure (Schuetz, 1967). In *Rana* this initial inhibitor-sensitive phase is followed by an inhibitor-insensitive period that precedes GVBD. On the basis of such responses, it is thought that the onset of the inhibitor-insensitive phase occurs when sufficient MPF is formed to induce GVBD and that MPF acts to break down the nuclear envelope, perhaps by an enzymatic process that does not require protein synthesis. The extent to which protein synthesis and MPF control GVBD in other species, particularly mammalian and even starfish oocytes, remains controversial (Guerrier *et al.*, 1983). There appears to be a general consensus, however, that protein synthesis or phosphorylation is involved in meiotic events occurring after germinal vesicle breakdown (Clarke and Masui, 1983).

With amphibian oocytes, inhibitors are typically added at or near the time of hormone or culture initiation. If similar studies are performed using starfish and mammalian oocytes, GVBD is not prevented, and it is generally assumed that protein synthesis is not involved in GVBD. Failure of protein synthesis inhibitors to prevent GVBD may, however, be dependent on the time of inhibitor addition. In rat follicle-enclosed GV–stage oocytes preincubated in inhibitors for several hours before release of oocytes into culture medium, GVBD was found to be reversibly inhibited (Ekholm and Magnusson, 1979). These workers concluded that the resumption of meiosis is dependent on proteins with a high turnover rate that exist in the oocyte before meiosis reinitiation. Other studies have demonstrated that mammalian somatic cells at certain stages of the cell cycle contain MPF-like activity as assessed when their cytoplasm is injected into amphibian oocytes (Sunkara *et al.*, 1979). Likewise, amphibian somatic cells also contain MPF that is active in the oocyte (Wasserman and Smith, 1978). It seems likely, therefore, that mammalian oocytes contain MPF (Balakier, 1978). Thus, such cytoplasmic factors appear to be relatively nonspecies specific, have been conserved unchanged through the evolutionary process, and function during mitotic and meiotic cell-division cycles. These ideas are consistent with the remarkable similarity in the meiotic process in most organisms.

The apparent lack of protein synthesis before GVBD in mammalian oocytes may be due to the specific experimental conditions; i.e., the fact that GVBD occurs relatively rapidly after oocytes are released into culture media and/or that insufficient time is allowed for the inhibitor to enter the oocytes. Alternatively, sufficient MPF already may have been synthesized and possibly stored in cytoplasmic organelles before the release of oocytes into the culture media. It is particularly relevant that amphibian oocytes, which exhibit steroid-induced protein synthesis and in which protein synthesis inhibitor treatment has been delayed, resemble mammalian oocytes undergoing spontaneous GVBD.

An additional possibility is that oocytes may be exposed to sufficient maturation-inducing hormones at some time before their isolation and *in vitro* culture. Support for such a hypothesis comes from a number of studies. Sampson and Schuetz (1979) exposed *Rana* oocytes to both progesterone and cycloheximide in an attempt to prevent GVBD. These oocytes, after simple washing and placement in fresh medium, proceeded to mature spontaneously in the absence of added progesterone and cycloheximide. Reversible inhibition of the progesterone-induced cytoplasmic calcium surge by cycloheximide and propionate has also been reported in urodele oocytes (Vilain *et al.*, 1980). These results clearly suggest that the oocytes are imprinted in some way to reinitiate the maturation process. Other studies using starfish and amphibian oocytes demonstrate that progesterone and 1-MA have cumulative subthreshold effects on the process of oocyte maturation. Thus, stable intermediate states could be induced such that less hormone is required to trigger maturation in oocytes previously exposed to hormone than in nonhormone-exposed oocytes (Nemoto, 1981; Marot *et al.*, 1977). There are two important questions: How long do such memory or intermediate states persist under *in vivo* conditions? What are the underlying biochemical mechanisms? The evidence suggests that these intermediate states vary and are short lived depending on the source of oocytes. Intracellular calcium release and the rate of hormone transfer and metabolism (inactivation) may all be involved in estabishing the duration of such memory.

Significantly, maturational changes seen in various species after GVBD are dependent to varying degrees on the mixing of nuclear and cytoplasmic materials. In starfish oocytes (Guerrier *et al.*, 1983), mechanical rupture of the germinal vesicle is sufficient to complete many but not all aspects of oocyte maturation, whereas in amphibian oocytes physically tearing the nuclear envelope leads to little or no detectable cytoplasmic maturation. In starfish oocytes, it has not been established whether MPF is formed during the process of nuclear envelope rupture, possibly as the result of increased intracellular calcium ions during oocyte manipulation. Coordinate changes in both the nucleus and cytoplasm thus appear to be required for the oocyte to mature after GVBD. Experiments involving enucleation and cytoplasmic transfer to germinal vesicle-stage oocytes have provided particularly critical information as to the hormonal site of action and the relative roles of the nucleus and cytoplasm in both oocyte and sperm differentiation after fertilization. On the basis of such studies, it has been possible to define germinal vesicle-dependent and -independent cytoplasmic maturation (Fig. 18).

Significantly, protein synthesis inhibitors have dichotomous effects (stimulatory or inhibitory) on the oocyte meiotic maturation process, depending on the meiotic stage at the time of treatment. In mice, inhibitor treatment of oviductal metaphase II oocytes stimulates activation and considerable parthenogenetic development (Siracusa *et al.*, 1978; Clarke and Masui, 1983). In amphibians, Schuetz and Samson (1979) observed activational changes in the cortex of progesterone-matured oocytes when cycloheximide treatment was delayed so as not to interfere with GVBD. Such data suggest that cytoplasmic proteins produced during hormone-induced maturation act in two ways: (1) to mediate cellular differentiation, or (2) to restrain or prevent cellular differentiation. Interestingly, it has been demonstrated that a number of germinal vesicle proteins shed into the cytoplasm during maturation reaccumulate in embryonic nuclei at characteristic stages of development (Dreyer and Hausen, 1983).

Cyclic nucleotides (cAMP, cGMP) also appear to play an important role in controlling GVBD in oocytes from many species. However, considerable confusion exists concerning their cellular source (follicle cells, oocytes), mode of action, and the relative importance of increasing or decreasing lev-

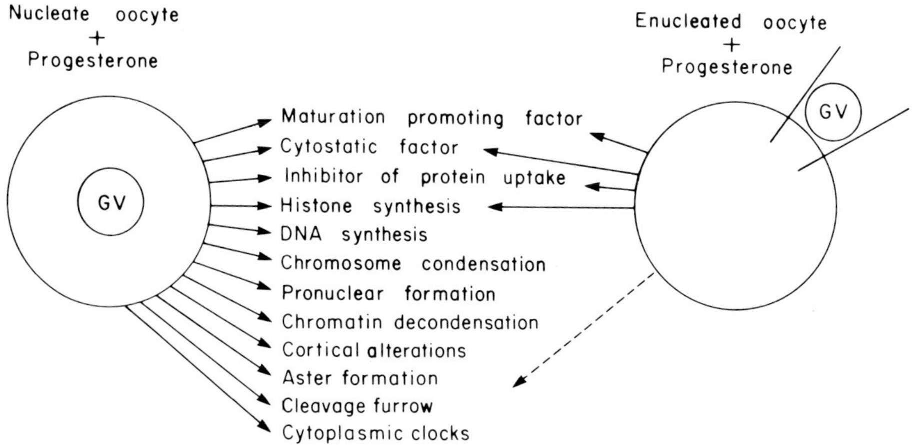


**Figure 18.** Nuclear–cytoplasmic interactions in oocyte maturation. This simplified diagram depicts various types of biological activities (center) detected in the cytoplasm of intact (left side) or enucleated (right side) oocytes after exposure to progesterone. The various biological activities can be characterized as being GV dependent (solid arrows on left only) or as independent (solid arrows on both sides). Incomplete or partial nuclear dependency is indicated by a broken arrow. (From Schuetz, 1979).

els of these nucleotides in preventing or initiating GVBD and other aspects of follicular differentiation. These topics and the interaction of steroids and cyclic nucleotides are discussed in greater detail in Chapter 6 and are reviewed by Maller, 1983.

## 9. Coordination of Ovulation and Oocyte Maturation

Ovulation is defined as the process by which oocytes are expelled from the follicular and/or ovarian environment and serves as a mechanism for transferring gametes to new environments where further differentiation and/or fertilization take place. In most species, ovulatory and oocyte maturation events are closely synchronized in the ovarian follicle by an appropriate gonadotropic hormone stimulus. How this synchronization is brought about at the local level, however, is poorly understood—in large part because ovulatory events have been studied independently of oocyte changes. Extensive studies in a wide range of species clearly demonstrate that complex processes that involve gonadotropins, a variety of ovarian hormones, enzymes, and smooth muscle cells interacting locally within the somatic elements of the follicle occur in preparation for ovulation. Excellent reviews of various aspects of these follicular proccesses are available and should be consulted for further details (e.g., Hisaw, 1947; Espey, 1980; Strickland and Beers, 1979; DiZerega and Hodgen, 1981). Studies of the control of oocyte maturation, however, have provided new insights as to how these two processes are integrated as well as evidence for some common cellular and molecular mechanisms linking ovulation and maturation—at least in starfish and amphibians.

In starfish, induction of ovulation and of oocyte maturation is mediated by a single hormone, 1-MA. In amphibians, direct involvement of particular steroid hormones in ovulation was suggested from *in vitro* studies even before oocyte maturation was investigated. Examination of data obtained from ovulation and maturation experiments demonstrates striking structural similarities in the steroids that induce both events.

As a result of locally produced or exogenously added maturation-inducing steroids (progesterone), a structural reorganization of ovarian follicle somatic and germinal elements appears to be a particularly crucial aspect in preparing the amphibian follicle and oocyte for ovulation (Schuetz, 1974) (see Fig. 19). In the unstimulated ovarian follicle, follicle cells are tightly attached to the oocyte by a large number of long macrovilli that extend through the vitelline envelope. Ovulated oocytes are typically devoid of all follicle cells; thus, it is apparent that physical separation of these two cell types precedes normal ovulation. Progesterone treatment ruptures these intercellular connections, at least partially through the action of cytoplasmic MPF produced in the oocyte itself. In addition to such changes involving the entire surface interface between the oocyte and fol-

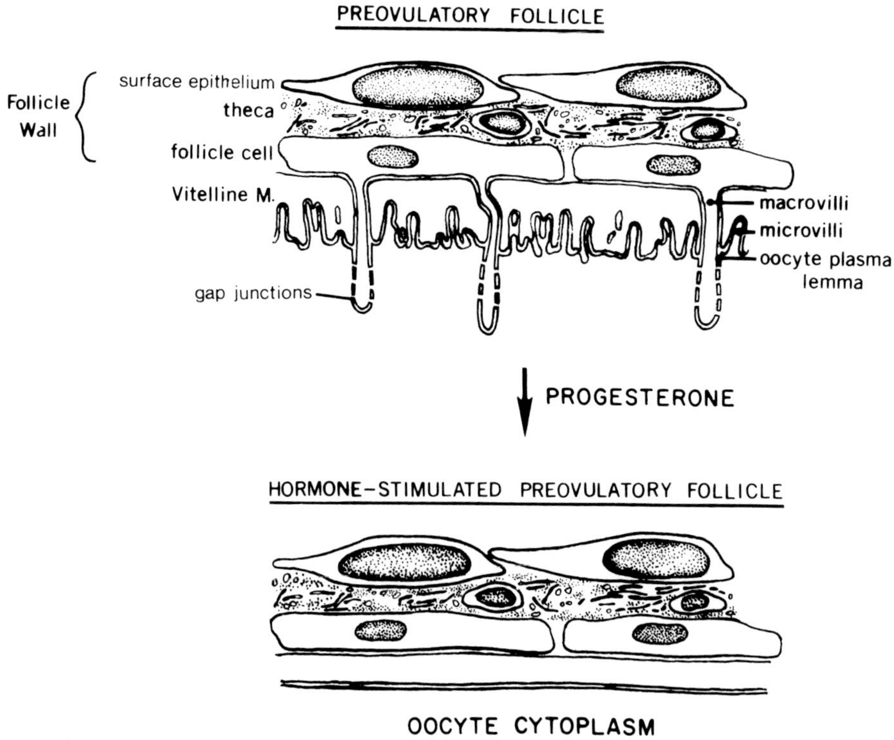


**Figure 19.** Diagram illustrating the multifaceted morphogenetic effects of hormones on the association between the somatic cells and oocyte within the amphibian ovarian follicle. Pituitary gonadotropic hormones are thought to stimulate the local intrafollicular production of progesterone, which in turn acts to interrupt—across the vitelline envelope—the macrovillus and gap junction connections between follicle cells and the oocyte cytoplasm. Progesterone also interrupts the physical interdigitation of the oocyte microvilli with the vitelline envelope and markedly alters the transport properties of the plasmalemma. The result of these transformations is that the vitelline envelope becomes physically detached from the oocyte and follicle wall in preparation for ovulation and fertilization.

licle wall, local structural and biochemical changes of a different nature occur at the predetermined site of follicle rupture. It was concluded that lytic substances emanating from the surface epithelium covering the preovulatory follicles have an important role in effecting local digestion of the wall of the rabbit follicle (Bjersing and Cajander, 1974; Bjersing, 1982).

In mammals, a progressive dissociation of cumulus cells from each other and the oocyte occurs in response to gonadotropin stimulation during the periovulatory period (Gilula *et al.*, 1978). Cumulus cells also secrete hyaluronic acid, mucify, remain associated with the oocyte through the ovulatory process, and are attached to the oocyte in the oviduct for a short period. In studying the process of cumulus cell mucification, N. Dekel and Phillips (1980) concluded that cAMP rather than steroids or prostaglandin induces this process. Thus, it appears that gonadotropic or maturation hormones have multiple and sequential effects on somatic and germinal elements of the follicle. Comparison of amphibian and starfish data also clearly demonstrates that different hormones mediate many of the same events in different species. Furthermore, with increasingly complex follicular structure, the evidence suggests that more intricate cellular and molecular interactions are involved in follicular differentiation.

In amphibians (Schuetz, 1971a), it has been demonstrated that a high incidence of steroid-induced maturation and structural reorganization of the follicle is associated with a low incidence of ovulation in the same follicle. A similar situation was observed in fish (Iwamatsu, 1978). By contrast, in pituitary hormone-treated follicles, the incidence of ovulation is typically much higher than that induced by the steroids alone. As these studies were carried out in sister follicles from individual animals, the results indicate that gonadotropic effects on ovulation cannot be explained entirely by the secretion of progesterone or other maturation-inducing hormones. Achievement of ovulation thus appears to require additional factors; considerable evidence suggests that prostaglandins might serve this function.

In a comprehensive review of ovulation in mammals, Espey (1980) concluded that ovulation involves a series of cellular and hormonal interactions that resemble a localized inflammatory response requiring both steroidal hormones and prostaglandins (Fig. 20). Attempts at sorting out these various interrelationships have been largely based on indirect observations and have been hindered by the lack of models that permit evaluation of what different cellular elements contribute to the follicular processes and in what way. This problem can be studied partially by determining the secretory activity of isolated follicular components as previously described for chickens. However, this approach does not permit investigation of the functional role of these cells or of their secretions on the oocyte.

In *Rana*, however, it is possible to remove selectively different layers of the follicle wall and subsequently assess and relate biological responsiveness (ovulation, oocyte maturation) and biochemical responsiveness (steroid production) of these follicular components to various hormonal stimuli (Schuetz and Lessman, 1982; Lin and Schuetz, 1984) (see Table I). Using this approach, it was demonstrated that neither ovulation nor follicle contractions occurs in the absence of the follicular epithelium and that progesterone production is dependent on the presence of follicle cells. Thus, the ovulatory, steroidogenic, and maturation responses are all dependent on the follicle wall, and there appears to be a segregation of function(s) among the different cellular elements of the follicle. In recent years, methods for perfusing mammalian ovaries have been developed such that ovulation can be induced *in vitro* (Janson *et al.*, 1982; Kobayaski *et al.*, 1984). *In vitro* ovulation from cultured mammalian follicles has also been reported (Talbot and Chacon, 1982). These approaches should provide important information on synchronization of follicular and oocyte differentiation.

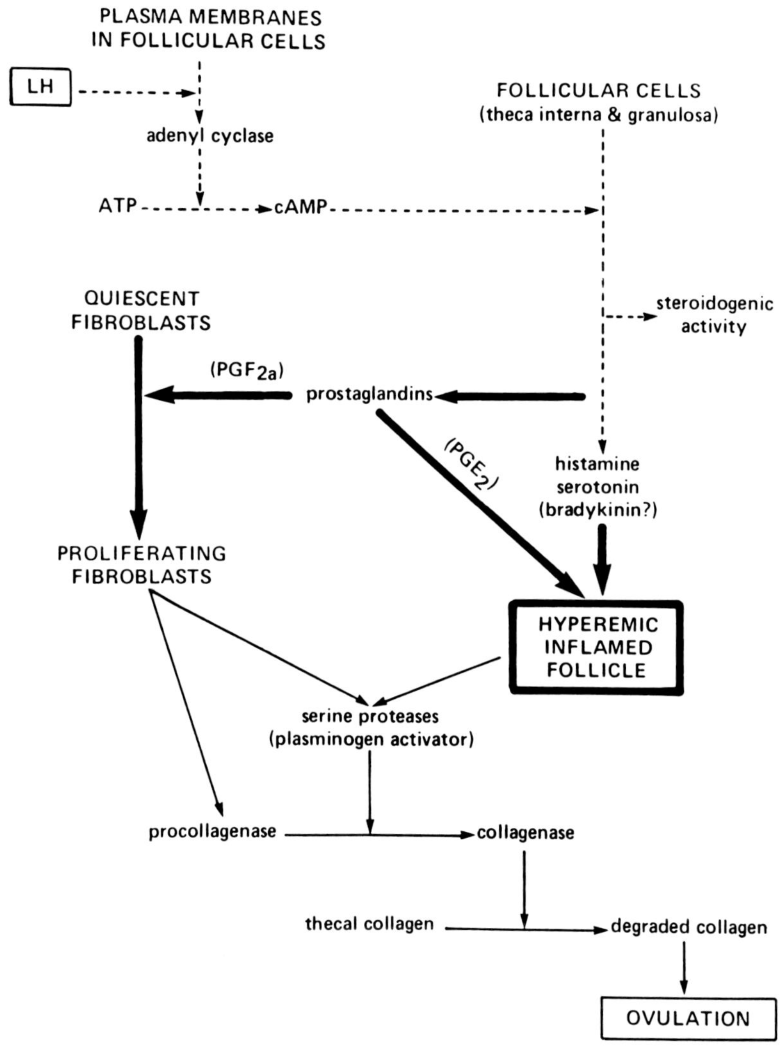


**Figure 20.** Simple model of the ovulatory process. (- - - - -) Reactions taking place during the first hours of the ovulatory process. (———) Hypothetical reactions during the intermediate stages of the process. (———) Final reactions, which culminate in ovulation. (From Espey, 1980.)

Under normal conditions in most species, oocyte maturation and ovulation occur at predictable times and in a particular sequence after gonadotropic stimulation. Natural or experimental dissociation of oocyte maturation and ovulation occurs and provides interesting clues concerning the local intrafollicular cellular and molecular mechanisms regulating these events. In many species, including most mammals, meiosis reinitiation

**Table I.** Experimentally Separable Follicle Components[a]

| Definition | Surface epithelium | Theca | Follicle cells | Vitelline envelope | Oocyte |
|---|---|---|---|---|---|
| | Follicular constituents present | | | | |
| Intact follicle | + | + | + | + | + |
| Theca intact follicle | | + | + | + | + |
| Defolliculated oocyte | | | + | + | + |
| Denuded oocyte | | | | + | + |
| Demembranated oocyte | | | | | + |

[a]Various types of follicular components that can be isolated or collected after microdissection and/or experimental treatment (enzymatic, calcium-free medium washes) of amphibian ovarian follicles.

(GVBD) precedes ovulation. Interestingly, in a few mammalian species (e.g., dog, fox), oocytes are normally ovulated in the germinal vesicle stage.

On the basis of the state of differentiation of the somatic follicle cells and of the oocyte in ovulated or unovulated oocytes, we can define different types of relationships (Fig. 21). In pigs, precocious induction of ovulation by exogenous hCG injections leads to the expulsion of germinal vesicle-stage oocytes (Hunter *et al.*, 1979) surrounded by undifferentiated cumulus cells. Cumulus cell mucification does not appear to be required for ovulation. Inhibition of gonadotropin-induced oocyte maturation (GVBD and meiotic progression) but not of ovulation has been reported (Tyler *et al.*, 1981). Likewise, during the normal estrous or reproductive cycle of many mammals, more follicles develop than are normally ovulated. Such nonovulated oocytes exhibit signs of having undergone reinitiation of meiosis, normal cumulus cell transformations, and eventually atresia. The number of such entrapped transformed oocyte complexes appears to be related to the amount and type of gonadotropic hormone used to stimulate follicular growth and maturation. See Donahue (1972) for a review of earlier mammalian literature concerning coordination of ovulation and maturation.

In starfish, many immature oocytes still surrounded by follicle cells may be ovulated. Ovulation and maturation, in the strictest sense, thus occur for these oocytes after the follicles are expelled from the ovary. In other instances, follicle breakdown and GVBD occur within the ovary.

Strikingly, it was recently demonstrated that ovulation and oocyte maturation are induced by hypophysectomized rats injected with a gonadotropin-releasing hormone analogue (GnRHa) (Ekholm *et al.*, 1982). The ovulatory requirement for gonadotropins can be circumvented with such treatment. Furthermore, both GnRHa and luteinizing hormone (LH) stimulate a marked increase in ovarian levels of prostaglandin (PGE). Pretreatment of such animals with prostaglandin synthetase inhibitor (i.e., indomethacin) inhibits PGE accumulation and ovulation but not oocyte GVBD.

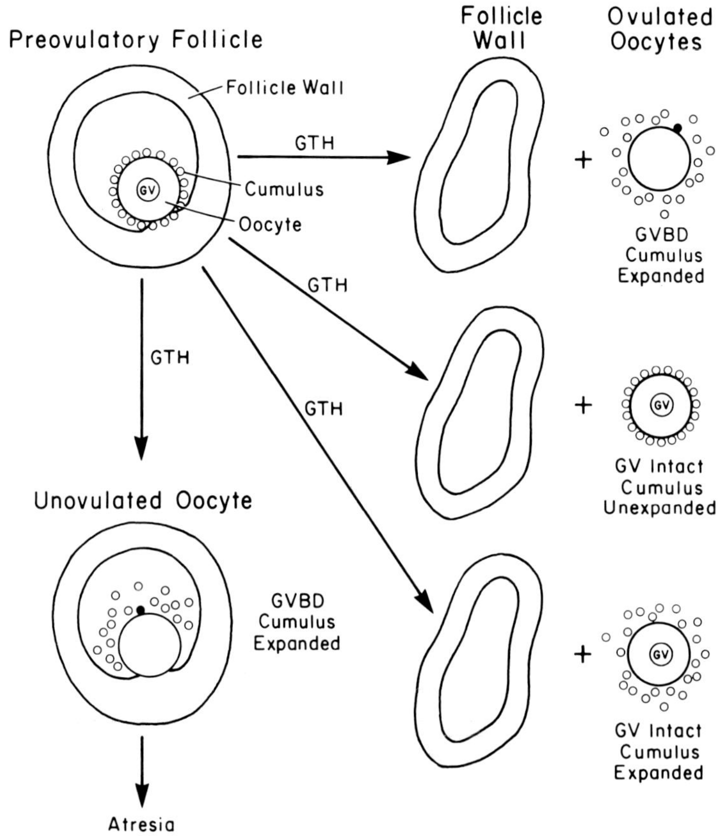


**Figure 21.** Diagram showing various types of ovarian (unovulated) and oviductal (ovulated) cumulus–oocyte complexes observed after gonadotropic hormone (GTH) stimulation of preovulatory ovarian follicles. Primarily based on data obtained in mammals. Results show that either cumulus cell or oocyte transformations can occur pre- or post-ovulation.

These results strongly suggest that GnRHa- and LH-induced ovulation—but not maturation—are mediated by prostaglandins. Since the induction of oocyte maturation and cumulus expansion by GnRHa can be induced in cultured ovarian follicles (Hillensjo and LeMaire, 1980), the action of the GnRHa appears to be directly on the ovarian follicle and may involve steroidogenesis. It is therefore of interest that GnRHa has been reported to stimulate, although less effectively than LH, progesterone production by cultured rat granulosa cells (Clark, 1982). Others (Jones and Hsueh, 1981) report that GnRHa inhibits progesterone production in cultured rat granulosa cells by stimulating the production of 20$\alpha$-OH progesterone. More recently, it was shown that ovulation of immature germinal vesicle-stage amphibian oocytes can be induced by appropriate treatment of follicles with certain prostaglandins (Schuetz, unpublished data). Earlier studies in fish (Jalabert and Szollosi, 1975) demonstrated that prostaglandins stimulate ovulation in cultured fish ovarian follicles in which maturation was induced. These investigators concluded that the effects of the prostaglandins are probably due to the induction of contractions in theca layer smooth muscle cells. In addition, contractions of smooth muscle cells at the base of the mammalian ovarian follicle have been directly implicated in the ovulatory process (Martin and Talbot, 1981a,b; Talbot and Chacon, 1982). To date, there is no evidence that prostaglandins are involved in ovulation of starfish oocytes, although contractile elements have been described in follicle cells. Thus, it appears in many species, that—whereas prostaglandins alone or in combination with maturation-inducing hormones have little or no direct effect on the oocyte maturation process—they have critical roles in preparing the follicle for ovulation and in mediating part of the ovulatory process. The relative amounts of locally produced hormones, which act on the different components of the follicle simultaneously or in a particular sequence, appear to determine the stage of oocyte and cumulus differentiation at the time of ovulation. Complex alterations in steroid and prostaglandin metabolic pathways appear to be closely correlated with dissociation of the ovulatory and oocyte maturation processes.

## 10. Postovarian Oocyte Maturation and Fertilization Mechanisms

Although oocyte growth and meiotic reinitiation occur within the ovarian follicle, many significant aspects of oocyte maturation continue or take place after ovulation. In fish, amphibians, and mammals, oocytes characteristically exist for a period in the coelomic cavity before being transported through the oviduct to the uterus or to the external environment. Oocyte–somatic cell interactions occur during this time. In starfish and other marine organisms in which the oocytes or follicles are extruded directly into seawater, there is little or no contribution by any reproductive tissues to the developing oocyte. Even in this circumstance, however, seawater has a dramatic influence on follicular or oocyte processes, as evidenced by the incidence of spontaneous maturation that occurs after liberation from the ovary of follicles with germinal vesicle-stage oocytes.

A clear functional relationship between oviductal secretions and the oocyte maturation process in amphibians was recently established. Typically, it is difficult or impossible to fertilize body cavity oocytes using standard procedures of *in vitro* fertilization. Full fertilizability appears to be attained after exposure of coelomic oocytes to oviductal secretions, which either condition the vitelline envelope or form the various jelly layers thought to be required for normal fertilization (Fig. 22). Vitelline coat conditioning is apparently mediated by oviductal secretions *(pars recta)*; this

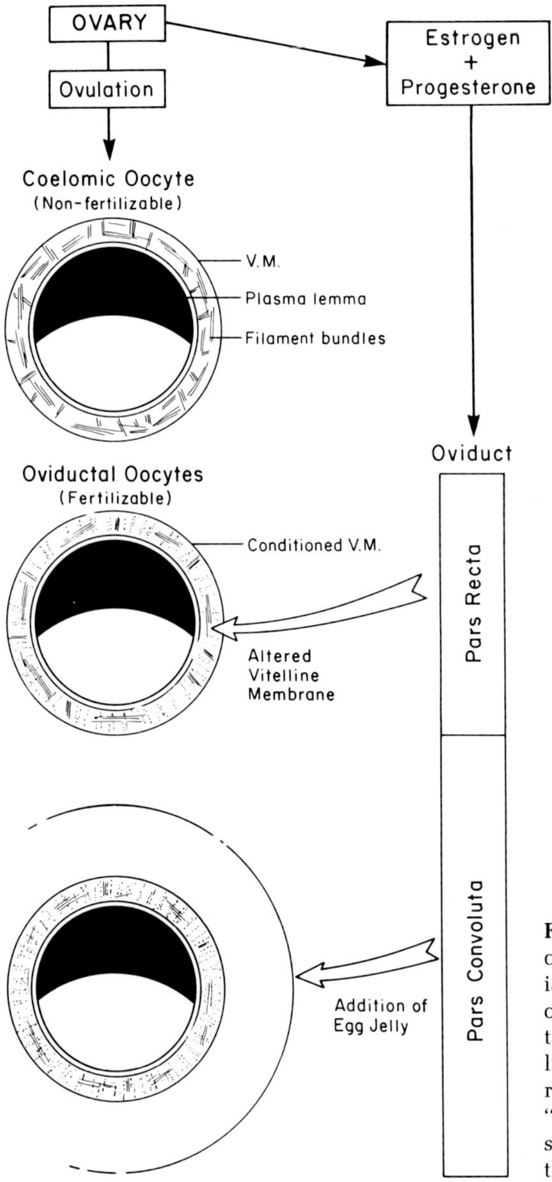


**Figure 22.** Diagram illustrating changes in fertilizability of amphibian oocytes at various stages after ovulation and the role of the oviductal secretions in attainment of fertilizability. Secretions from the pars recta region of the oviduct alter (or "condition") the vitelline envelope such that it is more readily penetrated by sperm.

process results in a softening and structural reorganization of the membrane. Such changes appear to enhance the effectiveness of sperm acrosomal enzymes and the ability of sperm to penetrate the vitelline coat (Yoshizaki and Katagiri, 1981; Barbieri, 1982; Grey *et al.*, 1977). Hardening of the zona pellucida occurs in mammalian ovarian or oviductal oocytes is mediated in part independent of the cortical reaction by an ovoperoxidasecatalyzed oxidation (Schmell and Gulyas, 1980; DeFelici and Siracusa, 1982). Zonal hardening may influence the fertilization process.

The mammalian oviduct is also a secretory organ; however, it remains to be determined whether these secretions play any significant role in fertilization or development of the oocyte (Greenwald, 1979).

In addition to the normal events of fertilization and early development that occur in oviductal oocytes, germ cells during this period are particularly fragile and susceptible to a variety of influences that can drastically alter the gamete's ability to be fertilized, or to develop following parthenogenetic activation (Fig. 23). Extensive clinical, genetic, and experimental data have established that a large proportion of developmental abnormalities, impaired fertility, and fetal and embryonic wastage in mammals can be linked to oviductal events that affect female gametes embedded in cumulus cells. As a result of meiotic errors, aberrations of chromosomal number constitute a considerable proportion of genetic abnormalities and

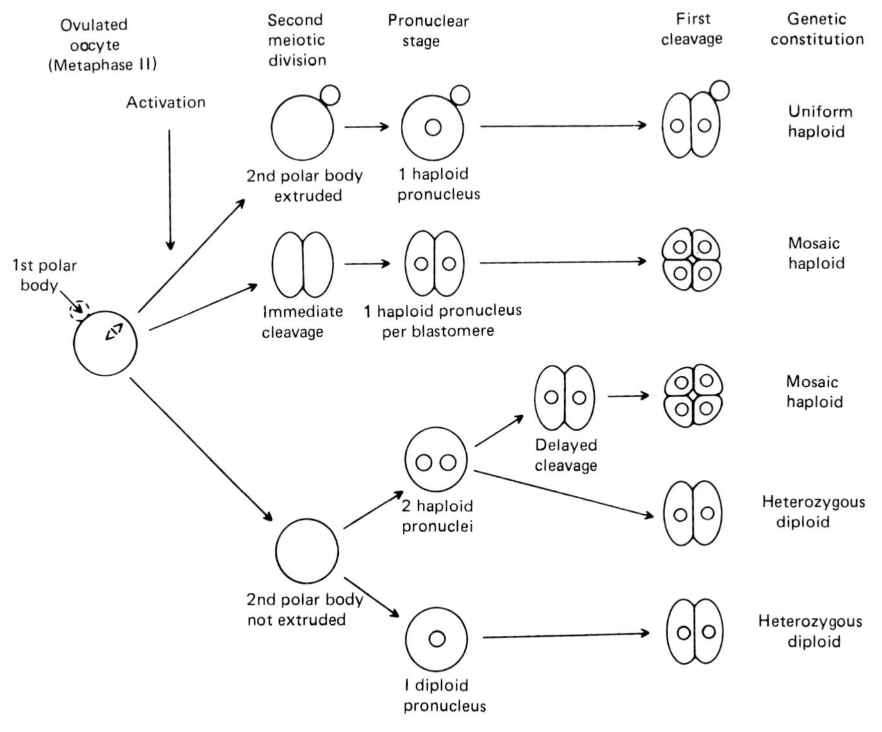


**Figure 23.** Classification of parthenogenones, their possible routes of development, and their genetic constitution after activation of unfertilized mouse eggs. Normal meiotic division gives rise to polar body extrusion, leaving one haploid pronucleus. When second polar body extrusion is suppressed, eggs either undergo immediate cleavage or develop two haploid pronuclei or one diploid pronucleus. (From Kaufman and Sachs, 1976.)

form the basis of triploidy, mosaics, monosomy, and trisomy. The latter can produce defects such as Down's, Klinefelter's, and Turner's syndromes. In humans, it is estimated that in 50% of all spontaneous abortions, abnormal karyotypes are present, of which approximately 50% are trisomic (Turpin and Lejune, 1969; Donahue, 1974; Golbus, 1981).

Mammalian and amphibian oocytes rapidly lose (by aging or overripening) their capacity to form normal embryos after ovulation and maturation. Under certain conditions, this decreased competence can be linked to events occurring in the ovarian follicles (Austin, 1970; Adams, 1970; Mikamo and Hamaguchi, 1975; Talbert, 1978; Peluso, 1976; Butcher, 1975). Convincing evidence for such degenerative changes comes from experiments in which mating or artificial insemination are increasingly delayed beyond ovulation. Such manipulation results in a progressive decrease in the number of offspring and an increase in the incidence of abnormal embryos and fetuses. Many of these abnormalities have been linked to meiotic and/or fertilization defects such as multiple sperm entry, suppression of second polar body, and disintegration of the chromosomal complement. Thus, the block to polyspermy and the functions of the second meiotic spindle appear to be altered by aging changes in oocytes. Cytological and ultrastructural studies have confirmed that morphological changes in the oocyte cortex cytoplasm and nucleus are also associated with the aging process (Longo, 1974; Szollosi, 1975). Interestingly, it has been hypothesized that an interaction between the rate of hormonally controlled meiosis and the timing of chiasmata terminalization could explain why the incidence of Down's syndrome increases with maternal age (Crowley *et al.*, 1979).

Until recently, it was thought that rat oviductal oocytes remain in second meiotic arrest during *in vitro* culture and fertilization experiments (Toyoda and Chang, 1974). However, studies from two laboratories demonstrate that a very high incidence of spontaneous parthenogenetic activation (second polar body formation) occurs when such oocytes are cultured in various media (Zeilmaker and Verhamme, 1974; Keefer and Schuetz, 1982). That this phenomenon has not been documented extensively may be partly due to the fact that the first polar body of rat oocytes disintegrates soon after its formation, making identification of the true state of meiosis difficult (Table II).

Cellular and molecular mechanisms underlying spontaneous or induced parthenogenetic development are unknown but are relevant to understanding oviductal functions and the basic mechanisms of fertilization (Kaufman, 1981, 1983). It is now recognized that a range of stimuli applied to oviductal oocytes *in vivo* and *in vitro* leads to parthenogenetic activation (Graham, 1974). Depending upon the type of treatment, activated oocytes proceed to various preimplantation or even postimplantation stages of development (Kaufman *et al.*, 1977). Thus, there appear to be gradations in the type of parthenogenetic activation that will occur *in vivo*.

**Table II.** Proportion of Oocytes (%) Remaining in Metaphase II Both *In Vitro*[a] and *In Vivo*[b]

| | Number of animals | Number of oocytes | Proportion in metaphase II | Proportion with second polar body |
|---|---|---|---|---|
| *In vivo* oocytes (15–17 hr post-hCG) | 14 | 318 | 95.1 ± 4.3 | 1.9 ± 1.7 |
| *In vitro*-cultured oocytes | 42 | 1297 | 1.3 ± 1 | 94.2 ± 3 |
| *In vivo* oocytes (21–22 hr post-HCG) | 10 | 189 | 91.4 ± 6 | 7 ± 6.5 |

[a]In Toyoda and Chang's medium at 37°C in 5% $CO_2$ plus air for 4–5 hr.
[b]From Keefer and Schuetz (1982).

Activating stimuli, which interrupt second meiotic metaphase arrest, include anesthetics, i.e., ether, nitrous oxide, and avertin. Under *in vitro* conditions, certain enzymes (hyaluronidase, pronase), ethanol, and benzyl alcohol are capable of activating oviductal mouse oocytes as are protein synthesis inhibitors such as cycloheximide and puromycin (Siracusa *et al.*, 1978). Heat or cold shock are also effective stimulators of activation *in vivo* or *in vitro*. Typically, these activating agents are most effective on oviductal oocytes at increasing periods after ovulation (i.e., aging eggs).

How all these changes are brought about is unknown. However, it is generally assumed that any or all of these stimuli or other environmental agents may produce alterations directly on the oocyte or indirectly through the local biochemical microenvironment within the oviduct such that normal development of the egg and embryo is altered. Keefer and Schuetz (1982) concluded that rapidly changing oviductal conditions after sacrifice of animals play an important role in determining whether parthenogenesis will occur *in vitro*. Interestingly, elevated levels of potassium exist in the ampulla fluid of mouse oviducts and are apparently secreted from the oviductal epithelium (Borland *et al.*, 1977). In starfish, increased polyspermic fertilization of aging oocytes has been linked to a deterioration in the membrane activation potential (Miyazaki and Hirai, 1979). In amphibians, Smith and Ecker (1970) concluded that the oviduct environment prevents pseudogastrulation from occurring in mature oocytes.

In most species, considerable evidence suggests that parthenogenesis-induced activation of oocytes is partially mediated through an increase in cytoplasmic levels of free calcium (Steinhardt *et al.*, 1977; Cuthbertson *et al.*, 1981; Whittingham, 1980). This occurs, as does sperm-induced activation, either by altering the cell membrane to allow entry of external calcium or by release of endogenous stores of bound calcium (Ridgeway *et al.*, 1977; Jaffee, 1983). In amphibian oocytes, calcium ions rapidly inactivate the cytoplasmic cytostatic factor that is thought to maintain the chromosomes in metaphase II arrest (Meyerhof and Masui, 1977).

In view of the multiplicity of events (nuclear and cytoplasmic maturation, meiosis, fertilization, activation, and cleavage) occurring during the oviductal period, one would anticipate that the chemical and physiological state of the oviductal microenvironment would play a key role in affecting the gametes and be well understood. However, the nature and physiology of the oviductal environment during this critical time are ill defined and await further detailed analysis and experimentation.

The stage of oocyte maturation at the time of ovulation and fertilization has a profound effect on the nature and course of sperm-egg interactions and subsequent developmental events and provides insight into the maturation process itself. As previously discussed, the maturational state of ovulated oocytes is closely linked to the local hormone–cellular interactions that synchronize the two major follicular processes involving oocytes: ovulation and maturation. In order to understand these events, it is appropriate to consider cytoplasmic and nuclear maturation as partially independent and interdependent components of the overall process of oocyte maturation.

In the past, the term cytoplasmic maturation, as contrasted to nuclear maturation was generally thought to occur after GVBD and has been used to describe the oocyte's ability to interact with sperm and give rise to a zygote. However, it is increasingly apparent that sperm–egg interaction and zygote formation are readily separable events; that is, fertilization can occur before classical cytoplasmic maturation (Schuetz, 1975). Thus, it is necessary to consider these as two different processes. In fact, the questions of how and when oocytes acquire the ability to interact with the sperm have not been critically addressed in many species.

## 11. Oocyte Maturation and the Block to Polyspermy

Initiation of normal development is dependent on the fusion of a single male pronucleus with the female pronucleus. Entry of excess sperm into the oocyte cytoplasm typically results in abnormal and usually limited cleavage and development. Restricting the number of sperm that enter the oocyte thus is an important component of fertilization, and a variety of structural and functional mechanisms (blocks to polyspermy) exist within the gamete to mediate this reaction (Dale and Monroy, 1981).

The best studied and most generally recognized mechanism for excluding excess sperm involves the discharge of cortical granule contents from the oocyte cytoplasm followed by the structural and functional transformation of the vitelline envelope (fertilization envelope formation) that prevent additional sperm from reaching the oocyte (Fig. 24). This chemically mediated "slow" block to polyspermy is normally triggered by the initial sperm that enters the oocyte and can be replicated by an appropriate

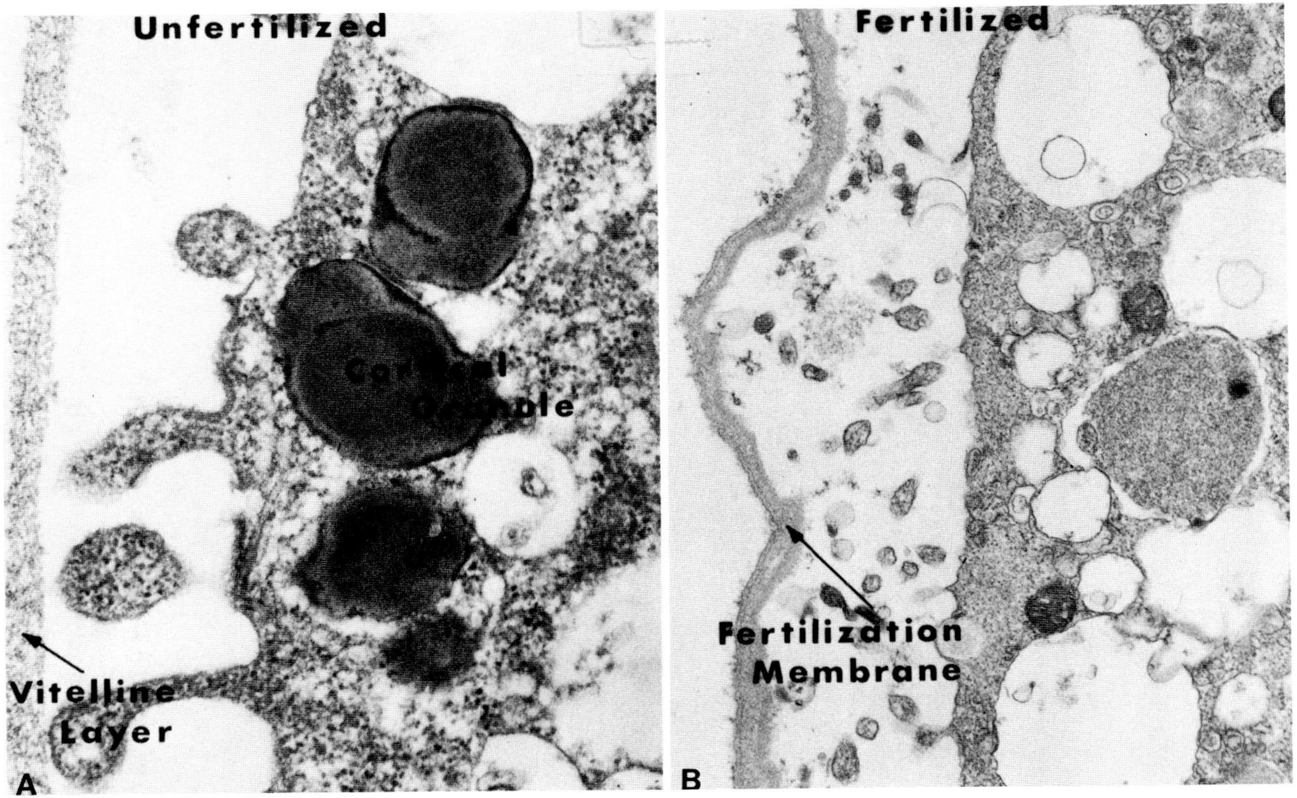


**Figure 24.** Morphological characteristics of the cortex and vitelline envelope before (A) and after (B) fertilization of starfish oocytes. After cortical granule exocytosis, the perivitelline space becomes filled with membranous structures, and the vitelline layer becomes elevated and transformed into the fertilization envelope and the cortical granule material. Note also the changes in the villus projections from the oocyte surface. (From A. W. Schuetz and F. Longo, unpublished observations.)

activation stimulus such as pricking with a glass needle or calcium ionophore (A23187) treatment. Local release of intracellular calcium appears to be an essential requirement of the activation stimulus that leads to propagation of the activation wave around the periphery of the oocyte cortex and to progressive exocytotic discharge of cortical granules (consult reviews by Moscona and Monroy, 1978; Schuel, 1978; and Yanagimachi, 1981).

An electrically mediated fast cortical block to polyspermy may also play a role in controlling the incidence of sperm penetration—at least in some species, including sea urchin and starfish (Jaffe, 1976; Miyazaki, 1979; Miyazaki and Hirai, 1979; Dale and Monroy, 1981). The existence of such a block is questioned by some investigators. Strikingly, recent evidence demonstrates that the final stages of the block to polyspermy develop late in the differentiation of some oocytes. This process is linked to meiosis reinitiation (GVBD) and the stimulus (hormonal or nonhormonal) that controls this event.

The absence (or incomplete development) of the polyspermy block is readily demonstrated by polyspermic fertilization of precociously ovulated (GV stage) oocytes in pigs Hunter *et al.*, 1979) and germinal vesicle-stage ovarian oocytes of starfish (Schuetz, 1975; Schuetz and Longo, 1981), sea urchins (Franklin, 1965), and amphibians (Katagiri, 1974; Elinson, 1977). Binor and Wolfe (1979) found little evidence of sperm penetration in zona-free germinal-vesicle stage mouse oocytes, although there was massive binding of spermatozoa to the plasmalemma. Interestingly, after GVBD, polyspermic fertilization was obtained in such oocytes.

A critical question therefore is why do GV stage or incompletely mature oocytes exhibit polyspermic fertilization or how does the maturation process make the polyspermy block fully operational? Our understanding of the mechanisms involved in these transformations is poorly delineated, although it is clear that such changes can occur rapidly in response to hormone treatment. In starfish, monospermic fertilization occurs during the interval between GVBD and first polar body formation, whereas in mammals and amphibians fertilization typically takes place after oocytes arrest at the second meiotic metaphase. Addition of 1-MA to immature (ovarian) starfish oocytes rapidly transforms ($\sim$30 min) the cell such that polyspermic fertilization essentially terminates with GVBD. Structural changes occurring in the oocyte cortex during maturation are thought to be of major importance in establishing a functional block to polyspermy (Vacquier, 1981). Although they are involved in the block to polyspermy, cortical granules as well as the zona pellucida are characteristically formed in the ovary before this time period. Thus, it is important to distinguish between structural and physiological mechanisms involved in induction of the cortical reaction.

In immature GV-stage amphibian oocytes, microvilli from the oocyte cell membrane penetrate and/or attach to the vitelline envelope. These interconnections typically break down after exposure of oocytes to progesterone (Schuetz, 1972b); unless these structural changes occur, the tight physical attachment between the oocyte and vitelline envelope persists and probably prevents elevation of the fertilization envelope, formation of the perivitelline space, and rotation of the oocyte after an activation stimulus. Limited ultrastructural evidence in mammals suggests that the migration of cortical granules beneath the plasmalemma occurs during final stages of oocyte maturation.

An important aspect of the cortical reaction is that the response is propagated from the site of sperm entry in mature oocytes but not in immature oocytes. The onset of the activation response during steroid-induced maturation was examined in *Rana* oocytes and was found to vary depending on the type of activating stimulus used (Fig. 25). Treatment of maturing oocytes (after but not before GVBD) with calcium ionophore (A23187) induced typical activational responses, including fertilization membrane elevation, cortical granule breakdown, as well as a dramatic capping reaction, during which the melanin granules became concentrated into a small nipple in the animal pole region. (Schroeder and Strickland, 1974). By contrast, in oocytes pricked with a glass needle, a typical activation response

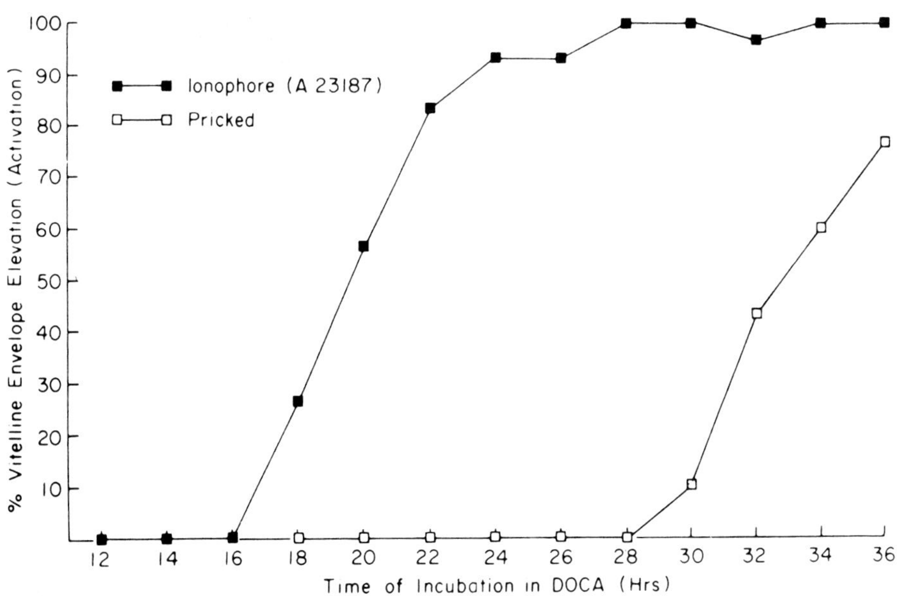


**Figure 25.** Timing of activation of *Rana* oocytes in response to ionophore and to pricking after *in vitro* steroid induction of maturation (1 $\mu$g/ml). Oocytes were treated wtih ionophore A23187 (1 $\mu$g/ml) or pricked with a glass needle at each designated time. After 15 min, the percentage of oocytes exhibiting vitelline envelope elevation was determined. (From Belanger and Schuetz, 1975.)

did not develop until many hours later. The differential onset of activation in response to these activating stimuli appears to be related to the fact that the ionophore acts on the entire oocyte surface, whereas pricking with a needle acts at only one point. Thus, the inability of pricking to induce the activation response immediately is attributed to failure to propagate around the oocyte surface. This interpretation is also consistent with the data of Elinson (1977), whose studies demonstrated that sperm enter but do not activate the oocyte when fertilization is carried out in incompletely matured oocytes.

Striking evidence that a block to polyspermy develops during maturation in amphibian oocyte cytoplasm comes from experiments (Katagiri, 1974) in which the the vitelline envelope of coelomic oocytes was removed before *in vitro* fertilization. Under these conditions, polyspermy was readily obtained for a period of time after the induction of maturation. With increasing maturation and in the absence of oviductal secretions, the block to polyspermy developed such that oocytes no longer exhibited polyspermic fertilization by metaphase II.

A similar maturation-dependent block to polyspermy appears to develop in starfish oocytes, as evidenced by the response of oocytes to sperm or ionophore (A23187) at various stages of maturation. Significantly, the activation responses develop slowly, as evidenced by the rate of elevation of the fertilization envelope in fertilized polyspermic immature oocytes. Thus, complete elevation of the fertilization envelope appears to result from the cumulative effects of multiple fertilizing sperm inducing local discharge of cortical granules and separation of the vitelline envelope from the oocyte. An inability of the activation stimulus to propagate may therefore explain why polyspermic fertilization occurs before the maturation process is completed (see also Hirai *et al.*, 1981). This possibility is also consistent with observations that local injections of calcium in the incompletely matured oocytes leads to localized cortical granule discharge (Hollinger and Schuetz, 1976). DeFelice and Dale (1979) also observed a delay in fertilization envelope elevation in immature sea urchin oocytes as well as in a high incidence of polyspermy.

## 12. Cytoplasmic Control of Sperm Nuclear Differentiation

Even after sperm have entered the oocyte cytoplasm (monospermic or polyspermic fertilization), fusion of the two genomes such that development will be initiated is not insured. Such sperm must undergo a sequence of nuclear differentiation steps (nuclear membrane breakdown, chromatic decondensation, nuclear swelling, DNA synthesis, pronuclear formation, and aster formation) before fusion of the male and female genomes can occur (Longo, 1981). These changes all occur in fully mature oocytes, but not in immature oocytes. Thus, the maturational state of the oocyte cytoplasm essentially controls the differentiation of the sperm nucleus and other cytoplasmic constituents (centrioles) (Heidemann and Kirschner, 1978). This fact is readily appreciated from studies in which fertilization or sperm injection is carried out at various stages of oocyte meiotic maturation. Sperm nuclear differentiation does not proceed in germinal vesicle-stage starfish (Schuetz and Longo, 1981), amphibian (Katagiri, 1980), or mammalian (Niwa and Chang, 1975; Balakier and Tarkowski, 1980) oocytes. Furthermore, after hormonally induced maturation, the incorporated sperm and zygote undergo different kinds of differentiation, depending on the stage of oocyte maturation or activation and on whether there is monospermic or polyspermic fertilization. For example, in fertilized immature starfish oocytes, polyspermic fertilization occurs, but the sperm are not readily detectable at the light microscopic level and remain dormant and undifferentiated within the oocyte cytoplasm. After oocyte exposure to 1-MA, however, sperm nuclear differentiation proceeds, asters form, and cleavage ensues, but only after the oocyte germinal vesicle has disintegrated (Fig. 26).

Significantly, the transformations in the sperm are dependent on maturational changes produced in the oocyte cytoplasm both before and after the breakdown of the germinal vesicle. This fact is elegantly demonstrated in amphibian oocytes, in which the germinal vesicle is surgically removed before hormonal treatment or before the time of GVBD in hormone-treated oocytes (Katagiri and Moriya, 1976). In such oocytes, incorporated sperm do not form pronuclei and asters, nor do they undergo DNA synthesis after progesterone exposure; by contrast, injection of germinal vesicle materials into similarly treated oocytes restores sperm-transforming activity to the cytoplasm. Such experiments demonstrate that the germinal vesicle contributes some factor(s) to the oocyte cytoplasm required for the sperm to differentiate and participate in the developmental process. Studies in mammalian oocytes indicate that reduction of disulfide bonds is involved in the process of sperm nuclear decondensation in the cytoplasm (Calvin & Bedford, 1971; Perrault *et al.*, 1984).

An activation stimulus also appears to be required for oocytes to induce aster and pronuclear formation and to initiate DNA synthesis in the sperm nucleus. Sperm entry into immature cytoplasm does not activate the cytoplasm. Thus, when such oocytes are later matured, a separate activation stimulus is required to complete transformation of the sperm nucleus. In a fully mature oocyte, when fertilization takes place, the activation stimulus is sufficient to complete all aspects of sperm nuclear differentiation, presumably as a result of the release of stored intracellular calcium ions. Katagiri and Moriya (1976) further concluded that pronucleus-inducing activity present in the cytoplasm is dependent on protein synthesis that occurs after GVBD. Significantly, mechanical breakdown of the amphibian germinal vesicle or injection of GV material into immature oocytes is inadequate to produce aster formation in incorporated sperm. Such results

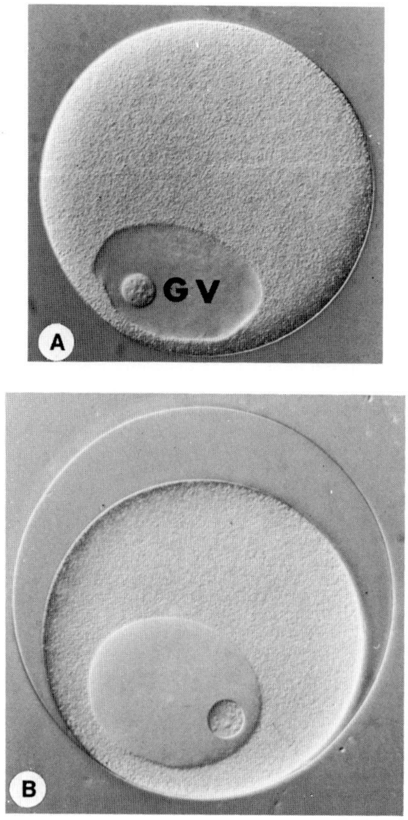


**Figure 26.** Maturation dependence of sperm nuclear differentiation in the starfish oocyte. (A) Immature (GV-intact stage) starfish oocyte before insemination. Nomarski optics. (B) Immature oocyte inseminated and incubated for 180 min. The fertilization envelope is elevated, and the cytoplasm remains clear and homogeneous. Nomarski optics. (C) Oocyte fertilized in GV-intact stage and treated with 1-MA. Note absence of GV and numerous decondensed sperm (pronuclei) in the cytoplasm. Nomarski optics. (Parts A, B, and C from Schuetz and Longo, 1982.) (D) Multiple sperm asters in the cytoplasm of oocyte fertilized in GV stage and treated with 1-MA. Polarized light microscopy. (E) Abnormal cleavage after maturation induction of polyspermic starfish oocytes. Nomarski optics. GV, germinal vesicle; 1-MA, 1-methyladenine.

clearly demonstrate that a degree of nuclear–cytoplasmic interaction is required before germinal vesicle materials are capable of becoming physiologically active in the cytoplasm (refer to Fig. 18). The results of these experiments demonstrate that full maturation of the oocyte requires hormone-induced cytoplasmic factors from two sources: those originating in the cytoplasm without nuclear involvement and those present in the cytoplasm but originating from the germinal vesicle or secondarily as a result of further interactions of nuclear and cytoplasmic components. Thus, it is

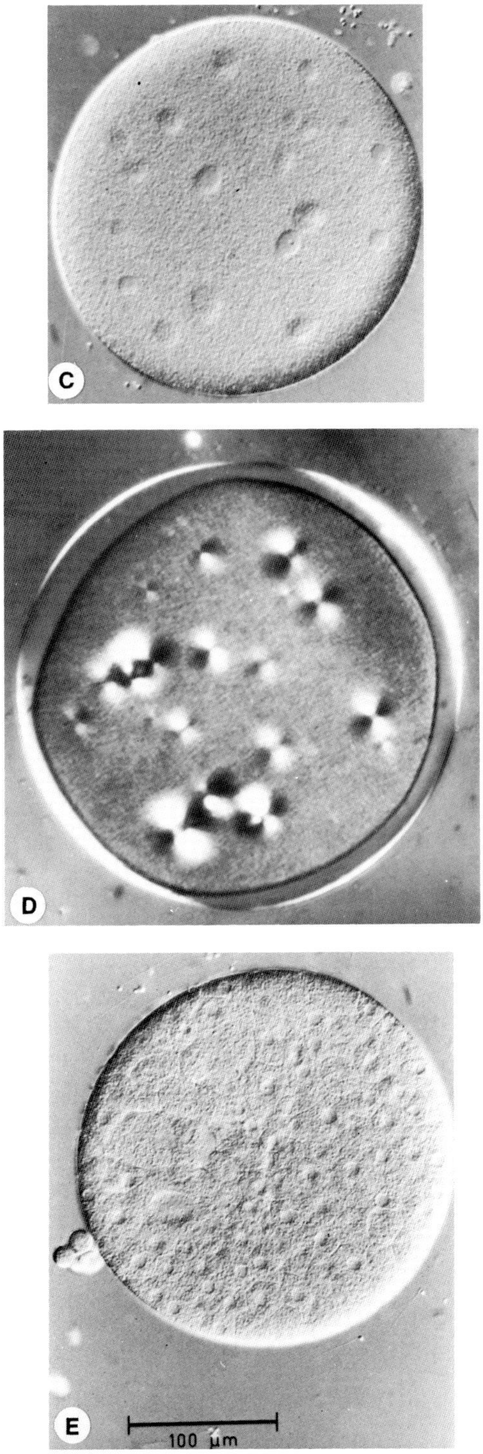


**Figure 26.** (*Continued*)

apparent that complete sperm nuclear differentiation is dependent on sequential changes in the cytoplasm (Fig. 18).

The early studies discussed above were all performed using intact oocytes; it was recently demonstrated, however, that cell-free preparations of cytoplasm taken from activated *Rana* oocytes induce demembranated sperm nuclei to undergo a cycle of nuclear (pronuclear) envelope formation, chromatin decondensation, nuclear swelling, DNA synthesis, and chromosomal condensation (Lohka and Masui, 1983) (Figs. 27 and 28). Sperm nuclear chromatin becomes dispersed within 6 min of being added to cytoplasmic fractions. Interestingly, soluble and particulate cytoplasmic constituents have separate effects on sperm nuclei. However, both are required to complete *in vitro* sperm nuclear differentiation. The morpho-

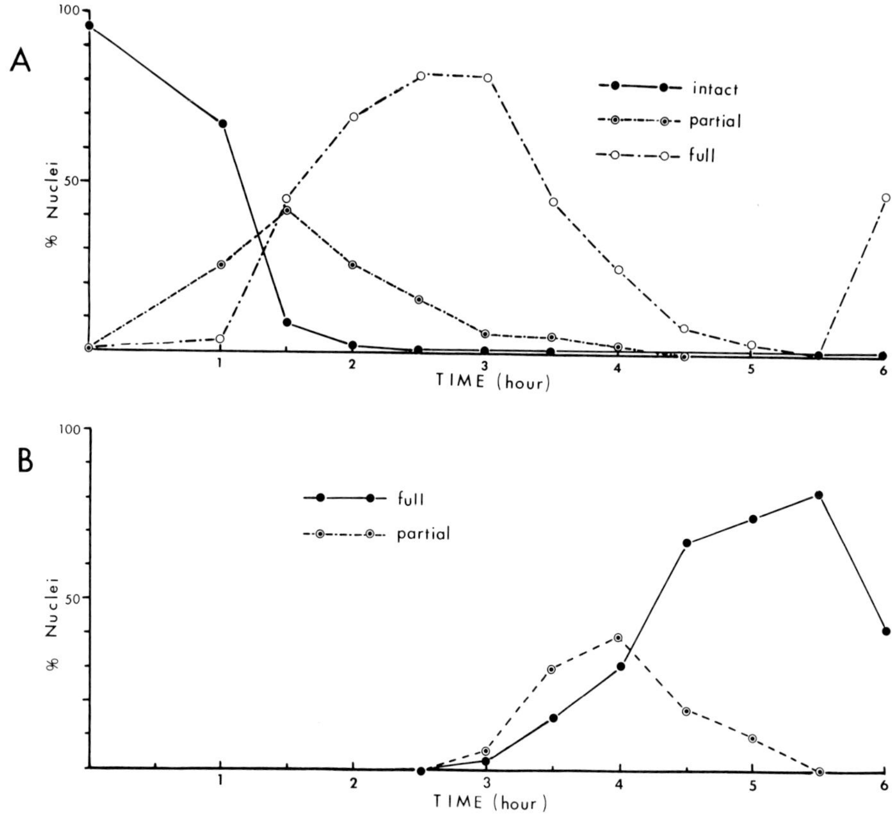


**Figure 27.** Time course of the changes in *Xenopus* sperm nuclear morphology in cell-free preparations from activated *Rana* oocytes. (A) Decondensation of sperm nuclei. (B) Chromosomal condensation by decondensed sperm nuclei. Partially recondensed nuclei have a metaphaselike appearance. (From Lohka and Masui, 1983. Copyright 1983 by the American Association for the Advancement of Science.)

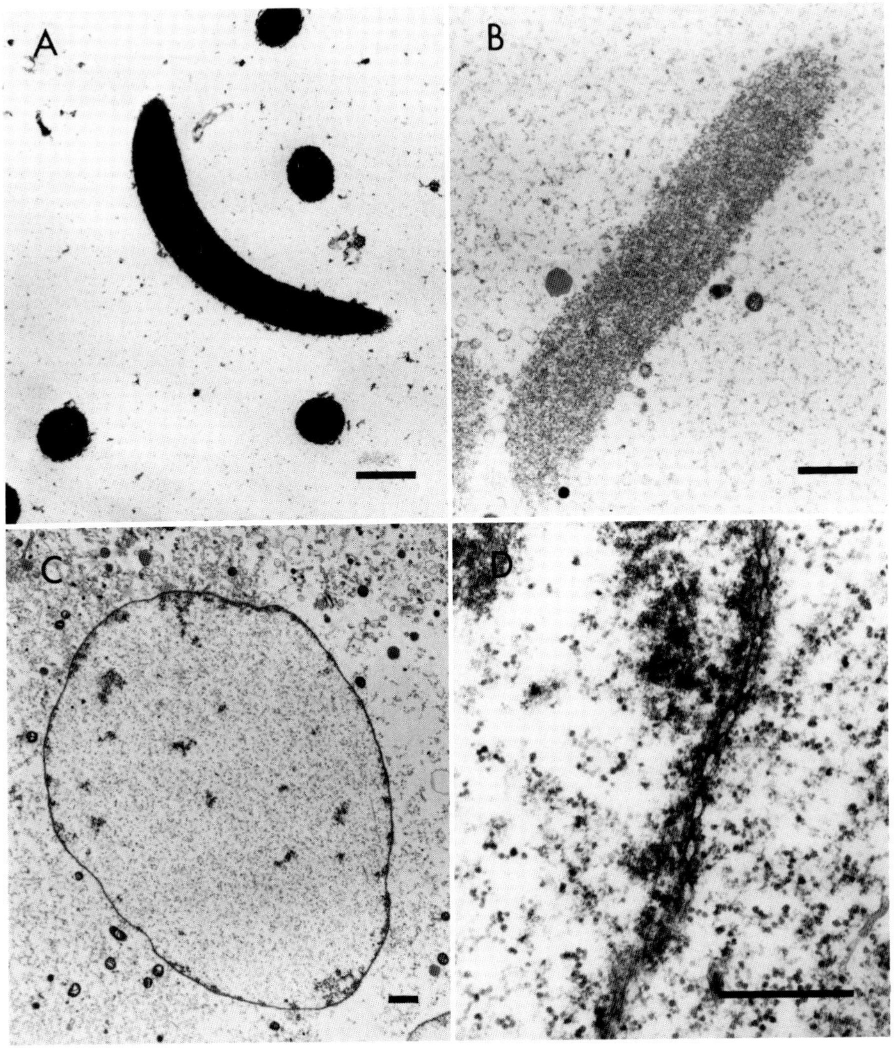


**Figure 28.** Ultrastructure of *Xenopus laevis* sperm nuclei in cytoplasmic preparations. Sperm nucleus (A) before addition to cytoplasmic preparation (scale bar: 1.0 μm); (B) 5 min after addition to cytoplasmic preparation (scale bar: 1.0 μm); and (C) 180 min after addition to cytoplasmic preparation (scale bar: 1.0 μm). (D) Nuclear envelope surrounding decondensed chromatin at 120 min of incubation (scale bar: 0.5 μm). (From Lohka and Masui, 1983. Copyright 1983 by the American Association for the Advancement of Science.)

logical and biochemical changes produced *in vitro* are remarkably similar to those that occur *in vivo* and may provide important information on how these events are controlled by the oocyte cytoplasm. Other workers (Eng and Metz, 1980) have found that a high-molecular-weight fraction from homogenates of sea urchin eggs causes sperm head decondensation. Glycosaminoglycan sulfate from sea urchin eggs was also found capable of decondensing human sperm nuclei (Delgado *et al.*, 1982).

Genetic factors participate as well in controlling sperm differentiation and subsequent development. Wakasugi (1974) described the existence of genetically determined incompatibility between sperm and egg after fertilization of oocytes from DDK strain mice. The defect is characterized by embryonic death at the morula–blastocyst and pre-egg cylinder stages as well as by abnormal trophoblast formation.

Cytoplasmic factors, presumably of maternal origin, also play a role in controlling subsequent development in many species (Malacinski and Brothers, 1974) (see Chapter 13). Embryonic development of cultured mouse embryos, obtained from certain strains of mice, typically exhibit arrested development at the two-cell stage. Muggleton-Harris *et al.* (1982) demonstrated that the two-cell block to cleavage can be circumvented by injection of cytoplasm from embryos that do not exhibit this block. A considerable degree of cross-species fertilization occurs. However, only a limited degree of development results (Elinson, 1977; Thadani, 1980). Significantly, cytoplasmic factors do not appear to be unlimited, since in polyspermic oocytes not all sperm nuclei undergo the same degree of differentiation. This suggests that only a limited amount of these cytoplasmic factors is available to act on the incorporated sperm. Lack of such cytoplasmic factors has been hypothesized to explain why oocytes undergoing spontaneous maturation develop poorly after fertilization (Thibault, 1977).

## 13. Oocyte Maturation and Cytoplasmic Clocks

It is readily apparent to anyone who has visually observed living gametes undergoing meiotic maturation, fertilization, or early development that these progressive transformations are highly synchronized under standardized conditions; as a result, specific criteria of differentiation (polar body formation, Met I, Met II, cleavage, gastrulation, morulae) occur at predictable times characteristic of each species. Thus, within a few minutes, thousands of starfish oocytes can be observed to undergo GVBD some 30 min after exposure to 1-MA and subsequently to form the first and second polar bodies and female pronucleus. In amphibians, the first 10 to 11 cleavages after fertilization are very rapid and synchronous within equatorial and vegetal regions until the mid-blastula stage is reached. Subsequently, the cleavage processes become quite asynchronous (Newport and Kirschner, 1982a,b). The regularity of these early events is indicative of some type of underlying developmental program or clock that appears to measure time flow for particular types of biochemical and/or morphological events (reviewed by Satoh, 1982).

Recent experiments, primarily in amphibians and starfish, have provided new insights as to how some of these cellular events are initiated, controlled, localized, and linked to the hormonal stimulus for maturation in oocytes and the cleavage cycle. Evidence for, and characterization of, cytoplasmic clocks has largely resulted from the development and use of a number of very elegant yet basically simple techniques for quantitative measurement of the mechanical properties (i.e., rigidity, turgidity, tension, resistance to deformation, contractility) of the cell surface.

In the unfertilized amphibian egg, the cell cycle is arrested. However, after fertilization, a rounding up of the animal hemisphere and increased stiffness and reactivity of the cortex to cleavage furrow-inducing substance occur synchronously with the cleavage cycle (Sawai, 1979). When precleavage-stage zygotes are surgically divided into nucleated and enucleated halves, it is further observed that enucleated fragments exhibit all these cyclic changes seen in intact nucleated fragments. Thus, it has been concluded that changes in the cortex that are necessary for cleavage furrow formation are brought about by cyclic change in the cytoplasm independent of the nucleus. Hara *et al.* (1980) further analyzed this cytoplasmic clock (surface contraction wave) in relationship to the division cycle; these workers showed that the contraction wave is initiated in the animal pole and is propagated in a circular manner to the equator of the gamete. Inhibition of cleavage by antimitotic drugs (e.g., colchicine, vinblastine) did not prevent the periodic contractile waves from occurring at expected times of division. Parthenogenetically activated (pricked) oocytes also exhibited cell-cycle changes in the cortex without cleavage furrow formation. This was attributed to the lack of a functional spindle centriole supplied by the sperm. It was concluded that persistent contractile cycles operate in the absence of any nuclear material (paternal or maternal) or of the sperm centriole.

Thus, the primary reason for periodic cortical changes being called cytoplasmic clocks is the fact that these events continue to occur in the absence of the nucleus. Such cytoplasmic clocks have been seen in sea urchins, amphibians, and starfish oocytes.

Although prior studies consistently showed that the existence of cytoplasmic clocks is independent of the nucleus, recent data obtained using starfish GV oocytes induced to mature with oocyte maturation hormone (1-MA) mandate a modification of this view (Nemoto *et al.*, 1980; Yamamato and Yoneda, 1983). Germinal vesicle-stage oocytes are extremely rigid and exhibit no evidence of periodic changes in stiffness. However, after addition of 1-MA to oocytes, a marked and rapid decrease in surface rigidity occurs that precedes disintegration of the germinal vesicle. This low level of tension is maintained except for two rapidly rising and declining peaks associated with formation of the first and second polar bodies (Fig. 29). Similar periodic changes result when reinitiation of meiosis is triggered with the use of dithiothreitol (DTT). At least two effects of 1-MA can be defined under these conditions: (1) reduction in surface tension, and (2) initiation of two contraction–relaxation peaks in association with the formation of

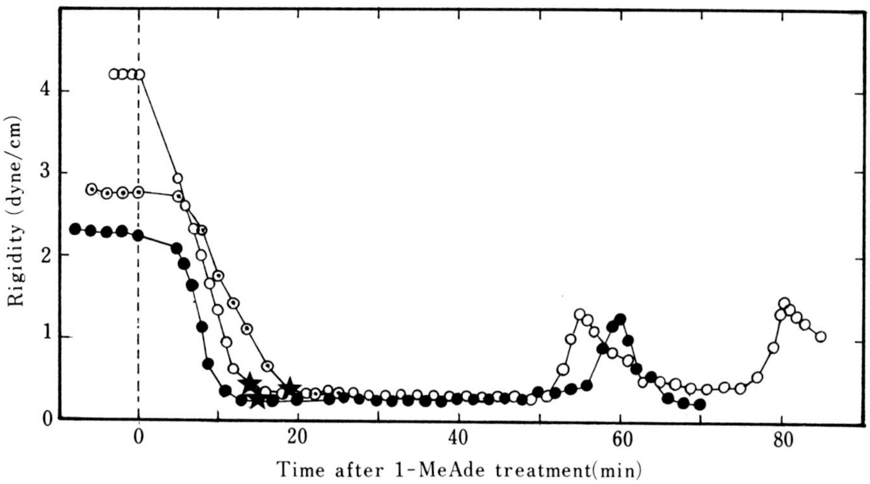


**Figure 29.** Effects of 1-methyladenine (1-MeAde, $10^{-6}$ M) on the stiffness of oocytes of *Asterina pectinifera*. Stars indicate germinal vesicle breakdown (GVBD). Sharp peaks in the stiffness at about 60 and 80 min are due to the first and second meiotic divisions, respectively. (From Nemoto *et al.*, 1980.)

the polar bodies. However, when nucleated and enucleated oocyte fragments are made and exposed to 1-MA, the cycle of tension in such oocytes varies drastically, depending on whether the enucleation is carried out before or after germinal vesicle breakdown (Fig. 30). Timing of enucleation is crucial to a determination of whether the nucleus functions in maintaining the periodic cytoplasmic cycle. Removal of the GV as close as 5 min before 1-MA-induced GVBD is not sufficient to restore cycle activity. Anucleate fragments obtained soon after GVBD exhibit the same responses to 1-MA as seen in intact oocytes except that the decline in tension following the second peak (associated with second polar body) is much more gradual. By contrast, enucleation of germinal vesicle-stage oocytes exhibits a typical rapid decline in surface tension, but this is not followed by the two contraction–relaxation peaks in tension. Rather, a gradual increase in tension is observed. In addition to undergoing a decrease in cortical tension, anucleate fragments treated with 1-MA are fertilizable and exhibit elevation of the fertilization envelope. Thus, certain aspects of cytoplasmic maturation can proceed in the absence of the nucleus, whereas other aspects are dependent on the germinal vesicle. Yamamato and Yoneda (1983) conclude that the diffusion of germinal vesicle materials is requisite in some way to

**Figure 30.** Effects of the nucleus on tension at the surface of the starfish oocyte. (A) Procedure for bisection of maturing oocyte (a) and immature oocyte (b). (B) (a,b) Tension at the surface of the oocyte fragment obtained from two maturing oocytes after GVBD. Nucleate (○) and nonnucleate (●) fragments from the same oocyte. (c,d) Tension at the surface of nucleate (○) and

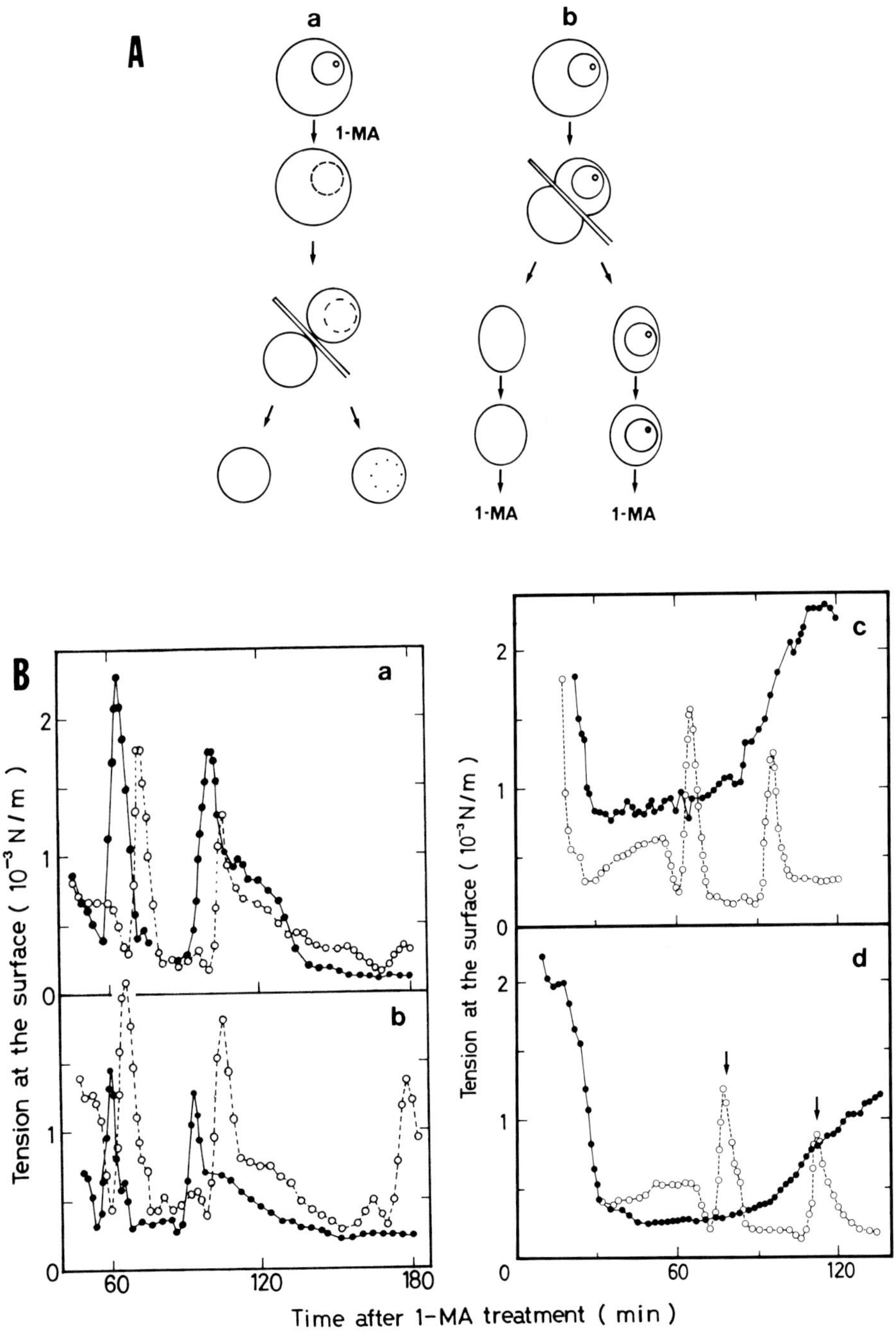


nonnucleate (●) fragments obtained from a single immature oocyte. (c) Treated with 1-MA 26 min. after the bisection. (d) Treated with 1-MA 73 min. after the bisection. Arrows indicate the time of polar body formation in the nucleate fragment. GVBD, germinal vesicle breakdown; 1-MA, 1-methyladenine. (From Yamamoto and Yoneda, 1983.)

the initiation of the cytoplasmic contractile cycles. Strikingly, the number of peaks of contraction–relaxation during maturation is directly correlated with the number of cell divisions that occur before meiotic arrest. It would be of interest to know whether other oocytes arrested at metaphase I or II have one or two such peaks.

On the basis of such studies, it appears that our designation of cytoplasmic clocks must be tempered by our understanding of the changing role of the nucleus at different stages of female gamete growth and maturation. Obviously, the results indicate that the greatly enlarged tetraploid meiotic (germinal vesicle) nucleus contains something that the cell nucleus at later stages of gamete or embryo development does not. Possibly, the material released from the germinal vesicle persists without undergoing metabolic degradation through activation and the initial cleavage stages. The nature of the germinal vesicle factor is unknown, but considerable evidence suggests that it may be related to a contractile element in the cortical cytoplasm of most oocytes.

Although it has not yet been demonstrated that the germinal vesicle contributes to the cytoplasmic clock in amphibian oocytes, it is interesting that GV material is required for the contractile responses seen in cycloheximide-treated oocytes (Fig. 31). Cycloheximide treatment of progesterone-matured oocytes was found to induce dramatic contractile changes in the cortex first seen in the region of GV breakdown in the animal pole. Significantly, these changes were invariably dependent on GVBD. Hormone-treated oocytes that did not mature never exhibited these changes, and the alterations were observed only after the germinal vesicle had disintegrated. Germinal vesicle contents were found to be necessary for this reaction, since enucleated oocytes underwent cytoplasmic maturation after progesterone treatment but did not respond to cycloheximide with cortical changes. Reinjection of GV materials partially restored the responsiveness to cycloheximide. It was concluded that proteins synthesized after GV breakdown are normally involved in suppressing the cortical changes.

These studies raise the question: Why doesn't the cell cycle clock associated with meiosis reinitiation continue through the mitotic cell division cycles associated with early embryogenesis? Possibly there are different components to a common clock mechanism or, alternatively, there may be different types of cell-cycle clocks. It is of particular interest that different

**Figure 31.** Effects of the nucleus on cortical responses of the amphibian oocyte. (A) Hormone and nuclear–cytoplasmic interactions in cortical differentation of amphibian oocytes. Cycloheximide induces cortical contractions in maturing (progesterone-induced) oocytes subsequent to germinal vesicle breakdown (GVBD). (B) Cycloheximide-treated enucleated oocyte after steroid induction of cytoplasmic maturation. Such oocytes were enucleated and treated with progesterone and cycloheximide (6 hr postprogesterone). Note the smooth surface of the oocyte and lack of typical cortical alterations. (From Schuetz and Samson, 1979.)

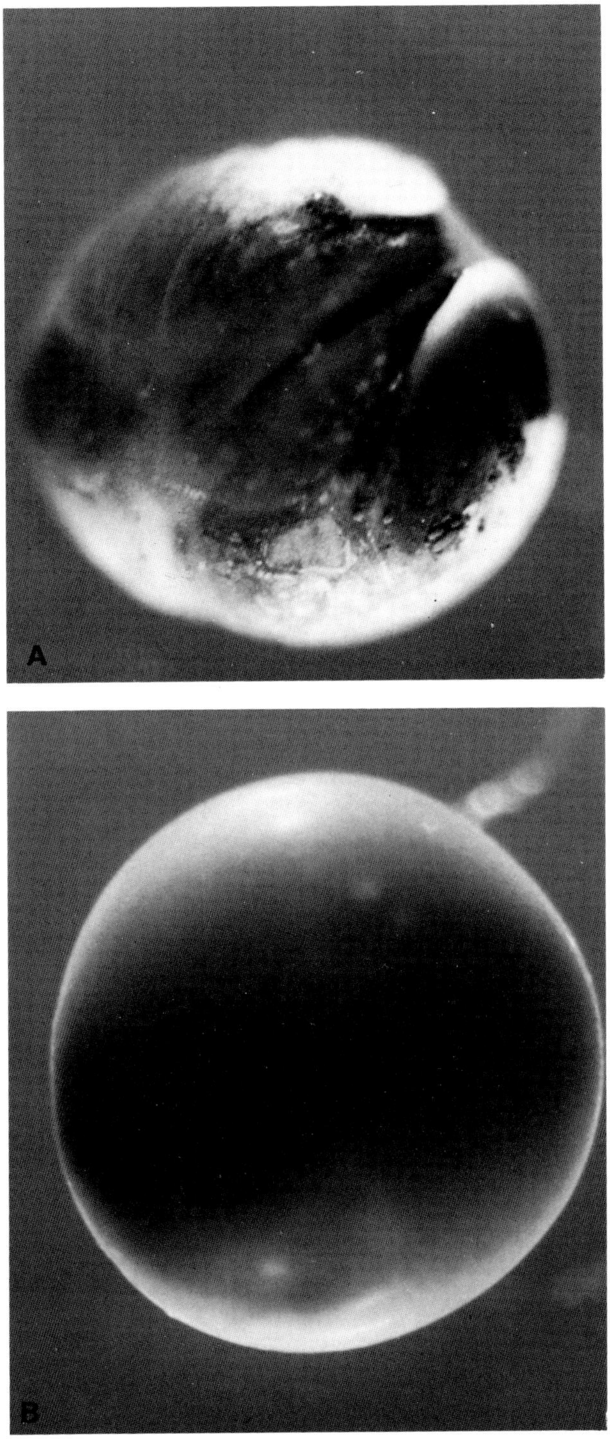
A
B

stimuli (e.g., hormones, sperm, pricking) trigger activation of cytoplasmic clocks in GV-stage oocytes and oocytes at later stages of differentiation. Clearly, the cytoplasmic clock during early embryogenesis does not appear to require a nuclear component, whereas such a component contributes to meiotic cycle events. Perhaps different factors in the cytoplasm control (i.e., inhibit, stimulate) the cytoplasmic clocks in meiotic and mitotic cells or during the transition between these two types of cell division. It remains to be determined whether and how cytoplasmic MPF, cytostatic factor, hormones, or some other factor controls the cytoplasmic clocks in gametes and embryos.

## 14. Concluding Remarks

Oocyte differentiation and attainment of full developmental competence in a wide range of species are inextricably linked to changes in the local intra- and extraoocytic microenvironment during periods of intra- and extraovarian existence. Within the gonads, development of each gamete is largely mediated through formation of the ovarian follicle, which undergoes progressive cellular and molecular differentiation. Intra- as well as extrafollicular fluctuations in the secretion of a variety of hormones, particularly steroids from different cellular components, appear to play a key role. In many cases, these molecules exert—as a result of their local and/or peripheral actions—a direct effect in bringing about growth and differentiation of the oocyte and follicle. Experimental evidence suggests that abnormal aspects of follicular, oocyte, and embryo development (i.e., atresia, teratogenesis, sexual differentiation, genetic defects, aging, parthenogenesis) are linked to alterations in the local environment. Many such dysfunctions appear to result from changes in somatic cells, their secretory products, and/or the physical relationship between somatic and germ cells. Thus, further characterization and quantification of these secretory products over the course of folliculogenesis are needed in order to understand fully folliculogenesis, oocyte maturation, and developmental events. The progressive differentiation of the somatic cell–oocyte units involves alterations of genomic processes in a variety of cells. The molecular basis for genomic regulation in oocytes and somatic cells remains largely undetermined. However, the evidence indicates that hormone-induced intermediate factors, including proteins, act to mediate or repress cellular differentiation. Significantly, oocyte maturation also involves extracellular components, including the vitelline envelope and egg jelly layers, which are produced within or outside the ovary by adjacent somatic tissues.

A central tenet of molecular embryology is that cellular change is mediated through the sequential nuclear synthesis of RNA molecules transferred to the cytoplasm. However, extensive structural and functional differentiation of the oocyte occurs at certain stages without direct nuclear participation. Thus, cytoplasmic control of the nucleus, nuclear control of the cytoplasm, as well as nucleocytoplasmic interactions are all important in certain aspects of the maturation and fertilization processes.

A particularly striking feature of the overall process of gamete differentiation is the repeated involvement of intra- and extracellular calcium ions in mediating cellular change. Calcium ions have been linked directly to synthesis and secretion of maturation-inducing hormones by somatic cells, induction of intraoocyte cytoplasmic maturation factor(s) by hormones, sperm–egg fusion, induction of cortical granule discharge, propagation of the block to polyspermy, interruption of meiotic arrest, inactivation of cytostatic factors, sperm nuclear differentiation, DNA synthesis, and cortical contractions. It remains to be determined to what extent these changes occur as a result of cyclic alterations in free and bound calcium ions localized in single or multiple compartments of the oocyte and somatic cells.

Strikingly, many of the basic mechanisms that regulate oocyte maturation in most species are similar despite wide variations in size, composition, timing, and duration of specific events. The available information provides a limited sketch of the process of oocyte maturation. Much remains to be accomplished before we can understand how oocytes function and simultaneously undergo construction and restructuring in order to acquire the capacity to generate the species.

ACKNOWLEDGMENTS. Financial support provided by NIH (RO1.HD 16429) and NSF (PCM-82-02818) during the preparation of this manuscript is gratefully acknowledged. I am particularly grateful to J. M. Schuetz for her contributions in editing the manuscript.

## References

Adams, C. E., 1970, Aging and reproduction in the female mammal with particular reference to the rabbit, *J. Reprod. Fertil. Suppl.* **12**:1.

Austin, C. R., 1970, Aging and reproduction: Preovulatory deterioriation of the egg, *J. Reprod. Fertil. Suppl.* **12**:39.

Bagnell, C. A., Mills, T. M., Costoff, A., and Mahesh, V. B., 1982, A model for the study of androgen effects on follicular atresia and ovulation. *Biol. Reprod.* **27**:903.

Bahr, J. M., Wang, S. C., Huang, M .Y., and Calvo, F. O., 1983, Steroid concentrations in isolated theca and granulosa layers of preovulatory follicles during the ovulatory cycle of the domestic hen, *Biol. Reprod.* **291**:326.

Balakier, H., 1978, Induction of maturation in small oocytes from sexually immature mice by fusion with meiotic or mitotic cells, *Exp. Cell Res.* **112**:137.

Balakier, H., and Tarkowski, A. I. L., 1980, The role of germinal vesicle karyoplasm in the development of male pronucleus in the mouse, *Exp. Cell Res.* **128**:79.

Barbieri, F. D., 1982, Biochemical mechanisms of fertilization, In: *Physiopathology of Hypophysical Disturbances and Diseases of Reproduction* (A. F. DeNicola, J. Bloquier, and R. J. Soto, eds.), p. 235, Alan R. Liss, New York.

Barrington, E. J. W., 1979, *Hormones and Evolution*, Academic Press, New York.

Baulieu, E. E., Godeau, F., Schorderet, M., and Schorderet-Slatkine, S., 1978, Steroid induced meiotic division in *Xenopus laevis* oocytes: Surface and calcium, *Nature* **275**:593.

Belanger, H. M., and Schuetz, A. W., 1975, Precocious induction of activation responses in amphibian oocytes by divalent ionophore A-23187, *Dev. Biol.* **45**:378.

Bellé, R., Boyer, J., and Ozon, R., 1982, Carbon dioxide reversibly inhibits meiosis of *Xenopus laevis* oocyte and the appearance of the maturation promoting factor. *Dev. Biol.* **90**:315.

Biggers, J. D., and Schuetz, A. W. (eds.), 1972, *Oogenesis*, University Park Press, Baltimore.

Biggers, J. D., Whittingham, D. G., and Donahue, R. P., 1967, The pattern of energy metabolism in the mouse oocyte and zygote. *Proc. Natl. Acad. Sci. USA* **58**:560.

Bigl, C. H., and Greenwald, G. S., 1981, Acute gonadotrophin deprivation: (1) A model for the study of follicular atresia, *Biol. Reprod.* **24**:913.

Binor, Z., and Wolfe, D. P., 1979, *In vitro* maturation and penetration of mouse primary oocytes after removal of the zona pellucida, *J. Reprod. Fertil.* **56**:309.

Bjersing, L., 1982, Maturation, morphology and endocrine function of the ovarian follicle, *Adv. Exp. Med. Biol.* **147**:1.

Bjersing, L., and Cajander, S., 1974, Ovulation and the mechanisms of follicle rupture. V. Ultrastructure of tunica albuginea and theca externa of rabbit graafian follicles prior to induced ovulation, *Cell Tissue Res.* **153**:15.

Borland, R. M., Hazra, S., Biggers, J. D., and Lechene, C. P., 1977, The elemental composition of the environments of the gametes and preimplanation embryo during the initiation of pregnancy, *Biol. Reprod.* **16**:147.

Browne, C. L., Wiley, H. S., and Dumont, J. N., 1979, Oocyte follicle cell gap junctions in *Xenopus laevis* and the effects of gonadotropin on their permeability, *Science* **203**:182.

Butcher, R. L., 1975, The role of intrauterine environment and intrafollicular aging of the oocyte on implantation rates and development, in: *Aging Gametes* (R. J. Blandau, ed.), p. 201, Karger, Basel.

Byskov, A. G., 1978, Follicular atresia, in: *The Vertebrate Ovary* (R. E., Jones, ed.), p. 533, Plenum, New York.

Calvin, H. I., and Bedford, J. M., 1971, Formation of disulfide bonds in the nucleus and accessory structures of mammalian spermatozoa during maturation in the epididymis, *J. Reprod. Fertil. (Suppl.)* **13**:65.

Cameron, I. L., Lum, J. B., Nations, C., Asch, R. H., and Silverman, A. Y., 1983, Assay for characterization of human folliclular oocyte maturation inhibitor using Xenopus oocytes, *Biol. Reprod.* **28**:817.

Carnegie, J. A., and Tsang, B. S., 1983, Follicle-stimulating hormone-regulated granulosa cell steroidogenesis: Involvement of calcium–calmodulin system, *Am. J. Obstet. Gynecol.* **145**:223.

Carson, R. S., Findlay, J. K., Clarke, I. J., and Burger, H. G., 1981, Estradiol, testosterone,, and androstenedione in ovine follicular fluid during growth and atresia of ovarian follicles, *Biol. Reprod.* **24**:105.

Channing, C. P., Anderson, L. D., Hoover, D. J., Kolena, J., Osteen, K. G., Pomerantz, S.H., and Tanabee, K., 1982, The role of non-steroidal regulators in control of oocyte and follicular maturation, *Rec. Prog. Horm. Res.* **38**:331.

Chen, Y-T., Mattison, D. R., Feigenbaum, L., Kukui, H., and Schulman, J. D., 1981, Reduction in oocyte number following preantral exposure to a diet high in galactose, *Science* **214**:1145.

Clark, M. R., 1982, Stimulation of progesterone and prostaglandin E accumulation by luteinizing hormone releasing hormone (LHRH) and LHRH analogs in rat granulosa cells, *Endocrinology* **110**:146.

Clarke, H. J., and Masui, Y., 1983, The induction of reversible and irreversible chromosome decondensation by protein synthesis inhibition during meiotic maturation of mouse oocytes, *Dev. Biol.* **97**:291.

Clavert, J., 1962, Symmetrization of the egg of vertebrates, in: *Advances in Morphogenesis,*

Vol. II (M. Abercrombie and J. Brachet, eds.), p. 27, Academic Press, New York and London.

Cloud, J., and Schuetz, A. W., 1973, Spontaneous maturation of starfish oocytes: Role of calcium ions and follicle cells, *Exp. Cell Res.* **79**:446.

Cloud, J., and Schuetz, A. W., 1977, Interaction of progesterone with all or isolated portions of the amphibian *(Rana pipiens)* oocyte surface, *Dev. Biol.* **60**:359.

Colonna, R., and Mangia, F., 1983, Mechanisms of amino acid uptake in cumulus-enclosed mouse oocytes, *Biol. Reprod.* **28**:797.

Crosby, I. M., Osborn,, J. C., and Moor, R. M., 1981, Follicle cell regulation of protein synthesis and developmental competence in sheep oocytes, *J. Reprod. Fertil.* **62**:575.

Crowley, P. H., Gulati, D. K., Hayden, T. L., Lopez, P., and Dyer, R., 1979, A chiasma-hormonal hypothesis relating Down's syndrome and maternal age, *Nature* **280**:417.

Cuthbertson, K. S. R., Whittingham, D. G., and Cobbold, P. H., 1981, Free $Ca^{2+}$ increase in exponential phases during mouse oocyte activation, *Nature* **295**:754.

Dale, B., and Monroy, A., 1981, How is polyspermy prevented? *Gamete Res.* **4**:151.

Davison, E. H., 1968, *Gene Activity in Early Development*, Academic Press, New York.

DeFelice, L. J., and Dale, B., 1979, Voltage response to fertilization and polyspermy in sea urchin eggs and oocytes, *Dev. Biol.* **72**:327.

DeFelici, M., and Siracusa, G., 1982, "Spontaneous" hardening of the zona pellucida of mouse oocytes during in vitro culture, *Gamete Res.* **6**:107.

Dekel, N., Kraicer, P. F., Phillips, D. M., Sanchez, R. S., and Segal, S. J., 1978, Cellular associations in the rat oocyte cumulus-cell complex. Morphology and ovulatory changes, *Gamete Res.* **1**:47.

Dekel, N., and Beers, W. H., 1978, Rat oocyte maturation *in-vitro*. Relief of cyclic AMP inhibition by gonadotrophins, *Proc. Natl. Acad. Sci. USA* **75**:4369.

Dekel, N., and Beers, W. H., 1980, Development of the rat oocyte in vitro. Inhibition and induction of maturation in the presence or absence of the cumulus oophorus, *Dev. Biol.* **75**:247.

Dekel, N., and Phillips, D. M., 1980, Cyclic AMP, prostaglandin $E_2$ and steroids: Possible mediators in the rat cumulus oophorus mucification, *Biol. Reprod.* **22**:289.

Delgado, N. M., Reyes, R., Huaccya, L., Carranco, A., Merchant, H., and Rosado, A., 1982, Decondensation of human sperm nuclei by glycosaminoglycan sulfate from sea urchin egg, *J. Exp. Zool.* **224**:2157.

Detlaff, T. A., Nikitina, L. A., and Stroeva, O. G., 1964, The role of the germinal vesicle in oocyte maturation in anurans as revealed by the removal and transplantation of nuclei, *J. Embryol. Exp. Morphol.* **12**:851.

DiBerardino, M., 1980, Genetic stability and modulation of metazoan nuclei transplanted into eggs and oocytes, *Differentiation* **17**:17.

DiZerega, G. S., and Hodgen, G. D., 1981a, Luteal phase dysfunction infertility: A sequel to aberrant folliculogenesis, *Fertil. Steril.* **35**:489.

DiZerega, G. S., and Hodgen, G. D., 1981b, Folliculogenesis in the primate ovarian cycle, *Endocrinol. Rev.* **2**:27.

Donahue, R. P,. 1972, The relation of oocyte maturation to ovulation in mammals, in: *Oogenesis* (J. D. Biggers, and A. W. Schuetz, eds.), p. *413*, University Park Press, Baltimore.

Donahue, R. P., 1974, Normal and abnormal chromosomal behavior in the meiotic divisions of mammalian oocytes, in : *Aging Gametes* (R. J., Blandau, ed.), p. 58, Karger, Basel.

Donaldson, E. M., and Hunter, G. A., 1982, Sex control in fish with particular reference to salmonids, *Can. J. Fisheries Aquatic Sci.* **39**:99.

Dreyer, C. and Hausen, P., 1983, Two-dimensional gel analysis of the fate of oocyte nuclear proteins in the development of *Xenopus laevis*. *Dev. Biol.* **100**:412.

Dumont, J. N,. 1972, Oogenesis in *Xenopus laevis* (Daudin). I. Stages of oocyte development in laboratory maintained animals, *J. Morphol.* **136**:153.

Edwards, R. G., 1974, Follicular fluid, *J. Reprod. Fertil.* **37**:189.

Ekholm, C., and Magnusson, C., 1979, Rat oocyte maturation: Effects of protein synthesis inhibitors, *Biol. Reprod.* **21**:1298.

Ekholm, C., Clark, M. R, Magnusson, C., Isaksson, O., and LeMaire, W. J., 1982, Ovulation induced by a gonadotropin releasing hormone analog in hypophysectomized rats involves prostaglandins, *Endocrinology* **110**:288.

El-Etr, M., Schorderet-Slatkine S., and Baulieu, E., 1979, Meiotic maturation in *Xenopus laevis* oocytes initiated by insulin, *Science* **205**:1397.

Elinson, R. P., 1977, Fertilization of immature frog eggs: Cleavage and development following subsequent activation, *J. Embryol. Exp. Morphol.* **37**:187.

Eng, L. A., and Metz, C.B., 1980, Sperm head decondensation by a high molecular weight fraction of sea urchin egg homogenate, *J. Exp. Zool.* **212**:159.

Eppig, J. J., 1978, Granulosa cell deficient follicles: Occurrence, structure, and relationship to ovarian teratocarcinogenesis in strain Lt/Sv mice, *Differentiation* **12**:111.

Eppig, J. J., 1982a, The relationship between parthenogenetic embryonic development and cumulus cell–oocyte intercellular coupling during oocyte meiotic maturation, *Gamete Res.* **5**:229.

Eppig, J. J, 1982b, The relationship between cumulus cell–oocyte coupling, oocyte meiotic maturation and cumulus expansion, *Dev. Biol.* **89**:268.

Erickson, G. F., 1978, Normal ovarian function, *Clin. Obstet. Gynecol.* **21**:31.

Espey, L. L., 1980, Ovulation as an inflammatory reaction: A hypothesis, *Biol. Reprod.* **22**:73.

Faddy, M. J., Jones, E. C., and Edwards, R. G., 1976, An analytical model for ovarian follicle dynamics, *J. Exp. Zool.* **197**:173.

Feder, H. H., 1981, Essentials of steroid structure. Nomenclature, reactions, biosynthesis and measurements, in: *Neuroendocrinology of Reproduction: Physiology and Behavior*, (N. T. Adler, ed.), p. 19, Plenum, New York.

Fleming, A. D., and McGaughey, R. W., 1982, A progesterone binding component in porcine follicular fluid, *J. Endocrinol.* **94**:69.

Fortune, J. E., 1983, Steroid production by Xenopus ovarian follicles at different developmental stages, *Dev. Biol.* **99**:502.

Fortune, J. E., and Tsang, P. C., 1981, Production of androgen and estradiol-17$\beta$ by Xenopus ovaries treated with gonadotropins *in-vitro*, *Gen. Comp. Endocrinol.* **43**:234.

Fortune, J. E,. and Vincent, S.E., 1983, Progesterone inhibits the induction of aromatase in rat granulosa cells *in vitro*, *Biol. Reprod.* **28**:1078.

Franchimont, P., and Channing, C. P., 1981, *Intragonadal Regulation of Reproduction*, Academic Press, London.

Franklin, L. E., 1965, Morphology of gamete membrane fusion and of sperm entry into oocytes of the sea urchin, *J. Cell Biol.* **25**:81.

Gallien, L., 1974, Intersexuality, in: *Physiology of the Amphibia*, Vol. II (B. Lofts, ed.), p. 523, Academic Press, New York.

Gemmell, R. T., and Stacy, B. D., 1979, Ultrastructural studies of secretory granules in the corpus luteum of the sheep during the estrous cycle, *Biol. Reprod.* **11**:447.

Gilula, N. B., Epstein, M. L., and Beers, W. H., 1978, Cell–cell communication and ovulation: A study of the cumulus–oocyte complex, *J. Cell Biol.* **78**:58.

Goad, L. J., Rubinstein, I., and Smith, A. G., 1972, The sterols of echinoderms, *Proc. R. Soc. Lond. B* **180**:223.

Golbus, M. S., 1981, Chromosome aberrations and mammalian reproduction, in: *Fertilization and Embryonic Development In Vitro* (L. Mastroianni and J. D. Biggers, eds.), p. 258, Plenum Press, New York.

Graham, C. F., 1974, The production of parthenogenetic mammalian embryos and their use in biological research, *Biol. Rev.* **49**:399.

Greenwald, G. S., 1969, Endocrinology of oviductal secretions, in: *The Mammalian Oviduct* (E. S. F. Hatez and R. J. Blandau, eds.), p. 183, University of Chicago Press, Chicago.

Grey, R. D., Working, P. K., and Hedrick, J. L., 1977, Alteration of structure and penetrability of the vitelline envelope after passage of eggs from coelom to oviduct in *Xenopus laevis*, *J. Exp. Zool.* **201**:73.

Guerrier, P., Meijer, L., Moreau, M., and Longo, F. J., 1983, Hormone independent GVBD induces cytoplasmic maturity in the starfish oocyte, *J. Exp. Zool.* **226**:303.

Gurdon, J. B., 1981, Molecular mechanisms in the control of gene expression during development, *Biochem. Soc. Trans.* **9(1)**:13.

Gurdon, J. B., DeRobertis, E. M., and Partington, G., 1976, Injected nuclei in frog oocytes provide a living cell system for the study of transcriptional control, *Nature* **260**:116.

Gurdon, J. B., Laskey, R.A., DeRobertis, E. M., and Partington, G. A., 1979, Reprogramming of transplanted nuclei in amphibia, *Int. Rev. Cytol. Suppl.* **9**:161.

Hanocq-Quertier, J., and Baltus, G., 1981, Induction of meiosis of xenopus laevis oocytes by mianserine, *Gamete Res.* **4**:49.

Hara, K., Tydeman, P., and Kirschner, M., 1980, A cytoplasmic clock with the same period as the division cycle in Xenopus eggs, *Proc. Natl. Acad. Sci. USA* **77**:462.

Harris, G. W., and Edwards, R. G., 1970, A discussion on determination of sex, *Philos. Trans. R. Soc. Lond. B. Biol. Sci.* **259**:1.

Hay, M. F., Cran, D., and Moor, R. M., 1976, Structural changes occurring during atresia in sheep ovarian follicles, *Cell Tissue Res.* **169**:515.

Heidemann, S. R., and Kirschner, M. W., 1978, Induced formation of asters and cleavage furrows in oocytes of *Xenopus laevis* during *in vitro* maturation, *J. Exp. Zool.* **204**:431.

Heller, D. T., and Schultz, R. M., 1980, Ribonucleoside metabolism by mouse oocytes: Metabolic cooperativity between fully grown oocyte and cumulus cells, *J. Exp. Zool.* **214**:355.

Hillensjo, T., and LeMaire, W. J., 1980, Gonadotrophin releasing hormone agonists stimulate meiotic maturation of follicle enclosed rat oocytes *in vitro*, *Nature* **287**:145.

Hillensjo, T. H., Pomerantz, S. H., Schwartz-Kripner, H., Anderson, L. D., and Channing, C. P., 1980, Inhibition of cumulus cell progesterone secretion by low molecular weight fractions of porcine follicular fluid which also inhibits oocyte maturation, *Endocrinology* **106**:584.

Hillier, S. G., and Ross, G. T., 1979, Effects of exogenous testosterone on ovarian weight, follicular morphology and intraovarian progesterone concentration in estrogen primed hypophysectomized immature female rats, *Biol. Reprod.* **20**:261.

Hirai, S., Nagahama, Y., Kishimoto, T., and Kanatani, H., 1981, Cytoplasmic maturity revealed by the structural changes in incorporated spermatozoa during the course of starfish oocyte maturation, *Dev. Growth Diff.* **23(5)**:465.

Hirai, S., Goascogne, C. S., and Baulieu, E. E., 1983, Induction of germinal vescile breakdown in *Xenopus laevis* oocytes: Response of denuded oocytes to progesterone and insulin, *Dev. Biol.* **100**:214.

Hisaw, F. L., 1947, Development of the graafian follicle and ovulation, *Physiol. Rev.* **27**:95.

Hollinger, T. G., and Schuetz, A. W., 1976, Cleavage and cortical granule breakdown in *Rana pipiens* oocytes induced by direct microinjection of calcium, *J. Cell Biol.* **71**:395.

Huchon, D., Crozet, N., Cantenot, N., and Ozon, R., 1981, Germinal vesicle breakdown in the *Xenopus laevis* oocyte: Description of a transient microtubular structure, *Reprod. Nutr. Dev.* **21**:135.

Hunter, R. H. F., Cook, B., and Baker, T. G., 1976, Dissociation of response to injected gonadotrophins between the graafian follicle and oocyte in pigs, *Nature* **260**:156.

Hunter, G. A., Donaldson, E. M., Goetz, F. W., and Edgell, P. R., 1982, Production of all female and sterile coho salmon, and experimental evidence for male heterogamity, *Trans. Am. Fisheries Soc.* **111**:367.

Ikegami, S., 1974, Role of asterosaponin A in starfish spawning induced by gonad stimulating substance and 1-methyladenine, *J. Exp. Zool.* **198**:359.

Ingram, D. L., 1962, Atresia, in: *The Ovary*, Vol. 1 (S. Zuckerman, A. M. Mandl, and P. Eckstein, eds.), p. 247, Academic Press, New York.

Iwamatsu, T., and Ohta, T., 1977, Fine structure of loach oocytes during maturation in vitro, *Dev. Growth Differen.* **19**:213.

Iwamatsu, T., 1978, Studies on oocyte maturation of the medaka, Oryzias latipes V. on the structure of steroids that induce maturation *in vitro*, *J. Exp. Zool.* **204**:401.

Jaffe, L. A., 1976, Fast block to polyspermy in sea urchin eggs electrically mediated, *Nature* (London) **261**:68.

Jaffe, L., 1983, Sources of calcium in egg activation: A review and hypothesis. *Dev. Biol.* **99**:265.

Jagiello, G., and Vogel, H. J. (eds.), 1981, *Bioregulators of Reproduction*, Academic Press, New York.

Jalabert, B., and Szollosi, D., 1975, *In vitro* ovulation of trout oocytes: Effect of prostaglandins on smooth muscle-like cells of the theca, *Prostaglandins* **9:**765.

Janson, P. O., LeMaire, W. J., Kallfelt, B., Holmes, P. V., Cajander, S., Bjersing, L., Wiquist, N., and Ahren, K., 1982, The study of ovulation in the isolated perfused rabbit ovary. I. Methodology and pattern of steroidogenesis, *Biol. Reprod.* **26:**456.

Jones, P. B. C., and Hsueh, A. J. W., 1981, Direct stimulation of ovarian progesterone metabolizing enzymes by gonadotropin-releasing hormone in cultured granulosa cells, *J. Biol. Chem.* **256:**1248.

Jones, R. E. (ed.), 1978, *The Vertebrate Ovary*, Plenum Press, New York.

Katagiri, C. H., 1974, A high frequency of fertilization in premature and mature coelomic toad eggs after enzymatic removal of vitelline membranes, *J. Embryol. Exp. Morphol.* **31:**573.

Katagiri, C., 1980, Appearance of pronucleus-inducing activity during hormonally induced meiotic maturation in toad oocytes, *Dev. Growth Diff.* **22:**237.

Katagiri, C., and Moriya, M., 1976, Spermatozoan response to the toad egg matured after removal of the germinal vesicle, *Dev. Biol.* **50:**235.

Kaufman, M. H., 1978, The experimental production of mammalian parthenogenetic embryos, in: *Methods in Mammalian Reproduction* (J. C. Daniel, Jr., ed.), p. 21, Academic Press, New York.

Kaufman, M. H., 1981, Parthenogenesis: A system facilitating understanding of factors that influence early mammalian development, in: *Progress in Anatomy*, (Harrison, R. J., ed), Vol. 1, p. 1, Cambridge University Press, London.

Kaufman, M. H., 1983, Early mammalian development: Partenogenetic studies, in: *Developmental and Cell Biology Series* (P. W. Barlow, P. B., Green, and C. C. Wylie, eds.), Cambridge University Press, London.

Kaufman, M. H., and Sachs, L., 1976, The complete preimplantation development of parthenogenetic mouse embryos, *J. Embryol. Exp. Morphol.* **35:**179.

Kaufman, M. H., Barton, S. C., and Surani, M. A. H., 1977, Normal postimplantation development of mouse parthenogenetic embryos to the forelimb bud stage, *Nature* **265:**53.

Kanatani, H., 1964, Spawning of starfish: Action of gamete-shedding substances obtained from radial nerves, *Science* **146:**1177.

Keefer, C. L., and Schuetz, A. W., 1982, Spontaneous activation of ovulated rat oocytes during *in vitro* culture, *J. Exp. Zool.* **224:**371.

Kobayashi, Y., Kitai, H., Santulli, R., Wright, K. H., and Wallach, E. E., 1984, Influence of calcium and magnesium deprivation on ovulation and ovum maturation in the perfused rabbit ovary, *Biol. Reprod.* **31:**287.

Korn, L. J., and Gurdon, J. B., 1981, The reactivation of developmentally inert 5S genes in somatic nuclei injected into *Xenopus* oocytes, *Nature* **289:**461.

Larsen, W. J., Tung, H. M., and Polking, C. M., 1982, Response of granulosa cell gap junctions to human chorionic gonadotropin (hCG) at ovulation, *Biol. Reprod.* **25:**1119.

Lessman, C. A., and Schuetz, A. W., 1981, Role of the follicle wall in meiosis reinitiation induced by insulin in *Rana pipiens* oocytes, *Am. J. Physiol.* **241:**E51.

Lessman, C. A., and Schuetz, A. W., 1982, Insulin induction of meiosis of *Rana pipiens* oocytes: Relation to endogenous progesterone, *Gamete Res.* **6:**95.

Lin, Y. W. P., and Schuetz, A. W., 1983, *In vitro* estrogen modulation of pituitary and progesterone induced oocyte maturation in *Rana pipiens*, *J. Exp. Zool.* **226:**281.

Lin, Y. W. P., and Schuetz, A. W., 1984, Intrafollicular action of estrogen in regulating pituitary induced ovarian progesterone synthesis and oocyte maturation in *Rana pipiens*: Temporal relationships and locus of action, *Gen. Comp. Endocrinol.*, in press.

Lohka, M. J., and Masui, Y., 1983, Formation in vitro of sperm pronuclei and mitotic chromosomes induced in amphibian ooplasmic components, *Science* **220:**719.

Longo, F. J., 1974, Ultrastructural changes in rabbit eggs aged *in vivo*, *Biol. Reprod.* **11:**22.

Longo, F. J., 1981, Regulation of pronuclear development, in: *Bioregulators of Reproduction* (G. Jagiello and H. J., Vogel, eds.), p. 529, Academic Press, New York.

Longo, F. J., So. F., and Schuetz, A. W., 1982, Meiotic maturation and the cortical granule reaction in starfish eggs, *Biol. Bull.* **163**:465.

Løvtrup, S., Landstrom, U., and Løvtrup-Rein, H., 1978, Polarities, cell differentiation and primary induction in the amphibian embryo, *Biol. Rev.* **73**:1.

Mahajan, D. K., and Little, A. B., 1978, Specific cortisol binding protein in porcine follicular fluid, *Biol. Reprod.* **18**:834.

Malacinski, G. M., and Brothers, A. J., 1974, Mutant genes in the Mexican Axolotl, *Science* **184**:1142.

Maller, J. L., 1983, Interactions of steroids with the cyclic nucleotide system in amphibian oocytes, *Adv. Cyclic Nucleotide Res.* **15**:295.

Manes, M. E., and Barbieri, F. D., 1976, Symmetrization in the amphibian egg by disrupted sperm cells, *Dev. Biol.* **53**:138.

Marot, J., Belle, R., and Ozon, R., 1977, In vitro facilitation of *Xenopus* oocyte maturation by subthreshold doses of progesterone, *Dev. Biol.* **59**:91.

Martin, G. G., and Talbot, P., 1981a, The role of follicular smooth muscle cells in hamster ovulation, *J. Exp. Zool.* **216**:469.

Martin, G. G., and Talbot, P., 1981b, Drugs that block smooth muscle contraction inhibit *in-vivo* ovulation in hamsters, *J. Exp. Zool.* **216**:483.

Masui, Y., and Clarke, H. J., 1979, Oocyte maturation, *Int. Rev. Cytol.* **57**:185.

Mattison, D. R., and Ross, G. T., 1983, Laboratory methods for evaluating and predicting specific reproductive dysfunctions: Oogenesis and ovulation, in: *Methods for Assessing the Effects of Chemicals on Reproductive Functions* (V. B. Vouk and P. J. Sheehan, eds.), pp. 217, John Wiley and Son, New York.

McGaughey, R. W., 1977, The culture of pig oocytes in minimal medium, and the influence of progesterone and estradiol-17$\beta$ in meiotic maturation, *Endocrinology* **100**:39.

McNatty, K. P., 1978, Follicular fluid, in: *The Vertebrate Ovary* (R. E. Jones, ed), p. 215, Plenum Press, New York.

McNatty, K. P., Smith, D. M., Makris, A., Osathanondl, R., and Ryan, K. J., 1980, Steroidogenesis by the human oocyte–cumulus cell complex *in vitro*, *Steroids* **35**:643.

Meyerhoff, P. G., and Masui, Y., 1977, Ca and Mg control of cytostatic factors from *Rana pipiens* oocytes which cause metaphase and cleavage arrest. *Dev. Biol.* **61**:214.

Mikamo, K., and Hamaguchi, H., 1975, Chromosomal disorders caused by preovulatory overripeness of oocytes, in: *Aging Gametes* (R. J. Blandau, ed.), p. 72, Karger, Basel.

Miller, B. G., and Armstrong, D. T., 1982, Infertility in superovulated immature rats: Role of ovarian steroid hypersecretion, *Biol. Reprod.* **26**:861.

Miyazaki, S., 1979, Fast polyspermy block and activation potential. Electrophysiological bases for this change during oocyte maturation of starfish, *Dev. Biol.* **70**:341.

Miyazaki, S-I., and Hirai, S., 1979, Fast polyspermy block and activation potential. Correlated changes during oocyte maturation of a starfish, *Dev. Biol.* **70**:327.

Moen, T. L., and Namenwirth, M., 1977, The distribution of soluble proteins along the animal–vegetal axis of frog eggs, *Dev. Biol.* **58**:1.

Monroy, A., Parisi, E., and Rosati, F., 1983, On the segregation of the germ and somatic cell lines in the embryo, *Differentiation* **23**:179.

Moor, R. M., and Trounson, A. O., 1977, Hormone and follicular factors affecting maturation of sheep oocytes *in-vitro* and their subsequent developmental capacity, *J. Reprod. Fertil.* **49**:101.

Moor, R. M., Polge, C., and Willadsen, S. M., 1980, Effect of follicular steroids on the maturation and fertilization of mammalian oocytes, *J. Embryol. Exp. Morphol.* **56**:319.

Moscona, A. A., and Monroy, A. (eds.), 1978, *Current Topics in Development Biology*, Vol. 12: *Fertilization*, Academic Press, New York.

Muggleton-Harris, A., Whittingham, D. G., and Wilson, L., 1982, Cytoplasmic control of preimplantation development *in vitro* in the mouse, *Nature* **299**:460.

Naftolin, F., and Butz, E., 1981, Sexual dimorphism, *Science* **211**:1263.

Nemoto, S. I., 1982, Nature of the 1-methyladenine-requiring phase in maturation of starfish oocytes, *Dev. Growth Dif.* **24(5)**:429.

Nemoto, S. I., Yoneda, M., and Uemura, I., 1980, Marked decrease in the rigidity of starfish oocytes by 1-methyladenine, *Dev. Growth Dif.* **22(3):**315.

Newport, J., and Kirschner, M., 1982, A major developmental transition in early Xenopus embryos. I. Characterization and timing of cellular changes at the midblastula stage, *Cell* **30:**675.

Newport, J., and Kirschner, M., 1982, A major developmental transition in early Xenopus embryos. II. Control of the onset of transcription, *Cell* **30:**687.

Nicosia, S. V., Wolf, D. P., and Inoue, M., 1977, Cortical granule distribution and cell surface characteristics in mouse eggs, *Dev. Biol.* **57:**56.

Niwa, K., and Chang, M. C., 1975, Fertilization of rat eggs *in vitro* at various times before and after ovulation with special reference to fertilization of ovarian oocytes matured in culture, *J. Reprod. Fertil.* **43:**435.

Ohno, S., and Smith, J. B., 1964, Role of fetal follicular cells in meiosis of mammalian oocytes, *Cytogenetics* **3:**324.

Parry, D. M., Willcox, D. L., and Thorburn, G. D., 1980, Ultrastructural and cytochemical study of the bovine corpus luteum, *J. Reprod. Fertil.* **60:**349.

Pasteels, J. J., 1961, The morphogenetic role of the cortex of the amphibian egg, in: *Advances in Morphogenesis*, Vol. III (M. Abercrombie and J. Brachet, eds.), p. 363, Academic Press, New York and London.

Pedersen, T., and Peters, H., 1968, Proposal for a classification of oocytes and follicles in the ovary of the mouse, *J. Reprod. Fertil.* **17:**555.

Peluso, J. J., 1976, Aging of mammalian ova, in: *Aging and Reproductive Physiology*, Vol. 2 (E. S. E., Hafez, ed.), p. 35, Ann Arbor Science, Ann Arbor, Michigan.

Perrault, S. D., Wolff, R. A., and Zirkin, B. R., 1984, The role of disulfide bond reduction during mammalian sperm nuclear decondensation *in vivo*, *Dev. Biol.* **101:**160.

Porter, R., and Whelan, J., 1983, *Molecular Biology of Egg Maturation*, CIBA Foundation Symposium 98, Pitman, London.

Readhead, C., Kaufman, M. H., Schuetz, A. W., and Abraham, G. E., 1979, Relationship between steroidogenesis and oocyte maturation in rat graafian follicles cultured *in vitro*, in: *Ovarian Follicular and Corpus Luteum Function* (C. P. Channing, J. Marsh, and W. A. Sadler, eds.), p. 293, Plenum Press, New York.

Reinboth, R., 1980, Can sex inversion be environmentally induced?, *Biol. Reprod.* **22:**49.

Ridgeway, E. B., Gilkey, J. C., and Jaffe, L. F., 1977, Free calcium increases explosively in activating medaka eggs, *Proc. Natl. Acad. Sci. USA* **74:**623.

Riley, R., Bennet, M. D., and Flavell, R. B., 1977, A discussion of the meiotic process, *Philos. Trans. R. Soc. Lond.* **277:**183.

Robinson, K. R., 1979, Electrical currents through full grown and maturing *Xenopus* oocytes, *Proc. Natl. Acad. Sci. USA* **76:**837.

Rolland, E., Van Hall, E. V., Hillier, S. G., McNatty, K. P., and Schoemaker, J. (eds.), 1982, *Follicular Maturation and Ovulation*, Excerpta Medica, Amsterdam.

Sakakura, T., Taguchi, O., Kojema, A., and Nishizuka, Y., 1979, Rapid oocyte death prior to and at puberty in neonatally thymectomized mice, in: *Ovarian Follicular and Corpus Luteum Function* (C. P. Channing, J. Marsh, and W. A. Sadler, eds.), p. 315, Plenum Press, New York.

Samson, G. A., and Schuetz, A. W., 1979, Progesterone induction of oocyte maturation in *Rana pipiens:* Reversibility of cycloheximide inhibition, *J. Exp. Zool.* **208:**213.

Sandor, T., 1980, Steroids in invertebrates, in: *Advances in Invertebrate Reproduction* (W. H. Clark, Jr. and T. S. Adams, eds.), p. 127, Elsevier North-Holland, New York.

Satoh, N., 1982, Timing mechanisms in early embryonic development, *Differentiation* **22:**156.

Sawai, T., 1979, Cyclic changes in the cortical layer of non-nucleated fragments of the newts' egg, *J. Embryol. Exp. Morphol.* **51:**183.

Schmell, E. D., and Gulyas, B. J., 1980, Ovoperoxidase activity in ionophore treated mouse eggs. II. Evidence of the enzymes' role in hardening the zona pellucida, *Gamete Res.* **3:**279.

Schoenmakers, H. J. N., 1980, The possible role of steroids in vitellogenesis in the starfish,

*Asterias rubens*, in: *Advances in Invertebrate Reproduction* (W. H. Clark, Jr. and T. S. Adams, eds.), p. 127, Elsevier North-Holland, New York.

Schreiber, J. R., and Hsueh, A. J. W., 1979, Progesterone "receptor" in rat ovary. *Endocrinology* **105:**915.

Schroeder, T. E., 1981, Microfilament-mediated surface change in starfish oocytes in response to 1-methyladenine: Implications for identifying the pathway and receptor sites for maturation-inducing hormones, *J. Cell Biol.* **90:**362.

Schroeder, T. E., and Stricker, S. A., 1983, Morphological changes during maturation of starfish oocytes: Surface ultrastructure and cortical action, *J. Cell Biol.* **98:**373.

Schroeder, T. E., and Strickland, P. L., 1974, Ionophore A-23187, calcium, and contractility in frog eggs, *Exp. Cell Res.* **83:**139.

Schuel, H., 1978, Secretory functions of egg cortical granules in fertilization and development. A critical review, *Gamete Res.* **1:**299.

Schuetz, A. W., 1967, Action of hormones in germinal vesicle breakdown in frog (*Rana pipiens*) oocytes, *J. Exp. Zool.* **166:**347.

Schuetz, A. W., 1971a, *In vitro* induction of ovulation and oocyte maturation in ovarian follicles of *Rana pipiens*. Effects of steroidal and non-steroidal hormones, *J. Exp. Zool.* **178:**377.

Schuetz, A. W., 1971b, Effects of precocious ovulation and mating on pregnancy in prepubertal rats, *Endocrinology* **89:**1325.

Schuetz, A. W., 1972a, Hormones and follicular functions, in: *Oogenesis* (J. D. Biggers, and A. W. Schuetz, eds.), p. 479, University Park Press, Baltimore.

Schuetz, A. W., 1972b, Induction of structural alterations in the preovulatory amphibian ovarian follicle by hormones, *Biol. Reprod.* **6:**67.

Schuetz, A. W,. 1972c, Estrogen and ovarian follicular functions in *Rana pipiens*, *Gen. Comp. Endocrinol.* **18:**32.

Schuetz, A. W., 1974, Role of hormones in oocyte maturation, *Biol. Reprod.* **10:**150.

Schuetz, A. W., 1975, Cytoplasmic activation of starfish oocytes by sperm and divalent ionophore A 23187, *J. Cell Biol.* **68:**86.

Schuetz, A. W., 1979a, Gametogenic processes and their relationship to normal and abnormal conceptus development, *J. Animal Sci. (Suppl. II)* **49:**1.

Schuetz, A. W., 1979b, The somatic–germ cell complex: interaction and transformations in relation to oocyte maturation, in: *Ovarian Follicular and Corpus Luteum Function* (C. P. Channing, J. Marsh, and W. A. Jedler, eds.), p. 307, Plenum Press, New York.

Schuetz, A. W., and Cloud, J. G., 1978, Cellular, hormonal and ionic interactions related to induced and spontaneous oocyte maturation, *Ann. Biol. Anim. Biochem. Biophys.* **213:**493.

Schuetz, A. W., and Dubin, N., 1981, Progesterone and prostaglandin secretion by ovulated rat cumulus cell–oocyte complexes, *Endocrinology* **108:**457.

Scheutz, A. W., and Lessman, C. A., 1982, Evidence for follicle wall involvement in ovulation and progesterone production by frog (*Rana pipiens*) follicles in vitro, *Differentiation* **22:**79.

Schuetz, A. W., and Longo, F., 1981, Hormone–cytoplasmic interactions controlling sperm nuclear decondensation and male pronuclear development in starfish oocytes, *J. Exp. Zool.* **215:**107.

Schuetz, A. W., and Rock, J., 1982, Stimulatory and inhibitory effects of human follicular fluid on amphibian oocyte maturation and ovulation *in-vitro*, *Differentiation* **21:**41.

Schuetz, A. W., and Samson, D., 1979, Nuclear requirement for post-maturational cortical differentiation of amphibian oocytes: Effects of cycloheximide, *J. Exp. Zool.* **210:**307.

Schuetz, A. W., Wallace, R. A., and Dumont, J. M., 1974, Steroid inhibition of protein incorporation by isolated amphibian oocytes, *J. Cell Biol.* **61:**26.

Schuetz, A. W., Hollinger, T. G., Wallace, R. A., and Samson, D., 1977, Inhibition of $^3$H-vitellogenin uptake by isolated amphibian oocytes injected with cytoplasm from progesterone treated mature oocytes, *Dev. Biol.* **58:**428.

Schuetz, A. W., Selman, K., and Samson, D., 1978, Alterations in growth, function and composition of *Rana pipiens* oocytes and follicles associated with microsporidian parasites, *J. Exp. Zool.* **204:**81.

Schultz, R. M., and Wasserman, P. M., 1977, Biochemical studies of mammalian oogenesis: Protein synthesis during oocyte growth and meiotic maturation, *J. Cell Sci.* **24**:167.

Schultz, R. M., Montgomery, R. Ward-Bailey, P. F., and Eppig, J. J., 1983, Regulation of oocyte maturation in the mouse: Possible roles of intercellular communication, cAMP and testosterone, *Dev. Biol.* **95**:294.

Shalgi, R., Dekel, N., and Kraicer, P. F., 1979, The effect of LH on the fertilizability and developmental capacity of rat oocytes matured *in-vitro, J. Reprod. Fertil.* **55**:429.

Siracusa, G., Whittingham, D. G., Molinaro, M., and Vivarelli, E., 1978, Parthenogenetic activation of mouse oocytes induced by inhibitors of protein synthesis, *J. Embryol. Exp. Morphl.* **43**:157.

Slack, J. M. W., 1983, *From Egg to Embryo, Determinative Events in Early Development*, Cambridge University Press, London.

Smith, L. D., and Ecker, R. E., 1970, Uterine suppression of biochemical and morphogenetic events in *Rana pipiens, Dev. Biol.* **22**:622.

Snyder, B. W., and Schuetz, A. W., 1973, Effect of cyanoketone on gonadotropin and steroid induced events related to ovarian follicular maturation in *Rana pipiens, J. Exp. Zool.* **183**:333.

Spiegel, J., Jones, E., and Snyder, B. W., 1978, Estradiol-17β interference with meiotic maturation in Rana pipiens ovarian follicles: Evidence for inhibition of 3β-hydroxysteroid dehydrogenase, *J. Exp. Zool.* **204**:187.

Steinhardt, R., Zucker, R., and Schatten, G., 1977, Intracellular calcium release at fertilization in the sea urchin egg, *Dev. Biol.* **58**:185.

Stevens, L. C., and Varnum, D. S., 1974, The development of teratomas from parthenogenetically activated ovarian mouse eggs, *Dev. Biol.* **37**:369.

Strauss, W. F., and Meyer, R. K., 1962, Neural timing of ovulation in immature rats treated with gonadotropin, *Science* **137**:860.

Strickland, S., and Beers, W. H., 1979, Studies on the enzymatic bases and hormonal control of ovulation, in: *Ovarian Follicular Development and Function*, (A. R. Midgley, and W. A. Sadler, eds.), p. 143, Raven Press, New York.

Sunkara, P. S., Wright, D. A., and Rao, P. N., 1979, Mitotic factors from mammalian cells induce germinal vesicle breakdown and chromosome condensation in amphibian oocytes, *Proc. Natl. Acad. Sci. USA* **76**:2799.

Szollosi, D., 1975, Mammalian eggs aging in the fallopian tubes, in: *Aging Gametes*, (R. J., Blandau, ed.), Karger, Basel.

Talbot, P., and Chacon, R. S., 1982, *In vitro* ovulation of hamster oocytes depends on contraction of follicular smooth muscle cells, *J. Exp. Zool.* **224**:409.

Talbert, G. B., 1978, Effect of aging of the ovaries and female gametes in reproductive capacity, in: *The Aging Reproductive System*, Vol. 4: *Aging* (E. L. Schneider, ed.), p. 59, Raven Press, New York.

Taya, K., and Greenwald, G. S., 1980, *In vitro* and *in vivo* ovarian steroidogenesis in the long term hypophysectomized hamster, *Endocrinology* **106**:1093.

Thadani, V. M., 1980, A study of hetero-specific sperm–egg interactions in the rat, mouse and deermouse using *in vitro* fertilization and sperm injection, *J. Exp. Zool.* **212**:435.

Thibault, C., 1977, Are follicular maturation and oocyte maturation independent processes?, *J. Reprod. Fertil.* **51**:1.

Thibier, C., Mulner, O., and Ozon, R., 1982, In vitro effects of progesterone and estradiol-17β on choleragen activated *Xenopus* oocyte adenylate cyclase, *J. Steroid Biochem.* **17**:191.

Toyoda, Y., and Chang, M. C., 1974, Fertilization of rat eggs *in vitro* by epididymal spermatozoa and the development of eggs following transfer, *J. Reprod. Fertil.* **36**:9.

Tsafriri, A., Dekel, N., and Bar-Ami, S., 1982, The role of oocyte maturation inhibitor in follicular regulation of oocyte maturation, *J. Reprod. Fertil.* **64**:541.

Tsang, B. K., and Carnegie, T. A., 1983, Calcium requirement in the gonadotrophic regulation of rat granulosa cell progesterone production, *Endocrinology* **113**:763.

Turpin, R., and Lejune, J., 1969, *Human Afflictions and Chromosomal Aberrations*, Pergamon Press, Oxford.

Tyler, J. P. P., Matson, P. L., Collins, W. P., and Dukes, M., 1981, Dissociation of oocyte maturation and ovulation in mice pretreated with a derivative of dihydropyridozinone, *J. Reprod. Fertil.* **62**:455.

Uno, V., and Hoshi, M., 1978, Separation of the sperm agglutinin and the acrosome reaction-inducing substances in egg jelly of starfish, *Science* **200**:58.

Vacquier, V. D., 1981, Dynamic changes of the egg cortex, *Dev. Biol.* **84**:1.

Vilain, J. P., Moreau, M., and Guerrier, P., 1980, Uncoupling of oocyte follicle cells triggers reinitiation of meiosis in amphibian oocytes, *Dev. Growth and Diff.* **22(4)**:687.

Wakasugi, N., 1974, A genetically determined incompatibility system between spermatozoa and eggs leading to embryonic death in mice, *J. Reprod. Fertil.* **41**:85.

Wallace, R., and Misulovin, Z., 1978, Long term growth and differentiation of *Xenopus* oocytes in a defined medium, *Proc. Natl. Acad. Sci. USA* **75**:5534.

Wasserman, W. J., and Smith, L. D., 1978, The cyclic behavior of a cytoplasmic factor controlling nuclear membrane breakdown, *J. Cell Biol.* **78**:R15.

Whittingham, D. G., 1979, In vitro fertilization, embryo transfer and storage, *Br. Med. Bull.* **35**:105.

Whittingham, D. G., 1980, Parthenogenesis in mammals, in: *Oxford Reviews of Reproductive Biology* Vol. 2, (D. A. Finn, ed), p. 205, Oxford University Press, Oxford.

Willcox, D., 1983, Two binding proteins for progesterone in the bovine corpus luteum, *Biol. Reprod.* **29**:487.

Willcox, D. L., and Alison, M. R., 1982, Release of protein which binds progesterone from the bovine corpus luteum, *J. Endocrinol.* **92**:57.

Willcox, D. L., Parry, D. M., and Morris, T. F., 1983, Storage of progesterone in secretory granules, *Spec. Sci. Technol.* **6**:3.

Wilson, E. B., 1925, *The Cell in Development and Heredity*, Macmillan, New York.

Yamamato, K., and Yoneda, M., 1983, Cytoplasmic cycle in meiotic division of starfish oocytes, *Dev. Biol.* **96**:166.

Yamamato, T., 1969, Sex differentiation, in: *Fish Physiology* Vol. III, (W. J. Hoar, and D. J. Randall, eds.), p. 117, Academic Press, New York.

Yamamato, T., and Kajishima, T., 1968, Sex hormone induction of sex reversal in the goldfish and evidence for male heterogamety, *J. Exp. Zool.* **168**:215.

Yanagimachi, R., 1981, Mechanisms of fertilization in mammals, in: *Fertilization and Embryo Development In Vitro* (L. Mastroianni and J. D. Biggers, eds.), p. 81, Plenum Press, New York.

Yanagimachi, R., Yanagimachi, H., and Roger, B. J., 1976, The use of zona free animal ova as a test system for the assessment of the fertilizing capacity of human spermatozoa, *Biol. Reprod.* **15**:471.

Yanagishita, M., Rodbard, D., and Hascall, V. C., 1979, Isolation and characterization of proteoglycans from porcine ovarian follicular fluid, *J. Biol. Chem.* **254**:911.

Yoshizaki, N., and Katagiri, C. H., 1981, Oviductal contribution to alteration of the vitelline coat in the frog, *Rana japonica*, an electron microscopic study, *Dev. Growth Dif.* **23(5)**:495.

Young, G., Kagawa, H., and Nagahama, Y., 1983, Evidence for a decrease in aromatase activity in the ovarian granulosa cells of amago salmon (Oncorhynchus rhodurus) associated with final oocyte maturation, *Biol. Reprod.* **29**:310.

Zeilmaker, G. H., and Verhamme, C. M. P. M., 1974, Observations in rat oocyte maturation *in vitro*. Morphology and energy requirements, *Biol. Reprod.* **11**:145.

Zuckerman, S., and Weir, B. J. (eds.), 1977, *The Ovary*, Academic Press, New York.

Chapter 2

# Vitellogenesis in Insects

JOHN H. POSTLETHWAIT and FRANCO GIORGI

## 1. Strategies for Provisioning the Egg

Fertilized eggs are the crossroads of development. They represent both the origin and the biological goal of each individual organism. Eggs provide in two ways for the embryo they enclose: They contain developmental instructions given by the mother to direct the initial phases of embryogenesis (see Chapters 11–13), and they are usually provisioned with nutritive substances to support the embryo until it can obtain its own food. This chapter concentrates on a portion of the second aspect of oogenesis—how storage materials come to reside in insect eggs.

There are two basic strategies for delivering nutrients to the developing embryo. In what can be called the "catering" strategy, the embryo stays inside the mother, who continuously provides nutrients via her bloodstream until the embryo becomes mature enough to eat for itself. This strategy is used by mammals. In the second approach—the "box-lunch" strategy—the mother packages nutrients **(yolk)** into the egg that the embryo uses after the mother lays the egg. This mechanism is employed by most invertebrates and many vertebrates.

The box-lunch strategy of provisioning the egg is called **vitellogenesis,** or yolk deposition. There are two general mechanisms of vitellogenesis: The first, in which the egg produces its own yolk, is called **autosynthetic vitellogenesis;** the second, in which the egg obtains its yolk from an extraovarian source, is called **heterosynthetic vitellogenesis.** In the latter strategy, yolk precursors are transferred to the oocyte via the maternal blood stream. Many lower invertebrates have opted for the autosynthetic approach (Boyer, 1972; Huebner and Anderson, 1976), including the primitive insects, the Apterygota (Bilinski, 1976, 1979). More advanced insects and the vertebrates use the heterosynthetic mechanism.

The tactical aspects of the heterosynthetic strategy are remarkably

JOHN H. POSTLETHWAIT • Department of Biology, University of Oregon, Eugene, Oregon 97403 FRANCO GIORGI • Istituto di Istología e Embriología, 56100 Pisa, Italy.

similar in different types of organisms. Usually a hormone stimulates a digestive organ, such as the liver of vertebrates or the fat body of insects, to secrete a specific protein into the blood. This protein, called **vitellogenin (Vg),** is taken from the circulation and incorporated into yolk granules in the oocyte. Post-translational modifications may accompany this process, altering the Vg to become **vitellin (Vn),** the mature yolk protein. Vn is then used by the embryo to support development.

The Vn peptides thus have a complicated ontogeny, which includes hormonally stimulated gene activity, secretion into the blood, specific recognition and uptake by the oocyte, packaging into yolk granules, and—finally—proteolysis in the embryo. Each of these steps has been intensively investigated in insects as models for our larger understanding of the mechanisms of endocrine action, gene regulation, peptide secretion, protein sequestration, and cell interactions in development. This chapter attempts to indicate the current status of our understanding of insect vitellogenesis in terms of these general processes. A comprehensive literature review is not intended. Instead, examples have been chosen to illustrate important general features. Since the details of vitellogenesis differ in each insect species, we have concentrated on a few species and have had to ignore many interesting and important studies on other insects. We first discuss the development and structure of ovaries and yolky oocytes and then the origin of yolk and how its production is regulated at developmental and molecular levels. Finally, we discuss how yolk is used during development.

## 2. How Insects Make Eggs

### 2.1. Development and Structure of the Female Reproductive System

A central idea in developmental biology is the continuity of life expressed in the germ cells: An egg is partitioned into cells, a few of which will again develop into eggs so that the cycle may be repeated. The germ cells—egg and sperm—arise from **pole cells,** which segregate from the rest of the single cell-layered embryo very early in development (Sonnenblick, 1941). In insects, pole cells originate from cells that come to contain the **polar granules**—particles of protein and RNA deposited in the posterior ooplasm (Fig. 1) by the mother (Mahowald, 1962, 1971). The polar granules change in form but remain associated with the nuclei of pole cells throughout development and function as germ cell determinants (see Chapter 13).

As development progresses, pole cells migrate to the interior of the embryo and become associated with a group of mesodermal cells to form the ovary. During the larval stages, the roughly spherical ovary increases in cell number, but differentiation commences only during the pupal stage

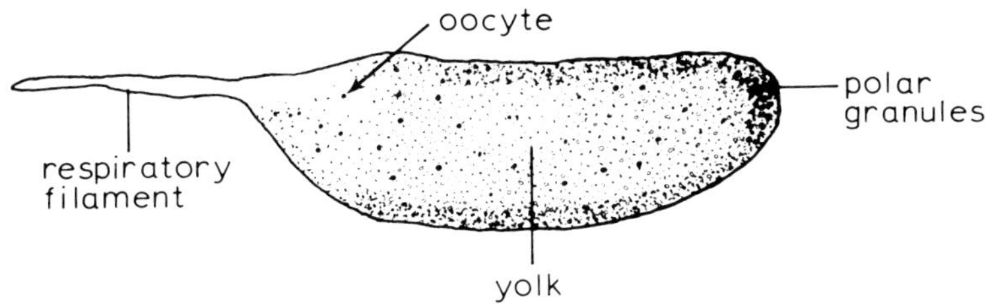


**Figure 1.** Insect egg. Oocyte refers to an oocyte nucleus. (After King, 1970.)

as cylindrical pillars of cells become evident in the ovary. These pillars develop into **ovarioles,** which are linear arrays of progressively more mature ovarian follicles (Fig. 3). Follicles in insects consist of the **oocytes** surrounded by an epithelial layer of **follicle cells,** sometimes associated with nutritive cells called **nurse cells (trophocytes).** During differentiation, the ovary unites with the oviduct, which grows anteriorly from the primordium of the external genitalia to complete the female reproductive system (Fig. 2).

Ovarioles are divided into several zones (Fig. 3). The anteriormost zone includes the **terminal filament,** which attaches the ovariole to the inner body wall. Zone 2 contains the anterior of the **germarium,** a region of active cell division consisting of both prefollicular cells and differentiating early oocytes. In the posterior germarium, the oocytes become sur-

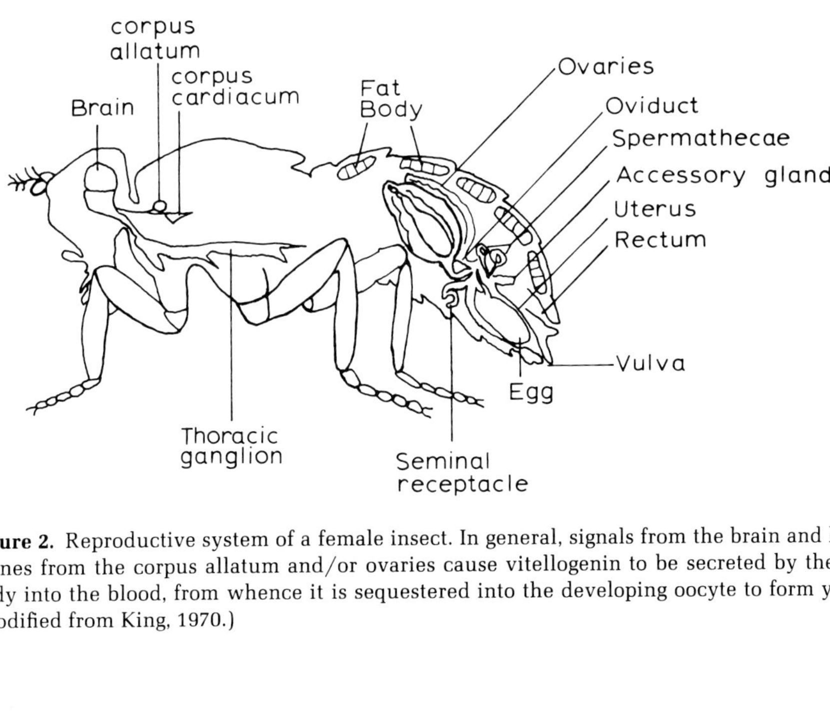


**Figure 2.** Reproductive system of a female insect. In general, signals from the brain and hormones from the corpus allatum and/or ovaries cause vitellogenin to be secreted by the fat body into the blood, from whence it is sequestered into the developing oocyte to form yolk. (Modified from King, 1970.)

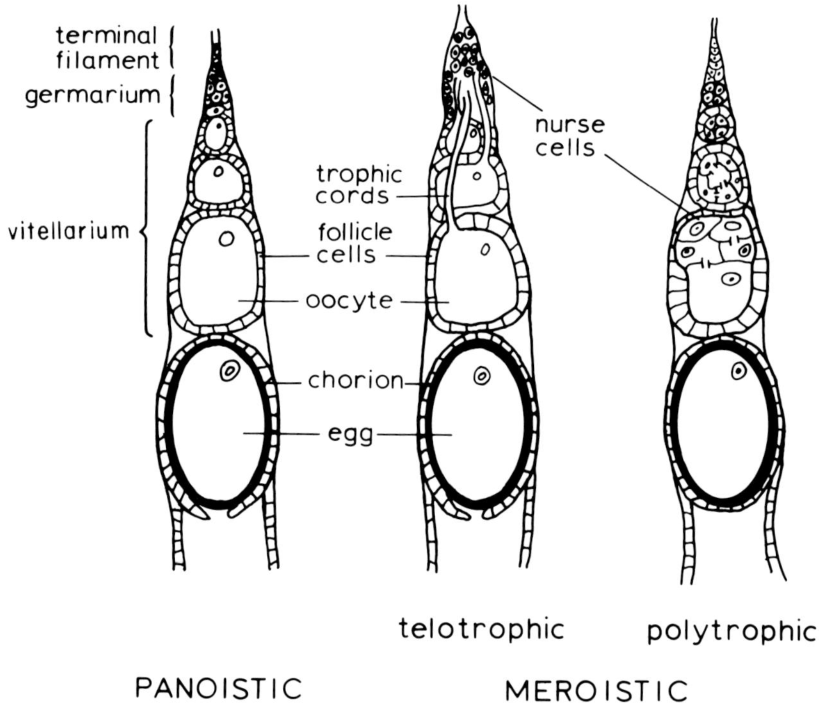


**Figure 3.** Three types of insect ovarioles. Ribosomes and other materials are produced by the oocyte nucleus in panoistic ovarioles but by nurse cells in meroistic ovarioles. In telotrophic ovarioles, the nurse cells are situated in the anterior of the ovariole and connected to the oocyte by slender trophic cords, but in polytrophic ovarioles, the nurse cells are close to the oocyte within the follicle.

rounded by a layer of follicle cells. The follicle then moves to the **vitellarium,** it increases in size, accumulates yolk, and becomes chorionated. Finally, the oocyte is released into the oviduct as a fully mature egg.

## 2.2. Approaches to Yolk Storage

The egg and its embryo represent a vulnerable stage in the life of an organism. Being immobile, the embryo can neither flee to escape predators nor seek protection from harsh environmental conditions. In most cases, a premium is set on completing embryonic development as rapidly as possible to form a mobile, self-feeding stage. Many new proteins must be produced to turn the single very large egg cell into a well-organized, many-celled creature. This process requires millions of ribosomes. The number of ribosomal RNA (rRNA) genes present in the genome of the single egg nucleus is simply not great enough to provide the necessary number of ribosomes in a biologically convenient time. Insects have devised two basic strategies in order to deliver a large number of rRNA molecules in a very short period of time (Fig. 3). Both strategies involve increasing the number of rRNA genes that deliver rRNA molecules to the oocyte, but they differ in the method used.

In one strategy, the number of rRNA genes in the oocyte nucleus is vastly and selectively amplified (see Cave, 1982). Since all the ribosomes come from the oocyte, this is called the **panoistic** mechanism. It is found in roaches and crickets, as well as certain other groups. In the second strategy—the **meroistic** mechanism—ribosomes are supplied to the oocyte not by the oocyte itself, but rather by several nurse cells. In meroistic ovaries the number of rRNA genes per nurse cell is increased, but most of the other genes are increased in number as well to form polytene chromosomes, that are similar to—but smaller and less coherent than—the polytene chromosomes found in larval salivary gland cells. In contrast to the nurse cells, the oocyte nucleus remains diploid until meiosis in meroistic ovaries, and it is not synthetically very active.

Meroistic ovaries themselves are of two subtypes. In one, the nurse cells are directly associated with the oocyte, and both cell types are surrounded by a layer of follicle cells. These are called **polytrophic** meroistic ovaries to indicate that many cells feed the oocyte. In the second subtype, called **telotrophic** meroistic ovarioles, the cluster of trophic cells remains at the tip of the ovary, as the name suggests, connected to the oocyte as it moves down the ovariole through ever-lengthening slender **trophic cords.**

### 2.2.1. Vitellogenesis in Panoistic Ovarioles

Each ovariole in the ovary of a panoistic insect matures a single oocyte at a time (Fig. 3). In the roach *Blatella germanica*, for example, the terminal oocyte increases in volume 180-fold over a period of 10 days, while the penultimate oocyte increases only threefold. The presence of a terminal oocyte inhibits the penultimate oocyte from developing (Pratt and Davey, 1972; Gillott and Eliott, 1976). This could be accomplished by the terminal oocyte inhibiting the penultimate oocyte either directly or indirectly via a neural or humoral mechanism. In the stick insect *Clitumnus extradentatus*, removal of the terminal follicle results in the resumption of vitellogenic growth in the penultimate follicle, but only if the tissue connecting the two adjacent follicles remains intact (Mesnier, 1980). This suggests that the interfollicular tissue notifies the penultimate oocyte about the state of the terminal oocyte directly. In a cockroach, on the other hand, a humoral factor from the terminal follicle regulates the growth of the penultimate follicle (Maa and Bell, 1977).

### 2.2.2. Vitellogenesis in Telotrophic Meroistic Ovarioles

In telotrophic meroistic ovarioles, nurse cells located at the anterior tip of the ovariole help deliver ribosomes and other substances to the oocyte through fine cytoplasmic cords (Fig. 3). Telotrophic ovaries are found among the true bugs and some beetles. In contrast to most panoistic ovarioles, many oocytes can undergo vitellogenesis simultaneously in telotrophic ovarioles.

The nutritive role of the apical trophocytes is readily demonstrated by following the time course of [$^3$H]uridine incorporation into ribonucleoprotein by autoradiography (Buning, 1972; MacGregor and Stebbings, 1970; Ullmann, 1973; Mays, 1972). Label is first detected in the nuclei of the apical trophocytes and with longer incubation times becomes visible in the trophic cords and finally in the oocyte cytoplasm. The transfer can be proven by ligation experiments: In ovarioles ligated to prevent cytoplasmic flow, the label accumulates on the trophocyte side of the ligation (Zinsmeister and Davenport, 1971; Davenport, 1976). The label in the oocyte nucleus is at all times very small, suggesting that it is not very active in RNA synthesis.

What causes the transport of materials down the trophic cords? The use of intracellular fluorescent probes has shown that the trophocytes and the oocyte form a cytoplasmic continuum (Huebner, 1981). The persistence of this continuity is ensured by a massive system of microtubules packed in the trophic cords (MacGregor and Stebbings, 1970). Agents such as colchicine and vinblastine, which are known to interfere with the assembly of the microtubular apparatus, decrease both the patency (i.e., openness) of the cords as well as the amount of [$^3$H]uridine finding its way to the oocyte (Stebbings, 1971). Whether the microtubules themselves play a role in the actual transport of materials down the cord or merely maintain the elongate structure of the cords is as yet unknown.

One possibility for the motive force in material transport down the nutritive cords is a difference in electrical potential (see DeLoof, 1983 and Chap. 8 for detailed discussion). **Juvenile hormone,** a sesquiterpenoid hormone that maintains larval insects in an immature form and also stimulates reproduction in adults, enhances the difference in electric potential between the oocyte and the trophocytes. This electrical gradient is thought to be sufficient to provide much of the driving force for the electrophoresis of trophocyte-produced ribosomes back to the oocyte (Woodruff and Telfer, 1980).

### 2.2.3. Vitellogenesis in Polytrophic Meroistic Ovarioles

Polytrophic ovarioles, which are found in flies, bees, and moths, contain follicles oriented in the ovariole such that the nurse cells are always anterior, the oocyte posterior, and the follicle cells surround both (Fig. 3). The oocyte itself is also polarized: The nucleus is located near the nurse cells. Whereas the anterioposterior axis of the egg (and eventually its embryo) corresponds to that of the mother, the dorsoventral axis does not. Fly follicles are oriented to have the notch that indicates the future hatching site of the embryo centered on a site within the ovary rather than on the mother's dorsal side (Kleine-Schonnefeld and Engels, 1981).

Follicular development commences in the germarium, in which a germ line cell undergoes mitosis several times (four in *Drosophila*) with incomplete cytokinesis to produce an interconnected set of cells called **cystocytes** (16 cells in *Drosophila*), one of which will become the oocyte (Brown and King, 1964). The cystocyte cluster migrates down the germarium and is enclosed by mesodermally derived cells destined to become follicle cells. The cystocyte cluster is reshaped to cause the oocyte to become the most posterior cell, and the follicle cell sheath becomes a single cell-layered epithelium (see King, 1970).

After the follicle leaves the germarium and enters the vitellarium, the oocyte begins to enlarge. The follicle cells then migrate posteriorly to surround the oocyte with a columnar epithelium, leaving a thin squamous epithelium around the anterior nurse cells. The oocyte continues to grow, sequesters yolk from the hemolymph and ribosomes from the nurse cells. The nurse cells become smaller and finally disappear as they collapse and dump their contents into the oocyte. As the follicle matures, the follicle cells synthesize and secrete first a product that is endocytosed by the oocyte and, subsequently, two egg coverings—the **vitelline envelope** and the **chorion,** or egg shell (Quattropani and Anderson, 1969; Cummings *et al.*, 1971) to form a mature oocyte that is ready for fertilization.

## 2.3. The Goal of Vitellogenesis: The Mature Egg

The insect egg forms a link between two generations of self-sufficient individuals: the mother and the feeding first-stage larva. To carry out this task, the oocyte must perform a number of functions, each with more or less obvious structural correlates. First, the egg provides developmental instructions both in its cytoplasmic architecture and in the genetic content of the oocyte nucleus. The egg contains a thin layer of cytoplasm below the oocyte plasma membrane. Embedded in this cytoplasm at the posterior end of the egg are the polar granules, which we have seen are closely associated with the determinants for pole cells. If other localized substances exist that canalize the development of cells that include them, they are not obvious from structural studies (see also Chapter 13).

Second, the oocyte prepares for fertilization, the union of sperm and egg. The portal for sperm entry is the **micropyle,** a small cone at the egg's anterior. The act of fertilization releases the developmental programs stored within the egg. Third, the egg protects the developing embryo from physical injury, and accomplishes the mutually antagonistic functions of encouraging gas exchange while discouraging water loss. This is accomplished by the chorion, a proteinaceous decorated coat surrounding the egg cell. In some eggs, delicate chorionic appendages similar to a skin diver's snorkel are elaborated to aid in gas exchange.

And finally, the egg provides the growing embryo with enough nutritional substances to fuel the impressively rapid growth and differentiation of the embryo. This is accomplished mainly by materials packaged in **yolk spheres**—vesicles of protein, lipid, and carbohydrate—which are the most obvious cytological and biochemical constituents of the egg. The yolk spheres virtually pack the interior of the egg, forcing the cytoplasm into a thin film lying just below the surface of the egg (Fig. 1).

## 2.4. The Nature of Yolk in Insect Eggs

There are three kinds of yolk in insect eggs: protein yolk, lipid yolk, and glycogen granules (Fig. 4). Each of these types of yolk is physically separate and easily recognizable in the electron microscope.

### 2.4.1. Protein Yolk

Most of the protein reserves found in an oocyte are contained in yolk spheres (Fig. 4), which consist of three basic components: a limiting membrane, a superficial layer, and an electron-dense main body. The relative proportion and morphology of yolk sphere components varies with the stage in oogenesis: The main body of the yolk sphere is initially rather homogeneous in appearance, but in later stages of oogenesis it becomes granular, or hexagonally crystalline (Roth and Porter, 1964; King, 1970). Immunohistochemistry has shown that the main body consists of Vn; there is evidence to suggest that the superficial layer may be rich in glycoprotein (Giorgi and Jacob, 1977a; Favard and Favard-Sereno, 1969).

The mechanisms by which yolk spheres attain their final positions in the egg have been investigated by studying the distribution of radioactivity after pulses of radioactively labeled Vg and by interfering with the cytoskeleton by using colchicine (Gutzeit and Koppa, 1982; Koch and Spitzer, 1983; Hausman *et al.*, 1971). These studies have shown that in the fruit fly *Drosophila*, cytoplasmic streaming distributes the yolk spheres widely throughout the ooplasm, but in the silk moth *Cecropia*, older yolk spheres are firmly positioned and are simply pushed deeper into the ooplasm by the newly formed spheres.

**2.4.1.a. Evolution of Vitellin.** The major female-specific protein sequestered from the blood and stored in yolk spheres is defined as vitellin, and its soluble counterpart in the hemolymph is vitellogenin (Fig. 5). The vitellins in different insect groups share structural similarities. They are generally high-molecular-weight phospholipoglycoproteins that some-

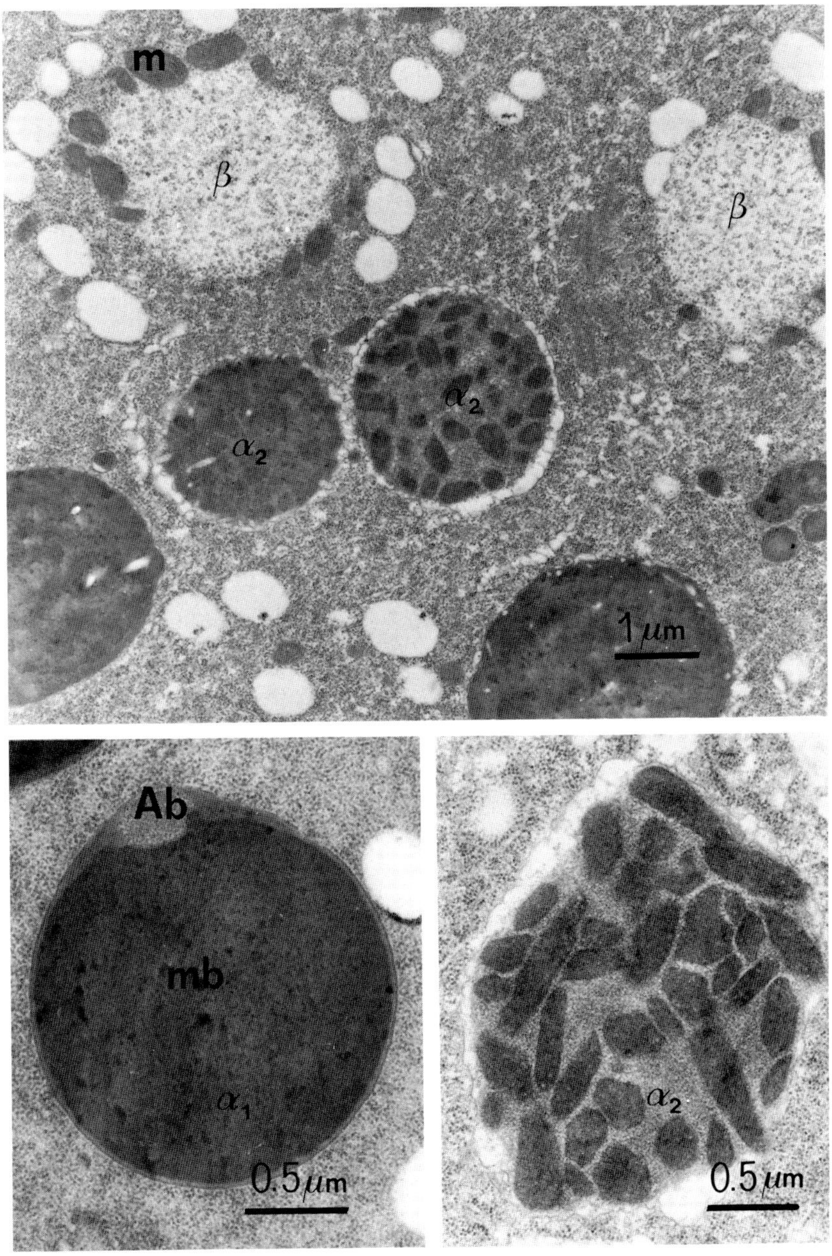


**Figure 4.** Electron micrographs of yolk from *Drosphila*. Ab, associated body; $\alpha_1$, immature protein yolk spheres; $\alpha_2$, mature protein yolk spheres; $\beta$, glycogen yolk; m, mitochondrion; mb, main body; unlabeled clear vesicles, lipid yolk.

times contain several polypeptides. Vgs and Vns from several species have been compared in a number of different ways (Hagedorn and Kunkel, 1979; Engelmann, 1979; Harnish and White, 1982). Depending on the size and amount of post-translational processing of their Vgs, insects have been clustered into three groups by Harnish and White (1982). Although this classification scheme may break down as more Vns are studied carefully, for now it represents a rather successful attempt to organize what otherwise would be just a list of data.

From the primary Vg translation products in probably all three groups of insects, a signal peptide (Blobel and Dobberstein, 1975) is cleaved cotranslationally (see Section 3.2.3). The Vg primary translation products of group 1 insects are $>200{,}000\ M_r$ in size. In the simplest cases such as the silk moth *Cecropia*, the primary product of 220 kd is cleaved in the fat body cell to yield two polypeptides—a large one of $180{,}000\ M_r$ and a small one of $47{,}000\ M_r$. Mundall and Law (1979) have proposed that after cleavage, the small peptide is encompassed by the large one, and thus the two peptides may represent different functional domains of the original primary translation product. Most insects belong to this group, including primitive insects, such as mayflies and roaches, and even some more advanced orders, such as bugs and moths.

Group 2 species have primary translation products of ~180,000, which is similar to the large Vn subunit of group 1. Except for the signal sequence, these Vgs are not cleaved before secretion. Group 2 includes the advanced order of bees and wasps and lower flies such as mosquitos.

Group 3 insects have Vn consisting of several polypeptides of ~50,000 each (Fig. 5), approximately the same size as the small fragment cleaved from Group 1 Vn. These polypeptides combine to form the native protein with a molecular weight of ~200,000 (Fourney *et al.*, 1982). Group 3 includes only the higher flies, such as fruit flies and blowflies (Bownes and Hames, 1978; Huybrechts and DeLoof, 1982).

Although insects can be divided into three distinct groups on the basis of Vg size, the amino acid composition of Vn from different species share some general characteristics. They contain large amounts of aspartic acid or asparagine, glutamic acid or glutamine, and leucine. However, they are generally low in histidine and methionine, and they usually have no cysteine. Immunological similarities have also been detected, but only within the family level, such as two locusts, or two kinds of blowflies. These data indicate that the Vn proteins are related, but not strongly conserved, suggesting that they are evolving at a rapid rate. This of course might be expected for a protein serving primarily a nutritive function with few further constraints to its structure.

Harnish and White (1982) suggested the following model for the evolution of the Vg genes in insects. The original gene may have encoded a protein with two domains: a large one of $\sim 160{,}000\ M_r$ and a small one of $\sim 50{,}000\ M_r$. This situation has been retained in contemporary primitive

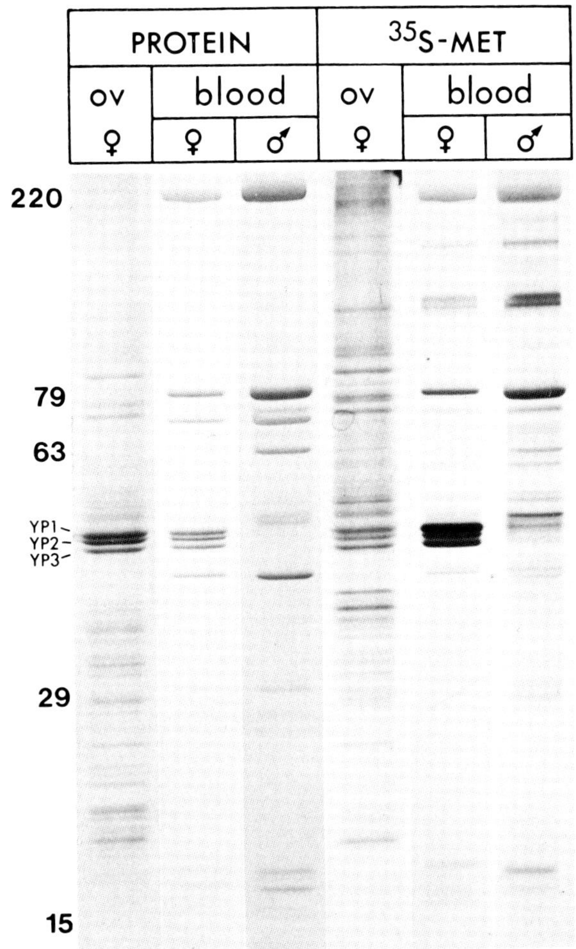


**Figure 5.** (Left) Sodium dodecylsulfate polyacrylamide gel electrophoresis of proteins from *Drosophila* ovaries and blood from females and males injected with [$^{35}$S]methionine. The yolk polypeptides (YP1, YP2, YP3), which constitute Vg, are found in the hemolymph and ovaries of sexually mature females, but not in the hemolymph of males. (Right) Autoradiography of the gel shows that YPs are rapidly being synthesized by females.

insects and even in some more advanced forms, such as moths. Groups 2 and 3 may have arisen by domain elimination, group 2 having lost the small domain, and group 3 having lost the large domain. The different functions of the large and small domains is unknown.

Along with these changes in gene structure, alterations in gene regulation apparently occurred. In the presumed primitive type, group 1, juvenile hormone is the main regulatory agent (Chen *et al.*, 1979), whereas in mosquitos (group 2), ecdysterone plays a significant role in regulation of Vg synthesis (Fallon *et al.*, 1974). Group 3 is more complex in that (1) both the ovarian follicle cells and the fat body synthesize Vg, and (2) there is evidence that both JH and an ecdysteroid may control Vg synthesis.

**2.4.1.b. Transformation of Vg to Vn.** The transformation of Vg to Vn in insects differs from that found in vertebrates (see Chapter 3). After uptake into the oocyte, frog Vg is cleaved to form the constituent polypeptides of Vn: phosvitin and lipovitellin. By contrast, in insects, the circulating Vg is taken into the oocyte without cleavage. For example, Vg and Vn are immunologically and electrophoretically identical in locusts, differing only in lipid content (Chen and Wyatt, 1981), and no major differences have been detected among the three YPs in *Drosophila* blood and ovary. The cleavage step in group 1 insects (Harnish *et al.*, 1982) occurs before secretion from the fat body cells, whereas analogous cleavage in frogs occurs only after uptake into the oocyte.

The carbohydrates, which are covalently attached to the yolk proteins in most insects, consist largely of mannose and glucosamine residues (Yamasaki, 1973; Gellissen *et al.*, 1976). These do not change substantially when the Vg enters the egg and becomes Vn (Mundall and Law, 1979; Kunkel *et al.*, 1980). Whether the carbohydrate moieties play a role in yolk protein recognition by the oocyte is not yet known.

Lipid is associated with some yolk proteins up to a level of about 10%. Unlike the carbohydrate component of vitellins, the lipid moiety (di- and triacylglycerol, cholesterol, and phospholipids) undergoes substantial change after Vg enters the oocyte, with some lipids being removed from the Vn and contributing to the overall lipid content of the oocyte.

**2.4.1.c. Other Proteins in Yolk Spheres.** Although pinocytosis preferentially removes Vg from the insect blood, proteins other than Vg can also be sequestered and packaged into yolk spheres. These nonvitellins are of three types: (1) blood proteins that are not female specific, (2) proteins made by the follicle cells, and (3) proteins made by the oocyte or nurse cells. The first two types enter the oocyte by pinocytosis, the third by fusion of vesicles from the Golgi apparatus.

Since exogenous proteins such as ferritin, peroxidase, or even the anionic polymer trypan blue can be taken into the oocyte after injection into the insect's hemocoel (Telfer, 1960; W. A. Anderson and Spielman, 1971; Mahowald, 1972a; Giorgi and Jacob, 1977b; Yonge and Hagedorn, 1977), it might be expected that yolk spheres would contain some blood proteins that enter by a nonspecific route, perhaps by entrapment in the pinocytotic vesicles. In fact, nearly all blood proteins can be detected in minor quantities in yolk, and some, such as **lipophorin** in the silkmoth *Cecropia*, occur in substantial quantities. Other blood proteins, such as **microvitellin** and **arylphorin** in *Cecropia*, may even be accumulated against a concentration gradient (Telfer *et al.*, 1981a; W. H. Telfer, personal communication), suggesting that there may exist a specific mechanism of uptake for them as well as for Vg.

There exists an interesting biological analogue to the scientist injecting foreign proteins into the hemolymph in order to study peptide uptake into the egg. In the grasshopper *Melanoplus*, certain proteins transferred to the female during copulation gain access to her hemolymph and are subsequently sequestered by the developing oocyte (Friedel and Gillott, 1977), a valiant attempt by the male to make a material contribution to his offspring.

The second type of prominent nonvitellin found in yolk granules is produced by the follicle cells, secreted into the space between the follicle cell and the oocyte, and sequestered by pinocytosis along the same route as Vg. These proteins appear to play an especially prominent role in the vitellogenesis of the Lepidoptera. One such protein is **paravitellin,** which is deposited both in yolk granules and in vesicles lying nearer the surface of the egg (Telfer *et al.*, 1981a; Shirk *et al.*, 1984). In silkmoths, paravitellin-like proteins seem to be used in embryogenesis before Vn, which in some cases persists well into the first instar (Telfer *et al.*, 1981a; Irie and Yamashita, 1980). Evidence for the importance of paravitellinlike proteins in embryogenesis comes from the finding that ovarian follicles of silkmoths transplanted into males will mature and are filled with paravitellinlike proteins, but no Vn. The male-matured eggs can be stimulated to develop parthenogenetically, and larvae can hatch even in the absence of Vg (Yamashita and Irie, 1980), perhaps due to the store of paravitellin deposited in the eggs. It is curious to consider that in *Drosophila*, the roles of paravitellin and vitellin may be filled by the same gene products, since the YPs are made by both the fat body and the follicle cells.

The third category of yolk sphere protein is germ-line derived and constitutes a small fraction of the total. The lysosomal enzyme acid phosphatase (ACPH) is made by the oocyte itself and can be demonstrated enzymatically or immunochemically in yolk spheres (Cone and Eschenberg, 1966; Hopkins and King, 1966; Giorgi and Jacob, 1977c). As in other cell types, ACPH in oocytes is first detected in the Golgi apparatus. As vitellogenesis commences, ACPH becomes associated with the yolk spheres, having been transferred via Golgi-derived vesicles. During embryogenesis, ACPH may help use the phosphate found in many Vns, or it may be involved in processing Vn during oocyte maturation (Sawicki and MacIntyre, 1978).

### 2.4.2. Lipid Yolk

In addition to the protein-containing yolk spheres, the ooplasm harbors a substantial quantity of lipid droplets (Fig. 4). The lipid droplets are not surrounded by a proper limiting membrane, although a surface coat of proteins seems to maintain their structural integrity; if this protein coat, which is visible in freeze-fracture analysis, is enzymatically removed, the integrity of the lipid droplets is destroyed (Liu and Davies, 1973; Wiemerslage, 1976).

Lipid yolk consists mainly of triglycerides and phopholipids, although substantial quantities of ecdysteroids, the insect steroid hormones, are also found (Van Handel, 1976; Monroe *et al.*, 1967; Lagueux *et al.*, 1981). No hormonal role has been found for the ecdysteroids in the oocyte.

Much of the glycerol found in lipid droplets originates outside the ovary, although some lipid may also come from within the ovary (King and Aggarawal, 1965). Lipid droplets first appear in the ooplasm near the surface of the egg, along with newly forming protein yolk spheres (Schreiner, 1977), and superficial droplets label first with radioactive palmitate, followed later by labeling of deeper strata (Przlecka and Dutkowski, 1965). The ovary itself makes ecdysteroids.

### 2.4.3. Carbohydrate Yolk

Besides the carbohydrate brought to the yolk granules attached to Vg, glycogen is found in mature eggs in the form of granules devoid of a limiting membrane but surrounded by mitochondria (Fig. 4). Glycogen yolk is distributed over the entire ooplasm (King, 1960; Giorgi and Deri, 1976). Glycogen begins to accumulate in oocytes later than either protein or lipid yolk, toward the end of the vitellogenic period (Mahowald, 1972a). Unlike most lipid and protein yolk, glycogen granules are mainly a result of endogenous synthetic activity in the oocyte (Ramamurty, 1968; Engels and Bier, 1967; Koch and Spitzer, 1982).

## 3. Regulation of Vitellogenin Synthesis and Secretion

### 3.1. Control of the Vitellogenic Process

In order for yolk to be deposited into oocytes, several processes must occur, each of which may be controlled separately. We shall focus here on the regulation of two of the most important steps: (1) transcription–translation–secretion of the Vg genes, and (2) sequestration of the Vg into the oocyte, its conversion into Vn, and its deposition into yolk granules. In general, insects use hormonal mechanisms to coordinate and regulate these processes. The endocrine circuits that regulate vitellogenesis will be discussed separately from the molecular mechanisms involved, since the precise endocrine pathways that control vitellogenesis differ among insect species. However, the molecular events triggered by the hormones are likely to be more or less similar among different insects. The different types of endocrine networks regulating vitellogenesis seem to parallel the three groups of insects described by Harnish and White (1982) on the basis of types of vitellin molecules (see Section 2.4.1.a).

### 3.1.1. Regulation of Vitellogenesis in Group 1 Insects

The general strategy of group 1 insects is to use the juvenile hormone (JH) to regulate Vg synthesis and secretion, as well as sequestration (Fig. 6). JH is the agent that is produced by the **corpus allatum** gland at the base of the insect brain and is present during larval stages to prevent metamorphosis and to maintain the expression of larval characters. JH disappears during metamorphosis but reappears in the adult to regulate reproduction in both males and females.

The most extensively studied Group 1 insects are locusts and cockroaches. The results for both species are in general similar; discussion here will focus on locusts. For a review of the extensive and important data on cockroaches, see Engelmann (1979).

In *Locusta migratoria*, Vg first appears in the blood of adult females about a week after emergence. By day 13, 60% of the protein secreted by the fat body is Vg. If the corpora allata are removed shortly after emergence, Vg fails to appear; Vg synthesis can be restored in allatectomized females by treatments with JH. These kinds of *in vivo* experiments show that JH is necessary to cause Vg to be synthesized and secreted by the fat body, but they do not prove that the action of the hormone is directly on the Vg-secreting cells. For that proof, organ culture experiments are required. When fat bodies from female locusts treated with the corpus allatum cytotoxic agent precocene are cultured *in vitro*, they synthesize and secrete Vg only after JH treatment. This evidence shows that the JH acts directly on the fat body cells themselves to induce both translation and secretion of Vg. The molecular aspects of these processes are discussed in the next section.

Although JH is necessary for Vg production by locust fat body cells, it is not sufficient (Dhadialla and Wyatt, 1983). Two factors influence the response of locust fat body cells to JH: age and sex. Adult females are unable to produce Vg in response to exogenously supplied JH analogue until about 4 days after emergence. This shows that a maturation process is necessary for the fat body to develop competence to respond to the hormone.

Adult males and larvae of both sexes have high levels of JH but no Vg synthesis. When larvae are treated with still higher amounts of a JH analogue, however, both male and female larvae can produce Vg. Although larval males can produce Vg when treated with a JH analogue, adult males do not. Apparently, during metamorphosis the Vg genes become insensitive to JH in the male fat body cells, although these cells do respond to JH in certain other ways (Gellissen and Wyatt, 1981). The mechanism that makes the male fat body Vg genes lose sensitivity to JH is not solely hormonal, since larval female fat body implanted into a male larva continues to respond to hormone after metamorphosis in the male host. There is thus a sex-specific regulatory mechanism inherent to the fat body cells that causes adult male cells to lose responsiveness to the hormone.

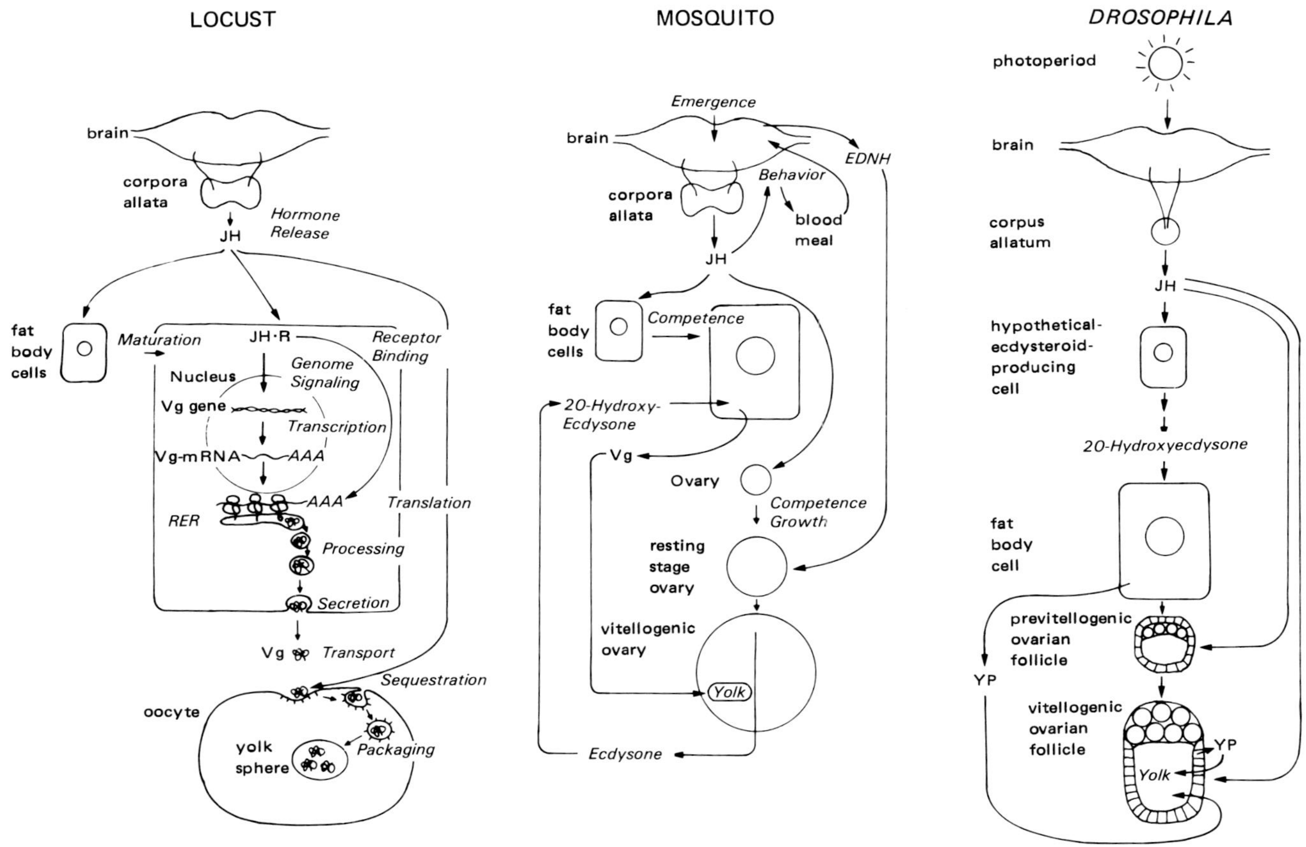


**Figure 6.** Endocrine networks proposed to regulate vitellogenesis in locusts (modified after Chen and Hillen, 1983), mosquitos (modified after Hagedorn, 1983), and *Drosophila* (modified after Postlethwait and Shirk, 1981). The proposed biochemical steps are shown only in the case of the locust, in which they have been studied the most thoroughly.

Control of vitellogenesis in locusts is more complex than a single triggering pulse of JH. Locusts go through reproductive cycles. After an active period of oogenesis, Vg production is turned off; it then recommences at the beginning of the next phase. Since JH induces Vg synthesis, the secretion of JH by the corpus allatum must be cyclic too. But what regulates the corpus allatum? Since either cutting the nerve between the brain and the corpus allatum or injection of anti-brain antibodies prevents corpus allatum activity, it can be concluded that a stimulatory signal arises from the brain to turn on the corpus allatum (Tobe and Chapman, 1977; Rembold *et al.*, 1980).

The factors that turn off the corpus allatum are less clear. To test whether the off switch is thrown by the ovary, ovaries were removed from reproductive females. These ovariectomized animals continued to produce Vg and accumulated high quantities in the blood (Chen and Wyatt, 1981). Apparently no signal from the ovary is needed to enable Vg synthesis to continue, but an ovarian factor is necessary to terminate Vg synthesis. How the ovary exerts its negative regulation on the corpus allatum is as yet unknown, but since the follicle cells of locust ovaries produce ecdysteroids (Goltzené *et al.*, 1978), it is possible that these steroids affect corpus allatum activity, even though only small amounts may escape into the hemolymph.

Whereas the regulation of Vg synthesis in most group 1 insects—locusts, cockroaches, beetles, and some Lepidoptera—seems to be under direct control of JH in the adult stage, the situation in other Lepidoptera is a bit different. In silk moths—the organisms in which Telfer (1954) first discovered and described Vg—the synthesis of Vg is precocious. It first appears in the blood just after the larva spins the cocoon. During the diapause of the pupa, Vg synthesis declines to low levels, but the amount of Vg in the blood is high, having accumulated during the earlier synthetic stage (Pan, 1971). Vg synthesis recommences during the development of the adult inside the pupal case, and yolk deposition into the oocytes begins just before the adult emerges. Animals from which the corpus allatum has been removed make Vg on schedule, and Vg synthesis is not affected by JH (Pan, 1977). Clearly, the regulation in silkmoths differs from other group 1 insects. It has been suggested that Vg synthesis is not controlled directly by hormones in silkmoths, but rather by an inherent developmental program in the fat body. It is also possible, however, that the endocrine events that occur during metamorphosis control Vg synthesis in silkmoths.

### 3.1.2. Regulation of Vitellogenesis in Group 2 Insects

This group includes Hymenoptera and lower flies, such as mosquitos. Vitellogenin regulation in honey bees seems to be similar to that of group 1 but is perhaps more complex. Recall that queen bees are reproductive but that female workers are not. Nevertheless, workers do have Vg in their blood (Engels, 1972). Removal of the corpus allatum from workers prevents Vg appearance, but this can be reversed by low doses of JH (Imboden *et al.*, 1976), as in locusts. The problem is that when similar experiments are performed with reproductive queens, Vg titers are not affected (Ramamurty and Engels, 1977). It has been suggested that Vg synthesis in bees may not be directly regulated by JH, but that JH affects in some way the factor that is the immediate controlling agent.

The regulation of vitellogenesis in mosquitos represents a dramatic departure from the mechanisms used by group 1 insects (for review, see Hagedorn, 1983). In mosquitos, Vg synthesis and secretion by the fat body are regulated by **20-hydroxyecdysone (20 HE),** the steroid-molting hormone of insects, rather than by JH as in group 1 insects. JH, however, plays a role in the maturation of both ovaries and fat body (Fig. 6).

The mosquito *Aedes aegypti* makes no Vg until after a meal of blood. Removal of ovaries blocks the increased rate of Vg synthesis, but 20-HE treatment can provoke Vg synthesis in ovariectomized animals. Coupled with the facts that mosquito ovaries can produce and secrete ecdysone, that ecdysone titers increase after a blood meal, and that fat body cells *in vitro* can respond to 20-HE by synthesizing and secreting Vg, Hagedorn (1983) has advanced the following hypothesis: The blood meal stimulates neurosecretory cells in the brain to produce the **egg development neurosecretory hormone (EDNH),** a peptide. EDNH enables the ovary to secrete ecdysone, which is then altered to 20-HE. The 20-HE acts on the fat body to stimulate Vg synthesis and secretion.

As in group 1 insects, JH plays a role in vitellogenesis, but it does not directly induce Vg synthesis in mosquitos as it does in locusts and cockroaches. Instead, JH in mosquitos plays a permissive role in Vg synthesis. Both the ovary and the fat body must be exposed to JH before they can respond to EDNH and subsequently to 20-HE. JH causes an early ovarian follicle to develop to the "resting stage," a developmental state in which the follicle is competent to respond to EDNH (Shapiro and Hagedorn, 1983). Likewise, JH causes a maturation process to occur in the fat body that permits it to become responsive to 20-HE (Flanagan and Hagedorn, 1977). Whether this maturation process is primarily due to the acquisition of receptor molecules for 20-HE is not yet known.

The parallels between the mechanisms regulating vitellogenesis in mosquitos and the endocrine control of the analogous process in birds and amphibians is striking (see Chapter 3). In both cases, neurosecretory peptide hormones from the brain (EDNH or follicle-stimulating hormone) pass in the circulation to the ovary. They stimulate the gonads to synthesize and secrete a steroid hormone (ecdysone or estrogen) into the blood. The steroid then goes to the major metabolic organ (fat body or liver) to cause Vg to be produced and secreted into the blood, whence it is sequestered into the oocytes to produce yolk.

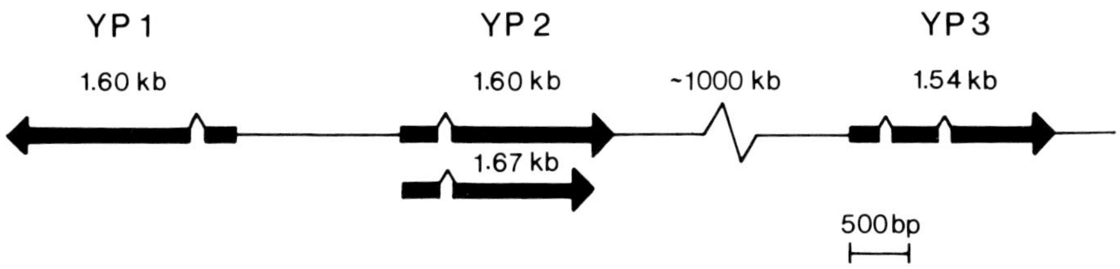


**Figure 7.** Transcript maps of the three yolk polypeptide genes from *Drosophila*. Brackets indicate introns, and arrows indicate the direction of transcription. bp, base pair; kb, kilobase. (Data from Hung and Wensink, 1981; Hovemann *et al.*, 1982.)

### 3.1.3. Regulation of Vitellogenesis in Group 3 Insects

The hormonal control of yolk deposition in the higher flies shares some of the features of insects of both group 1 and 2 (Fig. 6). As in the other insect groups, JH regulates Vg sequestration, and ecdysteroids are produced by the ovary. Furthermore, 20-hydroxyecdysone influences the rate of Vg synthesis as it does in mosquitos. The major improvisation of at least some group 3 insects is the production of Vg by the ovarian follicle cells (Brennan *et al.*, 1982) in addition to the more usual source, the fat body.

The analysis of vitellogenesis in the fruit fly *Drosophila* has used two technologies not yet readily available in many other insect species: mutations to block the process and cloned genes to probe for Vg messenger RNA. Despite these advantages, the precise endocrine circuitry driving vitellogenesis is not yet absolutely certain. This is largely due to the failure to obtain clear-cut induction of Vg synthesis from proven naive Vg-producing cells during tissue culture. The experiments to date have used living organisms manipulated surgically, chemically (Wilson *et al.*, 1983), or genetically. Thus the data must be interpreted with caution.

The synthesis of the three **yolk polypeptides (YPs)** that constitute Vg in *Drosophila* begins at about the time of eclosion in adult females, and deposition of yolk in oocytes starts in about 10 hr. Cloned probes for *Yp* genes can detect YP mRNA (see Fig. 8A) at about the time of eclosion (Hovemann and Galler, 1982), and message continues to accumulate up to about 24 hr (Barnett and Wensink, 1981). This increase in message parallels the increase of YP in the blood, arguing against translational control of YP synthesis from stored message.

Vitellogenesis can be blocked by decapitation within 10 min after eclosion in most females. Decapitation of older females results in the appearance of normal yolky oocytes. Hence, a signal from the head—presumably the brain—is required to initiate the vitellogenic process (Handler and Postlethwait, 1977).

The increase in the rate of YP synthesis and secretion and the increase in message can be blocked by removing the head and thorax from the abdo-

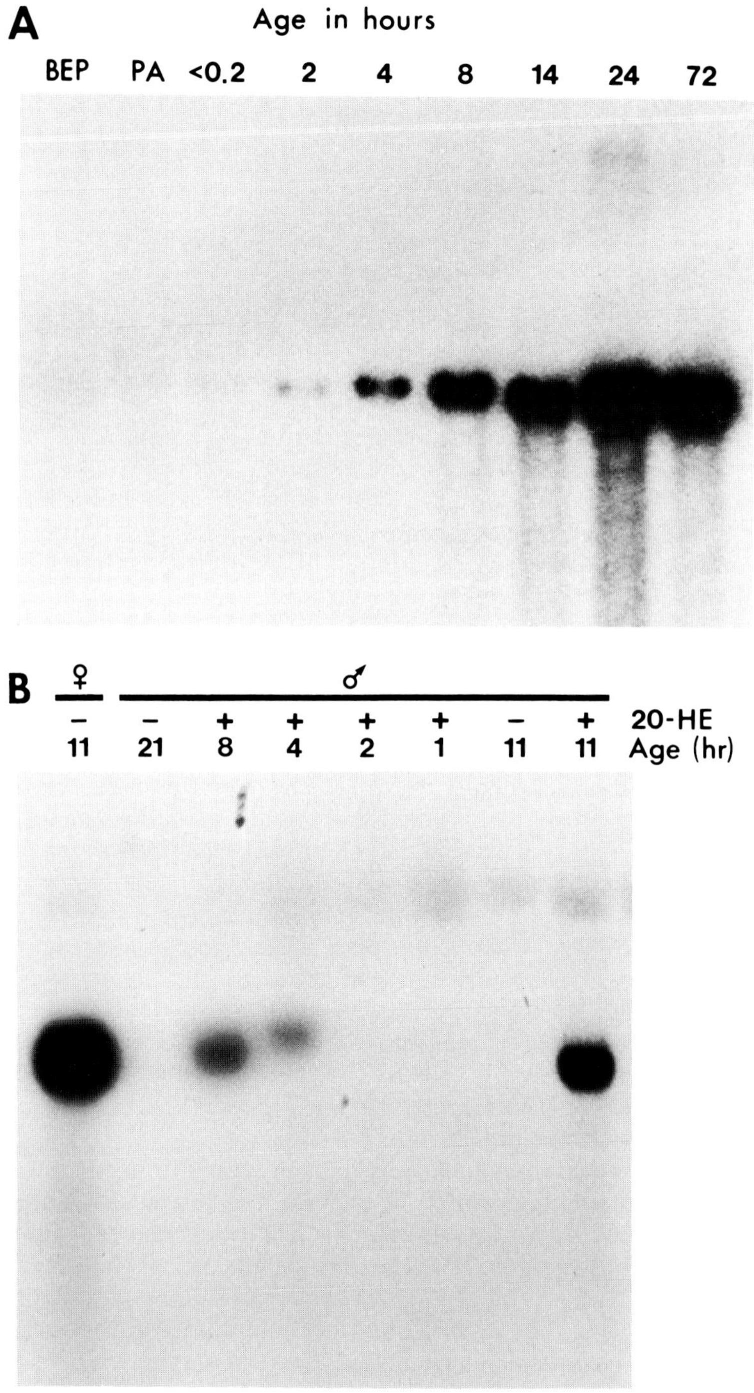
A
Age in hours
BEP PA <0.2 2 4 8 14 24 72
B
♀
♂
– – + + + + – + 20-HE
11 21 8 4 2 1 11 11 Age (hr)

mens shortly after eclosion (for review, see Bownes, 1982). The rate of YP synthesis in female abdomens isolated within 10 min of eclosion can be increased by applications of either 20-HE or juvenile hormone. A dose of 0.3 $\mu$l of a $10^{-6}$ M solution of 20-HE injected into isolated abdomens elicited a positive response (Jowett and Postlethwait, 1980). The effective dose of JH or its analogues is more difficult to determine, since they are delivered by topical application dissolved in acetone.

The question then became: Which tissue is responding to which hormone? When abdomens were isolated at eclosion, treated with hormones or control solutions, and then their fat bodies and ovaries separately put into culture in the presence of radiolabeled amino acid, it was found that YP synthesis by the fat body was stimulated by either 20-HE or JH, but that YP synthesis by the ovary was stimulated only by JH. Since vitellogenic follicles degenerate when cultured in an environment deficient in JH (Wilson, 1982), JH may stimulate ovarian Vg synthesis indirectly by supporting follicle maturation. Although both 20-HE and JH can stimulate YP synthesis, only JH causes YP to be sequestered into the oocytes to form yolk.

One intriguing finding in group 3 insects is that 20-HE injections stimulate the accumulation of YP transcript and its translation in males (Huybrechts and DeLoof, 1977, 1982; Postlethwait *et al.*, 1980; Shirk *et al.*, 1983). Substantially greater quantities of 20-HE are required for the males to respond than for isolated abdomens of females. In addition, *Drosophila* males seem to have only a slightly lower ecdysteroid content than that of females (Handler, 1982), although in other related flies the sex difference may be greater (Briers *et al.*, 1982). One possible explanation for these findings is that males may have fewer or different hormone receptor molecules and are hence less sensitive to the hormone. Another possibility is a fundamental difference in the chromatin structure of the YP genes in male and female fat body.

## 3.2. Molecular Mechanisms Controlling Vg Synthesis

Two basic approaches are available for investigating the regulation of expression of the Vg genes. The first method involves starting with mutant

**Figure 8.** *Drosophila* RNAs separated on agarose gels, blotted to nitrocellulose (Thomas, 1980), and hybridized to radiolabeled *Yp1* gene sequences as a probe for transcripts from the *Yp1* gene. (A) Equivalent amounts of total RNA from female flies of several different developmental ages. The amount of transcript increases rapidly during the first 8 hr after the fly emerges. BEP, brown-eyed pupae; PA, pharate adults. (B) Time course for the appearance of *Yp* gene transcript in males injected with 20-hydroxyecdysone (20-HE), $10^{-6}$ M final concentration. Transcript is first detected at 4 hr after treatment and continues to rise until at least 11 hr. Total RNA from 10 flies at each point.

animals that have observable defects in Vg gene regulation and finally identifying the structural defect in the DNA, RNA, or polypeptide responsible for the altered molecular physiology. This is more or less the traditional approach. It has become possible more recently to reverse the traditional approach. This second method uses cloned genes mutated in a defined fashion by the experimenter in the test tube; the mutant gene is reintroduced into an appropriate test system, such as a cell-free transcription system or even the living insect itself (Spradling and Rubin, 1982) in an attempt to investigate whether the known change in gene structure alters the physiology. The first method is the **mutant–animal-to-mutant-gene** approach, and the second is the **mutant-gene-to-mutant-animal** strategy. Work is now in progress in several laboratories to implement these two strategies in an effort to aid our study of gene regulation in vitellogenesis.

### 3.2.1. Genetic Variants of Vg Genes

In order to study mutants that affect the regulation of the Vg genes, it is first necessary to identify those genes. This was done in the classic manner in *Drosophila* by searching for variants in YP structure (Postlethwait, 1980; Postlethwait and Jowett, 1980). These genetic studies showed that the genes coding for YP1 and YP2 (the *YP1* and *YP2* genes) are closely linked on the X chromosome, and that the *Yp3* gene is also sex linked, but distant from the other two (Fig. 7). It is curious that at least one of the Vg genes in locusts has also been shown to be sex linked (Bradfield and Wyatt, 1983). Thus, the Vg genes are not only sex limited in their expression, but are also sex linked in location. This has led to the suggestion that there is functional significance to the sex-chromosomal localization of the Vg genes. Despite the phenomenon of dosage compensation (i.e., female *Drosophila* with two X chromosomes generally produce the same amount of a sex-linked gene product as do males with one X), the location of the Vg genes may play a role in their level of expression in the two sexes. The cytogenetic localization of the Vg genes in Lepidoptera may illuminate this hypothesis, since in moths, the female is the heterogametic sex.

The identification of the *Yp* gene loci aided geneticists in their search for mutants affecting *Yp* gene regulation. A number of mutants have been found that decrease the amount of a particular YP present in the blood or ovary of *Drosophila*. Only one of these mutations causes females to be sterile. The mutant *fs(1)1163* is deficient in the amount of YP1 found in the blood and eggs (Gans *et al.*, 1975; Bownes and Hames, 1978). This mutation alters a step in post-translational processing and will be discussed below. Another mutant, *(fs(1)12-1245)*, eliminates YP2 and has normal fertility. Since this mutant has normal amounts of *Yp2* gene transcript (C. Kunert and J. Postlethwait, unpublished observations), it is not likely to be a regulatory mutant in the classic sense. A third mutant *($Yp3^{RI09}$)* has reduced amounts of a YP3 with decreased mobility in SDS gels (Postlethwait and Shirk, 1981). This mutant has normal level of *Yp3* gene transcript and so may represent an unstable polypeptide. The search for possible regulatory mutants of the *Yp* genes continues.

### 3.2.2. The Structure of Cloned Vg Genes

Vg genes have been cloned from *Drosophila* (Barnett *et al.*, 1980), and from locusts (Wyatt *et al.*, 1981). Sequencing of the *Drosophila* genes has revealed structural features that may be involved in the regulation of the genes or in the function of the polypeptides.

Genomic clones for the three *Yp* genes in *Drosophila* were obtained by exploiting the fact that females but not males contain mRNA for YPs. Radiolabeled cDNA was made complementary to size-fractionated male and female poly(A)-containing RNAs. These cDNAs were used to screen a library of *Drosophila* genomic fragments that had been inserted into a variant of λ phage. Phages containing sequences homologous to female but not to male mRNA were isolated. In order to identify the polypeptides encoded by the cloned female-specific genes, mRNAs homologous to the phage were isolated from *Drosophila* females and translated in a cell-free translation system. The translation products were compared structurally to *bona fide* YPs. Phage were thus identified that contained *Yp1*, *Yp2*, and *Yp3* sequences. The genes were then cut from the phage and subcloned into the plasmid pBR322.

The first problem that was addressed with the cloned genes was their location in the genome. *Yp1* and *Yp2* are clearly very close together, since the series of clones that encoded YP1 and YP2 are overlapping (Fig. 7); only about 1200 base pairs separates the two genes. In order to localize the genes cytologically, radioactively labeled nucleic acid complementary to the *Yp* genes was hybridized *in situ* to salivary gland polytene chromosomes. The experiments showed that all three genes are on the X chromosome and that *Yp1* and *Yp2* are localized in region 9A, and *Yp3* about 1000 kilobases (kb) away, at 12B-C (Barnett *et al.*, 1980).

The next problem was to determine the number of copies of the *Yp* genes in the genome. It was found that the genes for the YPs are single copy, both in diploid DNA and in female fat body cells (Barnett and Wensink, 1981). This shows that no gene amplification occurs during the maturation of the fat body cells. This result contrasts to the situation found for the chorion genes in ovarian follicle cells, which become amplified several times in preparation for the massive synthesis of these proteins (Spradling and Mahowald, 1980). Apparently the polyploidization of the fat body cells coupled with the production of stable messages is enough to ensure the rapid synthesis of YPs without gene amplification.

Since the *Yp1* and *Yp2* genes are so close together and are coordinately controlled, it was of interest to determine the transcript maps of these two genes. YP mRNA was hybridized to the cloned genomic fragments bearing the *Yp* genes, and the hybrid molecules were digested with the single strand-specific nuclease S1. Unhybridized DNA was digested by S1 nuclease, leaving only those genomic regions protected from the nuclease in double-stranded DNA/mRNA hybrids. Analysis of the protected fragments indicated the message-encoding regions (Hung *et al.*, 1982). The two genes are transcribed divergently, and each has about an 80-base-pair intron in homologous locations. *Yp2* produces two transcripts, which differ slightly in size. It is not known yet whether both transcripts are made in both YP producing tissues. The third gene, *Yp3*, has two small introns.

The proximity of *Yp1* and *Yp2*, their coordinate expression, and their divergent orientation would lead one to expect that sequence signals necessary for regulation of individual genes might be present in two mirror image copies, one for each gene. To investigate these possibilities, both genes and the interval separating them have been sequenced (Hung and Wensink, 1981; Riddell *et al.*, 1981; Hovemann *et al.*, 1981) (see Table I).

**3.2.2.a. Sequences.** Most of the nucleotide sequences that are common to *Yp1* and *Yp2* are those features that are also conserved in many other eukaryotic genes. These include the TATA box, cap site, presumed ribosome-binding site, the consensus splice sequences at beginning and end of each intron, and the polyadenylation signal. These sequences are all probably involved in regulation of transcription or translation. The CCAAT box recognition sequence, which is often found 70 to 80 nucleotides 5′ from the mRNA starting position (Benoist *et al.*, 1980), was not found in the *Yp1* and *Yp2* genes.

**3.2.2.b. A Potential Loop/Stem.** Computer analysis of the DNA sequences revealed an unusual potential stem and loop structure at the 5′ end of each gene. In this structure, the TATA box, the capping site, the ribosomal RNA homology, and the translation initiation codon are at the four corners of the stem. Whether this stem and loop actually forms *in vivo* or plays a role in gene regulation is unknown.

**3.2.2.c. Hormone-Receptor Binding Homology.** A sequence 13 nucleotides long similar to a recently proposed consensus sequence for the progesterone–receptor binding site is located 100–150 base pairs 5′ to each gene. The roles of these sequences are yet to be demonstrated.

**3.2.2.d. Peptide Homologies.** The nucleic acid sequences were used to predict the amino acid sequences and secondary structures of the YP1 and YP2 polypeptides. The amino acid sequences are only 53% homologous, but the predicted secondary structures are nearly identical. This suggests that the two genes may have arisen by duplication and that the con-

**Table I.** Conserved Sequences in the *Yp1* and *Yp2* Genes

| Site | Gene | Location | Sequence |
|---|---|---|---|
| Putative DNAse-hypersensitive site | *Yp1* | −419 | AATATAAATCTACT |
| | *Yp2* | −350 | AAACAAAAGCTT . T |
| | Con | −325 ± 100 | AAAATAAAGCNNNT |
| Putative hormone-receptor binding site | *Yp1* | −101 | CTCAGCACAA . GTG |
| | *Yp2* | −149 | CTCATCAGAA . GTG |
| | Con | −250 | ATC.CCATTATCTGGTTGTA |
| CCAAT box | *Yp1* | | Not found |
| | *Yp2* | | Not found |
| | Con | −75 | GGCCAATCT |
| Hogness-Goldberg box | *Yp1* | −33 | TATATAAA |
| | *Yp2* | −30 | TATAAAAT |
| | Con | −25 ± 5 | TATATAAA |
| Capping site | *Yp1* | 0 | CCAGTTCA |
| | *Yp2* | 0 | GCATGCA |
| | Con | 0 | YCATTCR |
| Putative ribosome binding site | *Yp1* | +25 | TCGCATCCGCA |
| | *Yp2* | +21 | GTGTCTGAA |
| Initiation codon | *Yp1* | +59 | AACCATG |
| | *Yp2* | +52 | CACAATG |
| | Con | | CAYCATG |
| 5′ splicing site | *Yp1* | +278 | CT/GTGAGT |
| | *Yp2* | +286 | CT/GTAAGT |
| | Con | | AG/GTRAGT |
| 3′ splicing site | *Yp1* | +354 | CCTTTGTAG/ |
| | *Yp2* | +354 | ATCCCGCAG/ |
| | Con | | YNYYYNYAG/ |
| Termination codon | *Yp1* | +1452 | TAA |
| | *Yp2* | +1446 | TAG |
| | Con | | TRR |
| Poly-A recognition site | *Yp1* | +1604 | AATAAA |
| | *Yp2* | +1589 | AATAAA |
| | *Yp2* | +1665 | AATAAA |
| | Con | | AATAAA |
| 3′ end | *Yp1* | +1635 | |
| | *Yp2* | +1624 | |
| | *Yp2* | +1704 | |

Sequence data taken from Hung and Wensink (1981, 1983), Hovemann *et al.* (1981), Hovemann and Galler (1982), Shermoen and Beckendorf (1982), Benoist *et al.* (1980).
Con, consensus sequences for many eukaryotic genes; N, any nucleotide; R, purine; Y, pyrimidine.
A dot in a sequence indicates a space to improve homology.

served peptide structural features are important in either secretion, multimerization in the blood, sequestration, or yolk granule formation.

### 3.2.3. Regulation of the Vitellogenin Genes

The levels of control exerted on the Vg genes have been studied most extensively in locusts, cockroaches, and flies. The first question that has been asked is whether the hormones cause the increase in the secretion rate of Vg from the fat body by acting at the levels of transcription, translation, or secretion. The basic experiments are the same for each insect group: comparing newly radiolabeled protein or mRNA levels in a supposedly hormonally naive fat body cell population with that in the same cells after hormone treatments (Chinzei *et al.*, 1982; Engelmann, 1978; Shirk *et al.*, 1983). The general answer is that (depending on the organism) juvenile hormone or ecdysone causes the rate of label incorporation into Vg to increase due mainly to an increase in the amount of Vg mRNA present in the fat body cells. In locusts, Vg mRNA increases to about $10^6$ copies per cell, and 60% of the protein secreted by fat body cells is Vg (Chinzei *et al.*, 1982; Chen *et al.*, 1978).

The increase in the amount of mRNA can be detected as soon as 4 hr after the hormonal treatments by using cloned copies of the Vg genes as probes in RNA blot analyses, with Northern blots (Fig. 8B). This suggests that the hormones exert their effect on a process not too distant from the genes themselves. However, it has not yet been rigorously proved that the effect is directly on the Vg genes, i.e., that the hormones do not first induce the action of some regulatory gene, which only secondarily stimulates an increase in message levels. Nor has it been established that the effect on the amounts of message found after hormone application is due to a stimulation of transcription. It is still possible that the hormones could influence transcript stability. This possibility seems unlikely, however, because of the vast amount of transcript that appears in such a short time.

The juvenile hormone acts directly on the fat body of locusts by causing new synthesis of mRNA for Vg. Does the hormone act directly on the chromosomes at the site of the Vg genes, or is there a cytoplasmic component that mediates JH action? It has been shown that the fat body cytoplasm contains a protein that acts like a JH receptor with a high-affinity, low-capacity binding activity specific for the hormone (Engelmann, 1981; Roberts and Wyatt, 1983). The proposed JH receptor protein is also found in the nucleus; it is present at higher levels in the fat body of vitellogenic females than males. These workers argue that since its occurrence is correlated with the hormonal response of the tissue, the receptor helps transmit to the genome the fact that the hormone has arrived.

There is some evidence for control of Vg production at the level of translation. Locusts go through reproductive gonadotropic cycles, during which JH is high and then decreases. It has been found that when JH decreases after the first cycle and the rate of Vg synthesis decreases to a low level, the amount of mRNA for Vg fails to decrease to the same low level (Chinzei *et al.*, 1982). Since a substantial amount of untranslated Vg mRNA is stored in the cells between gonadotropic cycles, the JH must be exerting regulation on the translatability of Vg mRNA in addition to its rate of accumulation.

Further analysis of the activation of Vg genes awaits application of the tools of the molecular geneticist to the problem. The possibilities are exciting: Are there specific DNA sequences that ecdysone- or juvenile hormone-regulated genes have in common? What is the difference in chromatin conformation of the Vg genes in males and females? What sequences near the Vg genes are necessary for normal tissue- and stage-specific regulation? Can proteins be isolated that interact both with specific DNA sequences near the Vg genes and with the hormone? These questions and more will certainly be answered in the next few years, and the results will be of broad general interest and will aid our understanding of how genes regulate development.

### 3.3. Regulation of Vg Secretion

The mechanisms of Vg post-translational processing and secretion have not been studied as thoroughly as those of Vg transcription and translation. The Vgs have long been recognized to be altered after translation by the addition of phosphates, carbohydrates, and lipids, but the precise biochemical and cytological pathways involved are unknown.

An early step in the processing of Vg peptides is the cleavage of the **signal peptide,** a hydrophobic amino-terminal sequence of about 20 amino acids that helps thread the nascent secretory polypeptide through the membrane into the lumen of the endoplasmic reticulum (Blobel and Dobberstein, 1975; Warren *et al.*, 1979). In *Drosophila*, this is followed by modifications of the *Yp* to form several more negatively charged species, coincident with the addition of phosphate and carbohydrate moieties (Brennan and Mahowald, 1982; Minoo and Postlethwait, 1984).

A female sterile mutation *(fs(1)1163)* that decreases the amount of YP1 found in the blood of female *Drosophila* (Bownes and Hames, 1978) has been found to have normal levels of messenger RNA and wild-type levels of primary translation product, but to have a decreased rate of secretion into the blood. The mutant YP1 has a slightly altered mobility in two-dimensional gel electrophoresis and, thus, a defect in the amino acid sequence. This is associated with a buildup of processing intermediates not found in wild-type cells, which might cause the inhibition of secretion (Minoo and Postlethwait, 1984). Further work on the structure of the mutant polypeptide should help elucidate some of the structural requirements for normal secretion.

## 4. Regulation of Vitellogenin Sequestration

### 4.1. Mechanisms of Sequestration

After Vg has been translated, processed, and secreted, it is available for uptake into the oocyte (Fig. 9). When vitellogenesis actually com-

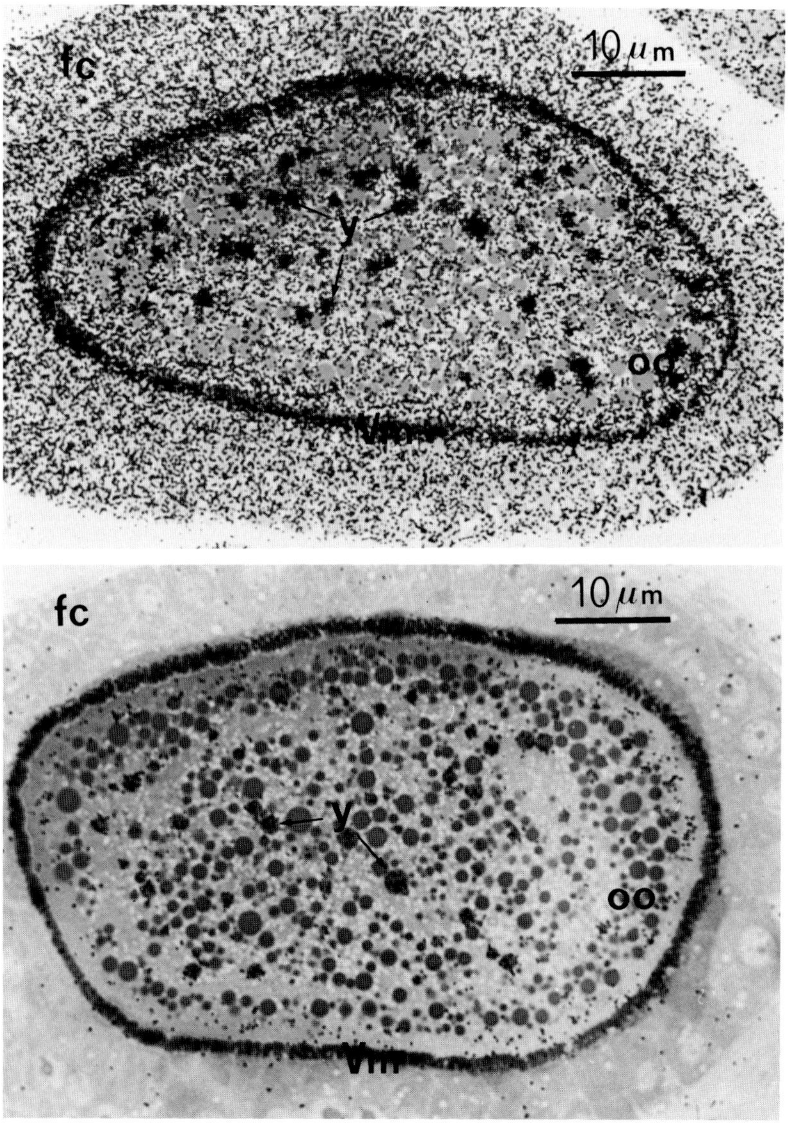


**Figure 9.** Sequestration of radiolabeled Vg in *Drosophila*. Females were injected with either [$^{3}$H]leucine (A) or [$^{3}$H]-Vg (B), incubated, and then prepared for autoradiography. Labeled leucine is incorporated throughout the ooplasm and in the cytoplasm of the follicle cells, but labeled Vg is confined to yolk granules and the cortical cytoplasm. fc, follicle cells; oo, ooplasm; Vm, vitelline membrane; y, yolk spheres.

mences, the follicle cells decrease their interactions with adjacent follicle cells but increase their contact with the oocyte. Large intercellular spaces form between adjacent follicle cells in a process called **patency.** Many junctions between follicle cells disappear (Huebner and Injeyan, 1981), follicle cells decrease in size (Davey, 1981), and the interfollicular cell spaces fill with proteoglycans (Telfer *et al.*, 1982). Presumably this happens so that Vg can more easily penetrate from the blood to the surface of the oocyte.

As these changes occur at the follicle cell interfaces, two important changes take place at the surface of the oocyte (Fig. 10). First, the oocyte sends numerous fingerlike projections across the perivitelline space to interdigitate with similar structures emerging from the follicle cells. These cell extensions make junctional contact (Huebner, 1981; Giorgi and Postlethwait, 1985), so that small-molecular-weight fluorescent probes—but not proteins—can cross from oocyte to follicle cell (Telfer *et al.*, 1982).

The second change in the oocyte surface is the appearance of **coated pits,** localized depressions in the plasma membrane that protrude toward the cortical ooplasm (Figs. 10 and 11). Coated pits are found widely among animal cells that are actively pinocytosing specific high-molecular-weight products. Coated pits consist of three layers: a hexagonal/pentagonal lattice meshwork on the cytoplasmic side of the pit; a fibrous coat on the luminal side; and—sandwiched between these two layers—a unit membrane that is continuous with the oocyte plasma membrane. The lattice forms a basket around the pit consisting of the protein **clathrin** (Kanaseki and Kadoka, 1969; Whyte and Ockleford, 1980). The fibrous material lining the luminal side of the pit consists of either the yolk precursors or their receptors anchored to the membrane.

The uptake of vitellogenin is specific (Ferenze, 1978; Ferenze *et al.*, 1981) and probably occurs via a phenomenon called **receptor-mediated endocytosis,** in which a membrane-enclosed receptor molecule specifically binds a target protein—a ligand—which is then brought into the cell via coated pits (see Roth and Woods, 1982). The nature of the receptor has been investigated in competent mosquito oocytes that, when cultured *in vitro*, can take up vitellogenin with a high affinity ($K_m = 10^{-7}$) and high specificity; exogenous proteins such as gamma globulin or bovine serum albumin (BSA) do not inhibit the uptake (Roth *et al.*, 1976). The role of the lattice in receptor-mediated endocytosis may be either to anchor the receptor/ligand complex to specialized membrane regions or to help generate the force that causes the pit to invaginate into the cytoplasm and become a vesicle. The latter seems possible, since clathrin polymerizes *in vitro* spontaneously into spheres—the shape of the pinocytotic vesicles.

The following picture emerges from static electron micrographs, but it must be remembered that the sequence is put together in the microscopist's mind, not from timed experiments that trace the history of a single coated pit. The coated pits seem to pinch off from the oocyte plasma membrane to form coated vesicles, which move to the cortical cytoplasm. The

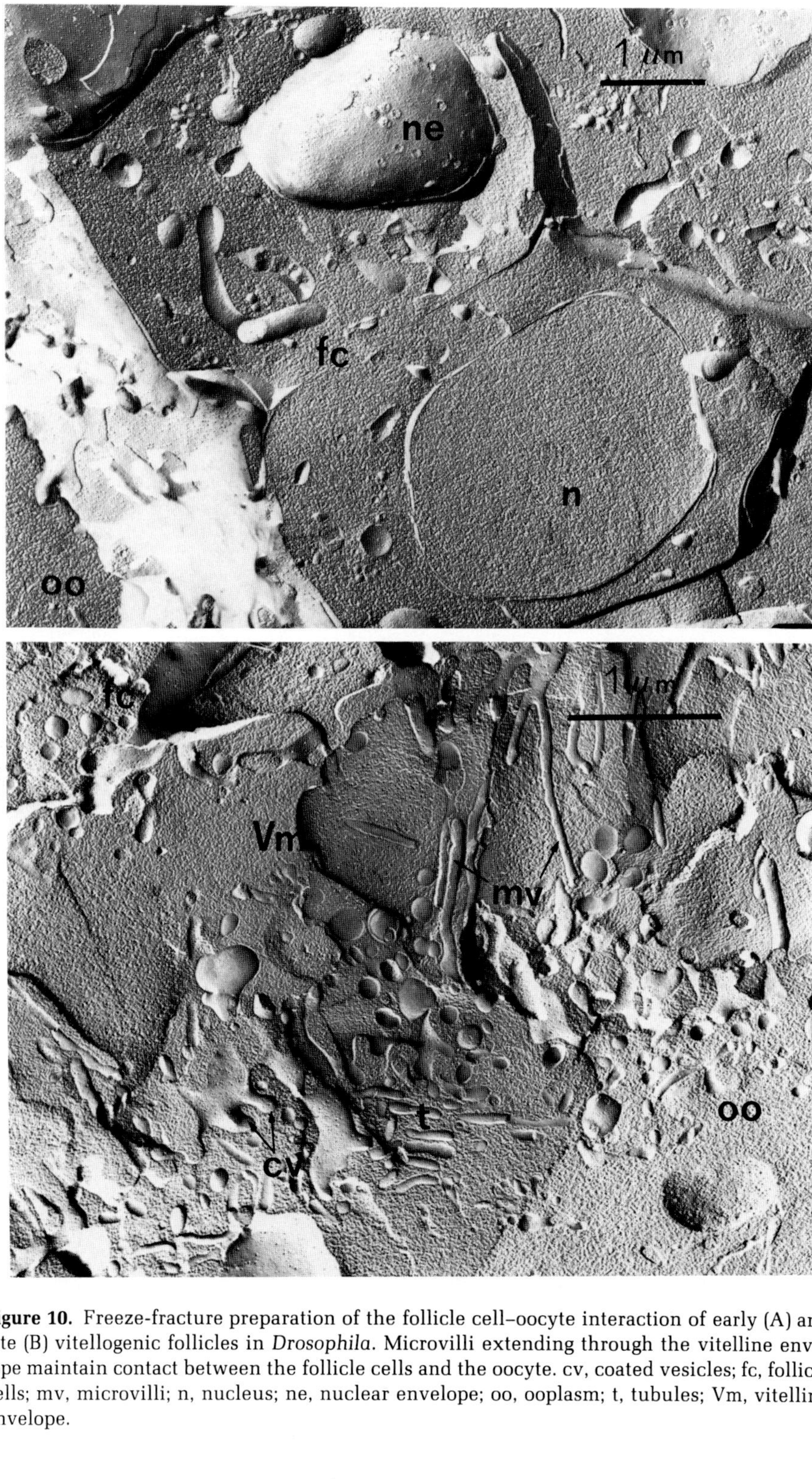


**Figure 10.** Freeze-fracture preparation of the follicle cell–oocyte interaction of early (A) and late (B) vitellogenic follicles in *Drosophila*. Microvilli extending through the vitelline envelope maintain contact between the follicle cells and the oocyte. cv, coated vesicles; fc, follicle cells; mv, microvilli; n, nucleus; ne, nuclear envelope; oo, ooplasm; t, tubules; Vm, vitelline envelope.

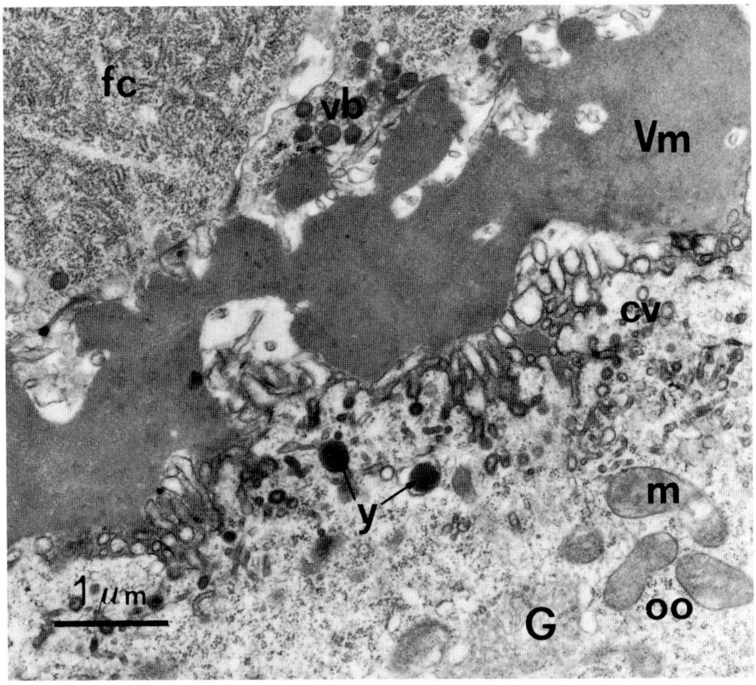


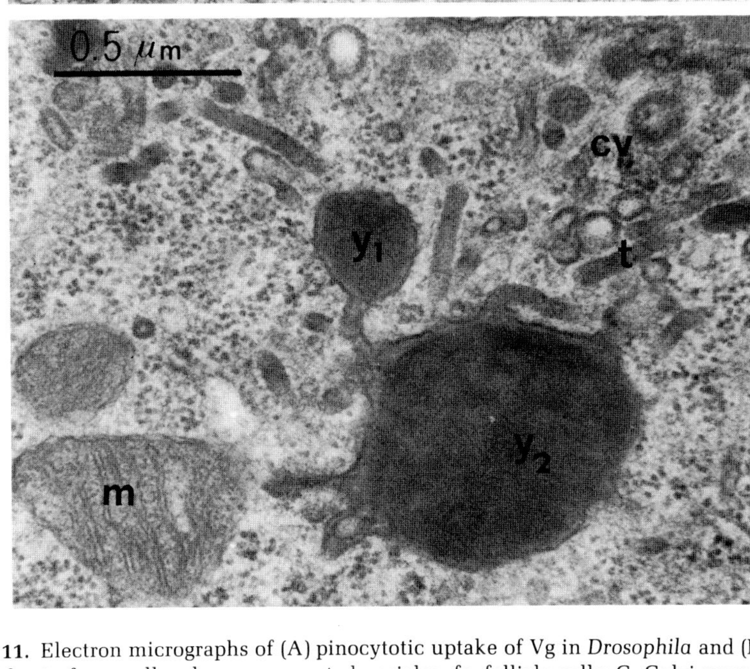


**Figure 11.** Electron micrographs of (A) pinocytotic uptake of Vg in *Drosophila* and (B) fusion of vesicles to form yolk spheres. cv, coated vesicles; fc, follicle cells; G, Golgi apparatus; m, mitochondrion; vb, vesicular bodies; Vm, vitelline envelope; y, y1, y2, yolk spheres.

latticework is then lost, and the fibrous material disappears to be replaced by a dense granular core. Vesicles then fuse to form large yolk spheres or tubular profiles.

During this uptake process, some materials are probably used and then reused. The lattice that disappears after internalization, as well as the vitellogenin receptor, are probably recycled several times or destroyed after use, since no protein occurs in a 1:1 ratio with Vg in insect oocytes (Telfer *et al.*, 1981a). Likewise, the excess membrane generated when vesicles fuse is likely to be reused, either directly by shunting vesicular or tubular structures to the membrane or, more likely, by disassembly and recycling the molecular constituents via the Golgi apparatus (Mahowald, 1972b; Giorgi, 1980).

## 4.2. Regulation of Sequestration

Insects could regulate yolk uptake either by controlling the Vg supply in the hemolymph, by manipulating the competence of the follicle to sequester it, or both. The first possibility—the regulation of Vg synthesis, secretion, transport, and stability—was considered above in Sections 3.2 and 3.3. The second possibility can be rephrased by asking: What governs the developmental program that culminates in the acquisition of Vg receptors and the coated pits mentioned above and the delivery of Vg through the sheet of follicle cells to the oocyte surface? The answers to this question involve both environmental and physiological mechanisms.

Among the environmental factors influencing yolk deposition, nutrition, and photoperiod have been studied the most intensely. Both parameters seem to exert their effect in most species via the endocrine system—particularly the activation of the corpus allatum, which is the source of juvenile hormone (Bell, 1971; Tobe and Chapman, 1979; Weaver and Pratt, 1981; Spielman and Wong, 1974; Khan *et al.*, 1982; Darjo, 1976; Allemand, 1976b).

Early investigators showed that the corpus allatum is important in yolk deposition (Vogt, 1943; Bodenstein, 1947) but were unable to tell whether the endocrine gland affects Vg concentration in the blood or its uptake into the oocyte. These two possibilities were distinguished by injecting Vg into animals from which the corpus allatum gland had been removed (Bell, 1969). These experiments showed that Vg itself is unable to stimulate uptake, but that vitellogenesis could be restored only by treating allatectomized animals with a juvenile hormone analogue.

Further evidence that juvenile hormone regulates Vg uptake has been demonstrated in experiments with mutant *Drosophila*. Homozygous *apterous-4* females are sterile; their follicles stop developing just before yolk deposition commences. This block can be overcome by transplanting mutant ovaries into wild-type hosts (King and Bodenstein, 1965), proving that a humoral factor—perhaps the Vg molecule or a hormone—regulates vitellogenesis. Treating mutant flies with a JH analogue also restores yolk deposition (Postlethwait and Weiser, 1973), thereby suggesting that the sterile mutant phenotype involves a JH insufficiency. But does the hormone act by increasing the amount of Vg in the blood or by stimulating Vg uptake in the ovary? When Vg levels were checked in mutant females, it was found that substantial, although lower than normal, amounts were present in some *apterous-4* stocks (Gavin and Williamson, 1976; Redfern and Bownes, 1982). Thus, in hormone-treated mutant flies, the hormone stimulates primarily the uptake of the preexisting Vg.

These conclusions were confirmed by ultrastructural examination of mutant ovaries both with and without exogenously added juvenile hormone (Tedesco *et al.*, 1981): The hormone causes previtellogenic oocytes to form microvilli and coated vesicles, the structural correlates to Vg uptake. Since this effect can occur as a result of JH treatments *in vitro* in mutant and wild-type ovaries (Giorgi, 1979), the general conclusion is that juvenile hormone acts directly on some cell type in the fly ovary to stimulate uptake of Vg from the blood.

The follicular cell type and the molecular level at which JH acts to effect Vg sequestration are unknown. One possibility is that JH induces the appearance of specific Vg receptors on the surface of the oocyte. Another effect of JH is to increase the patency of adjacent follicle cells (Abu-Hakima and Davey, 1977), which might also be the limiting factor in Vg uptake.

## 5. The Role of Yolk in Development

Yolk may play a role at two critical stages in the life of an insect. The first, of course, is during embryonic development. However, during the adult stage, yolk from unlaid eggs may be resorbed and reused by the female.

### 5.1. The Role of Yolk During Embryogenesis

It has long been assumed that yolk plays a nutritive role during the development of the embryo, but there are few critical experiments that prove this assumption. In *Drosophila*, the role of yolk spheres in embryonic determination has been tested by culturing developing embryos with and without yolk (Hadorn and Muller, 1974). Embryos just past the gastrula stage were cut in two and implanted into the body cavity of adult female flies for culture. From some of the eggs, the yolk mass was removed before the culture. After culture, the implants were recovered and were found to have developed the same inventory of larval structures whether yolk had been included or not. Hadorn and Muller (1974) concluded that the yolk as such is dispensable for differentiation at least after the gastrula stage. In these experiments, Vg was present in the host female hemolymph, which bathed the embryonic implant. Thus, one cannot infer that Vg is unnecessary—only that yolk is not required.

Experiments with silkmoths have gone even further to dispel the notion that Vn in yolk is necessary for embryonic life (Yamashita and Irie, 1980). Young ovaries were implanted into the body cavity of male moths, which make no Vg. In the male hosts, the oocytes matured but formed no Vg-containing yolk spheres. The male-matured eggs were then activated to develop parthenogenetically. The result was normal development, even though the eggs lacked Vg. Apparently, non-Vg proteins found their way into the yolk in sufficient quantities to support normal development, showing that Vg is not absolutely required for normal embryogenesis. These experiments also show that Vg is not needed to induce completion of oogenesis.

In *Drosophila*, whose Vg consists of three yolk polypeptides, a mutant is known that lacks YP2. Nevertheless, the stock has near-normal fertility. Thus, either the partially homologous YP1 and YP3 peptides have functions that overlap YP2, or the YP2 function is not needed for embryogenesis.

The finding that conjugated ecdysteroids and juvenile hormone can bind to Vg in locusts and cockroaches (Lagueux *et al.*, 1981; Sehnal, 1983) suggests that Vg may assist during transporation or storage of these hormones in the embryo and/or the adult female.

## 5.2. Utilization of Yolk During Embryogenesis

Although no unique role has actually been assigned to Vg or Vn, the patterns of utilization of yolk continue to convey the impression that it is used as a source of nutrients for the embryo.

Yolk spheres may not be absolutely essential for development, but they change morphologically and Vg largely disappears during embryogenesis (Takesue *et al.*, 1976; Irie and Yamashita, 1980; Bownes, 1982)—although in some cases not until just before hatching (McGregor and Loughton, 1974). Yolk accumulates centrally in the egg and is invaded by **vitellophages,** cells that originate from the cleavage nuclei and that fail to migrate to the egg's surface during blastoderm formation. Vitellophages appear to engulf yolk granules (Tanaka, 1977). During differentiation, the yolk mass becomes enclosed by the gut, indicating that the yolk possibly serves as a food supply.

## 5.3. Recycling of Yolk in Adult Females

If we were to limit our understanding of yolk to its role in the embryo, we would be restricting our appreciation of the adaptive strategies siezed by insects for reproduction. Insects not only store nutritional resources in eggs for embryonic use—they can also mobilize these nutrients in response to environmental stresses. Under such circumstances, a gravid female will resorb her vitellogenic oocytes and recycle the molecular constituents for her own use. The resorption of yolky oocytes is accomplished at least partially by follicle cells engulfing the deteriorating oocyte, and its regulation is likely to be from the withdrawal of hormonal support for the maturing oocyte (Wilson, 1982).

## 6. Conclusions

Although the outlines of the mechanisms governing the provisioning of the insect oocyte are now known, much remains to be learned. How do hormones select the Vg genes for expression? How does the structure of the Vg polypeptides determine their post-translational modifications and select them for secretion and recognition by the oocyte? How do hormones bring about the changes in cell behavior that result in selective uptake of Vg from the blood? Answers to these questions will not only solve some basic general problems in developmental and regulatory biology, but may have as a by-product the development of ecologically safer and better-targeted methods of restricting insect pests to environments not directly harmful to humans.

## References

Abu-Hakima, R., and Davey, K. G., 1977, The action of juvenile hormone on the follicle of Rhodnius prolixus: The importance of volume changes, *J. Exp. Biol.* **69**:33–44.

Allemand, R., 1976, Influence de modifications des conditions lumineuses sur les rythmes circadiens de vitellogenèse et d'ovulation chez Drosophila melanogaster, *J. Insect Physiol.* **22**:1075–1080.

Anderson, L. M., and Telfer, W. H., 1969, A follicle cell contribution to the yolk spheres of moth oocytes, *Tissue Cell* **1**:633–644.

Anderson, L. M., and Telfer, W. H., 1970, Trypan blue inhibition of yolk deposition—a clue to follicle cell function in the cecropia moth, *J. Embryol. Exp. Morphol.* **23**:35–52.

Anderson, W. A., and Spielman, A., 1971, Permeability of the ovarian follicle of *Aedes aegypti* mosquitoes, *J. Cell Biol.* **50**:201–221.

Barnett, T., and Wensink, P., 1981, Transcription and translation of yolk protein mRNA in the fat bodies of *Drosophila*, in: *Developmental Biology using Purified Genes*, Twenty-Third ICN–UCLA Symposium on Molecular and Cellular Biology (D. Brown, ed.), pp. 97–106, Academic Press, New York.

Barnett, T., Pachl, C., Gergen, J. P., and Wensink, P. C., 1980, The isolation and characterization of *Drosophila* yolk protein genes, *Cell* **21**:729–738.

Bast, R. E., and Telfer, W. H., 1976, Follicle cell protein synthesis and its contribution to the yolk of the cecropia moth oocyte, *Dev. Biol.* **52**:83–97.

Bell, W. J., 1969, Dual role of juvenile hormone in the control of yolk formation in *Periplaneta americana*, *J. Insect Physiol.* **15**:1279–1290.

Bell, W. J., 1971, Starvation induced oocyte resorption and yolk protein salvage in *Periplaneta americana*, *J. Insect Physiol.* **17**:1099–1111.

Benoist, C., O'Hare, K., Breathnach, R., and Chambon, P., 1980, The ovalbumin gene—sequence of putative control regions, *Nucleic Acids Res.* **8:**127–142.

Bilinski, S., 1976, Ultrastructural studies on the vitellogenesis of *Tetrodontophora bielanensis* Waga (Collembola), *Cell Tissue Res.* **168:**399–410.

Bilinski, S., 1979, Ultrastructural study of yolk formation in *Porcellio scaber* Latr. (Isopoda), *Cytobios* **26:**123–130.

Blobel, G., and Dobberstein, B., 1975, Transfer of proteins across membranes. I. Presence of proteolytically processed and unprocessed nascent immunoglobin light chains on membrane bound ribosomes of murine myeloma, *J. Cell Biol.* **67:**835–851.

Bodenstein, D., 1947, Investigations on the reproductive system of *Drosophila*, *J. Exp. Zool.* **104:**101–152.

Bownes, M., 1982, Hormonal and genetic regulation of vitellogenesis in *Drosophila*, *Q. Rev. Biol.* **57:**247–274.

Bownes, M., and Hames, B. D., 1978, Genetic analysis of vitellogenesis in *Drosophila melanogaster:* The identification of a temperature sensitive mutation affecting one of the yolk proteins, *J. Embryol. Exp. Morphol.* **47:**111–120.

Boyer, B. C., 1972, Ultrastructural studies of differentiation in the oocyte of the polyclad turbellarian *Prostheceraeus floridanus*, *J. Morphol.* **136:**273–296.

Bradfield, J. Y., and Wyatt, G., 1983, X-linkage of a vitellogenin gene in Locusta migratoria, *Chromosoma* **88:**190–193.

Brennan, M. D., and Mahowald, A. P., 1982, Phosphorylation of the vitellogenin polypeptides of *Drosophila melanogaster*, *Insect Biochem.* **12:**669–673.

Brennan, M. D., Warren, T. G., and Mahowald, A. P., 1980, Signal peptides and signal peptidase in *Drosophila melanogaster*, *J. Cell Biol.* **87:**516–520.

Brennan, M. D., Weiner, A. J., Goralski, T. J., and Mahowald, A. P., 1982, The follicle cells are a major site of vitellogenin synthesis in *Drosophila melanogaster*, *Dev. Biol.* **89:**225–236.

Briers, T., van Beek, E., and De Loof, A., 1982, Ecdysteroid activity during metamorphosis and in male and female adults of four blowfly species, *Comp. Biochem. Physiol.* **74:**521–524.

Brown, E. H., and King, R. C., 1964, Studies on the events resulting in the formation of an egg chamber in *Drosophila melanogaster*, *Growth* **28:**41–81.

Buning, J., 1972, Untersuchungen am Ovär *Bruchidius obtectus* Say. (Coleoptera-Polyphaga) zur Klärung des Oocyten wachstums in der Prävitellogenese, *Z. Zellforsch.* **128:**241–282.

Cave, M. D., 1982, Morphological manifestations of ribosomal DNA amplification during insect oogenesis, in: *Insect Ultrastructure*, Volume 1 (R. C. King and H. Akai, eds.), pp. 86–117, Plenum Press, New York.

Chen, T. T., and Hillen, L. J., 1983, Expression of the vitellogenin genes in insects, *Gamete Res.* **7:**179–196.

Chen, T. T., and Wyatt, G. R., 1981, Juvenile hormone control of vitellogenin synthesis in *Locusta migratoria*, in: *The Regulation of Insect Development and Behavior* (F. Sehnal and A. Zabza, eds.), pp. 535–536, Technical University of Wroclaw Press, Wroclaw, Poland.

Chen, T. T., Strahlendorf, P., and Wyatt, G. R., 1978, Vitellin and vitellogenin from locusts (Locusta migratoria): Properties and post-translational modification in the fat body, *J. Biol. Chem.* **253:**5325–5331.

Chinzei, Y., White, B. N., and Wyatt, G. R., 1982, Vitellogenin mRNA in locust fat body: Identification, isolation and quantitative changes induced by juvenile hormones, *Can. J. Biochem.* **24:**243–251.

Cone, M. V., and Eschenberg, K. M., 1966, Histochemical localization of acid phosphatase in the ovary of *Gerris remigis* (Hemiptera), *J. Exp. Zool.* **161:**337–352.

Cummings, M. R., Brown, N. M., and King, R. C., 1971, The cytology of the vitellogenic stages of oogenesis in *Drosophila melanogaster*. III. Formation of the vitelline membrane, *Z. Zellforsch*, **118:**482–492.

Darjo, A., 1976, Activité des corpora allata et controle photoperiodique de la maturation ovarienne chez *Locusta migratoria*, *J. Insect Physiol.* **22:**347–355.

Davenport, R., 1976, Transport of ribosomal RNA into the oocytes of the milkweed bug, *Oncopeltus fasciatus*, *J. Insect Physiol.* **22:**925–926.

Davey, K. G., 1981, Regulation of cell volume in follicle cells of *Rhodnius* by JH, in: *Juvenile Hormone Biochemistry* (G. E. Pratt and G. T. Brooks, eds.), pp. 328–342, Elsevier/North-Holland Biomedical Press, Amsterdam.

De Loof, A., 1983, The meroistic insect ovary as a miniature electrophoretic chamber, *Comp. Biochem. Physiol.* **74A**:3–9.

Dhadialla, T. S., and Wyatt, G. R., 1983, Juvenile hormone-dependent vitellogenin synthesis in *Locusta migratoria* fat body: Inductibility related to sex and stage, *Dev. Biol.* **96**:436–444.

Engelmann, F., 1978, Juvenile hormone regulated vitellogenin synthesis in an insect, in: *Comparative Endocrinology* (P. J. Gaillard and H. H. Boer, eds.), pp. 17–20, Elsevier/North-Holland Biomedical Press, Amsterdam.

Engelmann, F., 1979, Insect vitellogenin: Identification, biosynthesis and role in vitellogenesis, *Adv. Insect Physiol.* **14**:49–108.

Engelmann, F., 1981, Heterogeneity of juvenile hormone binding compounds in fat bodies of a cockroach, *Mol. Cell. Endocrinol.* **24**:103–112.

Engels, W., 1972, Quantitative Untersuchungen zum Dotterprotein-Haushalt der Honigbiene (*Apis mellifica*), *Arch. Entwicklungsmech. Org.* **171**:55–86.

Engels, W., and Bier, K., 1967, Zür Glycogen speicherung während der Oogenese und ihrer Vorzeitigen auslösung durch Blockierung der FNS versorgung (Untersuchungen an *Musca domestica*), *Arch. Entwicklungsmech. Org.* **158**:64–88.

Fallon, A. M., Hagedorn, H. H., Wyatt, G. R., and Laufer, H., 1974, Activation of vitellogenin synthesis in the mosquito *Aedes aegypti* by ecdysone, *J. Insect Physiol.* **20**:1815–1823.

Favard, P., and Favard-Sereno, C., 1969, Electron microscope study of polysaccharides in the amphibian oocyte, *J. Submicrosc. Cytol.* **1**:91–111.

Ferenze, H. J., 1978, Uptake of vitellogenin to developing oocytes of *Locusta migratoria*, *J. Insect Physiol.* **24**:273–278.

Ferenze, H. J., Lubzens, E. W., and Glass, H., 1981, Vitellin and vitellogenin incorporation by isolated oocytes of *Locusta migratoria migratorioides*, *J. Insect Physiol.* **12**:869–875.

Flanagan, T. R., and Hagedorn, H. H., 1977, Vitellogenin synthesis in the mosquito: The role of juvenile hormone in the development of responsiveness to ecdysone, *Physiol. Entomol.* **2**:173–178.

Fourney, R. M., Pratt, G. F., Harnish, D. G., Wyatt, G. R., and White, B. N., 1982, Structure and synthesis of vitellogenin and vitellin from *Calliphora erythrocephala*, *Insect Biochem.* **12**:311–321.

Friedel, T., and Gillott, C., 1977, Contribution of male-produced proteins to vitellogenesis in *Melanoplus sanguinipes*, *J. Insect Physiol.* **23**:145–151.

Gans, M., Audit, C., and Masson, M., 1975, Isolation and characterization of sex-linked female sterile mutants in *Drosophila melanogaster*, *Genetics* **81**:683–704.

Gavin, J. A., and Williamson, J. H., 1976, Juvenile hormone induced vitellogenesis in apterous-4, a nonvitellogenic mutant in *Drosophila melanogaster*, *J. Insect Physiol.* **22**:1737–1742.

Gellissen, G., and Wyatt, G. R., 1981, Production of lipophorin in the fat body of adult *Locusta migratoria*: Comparison with vitellogenin, *Can. J. Biochem.* **59**:648–654.

Gellissen, G., Wajc, E., Cohen, E., Emmerich, H., Applebaum, S. W., and Flossdorf, J., 1976, Purification and properties of oocyte vitellin from the migratory locust, *J. Comp. Physiol.* **108B**:287–301.

Gillott, C., and Elliott, R. H., 1976, Reproductive growth in normal, allatectomized, median neurosecretory cell cauterized and ovariectomized females of *Melanoplus sanguinipes*, *Can. J. Zool.* **54**:162–171.

Giorgi, F., 1979, In vitro induced pinocytotic activity by a juvenile hormone analogue in oocytes of *Drosophila melanogaster*, *Cell Tissue Res.* **203**:241–247.

Giorgi, F., 1980, Coated vesicles in the oocyte, in: *Coated Vesicles* (C. D. Ockleford and A. Whyte, eds.), pp. 135–177, Cambridge University Press, New York.

Giorgi, F., and Deri, P., 1976, Cytochemistry of late ovarian chambers of *Drosophila melanogaster*, *Histochemistry* **48**:325–334.

Giorgi, F., and Jacob, J., 1977a, Recent findings on oogenesis of *Drosophila melanogaster*, I.

Ultrastructural observations on the developing ooplasm, *J. Embryol. Exp. Morphol.* **38**:115–124.

Giorgi, F., and Jacob, J., 1977b, Recent findings on oogenesis of *Drosophila melanogaster*. II. Further evidence on the origin of yolk platelets, *J. Embryol. Exp. Morphol.* **38**:125–138.

Giorgi, F., and Jacob, J., 1977c, Recent findings on oogenesis of *Drosophila melanogaster*. III. Lysosomes and yolk platelets, *J. Embryol. Exp. Morphol.* **39**:45–57.

Giorgi, F., and Postlethwait, J. H., 1985, Development of gap junctions in normal and mutant ovaries of *Drosophila melanogaster*, *J. Morphol.*, in press.

Goltzené, F., Lagueux, M., Charlet, M., and Hoffmann, J. A., 1978, Follicle cell epithelium of maturing ovaries of Locusta migratoria: New biosynthetic tissue for ecdysone, *Hoppe Seylers Z. Physiol. Chem.* **359**:1427–1434.

Gutzeit, H. O., and Koppa, R., 1982, Time-lapse film analysis of cytoplasmic streaming during late oogenesis of *Drosophila*, *J. Embryol. Exp. Morphol.* **57**:101–111.

Hadorn, E., and Muller, G., 1974, Has the yolk an influence on the differentiation of embryonic Drosophila blastemas?, *Arch. Entwicklungsmech. Org.* **174**:333–335.

Hagedorn, H. H., 1983, The role of ecdysteroids in the adult insect, in: *Endocrinology of Insects* (G. Downer and H. Laufer, eds.), pp. 241–304, Alan R. Liss, New York.

Hagedorn, H. H., and Kunkel, J. G., 1979, Vitellogenin and vitellin in insects, *Annu. Rev. Entomol.* **24**:465–505.

Handler, A. M., 1982, Ecdysteroid titers during pupal and adult development in *Drosophila melanogaster*, *Dev. Biol.* **93**:73–82.

Handler, A. M., and Postlethwait, J. H., 1977, Endocrine control of vitellogenesis in *Drosophila melanogaster*: Effects of the brain and corpus allatum, *J. Exp. Zool.* **202**:389–402.

Harnish, D. G., and White, B. N., 1982, Insect vitellins: Identification, purification, and characterization from eight orders, *J. Exp. Zool.* **220**:1–10.

Harnish, D. G., Wyatt, G. R., and White, B. N., 1982, Insect vitellins: Identification of primary products of translation, *J. Exp. Zool.* **220**:11–19.

Hausman, S. J., Anderson, L. M., and Telfer, W. H., 1971, The dependence of cecropia yolk formation in vitro on specific blood proteins, *J. Cell Biol.* **48**:303–313.

Hopkins, C. R., and King, P.E., 1966, An electron microscope and histochemical study on the oocyte periphery in *Bombus terrestris* during vitellogenesis, *J. Cell Sci.* **1**:201–216.

Hovemann, B., and Galler, R., 1982, Vitellogenin in *Drosophila melanogaster*: A comparison of the YPI and YPII genes and their transcription products, *Nucleic Acids Res.* **10**:2261–2275.

Hovemann, B., Galler, R., Walldorf, U., Küpper, H., and Bautz, E. K. F., 1981, Vitellogenin in *Drosophila melanogaster*: Sequence of the yolk protein-1 gene and its flanking regions, *Nucleic Acids Res.* **9**:4721–4734.

Huebner, E., 1981, Nurse cell–oocyte interaction in the telotrophic ovarioles of an insect, *Rhodnius prolixus*, *Tissue Cell* **13**:105–125.

Huebner, E., and Anderson, E., 1976, Comparative spiralian oogenesis—Structural aspects: An overview, *Am. Zool.* **16**:315–343.

Huebner, E., and Injeyan, H., 1981, Follicular modulation during oocyte development in an insect: Formation and modification of septate and gap junctions, *Develop. Biol.* **83**:101–113.

Huebner, E., Tobe, S. S., and Davey, K. G., 1975, Structural and functional dynamics of oogenesis in *Glossina austeni*: Vitellogenesis with special reference to the follicular epithelium, *Tissue Cell* **7**:535–558.

Hung, M., and Wensink, P. C., 1981, The sequence of the *Drosophila melanogaster* gene for yolk protein, *Nucleic Acids Res.* **9**:6407–6419.

Hung, M., and Wensink, P. C., 1983, Sequence and structure conservation in yolk proteins and their genes, *J. Mol. Biol.* 164:481–489.

Hung, M., Barnett, T., Woolford, C., and Wensink, P. C., 1982, Transcript maps of Drosophila yolk protein genes, *J. Mol. Biol.* **154**:581–602.

Huybrechts, R., and De Loof, A., 1977, Induction of vitellogenin synthesis in male *Sarcophaga bullata*, *J. Insect Physiol.* **23**:1359–1362.

Huybrechts, R., and De Loof, A., 1982, Similarities in vitellogenin and control of vitellogenin synthesis within the genera *Sarcophaga, Calliphora, Phormia and Lucilia* (Diptera), *Comp. Biochem. Physiol.* **72B**:399–344.

Irie, K., and Yamashita, O., 1980, Changes in vitellin and other yolk proteins during embryonic development in the silkworm, *Bombyx mori, J. Insect Physiol.* **26**:811–817.

Jowett, T., and Postlethwait, J. H., 1980, The regulation of yolk polypeptide synthesis in *Drosophila* ovaries and fat body by 20-hydroxyecdysone and a juvenile hormone analog, *Dev. Biol.* **80**:225–234.

Kanaseki, T., and Kadota, K., 1969, The "vesicle in a basket." A morphological study of the coated vesicles isolated from the nerve endings of the guinea pig brain, with special reference to the mechanism of membrane movements, *J. Cell Biol.* **42**:202–220.

Khan, M. A., Koopmanschap, A. B., Privee, H., and DeKort, C. A. D., 1982, The mode of regulation of the corpus allatum activity during starvation in adult females of the Colorado potato beetle *Leptinotarsa decemlineata* (Say), *J. Insect Physiol.* **28**:791–796.

King, R. C., 1960, Oogenesis in adult *Drosophila melanogaster*. IX. Studies on the cytochemistry and ultrastructure of developing oocytes, *Growth* **24**:265–323.

King, R. C., 1970, *Ovarian Development in Drosophila melanogaster*. Academic Press, New York.

King, R. C., and Aggarwal, S. K., 1965, Oogenesis in *Hyalophora cecropia, Growth* **29**:17–83.

King, R. C., and Bodenstein, D., 1965, The transplantation of ovaries between genetically sterile and wild type *Drosophila melanogaster, Z. Naturforsch.* **206**:292–297.

Kleine-Schonnefeld, H., and Engels, W., 1981, Symmetrical pattern of follicle arrangement in the ovary of *Musca domestica* (Insecta, Diptera), *Zoomorphology* **98**:185–190.

Koch, E. A., and Spitzer, R. H., 1982, Autoradiographic studies of protein and polysaccharide synthesis during vitellogenesis in *Drosophila melanogaster, Cell Tissue Res.* **224**:315–333.

Koch, E. A., and Spitzer, R. H., 1983, Multiple effects of colchicine on oogenesis in Drosophila: Induced sterility and switch of potential oocyte to nurse cell developmental pathway, *Cell Tissue Res.* **228**:21–32.

Kunkel, J. G., Shepard, G. L., McCarthy, R.A., Ethier, D. B., and Nordin, J. H., 1980, Concanavalin A reactivity and carbohydrate structure of *Blattella germanien* vitellin, *Insect Biochem.* **10**:703–714.

Lagueux, M., Harry, P., and Hoffmann, J. A., 1981, Ecdysteroids are bound to vitellin in newly laid eggs of *Locusta, Mol. Cell. Endocrinol.* **24**:325–338.

Liu, T. P., and Davies, D. M., 1973, Intramitochondrial transformation during lipid vitellogenesis in oocytes of a black fly, *Simulus vittatus* Zetterstedt (Diptera: S-muliidae), *Int. J. Insect Morphol. Embryol.* **2**:233–245.

Maa, W. C. J., and Bell, W. J., 1977, An endogenous component of the mechanism controlling the vitellogenic cycle in the American cockroach, *J. Insect Physiol.* **23**:895–897.

MacGregor, H. C., and Stebbings, H., 1970, A massive system of microtubules associated with cytoplasmic movement in telotrophic ovarioles, *J. Cell Sci.* **6**:431–449.

Mahowald, A. P., 1962, Fine structure of pole cells and polar granules in *Drosophila melanogaster, J. Exp. Zool.* **151**:201–207.

Mahowald, A. P., 1971, Polar granules of Drosophila. III. The continuity of polar granules during the life cycle of *Drosophila, J. Exp. Zool.* **176**:329–344.

Mahowald, A. P., 1972a, Ultrastructural observations on oogenesis in *Drosophila, J. Morphol.* **137**:29–48.

Mahowald, A. P., 1972b, Oogenesis, in: *Developmental Systems: Insects*, Vol. 1 (S. J. Counce and C. H. Waddington, eds.), pp. 1–47, Academic Press, New York.

Mays, U., 1972, Autoradiographische Untersuchungen zum Stofftransport von den Narrzellen zur Oocyte der Feuerawnze *Pyrrhocoris apterus*. L. *Z. Zellforsch. Mikrosk. Anat.* **123**:395–410.

McGregor, D. A., and Loughton, B. G., 1974, Yolk protein degradation during embryogenesis of the African migratory locust, *Can. J. Zool.* **52**:907–917.

Mesnier, M., 1980, Study of oocyte growth within the ovariole in a stick-insect, *Clitumnus extradentatus, J. Insect Physiol.* **26**:59–65.

Minoo, P., and Postlethwait, J. H., 1984, Biosynthesis of *Drosophila* yolk polypeptides, *Arch. Insect Biochem. Physiol.* (in press).

Monroe, R. E., Hopkins, T. L., and Valder, S. A., 1967, Metabolism and utilization of cholesterol-4$^{14}$C for growth and reproduction of aseptically reared houseflies, *Musca domestica* L, *J. Insect Physiol.* **13:**219–233.

Mundall, E. C., and Law, J. H., 1979, Physical and chemical characterization of vitellogenin from the hemolymph and eggs of the tobacco hornworm *Manduca sexta, Comp. Biochem. Physiol.* **63B:**459–468.

Pan, M. L., 1971, Synthesis of vitellogenin in Cecropia silkworm, *J. Insect Physiol.* **17:**677–678.

Postlethwait, J. H., 1980, Genetic analysis of Drosophila vitellogenesis: A molecular weight variant of yolk polypeptide-2 in *Drosophila simulans, Mol. Gen. Genet.* **178:**253–260.

Postlethwait, J. H., and Jowett, T., 1980, Genetic analysis of the hormonally regulated yolk polypeptide genes in *D. melanogaster, Cell* **20:**671–678.

Postlethwait, J. H., and Shirk, P. D., 1981, Genetic and endocrine regulation of vitellogenesis in *Drosophila Am. Zool.* **21:**687–700.

Postlethwait, J. H., and Weiser, K., 1973, Vitellogenesis induced by juvenile hormone in the female sterile mutant apterous-four in *Drosophila melanogaster, Nature New Biol.* **244:**284–285.

Postlethwait, J. H., Bownes, M., and Jowett, T., 1980, Sexual phenotype and vitellogenin synthesis in *Drosophila melanogaster, Develop. Biol.* **79:**379–387.

Pratt, G. E., and Davey, K. G., 1972, The corpus allatum and oogenesis in *Rhodnius prolixus* (Stal). III. Effect of mating, *J. Exp. Biol.* **56:**223–237.

Przelecka, A., and Dutkowski, A., 1965, Autoradiographic investigation of incorporation of fatty acids into the lipids of insect ovarioles, *Bull. Acad. Pol. Sci.* [*Biol.*] **13:**573–575.

Quattropani, S. L., and Anderson, E., 1969, The origin and structure of the secondary coat of the egg of *Drosophila melanogaster, Z. Zellforsch.* **95:**495–510.

Ramamurty, P. S., 1968, The route of haemolymph protein transport into the oocytes of *Panorpa communis* (Mecoptera—Insects), *J. Anim. Morphol. Physiol.* **15:**188–190.

Ramamurty, P. S., and Engels, W., 1977, Occurrence of intercellular bridges between follicle epithelial cells in the ovary of *Apis mellifica* queens, *J. Cell Sci.* **24:**195–202.

Redfern, C. P. F., and Bownes, M., 1982, Yolk polypeptide synthesis and uptake in ovaries of the *apterous-4* mutant of *Drosophila melanogaster, Mol. Gen. Genet.* **185:**181–185.

Rembold, H., Eder, J., and Ulrich, G. M., 1980, Inhibition of ullatotropic activity and ovary development in *Locusta migratoria* by anti-brain antibodies, *Z. Naturforsch. C.* **35:**1117–1119.

Riddell, D. C., Higgins, M. J., McMillan, B. J., and White, B. N., 1981, Structural analysis of the three vitellogenin genes in *Drosophila melanogaster, Nucleic Acids Res.* **9:**1323–1338.

Roberts, P. E., and Wyatt, G. R., 1983, Juvenile hormone binding by components of fat body cytosol from vitellogenic locusts, *Mol. Cell. Endocrinol.* **31:**53–69.

Roth, T. F., and Porter, K. R., 1964, Yolk protein uptake in the mosquito *Aedes aegypti* L, *J. Cell Biol.* **20:**313–332.

Roth, T. F., and Woods, J. W., 1982, Fundamental questions in receptor-mediated endocytosis, in: *Differentiation and Function of Hematopoietic Cell Surfaces,* (V. P. Marchesi and R. C. Gallo, eds.) pp. 163–181, Alan R. Liss, New York.

Roth, T. F., Cutting, J. A., and Allas, S. B., 1976, Protein transport: A selective membrane mechanism, *J. Supramol. Struct.* **4:**527–548.

Sawicki, J. A., and MacIntyre, R. J., 1978, Localization at the ultrastructural level of maternally derived enzyme and determination of the time of paternal gene expression for acid phosphatase-1 in *Drosophila melanogaster, Dev. Biol.* **63:**47–58.

Schreiner, B., 1977, Vitellogenesis in the milkweed bug, *Oncopeltus fasciatus* Dallas (Hemiptera). A light and electron microscopic investigation, *J. Morphol.* **151:**35–80.

Shermoen, A. W., and Beckendorf, S. K., 1982, A complex of interacting DNAase I-hypersensitive sites near the Drosophila glue protein gene, *Sgs4, Cell* **29:**601–607.

Shirk, P. D., Bean, D., Millemann, A., and Brookes, V. J., 1984, Identification, synthesis and characterization of the yolk polypeptides of *Plodia interpunctella* (submitted for publication).

Sonnenblick, B. P., 1941, Germ cell movements and sex differentiation of the gonads in the *Drosophila* embryo, *Proc. Natl. Acad. Sci. USA* **27**:484–489.

Spielman, A., and Wong, J., 1974, Dietary factors stimulating oogenesis in *Aedes aegypti*, *Biol. Bull.* **147**:433–442.

Spradling, A., and Rubin, G. M., 1982, Transposition of cloned P elements into Drosophila germ line chromosomes, *Science* **218**:341–347.

Spradling, A. C., and Mahowald, A. P., 1980, Amplification of genes for chorion proteins during oogenesis in *Drosphila melanogaster*, *Proc. Natl. Acad. Sci. USA* **77**:1096–1100.

Stebbings, H., 1971, Influence of vinblastine sulphate on the development of microtubules and ribosomes in telotrophic ovarioles, *J. Cell Sci.* **8**:111–125.

Takesue, S., Keino, H., and Endo, K., 1976, Studies on the yolk granules of the silkworm, *Bombyx mori* L.: The morphology of diapause and nondiapause eggs during early developmental stages, *Arch. Entwichlungsmech. Org.* **180**:93–105.

Tanaka, A., 1977, Immunohistochemical studies of vitellogenin during embryogenesis in the cockroach *Blattella germanica*, *J. Embryol. Exp. Morphol.* **38**:49–62.

Tata, J. R., 1979, Vitellogenesis: A versatile model for hormonal regulation of gene expression, *Rec. Prog. Horm. Res.* **35**:47–95.

Tedesco, J. L., Courtright, J. B., and Kumaran, A. K., 1981, Ultrastructural changes induced by juvenile hormone analogue in oocyte membranes of apterous 4 *Drosophila melanogaster*, *J. Insect Physiol.* **27**:895–902.

Telfer, W. H., 1954, Immunological studies of insect metamorphosis. II. The role of a sex-limited blood protein in egg formation by the cecropia silkworm, *J. Gen. Physiol.* **37**:539–558.

Telfer, W. H., 1960, The selective accumulation of blood proteins by the oocytes of saturniid moths, *Biol. Bull.* **118**:338–351.

Telfer, W. H., Rubenstein, E. C., and Pan, M. L., 1981a, How the ovary makes yolk in *Hyalophora*, in: *Regulation of Insect Development and Behavior* (F. Sehnal, A. Zabza, J. J. Menn, and B. Cymborowski, eds.), pp. 637–654, Wroclaw Technical University, Wroclaw, Poland.

Telfer, W. H., Woodruff, R. L., and Huebner, E., 1981b, Electrical polarity and cellular differentiation in meroistic ovaries, *Am. Zool.* **21**:675–686.

Telfer, W. H., Huebner, E., and Smith, D. S., 1982, The cell biology of vitellogenic follicles in *Hyalophora* and *Rhodnius*, in: *Insect Ultrastructure*, Vol. I (R. C. King and H. Akai, eds.), pp. 118–149, Plenum Press, New York.

Thomas, P. S., 1980, Hybridization of denatured RNA and small DNA fragments transferred to nitrocellulose, *Proc. Natl. Acad. Sci. USA* **77**:5201–5205.

Tobe, S. S., and Chapman, C. S., 1977, Decay in juvenile hormone biosynthesis by insect corpus allatum after nerve transection, *Nature* **268**:728–730.

Tobe, S. S., and Chapman, C. S., 1979, The effects of starvation and subsequent feeding on juvenile hormone synthesis and oocyte growth in *Schistocerca americana gregaria*, *J. Insect Physiol.* **25**:701–708.

Ullmann, S. L., 1973, Oogenesis in *Tenebrio molitor*: Histological and autoradiographical observations on pupal and adult ovaries, *J. Embryol. Exp. Morphol.* **30**:179–217.

Van Handel, E., 1976, The chemistry of egg maturation in the unfed mosquito *Aedes atropalpus*, *J. Insect Physiol.* **22**:521–522.

Vogt, M., 1943, Zur Produktion und Bedeutung metamorphosefordernder hormone während der Larvenentwicklung von *Drosophila*, *Biol. Zentralbl.* **63**:395–446.

Warren, T., Brennan, M., and Mahowald, A., 1979, Two processing steps in maturation of vitellogenin polypeptides in *Drosophila melanogaster*, *Proc. Natl. Acad. Sci. USA* **76**:2848–2852.

Weaver, R. J., and Pratt, G. E., 1981, Effects of starvation and feeding upon corpus allatum activity and oocyte growth in adult female *Periplaneta americana*, *J. Insect Physiol.* **27**:75–83.

Whyte, A., and Ockleford, C. D., 1980, Structural aspects of coated vesicles at the molecular level, in: *Coated Vesicles* (C. D. Ockleford and A. Whyte, eds.), pp. 283–302, Cambridge University Press, New York.

Wiemerslage, L. J., 1977, Lipid droplet formation during vitellogenesis in the cecropia moth, *J. Insect Physiol.* **22**:41–50.

Wilson, T. G., 1982, A correlation between juvenile hormone deficiency and vitellogenic oocyte degeneration in *Drosophila melanogaster, Arch. Entwicklungsmech. Org.* **191**:257–263.

Wilson, T. G., Landers, M. H., and Happ, G. M., 1983, Precocene I and II inhibition of vitellogenic oocyte development in *Drosophila melanogaster, J. Insect Physiol.* **29**:249–254.

Woodruff, R. I., and Telfer, W. H., 1980, Electrophoresis of proteins in intercellular bridges, *Nature* **286**:84–86.

Wyatt, G. R., Locke, J., Bradfield, J. Y., White, B. N., and Deeley, R. G., 1981, Molecular cloning of vitellogenin gene sequences from *Locusta migratoria*, in: *Juvenile Hormone Biochemistry* (G. E. Pratt and G. T. Brooks, eds.), pp. 299–307, Elsevier/North-Holland Biomedical Press, Amsterdam.

Yamasaki, K., 1973, Characterization and partial purification of a mannan-like polysaccharide in the eggs of *Locusta migratoria, Insect Biochem.* **3**:79–90.

Yamashita, O., and Irie, K., 1980, Larval hatching from vitellogenin-deficient eggs developed in male hosts of the silkworm, *Nature* **283**:385–386.

Yonge, C., and Hagedorn, H. H., 1977, Dynamics of vitellogenin uptake in *Aedes aegypti* as demonstrated by trypan blue, *J. Insect. Physiol.* **23**:1199–1203.

Zinsmeister, P. P., and Davenport, R., 1971, RNA and protein synthesis in the cockroach ovary, *J. Insect Physiol.* **17**:29–34.

Chapter 3

# Vitellogenesis and Oocyte Growth in Nonmammalian Vertebrates

ROBIN A. WALLACE

## 1. Introduction

The outstanding characteristic of most vertebrate eggs is their large size. These specialized cells have grown so huge that many can be appreciated without instrumental aids as part of our manifest environment—from the amphibian eggs that grace rural ponds to the omelets and caviar that complement our cuisine. The egg proper of the domestic chicken averages 32 mm in diameter (Romanoff, 1943), for instance, but even this size is exceeded by a number of living vertebrates, such as the coelacanth, with eggs measuring 85–90 mm (Anthony and Millot, 1972). The chicken egg acquires about 99% of its size during the 6 days before laying (Romanoff and Romanoff, 1949), when approximately 1 g protein is accumulated per day (Cutting and Roth, 1973). Thus, pronounced growth occurs in these single cells over a relatively short period of time.

The major event during this growth period is the accumulation within the oocyte of what is generally termed yolk, and it is the general aspects of protein yolk formation or **vitellogenesis** to which this chapter is addressed. The overall process is discussed as it is known in nonmammalian vertebrates. Further details on any particular aspect can be found in several recent reviews (Ryffel, 1978; Wallace, 1978, 1983; Chan *et al.*, 1979; Knowland and Westley, 1979; Tata and Smith, 1979; Wahli *et al.*, 1981; Wallace and Selman, 1981; Ho *et al.*, 1982b; Shapiro, 1982; Wiegand, 1982; Wallace *et al.*, 1983). The scope of this chapter is limited to the cellular and molecular events that occur in the ovary and liver and does not address the environmental regulation of ovarian recrudescence, sexual behavior, or the complex interplay among hypothalamic, hypophyseal, and ovarian hor-

---

ROBIN A. WALLACE • C. V. Whitney Laboratory, St. Augustine, Florida 32086.

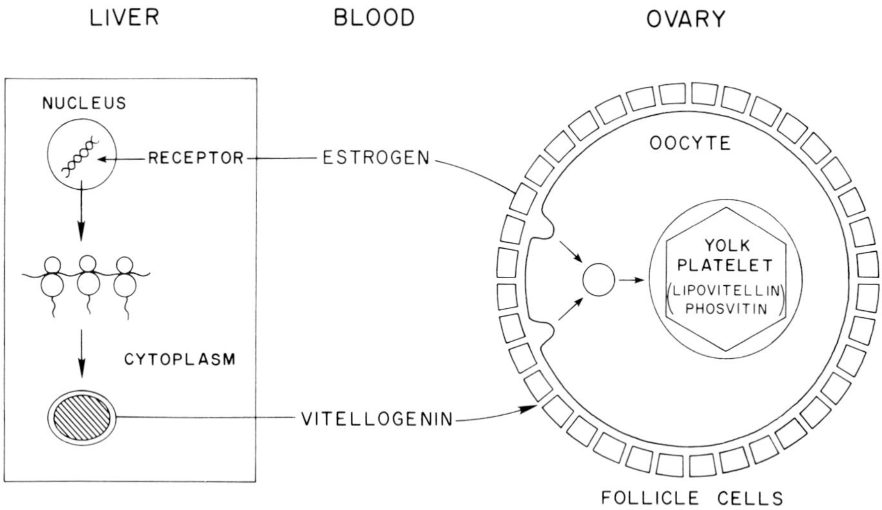


**Figure 1.** Scheme depicting the hepatic–ovarian relationship in a vitellogenic female vertebrate; see text for details. (Adapted from Tata and Smith, 1979.)

mones during reproductive periods (see, e.g., Chapter 1 of this volume and Richter and Goos, 1982).

A highly simplified model of the vitellogenic process is provided in Fig. 1, which represents the most common features found among nonmammalian vertebrates. Estrogen produced by the ovary in response to gonadotropin is introduced into the bloodstream and regulates the hepatic synthesis and secretion of **vitellogenin,** a large lipoglycophosphoprotein. Also as a response to gonadotropin, vitellogenin in turn is selectively sequestered from the bloodstream by growing oocytes and is proteolytically processed within the oocyte into accruing yolk proteins, of which **lipovitellin** and **phosvitin** are the most familiar. These yolk proteins comprise 80–90% of the total protein in amphibian eggs (Benbow *et al.*, 1975; Callen *et al.*, 1980), for instance, and are derived exclusively from vitellogenin (Wallace *et al.*, 1972). Thus, ovarian sequestration of vitellogenin secreted by the liver is predominantly responsible for oocyte growth. Additional proteins or physiological processes contribute to oocyte growth in birds and some fish and are discussed below together with vitellogenin-related events.

## 2. Vitellogenin and Vitellogenin Genes

Vitellogenin was first obtained in crude form from any animal in 1935, when Laskowski diluted the serum of laying hens with water and obtained a precipitate. He called this material Serumvitellin, because of the prevailing evidence indicating similarities between yolk and blood constituents in the laying hen. McIndoe (1959) later showed that Laskowski's Serumvitellin could be resolved into at least two components in the ultracentrifuge. During the subsequent decade, the view prevailed that the yolk proteins of the chicken *(Gallus domesticus)* could be found in the bloodstream noncovalently bound together as a complex. What were thought to be individual yolk proteins, and particularly phosvitin, were extracted and identified as part of this complex (McCully and Common, 1961; Mok *et al.*, 1961; Heald and McLachlan, 1963, 1964; Beuving and Gruber, 1971; Culbert and McIndoe, 1971). Meantime, a lipophosphoprotein was identified and isolated from the blood of vitellogenic females and estrogen-treated males of the amphibian, *Xenopus laevis*, but—unlike the chicken complex—it could not be dissociated by relatively mild procedures into yolk proteins (Wallace, 1970). The name vitellogenin, proposed by Pan *et al.* (1969) as a generic term for female-specific, bloodborne yolk precursors in insects (see Chapter 2), was adopted for the *X. laevis* protein. It was characterized in its native state as a large lipophosphoprotein (460,000 $M_r$) with a lipid and protein-phosphorus content of 12% and 1.4%, respectively; one atom of calcium is associated with every protein-phosphorus group (Wallace, 1970). *Xenopus laevis* vitellogenin was later shown to be a glycoprotein as well (Ansari *et al.*, 1971; Redshaw and Follett, 1971) and to contain nonstoichiometric amounts of biliverdin (Redshaw *et al.*, 1971).

Until this point, no special precautions against proteolysis had been taken by any workers during isolation of *X. laevis* vitellogenin or the chicken complex. The next major advance came when it was shown by gel electrophoresis under denaturing conditions that both *X. laevis* vitellogenin and the chicken complex were composed of a heterogeneous collection of peptides, but that when appropriate measures were taken to avoid proteolysis during the isolation procedure, a major peptide band was found, within a range of 200,000–240,000 $M_r$, depending on the animal used and the investigative conditions (Bergink, 1973; Bergink and Wallace, 1974; Bergink *et al.*, 1974; Deeley *et al.*, 1975). The native vitellogenin molecules in each species are thus composed of two large polypeptides and some lipid. In retrospect, it appears that when proteolytically nicked, the native chicken vitellogenin could be readily disassembled into yolklike proteins, whereas the *X. laevis* vitellogenin could not. Vitellogenins have now been identified or isolated from nearly all vertebrate classes, including agnathans and reptiles; in general, properties similar to those of the *X. laevis* or chicken vitellogenins have been observed (Ho *et al.*, 1980; Yu *et al.*, 1981); see Wiegand (1982) for the extensive literature on fish.

Wahli and associates isolated from *X. laevis* liver a 6.3-kilobase (kb) vitellogenin mRNA fraction from which 21 cDNA sequences were transcribed and cloned in *Escherichia coli* (Wahli *et al.*, 1976, 1978). Restriction endonuclease mapping, crosshybridization, heteroduplex mapping in the electron microscope, and heteroduplex melting experiments with these cloned DNAs unambiguously and surprisingly demonstrated that they could be divided into four categories, revealing the expression of four different genes (Wahli *et al.*, 1979). Specifically, the cloned DNAs fell into two main groups of sequences (A and B) that differ in approximately 20% of their nucleotides. Each main group in turn contains two subgroups (A1 and A2, B1 and B2) that have about 5% sequence divergence (Fig. 2). Wahli *et al.* (1979) therefore suggested that *X. laevis* vitellogenin is encoded by a small family of related genes. R-loop and RNA–DNA hybrid analysis in the electron microscope using DNA fragments containing genes A1 and A2 indicated that the 6-kb coding regions are interrupted 33 times by intervening sequences of DNA **(introns),** resulting in total gene lengths of 21 kb for gene A1 and 16 kb for gene A2 (Wahli *et al.*, 1980). The introns apparently interrupt structural sequences **(exons)** at homologous positions, thus indicating the close relatedness of the two genes. Similar results have been found for genes B1 and B2 (Germond *et al.*, 1983). Genes A1 and B1 have also been found to be linked (Wahli *et al.*, 1982), suggesting that the A and B sequences arose first by the duplication of an ancestral vitellogenin gene and that the two subgroups of each sequence diverged after genomic duplication, which is postulated to have occurred in *X. laevis* 30 million years ago (Bisbee *et al.*, 1977). Another fascinating recent discovery is that many additional silent, or nonexpressible, vitellogeninlike genes may also exist in *X. laevis* (Tata *et al.*, 1980).

The four types of mRNA molecules from which the various cDNAs were transcribed have since been isolated; each has been found with a reticulocyte lysate to encode a 200,000-$M_r$ polypeptide, which is immunoprecipitable by polyclonal vitellogenin antibody (Felber *et al.*, 1980). Wiley and Wallace (1978) also obtained evidence for at least three vitellogenin polypeptides when vitellogenin was denatured and analyzed electrophoretically under conditions of higher resolution (197,000, 188,000, and 182,000 $M_r$, present in a 2:2:1 ratio, respectively). Multiple high-molecular-weight vitellogenin polypeptides, apparently not generated by proteolysis,

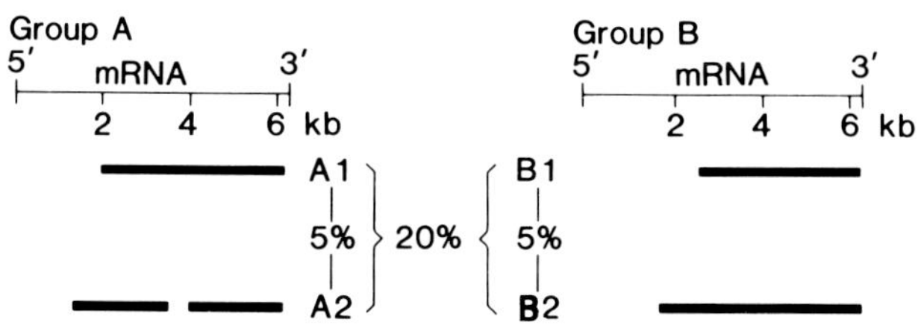


**Figure 2.** Relationship between the different cloned vitellogenin cDNAs in *Xenopus laevis*. cDNAs can be organized into two main groups of sequences, A and B, which differ in about 20% of their nucleotides. Each main group contains two subgroups (A1 and A2, or B1 and B2), which differ in about 5% of their sequence. A scale is provided for the mRNA maps; and the horizontal black bars below indicate the portions of the four mRNAs that have been cloned as cDNAs. (From W. Wahli *et al.*, 1981, *Science* **212:** 298–304. Copyright 1981 by the American Association for the Advancement of Science.)

have also been observed in the goldfish, *Carassius auratus* (Hori *et al.*, 1979; de Vlaming *et al.*, 1980), and in the axolotl, *Ambystoma mexicanum* (May *et al.*, 1981). In none of these studies, however, has a difference in the primary sequence of the various vitellogenin polypeptides been established.

Even though only one vitellogenin gene has been described thus far for the chicken (e.g., Wilks *et al.*, 1981; Protter *et al.*, 1982), the analysis of the vitellogenin protein is much more advanced. Three polypeptides have now been resolved and isolated by preparative electrophoresis and designated vitellogenin I, II, and III; the molecular weights were estimated to be 190,000, 180,000 and 170,000, respectively (Wang and Williams, 1980; Wang *et al.*, 1983). Vitellogenins I and II were each found to be associated with 116 moles of phosphate, whereas III contains only 44 moles of phosphate. Of greater importance, however, are indications that the three polypeptides differ with respect to amino acid analysis, peptide mapping patterns, and immunological crossreactivity. Thus, the three vitellogenin polypeptides reflect at least three different vitellogenin genes expressed in the chicken. When synthesized and secreted into the bloodstream under conditions of maximum stimulation, the ratio of the vitellogenin I, II, and III polypeptides was found to be 0.33:1.0:0.08, respectively. Since native chicken vitellogenin is composed of two polypeptides (Deeley *et al.*, 1975), it would be interesting to know whether it is a homo- or heterodimer, but such information is thus far lacking for any of the animals in which multiple vitellogenin polypeptides have been described. Recent information for the quail, *Coturnix coturnix*, also indicates vitellogenin polypeptides similar to those found in the chicken (Gibbins *et al.*, 1981).

## 3. Liver-Related Events

### 3.1. Vitellogenin Synthesis and Secretion

After it was simultaneously established in several laboratories that the liver of *X. laevis* is the only site of vitellogenin synthesis and secretion and that the process is normally controlled in females and can be initiated to a pronounced extent in males by estrogen (Follett *et al.*, 1968; Munday *et al.*, 1968; Rudack and Wallace, 1968; Wallace and Dumont, 1968; Wallace and Jared, 1968a,b, 1969), many laboratories began to examine vitellogenin synthesis as a model for steroid regulation of eukaryotic gene expression. By now, all types of nonmammalian vertebrates, from hagfish (Yu *et al.*, 1981), sharks (Craik, 1978c), and other fish (Wiegand, 1982) to turtles (Ho *et al.*, 1981) and birds (Bergink *et al.*, 1974; Gibbins and Robinson, 1982) have been shown to synthesize and secrete vitellogenin as a response to estrogen (e.g., Fig. 3). In general, estrogens and particularly estradiol-17$\beta$ have been found to be the most effective steroids *in vivo* (Redshaw *et al.*, 1969; Sundararaj and Nath, 1981; van Bohemen *et al.*, 1982), although a weak effect

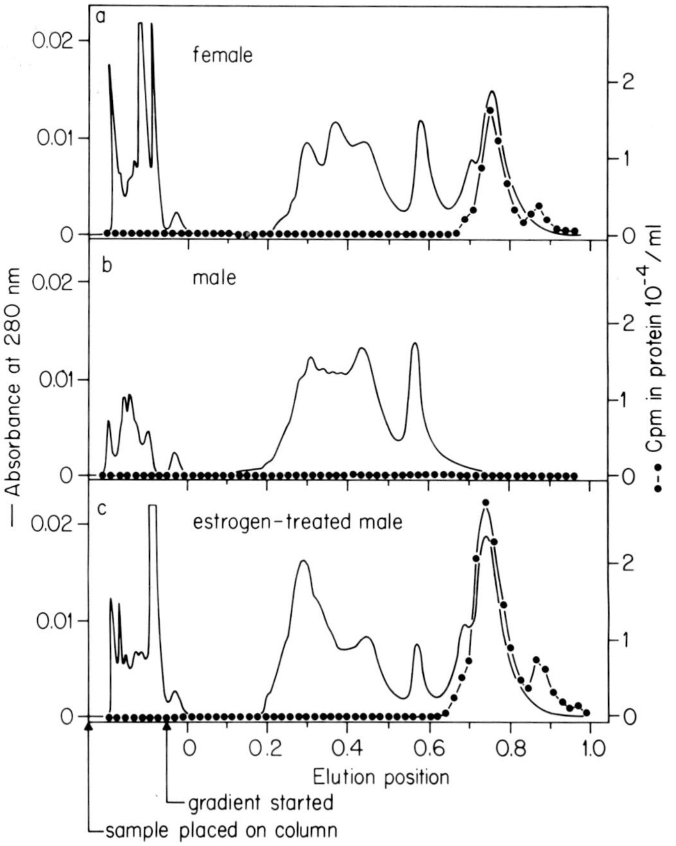


**Figure 3.** DEAE-cellulose chromatography of approximately 100 $\mu$l plasma from an individual (a) female, (b) male, and (c) estrogen-treated male *Fundulus heteroclitus*, each previously injected with [$^{32}$P]orthophosphate. Effluent fractions were monitored for absorbance at 280 nm or [$^{32}$P]protein (— ● —). Vitellogenin appears in the female and estrogen-treated male as a UV-absorbing labeled doublet, with a minor and major component eluting at positions 0.69 and 0.75, respectively. A minor non-UV-absorbing labeled component also appears at position 0.86; this phosvitinlike material represents a degradation product of vitellogenin. (From Selman and Wallace 1983.)

of androgens has also been noted for some fish (Hori *et al.*, 1979; Le Menn *et al.*, 1980). The hormonal requirements for vitellogenin synthesis have been best elucidated for *X. laevis* by using small liver pieces previously incubated in a defined medium for 1–2 weeks in order to remove traces of endogenous hormones. Such liver pieces do not synthesize and secrete either vitellogenin or the glucocorticoid-dependent proteins serum albumin and fibrinogen. Thyroxine was found to potentiate the ability of dexamethasone to reinduce the synthesis and secretion of serum albumin or fibrinogen and was required for the induction of vitellogenin synthesis and secretion by estradiol-17$\beta$ (Wangh, 1982; Wangh and Schneider, 1982).

These results help explain the acquisition of competence by the *X. laevis* liver to respond to estrogen during late metamorphosis (Knowland, 1978; Huber *et al.*, 1979; May and Knowland, 1980). Similar studies on neotenous animals such as *A. mexicanum* would be helpful to ascertain whether the thyroxine dependency of the estrogen effect is a general phenomenon. Also, the observations indicating a pituitary-mediated estrogen response in reptiles (Ho *et al.*, 1982a) need to be verified in organ or tissue cultures (Ho *et al.*, 1982b).

Another effect of estrogen on uninduced (male or immature female) hepatocytes of both the chicken and *X. laevis* is to increase irreversibly the number of estrogen receptors, thus enhancing responsivity (Mester and Baulieu, 1972; Lazier and Haggerty, 1979; Westley and Knowland, 1979; Hayward *et al.*, 1980); this effect is not observed in *A. mexicanum* (May *et al.*, 1981). The adult female is thus particularly sensitive to low levels of circulating estrogen and responds by synthesizing large vitellogenin transcripts within the nuclei of hepatic parenchymal cells. These large transcripts contain many intervening sequences, which are nonsequentially spliced out before translocation of the mature vitellogenin mRNAs to the cytoplasm (Ryffel *et al.*, 1980). Under conditions of chronic estrogen exposure, cytoplasmic vitellogenin mRNAs appear to be very stable and are therefore probably translated many thousands of times subsequent to ribosome and membrane attachment (Shapiro *et al.*, 1976; Shapiro, 1982).

Pulse-chase experiments, combined with subcellular fractionation of estrogen-treated female *X. laevis* liver explants, have indicated that vitellogenin polypeptides are completely synthesized in the rough endoplasmic reticulum before modification by phosphorylation or glycosylation (Gottlieb and Wallace, 1981, 1982; see Fig. 4). Approximately 30% of the phosphate residues is covalently attached to vitellogenin in the rough endoplasmic reticulum after termination of polypeptide synthesis; the remainder of the phosphate is incorporated during intracellular translocation through membrane compartments comprising the smooth microsome subfraction (smooth endoplasmic reticulum, Golgi apparatus, secretory elements). Carbohydrate incorporation, including tunicamycin-sensitive core glycosylation, appears to occur entirely within a smooth membrane compartment—a significant departure from currently accepted descriptions of glycoprotein synthesis based on mammalian models. An apparently different pattern of post-translational modification has been found for the chicken, in which the stages of vitellogenin biosynthesis follow the pattern: polypeptide synthesis → glycosylation → phosphorylation → secretion (Wang and Williams, 1982). Although subcellular fractionation was not done, data indicating that 94–98% of the intracellular vitellogenin is unphosphorylated (Wang and Williams, 1982), together with the observation that [$^{32}$P]vitellogenin appears in plasma with a lag of only several minutes after [$^{32}$P]orthophosphate administration to estrogen-treated roosters (Bergink *et al.*, 1973), argue for vitellogenin phosphorylation occurring immediately before secretion in the chicken. Lipidation of vitellogenin in

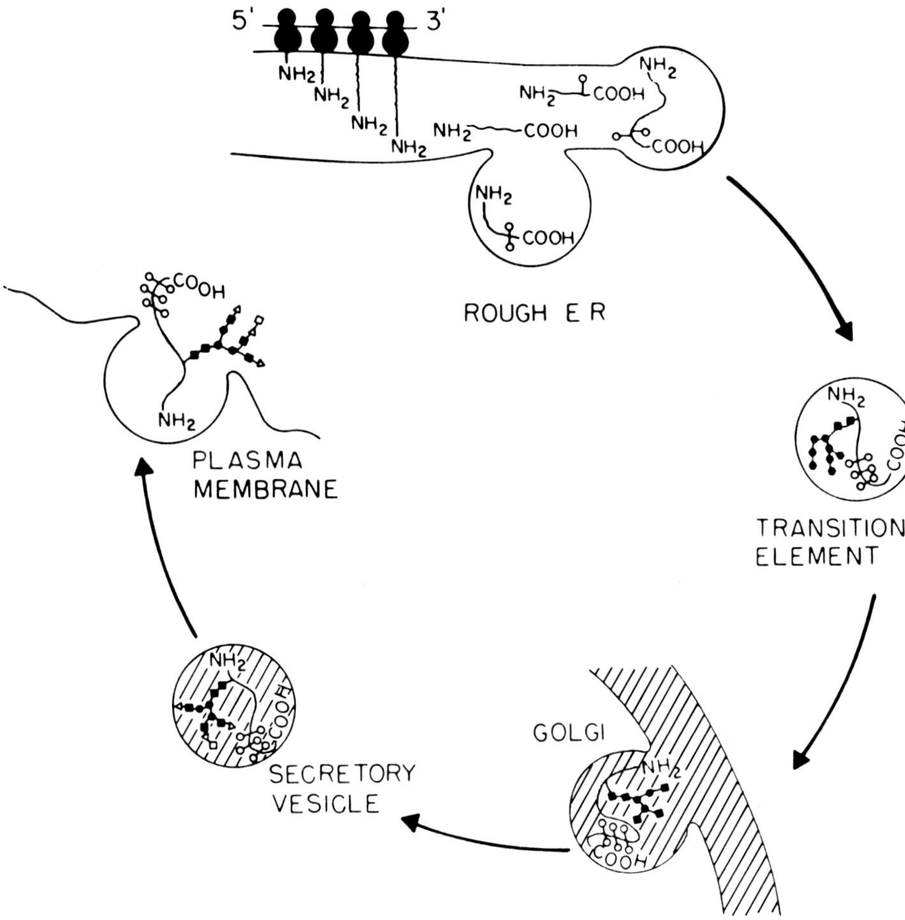


**Figure 4.** Scheme indicating the post-translational modification and secretion of vitellogenin in hepatocytes of *Xenopus laevis.* The intact newly synthesized polypeptide is partially phosphorylated (○) after its release into the lumen of the rough endoplasmic reticulum. The subsequent segregation of vitellogenin within the budding smooth-surfaced regions of the rough endoplasmic reticulum (referred to as transition elements) is inferred from the ultrastructure of these hepatocytes. Additional phosphorylation occurs within its COOH-terminal region during this vesicular transit to the Golgi apparatus. The kinetic evidence would indicate that by the time vitellogenin is being packaged into secretory vesicles, phosphorylation has ceased. A mannose-containing oligosaccharide (●) is also linked to vitellogenin by an *N*-glycosidic bond between asparagine and *N*-acetylglucosamine (■), apparently within the transition elements. After transfer to vitellogenin, the oligosaccharide undergoes an enzymatic removal of several, but not all mannosyl groups. This trimming step is relatively slow and is thought to occur in the Golgi. The sequential addition of *N*-acetylglucosamine, galactose (△), and scialic acid (□) follows trimming, with the last two steps occurring just before secretion. (From Gottlieb, 1981.)

the liver is much more difficult to study because of the noncovalent nature of the association. Nevertheless, a beginning has been made by Smith *et al.* (1978), who found that prior administration of estradiol-17$\beta$ to *X. laevis* markedly stimulates cholesterol and fatty acid synthesis by liver extracts. Most of the lipids and cholesterol synthesized by liver slices were associated with membrane fractions, whereas *in vivo* about 15% of the synthesized lipid and cholesterol could be chased into vitellogenin in the bloodstream. Unfortunately, these results tell us little about the mechanism or site of vitellogenin lipidation in the liver or the extent to which lipid exchange or sequestration can occur once vitellogenin is secreted into the bloodstream. Since vitellogenin represents the major source of lipid in most nonavian eggs, further studies on this significant problem are needed.

Initiation of vitellogenin synthesis and secretion by estrogen is a reversible process that does not require DNA synthesis (Green and Tata, 1976) or cytodifferentiation, since the response occurs within a few hours (Wangh and Knowland, 1975). Nevertheless, the cytoplasm of hepatocytes from vitellogenic females or estrogen-treated males generally appears much more complex than that from males or immature females (see Fig. 5; see also Herbener *et al.* (1983) for a morphometric analysis in *Rana pipiens*). Loss of glycogen and an extensive elaboration of the endoplasmic reticulum together with Golgi elements in particular highlight the observable changes; in addition, an increase in enzymes involved in lipidation (Philipp and Shapiro, 1981) and an increase in new ribosomes and ribosomal proteins (Shapiro, 1982) take place. However, the mechanism of vitellogenin packaging and secretion has not yet been decisively established. Despite evidence that all liver parenchymal cells participate in the vitellogenic response in the chicken (Wachsmuth and Jost, 1976) and *X. laevis* (Wangh *et al.*, 1979), this has been disputed at least for the chicken (Lin and Chan, 1981). Several areas for future research appear to include (1) morphological identification of the secretory pathway for vitellogenin and a correlation of this pathway with existing biochemical studies, (2) an indication of other enzymatic components necessary for the vitellogenic response and elucidation of whether such enzymes are directly or indirectly regulated by estrogen, and (3) a resolution of whether all hepatic parenchymal cells in various animals participate in vitellogenin synthesis and secretion.

### 3.2. Other Egg-Related Proteins Synthesized and Secreted by the Liver

In terms of dry weight, the major yolk-related protein in birds is not vitellogenin, but rather a very-low-density lipoprotein (VLDL), which was first observed in the plasma of laying hens by Schjeide (1954) and identified by McIndoe (1959) as a component of Laskowski's (1935) Serumvitellin.

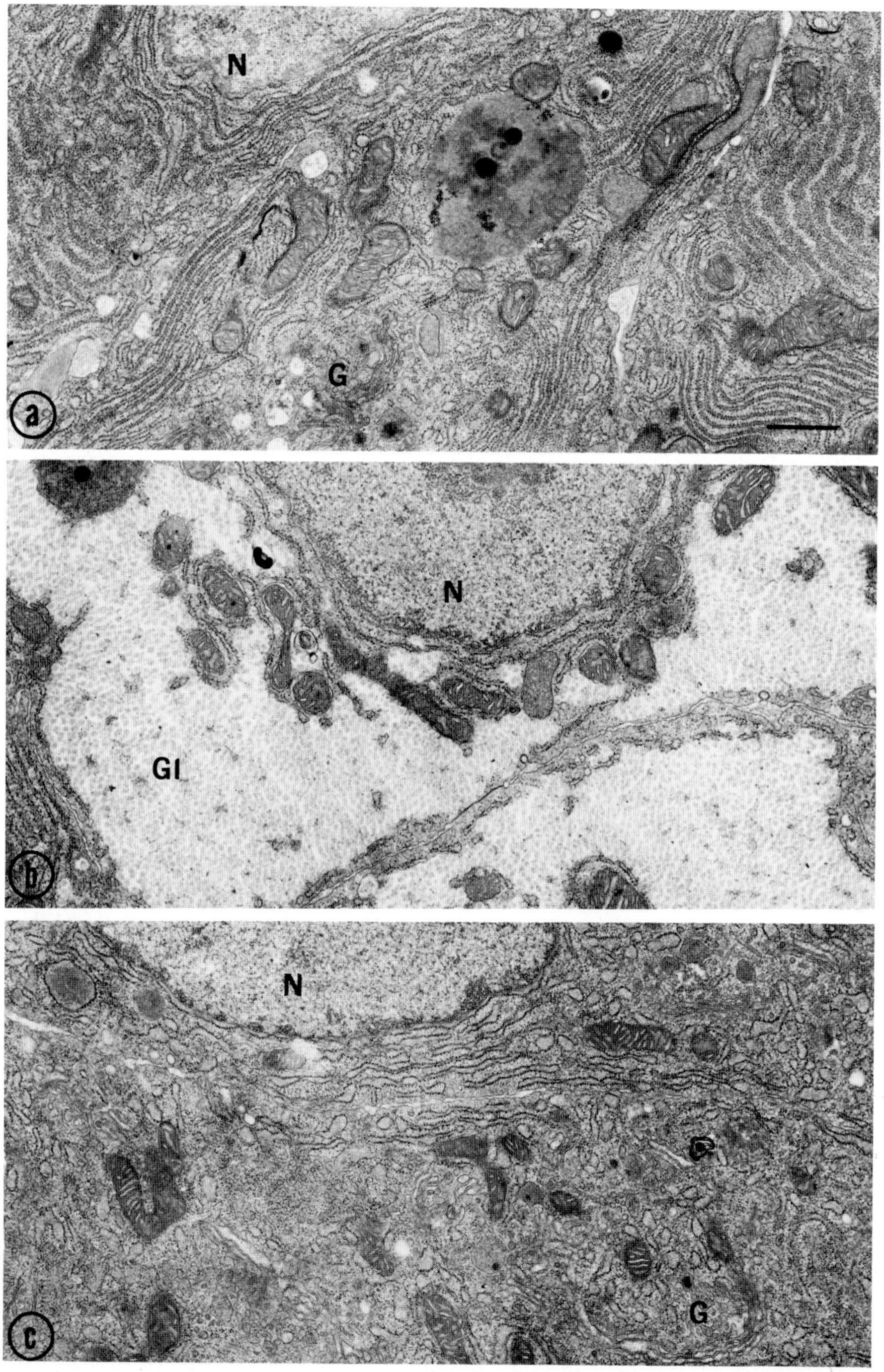


**Figure 5.** Hepatocytes from (a) female, (b) male, and (c) estrogen-treated male *Fundulus heteroclitus*. Hepatocytes from both female and estrogen-treated male contain considerable rough endoplasmic reticulum and large Golgi complexes. The male hepatocytes display large stores of glycogen. A secondary lysosome is visible in the female hepatocyte. G, Golgi complex; Gl, glycogen; N, nucleus. Scale bar: 1 $\mu$m. (From Selman and Wallace, 1983.)

VLDL contains 82–88% lipid associated with protein (Hillyard *et al.*, 1956) and is synthesized and secreted by the liver in large amounts in response to estrogen (Urist and Schjeide, 1961; Luskey *et al.*, 1974; Chan *et al.*, 1976). Unlike vitellogenin, however, it is normally found in somewhat less lipidated form as a normal blood constituent of pullets and roosters (Hillyard *et al.*, 1972; Yu and Marquardt, 1973; Capony and Williams, 1980). When particular care was taken to avoid proteolysis, chicken VLDL was shown to be made up of two apoproteins, now known as apo-VLDL-B and apo-VLDL-II, with molecular weights of approximately 350,000 and 9444, respectively (D. L. Williams, 1979). The small apo-VLDL-II peptide has been isolated and sequenced from both the chicken and the emu (Dopheide and Inglis, 1974, 1976; Jackson *et al.*, 1977); the signal peptide of chicken apo-VLDL-II has also been sequenced (Chan *et al.*, 1980). The secretory pathway of apo-VLDL-II has been studied by immunoelectron microscopy, the principal conclusion being that lipidation occurs in the Golgi apparatus before secretion (Lin and Chan, 1982). The molar ratio of apo-VLDL-B to apo-VLDL-II in the plasma of estrogen-treated roosters was found to be 1:32, indicating that these two apoproteins represent 54% and 46%, respectively, of the total VLDL protein. Unlike apo-VLDL-II (Blue and Williams, 1981), apo-VLDL-B also appears to be synthesized and secreted by nonhepatic tissues, notably the kidney and the entire length of the intestine (Blue *et al.*, 1980), but levels of synthesis and secretion are not modulated by estrogen at these sites (Blue, 1981).

Other estrogen-regulated components synthesized by the liver and found in yolk are the vitamin-carrying proteins (Clagett, 1971; Muniyappa and Adiga, 1981; Adiga and Murty, 1983). The kinetics of synthesis and secretion of these proteins as a response to estrogen are somewhat different from that observed for vitellogenin, and the amount of riboflavin- or thiamin-binding proteins produced by the liver is about three orders of magnitude less (Murthy and Adiga, 1978; Muniyappa and Adiga, 1980), but they nevertheless appear to be indispensable for embryonic development and survival (Clagett, 1971; Adiga and Murty, 1983). Since trace amounts of riboflavin-carrying protein are found in both the yolk and white of the egg, the latter being secreted by the oviduct and swallowed by the late chick embryo (Rhodes *et al.*, 1959; Ostrowski, 1968), the relative importance of the yolk component to overall embryonic development is unclear. Two minor estrogen-regulated polypeptides have also recently been identified in *X. laevis* hepatocyte cultures (Wangh *et al.*, 1979) and may represent similar vitamin-carrying proteins.

## 4. Blood-Related Events: Kinetics of Vitellogenin and VLDL Turnover

Animals that have a seasonal breeding pattern generally also show a seasonal variation of vitellogenin levels in the bloodstream (Giorgi *et al.*,

1982); an example for a population of trout, *Salmo gairdneri*, which spawn in January/February, is provided in Fig. 6a. Other patterns exist, however, such as that found for the turtle, *Chrysemys picta* (Fig. 6b). In this case, ovulation occurs in June, and subsequent vitellogenesis in the fall is interrupted by a winter hibernation period before continuing in the spring. Peak blood levels of vitellogenin in these two animals range from 13 to 40 mg/ml.

Quite a different situation exists for the elasmobranch, *Scyliorhinus canicula*, in which oviposition continues for 10–12 months of the year. A very low vitellogenin level (0.4 mg/ml) displays little seasonal variation, so that vitellogenin turnover in the bloodstream remains fairly constant (Craik, 1978a). At 7°C, the plasma half-life of vitellogenin was found to be about 9 days, reflecting a slow secretion by the liver and uptake by the ovary throughout the year (Craik, 1978b).

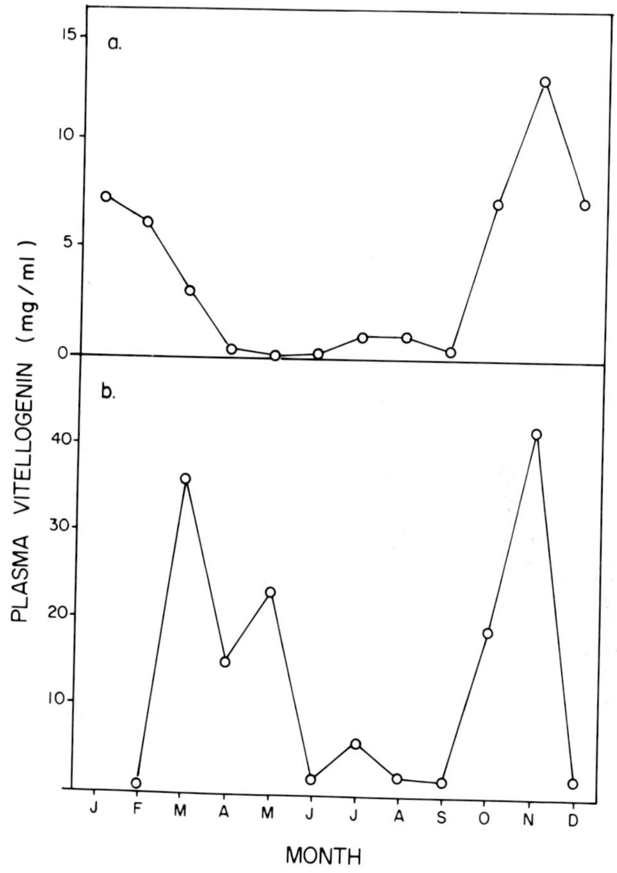


**Figure 6.** Monthly variation of vitellogenin plasma levels in (a) rainbow trout, *Salmo gairdneri*, and (b) painted turtle, *Chrysemys picta*. (Drawn from data provided by van Bohemen and Lambert, 1981 and Gapp *et al.*, 1979 b.)

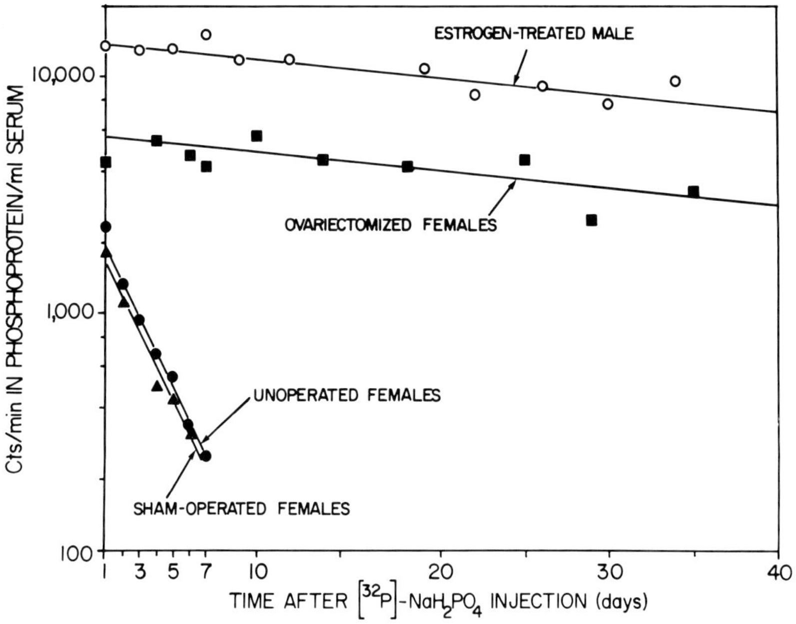


**Figure 7.** Turnover of vitellogenin in the bloodstream of *Xenopus laevis* under different physiological conditions. Males and females were injected with estrogen and gonadotropin, respectively, on day 0 in order to stimulate vitellogenin production in both groups. On day 8, some of the females were injected with [$^{32}$P]orthophosphate in order to label specifically and exclusively any vitellogenin secreted into the bloodstream during the subsequent 20 hr. On various days thereafter, the amount of [$^{32}$P]vitellogenin present in the serum was measured. Each point represents the average value derived from three animals. (From Wallace and Jared, 1968b.)

Laboratory-maintained *X. laevis* also display a low level of vitellogenic activity throughout the year (Keem *et al.*, 1979), apparently reflecting a continuous food supply (Wallace and Bergink, 1974); thus, they resemble the situation found for *S. canicula.* Vitellogenesis can be stimulated in the intact animal by exogenous gonadotropin (Wallace *et al.*, 1970), however, under which condition the physiological half-life of vitellogenin was found to be somewhat less than 2 days at 20°C (Wallace and Jared, 1968b), representing a turnover of 7.7 mg/day (Wallace and Jared, 1969). The corresponding vitellogenin half-life for estrogen-treated males and ovariectomized females was 40 days, reflecting normal blood protein turnover, whereas the turnover in sham-operated females was indistinguishable from that in unoperated, vitellogenic females (Fig. 7). These results indicate that vitellogenin undergoes a slow turnover in the bloodstream typical of circulating proteins in the absence of an ovary, whereas a functioning ovary normally serves to remove vitellogenin from the circulation in vitellogenic females.

A more direct test of this notion was performed by injecting labeled vitellogenin and blood proteins into vitellogenic females and simultaneously following the course of specific activity in the blood and ovarian protein. The results indicate that vitellogenin leaves the circulation and becomes associated with ovarian protein much more rapidly than the other two proteins tested (Fig. 8a). At the beginning of the experiment, when the relative amounts of the three labeled components in the bloodstream are similar, the incorporation of vitellogenin into ovarian protein appears to be about 25–50 times more rapid than the other blood proteins (Fig. 8b). It is important to note that the blood proteins other than vitellogenin do become associated with ovarian protein to some extent during the short-term course of the experiment. These results, in agreement with long-standing observations on the uptake of autologous and heterologous macromolecules into growing oocytes, indicate that all macromolecular components of the bloodstream are taken up into the recrudescent ovary and that this process is particularly pronounced for vitellogenin.

It has generally been assumed that vitellogenin is absent from the circulation before estrogen induction, but this notion may have to be reevaluated in the light of recent evidence by Blue and Williams (1981), who found circulating levels of vitellogenin in several roosters, ranging from 4 to 100 ng/ml, using a very sensitive radioimmunoassay. With a somewhat less sensitive radioimmunoassay, Redshaw and Follett (1976) found that circulating levels of vitellogenin in pullets were undetectable (<50 ng/ml) until 16 weeks of age. The concentration was then found to rise rapidly

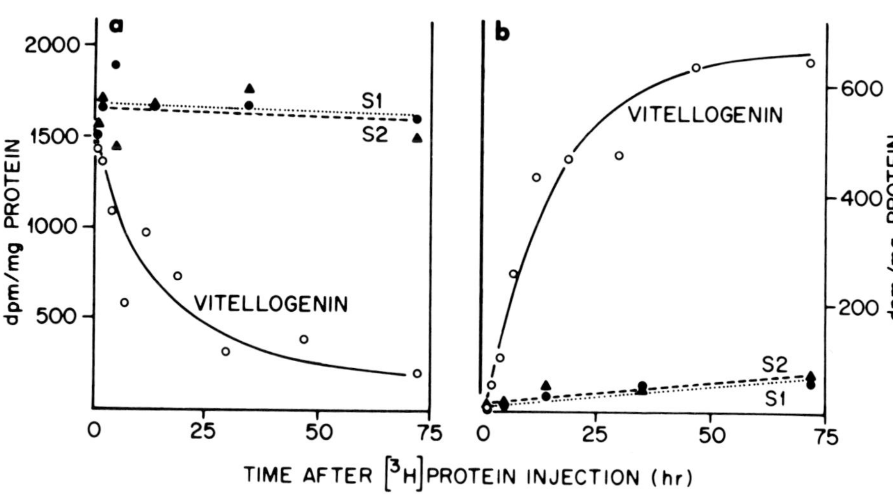


**Figure 8.** Loss of labeled protein from the blood (a) and its incorporation into the ovary (b) of *Xenopus laevis* after the injection of equivalent amounts of [$^3$H]vitellogenin or serum protein fractions [$^3$H]-S1 or [$^3$H]-S2 into gonadotropin-treated females. The various injected proteins were initially labeled *in vivo* and chromatographically isolated. Each point represents the average value derived from two animals. (From Wallace and Jared, 1969.)

until, at 20 weeks, it plateaus at levels of 20–25 mg/ml. Egg laying then commences at 21–23 weeks, and the physiological half-life for vitellogenin in the serum of laying hens is reduced to about 1.0 day. In older hens, vitellogenin levels of 10–20 mg/ml are generally recorded, but strain differences can sometimes be an important factor. The circulating level of VLDL in laying hens has not been directly measured, but judging from older data (Schjeide, 1954) it can be estimated to be also around 20 mg/ml at the onset of egg production.

## 5. Ovary-Related Events

### 5.1. Passage of Macromolecular Materials through the Follicular Epithelium

Once oogonia give rise to oocytes within the ovary, they become invested by a layer of follicle cells and connective tissue, thereby forming a follicle. Although the organization of the growing follicle differs to some extent among the various types of vertebrates, in general an acellular investment (primary envelope, vitelline envelope, chorion, zona pellucida, zona radiata) forms immediately adjacent to the oocyte and is penetrated by microvilli from the oocyte and also by processes from an overlying layer of follicle (granulosa) cells. In turn, this cellular layer is invested by a basal lamina and a collagenous matrix through which a vascular network courses. The matrix layer, most frequently termed the **theca,** also contains fibroblasts and sometimes specialized steroid-secreting cells and is externally limited by a squamous and impermeable ovarian epithelium (Fig. 9); see Dumont and Brummett (1978) for *X. laevis* and Chapter 5, this volume, for a detailed discussion of vertebrate egg envelopes.

Injection of labeled vitellogenin or macromolecular tracers into a variety of animals known to be vitellogenic has invariably demonstrated that such materials freely penetrate throughout the follicle primarily by an intercellular route (Neaves, 1972; Brummett and Dumont, 1977; Dumont, 1978; Selman and Wallace, 1982a,b, 1983). Circulating materials appear to leave the perifollicular capillaries both by intracellular (pinocytotic) transport across endothelial cells and by passage between adjacent endothelial cells. Substances with a diameter of up to at least 10 nm (such as ferritin or native vitellogenin) can then pass from the theca through a basal lamina and penetrate patent intercellular channels of the follicular epithelium and the pore canals of the vitelline envelope. Thus, bloodborne proteins have ready access to the growing oocyte during vitellogenesis (Fig. 9).

A major unstudied question concerns the extent to which the above conditions prevail throughout the growth of the oocyte in any one type of animal. The precise timing of follicle vascularization as well as the extent of follicular permeability to blood proteins may prove to have important

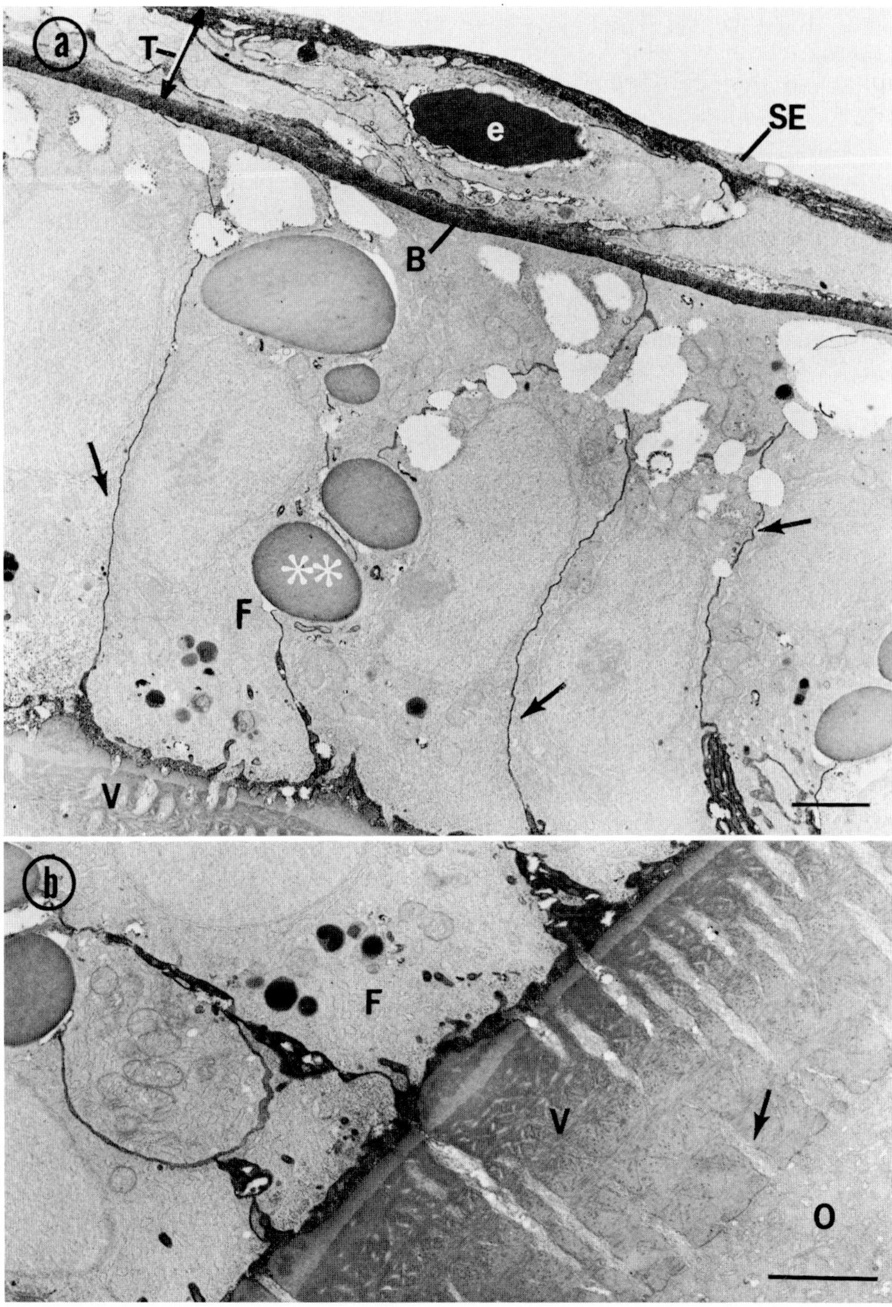


**Figure 9.** Vitellogenic follicle from *Fundulus heteroclitus* 5 hr after injection of horseradish peroxidase. (a) Electron-dense peroxidase reaction product seen within the extracellular components of the theca (T), the basal lamina (B), overlying the follicular epithelium (F), and the intercellular channels between the follicle cells (arrows). SE, surface epithelium; e, erythro-

regulatory implications. In the silkmoth, for instance, the formation of an occlusion zone, consisting of tight junctions joining adjacent follicle cells, appears to prevent nutrients from reaching the oocyte, thus effectively terminating vitellogenesis (Rubenstein, 1979). Analogous situations may also occur in nonmammalian vertebrates, since recent evidence (see Section 6) indicates that once oocytes above a certain size are removed from their follicles and provided with a saturating concentration of vitellogenin *in vitro*, they can grow at a maximum rate to sizes larger than normally found *in vivo*.

### 5.2. Selective Binding and Uptake of Vitellogenin by the Growing Oocyte

Observations on selective vitellogenin incorporation *in vivo* for *X. laevis* were confirmed and extended once culture procedures were developed for isolated oocytes. Vitellogenin uptake *in vitro* was found to be saturable at an external concentration of about 3 mg/ml, and at this concentration about 20–50 times as many vitellogenin macromolecules were incorporated per unit time as other proteins tested (Table I). Selective vitellogenin uptake was found relative both to plasma proteins normally present in the bloodstream (Wallace *et al.*, 1970) as well as to heterologous macromolecules (Wallace and Jared, 1976). However, the animal source of vertebrate vitellogenin is relatively unimportant (Wallace *et al.*, 1980a). Thus, a receptor-mediated process appears to be involved in vitellogenin uptake, and the ligand attachment site is relatively conserved among nonmammalian vertebrates.

**Table I.** Relative Uptake of Various Proteins by *Xenopus laevis* Oocytes *In Vitro*

| Protein | $M_r$ | Uptake | |
|---|---|---|---|
| | | ng/mm² per hr | fmoles/mm² per hr |
| Ferritin | 465,000 | 0.7 | 2 |
| Hemoglobin | 64,000 | 0.3 | 4 |
| Serum albumin | 68,000 | 0.3 | 5 |
| Vitellogenin | 460,000 | 49.2 | 107 |

From Wallace and Jared, 1976. Adapted from *The Journal of Cell Biology*, 1976, vol. 69, pp. 345–351 by copyright permission of The Rockefeller University Press.

cyte within perifollicular capillary; (**), fibrillar appendage of the vitelline envelope (V). (b) Long microvillar processes (arrow) extending from the oocyte surface (O) can be seen within pore canals of the vitelline envelope (V) and extend into reaction product-filled channels between follicle cells (F). Scale bars: 2 μm. (From Selman and Wallace, 1983.)

The only existing study of a similar nature on another animal is that of Campbell and Jalabert (1979) for the trout, *S. gairdneri*. However, their data indicate that alkylated $^{14}C$-labeled vitellogenin is incorporated three to six times faster than alkylated $^{14}C$-labeled bovine serum albumin (BSA) on a weight basis, depending on the method of oocyte preparation. Assuming molecular weights of ~400,000 and 68,000 for vitellogenin and BSA, respectively, the results of Campbell and Jalabert (1979) would indicate that on a molar basis vitellogenin is sequestered at only 0.5–1.0 times the rate of BSA; i.e., no selective uptake into oocytes occurs. Furthermore, vitellogenin uptake is saturable in the *X. laevis* system (Wallace and Jared, 1976), but the data provided by Campbell and Jalabert (1979) indicate that this is not the case for *S. gairdneri*. Finally, the absolute vitellogenin uptake rates reported (1.9–4.4 ng/mm$^2$ per hr) are 4–9% of those found for *X. laevis* oocytes. Something is clearly amiss in the trout system. Either the oocytes used were nonvitellogenic, the receptor mechanism was damaged by the enzymatic procedures used to isolate the oocytes, the alkylation process destroyed ligand recognition, intact vitellogenin was not prepared and the ligand was missing, or some other problem was responsible.

Preliminary studies on the interaction of *X. laevis* vitellogenin with its cell surface receptor have also begun (Opresko *et al.*, 1981). At 0°C, vitellogenin binding to intact oocytes is saturable, reaching equilibrium by 4 hr, and nonspecific binding is generally found to be 1% or less at a labeled vitellogenin concentration of 1 mg/ml (2.2 nM). Scatchard plot analyses of 1.0-mm-diameter oocytes maintained at 0°C reveal a single class of surface-binding sites whose number varies from 2 to 28 $\times 10^{10}$/oocyte, depending on the animal under study and assuming univalent binding. Correlation between surface receptor number and internalization of vitellogenin at 20°C indicates that the number of vitellogenin receptors is inversely related to the turnover rate of the receptor at the surface and positively correlated with the rate of receptor insertion into the membrane (Wallace *et al.*, 1983). Vitellogenin sequestration could also occur for at least 17 hr in the presence of sufficient cycloheximide to inhibit protein synthesis by 99%, suggesting that vitellogenin receptor synthesis is not directly coupled to receptor uptake (Wallace and Ho, 1972). Preincubation of oocytes with unlabeled vitellogenin does not alter the number of cell-surface receptors, so that vitellogenin receptor appears to be continuously internalized, regardless of occupancy (Opresko *et al.*, 1981). This lack of downregulation of the vitellogenin receptor also suggests that occupied receptors do not induce the formation of endocytic vesicles, but instead are internalized at preformed sites. These characteristics place the vitellogenin receptor among the class II type of receptors as defined by Kaplan (1981).

Membrane preparations have also been made from cortices derived from chicken oocytes and have been used to study vitellogenin binding (Yusko *et al.*, 1981). In this case, vitellogenin and phosvitin, a yolk protein proteolytically derived from vitellogenin, both bound to oocyte membranes with $K_D$ values of $\sim 6 \times 10^{-7}$ M but showed little affinity for erythrocyte membranes. Both proteins also competitively inhibit the binding of each other, suggesting that chicken vitellogenin binds to the oocyte surface by means of its phosvitin moiety. Preliminary identification of the chicken vitellogenin receptor has been achieved by extracting membrane preparations with Triton X-100 and passing the extract either through a gel exclusion or phosvitin-affinity column. A large (500,000 $M_r$) complex found to bind phosvitin specifically yielded primarily a 116,000-$M_r$ subunit when analyzed by sodium dodecylsulfate polyacrylamide gel electrophoresis (SDS-PAGE). A similar peptide was obtained from the extract-treated affinity column by eluting with phosvitin and hence apparently represents the vitellogenin receptor (Woods and Roth, 1979, 1980).

The enormous amount of membrane material that can be obtained from chicken oocytes would appear to make this system ideal for biochemical characterization of both the vitellogenin receptor and the presumed VLDL receptor present in avian oocytes. Preparation of antibodies to the purified receptors should also enable interesting immunocytochemical studies. Unfortunately, adequate culture procedures have not yet been, and may never be, developed for the relatively huge chicken oocyte, so that physiological questions related to vitellogenin and VLDL receptor turnover and cooperativity during oocyte growth probably cannot be addressed with this system. Nevertheless, an interesting genetic variant, which suppresses egg laying, has been described in the chicken. Approximately one-half the daughters from carrier males either completely fail to lay eggs or lay only a few widely spaced eggs during their reproductive period (Jones *et al.*, 1975). Affected females also exhibit extremely hyperlipemic sera, and ultracentrifugal analysis (Schjeide *et al.*, 1976) indicates abnormally high concentrations of vitellogenin and VLDL. The yolk of affected females, however, is relatively deficient in the major yolk components when compared with a protein (albumin) presumably taken up nonselectively (Schjeide *et al.*, 1976). The mutation may thus affect a part of the specific uptake mechanism in the ovary.

## 5.3. Protein Processing and Translocation within the Vitellogenic Oocyte

Examination of the vitellogenic oocyte surface by transmission electron microscopy invariably shows extensive numbers of coated pits and vesicles (Fig. 10), suggesting that an endocytic mechanism is involved in vitellogenin sequestration; (see Roth *et al.*, 1976, for a particularly pronounced example in the chicken). However, vitellogenin has been transiently observed in, but has never been isolated as such from, the ovary or oocyte. Rather, most of the total protein in full-grown oocytes is found in membrane-enclosed yolk granules, platelets, or spheres containing yolk

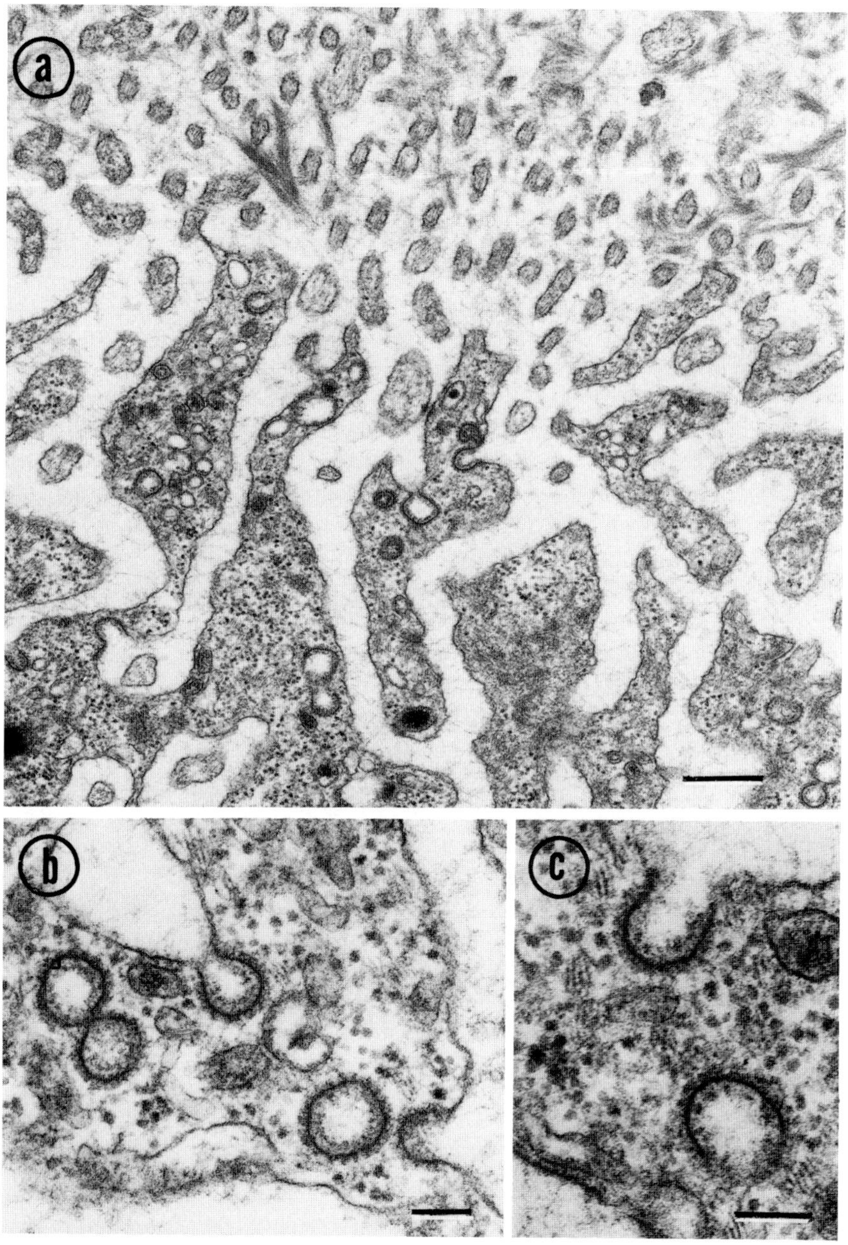


**Figure 10.** Pinocytotic pits and vesicles at the highly infolded surface of a vitellogenic *Xenopus laevis* oocyte. (a) Overview of the surface, with wispy vitelline envelope (near top) overlying the oocyte. (b) A portion of the infolded surface showing clathrin-coated pits and vesicles. (c) Somewhat higher magnification of two pinocytotic pits. Particles lining the lumenal surface have approximately the same dimension as native vitellogenin macromolecules (10 nm). Scale bars: 500 nm in (a) and 100 nm in (b) and (c). (From Wallace *et al.*,1983.)

proteins such as lipovitellin and phosvitin (see below). Vitellogenin from *X. laevis* has been found to have essentially the same amino acid profile as the combined yolk proteins (Redshaw and Follett, 1971) and to crossreact with antibody prepared against yolk protein (Wittliff and Kenney, 1972). Labeled vitellogenin is also proteolytically processed into lipovitellins and phosvitins after being sequestered by oocytes either *in vivo* or *in vitro* (Fig. 11b). Despite this processing, the total labeling found in the oocyte remains constant over at least a 2-week period, indicating that further proteolysis or degradation does not occur (Opresko *et al.*, 1980). Measurement of the protein synthesized or sequestered by the *X. laevis* oocyte has established that at least 99% of the yolk protein is derived from vitellogenin (Wallace *et al.*, 1972). Vitellogenin sequestration thus represents the predominant mechanism for oocyte growth in this animal. It is important that the relative contribution of vitellogenin to oocyte growth in other animals also be determined at some time, particularly for those animals in which other yolk-forming processes have been implicated.

In contrast to the fate of sequestered vitellogenin, that injected into oocytes is found to be simply degraded (Fig. 11C) with a half-life of 6–14 hr (Wallace and Hollinger, 1979). These results indicate that vitellogenin is taken up by oocytes and converted into yolk proteins within a compartment different from and inaccessible to vitellogenin injected into the oocyte cytoplasm. A combined autoradiographic and ultrastructural study has confirmed that this compartment is an endocytic system (Brummett and Dumont, 1977).

Pulse-chase experiments with labeled proteins *in vitro* followed by cell fractionation and gel electrophoretic analyses have been used to elucidate the dynamic aspects of this endocytic system and hence to construct a working model for endocytic compartmentation of protein in *X. laevis* oocytes (Fig. 12). Vitellogenin apparently binds to specific receptors on the oocyte surface, although the site of initial binding is unknown. Internalization occurs at preformed membrane sites known as **coated pits,** which invaginate into the oocyte cortex giving rise to coated vesicles. This process takes place continuously, whether or not vitellogenin receptors are occupied, and nonbound macromolecules (blood proteins, heterologous tracers) may also be adventitiously trapped during membrane internalization. Morphological evidence indicates that sequestered material is subsequently seen in uncoated vesicles or endosomes (Dumont, 1978). In the absence of vitellogenin, these fuse directly with yolk platelets. Alternatively, endosomes containing occupied vitellogenin receptors associate with one another and condense, giving rise to transitional yolk bodies. These in turn fuse with yolk platelets only after proteolytic processing of vitellogenin. Observations on protein translocation in the absence or presence of vitellogenin are reconciled by assuming that proteolytic processing of vitellogenin serves to sever the vitellogenin–receptor complex. Thus, the transitional yolk body would contain unoccupied receptors after vitellogenin

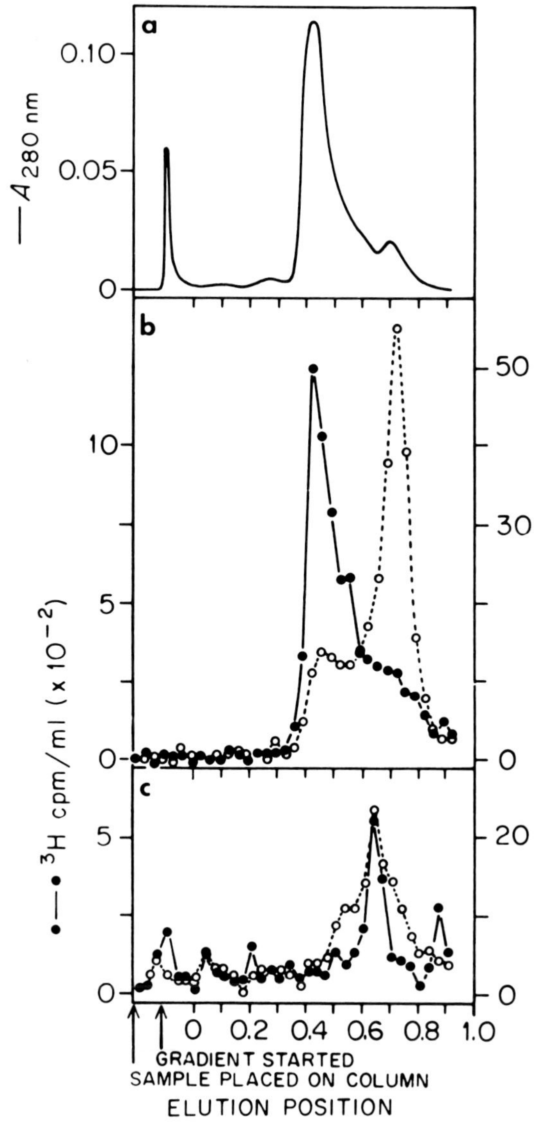


**Figure 11.** TEAE-cellulose chromatography of *Xenopus laevis* oocyte extracts containing approximately 5 mg undenatured protein derived from 60 vitellogenic oocytes. (a) Absorbance trace indicating the elution position of the yolk proteins lipovitellin and phosvitin (at elution positions 0.42 and 0.72, respectively). (b) Distribution of radioactivity in saline-injected oocytes 11 hr after sequestering approximately 88 ng [$^3$H, $^{32}$P]vitellogenin, previously labeled *in vitro* by incubating liver slices from estrogen-treated males with [$^3$H]leucine and [$^{32}$P] orthophosphate. The [$^3$H, $^{32}$P]vitellogenin, which normally elutes at position 0.64, has been processed into lipovitellin and phosvitin containing the expected relative abundance of $^3$H- and $^{32}$P-labeling. (c) Distribution of radioactivity in oocytes 11 hr after injecting 84 ng of [$^3$H, $^{32}$P] vitellogenin. Less than one-half the injected [$^3$H, $^{32}$P]vitellogenin is still present, no conversion to lipovitellin or phosvitin is indicated, and heterogeneously labeled components appear throughout the rest of the chromatograph. (From Dehn and Wallace, 1973. Reproduced from *The Journal of Cell Biology*, 1980, vol. 58, pp. 721–724 by copyright permission of the Rockefeller University Press.)

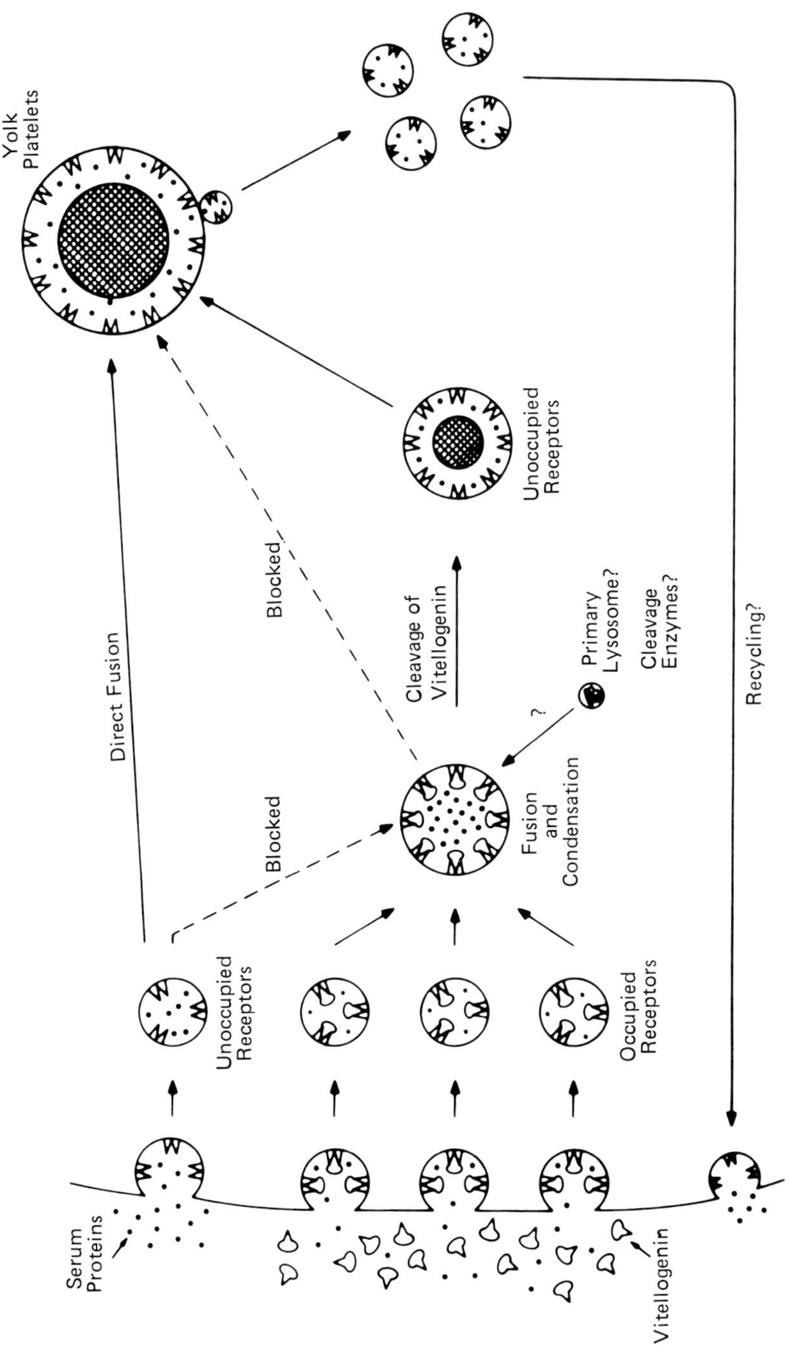


**Figure 12.** Current working model for postendocytic compartmentation of protein in *Xenopus laevis* oocytes. (From L. Opresko, *et al.*, 1980.)

processing and would fuse with a yolk platelet, as occurs in the absence of vitellogenin. A corollary of this assumption is that receptor occupancy can act as a transmembrane signal that directs the postendocytic compartmentalization of protein (Opresko *et al.*, 1980).

An additional interesting feature of sequestration of serum albumin, as well as that of other nonspecifically incorporated macromolecules, is that, unlike vitellogenin, these substances are removed from the oocyte with time (Opresko *et al.*, 1979, 1980). The loss occurs with biphasic kinetics, and it remains to be evaluated to what extent processes such as proteolysis or secretion during membrane recycling are responsible. Nevertheless, at least two mechanisms for vitellogenin accrual have now been identified: (1) selective vitellogenin uptake, and (2) turnover of incorporated materials other than vitellogenin.

Endocytic internalization of external protein generally leads to digestive sequelae mediated by lysosomes (Kaplan, 1981). Although sedimentable acid hydrolases have been found in full-grown oocytes and eggs (Decroly *et al.*, 1979; Steinert and Hanocq, 1979), typical lysosomes have not been detected in healthy vitellogenic oocytes (Dehn and Wallace, 1973; Wallace and Hollinger, 1979), and sequestered vitellogenin is not digested (Opresko *et al.*, 1980). Proteolytic processing of vitellogenin into yolk polypeptides does occur in transitional yolk bodies, however, and loss of serum albumin takes place in both transitional yolk bodies and yolk platelets (Opresko *et al.*, 1980). These cellular compartments may thus represent modified secondary lysosomes that contain a vitellogenin-processing hydrolase. This hydrolase may also be responsible for digesting nonvitellogenin protein incorporated adventitiously, such as serum albumin. Processed vitellogenin forms an insoluble matrix of yolk protein in most vertebrates and by this agency may be protected from further proteolysis, whereas soluble macromolecules remain susceptible to degradation. Insoluble yolk protein is also unavailable to any membranous elements that pinch off from forming yolk platelets, whereas materials that remain soluble may be translocated back to the oocyte surface by means of membrane recycling.

### 5.4. Morphology of Yolk Protein

Light microscopic study shows that in most anamniotic vertebrates, yolk protein is present in numerous structures that have a somewhat geometric shape, traditionally termed **yolk platelets** (Fig. 13a). Small nascent platelets appear peripherally in the growing oocyte (Massover, 1971c) and, as new platelets accumulate, the older platelets migrate inward and increase in size (Wittek, 1952). The size increase of centripetally migrating platelets appears to be due both to the continued accrual of yolk materials

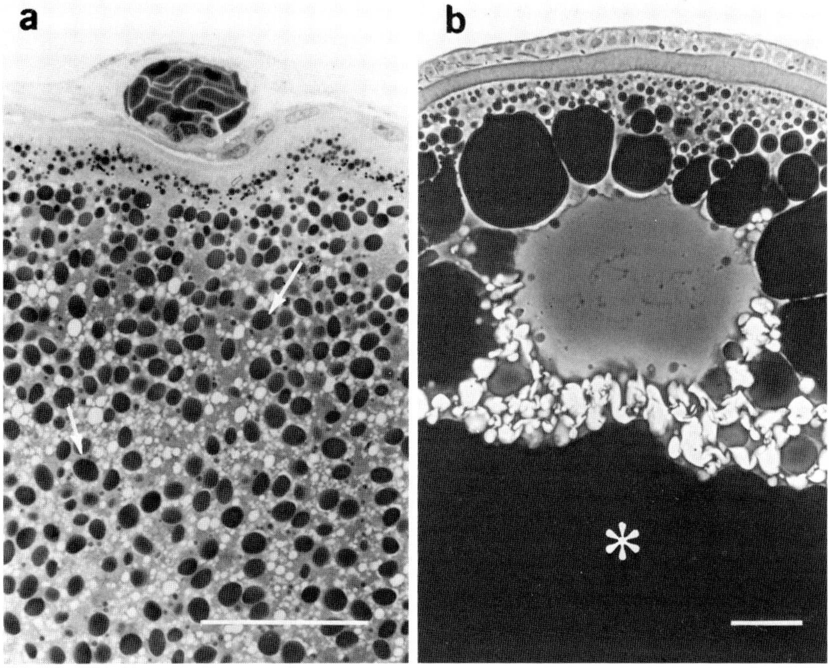


**Figure 13.** Light micrographs of sections from vitellogenic follicles of (a) *Xenopus laevis* and (b) *Cyprinodon variegatus*. In *X. laevis*, yolk protein is found in numerous yolk platelets of irregular shape and distributed through the oocyte cytoplasm (see arrows). Platelets tend to be somewhat larger toward the interior of the oocyte; small dark spheres near the oocyte surface are cortical graunules and pigment. By contrast, in *C. variegatus*, small yolk spheres form near the periphery of the oocyte and fuse with one another as they move toward the interior, eventually producing a central core of fluid yolk protein (*). Scale bars: 50 μm. (Courtesy of Dr. Kelly Selman.)

(Wallace *et al.*, 1972) and to the fusion of platelets with one another (Karasaki, 1963a). The net result of these processes is that the mature egg contains a heterodisperse size range of platelets. In *Rana esculenta*, for instance, yolk platelets range in size up to 50 μm in diameter and have a bimodal frequency distribution with peaks at 2 μm and 35 μm (Panijel, 1950).

In 1957, Wischnitzer was the first to observe with the electron microscope that the central region (main body) of amphibian platelets has a crystalline matrix surrounded by amorphous material, termed a superficial layer by Karasaki (1963a), and by a membrane; since that time, similar observations have been made for virtually all anamniotic taxa. About 98% of the total platelet protein was found to be associated with the main body in *R. pipiens* (Wallace, 1963a). In analyzing Fourier transforms of a series of electron micrographs of negatively stained crystalline fragments from *X. laevis*, Ohlendorf *et al.* (1978) concluded that the crystals are made up of an orthorhombic array of lipoprotein complexes, measuring ~5.5 × 11.5 × 25 nm. Each complex, in turn, consists of two subunits that most likely are not identical (Lange *et al.*, 1983). A similar orthorhombic lattice has now been found in other amphibians and representative members of the bony fishes by electron diffraction and freeze-fracture analysis (Lange, 1981a; Lange *et al.*, 1982, 1983). In cyclostomes, however, the crystal-like main body of the yolk platelets is organized according to a monoclinic lattice of possibly symmetric dimers (Lange and Richter, 1981; Lange, 1982). Thus, two basic types of yolk platelet architecture, illustrated in Fig. 14, have survived 400 million years since the separation of these two groups of anamniotes. The physiological and evolutionary constraints that have preserved the two types of crystalline structure over this period are yet to be elucidated. X-ray microanalysis of platelet cryosections has indicated the presence of chloride and of a variety of cations (Lange, 1981b, 1982; Lange *et al.*, 1983). Such results have led Lange to suggest that the crystalline architecture represents a device for electrolyte storage in those animals that lay eggs in freshwater (Lange, 1982), although the presence of yolk platelet crystals in hagfish eggs and their lack in the eggs of anadromous salmon are not in keeping with this interpretation.

In the eggs of some vertebrates, particularly in marine teleosts, the yolk protein is not found in relatively insoluble platelets; rather, it accumulates in fluid-filled **yolk spheres** or **yolk globules** (Grodziński, 1954; Wallace and Selman, 1981). These structures may either maintain their integrity throughout oocyte growth (K. Yamamoto, 1957b) or fuse centripetally (Fig. 13b), eventually forming a continuous mass of fluid yolk. This process can occur relatively soon after the initial formation of yolk spheres at the periphery, as in euryhaline sticklebacks (Wallace and Selman, 1981), during the later stages of vitellogenesis, as in the sheepshead minnow (Fig. 13b), or during postvitellogenic maturation, as in those marine teleosts that spawn pelagic eggs (K. Yamamoto, 1957a; Oshiro and Hibiya, 1982), conferring on such eggs their characteristic transparency.

The yolk of chicken eggs appears to be composed primarily of large, osmotically active yolk spheres (Grodziński, 1951), which are membrane limited and generally contain numerous osmiophilic subdroplets (Bellairs, 1961, 1967). These subdroplets have been interpreted by Schjeide *et al.* (1970) to be the granules obtained when chicken yolk is fractionated by centrifugation, and hence are homologous with the crystalline main body of the yolk platelet found in anamniotes (Wallace, 1963b). Although the subdroplets have been found to contain closely packed particles ~3–6 nm in diameter, they do not have a crystalline structure (Bellairs, 1967). The matrix of the yolk sphere in which the subdroplets are embedded presumably is derived from bloodborne protein other than vitellogenin, particularly VLDL.

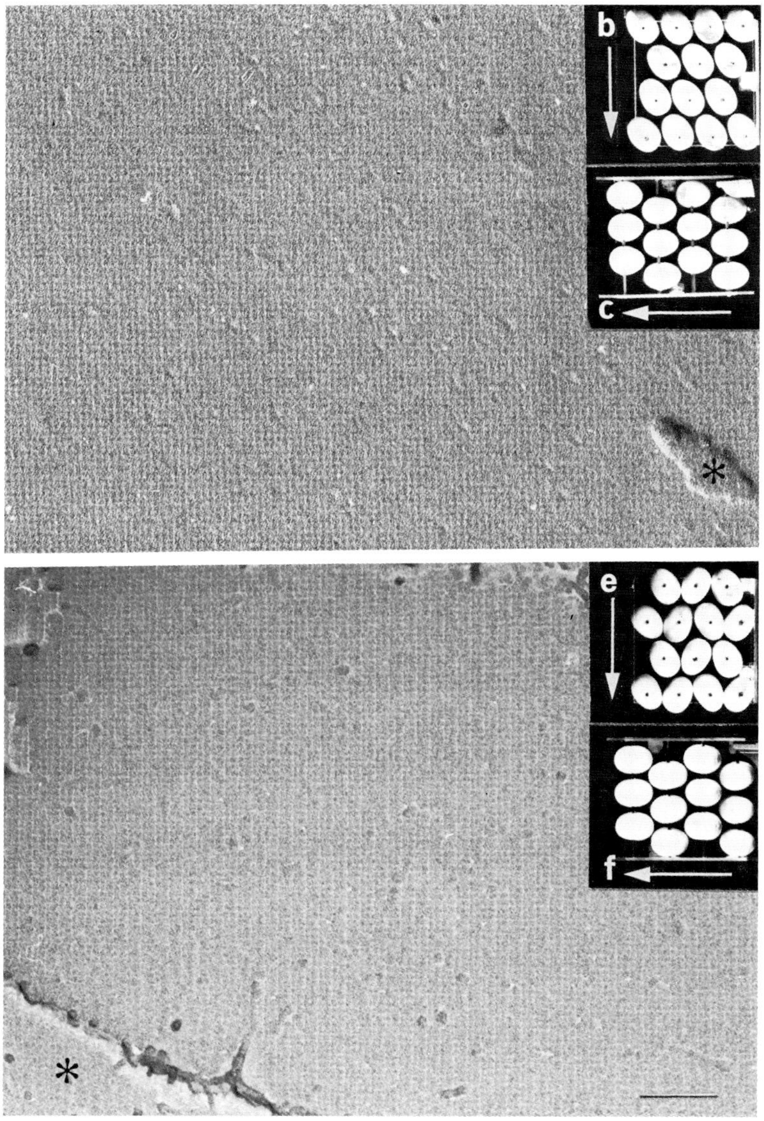


**Figure 14.** Corresponding freeze-fracture (a, d) and model views (b, c, e, f) of the lipovitellin dimers in a cyclostome, *Myxine glutinosa* (a–c) and an amphibian, *Xenopus laevis* (d–f). The oblong lipovitellin dimers possibly are symmetric in cyclostomes (Lange and Richter, 1981; Lange, 1982) and are composed of heterodimers of two different types in *X. laevis* (Lange *et al.*, 1983). In both cases, they form monomolecular layers in the crystals with different kinds of packaging. The freeze-fracture micrographs portray end-on views of the oblong lipovitellin dimers from a surface of such monolayers (asterisks mark exposure of the next surface layer). The model photographs are projected crystal views perpendicular to the freeze-fracture views (arrows give view lines) and show the suggested difference in dimer packing from a top-side view (b, e) and a similar aspect in both cases from a lateral view (c, f). Phosvitins probably occupy most of the interstices between the lipovitellin dimers. The models give only a rough representation of the structure and do not take into account certain established details in the contact zones of lipovitellin dimers. Scale bar: 100 nm. (Courtesy of Dr. Rainer H. Lange.)

## 5.5. Biochemistry of Yolk Protein

The water-insoluble platelets from anamniotic eggs or the granules from bird eggs can be dissolved in high-ionic-strength solutions, yielding two types of yolk protein, the lipovitellins and phosvitins; these can be separated from one another by cellulose chromatography or ammonium sulfate fractionation (Wallace, 1965; Wallace *et al.*, 1966). In their native state, lipovitellins are large lipoprotein dimers (~400,000 $M_r$) that dissociate into monomeric subunits (~200,000 $M_r$) under alkaline conditions (Wallace, 1965). By contrast, phosvitins are relatively small phosphoproteins (30,000–40,000 $M_r$), which are highly phosphorylated (~10% protein–phosphorus) and in which blocks of serine represent about one-half the total amino acids (Taborsky, 1974; Wallace, 1978). The stoichiometric ratio between the native lipovitellin dimer and phosvitin in *R. pipiens* platelets is 1:2 (Wallace, 1963b). In *X. laevis*, monomeric lipovitellin has been found to yield further a large peptide, apo-LV-1 (~120,000 $M_r$) and a small peptide apo-LV-2 (~31,000 $M_r$), under denaturing conditions, whereas phosvitin remains unchanged in denaturing solvents; see also Franzen *et al.* (1968) for chicken lipovitellin. Thus, each delipidated vitellogenin polypeptide (~200,000 $M_r$) appears to be cleaved within the oocyte in two places, yielding apo-LV-1, apo-LV-2, and phosvitin (Bergink and Wallace, 1974), unlike the scheme recently propounded by Tata (1976) and Tata and Smith (1979) indicating two phosvitin molecules derived from each vitellogenin polypeptide.

The further complexity of the situation for *X. laevis* was first indicated when Ohlendorf *et al.* (1977) discovered at least one additional low-molecular-weight apolipovitellin polypeptide in *X. laevis* platelets. Wiley and Wallace (1981) therefore reinvestigated the proteins present in the *X. laevis* platelet using higher-resolution analytical procedures and found three apo-LV-1 polypeptides (121,000, 116,000, and 111,000 $M_r$), three apo-LV-2 polypeptides (34,000, 31,500, and 30,500 $M_r$), and two phosvitins (34,000 and 33,000 $M_r$). In addition, two small phosvitinlike peptides, termed phosvettes, were isolated and found to have molecular weights of 19,000 and 13,500. The observations that at least three vitellogenin polypeptides are secreted by the liver (Wiley and Wallace, 1978) and the stoichiometry of the various platelet proteins found led these workers to propose that two of the vitellogenin polypeptides are cleaved in two places, whereas the third vitellogenin polypeptide is cleaved at an additional site in the phosvitin domain yielding the two phosvettes (Wiley and Wallace, 1981). Since four different vitellogenin genes and four vitellogenin mRNAs have been found in *X. laevis* to date (Wahli *et al.*, 1979; Felber *et al.*, 1980), a second interpretation of the available information is that one each of the apo-LV-1 and apo-LV-2 components has yet to be resolved into a doublet and that each phosvette is derived from a different, less phosphorylated, parent vitellogenin similar to the chicken vitellogenin III polypeptide. As yet, no sequence analysis or peptide mapping has been performed on the *X. laevis* proteins to resolve this question.

Preparative procedures for the constituents of chicken yolk typically begin with a brief centrifugation of partially diluted yolk to obtain a granule fraction and a supernatant fraction. Upon further extensive centrifugation, the supernatant yields both a gel-like floating material, a low-density fraction (LDF), and a subnatant of soluble proteins, the so-called livetins (MacKenzie and Martin, 1967). The granule fraction, approximately 20% of the yolk solids, contains two lipovitellins ($\alpha$ and $\beta$) in equal amounts, phosvitin, and a variable amount of low-density lipoprotein, which is most likely a contaminant from the supernatant material (Burley and Cook, 1961). The native (dimer) $\alpha$- and $\beta$-lipovitellins seem to have a similar size ($\sim$400,000 $M_r$, Bernardi and Cook, 1960b) and lipid distribution, primarily phospholipid (Martin *et al.*, 1963), but differ from one another in electrophoretic migration (Bernardi and Cook, 1960a), chromatographic behavior (Burley and Cook, 1961), dissociative behavior (Burley and Cook, 1962), protein-phosphorus content (Wallace, 1965), and polypeptide composition (Bergink *et al.*, 1974). Partial proteolysis experiments have indicated that vitellogenin II gives rise to polypeptides in both $\alpha$- and $\beta$-lipovitellin, whereas vitellogenin I gives rise to only $\alpha$-lipovitellin polypeptides (Wang and Williams, 1980). Phosvitin has also been resolved into two components by a variety of procedures; both components contain 10% protein-phosphorus but have slightly different amino acid compositions (Connelly and Taborsky, 1961; Clark, 1970) and sizes (40,000 and 36,000 $M_r$, Taborsky and Mok, 1967). Judging from their phosphorus contents, the two phosvitins appear to be derived from the more highly phosphorylated vitellogenins I and II (Wang and Williams, 1980). None of the yolk peptides has yet been associated with vitellogenin III; nevertheless, the available evidence for the chicken reinforces the possibility that multiple vitellogenin polypeptides are processed into multiple yolk peptides within the growing oocyte (de Vlaming *et al.*, 1980; Wang and Williams, 1980; Wiley and Wallace, 1981) and mitigates against the notion that a single vitellogenin precursor gives rise to several types of yolk protein by differential proteolytic cleavage (Gordon *et al.*, 1977).

The floating material, or LDF, contains 89% lipid (primarily neutral lipid) and is the major component of chicken yolk, making up approximately 70% of the yolk solids (Martin *et al.*, 1963). It is the unique component of avian yolk and is presumably the nonsubdroplet constituent of yolk spheres. Native LDF in aqueous solvents is polydisperse, with a molecular weight of 5–17 $\times$ $10^6$ (Evans *et al.*, 1975). The apoprotein associated with LDF has been fractionated into several partially characterized peptides (Burley, 1975). The relative concentrations of amino acids present in plasma VLDL and yolk LDF apoproteins are similar, and the two components in native form are immunologically identical (Hillyard *et al.*, 1972), so that plasma VLDL is apparently the precursor for yolk LDF. Plasma

VLDL is comprised of only two polypeptide species (D. L. Williams, 1979); whether the more numerous peptides found in LDF apoprotein are formed by proteolytic processing within the oocyte or during the lengthy manipulations of isolation is still uncertain.

The livetins, approximately 10% of the total yolk solids, are a heterogeneous collection of proteins shown conclusively by several investigators to be related to the normal serum proteins (J. Williams, 1962; Mok and Common, 1964). Albumin, $\alpha_2$-glycoproteins, transferrin, and $\gamma$-globulins are the most abundant components of the livetin fraction. Neither their subcellular location within egg yolk nor the extent to which they may be partially degraded has been clarified.

Changes in relative proportions of the three major classes of yolk components—granules, LDF, livetins—have also been observed to occur during oocyte growth (MacKenzie and Martin, 1967). The granule proteins and livetins are the predominant constituents of early ($\leq$9-mm diameter) oocytes. During subsequent growth, the relative proportion of LDF increases until it reaches almost 80% of the total yolk solids in 20-mm oocytes; thereafter, the proportion declines toward values found for the fully developed oocyte (70%). Values for the granule proteins and livetins change coordinately with each other and conversely with LDF. Since the concentration of blood precursors is constant throughout oocyte growth, these results would imply that (1) there are two different physiological mechanisms for the selective incorporation of vitellogenin and VLDL, (2) their functioning is not strictly coordinated during oocyte growth, and (3) the serum proteins are adventitiously acquired primarily through the mechanism whereby vitellogenin is sequestered.

Although several recent electrophoretic studies on teleost yolk proteins have yielded results consistent with the data obtained for lipovitellins and phosvitins in other species (Hara and Hirai, 1978; de Vlaming *et al.*, 1980; Lange *et al.*, 1983), the yolk proteins isolated from teleosts have generally been found to have atypical properties. The native lipovitellins are frequently numerous, are highly lipidated, contain very little or no protein-phosphorus, and—unlike all other lipovitellins—are generally soluble in low-ionic-strength solutions (Jared and Wallace, 1968). The phosvitins are also frequently numerous and heterogeneous (Mano and Lipmann, 1966; Jared and Wallace, 1968), have lower molecular weights (Schmidt *et al.*, 1965; Wallace *et al.*, 1966; Markert and Vanstone, 1971; de Vlaming *et al.*, 1980), contain widely divergent amounts of protein-phosphorus (Barman *et al.*, 1964; Mano and Lipmann, 1966), or sometimes appear to be almost or completely absent from yolk preparations (Jared and Wallace, 1968; Craik, 1982). Some of the native proteins derived from yolk preparations also seem to have the characteristics of neither lipovitellin nor phosvitin (Jared and Wallace, 1968), and at least one of these—the $\beta$ component from salmonid yolk—has been found to contain neither lipid nor protein-phosphorus, to have a molecular weight of only ~30,000, and to be derived from the blood of sexually maturing females; that is, it is not a livetin or normal blood protein (Markert and Vanstone, 1971). These anomalies originally led us (Jared and Wallace, 1968) as well as others (Mano and Lipmann, 1966) to postulate that teleost yolk proteins are phosphorylated in either a deficient or heterogeneous manner. However, more recent observations have suggested a different conclusion.

Many teleost vitellogenins appear to contain about one-half the amount of protein-phosphorus found in that of other vertebrates; i.e., the phosvitin domain in teleost vitellogenins may be considerably smaller (de Vlaming *et al.*, 1980). Teleost vitellogenins also are very susceptible to proteolysis (Hickey and Wallace, 1974; Wallace and Selman, 1982). Finally, after the conclusion of vitellogenesis, many teleost oocytes undergo hydration to a variable extent during meiotic maturation—a process that can be achieved *in vitro* simply by adding steroid hormone to a saline medium containing postvitellogenic oocytes (Wallace and Selman, 1981). During the hydration process, the yolk proteins present in postvitellogenic oocytes have been found to undergo further proteolysis both *in vitro* and *in vivo* (see Fig. 15; compare especially lanes B and b with C and c). Craik (1982) has even found that the protein-phosphorus content of the ovary decreases approximately 20-fold during hydration in plaice, a marine spawner in which oocyte swelling is particularly pronounced. These observations indicate that additional proteolysis of yolk peptides and perhaps dephosphorylation of smaller than average phosvitin molecules occur during the final stage of egg formation in at least some teleosts. Analysis of egg proteins in such animals will therefore show numerous atypical yolk peptides.

## 5.6 Other Processes Ascribed to Vitellogenesis or Related to Oocyte Growth

In amphibian eggs, yolk platelets make up 80–90% of the total protein (Benbow *et al.*, 1975; Callen *et al.*, 1980); in other animals, such as the chicken, yolk proteins make up essentially all the dry weight of the fully developed egg (MacKenzie and Martin, 1967). We have therefore concentrated on those processes involved in yolk protein formation as the major events contributing to oocyte growth in nonmammalian vertebrates. Over the years, numerous reports have appeared documenting the unusual (for a single cell) accumulation in the growing oocyte of other materials, such as lipid, glycogen, histones, and maternal RNA transcripts, all of which appear to be important for successful embryogenesis after the egg is fertilized. Occasionally, certain materials accumulate to an extent normally associated with yolk protein, such as may be the case for wax ester, which makes up 37% of the ovarian dry weight in the opaline gourami, *Trichogaster cosby* (Sand *et al.*, 1969). Presumably, this material is found in devel-

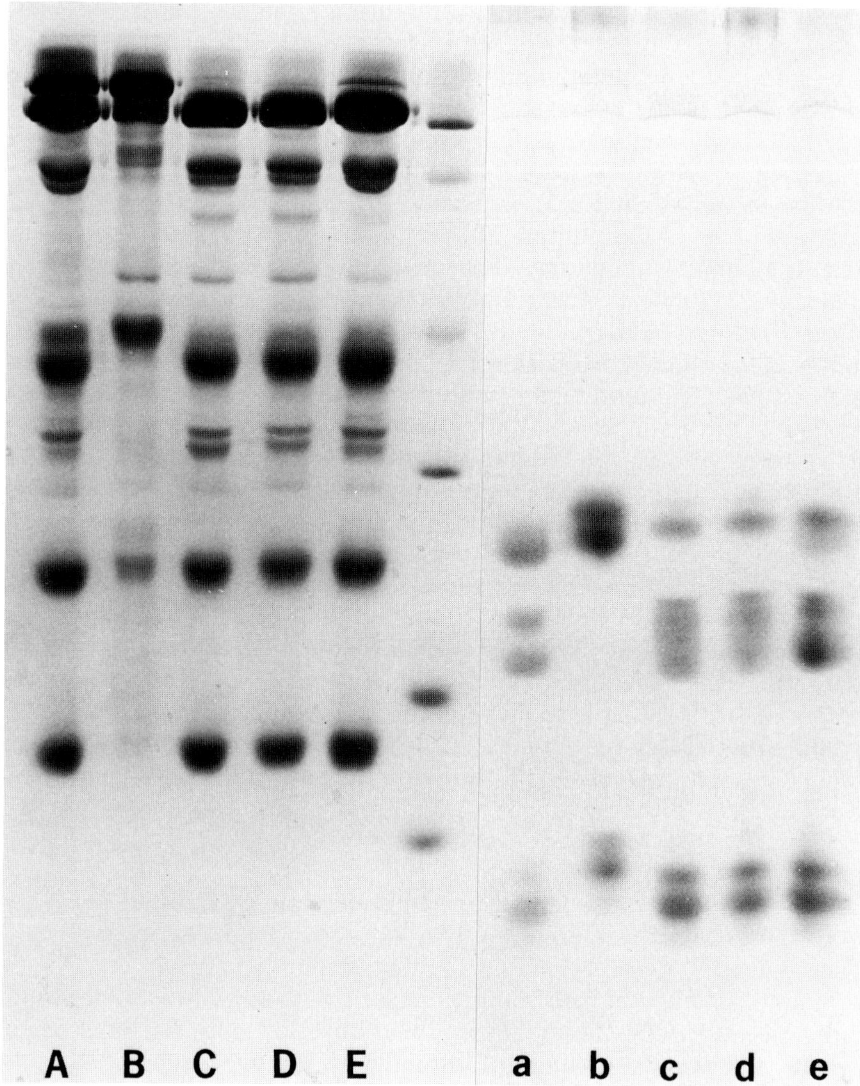


**Figure 15.** Sodium dodecylsulfate electrophoresis on 12% polyacrylamide slab gels of yolk proteins present in extracts of *Fundulus heteroclitus* ovary (A, a), postvitellogenic prematurational follicles (B, b; avg diameter, 1.33 mm), postmaturational follicles matured *in vivo* (C, c; avg diameter, 1.79 mm) or *in vitro* (D, d; avg diameter, 1.67 mm), and ovulated eggs (E, e; avg diameter = 1.75 mm). Lanes A–E represent samples containing approximately 50 μg protein and are stained with Coomassie blue, as is the middle lane indicating molecular-weight standards (92,500, 66,200, 45,000, 31,000, 21,000, and 14,400 from top to bottom, respectively). Lanes a–e represent samples containing approximately 100 μg protein electrophoresed under identical conditions but stained with Stains-all to indicate phosphoproteins (phosvitins), which do not stain with Coomassie blue. The ovary contains both vitellogenic follicles (≤1.35 ± 0.05 mm diameter) as well as maturational stages (1.35 ± 0.05- to ~1.8 diameter); hence the proteins present in B, b and C, c should be represented in A, a. During meiotic

oping oocytes, since the eggs of *Trichogaster* species characteristically float in freshwater (Breder and Rosen, 1966), but this point needs clarification.

Other processes besides vitellogenin uptake have been ascribed to the formation of yolk materials, one of the most entrenched notions in the prevailing literature being the synthesis of "primary" or "intravesicular" yolk within teleost oocytes before vitellogenin uptake. These yolk vesicles contain periodic acid-Schiff (PAS)-positive material and increase in size and number as they move peripherally within the oocyte. In an autoradiographic study of yolk formation in the zebrafish, *Brachydanio rerio*, Korfsmeier (1966) demonstrated that yolk vesicles rapidly incorporate both [$^3$H]histidine and [$^3$H]glucose, and he suggested that they contain a glycoprotein that is endogenously synthesized by the oocyte. Subsequently, electrophoretic studies by te Heesen and Engels (1973) identified a major glycoprotein component in extracts of ovaries containing primarily yolk vesicle-stage oocytes. As this glycoprotein component was not found in extracts of liver or serum, these workers also concluded that it is synthesized endogenously. However, a series of studies reported a number of years ago and not generally appreciated indicate that yolk vesicles eventually give rise to cortical alveoli in the growing oocyte. (T. S. Yamamoto, 1955; Osanai, 1956; K. Yamamoto, 1956a; Malone and Hisaoka, 1963). These structures, which are found in variable size and number in the egg cortex, fuse with the plasma membrane and release their glycoprotein contents into the perivitelline space during the cortical reaction at fertilization; hence they are not yolk in a strict sense but are analogous to the cortical granules that have been extensively studied in the oocytes of numerous invertebrate and vertebrate species (T. Yamamoto, 1961; Guraya, 1982).

One of the most influential observations made on amphibian oocytes was that by Ward (1962), who described enlarging crystalline inclusions within the mitochondria of *R. pipiens* tadpole oocytes. These inclusions were interpreted at the time as representing progenitors of the crystals formed within the oocyte yolk platelets. Intramitochondrial yolk crystals were then observed and similarly interpreted by many others (see Massover, 1971a and references cited therein). Massover (1971b) also provided evidence that yolk crystals in *Rana catesbeiana* oocytes were extruded

---

maturation, yolk polypeptides formed during vitellogenesis—notably those with molecular weights of 120,000 and 45,000 in B and 28,000 and 14,000 in b—disappear, and new lower-molecular-weight bands appear in C and c. A comparison of follicles matured *in vivo* (C, c) and *in vitro* (D, d) according to the procedures of Wallace and Selman (1978) reveals no fundamental differences, indicating that the proteolytic processes that occur during maturation *in vivo* can be achieved *in vitro* simply by adding deoxycorticosterone to postvitellogenic follicles such as those used for lanes B and b. Lanes E and e represent ovulated eggs and indicate no fundamental differences from those (C and c, D and d) representing postmaturational, preovulatory follicles; thus the polypeptides indicated are derived from oocytes or eggs and not from associated follicular tissues. (R. A. Wallace, previously unpublished data.)

from mitochondria to form single-membrane-bound hexagonal crystalloids. As evidence accumulated, however, it soon became apparent that the occurrence of such structures is limited to ranid species of frogs (*R. catebeiana*, *R. clamitans*, *R. esculenta*, *R. japonica*, *R. nigromaculata*, *R. oranativenris*, *R. pipiens*; see also Karasaki, 1963b), and their relative abundance does not exceed 1% of the total crystalline inclusions in the egg (Karasaki,1963b). Even more importantly, intramitochondrial yolk crystals have a different matrix structure and periodicity (Lanzavecchia, 1968; Spornitz, 1972) and protein content (R. T. Ward, unpublished observations) than that of yolk platelet crystals. Their occurrence and function thus remain an enigma.

Another consistent observation made over the years by a number of investigators is the presence of so-called multivesicular bodies in growing oocytes (Balinsky and Devis, 1963; Hope *et al.*, 1964; Lanzavecchia, 1968; Spornitz and Kress, 1971, 1973; Ward, 1978a; Kress, 1982). In general, such structures have been observed in early oocytes before the formation of the bulk of yolk by endocytosis, although in *Discoglossus pictus* they appear to persist in the region of the animal hemisphere throughout vitellogenesis (Andreuccetti and Campanella, 1982). In this frog species, the animal hemisphere atypically appears as a slightly depressed disc in the vitellogenic oocyte and gives rise in the egg to an animal dimple, where sperm entrance occurs (Andreuccetti and Campanella, 1980). Multivesicular bodies in growing oocytes have generally been interpreted as participating in vitellogenesis, since crystalline structures appear to accumulate in them gradually as oocyte growth progresses; an extension of this interpretation has been the argument that endogenous yolk synthesis occurs in the oocyte, at least during early stages of oocyte growth.

In view of the observations that (1) multivesicular bodies appear to predominate in oocytes immediately before the onset of yolk protein accumulation, except in the animal hemisphere of *D. pictus*, where a diminished vitellogenesis occurs; (2) endocytosis has never been shown to be completely absent in oocytes when multivesicular bodies are particularly numerous; (3) multivesicular bodies have generally been interpreted in other cells as representing the major depot to which endocytosed molecules are delivered (Holtzman, 1976); and (4) quantitative data indicate that at least 99% of the yolk platelet protein in *X. laevis* is derived from an endocytic process (Wallace *et al.*, 1972), an alternative interpretation of the role of multivesicular bodies would seem to be warranted. The normal course of endocytic protein uptake by a cell leads to digestion of incorporated protein by lysosomes (Kaplan, 1981), except in vitellogenic oocytes. Since the previtellogenic follicle contains a capillary bed (Dumont, 1972), any circulating vitellogenin or other protein filtering through to the oocyte surface would, under normal circumstances, provoke the appearance of secondary lysosomes (i.e., multivesicular bodies). According to this interpretation, the onset of vitellogenesis is not necessarily triggered by the availability of vitellogenin or insertion of vitellogenin receptors into the oocyte surface but rather by cessation of normal lysosomal synthesis. Consistent with this interpretation (Giorgi, 1980; Giorgi *et al.*, 1976), the forming yolk platelet or sphere is viewed as a modified secondary lysosome containing only the enzymatic apparatus for processing the vitellogenin polypeptide into yolk proteins but not those enzymes necessary for complete digestion. Quantitation of vitellogenin access, vitellogenin receptors, and lysosomal enzymes during the transition from previtellogenesis into vitellogenesis is needed in order to resolve these possibilities.

Finally, in most nonmammalian vertebrates, vitellogenesis ceases *in vivo* once oocytes reach their fully developed size, and such oocytes eventually undergo meiotic maturation and ovulation after appropriate hormonal stimulation (Masui and Clarke, 1979) (see Chapters 1 and 6 of this volume). In many organisms, particularly marine teleosts, a further rapid increase in size of the oocyte also occurs due to hydration during meiotic maturation, as was first described *in vivo* by Fulton (1898) and subsequently achieved *in vitro* (Wallace and Selman, 1981 and references cited therein). A dramatic example of this process *in vitro* is shown for the oocytes of the cunner in Figure 16. In this case, there is a 4.3-fold increase in volume over 20 hr after the addition of steroid to the medium, so that approximately 77% of the final egg volume is formed during maturation. This process is especially pronounced among marine teleosts that spawn pelagic eggs and renders the eggs buoyant in seawater.

## 6. Vitellogenesis and Oocyte Growth *In Vitro*

Several years ago, a procedure was developed for growing amphibian oocytes *in vitro*, which involved placing oocytes dissected from their follicles into a nutrient medium containing vitellogenin (Wallace and Misulovin, 1978; Wallace *et al.*, 1978, 1980b). Early- to mid-vitellogenic stage oocytes obtained from gonadotropin-stimulated females were found to grow *in vitro* in the absence of added serum or gonadotropin at rates similar to or exceeding those found for the most rapidly growing oocytes in intact gonadotropin-stimulated females (e.g., curves a–c in Fig. 17). Insulin was the only hormone present; its deletion reduced the growth date by an average of 25%. In contrast, the addition of serum increased the growth rate by about 30%. the most important variable, however, proved to be the external vitellogenin concentration, with growth rates that could be reduced to zero simply by lowering the concentration to $<2$ mg/ml.

With continued experience, it also became apparent that oocyte growth *in vitro* never slowed down once oocytes reached a diameter of 1.1 mm, as occurs *in vivo* (curves b and c in Fig. 17). Another consistent observation was that oocytes obtained from unstimulated females grew equally well over the long term *in vitro* as did those from gonadotropin-stimulated

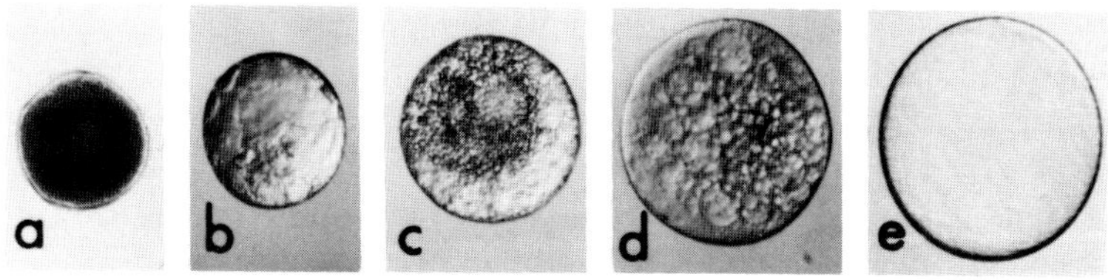


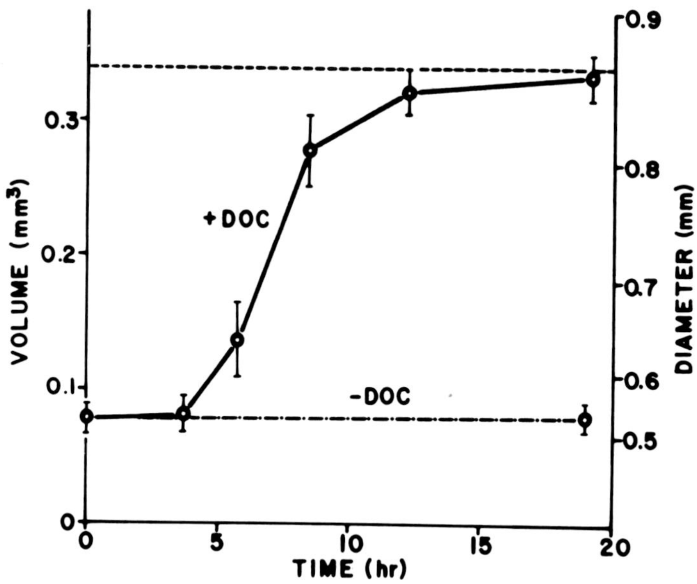


**Figure 16.** Size change that occurs in the ooyctes of the cunner, *Tautogolabrus adspersus*, during maturation. The largest follicles present in the ovary were dissected out and incubated at 20°C in the presence (———) or absence (—·—) of deoxycorticosterone according to the procedures of Wallace and Selman (1978). Each point represents the average size of 10 follicles ±SD. The dashed horizontal line at 0.87 mm indicates the average size of 10 eggs collected from the ovarian lumen. Photomicrographs at top depict the appearance of individual follicles with diameters of (a) 0.50 mm, (b) 0.55 mm, (c) 0.67 mm (germinal vesicle apparent near center), (d) 0.82 mm (germinal vesicle approaching animal poles, and (e) 0.85 mm, in which the germinal vesicle is no longer apparent and the oocyte has clarified. (From Wallace and Selman, 1981.)

females (Wallace and Misulovin, 1978; Wallace *et al.*, 1980b). It thus appeared that once oocytes were removed from the confines of the follicle and placed in culture, any shackles to growth were somehow either removed or overcome. An extreme test of this hypothesis was then made by placing full-grown oocytes from unstimulated females in culture, with the result that most oocytes resumed growth at remarkable rates; some even doubled in volume over a 3-week period (e.g., curves d and e in Fig.

17). Thus, the maximum size reached by oocytes in the ovary is not an intrinsic property of the oocyte but is imposed extrinsically by the ovarian environment (Wallace *et al.*, 1981).

Further analysis and comparison of oocyte growth *in vivo* and *in vitro* (Wallace, 1983) has led to the conclusion that the *X. laevis* oocyte appears capable of sustained growth and differentiation in the absence of external cues if it is simply removed from its somatic environment and is provided with the appropriate nutrients, including a saturating concentration of vitellogenin. *In vivo*, on the other hand, the somatic tissues of the host are used by the growing oocyte for its nutritional and hormonal milieu—a milieu over which it has a measure of control. During vitellogenic growth, investing follicle cells respond to circulating gonadotropin by producing estrogen and by permitting consequently synthesized vitellogenin to reach the oocyte surface. Vitellogenin access appears to be provided by patent channels between follicle cells. However, once the growing oocyte acquires the competence to resume meiosis and become an egg (which it can acquire automatically *in vitro*), the follicle cells appear to restrict vitellogenin uptake and to alter steroid synthesis by producing progesterone in order to accommodate the full-grown oocyte (Wallace, 1983). Although gonadotropin levels have been shown to regulate the amount of steroid synthesis, the manner in which follicle cells respond to circulating gonadotropin(s) appears to be ultimately determined by the developing oocyte. Although the mechanism of this influence remains to be elucidated, the evidence for gap junctions between the oocyte and follicle cells (Browne *et al.*, 1979; see also Fig. 18) could be reinterpreted to indicate that signals can be passed from the oocyte to the follicle cells rather than vice versa.

## 7. Epilogue

Vitellogenin as a generic term was first used for a female-specific protein present in an insect (Pan *et al.*, 1969); since that time, bloodborne proteins having a similar function as a precursor for yolk protein have been described in several animal phyla. Thus far, there is little biochemical evidence to indicate that the vitellogenins or yolk proteins described here for vertebrates have any relationship other than a functional one to those described for invertebrates. Nevertheless, a survey of vitellogenin coding sequences in the genomic DNA of insect, amphibian, and avian species by hybridization to cDNA probes suggests a high degree of conservation of these coding sequences (James *et al.*, 1982). Extensive crosshybridization of the expressed locust vitellogenin gene with *X. laevis* and chicken sequences indicates that at least part of this gene has been conserved for 500 million years (Tata *et al.*, 1983). A phosvitin peptide has not as yet been described for any invertebrate species, however, and may therefore represent a vitellogenin domain unique to vertebrates.

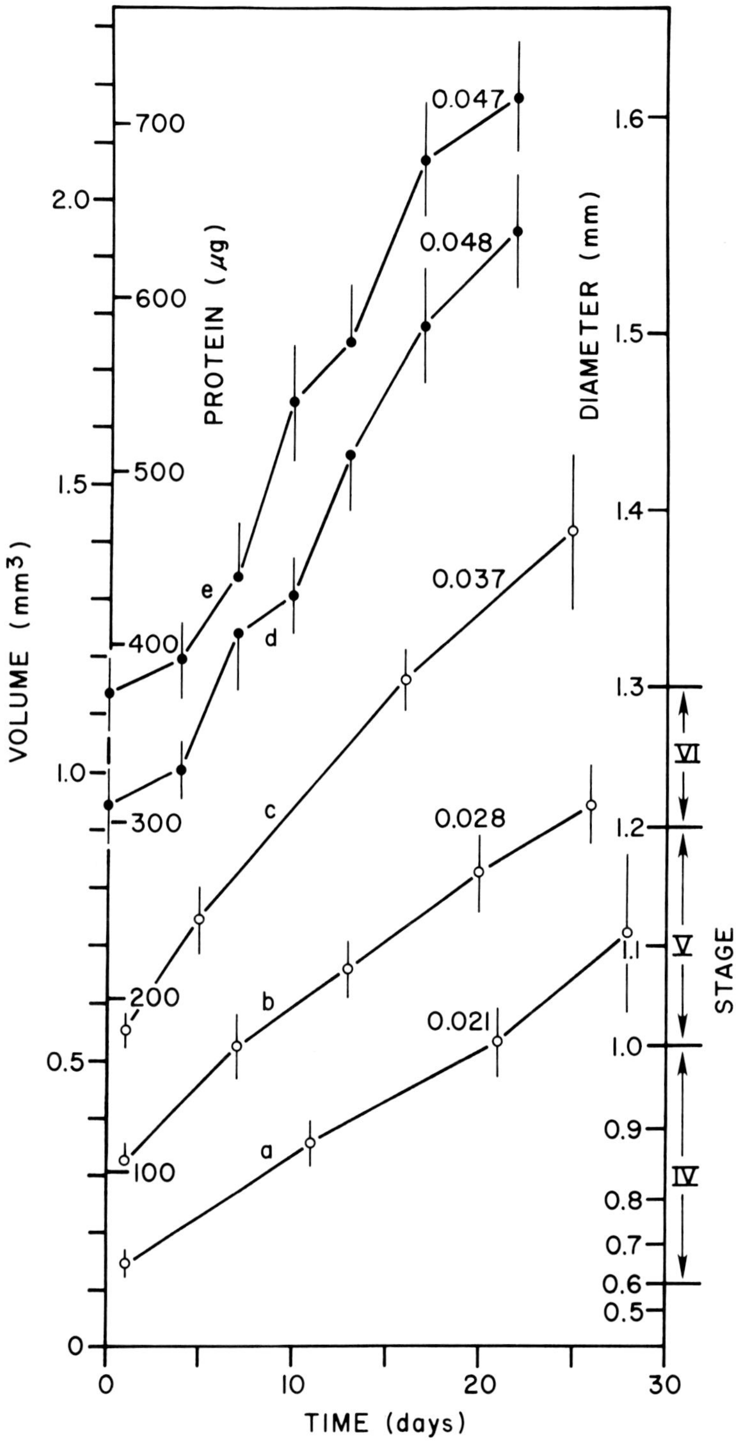
0.047
0.048
0.037
0.028
0.021
a
b
c
d
e
700
600
500
400
300
200
100
PROTEIN (μg)
VOLUME (mm3)
2.0
1.5
1.0
0.5
0
DIAMETER (mm)
1.6
1.5
1.4
1.3
1.2
1.1
1.0
0.9
0.8
0.7
0.6
0.5
STAGE
VI
V
IV
0
10
20
30
TIME (days)

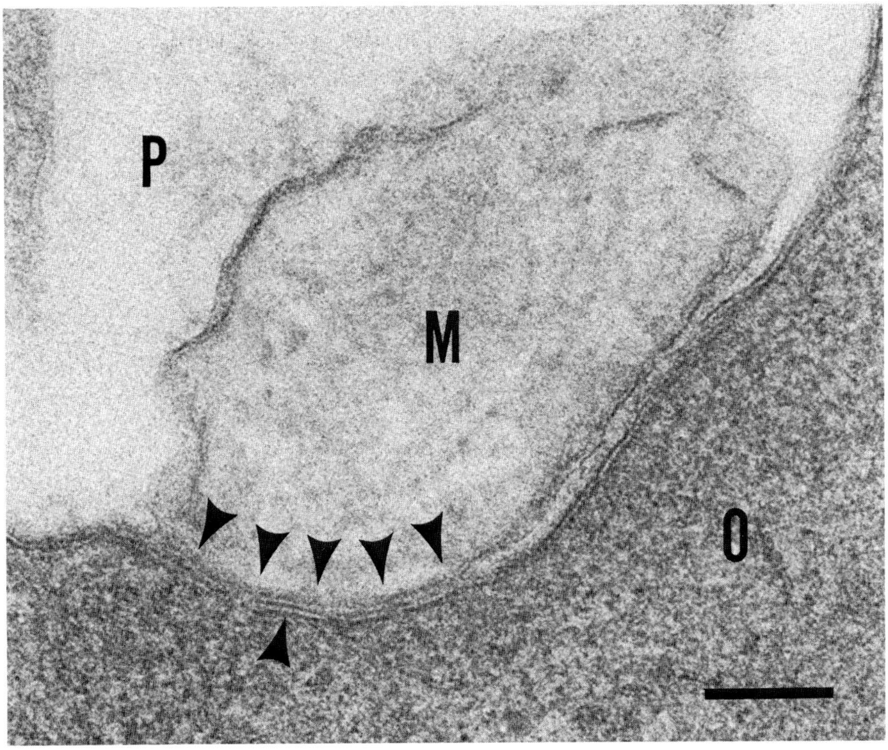


**Figure 18.** Gap junctionlike contact (upper set of arrowheads) between macrovillus (M) of follicle cell and surface of a large *Hyla cinerea* oocyte (O). The trilaminar appearance shown by both the plasmalemma of the macrovillus and the oolemma changes to a pentalaminar appearance (lower arrowhead) in the focal region of this intercellular junction. Scale bar: 100 nm. (Courtesy of Dr. William H. Massover.)

**Figure 17.** Progression of oocyte size with time *in vitro*. *Xenopus laevis* oocytes were isolated from individual females and placed in culture with vitellogenin in the presence (●) or absence (○) of serum. The points (±SD) represent the average volume or protein content found for 8–10 oocytes; these parameters, indicated on the left ordinate, are interchangeable (Wallace and Misulovin, 1978). The right ordinate indicates both the oocyte diameter corresponding to the volume measurements and the staging system described by Dumont (1972). Full-grown (stage VI) oocytes generally have a diameter of 1.2–1.3 mm. The numbers associated with each growth curve refer to the average growth rates ($mm^3$ $day^{-1}$). For comparison, maximum growth rates of 0.008 and 0.025 $mm^3$ $day^{-1}$ have been found for oocytes *in vivo* in unstimulated and gonadotropin-stimulated females, respectively (Keem *et al.*, 1979). (From Wallace, 1983.)

The most primitive animal for which a vertebrate-type vitellogenin–yolk system has been documented is the hagfish (Yu *et al.*, 1981; Lange and Richter, 1981). Most other types of nonmammalian vertebrates also seem to have estrogen-inducible vitellogenic systems, although information on elasmobranchs, reptiles, and birds other than the chicken is still relatively scant. The oocytes of some lower vertebrates apparently have a greatly reduced capacity for yolk protein accumulation. In particular, published pictures of the relatively tiny (0.25-mm diameter) full-grown oocytes from the viviparous seaperch, *Cymatogaster aggregata*, seem to indicate that they do not advance beyond the yolk vesicle stage, i.e., true vitellogenesis never takes place (Turner, 1938; Wiebe, 1968). If this assumption is correct, it would be interesting to explore whether these animals have lost the capacity to make vitellogenin—and, if so, why—or whether sequestration itself is deficient.

Although polymorphic fibrillar material present in mammalian, particularly rodent, eggs has been referred to as yolk (e.g., Nilsson, 1980), membrane-enclosed yolk inclusions suggestive of a lipovitellin–phosvitin complex have not been described in placental mammals and, at least in the mouse, changes in serum proteins indicative of vitellogenin synthesis have not been observed (Urist and Schjeide, 1961). The apo-VLDL-B and vitamin-carrying proteins described for the chicken apparently have mammalian homologues (Chapman *et al.*, 1977; Adiga and Murty, 1983), but there is no evidence to show that these two types of protein accumulate in either the chicken or mammalian oocyte after delivering lipid and vitamins, respectively. Micropinocytosis has been indicated at the surface of the guinea pig oocyte (Anderson and Beams, 1960); Anderson (1972) demonstrated the endocytic incorporation of horseradish peroxidase by the developing mouse oocyte. The fate of all incorporated external protein, however, would appear to be digestion via secondary lysosomes rather than accrual in storage granules (Anderson, 1972). The eggs of monotremes are known to be relatively large and to contain considerable quantities of yolk (e.g., Gatenby and Hill, 1924); an investigation of these forms would therefore be desirable. Evidence to date, however, seems to indicate that vitellogenin genes were either silenced or lost in the transition from prototherian to therian mammals.

## References

Adiga, P. R., and Murty, C. V. R., 1983, Vitamin carrier proteins during embryonic development in birds and mammals, in: *Molecular Biology of Egg Maturation* (Ciba Foundation Symposium 98) (R. Porter and J. Whelan, eds.), pp. 111–136, Pitman, London.

Anderson, E., 1972, The localization of acid phosphatase and the uptake of horseradish peroxidase in the oocyte and follicle cells of mammals, in: *Oogenesis* (J. D. Biggers and A. W. Schuetz, eds.), pp. 87–117, University Park Press, Baltimore.

Anderson, E., and Beams, H. W., 1960, Cytological observations on the fine structure of the guinea pig ovary with special reference to the oogonium, primary oocyte and associated follicle cells, *J. Ultrastruct. Res.* **3**:432–446.

Andreuccetti, P., and Campanella, C., 1980, Origin and cytochemistry of the animal dimple granules in *Discoglossus pictus* (Anura) eggs, *J. Embryol. Exp. Morphol.* **56**:239–252.

Andreuccetti, P., and Campanella, C., 1982, Regional differences in the pattern of vitellogenesis in the painted frog, *Discoglossus pictus, Tissue Cell* **14**:681–690.

Ansari, A. Q., Dolphin, P. J., Lazier, C. B., Munday, K. A., and Akhtar, M., 1971, Chemical composition of an estrogen-induced calcium-binding glycolipophosphoprotein in *Xenopus laevis, Biochem. J.* **122**:107–113.

Anthony, J., and Millot, J., 1972, Première capture d'une femelle de Coelacanthe en état de maturité sexuelle, *C. R. Acad. Sci. Ser. D* **274**:1925–1926.

Balinsky, B. I., and Devis, R. J., 1963, Origin and differentiation of cytoplasmic structures in the oocytes of *Xenopus laevis, Acta Embryol. Morphol. Exp.* **6**:55–108.

Barman, T. E., Bai, N. K., and Thoai, N. V., 1964, Studies on herring-egg phosphoprotein, *Biochem. J.* **90**:555–558.

Bellairs, R., 1961, The structure of the yolk of the hen's egg as studied by electron microscopy I. The yolk of the unincubated egg. *J. Biochem. Biophys. Cytol.* **11**:207–225.

Bellairs, R., 1967, Aspects of the development of yolk spheres in the hen's oocyte, studied by electron microscopy, *J. Embryol. Exp. Morphol.* **17**:267–281.

Benbow, R. M., Pestell, R. Q. W., and Ford, C. C., 1975, Appearance of DNA polymerase activities during early development of *Xenopus laevis, Dev. Biol.* **43**:159–174.

Bergink, E. W., 1973, Synthese van dooiereiwitten door de haan, Ph.D. thesis, University of Groningen, Groningen, The Netherlands.

Bergink, E. W., and Wallace, R. A., 1974, Precursor–product relationship between amphibian vitellogenin and the yolk proteins, lipovitellin and phosvitin, *J. Biol. Chem.* **249**:2897–2903.

Bergink, E. W., Kloosterboer, H. J., Gruber, M., and AB, G., 1973, Estrogen-induced phosphoprotein synthesis in roosters. Kinetics of induction, *Biochim. Biophys. Acta* **294**:497–506.

Bergink, E. W., Wallace, R. A., Van de Berg, J. A., Bos, E. S., Gruber, M., and AB, G., 1974, Estrogen-induced synthesis of yolk proteins in roosters, *Am. Zool.* **14**:1177–1193.

Bernardi, G., and Cook, W. H., 1960a, An electrophoretic and ultracentrifugal study of the proteins of the high density fraction of egg yolk, *Biochim. Biophys. Acta* **44**:86–96.

Bernardi, G., and Cook, W. H., 1960b, Separation and characterization of the two high density lipoproteins of egg yolk, $\alpha$- and $\beta$-lipovitellin, *Biochim. Biophys. Acta* **44**:96–105.

Beuving, G., and Gruber, M., 1971, Isolation of phosvitin from the plasma of estrogenized roosters, *Biochim. Biophys. Acta* **232**:524–528.

Bisbee, C. A., Baker, M. A., Wilson, A. C., Hadji-Azimi, I., and Fischberg, M., 1977, Albumin phylogeny for clawed frogs (*Xenopus*), *Science* **195**:785–787.

Blue, M. L., 1981, The biosynthesis of major avian and human plasma apoproteins, Ph.D. thesis, State University of New York at Stony Brook, Stony Brook, New York.

Blue, M. L., and Williams, D. L., 1981, Induction of avian serum apolipoprotein II and vitellogenin by tamoxifen, *Biochem. Biophys. Res. Commun.* **98**:785–791.

Blue, M. L., Protter, A. A., and Williams, D. L., 1980, Biosynthesis of apolipoprotein B in rooster kidney, intestine, and liver, *J. Biol. Chem.* **255**:10048–10051.

Breeder, C. M., and Rosen, D. E., 1966, *Modes of Reproduction in Fishes*, Natural History Press, Garden City, New York.

Browne, C. L., Wiley, H. S., and Dumont, J. N., 1979, Oocyte–follicle cell gap junctions in *Xenopus laevis* and the effects of gonadotropin on their permeability, *Science* **203**:182–183.

Brummett, A. R., and Dumont, J. N., 1977, Intracellular transport of vitellogenin in *Xenopus* oocytes: An autoradiographic study, *Dev. Biol.* **60**:482–486.

Burley, R. W., 1975, Studies on the apoproteins of the major lipoprotein of the yolk of hen's eggs I. Isolation and properties of the low-molecular-weight apoproteins, *Aust. J. Biol. Sci.* **28**:121–132.

Burley, R. W., and Cook, W. H., 1961, Isolation and composition of avian egg yolk granules and their constituent $\alpha$- and $\beta$-lipovitellins, *Can. J. Biochem. Physiol.* **391**:1295–1307.

Burley, R. W., and Cook, W. H., 1962, The dissociation of $\alpha$- and $\beta$-lipovitellin in aqueous solution I. Effect of pH, temperature, and other factors, *Can. J. Biochem. Physiol.* **40:**363–372.

Callen, J. C., Tourte, M. Dennebouy, N., and Mounolou, J. C., 1980, Mitochondrial development in oocytes of *Xenopus laevis, Biol. Cell.* **38:**13–18.

Campbell, C. M., and Jalabert, B., 1979, Selective protein incorporation by vitellogenic *Salmo gairdneri* oocytes *in vitro, Ann. Biol. Anim. Biochim. Biophys.* **19:**429–437.

Capony, F., and Williams, D. L., 1980, Apolipoprotein B of avian very low density lipoprotein: Characteristics of its regulation in nonstimulated and estrogen-stimulated rooster, *Biochemistry* **19:**2219–2226.

Chan, L., Jackson, R. L., O'Malley, B. W., and Means, B. W., 1976, Synthesis of very low density lipoproteins in the cockerel. Effects of estrogen, *J. Clin. Invest.* **58:**368–379.

Chan, L., Snow, L. D., Jackson, R. L., and Means, A. R., 1979, Hormonal regulation of lipoprotein synthesis in the cockerel, in: *Ontogeny of Receptors and Reproductive Hormonal Action* (T. H. Hamilton, J. H. Clark, and W. A. Sadler, eds.), pp. 331–351, Raven Press, New York.

Chan, L., Bradley, W. A., and Means, A. R., 1980, Amino acid sequence of the signal peptide of apo VLDL-II, a major apoprotein in avian very low density lipoprotein, *J. Biol. Chem.* **255:**10060–10063.

Chapman, M. J., Goldstein, S., and Laudat, M. H., 1977, Characterization and comparative aspects of the serum very low and low density lipoproteins and their apoproteins in the chicken *(Galus domesticus), Biochemistry* **16:**3006–3015.

Clagett, C. O., 1971, Genetic control of the riboflavin carrier protein, *Fed. Proc.* **30:**127–129.

Clark, R. C., 1970, The isolation and comparison of two phosphoproteins from hen's egg, *Biochem. J.* **118:**537–542.

Connelly, C., and Taborsky, G., 1961, Chromatographic fractionation of phosvitin, *J. Biol. Chem.* **236:**1364–1368.

Craik, J. C. A., 1978a, Plasma levels of vitellogenin in the elasmobranch *Scyliorhinus canicula* L. (lesser spotted dogfish), *Comp. Biochem. Physiol.* **60B:**9–18.

Craik, J. C. A., 1978b, Kinetic studies of vitellogenin metabolism in the elasmobranch *Scyliorhinus canicula* L., *Comp. Biochem. Physiol.* **61A:**355–361.

Craik, J. C. A., 1978c, The effects of estrogen treatment on certain plasma constituents associated with vitellogenesis in the elasmobranch *Scyliorhinus canicula* L., *Gen. Comp. Endocrinol.* **35:**455–464.

Craik, J. C. A., 1982, Levels of phosphoprotein in the eggs and ovaries of some fish species, *Comp. Biochem. Physiol.* **72B:**507–510.

Culbert, J., and McIndoe, W. M., 1971, A comparison of the heterogeneity of phosvitin of egg-yolk and blood-plasma of the domestic fowl *(Gallus domesticus), Int. J. Biochem.* **2:**616–622.

Cutting, J. A., and Roth, T. F., 1973, Changes in specific sequestering of protein during transport into the developing oocyte of the chicken, *Biochim. Biophys Acta* **298:**951–955.

Decroly, M., Goldfinger, M., and Six-Tondeur, N., 1979, Biochemical characterization of lysosomes in unfertilized eggs of *Xenopus laevis, Biochim, Biophys. Acta* **587:**567–578.

Deeley, R. G., Mullinix, K. P., Wetekam, W., Kronenberg, H. M., Meyers, M., Eldridge, J. D., and Goldberger, R. F., 1975, Vitellogenin synthesis in the avian liver: Vitellogenin is the precursor of the egg yolk phosphoproteins, *J. Biol. Chem.* **250:**9060–9066.

Dehn, P. F., and Wallace, R. A., 1973, Sequestered and injected vitellogenin: Alternative routes of protein processing in *Xenopus* oocytes, *J. Cell Biol.* **58:**721–724.

de Vlaming, V. L., Wiley, H. S., Delahunty, G., and Wallace, R. A., 1980, Goldfish *(Carassius auratus)* vitellogenin: Induction, isolation, properties and relationship to yolk proteins, *Comp. Biochem. Physiol.* **67B:**613–623.

Dopheide, T. A. A., and Inglis, A. S., 1974, The amino acid sequence of a protein (apovitellinin I) from the low density lipoprotein of emu egg yolk, *Aust. J. Biol. Sci.* **27:**15–21.

Dopheide, T. A. A., and Inglis, A. S., 1976, Primary structure of apovitellinin I from hen egg yolk and its comparison with emu apovitellinin I, *Aust. J. Biol. Sci.* **29:**175–180.

Dumont, J. N., 1972, Oogenesis in *Xenopus laevis* (Daudin) I. Stages of oocyte development in laboratory maintained animals, *J. Morphol.* **136:**153–179.

Dumont, J. N., 1978, Oogenesis in *Xenopus laevis* (Daudin) VI. The route of injected tracer transport in the follicle and developing oocyte, *J. Exp. Zool.* **204:**193–217.

Dumont, J. N., and Brummett, A. R., 1978, Oogenesis in *Xenopus laevis* (Daudin) V. Relationships between developing oocytes and their investing follicular tissues, *J. Morphol.* **155:**73–97.

Evans, R. J., Flegel, C. J., and Bauer, D. H., 1975, Molecular sizes of egg yolk very low density lipoproteins fractionated by ultracentrifugation, *Poult. Sci.* **54:**889–895.

Felber, B. K., Maurhofer, S., Jaggi, R. B., Wyler, T., Wahli, W., Ryffel, G. U., and Weber, R., 1980, Isolation and translation *in vitro* of four related vitellogenin mRNAs of estrogen-stimulated *Xenopus laevis*, *Eur. J. Biochem.* **105:**17–24.

Follett, B. K., Nicholls, T. J., and Redshaw, M. R., 1968, The vitellogenic response in the South African clawed toad *(Xenopus laevis)* Daudin, *J. Cell Physiol. Suppl.* **72:**91–102.

Franzen, J. S., Bobik, C. M., Stuchell, R. N., and Lee, L., 1968, Polypeptide subunits of lipovitellin, *Arch. Biochem. Biophys.* **123:**127–132.

Fulton, T. W., 1898, On the growth and maturation of the ovarian eggs of teleostean fishes, *Fish. Bd Scotland 16th Annu. Rep.* **3:**83–134.

Gapp, D. A., Ho, S. M., and Callard, I. P., 1979, Plasma levels of vitellogenin in *Chrysemys picta* during the annual gonadal cycle: Measurement by specific radioimmunoassay, *Endocrinology* **3:**784–790.

Gatenby, J. B., and Hill, J. P., 1924, On an ovum of *Ornithorhynchus* exhibiting polar bodies and polyspermy, *Q. J. Microsc. Sci.* **68:**229–238.

Germond, J. E., ten Heggeler, B., Schubiger, J. L., Walker, P., Westley, B., and Wahli, W., 1983, Vitellogenin B2 gene in *Xenopus laevis:* Isolation, *in vitro* transcription and relation to other vitellogenin genes, *Nucleic Acids Res.* **11:**2979–2997.

Gibbins, A. M. V., van de Voort, F. R., and Braham, R., 1981, Comparative study of three different forms of both native and subunit vitellogenin identified in quail plasma, *Comp. Biochem. Physiol.* **70B:**731–738.

Gibbins, A. M. V., and Robinson, G. A., 1982, Comparative study of the physiology of vitellogenesis in Japanese quail, *Comp. Biochem. Physiol.* **72A:**149–152.

Giorgi, F., 1980, Coated vesicles in the oocyte, in: *Coated Vesicles* (C. J. Ockleford and A. Whyte, eds.), pp. 135–177, Cambridge University Press, Cambridge.

Giorgi, F., Bucci Innocenti, S., and Ragghianti, M., 1976, Osmium zinc iodide staining of Golgi elements in oocytes of *Triturus cristatus*, *Cell Tissue Res.* **172:**121–131.

Giorgi, F., Gobbetti, A., and Polzonetti-Magni, A., 1982, Variations in the vitellogenin titre during the reproductive cycle of *Rana esculenta* L., *Comp. Biochem. Physiol.* **72B:**501–506.

Gordon, J. I., Deeley, R. G., Burns, A. T. H., Paterson, B. M., Christmann, J. L., and Goldberger, R. F., 1977, *In vitro* translation of avian vitellogenin messenger RNA, *J. Biol. Chem.* **252:**8320–8327.

Gottlieb, T. A., 1981, The posttranslational modification and secretion of vitellogenin in *Xenopus laevis*, Ph.D. thesis, University of Tennessee, Knoxville, Tennessee.

Gottlieb, T. A., and Wallace, R. A., 1981, Intracellular phosphorylation of vitellogenin in the liver of estrogen-stimulated *Xenopus laevis*, *J. Biol. Chem.* **256:**4116–4123.

Gottlieb, T. A., and Wallace, R. A., 1982, Intracellular glycosylation of vitellogenin in the liver of estrogen-stimulated *Xenopus laevis*, *J. Biol. Chem.* **257:**95–103.

Green, C. D., and Tata, J. R., 1976, Direct induction by estradiol of vitellogenin synthesis in organ cultures of male Xenopus laevis liver, *Cell* **7:**131–139.

Grodziński, Z., 1951, The yolk spheres of the hen's egg as osmometers, *Biol. Rev.* **26:**253–264.

Grodziński, Z., 1954, The yolk of bitterling *Rhodeus amarus* Bl, *Folia Morphol. (Warsz.)* **13:**13–26.

Guraya, S. S., 1982, Recent progress in the structure, origin, composition, and function of cortical granules in animal egg, *Int. Rev. Cytol.* **78:**257–360.

Hara, A., and Hirai, H., 1978, Comparative studies on immunochemical properties of female-specific serum protein and egg yolk proteins in rainbow trout *(Salmo gairdneri)*, *Comp. Biochem. Physiol.* **59B**:339–343.

Hayward, M. A., Mitchell, R. A., and Shapiro, D. J., 1980, Induction of estrogen receptor and reversal of the nuclear–cytoplasmic receptor ratio during vitellogenin synthesis and withdrawal in *Xenopus laevis*, *J. Biol. Chem.* **255**:11308–11312.

Heald, P. J., and McLachlan, P. M., 1963, Isolation of phosvitin from the plasma of the laying hen, *Biochem. J.* **87**:571–576.

Heald, P. J., and McLachlan, P. M., 1964, The isolation of phosvitin from the plasma of oestrogen-treated immature pullet, *Biochem. J.* **92**:51–55.

Herbener, G. H., Feldhoff, R. C., and Fonda, M. L., 1983, A correlated morphometric and biochemical study of estrogen-induced vitellogenesis in male *Rana pipiens*, *J. Ultrastruct. Res.* **83**:28–42.

Hickey, E. D., and Wallace, R. A., 1974, A study of the vitellogenic protein in the serum of estrogen-treated *Ictalurus nebulosus*, *Biol. Bull.* **147**:481.

Hillyard, L. A., Entenman, C., and Chaikoff, I. L., 1956, Concentration and composition of serum lipoproteins of cholesterol-fed and stilbestrol-injected birds, *J. Biol. Chem.* **223**:359–368.

Hillyard, L. A., White, H. M., and Pangburn, S. A., 1972, Characterization of apolipoproteins in chicken serum and egg yolk, *Biochemistry* **11**:511–518.

Ho, S. M., L'Italien, J., and Callard, I. P., 1980, Studies on reptilian yolk: *Chrysemys* vitellogenin and phosvitin, *Comp. Biochem. Physiol.* **65B**:139–144.

Ho, S. M., Dawko, D., and Callard, I. P., 1981, Effect of exogenous estradiol-17$\beta$ on plasma vitellogenin levels in male and female *Chrysemys* and its modulation by testosterone and progesterone, *Gen. Comp. Endocrinol.* **43**:413–421.

Ho. S. M., Taylor, S., and Callard, I. P., 1982a, Effect of hypophysectomy and growth hormone on estrogen-induced vitellogenesis in the freshwater turtle, *Chrysemys picta*, *Gen Comp. Endocrinol.* **48**:254–260.

Ho, S. M., Kleis, S., McPherson, R., Heiserman, G. J., and Callard, I. P., 1982b, Regulation of vitellogenesis in reptiles, *Reprod. Biol. Reptiles Herpetol.* **38**:40–50.

Holtzman, E., 1976, *Lysosomes: A Survey*, Springer-Verlag, Vienna.

Hope, J., Humphries, A. A., and Bourne, G. H., 1964, Ultrastructural studies on developing oocytes of the salamander *Triturus viridescens* II. The formation of yolk, *J. Ultrastruct. Res.* **10**:547–556.

Hori, S. H., Kodama, T., and Tamahashi, K., 1979, Induction of vitellogenin synthesis in goldfish by massive doses of androgens, *Gen. Comp. Endocrinol.* **37**:306–320.

Huber, S., Ryffel, G. U., and Weber, R., 1979, Thyroid hormone induces competence for oestrogen-dependent vitellogenin synthesis in developing *Xenopus laevis* liver, *Nature* **278**:65–67.

Jackson, R. L., Lin, H. Y., Chan, L., and Means, A. R., 1977, Amino acid sequence of a major apoprotein from hen plasma very low density lipoproteins, *J. Biol. Chem.* **252**:250–253.

James, T. C., Bond, Y. M., Maack, C. A., Applebaum, S. W., and Tata, J. R., 1982, Evolutionary conservation of vitellogenin genes, *DNA* **1**:345–351.

Jared, D. W., and Wallace, R. A., 1968, Comparative chromatography of the yolk proteins of teleosts, *Comp. Biochem. Physiol.* **24**:437–443.

Jones, D. G., Briles, W. E., and Schjeide, O. A., 1975, A mutation restricting ovulation in chickens, *Poult. Sci.* **54**:1780.

Kaplan, J., 1981, Polypeptide-binding membrane receptors: Analysis and classification, *Science* **212**:14–20.

Karasaki, S., 1963a, Studies on amphibian yolk 1. The ultrastructure of the yolk platelet, *J. Cell Biol.* **18**:135–151.

Karasaki, S., 1963b, Studies on amphibian yolk 5. Electron microscopic observations on the utilization of yolk platelets during embryogenesis, *J. Ultrastruct. Res.* **9**:225–247.

Keem, K., Smith, L. D., Wallace, R. A., and Wolf, D., 1979, Growth rates of oocytes in laboratory-maintained Xenopus laevis, *Gamete Res.* **2**:125–135.

Knowland, J., 1978, Induction of vitellogenin synthesis in *Xenopus laevis* tadpoles, *Differentiation* **12**:47–51.

Knowland, J., and Westley, B., 1979, The control of vitellogenin synthesis by oestrogen in *Xenopus laevis* and conversion of vitellogenin into yolk proteins, in: *Maternal Effects in Development* (D. R. Newth and M. Balls, eds.), pp. 111–126, Cambridge University Press, Cambridge.

Korfsmeier, K. H., 1966, Zur Genese der Dottersystems in der Oocyte von *Brachydanio rerio*. Autoradiographische Untersuchungen, *Z. Zellforsch.* **71**:283–296.

Kress, A., 1982, Ultrastructural indications for autosynthetic proteinaceous yolk formation in amphibian oocytes, *Experientia* **38**:761–771.

Lange, R. H., 1981a, Highly conserved lipoprotein assembly in teleost and amphibian yolk-platelet crystals, *Nature* **289**:329–330.

Lange, R. H., 1981b, Are yolk phosvitins carriers for specific cations? Comparative microanalysis in vertebrate yolk platelets, *Z. Naturforsch.* **36c**:686–687.

Lange, R. H., 1982, The lipoprotein crystals of cyclostome yolk platelets (*Myxine glutinosa* L., *Lampetra planeri* [Block], *L. fluviatilis* [L.]), *J. Ultrastruct. Res.* **79**:1–17.

Lange, R. H., and Richter, H. P., 1981, A symmetric lipovitellin-phosvitin dimer in cyclostome yolk platelet crystals: Structural and biochemical observations, *J. Mol. Biol.* **148**:487–491.

Lange, R. H., Grodziński, Z., and Kilarski, W., 1982, Yolk-platelet crystals in three ancient bony fishes: *Polypterus bichir* (Polypteri), *Amia calva* L., and *Lepisosteus osseus* (L.) (Holostei), *Cell Tissue Res.* **222**:159–165.

Lange, R. H., Richter, H. P., Riehl, R., Zierold, K., Trandaburu, T., and Magdowski, G., 1983, Lipovitellin-phosvitin crystals with orthorhombic features: Thin-section electron microscopy, gel electrophoresis, and microanalysis in teleost and amphibian yolk platelets and a comparison with other vertebrates, *J. Ultrastruct. Res.* **83**:122–140.

Lanzavecchia, G., 1968, Problems of the yolk formation in amphibia, *Atti. Simp. Accad. Naz. Lincei.* **104**:101–109.

Laskowski, M., 1935, Darstellungsmethode des Serumvitellins, *Biochem. Z.* **278**:345–348.

Lazier, C. B., and Haggerty, A. J., 1979, A high-affinity estrogen-binding protein in cockerel liver cytosol, *Biochem. J.* **180**:347–353.

Le Menn, F., Rochefort, H., and Garcia, M., 1980, Effect of androgen mediated by the estrogen receptor of fish liver: Vitellogenin accumulation, *Steroids* **35**:315–328.

Lin, C. T., and Chan, L., 1981, Estrogen regulation of yolk and non-yolk protein synthesis in the avian liver, *Differentiation* **18**:105–114.

Lin, C. T., and Chan, L., 1982, Localization of apo-VLDL-II, a major apoprotein in very low density lipoproteins, in the estrogen-treated cockerel liver by immunoelectron microscopy, *Histochemistry* **76**:237–246.

Luskey, K. L., Brown, M. S., and Goldstein, J. L., 1974, Stimulation of synthesis of very low density liproproteins in rooster liver by estradiol, *J. Biol. Chem.* **249**:5939–5947.

MacKenzie, S. L., and Martin, W. G., 1967, The macromolecular composition of hen's egg yolk at successive stages of maturation, *Can. J. Biochem.* **45**:591–601.

Malone, T. E., and Hisaoka, K. K., 1963, A histochemical study of the formation of deutoplasmic components in developing oocytes of the zebrafish, *Brachydanio rerio*, *J. Morphol.* **112**:61–75.

Mano, Y., and Lipmann, F., 1966, Characteristics of phosphoproteins (phosvitins) from a variety of fish roes, *J. Biol. Chem.* **241**:3822–3833.

Markert, J. R., and Vanstone, W. E., 1971, Egg proteins of coho salmon (*Oncorhynchus kisutch*): Chromatographic separation and molecular weights of the major proteins in the high density fraction and their presence in salmon plasma, *Fish. Res. Bd. Can.* **28**:1853–1856.

Martin, W. G., Tattrie, N. H., and Cook, W. H., 1963, Lipid extraction and distribution studies of egg yolk lipoproteins, *Can. J. Biochem. Physiol.* **41**:657–666.

Massover, W. H., 1971a, Intramitochondrial yolk-crystals of frog oocytes. I. Formation of yolk-crystal inclusions by mitochondria during bullfrog oogenesis, *J. Cell Biol.* **48**:266–279.

Massover, W. H., 1971b, Intramitochondrial yolk-crystals of frog oocytes. II. Expulsion of intramitochondrial yolk-crystals to form single-membrane-bound hexagonal crystalloids, *J. Ultrastruct. Res.* **36**:603–620.

Massover, W. H., 1971c, Nascent yolk platelets of anuran amphibian oocytes, *J. Ultrastruct. Res.* **37**:574–590.

Masui, Y., and Clarke, H. J., 1979, Oocyte maturation, *Int. Rev. Cytol.* **57**:185–282.

May, F. E. B., and Knowland, J., 1980, The role of thyroxine in the transition of vitellogenin synthesis from noninducibility to inducibility during metamorphosis in *Xenopus laevis*, *Devl. Biol.* **77**:419–430.

May, F. E. B., Westley, B. R., and Knowland, J., 1981, Vitellogenin synthesis and characterization of the liver estrogen receptor in the neotenous salamander *Ambystoma mexicanum*, *Devl. Biol.* **82**:350–357.

McCully, K. A., and Common, R. H., 1961, Zone electrophoresis of protein-bound phosphorus of fowl's serum, *Can. J. Biochem. Physiol.* **39**:1451–1466.

McIndoe, W. M., 1959, A lipophosphoprotein complex in hen plasma associated with yolk production, *Biochem. J.* **72**:153–159.

Mester, J., and Baulieu, E. E., 1972, Nuclear estrogen receptor of chick liver, *Biochim. Biophys. Acta* **261**:236–244.

Mok, C. C., and Common, R. H., 1964, Studies of the livetins of hen's egg yolk II. Immunoelectrophoretic identification of livetins with serum proteins, *Can. J. Biochem.* **42**:1119–1131.

Mok, C. C., Martin, W. G., and Common, R. H., 1961, A comparison of phosvitins prepared from hen's serum and hen's egg yolk, *Can. J. Biochem. Physiol.* **39**:109–117.

Munday, K. A., Ansari, A. Q., Oldroyd, D., and Akhtar, M., 1968, Oestrogen-induced calcium-binding protein in *Xenopus laevis*, *Biochem. Biophys. Acta* **166**:748–751.

Muniyappa, K., and Adiga, P. R., 1980, Oestrogen-induced synthesis of thiamin-binding protein in immature chicks. Kinetics of induction, hormonal specificity and modulation, *Biochem. J.* **186**:201–210.

Muniyappa, K., and Adiga, P. R., 1981, Nature of thiamin-binding protein from chicken egg yolk, *Biochem. J.* **193**:679–685.

Murthy, V. S., and Adiga, P. R., 1978, Estrogen-induced synthesis of riboflavin-binding protein in immature chicks, *Biochim. Biophys. Acta* **538**:364–375.

Neaves, W. B., 1972, The passage of extracellular tracers through the follicular epithelium of lizard ovaries, *J. Exp. Zool.* **179**:339–364.

Nilsson, B. O., 1980, Comparative ultrastructure of the yolk material in preimplantation stages of the hamster, mouse and rat embryos, *Gamete Res.* **3**:369–377.

Ohlendorf, D. H., Barbarash, G. R., Trout, A., Kent, C., and Banaszak, L. J., 1977, Lipid and polypeptide components of the crystalline yolk system from *Xenopus laevis*, *J. Biol. Chem.* **252**:7992–8001.

Ohlendorf, D. H., Wrenn, R. F., and Banaszak, L. J., 1978, Three dimensional structure of the lipovitellin-phosvitin complex from amphibian oocytes, *Nature* **272**:28–32.

Opresko, L., Wiley, H. S., and Wallace, R. A., 1979, The origin of yolk-DNA in *Xenopus laevis*, *J. Exp. Zool.* **209**:367–376.

Opresko, L., Wiley, H. S., and Wallace, R. A., 1980, Differential postendocytotic compartmentation in Xenopus oocytes is mediated by a specifically bound ligand, *Cell* **22**:47–57.

Opresko, L., Wiley, H. S., and Wallace, R. A., 1981, Receptor-mediated binding and internalization of vitellogenin by Xenopus oocytes, *J. Cell Biol.* **91**:218a.

Osanai, K., 1956, On the ovarian eggs of the loach, *Lefua echigonia*, with special reference to the formation of the cortical alveoli, *Sci. Rep. Tohoku Univ. Ser. 4* **22**:181–188.

Oshiro, T., and Hibiya, T., 1982, *In vitro* yolk globule fusion of the oocytes in the plaice *Limanda yokohamae*, *Bull. Jpn. Soc. Sci. Fish.* **48**:391–399.

Ostrowski, W., Zak, Z., and Krawczyk, A., 1968, Riboflavin flavoprotein from egg yolk. Analytical and biophysical data, *Acta Biochem. Pol.* **15**:241–260.

Pan, M. L., Bell, W. J., and Telfer, W. H., 1969, Vitellogenic blood protein synthesis by insect fat body, *Science* **165**:393–394.

Panijel, J., 1950, L'organisation du vitellus dans les oeufs d'amphibiens, *Biochim. Biophys. Acta* **5**:343–357.

Philipp, B. W., and Shapiro, D. J., 1981, Estrogen regulation of hepatic 3-hydroxy-3-methylglutaryl coenzyme A reductase and acetyl-CoA carboxylase in *Xenopus laevis*, *J. Biol. Chem.* **256**:2922–2927.

Protter, A. A., Wang, S. Y., Shelness, G. S., Ostapchuk, P., and Williams, D. L., 1982, Isolation and characterization of a cDNA clone specific for avian vitellogenin II, *Nucleic Acids Res.* **10**:4935–4950.

Redshaw, M. R., and Follett, B. K., 1971, The crystalline yolk-platelet proteins and their soluble plasma precursor in an amphibian, *Xenopus laevis*, *Biochem. J.* **124**:759–766.

Redshaw, M. R., and Follett, B. K., 1976, Physiology of the egg yolk production by the fowl: The measurement of circulating levels of vitellogenin employing a specific radioimmunoassay, *Comp. Biochem. Biophys.* **55A**:399–405.

Redshaw, M. R., Follett, B. K., and Nicholls, T. J., 1969, Comparative effects of oestrogens and other steroid hormones on serum lipids and proteins in *Xenopus laevis* Daudin, *J. Endocrinol.* **43**:47–53.

Redshaw, M. R., Follett, B. K., and Lawes, G. J., 1971, Biliverdin: A component of yolk proteins in *Xenopus laevis*, *Int. J. Biochem.* **2**:80–84.

Rhodes, M. B., Bennett, M., and Fenney, R. E., 1959, The flavoprotein–apoprotein system of egg white, *J. Biol. Chem.* **234**:2054–2060.

Richter, C. J. J., and Goos, H. J. T. (eds.), 1982, *Reproductive Physiology of Fish*, Pudoc, Wageningen.

Romanoff, A. L., 1943, Growth of avian ovum, *Anat. Rec.* **85**:261–267.

Romanoff, A. L., and Romanoff, A. J., 1949, *The Avian Egg*, Wiley, New York.

Roth, T. F., Cutting, J. A., and Atlas, S. B., 1976, Protein transport: A selective membrane mechanism, *J. Supramol. Struct.* **4**:527–548.

Rubenstein, E. C., 1979, The role of an epithelial occlusion zone in the termination of vitellogenesis in *Hyalophora cecropia* ovarian follicles, *Dev. Biol.* **71**:115–127.

Rudack, D., and Wallace, R. A., 1968, On the site of phosvitin synthesis in *Xenopus laevis*, *Biochim. Biophys. Acta* **155**:299–301.

Ryffel, G. U., 1978, Synthesis of vitellogenin, an attractive model for investigating hormone-induced gene activation, *Mol. Cell. Endocrinol.* **12**:237–246.

Ryffel, G. U., Wyler, T., Muellener, D. B., and Weber, R., 1980, Identification, organization and processing intermediates of the putative precursors of Xenopus vitellogenin messenger RNA, *Cell* **19**:53–61.

Sand, D. M., Hehl, J. L., and Schlenk, H., 1969, Biosynthesis of wax esters in fish. Reduction of fatty acids and oxidation of alcohols, *Biochemistry* **10**:2536–2541.

Schjeide, O. A., 1954, Studies of the New Hampshire chicken embryo III. Nitrogen and lipid analyses of ultracentrifugal fractions of plasma, *J. Biol. Chem.* **211**:355–362.

Schjeide, O. A., Galey, F., Grellert, E. A., Lin, R. I., deVellis, J., and Mead, F. F., 1970, Macromolecules in oocyte maturation, *Biol. Reprod. Suppl.* **2**:14–43.

Schjeide, O. A., Briles, W. E., Holshouser, S., and Jones, D. G., 1976, Effect of "restricted ovulator" gene on uptake of yolk-precursor protein, *Cell Tissue Res.* **166**:109–116.

Schmidt, G., Bartsch, G., Kitagawa, T., Fujisawa, K., Knolle, J., Joseph, J., De Marco, P., Liss, M., and Haschemeyer, R., 1965, Isolation of a protein of high phosphorus content from the eggs of brown brook trout, *Biochem. Biophys. Res. Commun.* **18**:60–65.

Selman, K., and Wallace, R. A., 1982a, Oocyte growth in the sheepshead minnow: Uptake of exogenous proteins by vitellogenic oocytes, *Tissue Cell* **14**:555–571.

Selman, K., and Wallace, R. A., 1982b, The inter- and intracellular passage of proteins through the ovarian follicle in teleosts, in: *Reproductive Physiology of Fish* (C. J. J. Richter and H. J. T. Goos, eds.), pp. 151–154, Pudoc, Wageningen.

Selman, K., and Wallace, R. A., 1983, Oogenesis in *Fundulus heteroclitus*. III. Vitellogenesis, *J. Exp. Zool.* **226**:441–457.

Shapiro, D., 1982, Steroid hormone regulation of vitellogenin gene expression, *CRC Crit. Rev. Biochem.* **12**:187–203.

Shapiro, D. J., Baker, H. J., and Stitt, D. T., 1976, In vitro translation and estradiol-17β induction of *Xenopus laevis* vitellogenin messenger RNA, *J. Biol. Chem.* **251**:3105–3111.

Smith, D. F., Penning, T. M., Ansari, A. Q., Munday, K. A., and Akhtar, M., 1978, Oestrogen-induced cholesterol and fatty acid biosynthesis in *Xenopus laevis* liver during vitellogenic response, *Biochem. J.* **174**:353–361.

Spornitz, U. M., 1972, Some properties of crystalline inclusion bodies in oocytes of *Rana temporaria* and *Rana esculenta*, *Experientia* **28**:66–67.

Spornitz, U. M., and Kress, A., 1971, Yolk-platelet formation in oocytes of *Xenopus laevis* (Daudin), *Z. Zellforsch.* **117**:235–251.

Spornitz, U. M., and Kress, A., 1973, Ultrastructural studies of oogenesis in some European amphibians. II. *Triturus vulgaris*, *Z. Zellforsch.* **143**:387–407.

Steinert, G., and Hanocq, J., 1979, Ultrastructural localization of acid phosphatase activity in matured *Xenopus laevis* oocyte, *Biol. Cell.* **34**:247–254.

Sundararaj, B. I., and Nath, P., 1981, Steroid-induced synthesis of vitellogenin in the catfish, *Heteropneustes fossilis* (Bloch), *Gen. Comp. Endocrinol.* **43**:201–210.

Taborsky, G., 1974, Phosphoproteins, *Adv. Prot. Chem.* **28**:1–210.

Taborsky, G., and Mok, C. C., 1967, Phosvitin. Homogeneity and molecular weight, *J. Biol. Chem.* **242**:1495–1501.

Tata, J. R., 1976, The expression of the vitellogenin gene, *Cell* **9**:1–14.

Tata, J. R., and Smith, D. F., 1979, Vitellogenesis: A versatile model for hormonal regulation of gene expression, *Rec. Prog. Horm. Res.* **35**:47–95.

Tata, J. R., Baker B. S., and Deeley, J. V., 1980, Vitellogenin as a multigene family. Not all *Xenopus* vitellogenin genes may be in an "expressible" configuration, *J. Biol. Chem.* **255**:*6721–6726.*

Tata, J. R., James, T. C., Watson, C. S., Williams, J. L., and Wolffe, A. P., 1983, Hormonal regulation and expression of vitellogenin multigene family, in *Molecular Biology of Egg Maturation* (Ciba Foundation Symp. 98), (R. Porter and J. Whelan, eds.), pp. 96–110, Pitman, London.

te Heesen, D., and Engels, W., 1973, Elektrophoretische Untersuchungen zur Vitellogenese von *Brachydanio rerio* (Cyprinidae, Teleostoi), *Arch. Entwicklungsmech. Org.* **173**:46–59.

Turner, C. L., 1938, Histological and cytological changes in the ovary of Cymatogaster aggregatus during gestation, *J. Morphol.* **62**:351–373.

Urist, M. R., and Schjeide, O. A., 1961, The partition of calcium and protein in the blood of oviparous vertebrates during oestrus, *J. Gen. Physiol.* **44**:743–756.

van Bohemen, C. G., and Lambert, J. G. D., 1981, Estrogen synthesis in relation to estrone, estradiol, and vitellogenin plasma levels during the reproductive cycle of the female rainbow trout, *Salmo gairdneri*, *Gen. Comp. Endocrinol.* **45**:105–114.

van Bohemen, C. G., Lambert, J. G. D., Goos, H. J. T., and van Oordt, P. G. W. J., 1982, Estrone and estradiol participation during exogenous vitellogenesis in the female rainbow trout, *Salmo gairdneri*, *Gen. Comp. Endocrinol.* **46**:81–92.

Wachsmuth, E. D., and Jost, J. P., 1976, Localization of vitellogenin and serum albumin in hepatic parenchymal cells of normal and estradiol-treated immature chickens, *Biochim. Biophys. Acta* **437**:454–461.

Wahli, W., Wyler, T., Weber, R., and Ryffel, G. U., 1976, Size, complexity and abundance of a specific poly(A)-containing RNA of liver from male *Xenopus* induced to vitellogenin synthesis by estrogen, *Eur. J. Biochem.* **66**:457–465.

Wahli, W., Ryffel, G. U., Wyler, T., Jaggi, R. B., Weber, R., and Dawid, I. B., 1978, Cloning and characterization of synthetic sequences from the *Xenopus laevis* structural gene, *Dev. Biol.* **67**:371–383.

Wahli, W., Dawid, I. B., Wyler, T., Jaggi, R. B., Weber, R., and Ryffel, G. U., 1979, Vitellogenin in *Xenopus laevis* is encoded in a small family of genes, *Cell* **16**:535–549.

Wahli, W., Dawid, I. B., Wyler, T., Weber, R., and Ryffel, G. U., 1980, Comparative analysis of the structural organization of two closely related vitellogenin genes in X. Laevis, *Cell* **30**:107–112.

Wahli, W., Dawid, I. B., Ryffel, G. U., and Weber, R., 1981, Vitellogenesis and the vitellogenin gene family, *Science* **212**:298–304.
Wahli, W., Germond, J. E., ten Heggeler, B., and May, F. E. B., 1982, Vitellogenin genes *A1* and *B1* are linked in the *Xenopus laevis* genome, *Proc. Natl. Acad. Sci. USA* **79**:6832–6836.
Wallace, R. A., 1963a, Studies on amphibian yolk. III. A resolution of yolk platelet components, *Biochim. Biophys. Acta* **74**:495–504.
Wallace, R. A., 1963b, Studies on amphibian yolk. IV. An analysis of the main-body component of yolk platelets, *Biochim. Biophys. Acta* **74**:505–518.
Wallace, R. A., 1965, Resolution and isolation of avian and amphibian yolk-granule proteins using TEAE-cellulose, *Anal. Biochem.* **11**:297–311.
Wallace, R. A., 1970, Studies on amphibian yolk. IX. *Xenopus* vitellogenin, *Biochim. Biophys. Acta* **215**:176–183.
Wallace, R. A., 1978, Oocyte growth in nonmammalian vertebrates, in: *The Vertebrate Ovary* (R. E. Jones, ed.), pp. 469–502, Plenum Press, New York.
Wallace, R. A., 1983, Interactions between somatic cells and the growing oocyte of *Xenopus laevis*, in: *Current Problems in Germ Cells* (A. McLaren and C. C. Wylie, eds.), pp. 285–306, Cambridge University Press, Cambridge.
Wallace, R. A., and Bergink, E. W., 1974, Amphibian vitellogenin: Properties, hormonal regulation of hepatic synthesis and ovarian uptake, and conversion to yolk proteins, *Am. Zool.* **14**:1159–1175.
Wallace, R. A., and Dumont, J. N., 1968, The induced synthesis and transport of yolk proteins and their accumulation by the oocyte in *Xenopus laevis*, *J. Cell Physiol. Suppl.* **72**:73–89.
Wallace, R. A., and Ho, T., 1972, Protein incorporation by isolated amphibian oocytes. II. A survey of inhibitors, J. Exp. Zool. **81**:303–318.
Wallace, R. A., and Hollinger, T. G., 1979, Turnover of endogenous, microinjected, and sequestered protein in Xenopus oocytes, *Exp. Cell Res.* **119**:277–287.
Wallace, R. A., and Jared, D. W., 1968a, Estrogen induces lipophosphoprotein in serum of male *Xenopus laevis*, *Science* **160**:91–92.
Wallace, R. A., and Jared, D. W., 1968b, Studies on amphibian yolk. VII. Serum-phosphoprotein synthesis by vitellogenic females and estrogen-treated males of *Xenopus laevis*, *Can. J. Biochem.* **46**:953–959.
Wallace, R. A., and Jared, D. W., 1969, Studies on amphibian yolk. VIII. The estrogen-induced hepatic synthesis of a serum lipophosphoprotein and its selective uptake by the ovary and transformation into yolk platelet proteins in *Xenopus laevis*, *Dev. Biol.* **19**:498–526.
Wallace, R. A., and Jared, D. W., 1976, Protein incorporation by isolated amphibian oocytes. V. Specificity for vitellogenin incorporation, *J. Cell Biol.* **69**:345–351.
Wallace, R. A., and Misulovin, Z., 1978, Long-term growth and differentiation of Xenopus oocytes in a defined medium, *Proc. Natl. Acad. Sci. USA* **75**:5534–5538.
Wallace, R. A., and Selman, K., 1978, Oogenesis in *Fundulus heteroclitus*. I. Preliminary observations on oocyte maturation *in vivo* and *in vitro*, *Dev. Biol.* **62**:354–369.
Wallace, R. A., and Selman, K., 1981, Cellular and dynamic aspects of oocyte growth in teleosts, *Am. Zool.* **21**:325–343.
Wallace, R. A., and Selman, K., 1982, A new procedure for the isolation of intact vitellogenin from teleosts, in: *Reproductive Physiology of Fish* (C. J. J. Richter and H. J. T. Goos, eds.), p. 161, Pudoc, Wageningen.
Wallace, R. A., Jared, D. W., and Eisen, A. Z., 1966, A general method for the isolation and purification of phosvitin from vertebrate eggs, *Can. J. Biochem.* **44**:1647–1655.
Wallace, R. A., Jared, D. W., and Nelson, B. L., 1970, Protein incorporation by isolated amphibian oocytes. I. Preliminary studies, *J. Exp. Zool.* **175**:259–270.
Wallace, R. A., Nickol, J. M., Ho, T., and Jared, D. W., 1972, Studies on amphibian yolk. X. The relative roles of autosynthetic and heterosynthetic processes during yolk protein assembly by isolated oocytes, *Dev. Biol.* **29**:255–272.
Wallace, R. A., Misulovin, Z., Jared, D. W., and Wiley, H. S., 1978, Development of a culture medium for growing *Xenopus laevis* oocytes, *Gamete Res.* **1**:269–280.

Wallace, R. A., Deufel, R. A., and Misulovin, Z., 1980a, Protein incorporation of isolated amphibian oocytes. VI. Comparison of autologous and xenogeneic vitellogenins, *Comp. Biochem. Physiol.* **65B**:151–155.

Wallace, R. A., Misulovin, Z., and Wiley, H. S., 1980b, Growth of anuran oocytes in serum-supplemented medium, *Reprod. Nutr. Dév.* **20**:699–708.

Wallace, R. A., Misulovin, Z., and Etkin, L. D., 1981, Full-grown oocytes from *Xenopus laevis* resume growth when placed in culture, *Proc. Natl. Acad. Sci. USA* **78**:3078–3082.

Wallace, R. A., Opresko, L., Wiley, H. S., and Selman, K., 1983, The oocyte as an endocytic cell, in: *Molecular Biology of Egg Maturation* (Ciba Foundation symp. 98), (R. Porter and J. Whelan, eds.), pp. 228–248, Pitman, London.

Wang, S. Y., and Williams, D. L., 1980, Identification, purification and characterization of two distinct avian vitellogenins, *Biochemistry* **19**:1557–1563.

Wang, S. Y., and Williams, D. L., 1982, Biosynthesis of vitellogenins. Identification and characterization of nonphosphorylated precursors to avian vitellogenin I and vitellogenin II, *J. Biol. Chem.* **257**:3837–3846.

Wang, S. Y., Smith, D. E., and Williams, D. L., 1983, Purification of avian vitellogenin III: Comparison with vitellogenins I and II, *Biochemistry* **22**:6206–6212.

Wangh, L. J., 1982, Glucocorticoids act together with estrogens and thyroid hormones in regulating the synthesis and secretion of *Xenopus* vitellogenin, serum albumin, and fibrinogen, *Dev. Biol.* **89**:294–298.

Wangh, L. J., and Knowland, J., 1975, Synthesis of vitellogenin in cultures of male and female frog liver regulated by estradiol treatment *in vitro*, *Proc. Natl. Acad. Sci. USA* **72**:3172–3175.

Wangh, L. J., and Schneider, W., 1982, Thyroid hormones are corequisites for estradiol-17$\beta$ *in vitro* induction of *Xenopus* vitellogenin synthesis and secretion, *Dev. Biol.* **89**:287–293.

Wangh, L. J., Osborne, J. A., Hentschel, C. C., and Tilly, R., 1979, Parenchymal cells purified from *Xenopus* liver and maintained in primary culture synthesize vitellogenin in repsonse to estradiol-17$\beta$ and serum albumin in response to dexamethasone, *Dev. Biol.* **70**:479–499.

Ward, R. T., 1962, The origin of protein and fatty yolk in *Rana pipiens*. II. Electron microscopical and cytochemical observations of young and mature oocytes, *J. Cell Biol.* **14**:309–341.

Ward, R. T., 1978a, The origin of protein and fatty yolk in *Rana pipiens*. III. Intramitochondrial and primary vesicular yolk formation in frog oocytes, *Tissue Cell* **10**:515–524.

Ward, R. T., 1978b, The origin of protein and fatty yolk in *Rana pipiens*. IV. Secondary vesicular yolk formation in frog oocytes, *Tissue Cell* **10**:525–534.

Westley, B., and Knowland, J., 1979, Estrogen causes a rapid, large and prolonged rise in the level of nuclear estrogen receptors in *Xenopus laevis* liver, *Biochem, Biophys. Res. Commun.* **88**:1167–1172.

Wiebe, J. P., 1968, The reproductive cycle of the viviparous seaperch, *Cymatogaster aggregata* Gibbons, *Can. J. Zool.* **46**:1221–1234.

Wiegand, M. D., 1982, Vitellogenesis in fishes, in: *Reproductive Physiology of Fish* (C. J. J. Richter and H. J. T. Goos, eds.), pp. 136–146, Pudoc, Wageningen.

Wiley, H. S., and Wallace, R. A., 1978, Three different molecular weight forms of the vitellogenin peptide from *Xenopus laevis*, *Biochem. Biophys. Res. Commun.* **85**:153–159.

Wiley, H. S., and Wallace, R. A., 1981, The structure of vitellogenin. Multiple vitellogenins in *Xenopus laevis* give rise to multiple forms of the yolk proteins, *J. Biol. Chem.* **256**:8626–8634.

Wilks, A., Cato, A. C. B., Cozens, P. J., Mattaj, I. W., and Jost, J. P., 1981, Isolation and fine structure organization of an avian gene coding for the major estrogen-inducible mRNA, *Gene* **16**:249–259.

Williams, D. L., 1979, Apoproteins of avian very low density lipoprotein: Demonstration of a single high molecular weight apoprotein, *Biochemistry* **18**:1056–1063.

Williams, J., 1962, Serum proteins and the livetins of hen's egg yolk, *Biochem. J.* **83**:346–355.

Wischnitzer, S., 1957, The ultrastructure of yolk platelets of amphibian oocytes, *J. Biochem. Biophys. Cytol.* **3**:1040–1041.
Wittek, M., 1952, La vitellogénèse chez les amphibiens, *Arch. Biol. (Liége)* **63**:133–198.
Wittliff, J. L., and Kenney, F. T., 1972, Regulation of yolk protein synthesis in amphibian liver. I. Induction of lipovitellin synthesis by estrogen, *Biochim. Biophys. Acta* **269**:485–492.
Woods, J. W., and Roth, T. F., 1979, Selective protein transport: Characterization and solubilization of the phosvitin receptor from chicken oocytes, *J. Supramol. Struct.* **12**:491–504.
Woods, J. W., and Roth, T. F., 1980, Selective protein transport: Identity of the solubilized phosvitin receptor from chicken oocytes, *J. Supramol. Struct.* **14**:473–481.
Yamamoto, K., 1956a, Studies on the formation of fish eggs. VII. The fate of the yolk vesicle in the oocyte of the herring, *Clupea pallasii*, during vitellogenesis, *Annotnes. Zool. Jap.* **29**:91–96.
Yamamoto, K., 1956b, Studies on the formation of fish eggs. VIII. The fate of the yolk vesicle in the oocyte of the smelt, *Hypomesus japonicus*, during vitellogenesis, *Embryologia* **3**:131–138.
Yamamoto, K., 1957a, Studies on the formation of fish eggs. XI. The formation of a continuous mass of yolk and the chemical nature of the lipids contained in it in the oocyte of the flounder, *Liopsetta obscura*, *J. Fac. Sci. Hokkaido Univ. Ser. VI* **13**:344–351.
Yamamoto, K., 1957b, Studies on the formation of fish eggs. XII. On the nonmassed yolk in the egg of the herring, *Clupea pallasii*, *Bull. Fac. Sci. Hokkaido Univ.* **8**:270–277.
Yamamoto, T., 1961, Physiology of fertilization in fish eggs, *Int. Rev. Cytol.* **12**:361–405.
Yamamoto, T. S., 1955, Morphological and cytochemical studies on oogenesis in the freshwater fish, medaka, *Oryzias latipes*, *Jpn. J. Ichthyol.* **4**:170–181.
Yu, J. Y. L., and Marquardt, R. R., 1973, Effects of estradiol and testosterone on the immature female chicken *(Gallus domesticus)*. II. Immune diffusional patterns of yolk lipoproteins in the liver and plasma, *Comp. Biochem. Physiol.* **46B**:759–767.
Yu, J. Y. L., Dickhoff, W. W., Swanson, P., and Gorbman, A., 1981, Vitellogenesis and its hormonal regulation in the Pacific hagfish, *Eptatretus stouti* L., *Gen. Comp. Endocrinol.* **43**:492–502.
Yusko, S., Roth, T. F., and Smith, T., 1981, Receptor-mediated vitellogenin binding to chicken oocytes, *Biochem. J.* **200**:43–50.

Chapter 4

# Annulate Lamellae (Porous Cytomembranes): With Particular Emphasis on Their Possible Role in Differentiation of the Female Gamete

RICHARD G. KESSEL

## 1. Introduction/Background

Annulate lamellae (AL) is a term used to identify and characterize an organelle that is relatively widespread in occurrence, especially in the cytoplasm of the developing female gamete. AL are primarily cytoplasmic membranes. However, in a limited number of cell types, including oocytes, they are present within the nucleus and are called **intranuclear annulate lamellae (IAL).** Details about the structure, function, and morphogenesis of AL in the female gamete are considered in this chapter.

Although AL have been known to exist since their discovery about three decades ago, a functional role has not yet been established, and their significance has only been speculated upon. In addition to their rather abundant occurrence in both male and female germ cells at some period during differentiation, AL have also been described in a variety of somatic cells, embryonic cells, and plant cells, as well as in transformed or tumorigenic cells (reviewed by Kessel, 1983a,) and Kessel (1968d) noted earlier the frequent occurrence of AL in rapidly growing or differentiating cells. Subsequently, Kessel (1983a) noted that AL are especially abundant in cells in which there is a substantial interval or delay between transcription and the utilization of the transcripts in protein synthesis as in certain oocytes and spermatocytes; he has proposed that AL might be involved in events surrounding post-transcriptional regulation of gene expression.

Annulate lamellae bear a striking resemblance to the nuclear envelope (NE) in many structural details; both organelles consist of membranous as

RICHARD G. KESSEL • Department of Biology, University of Iowa, Iowa City, Iowa 52242.

well as nonmembranous elements (Figs. 1–5)*. In fact, AL often appear to have resulted from the organization of many segments or sheets of nuclear membrane into stacked, parallel arrays in the cytoplasm (Figs. 5 and 6). It is not surprising, perhaps, that several early investigators believed AL to be remnants of the NE that underwent disorganization during mitosis (for reviews, cf. Afzelius, 1955; Kessel, 1968d, 1973, 1983a), a notion that does not appear to be the case. Among the similarities between NE and AL are the following:

1. Both are composed of two membranes enclosing a compartment or space (cisterna); these two membranes are continuous at the margin of the many discontinuities in the membrane (Figs. 1–3). The discontinuities are called **pores,** and their appearance varies, depending on whether they are examined in surface or side views; that is, in perpendicular (Figs. 1, 2, 5 and 7) or tangential (Figs. 3 and 8) sections of the pores. The term **annulus** (plural: annuli) is used to characterize a pore in either the nuclear envelope or AL when it is tangentially sectioned and thus observed in surface view. In this case pores appear variously as rounded, circular, or angular discontinuities in the membranes (Figs. 3, 8, and 21). Central granules (Figs. 8 and 40) sometimes appear inside the pores of both AL and nuclear pores (reviewed in Kessel, 1973; Franke, 1974); they may be either single or multiple and measure approximately 20 nm in diameter.
2. In both organelles, pores contain a diffuse, nonmembranous material called the **pore-associated material,** which appears to have a rather complex structure. Sometimes, the pore-associated material is continuous with interlamellar material in the stacks of AL such that these interlamellar regions appear variously to be fibrillar or granular, or both, in composition and of considerable electron density (Figs. 5 and 6).
3. Occasionally, the membranes of both the NE and AL have ribosomes attached in certain places under certain conditions (Figs. 4, 9, and 11), and both organelles may exhibit continuity with the rough-surfaced endoplasmic reticulum (RER). The continuity between the AL and RER can be seen in Figs. 10, 16, 47, and 49. This continuity has led to the view that a close functional relationship exists between the two organelles. Less frequently, the AL may exhibit continuity with lamellae of agranular or smooth endoplasmic reticulum (SER) (Figs. 1 and 18).

Information on the function of AL has been difficult to secure. This is probably due, in part, to the initial impressions that annulate lamellae are

*All figures for this chapter can be found on pages 194–212.

**Table I.** Synonyms for Annulate Lamellae

| | Reference |
|---|---|
| Coarse fibrous component | McCullough (1952) |
| Fenestrated cisternae | Palade (1955); Schulz (1957) |
| Annulate lamellae | Swift (1956) |
| Periodic lamellae | Rebhun (1956) |
| Fenestrated lamellae | Pasteels *et al.* (1958) |
| Secondary membranes | Merriam (1959) |
| Pitted membranes | Balinsky and Devis (1963) |
| Porous cytomembranes | Kessel (1968d) |

largely restricted to oocytes and, in lesser amounts, to cancer cells. Another complication is the difficulty in isolating AL for biochemical analysis—a feat that has yet to be achieved. Oocytes are large cells, reaching a diameter of several millimeters in a differentiation process spanning a period of months or years in some species. Hence, oocytes should be an excellent source of AL. However, they also contain vast quantities of diverse organelles as well as formed elements such as lipid, carbohydrate, and proteinaceous yolk bodies. Furthermore, the female gamete is invested by various extracellular coats (e.g., vitelline envelope and follicular cells), thereby complicating the isolation of AL. These problems are compounded by the fact that AL appear to be cyclical or transitory in appearance in many cells including oocytes, and the timed sequence of this variability is not precisely known in most cases.

A variety of synonyms have been used to denote and describe annulate lamellae in the past, including those listed in Table I.

## 2. Distribution in Oocytes

Initial descriptions of annulate lamellae were most frequently based on observations of the fine structure of the cytoplasm of the female gamete—a feature that has probably resulted in these membranes being considered by some to be a rather specialized structure perhaps related to one or more specific events in oogenesis. It has now become apparent, however, that although AL may indeed be found in great abundance in developing female gametes of a wide variety of organisms (Table II), they can by no means be considered restricted in their distribution to these particular cells. In fact, a careful survey of the cell types in which AL have thus far been described indicates that many cell types can have them at some period in their developmental history, including such diverse cell types as epithelium, muscle, nerve, and secretory cells, as well as radiolarian and foraminiferan cells (reviewed by Kessel, 1983a).

**Table II.** Distribution of Annulate Lamellae in Oocytes

| | Reference |
|---|---|
| Planktic foraminifer | |
| *Hastigerina pelagica* (gametes, gamont) | Spindler and Hemleben (1982) |
| Cnidaria | |
| Hydrozoan jellyfish (Hydrozoa, Trachylina) | Kessel (1968c) |
| Platyhelminths | |
| *Gorgoderina attenuata* (trematode) | Koulish (1965) |
| *Hymenolepus diminuta* (cestode) | Lumsden (1965) |
| *Diplozoon paradoxum* (Trematoda, Monogenea) | Stranock (1979) |
| Acanthocephala | |
| *Moniliformis dubius* (spiny-headed worm) | Atkinson and Bryam (1976) |
| Annelida | |
| *Capitella jonesi* (polychete) | Eckelbarger and Grassle (1982) |
| *Promatoceros triqueter (serpulid worm)* | Gwynn *et al.* (1971) |
| *Spirorbus borealis* (serpulid worm) | Babbage and King (1970) |
| *Nereis diversicolor* (polychete) | Dhainaut (1966, 1973) |
| *Urechis caupo* (echiuroid worm) | Gould-Somero and Holland (1975) |
| Mollusca | |
| *Onithochiton quercinus* (polyplacophora, a chiton) | E. Sakker (personal communication) |
| *Mopalia mucosa* (an amphineuran) | Anderson (1969) |
| *Spisula soldissima (pelecypod)* | Rebhun (1956, 1961); Swift (1956) |
| *Otala lactea* (gastropod, pulmonate snail) | Swift (1956) |
| *Viviparus contectus* (gastropod) | Bottke (1973) |
| *Acmaea digitalis* (gastropod, limpet) | Kessel (1981a) |
| *Sepia officinalis* (cephalopod, a squid) | Dhainaut and Richard (1972) |
| *Ilyanassa obsoleta* (a gastropod) | Taylor and Anderson (1969) |
| Echinodermata | |
| *Thyone briareus* (holothuroid, sea cucumber) | Kessel (1964a, 1966b) |
| *Ophioderma panamensis* (ophiuroid, brittle star) | Kessel (1968b) |
| *Dendraster* (echinoid, sanddollar) | Merriam (1959) |
| *Arbacia punctulata, Strongylocentrotus,* and others (echinoid, sea urchins) | Afzelius (1957); Bal *et al.* (1968); Verhey and Moyer (1967); Sachs and Anderson (1970); Anderson (1970); Conway (1971); Conway and Metz (1974); Gross *et al.* (1960); McCullough (1952) |
| *Nemaster rubiginosa*, a crinoid | |
| Arthropoda Chelicerates | Dumont and Anderson (1971) |
| *Limulus polyphemus, Xiphosura* (horseshoe crab) | |
| *Ornithodorus moubata*, Isodoidea, Argasidae, (tick) | Aeschlimann and Hecker (1970) |
| Aquatic mandibulates | Kessel (unpublished observations) |
| *Uca pubilitor* (fiddler crab) | |
| *Artemia* (brine shrimp) | Anteunis (1964) |
| *Orchestia gammarella* (amphipod) | Meusy (1968) |
| *Oniscus asellus* (crustacean) | Beams and Kessel (1980) |
| *Triops cancriformis* (crustacea, notostraca) | Scanabissi and Trentini (1979) |

**Table II.** Distribution of Annulate Lamellae in Oocytes (*continued*)

| | Reference |
|---|---|
| Terrestrial mandibulates | |
| *Lithobius forficatus* L. (Myriapoda, Chilopoda) | Herbaut (1972) |
| *Polydesmus angustus* (Myriapoda, Diplopoda) | Petit (1973) |
| Insects | |
| *Anagasta kuhniella* (flour moth oocytes) | Cruickshank (1972) |
| *Libellula pulchella* (dragonfly) | Kessel and Beams (1969) |
| *Cordulia aenea* (dragonfly) | Halkka and Halkka (1977) |
| *Glossina* (fly) | Huebner *et al.* (1975) |
| *Drosophila* (fruit fly) | Mahowald (1962, 1963); Okada and Waddington (1959); Mahowald and Tiefert (1970) |
| *Nasonia vitripennis* and *Ophion luteus* (hymenoptera) | King and Richard (1968) |
| *Ophion luteus* (L.) and *Aparteles glomeratus* (L.) (parasitoid hymenopterans) | King and Fordy (1970) |
| *Xyleborus ferrugineus* (Fabr.), beetle | Chu *et al.* (1982) |
| Cephalochordata | |
| *Amphioxus lanceolatus* | Reverberi and DeLeo (1972) |
| Ascidians (Urochordata) | Kessel (1964b, 1965, 1968d); |
| *Ciona intestinalis*, *Sytela partita*, *Boltenia villosa*, and others | Everingham (1968); Mancuso (1964); Hsu (1967) |
| Teleosts | |
| *Salmo gairdneri* (rainbow trout) | Kessel (1981a) |
| Goldfish oocytes | Yamamoto and Onozato (1965) |
| Carp oocytes | Kessel (1968d) |
| *Acipenser stellatus* Pall. (sturgeon oocytes) | Aizenshtadt and Detlaff (1972) |
| Amphibians | |
| *Necturus maculosus* | Kessel (1963) |
| *Bufo bufo* | Humeau and Temple (1969); Kress (1982) |
| *Xenopus laevis* | Balinksy and Devis (1963); Dumont and Wallace (1972); Franke and Scheer (1970a,b, 1971) |
| *Ambystoma mexicanum* | Kress (1982) |
| *Cynops pyrrhogaster* (newt) | Imoh (1982) |
| *Triturus* | Wischnitzer (1970b); Humeau and Temple (1969); Kress (1982) |
| *Rana* sp. | Kessel (1969b); Wang and Hsü (1974); Ward and Ward (1975) |
| Birds | |
| Hen's oocyte | Bellairs (1967) |
| Reptiles | |
| Chamelion | Dunne (1965) |
| Turtle | Kessel (unpublished observations) |
| Mammals | |
| Chimpanzees | Barton and Hertig (1972) |
| Human (primary ooctye) | Baca and Zamboni (1967); Hertig and Adams (1967); Hertig (1968); Zamboni *et al.* (1972); Wartenberg and Stegner (1960); Tardini *et al.* (1960) |

(*continued*)

Table II. Distribution of Annulate Lamellae in Oocytes (*continued*)

| | Reference |
|---|---|
| Human (atretic) | Baker and Franchi (1967) |
| Hamster (atretic) | Weakley (1969) |
| Rabbit (pronuclei and early cleavages) | Zamboni and Mastroianni (1966a,b); Longo and Anderson (1969b); Merchant (1970) |
| Fertilized ova | |
| Hamster | Weakley (1969) |
| Rabbit | Zamboni and Mastroianni (1966a); Gulyas (1971, 1972) |
| Human | Zamboni *et al.* (1966b) |

## 3. Ooplasmic Position and Variability

Annulate lamellae commonly occur in stacked parallel array in oocytes (Figs. 5 and 6), but single lamellae and concentric lamellae (Figs. 1 and 13) have been noted as well. The spatial distribution of AL in oocytes does not appear to be consistent or rigid, since AL have been described in the perinuclear, intermediate, and peripheral regions of the ooplasm—sometimes in all three locations in the oocytes of the same organism. When AL are organized as stacks, the pores in adjacent lamellae are characteristically aligned and adjacent to each other. When stacks of AL are aligned adjacent to and parallel with the nuclear envelope, the pores in the nuclear envelope are also aligned with those in the AL (Figs. 1 and 5). Whereas stacks of AL are frequently aligned parallel to the nucleus (nuclear envelope), the stacked lamellae may also be aligned at right angles to the nuclear envelope. The Balbiani vitelline body or yolk nucleus of some oocytes may contain a stack of AL (e.g., Zamboni, 1971). In the chimpanzee oocyte (Barton and Hertig, 1972), one or more stacks of AL are located in a region consisting of a cluster of organelles known as the yolk nucleus or Balbiani body, but the remainder of the stacks of AL are distributed in the perinuclear cytoplasm.

In most oocytes, no quantitative information on the AL is available. However, Hertig and Adams (1967) and Hertig (1968) indicate that in the human oocyte only a single stack of AL is present, but in chimpanzees (Barton and Hertig, 1972) as many as five stacks of AL of 2–35 lamellae were observed in a single oocyte. However, Zamboni *et al.* (1972) cultured 159 human oocytes from 24 to 53 hr and studied them by means of serial sectioning for high-resolution microscopy. These workers indicated that AL were observed "only occasionally; they were present exclusively in meiotically inactive oocytes and only in those which displayed signs of degeneration or were anomalous."

Although both the number and length of individual lamellae are quite variable, the distance between adjacent lamellae in a stack of AL is rather constant at approximately 80–100 nm. The number of individual AL can range from one or only several to more than 100 in a single stack.

Variability in AL is not only represented in the number and length of lamellae as well as their location in the ooplasm, but the lamellae within stacks of AL can also exhibit variability in terminal connections. The stacks of AL are frequently connected with ribosome-covered membranes of the RER (Figs. 10 and 16). Less frequently, the AL may be continuous with smooth membranes of the SER. In still other cases, the lamellae extend or terminate as rounded, smooth-surfaced dilations (Fig. 6). In a few cases, ribosomes appear to be attached to the nonporous portion of the AL membranes (Figs. 9–11), but this situation appears to be rather rare. Possible variations in terminal connections of AL that have been noted rather infrequently are shown in Figs. 14 and 15; AL from small rainbow trout oocytes are shown, in which the ends of some individual lamellae may be connected with membranous rosettes having an electron-dense central region. The ends of other lamellae in a stack of AL may end in a complex of closely-spaced, smooth-surfaced membranous elements such as those shown in Fig. 15. The significance of these structural variations is not clear. Although not yet observed in oocytes, the AL in some cultured cells may extend as RER that, in turn, is continuous with SER (Kessel and Katow, 1984a). The membranes extending from individual AL in a stack occasionally appear to be fenestrated (Fig. 17), much like those described for the fenestrated Golgi saccules. The association of fenestrated membranes with AL such as those shown in Fig. 17 have been observed in both oocytes and nongerm cells, but with limited frequency.

The variability in terminal configurations of AL has led to a variety of functional postulates, including the possibilities that (1) AL may undergo vesiculation, (2) the RER or SER may form AL, or (3) AL may form RER or SER. These suggestions are explored in somewhat greater detail in Sections 5.2 and 5.3.

Relatively little information is available about AL in relationship to the developmental history of cells, particularly oocytes. What information is available tends to indicate that considerable variability exists with respect to the appearance and disappearance of AL in oocytes. For example, in *Rana pipiens* oocytes (Kessel, 1969b; Kessel and Subtelny, 1981), stacks of AL appear in the ooplasm before vitellogenesis and persist until the egg is fully grown. However, AL have been observed in synaptic germ cells of 6-month-old *Rana catebeiana* tadpoles by Wang and Hsü (1974), who indicate that AL are also present in 10- and 12-month-old tadpoles, but disappear in 15-month-old froglets. Annulate lamellae tend to appear before vitellogenesis, but are difficult to find in the ooplasm of full-grown dragonfly oocytes (Kessel and Beams, 1969). In a number of ascidian oocytes (e.g., *Ciona* and *Styela*), stacks of AL are observed in young previtellogenic oocytes and persists throughout vitellogenesis (Kessel, 1965, 1966a). During vitellogenesis, the ends of the lamellae extend as long ribosome-coated lamellae of RER. In nearly full-grown oocytes, however, stacks of AL are sparse, but stacks of RER are frequently observed. Annulate lamellae that appear in vitellogenic *Drosophila* oocytes persist in the cortical ooplasm of the mature egg and are present in early embryo and pole cells (Mahowald, 1962, 1963, 1968, 1971a,b; Okada and Waddington, 1959; Mahowald and Tiefert, 1970). In several amphibian oocytes (see Section 7.7), the breakdown of AL appears to be an initial event in maturation phenomena.

By contrast, in other oocytes, AL appear to be restricted to a shorter period of oogenesis. Electron microscopic studies have failed to detect AL in rabbit ovarian oocytes (Blanchette, 1961; Zamboni and Mastroianni, 1966a; Krauskopf, 1968a), but AL have been observed in fertilized eggs at 17 hr post coitum (Zamboni and Mastroianni, 1966b), at 22 hr after mating (Hadek and Swift, 1961; Longo and Anderson, 1969), and in blastomeres of the two-cell embryo (Krauskopf, 1968a,b). Gulyas (1971) pointed out that AL are not observed in rabbit follicular oocytes, unfertilized eggs, or fertilized eggs before and during formation of pronuclei. Thus, the developmental sequence of AL appears to be somewhat different in the rabbit as compared with other oocytes.

## 4. Comparative Features of Nuclear Envelope and Annulate Lamellae

The nuclear envelope characterizes and defines the eukaryotic cell. In addition to well-defined and apparently complex membrane interruptions—pores (Figs. 1–3 and 21)—the nonmembranous pore-associated material is intimately associated with the membranous interruption (Figs. 19 and 20). The pore-associated material extends through the pore and for a short, but somewhat variable distance into the surrounding nucleoplasm and cytoplasm (Figs. 19 and 20). It is not within the scope of this chapter to review all the details in the structure, relationships, and composition of the various granules and fibrils associated with nuclear pores or the various models that have appeared in an attempt to define the structural interrelationships of the pore. A number of review articles are available that investigate the nuclear envelope in considerable depth (e.g., Gouranton, 1969; Feldherr, 1972; Kaye and Johnston, 1973; Kessel, 1973; Franke, 1974; Maul, 1977a; Zbarsky, 1978; Harris, 1978).

The number of pores per square micrometer (pore frequency) in the nuclear envelope has been described to vary in different cell types from 0–3/$\mu m^2$, to more than 75/$\mu m^2$, whereas the relative nuclear surface area occupied by pores varies over a range of about 1–30% (Franke and Scheer, 1975).

Pore frequency is variable in both the NE and AL and can change during cell differentiation. Franke and Scheer (1970b) reported that the nuclear envelope pore frequency in lampbrush-stage oocytes of the newt *Triturus* is 68 ± 6.6, while it is 50 ± 6.4 in mature oocytes (see Chapter 9 for a discussion of the lampbrush stage). In general, in those cells in which the pore frequency in the nuclear envelope and AL has been compared, the pore frequency is higher in the AL—in some cases by a factor of 2 or 3 (Franke and Scheer, 1971). Pore configuration in the AL and NE may differ: It is more common to observe hexagonal arrays of pores when tangential sections of AL are observed than when nuclear pores are observed in similar profile (Fig. 8). However, orthogonal arrays of pores in AL are possible. The pore diameter in both the NE and AL ranges from about 60 to 100 nm.

Whereas many of the pores in both the NE and AL appear to be circular in outline, an octagonal symmetry has been reported by a number of investigators using both isolated and negatively stained preparations and by workers using high-resolution transmission electron microscopy. Although there is no universal agreement, octagonal symmetry (Abelson and Smith, 1968; Gall, 1967; Kessel, 1969a; Fisher and Cooper, 1967; Yoo and Bayley, 1967; Maul, 1971; Fabergé, 1973; Smith and Berlin, 1977) as well as spherical or oval shapes (Scheer, 1970; Ward and Ward, 1975) have been described for pores in the AL and the NE.

The pore-associated material of both NE and AL consists of a rather homogeneous, moderately electron-dense, nonmembranous material (Figs. 12 and 19). With sufficient magnification and resolution, some indication of subunit organization has been described (Kessel, 1968a,b; Kessel, 1969a; Kessel and Beams, 1969; Kessel, 1981a,b). A thin fibrillar component (~2–6 nm wide) and a small granular component (~6–14 nm in diameter) seem to be present in many cases within the pore-associated material of the AL and the nuclear pores as well (cf. Figs. 3, 20, 39, and 40). A more thorough discussion of the structure of the nuclear pore and nuclear pore models can be found elsewhere (e.g., Kessel, 1983a).

The information available about the composition of the pore complex is largely based on studies of NE pores using enzymatic digestion and extraction methods coupled with transmission electron microscopy. That the annular material (and central granule), including fibrils of nuclear pores, contains protein has been described (Merriam, 1961; DuPraw, 1965; Beaulaton, 1968a,b; Clérot, 1968; Koshiba *et al.*, 1970). Stick and Krohne (1982) have reported that the NE of *Xenopus laevis* contains what they call a "major architectural protein" resistant to extraction in high-salt buffer and nonionic detergent and characterized by a polypeptide molecular weight of 68,000. These investigators indicate that the polypeptide is a major karyoskeletal component of the pore complex. In a recent study concerned with the attempted isolation of the nuclear pore complex, Aaronson *et al.* (1982) concluded, "This evidence suggests to us that minor polypeptides or even presently undetected polypeptides are constituents of the pore complex and that its structure is, in fact, very complex." Other investigators have noted, on the basis of RNAse digestion or EDTA staining, that ribonucleic acid seems to be associated with the nuclear pores (Mentré, 1969; Bernhard, 1969; Monneron and Bernhard, 1969; Franke and Falk, 1970; Dhainaut, 1970a,c; Kessel, 1981b; Scheer, 1972, 1973). These results suggest the presence of ribonucleoprotein in the pore-associated material (for additional discussion, see Dhainaut, 1973; Franke and Scheer, 1974; Kessel, 1981a).

Clearly, the nuclear pores are important gateways in the translocation of macromolecules and particles of different size, including RNP. According to Matsuura *et al.* (1974), nucleolar RNP particles that contain rRNA and precursors of rRNA are pleomorphic (i.e., can assume various shapes) filamentous structures with nodular thickenings. Clawson and Smuckler (1982) have postulated that during translocation of ribonucleoprotein particles into the cytoplasm, localized regions of ribonucleoprotein "unfold"; these linear segments then react with a nucleoside triphosphatase, and the RNA chain is linearly translocated through the aqueous nuclear pore channel.

In addition, it was noted some time ago (Kessel, 1968a, 1969a) that small granular and fibrillar subunits of similar sizes are associated in the pores of NE and AL as well as in the nucleoli of certain oocytes (see also Matsuura *et al.*, 1974). Scheer (1972) found that isolated nuclear envelopes of *Xenopus* oocytes contain considerable amounts of membrane-attached RNA or RNP. In fact, about 9% of the nuclear RNA in mature *Xenopus* oocytes appears to be associated with the nuclear envelope. Scheer (1972) subjected manually isolated nuclear envelopes of mature *Xenopus* oocytes to biochemical analysis and found each nuclear pore complex to contain ribonucleoprotein with $41 \times 10^{-6}$ pg RNA. It appears that a considerable part of the nuclear envelope pore complex material is ribosomal RNP, but little information is available about the biochemical nature, labeling kinetics, or molecular weight of the nuclear pore complex RNA (Scheer, 1972, 1973; Franke and Scheer, 1974). It has been suggested that pre-rRNA molecules in the later stage of processing are enriched in isolated nuclear envelopes (cf. discussion in Franke and Scheer, 1975) and that processing and assembly stages of RNAs and associated proteins might be associated with the nuclear envelope—perhaps the pore complexes.

Ribosomes are attached to the outer membrane of the NE in some cells (Fig. 4), and they appear to be transitory stages in the movement of cytoplasmic ribosomes (Bach and Johnson, 1966; Smith *et al.*, 1969). It has been suggested that newly produced messenger is translated on nuclear envelope polyribosomes before appearing in association with polyribosomes attached to the RER (Avrameas and Bouteille, 1968; Leduc *et al.*, 1968; Avrameas, 1970). Fibrils and granules in the pore-complex material have been likened to initiator complexes consisting of mRNA and smaller ribosomal subunits (Joklik and Becker, 1965; Latham and Darnell, 1965; McConkey and Hopkins, 1965; Henshaw *et al.*, 1965). The pores of the NE may play a role in polyribosomal assembly (reviewed in Kessel, 1973, 1983a). It has also been suggested that polyribosomal formation begins at the outer annulus of the nuclear pore complex (Mepham and Lane, 1969; Jacob and Danieli, 1972; Danieli and Burighel, 1974; Hanzely and Olah, 1973) on the basis of variable associations of polyribosome chains with the annulus.

Consideration has also been directed to the possibility that the fibrils in the pore-associated material could represent informator complexes or informosomes, that is, mRNA complexed with nonribosomal informofer protein (cf. Samarina *et al.*, 1967, 1968; Infante and Nemer, 1968; Lukanidin *et al.*, 1972). Finally, further discussion has surrounded the possiblity that the nuclear pore complex might be involved in cleaving polycistronic 40–30 S RNA-containing RNP strands (Rogers, 1968) to smaller RNAs (Brown, 1967) or cleaving a polycistronic messenger into smaller messengers (Scherrer *et al.*, 1966; Scherrer and Marcaud, 1968). For example, it has been suggested that the nuclear constituents of various kinds (most commonly various kinds of RNAs are considered) could be channeled through the pores of the nuclear envelope and into the pores of the adjacent stacked AL, which could then be considered storage depots for material that might subsequently be dispersed into the peripheral ooplasm as the stacks of AL migrate to these regions and perhaps then undergo breakdown or vesiculation. It is once again of interest to point out the similarity in the pore-associated material of both the nuclear pores and AL pores and the fact that the pores in the two membrane systems tend to be aligned in precise register.

Central granules (also called intra-annular granules) were noted to be especially numerous in pores of germinal vesicle nuclear envelopes in amphibian oocytes at a time of intense RNA synthesis, and it has been found (Scheer, 1970) that actinomycin D, which blocks RNA synthesis, causes a reduction in frequency of central granules in nuclear pores (46–10% after 50-min treatment with antibiotic).

It seems clear that the pore structure, whether located in the NE or amplified in the cytoplasm in the form of stacks of AL, is identical regardless of its location. However, some variability apparently exists in size and in number of granular and fibrillar subunits. The frequency of pores in AL tends to be higher than that of the NE. As will be discussed in more detail subsequently, it has been proposed that the function of nuclear pores and AL pores may be similar, but that AL may represent an amplification of pore function—one that in some cells may be exclusively performed by nuclear pores, since AL appear to be lacking in some cells (Kessel, 1981a,b, 1982, 1983b).

Information about the chemical nature of the pores, either in the nuclear envelope or AL, is rather sparse. However, more extensive information is becoming available about the biochemical composition of the NE (cf. reviews by Scheer, 1972; Eckert *et al.*, 1972; Kessel, 1973; Franke, 1974; Maul, 1977a; Feldherr, 1972; Harris, 1978). On the basis of transmission electron microscopic cytochemistry, alkaline phosphatase activity has been described in association with the pore-associated material of nuclear pores, primarily in human neoplastic salivary gland tissue (Cutler *et al.*, 1974). Biochemical data (Franke *et al.*, 1970) also suggest alkaline phosphatase activity in the NE. Alkaline phosphatase activity has also been reported for the fibrogranular material (but not AL *per se*) that comprises the heavy bodies in sea urchin eggs (see Section 5.6). $Na^+$, $K^+$-stimulated $Mg^{2+}$-ATPase activity has been reported for AL on the basis of cytochemical techniques applied to transmission electron microscopy (Scheer and Franke, 1969) in oocytes of the newt. Adenosine triphosphatase activity, again based on transmission electron microscopic cytochemistry, has been described in the pore-complex material of both NE (Yasuzumi and Tsubo, 1966) and AL (Scheer and Franke, 1969). A preferential localization of lanthanum to the nuclear pore complexes of cultured cells was described recently by Shaklai and Tavassoli (1982), but no localization was found to be associated with pores of AL in these cells. Binding of negatively charged particles to the nuclear pore complex of frog ooctyes was described by Feldherr (1974). Binding was thought to be due to fixed positive charges associated with protein components of the annuli.

Annulate lamellae were early noted to be basophilic in their staining (Swift, 1956; Rebhun, 1956; Ruthmann, 1958), and the basophilia was shown to be ribonuclease sensitive. Other studies ranging from biochemical analysis to either enzymatic digestion and/or extraction have suggested that the pores in the AL (as with the NE) have ribonucleoprotein associated with them (Mentré, 1969; Franke and Falk, 1970).

In autoradiographic studies of *Drosophila* oocytes, Mahowald and Tiefert (1970) found that a period of active RNA synthesis occurs during a short period of vitellogenesis (stage 10A and early stage 10B). These workers were unable to specify the nature of this special RNA made by the oocyte at this time but suggested that it did not appear to be ribosomal. When stage 10B ooctyes were examined by transmission electron microscopy, Mahowald and Tiefert (1970) observed stacks of AL closely positioned adjacent to the nucleus as well as small projections of nucleoplasm adjacent to the AL.

Tritiated thymidine was observed by Gwynn *et al.* (1971) to be incorporated by stacks of cytoplasmic annulate lamellae in the serpulid worm, *Promatoceros*. However, in studies on rabbit zygotes, Maul and Cross (1974) found dense label over intranuclear AL but no significant label over the cytoplasmic AL after [$^3$H]thymidine incorporation.

## 5. Origin and Morphogenesis of Annulate Lamellae and Associations with Other Organelles

### 5.1. Nuclear Envelope

Because of the structural similarity between the nuclear envelope and annulate lamellae, together with the frequent position of stacks of AL adjacent to and parallel with the nuclear surface, as well as the actual continuity of the two membrane systems in some cases (Hertig, 1968), it is not surprising that many of the earlier investigators indicated that AL had their origin from the NE. The rather varied mechanisms proposed have included delamination, folding, and vesiculation (Swift, 1956; Bal *et al.*, 1968; Kessel, 1963).

An initial detailed study of the origin of AL was made on the oocytes of the neotenous salamander, *Necturus maculosus* (Kessel, 1963). In previtellogenic oocytes, the outer membrane of the nuclear envelope is often repeatedly folded into the adjacent cytoplasm, which may be filled with many vesicles, most of which appear in the form of long chains of attached vesicles. Annulate lamellae were described to form by the specialized fusion process of adjacent vesicles in the long vesicular chains. These initial studies were made on oocytes preserved only with a fixative of buffered osmium tetroxide. More recent investigations (cf. Kessel, 1983a) of these oocytes using a primary fixative of buffered glutaraldehyde, followed by secondary fixation in buffered osmium tetroxide, have shown that long membranous lamellae extend from the outer layer of the nuclear envelope into the surrounding ooplasm (Figs. 23 and 24). Vesicles are only sparsely observed. These long smooth-surfaced membranes appear to be elaborated by the outer membrane of the NE and released into the surrounding ooplasm, which progressively fills with many smooth, membranous lamellae that become more numerous in the distal or peripheral regions of the ooplasm (Figs. 23–25). These marked events in membrane elaboration in the *Necturus* ooctye occur before vitellogenesis. Thus, AL differentiation is a relatively early event in cytoplasmic oogenesis. When glutaraldehyde is the primary fixative for the *Necturus* oocyte, rounded expansions of the outer nuclear membrane are sometimes observed, but they are much less frequent than when osmium tetroxide is the primary and only fixative. Whereas single AL may be located adjacent to the nucleus, small stacks of AL are more commonly encountered in the peripheral ooplasm (Fig. 25). This finding suggests that the membrane elaborated by the NE may migrate into other areas of the ooplasm before differentiating into AL (i.e., before pore formation occurs). Such a condition is in contrast to reports in other oocytes that AL differentiate very close to the NE from a delamination or folding process. For example, Gulyas (1971) has indicated for rabbit pronuclear stages that vesicles presumed to be derived from the pronuclear envelopes accumulate in the cytoplasm and then give rise to AL, which are located at random sites near the pronuclei. Several other studies on various oocytes and other cell types have suggested that AL form or differentiate from vesicles that are budded from the outer layer of the NE (Kessel, 1964a; Harrison, 1966; Smith and Berlin, 1977).

Other studies interpreted that the annulate lamellae form by folding and detachment of nuclear envelope (in *Arbacia* oocytes, Bal *et al.*, 1968; in oocytes of a serpulid worm, Gwynn *et al.*, 1971; in human oocyte, Hertig, 1968). For example, the AL in *Arbacia punctulata* eggs are described as involving a folding of the nuclear envelope. In the mature *Arbacia* egg, the stacks of AL are scattered throughout the cytoplasm without preferential localization as described by Bal *et al.* (1968). These workers indicate that the formation of AL after vitellogenesis from folds that detach from the NE could result in the marked reduction in nuclear volume during maturation, ie., the germinal vesicle in *Arbacia* is reduced to 1/40th its original volume (Harvey, 1956).

### 5.2. Rough-Surfaced Endoplasmic Reticulum

In the past, a consistently observed feature was that annulate lamellae and endoplasmic reticulum are often connected or continuous. This led to the view that AL represent a form of ergastoplasm (Swift, 1956) or endoplasmic reticulum. Several studies, including those on developing oocytes, have interpreted variations in the associations between AL and ER to indicate that AL may originate from cisternae of the ER. This has been suggested, for example, in the case of *Nereis* oocytes (Dhainaut, 1973), *Ciona* oocytes (Mancuso, 1964), and human follicular oocytes (Baca and Zamboni, 1967). Zamboni *et al.* (1972) described parallel laminae in the cytoplasm of human oocytes. These laminae did not contain pores, and the most external lamina of each array was typically coated with or covered by ribosomes. Zamboni *et al.* (1972) thought these parallel membranes were a form of AL and believed that "these laminae, which were consistently associated with numerous ribosomes, may represent a form of transition or degradation of ergastoplasmic reticulum elements into annulate lamellae." Spindler and Hemleben (1982) indicated that during gametogenesis of a planktic foraminifer, *Hastigerina pelagica,* the AL were formed at Golgi complexes, where the ER forms vesicles that later transform into AL. Finally, Scheer and Franke (1972) postulated that AL cisternae originate as a degenerative form of ER.

By contrast, some investigators have postulated that AL may be an intermediate stage in the formation of some RER, i.e., that AL can transform into or give rise to RER. This suggestion has been especially prevalent in the case of some oocytes (Babbage and King, 1970; Fonzo and Esponda, 1974; Cruickshank, 1972) as well as in other cell types (e.g., Elliott and Arhelger, 1970; Jasper, 1976). Although not typically observed in oocytes, spermatogenic cells of the honeybee (Hoage and Kessel, 1978) and a number of somatic cell types show single pore complexes in the RER lamellae (Kessel, 1983a). The appearance of one or several pore complexes in such membranes suggests that perhaps pores can form *de novo* within extant membranes. Annulate lamellae and RER have been described (Kessel, 1982) in close relationship with the surface or lipid yolk droplets in the oocytes of the mollusc, *Acmaea digitalis* (Figs. 50–52).

### 5.3. Smooth-Surfaced Endoplasmic Reticulum

Several instances in which annulate lamellae and smooth-surfaced endoplasmic reticulum are continuous have been noted in oocytes as well as in cells other than oocytes. Observations that the SER and the AL may exhibit continuity perhaps have contributed to several reports that the SER gives rise to the AL. This has been the case for the cystocarp of an alga (Wetherbee *et al.*, 1974), cultured MO cells (DeBrabander and Borgers, 1975), and B16 melanoma cells (Hu, 1971). On the basis of ultrastructural changes occurring in *Rana catesbeiana* oocytes after heat treatment, Wang *et al.* (1980) suggested that stacks of AL move from the perinuclear position to a peripheral position in the ooplasm. In addition, AL were described to exhibit a transformation into SER, and with further heat treatment, neither AL nor the SER was observed.

### 5.4. Golgi Apparatus

There are several reports that annulate lamellae have their origin in the Golgi apparatus. In a study of cultured melanotic melanoma (H-19) cells, Maul (1970b) indicated "that annulate lamellae are transformed fenestrated cisternae of dictyosomes because of resemblance in size and arrangement of perforation." In the stellate sturgeon oocytes, the ends of AL were described as being expanded into large, rounded vesicle-like structures. Aizenshtadt and Detlaf (1972) reported that "budding" took place from the terminal expansions of the AL, and a similarity was noted between these blebs and surrounding vesicles of the Golgi system. Therefore, the authors thought that an intimate relationship, perhaps even a formative one, exists between the AL and adjacent Golgi complexes in the sturgeon oocytes at particular periods of development. In fact, these workers suggested that in maturing stellate sturgeon oocytes the annulate lamellae are, "functionally a system supplementary to the Golgi complexes." Baca and Zamboni (1967) also suggest a Golgi apparatus origin for the AL in a mammalian oocyte.

### 5.5 Mitochondria

Close relationships have been observed on occasion between annulate lamellae and other cytoplasmic constituents, including mitochondria. A close relationship between mitochondria and AL has been noted in the fertilized egg of the annelid, *Spirorbis* (Babbage and King, 1970), as well as in other cell types. In cases where this relationship has been noted, it has been felt that such relationships reflect a stage of formation of mitochondria from AL.

**Figure 1.** Nuclear envelope (NE) and annulate lamellae (AL) shown connected to smooth endoplasmic reticulum (ser), in young trout oocyte. ×48,000.

**Figure 2.** Perpendicular section showing two nuclear pores (arrows) in side view. Nucleus is toward top of figure, cytoplasm in lower part of figure. CH, chromatin; R, ribosome. ×100,000.

**Figure 3.** Tangential section of nuclear envelope showing nuclear pore in surface view. Note small granular and fibrillar subunits (arrows) comprising the pore-associated material. × 320,000.

**Figure 4.** Nuclear envelope with ribosomes (R) on outer nuclear membrane next to cytoplasm. ×100,000.

**Figure 5.** Nuclear envelope (NE) and part of a stack of annulate lamellae (AL) in echinoderm oocyte. ×32,800.

**Figure 6.** Stack of annulate lamellae from *Rana pipiens* oocyte. Note smooth membranous expansions (arrows) at end of some lamellae. ×32,800.

**Figure 7.** Portion of a stack of AL illustrating several pores (po) in side view and the nature and disposition of the pore-associated material. × 125,000.

**Figure 8.** Several surface views of AL pores (po) illustrating hexagonal packing of the annuli in this case. Note intra-annular granules (arrows) associated with at least two of the pores. ×125,000.

**Figure 9.** Annulate lamellae showing ribosomes (R) attached to the membranes. Note the many polyribosomes surrounding the AL that appear to be in contact (arrows) with the pore-associated material. ×100,000.

**Figures 10–12.** Fibrillar and granular nature (*) of the pore-associated and interlamellar material in annulate lamellae. R, ribosomes; ER, rough-surfaced endoplasmic reticulum. Note that polyribosomes are closely associated with the pore-associated material at arrows in Fig. 10. ×100,000.

**Figure 13.** Circular concentric rings of AL are present in some cells, such as the rainbow trout oocyte. ×23,600.

**Figure 14.** The ends of some AL extend so as to connect with rosettes of membranes surrounding a central electron-dense region (arrows). ×48,000.

**Figure 15.** Ends of some AL are connected to closely packed membranous lamellae (arrows). ×32,800.

**Figures 16–18.** Annulate lamellae from zebrafish oocytes. AL are connected to rough endoplasmic reticulum (ER) in Fig. 16, to fenestrated membranes (FM) in Fig. 17, and to smooth endoplasmic reticulum (SER) in Fig. 18. Figs. 16, 17, ×30,250; Fig. 18, ×62,250. (Courtesy of Dr. H. W. Beams.)

**Figures 19 and 20.** Nuclear envelope (NE) of full-grown *Rana pipiens* oocyte, showing nature of the pore-associated material (arrows), which is enlarged in Fig. 20 so as to show the extent and fibrogranular nature. Fig. 19, ×62,250; Fig. 20 ×125,000.

**Figure 21.** Freeze-fracture replica of the nuclear envelope showing nuclear pores in surface view. ×62,250.

**Figure 22.** Stack of annulate lamellae in *Styela* oocyte. Note the membranous vesicles (V) located between the lamellae. ×62,250.

**Figures 23 and 24.** Sections of *Necturus* oocyte nuclear envelope, illustrating continuity (arrows) of outer nuclear membrane with smooth lamellae that extend into the perniuclear ooplasm. N, nucleus. Fig. 23, ×62,250; Fig. 24, ×19,500.

**Figure 25.** Early stage in differentiation of annulate lamellae (AL) from long membranous lamellae (ML) in the peripheral ooplasm of the *Necturus* oocyte. ×19,500.

**Figure 26.** Stack of annulate lamellae in the animal pole ooplasm of a full-grown *Rana pipiens* oocyte (osmium tetroxide fixation). ×18,800.

**Figure 27.** Large cluster of smooth membranous vesicles (osmium tetroxide fixation) in animal pole cytoplasm of full-grown *Rana pipiens* oocyte that had been cultured for 12 hr in progesterone. Annulate lamellae are not observed in oocytes treated for this time interval. ×18,880.

**Figures 28–33.** Heavy bodies in developing oocytes of the purple sea urchin, *Strongylocentrotus purpuratus*. The heavy bodies consist of a fibrogranular material (*) and annulate lamellae (AL) and may be located in all regions of the ooplasm, but apparently originate near the nuclear envelope (NE). Fig. 28, ×2400; Fig. 29, ×14,000; Fig. 30, ×15,500; Fig. 31, ×24,000; Fig. 32, ×48,000; Fig. 33, ×30,250.

**Figure 34.** Portion of a fibrogranular body in peripheral ooplasm of previtellogenic dragonfly oocyte, showing fibrillar and granular subunits. ×125,000.

**Figure 35.** Stage in which five annulate lamellae (AL) have differentiated within a fibrogranular body. Note that lamellae in this case do not extend into surrounding ooplasm. ×23,6000.

**Figure 36.** Differentiation of annulate lamellae (AL) in a fibrogranular body of dragonfly oocyte. Areas of possible nascent pores (arrows) are identified. ×50,150.

**Figures 37 and 38.** Later stages in the differentiation of annulate lamellae within fibrogranular bodies of dragonfly oocytes. Fig. 37, ×17,400; Fig. 28, ×32,800.

**Figure 39.** Stage in differentiation of pores (po) in a single annulate lamella (AL) within a fibrogranular body. The fibrils and granules are denoted by unlabeled arrows. Ribosomes (R) are located around the fibrogranular body. ×120,000.

**Figure 40.** Surface view of pores (po) during differentiation of AL within fibrogranular body. Note that the centers of the pores contain variable numbers of small, electron-dense granules and fibrillar elements (arrows), a feature that suggests recruitment of the granular and fibrillar subunits from the adjacent material. ×138,000.

**Figures 41 and 42.** Stages in differentiation of annulate lamellae (AL) in fibrogranular bodies of dragonfly oocyte. Note several microtubules (arrows) that appear within the fibrogranular body and what appears to be a polyribosome (double arrows) also within the fibrogranular body. Note also the electron-lucent halo surrounding the microtubules. Fig. 41, ×62,250; Fig. 42, ×100,000.

**Figure 43.** Annulate lamellae in dragonfly oocyte. Note two areas (arrows) in which pore-associated material appears to be connected to adjacent polyribosomes (R). Short ribosome-covered membranes (ER) extend beyond the fibrogranular material in the case of some of the lamellae. ×62,250.

**Figure 44.** Stack of annulate lamellae in dragonfly oocyte. Note regions in which polyribosomes (R) are in contact with the fibrogranular body material or with pore regions (arrows). ×100,000.

**Figures 45–47.** Stages in differentiation of annulate lamellae (AL) in fibrogranular bodies of dragonfly oocytes. During oogenesis, the amount of fibrogranular material decreases as the annulate lamellae become less extensive. However, the rough-surfaced endoplasmic reticulum (ER) becomes more extensive. Fig. 45, ×62,250; Fig. 46, ×19,500; Fig. 47, ×62,250.

**Figure 48.** Portion of a fibrogranular body (FGB) in a spermatid of *Drosophila*. At this time, several annulate lamellae (AL) have differentiated within the FGB, and in these regions there is decreased density. Note also several microtubules (MT) in transverse section within the fibrogranular body. ×62,250.

**Figure 49.** Alignment of individual lamellae within a stack of annulate lamellae in a teleost oocyte. Note that all lamellae are continuous at each end with lamellae of rough-surfaced endoplasmic reticulum (ER). ×62,250.

**Figure 50–52.** Annulate lamellae (AL) are closely associated with localized areas of the periphery of lipid (L) droplets in certain snail oocytes (*Acmaea digitalis*). The annulate lamellae extend as rough-surfaced endoplasmic reticulum (ER). Fig. 50, ×48,000; Fig. 51, ×62,250; Fig. 52, ×78,750.

**Figure 53.** A fibrogranular body (FGB) in a sow bug oocyte is traversed in different regions by single annulate lamellae (AL) of different lengths. ×69,400.

**Figure 54.** Section of *Ciona intestinalis* oocyte illustrating intranuclear annulate lamellae (IAL) in the periphery of the nucleoplasm. Nuclear envelope (NE). ×17,400.

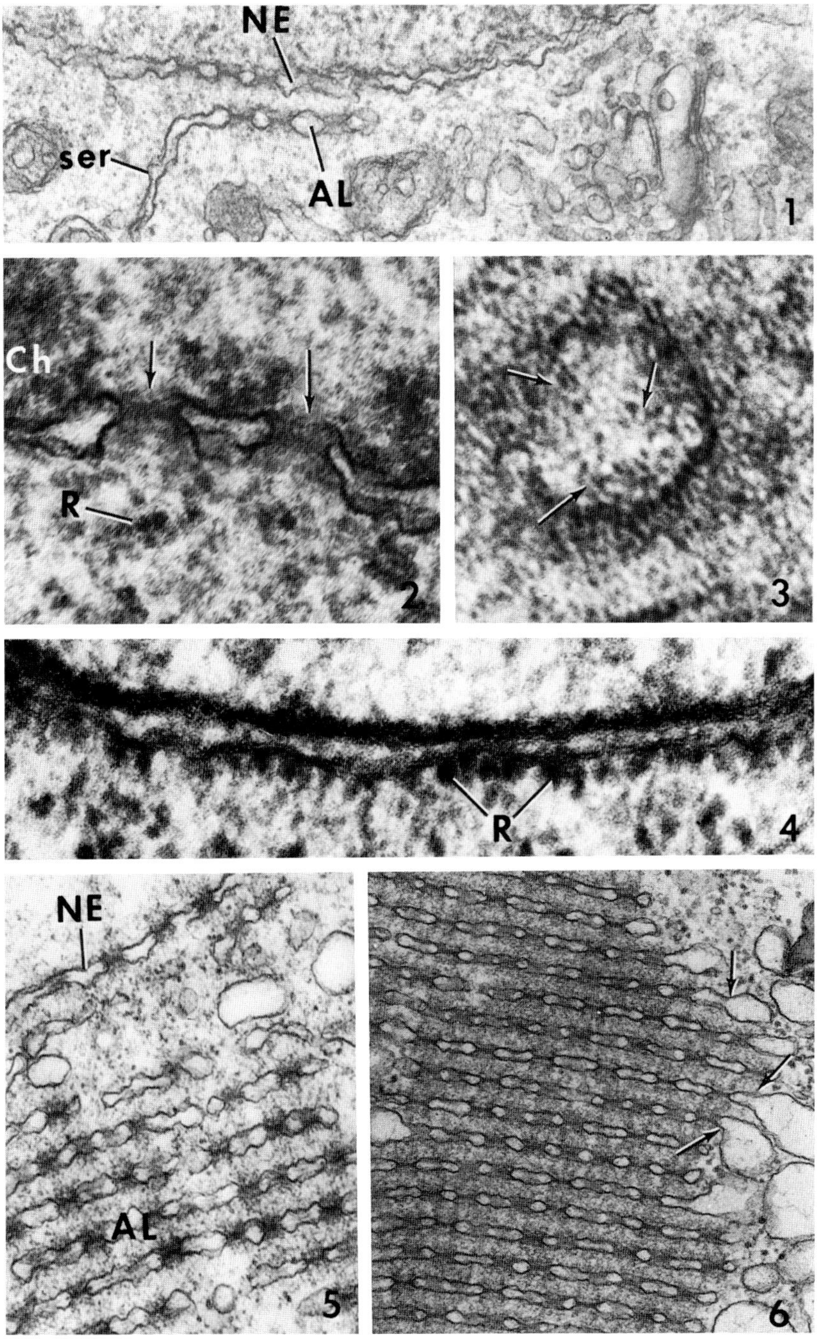


**Figures 1–6**

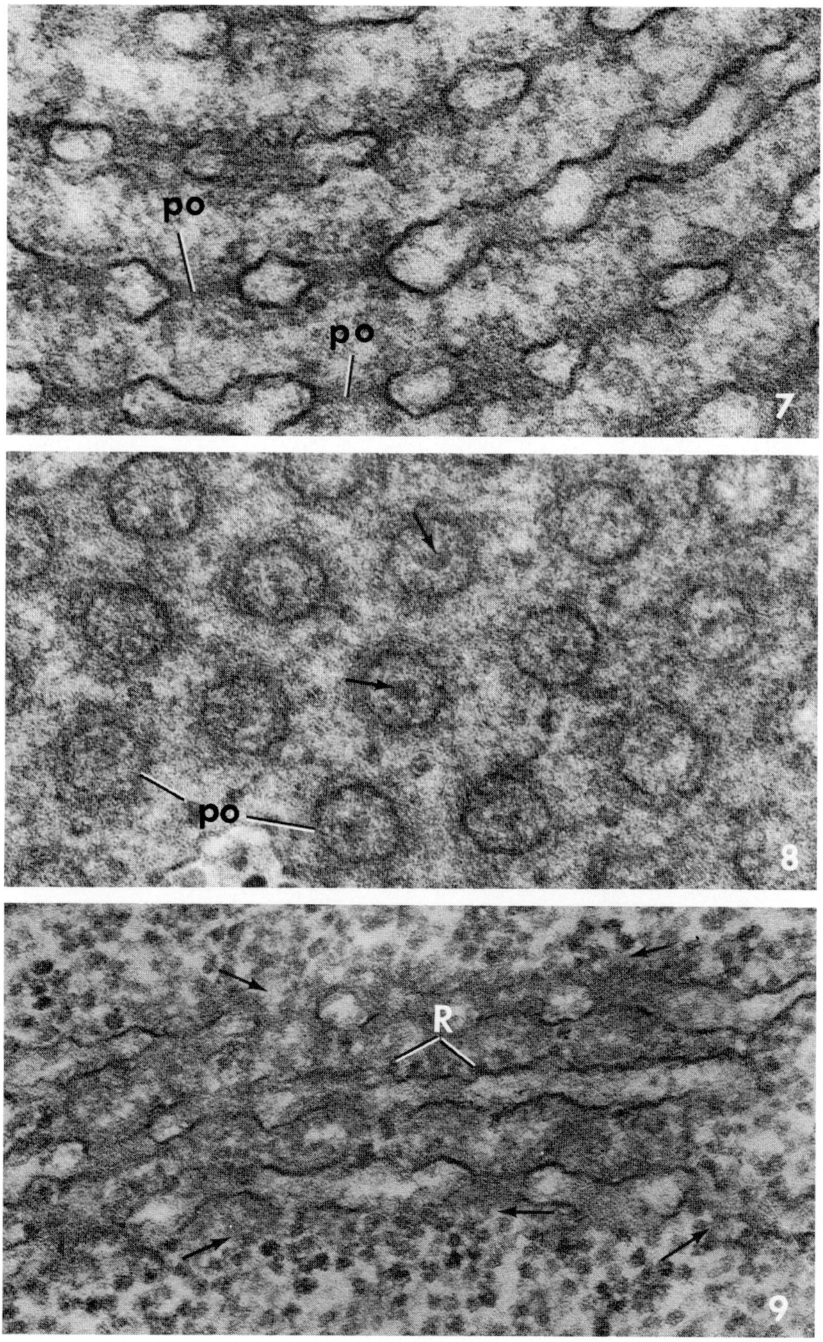


Figures 7–9

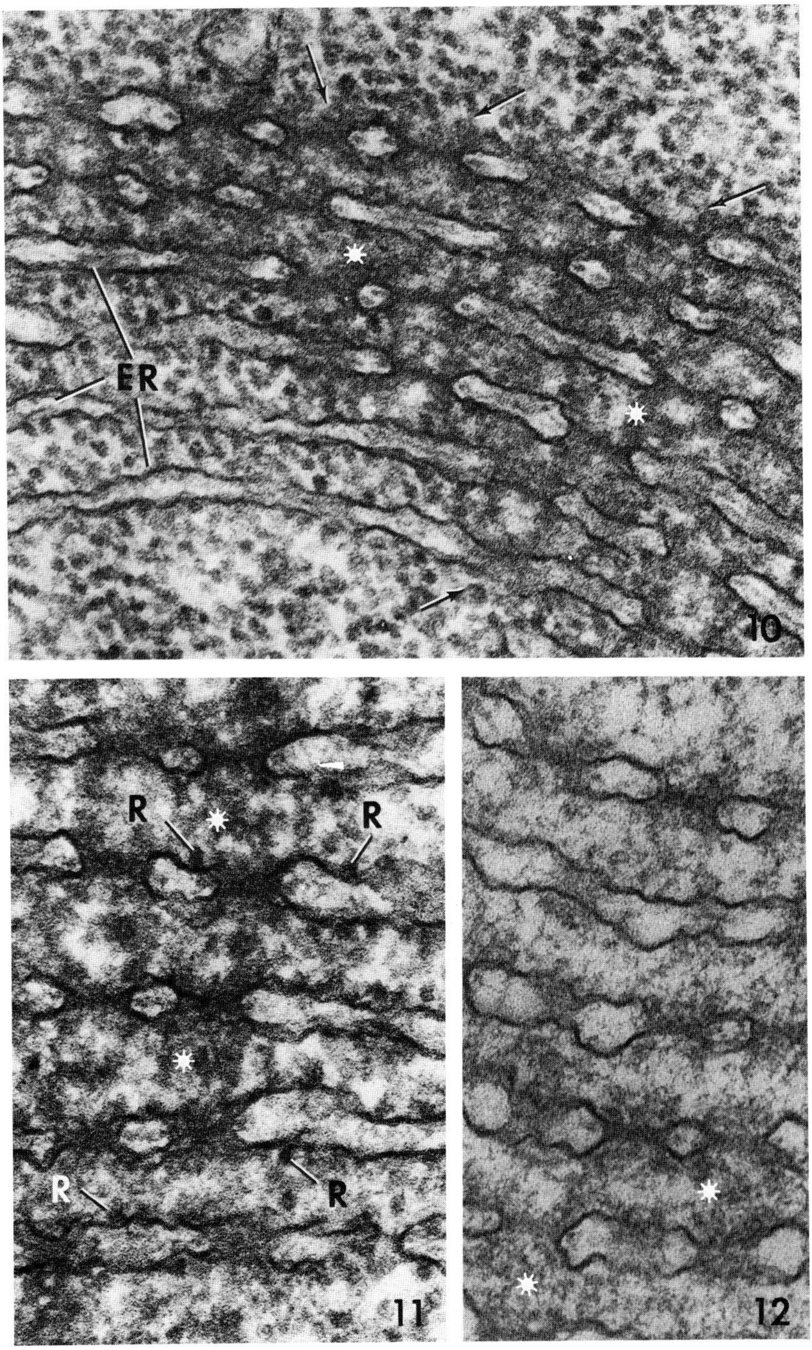


Figures 10–12

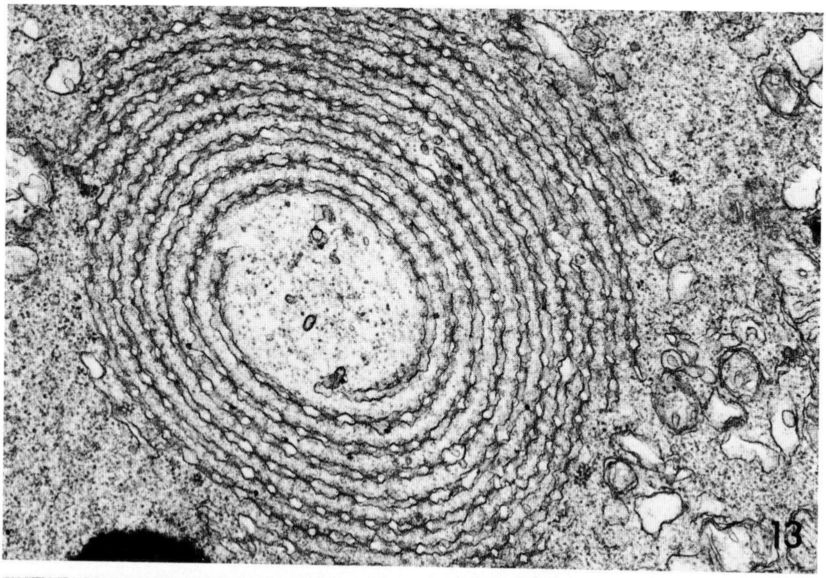


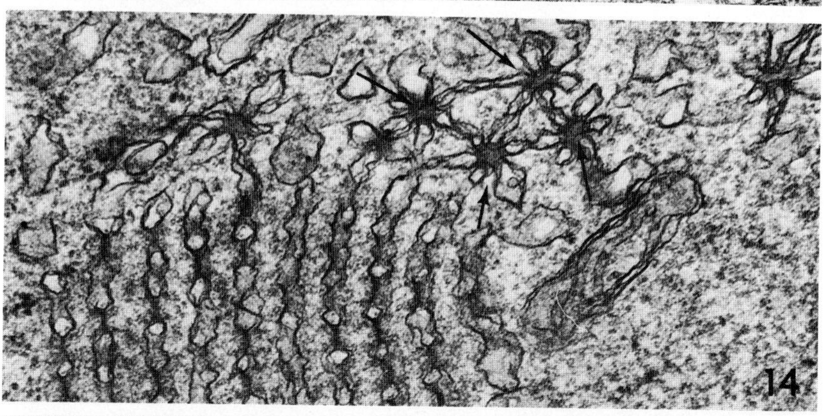


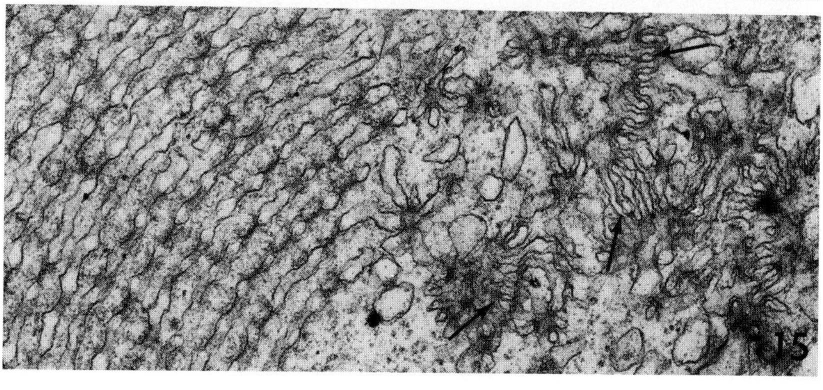


**Figures 13–15**

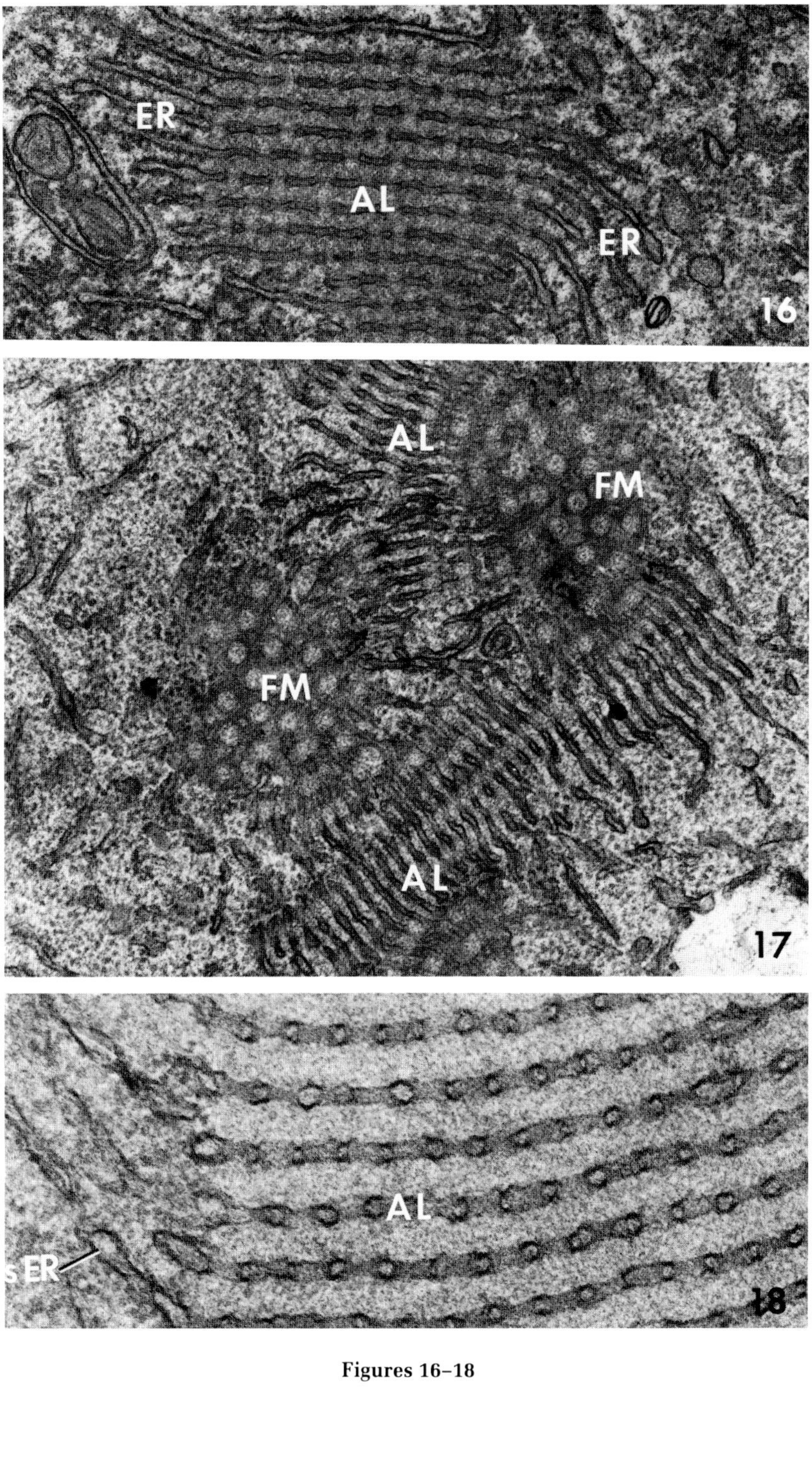


**Figures 16–18**

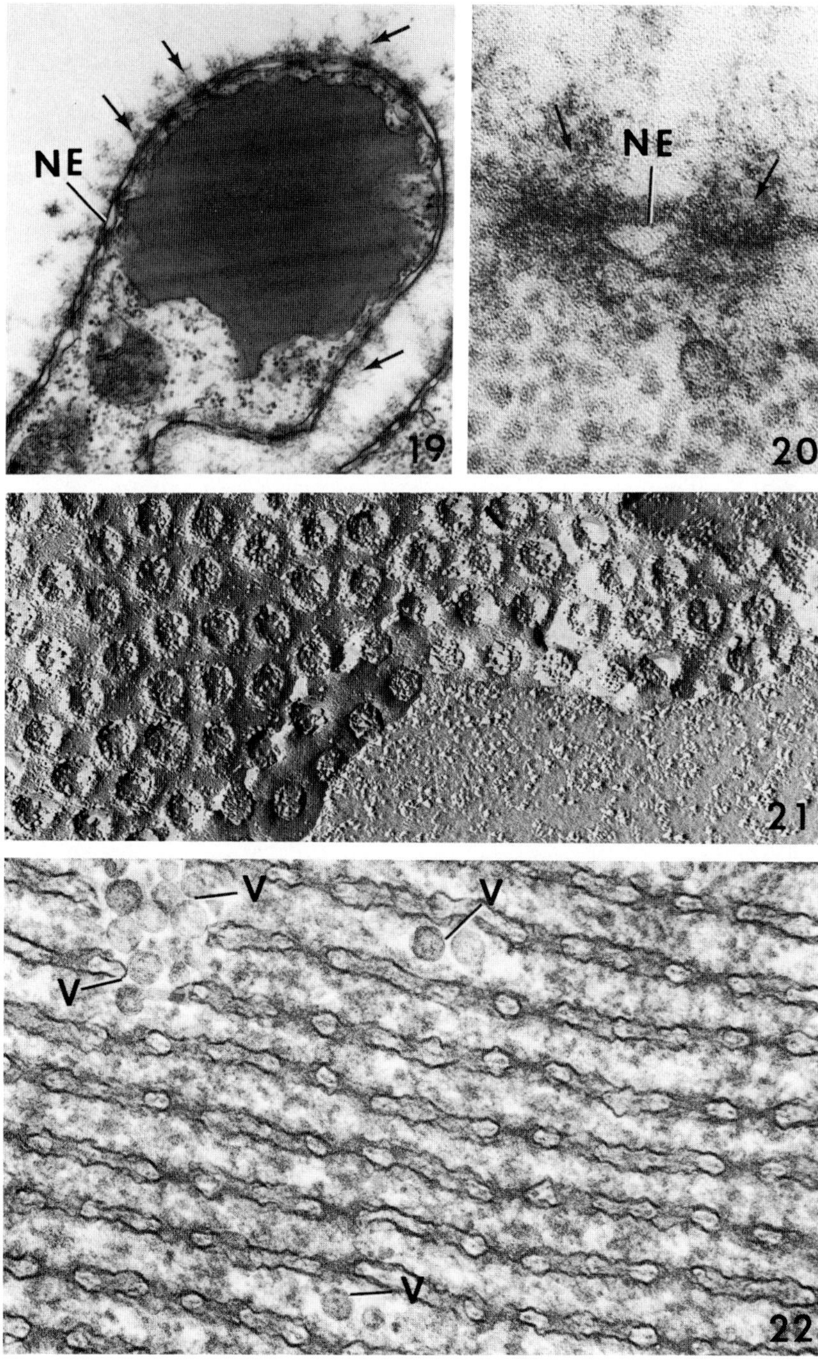


**Figures 19–22**

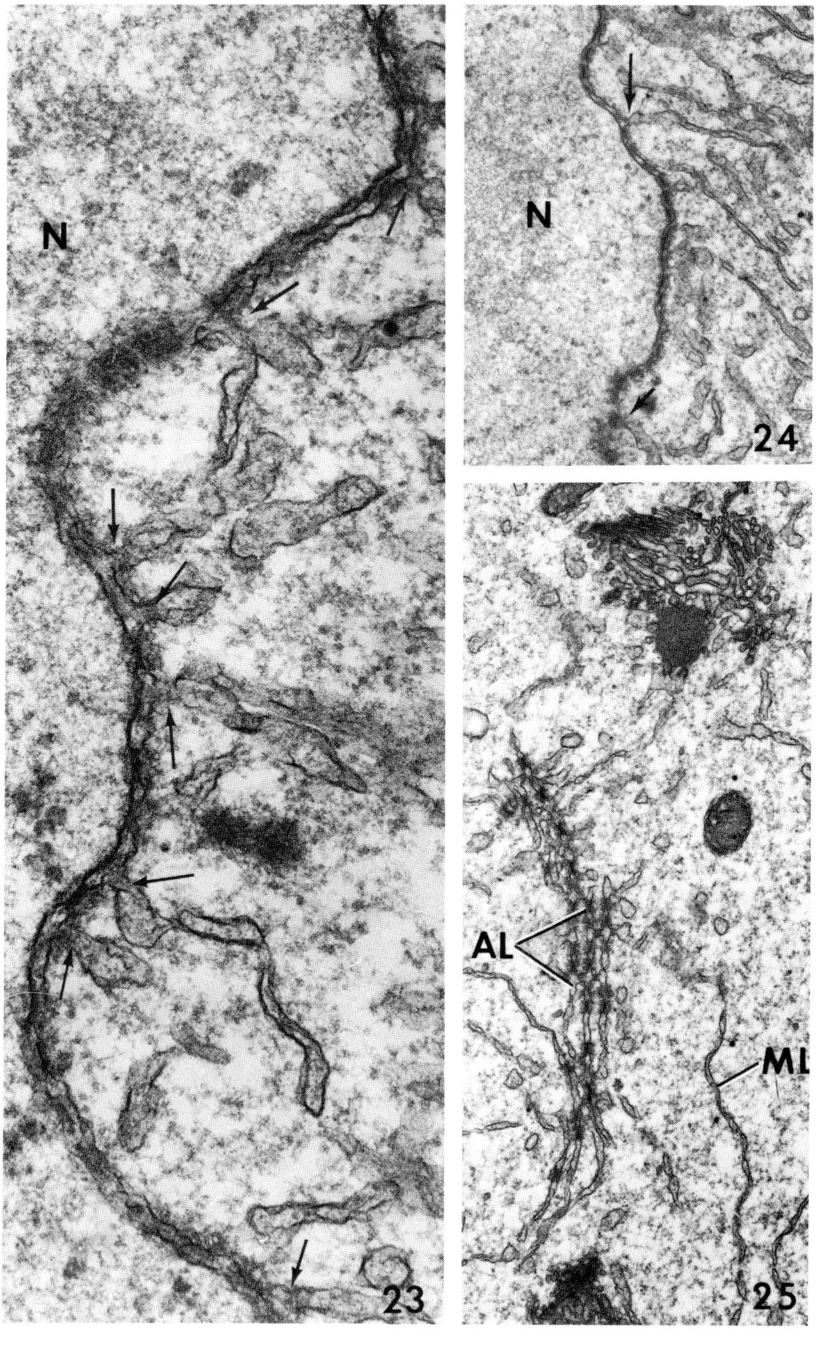


**Figures 23–25**

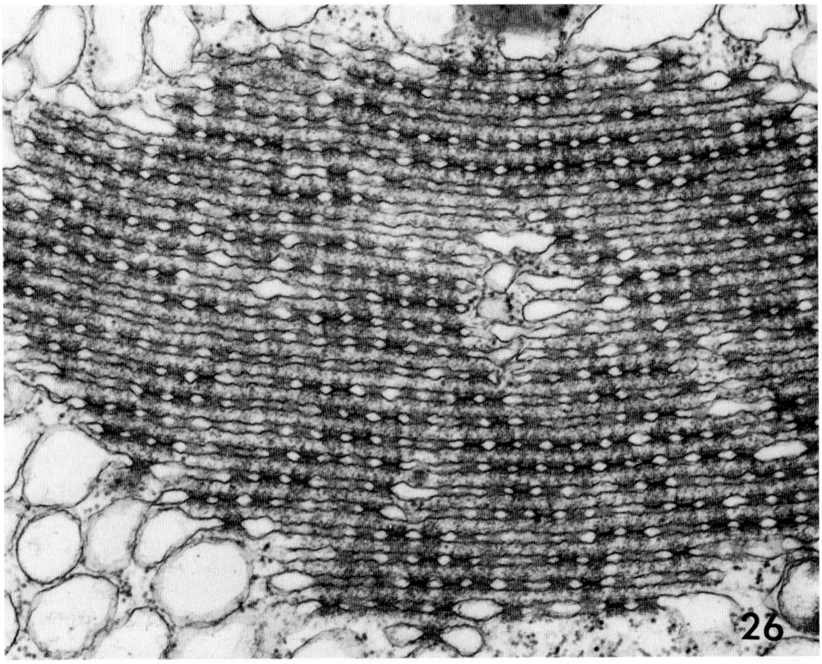


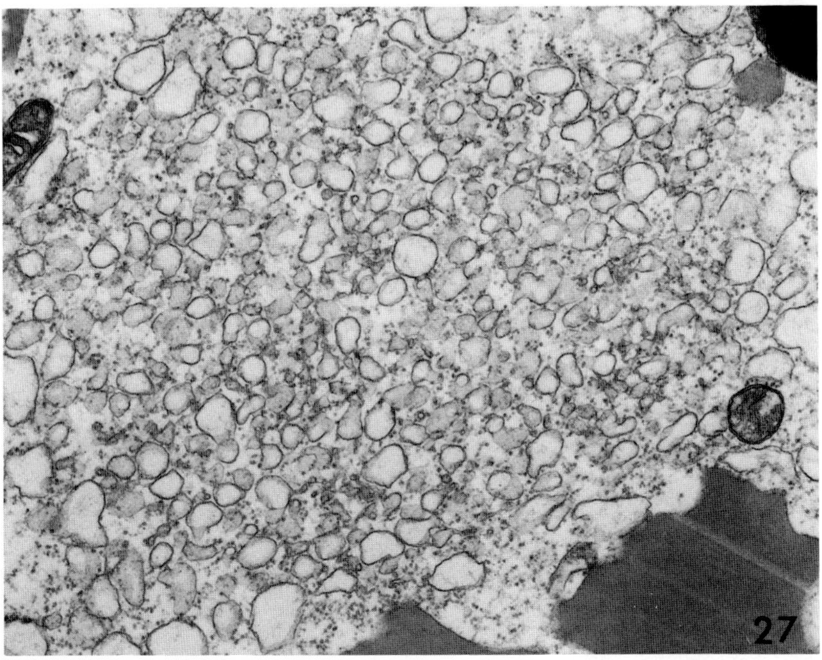


**Figures 26–27**

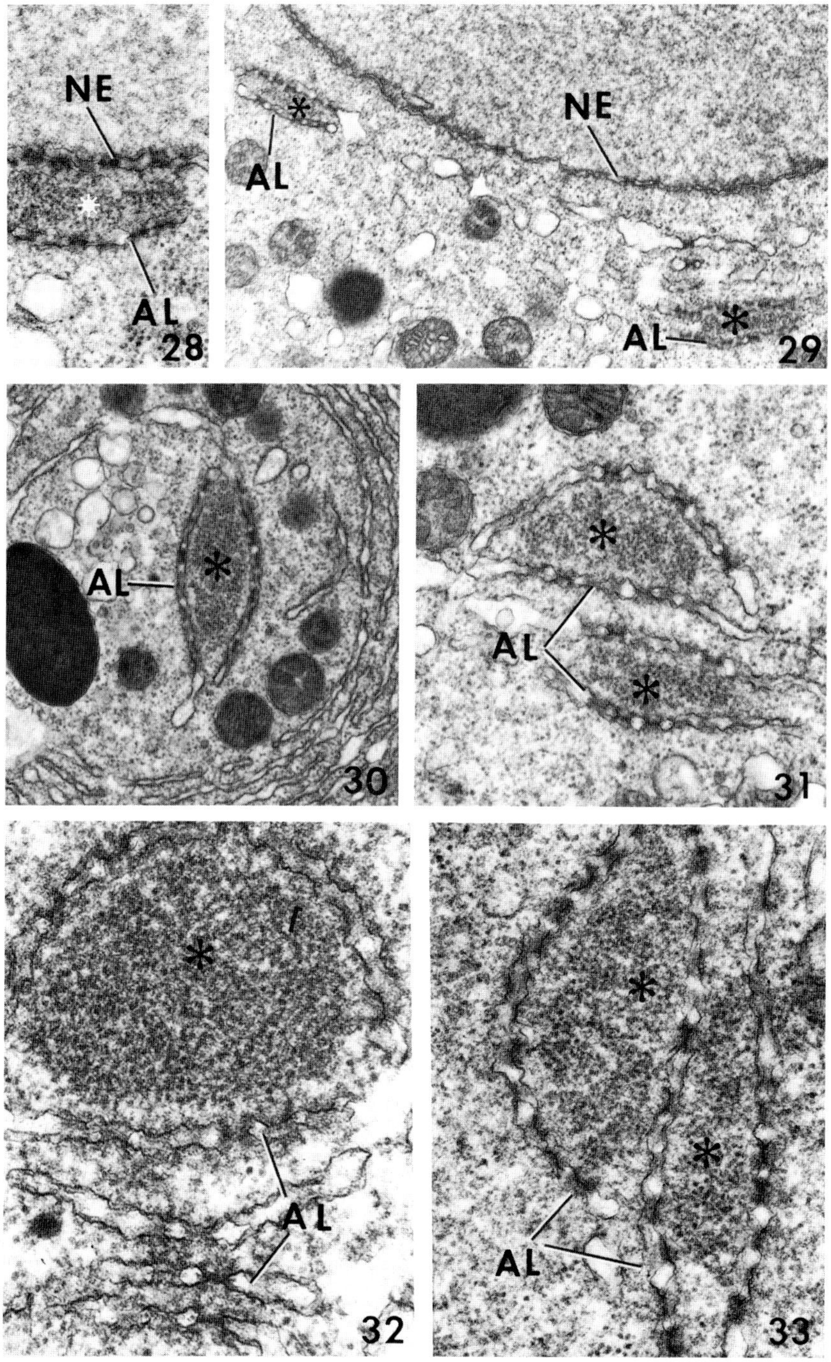


**Figures 28–33**

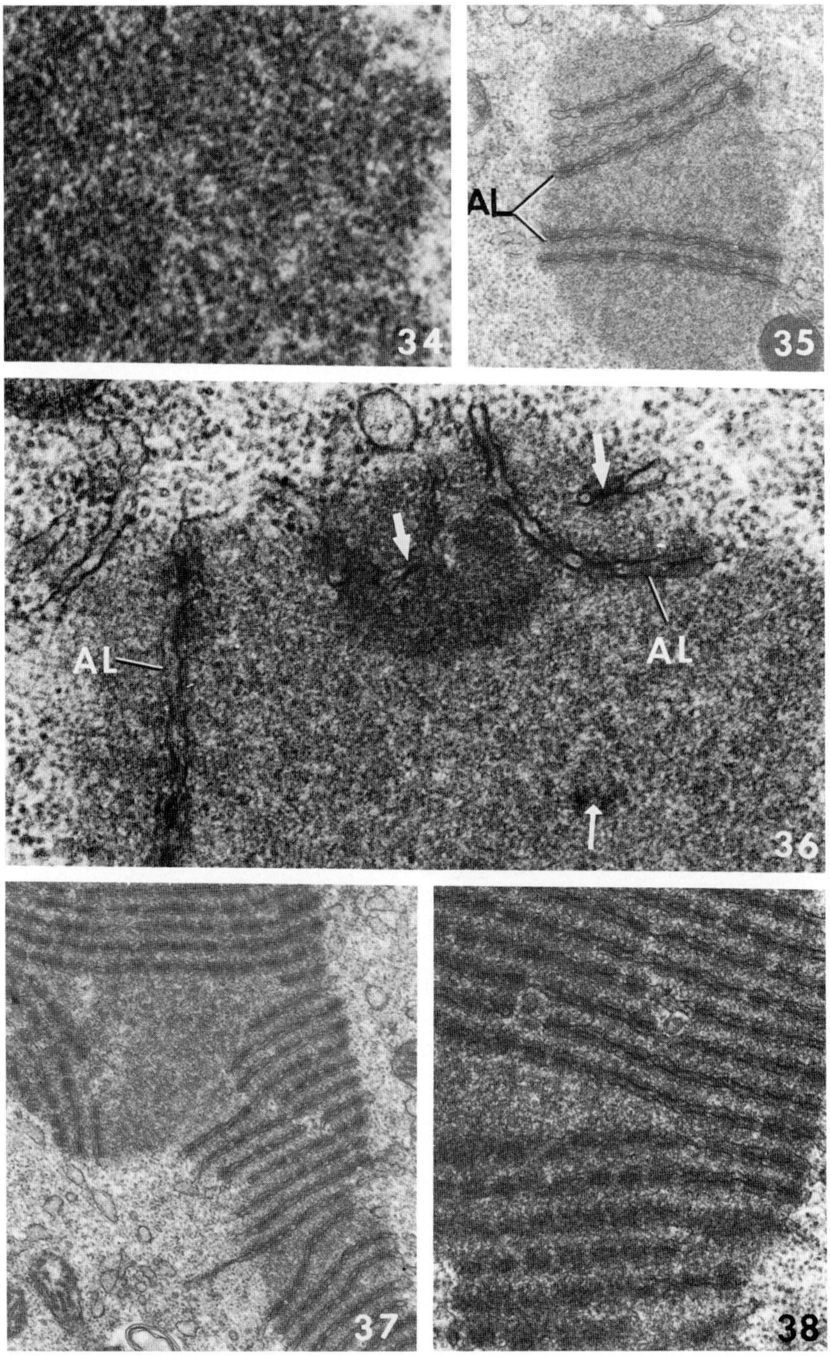


**Figures 34–38**

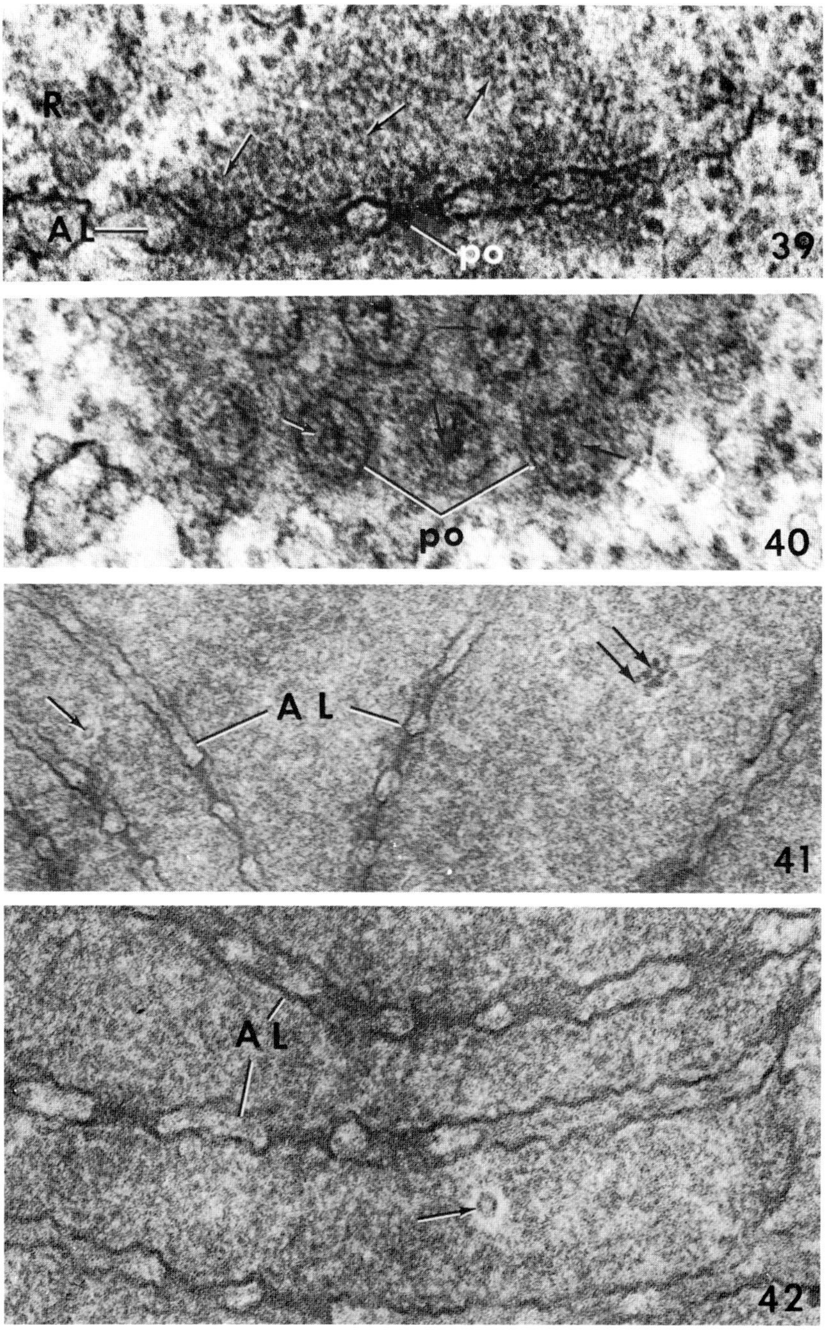


Figures 39–42

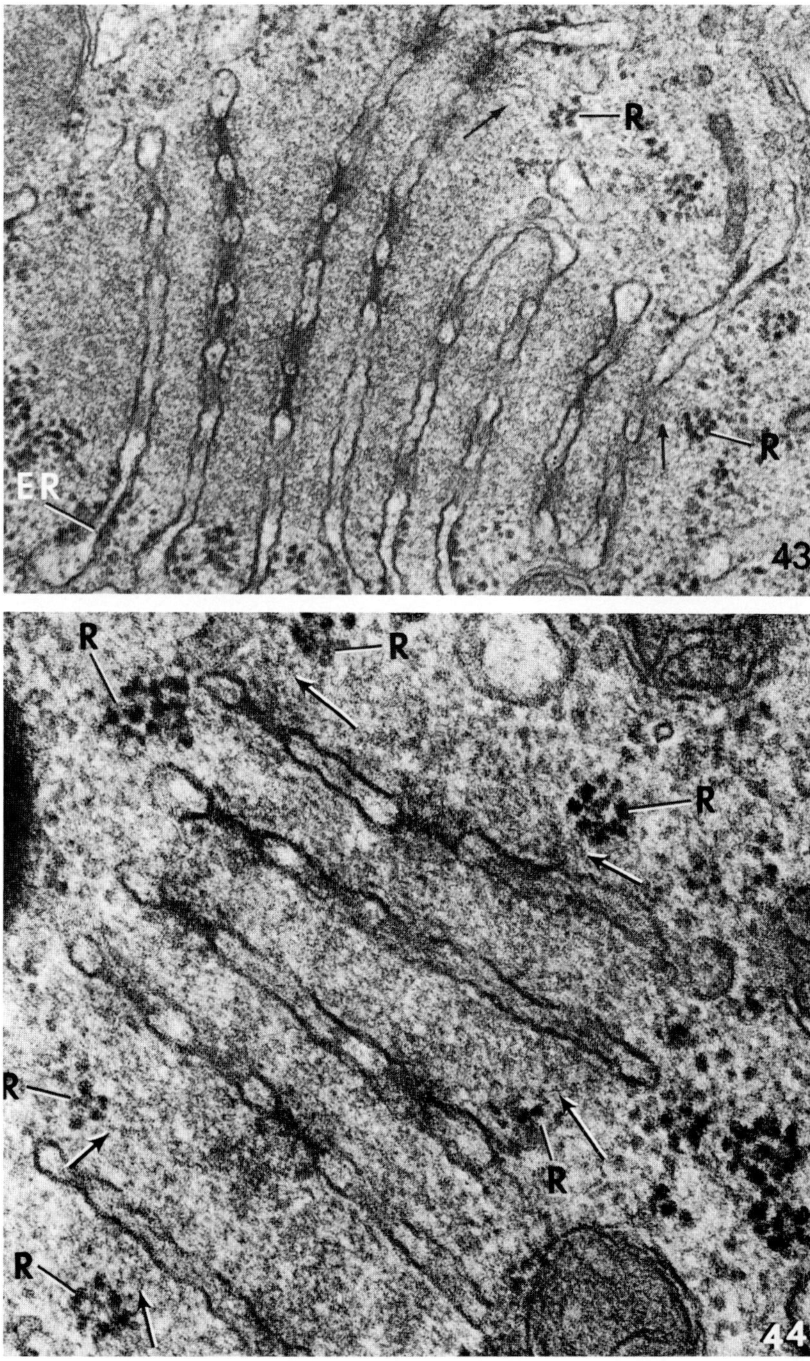


**Figures 43–44**

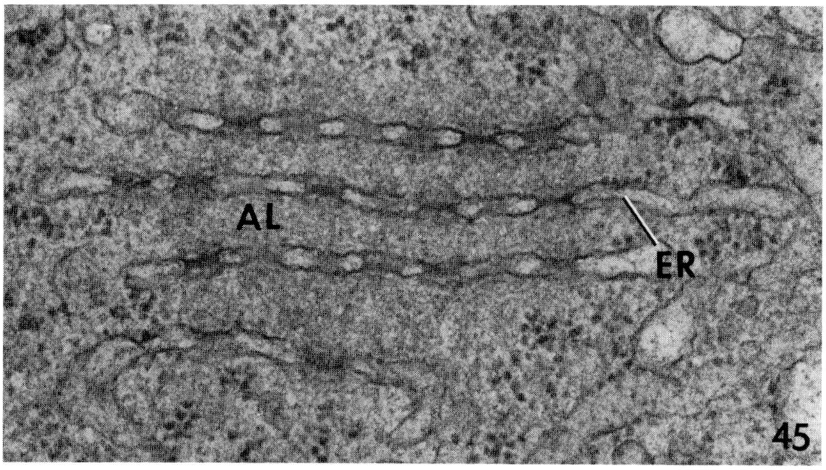


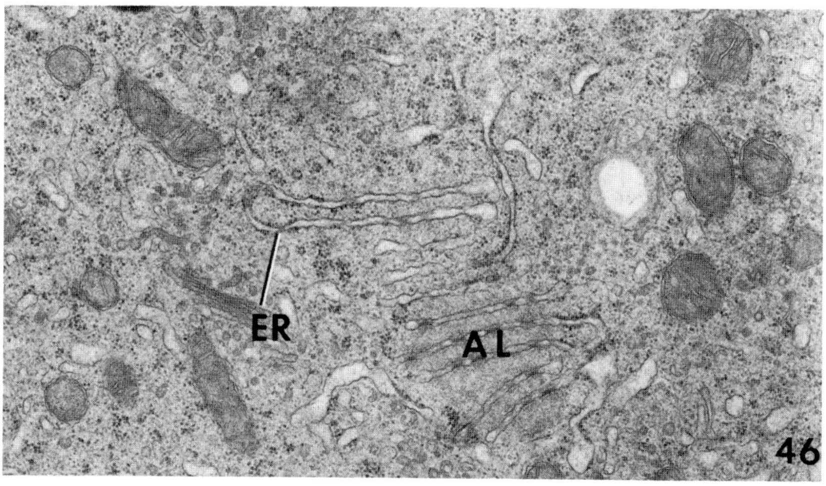


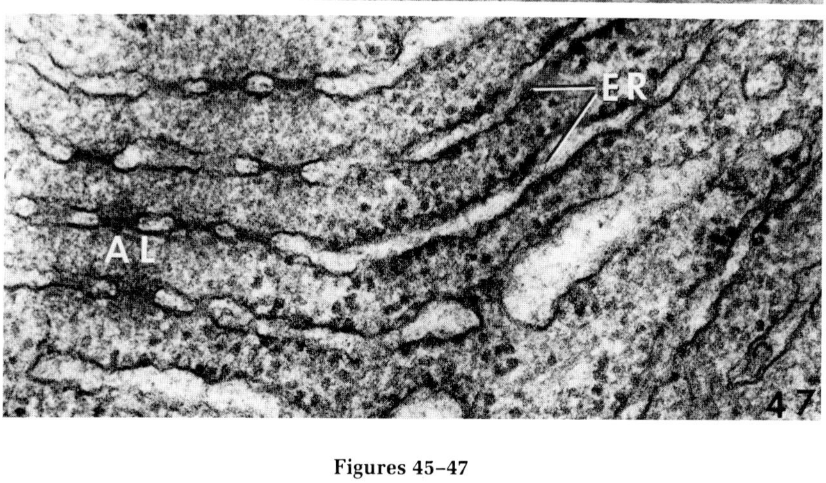


Figures 45–47

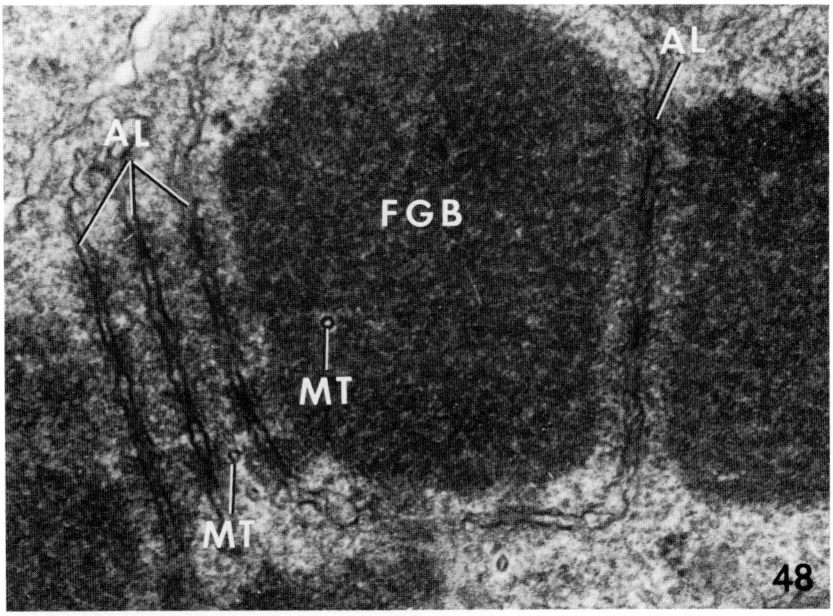


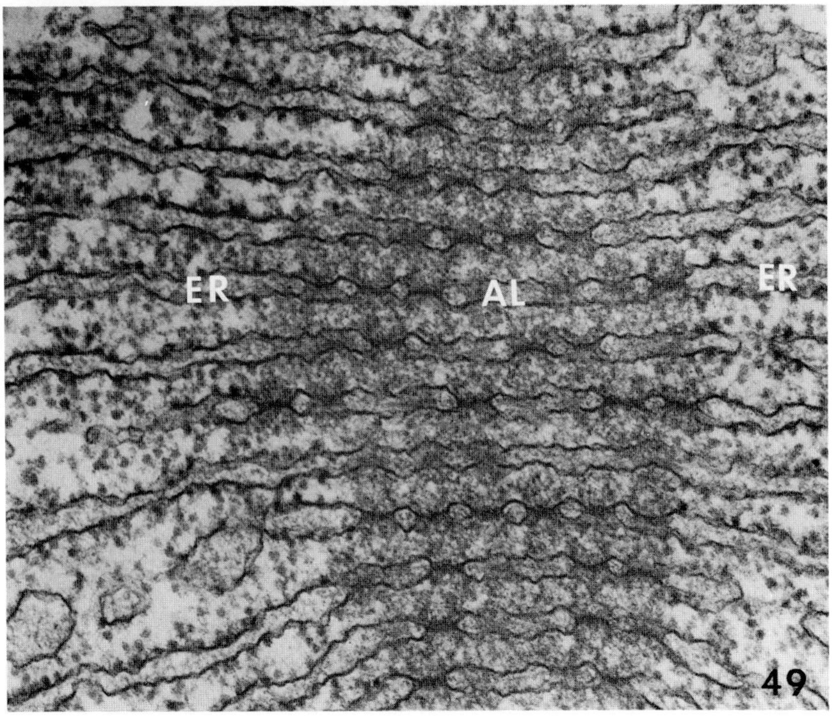


**Figures 48–49**

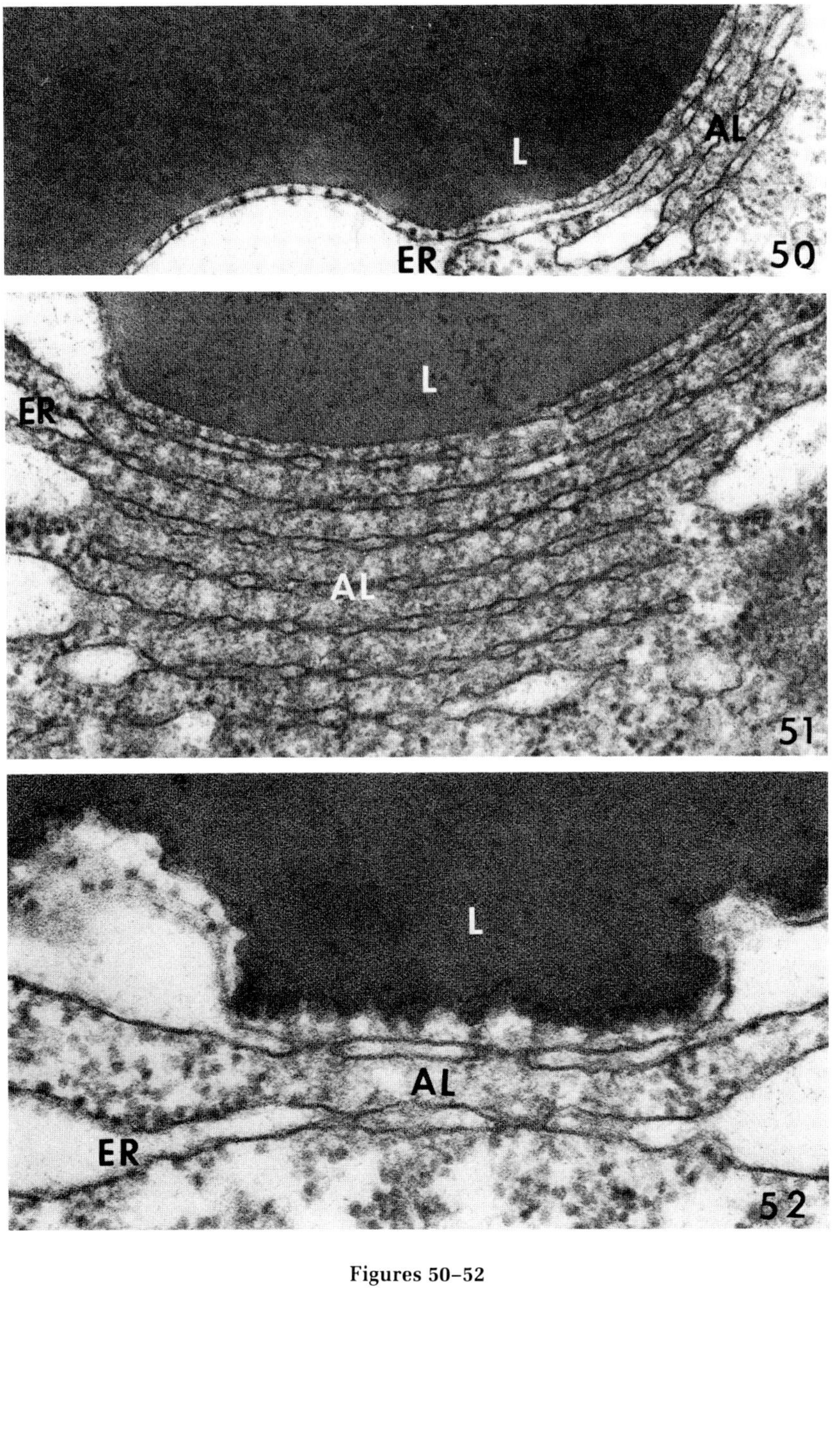


Figures 50–52

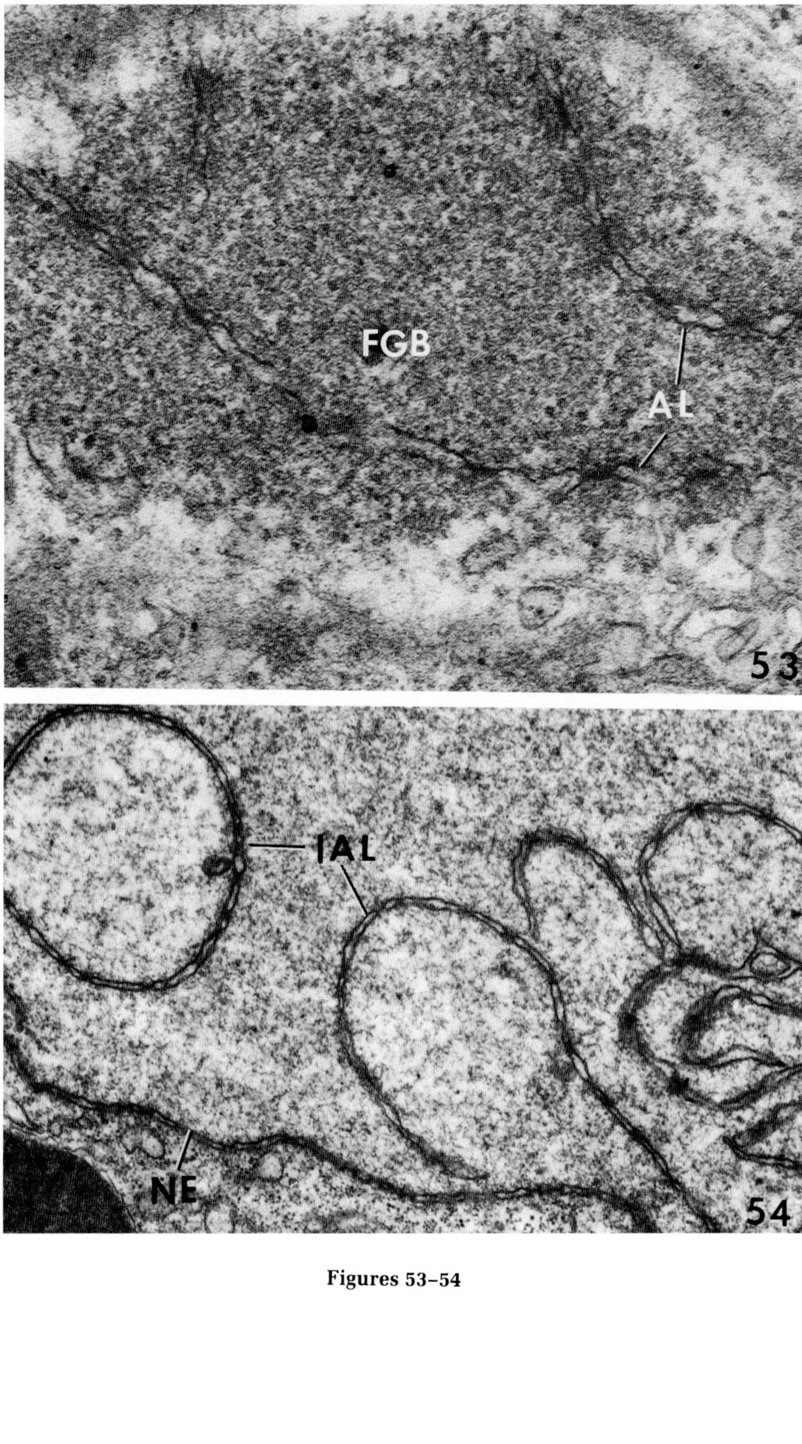


**Figures 53–54**

## 5.6 Heavy Bodies

One of the first examples in which annulate lamellae were reported to be associated with a distinct, electron-dense product in transmission electron micrographs was in the case of heavy bodies (Figs. 28–33). These structures were noted in a number of echinoderm oocytes (Afzelius, 1957) to consist of a diffuse, electron dense product that at much higher magnification appeared to contain closely packed filaments and granules (Harris, 1967a,b). This component of the heavy body was suggested to originate from the nucleus (particularly the nucleolus), and it was observed that single AL frequently were observed to border the fibrogranular material or to traverse it (Figs. 30–33). A number of investigators have since indicated that heavy bodies contain ribonucleic acid (Afzelius, 1957; Pasteels *et al.*, 1958; Sanchez, 1968; Conway and Metz, 1974; Conway, 1971), and a similarity between the fibrogranular material in the heavy body to ribosomal-like particles has been emphasized (Harris, 1967b).

## 5.7. Fibrogranular Bodies (Dense Cytoplasmic Masses)

In several different germ cells, a distinct cytoplasmic material can be observed that seems, at least in a number of cases, to control or exert an influence on the differentiation of AL. Two of the better documented systems include the dragonfly oocyte and the *Drosophila* spermatocyte/spermatid cells. Before discussing in some detail the sequence of events associated with the differentiation or formation of AL in relationship to a distinct cytoplasmic product, a brief summary of the form and nature of the product in different oocytes is warranted. Thus, oocytes frequently are observed to contain variable quantities of moderately electron-dense material in the cytoplasm. This electron-dense material often appears to be homogeneous; it is not membrane bound, and it varies greatly in size, shape, and position within the ooplasm. In some cases, it appears that the cytoplasmic material is formed by movement of nucleolar fragments through the nucleolar pores into the cytoplasm. A variety of names have been used to designate these cytoplasmic masses in germ cells. In these cases, terms such as "nucleolar or nuclear emissions" or "nuage" have been used to characterize the material. In some oocytes, the cytoplasmic masses seem to contain, when examined at very high magnification, small granular and fibrillar appearing subunits (Figs. 34, 39, and 40). As a result, the term "fibrogranular body" has been applied to this cytoplasmic material. Cytoplasmic masses of material are frequently observed also in germ cells to be closely associated with mitochondria, and this material has been called "intermitochondrial cement." In the past, it has not been very clear as to whether the cytoplasmic masses observed in different germ cells, especially oocytes, are similar, or if they represent a variety of macromolecular species. In only relatively recent times has information appeared about the composition of the material. Early cytological studies pointed out the similarity of the material to nucleolar constituents and suggested that the material originated by nucleolar emissions and nuclear-cytoplasmic exchanges. A ribonucleoprotein composition was postulated (see Kessel, 1973, 1983a for review).

In the dragonfly oocyte, large masses of basophilic staining material appear in the perinuclear ooplasm before vitellogenesis. The electron-dense masses consist of closely packed granules about 8–14 nm in diameter as well as very thin, short fibrils about 2–8 nm wide (Figs. 34, 39, and 40). These electron-dense masses, which contain both granular and fibrillar subunits, will be called **fibrogranular bodies.** Previous studies (Kessel and Beams, 1969) have provided evidence that the fibrogranular bodies arise from the nucleolar fragments that move through the nuclear pores and into the perinuclear ooplasm. These large electron-dense masses (fibrogranular bodies) then undergo fragmentation into smaller, but multiple bodies that move into the peripheral ooplasm. Initially, the fibrogranular bodies are several micrometers in either length or width, or both, and are readily visible in light microscope preparations. On the basis of their staining reactions (Kessel and Beams, 1969), these masses appear to contain ribonucleic acid and protein. It was subsequently considered that these bodies may represent long-lived gene products that are stored for a time in the cytoplasm and whose constituents might include mRNA (Kessel and Beams, 1969; Halkka and Halkka, 1975; Kessel, 1983a,c). In some dragonflies during the three years of nymphal development, gene products slowly accumulate in the cytoplasm of previtellogenic oocytes (Halkka and Halkka, 1975). The fibrogranular bodies may represent a storage form of these gene products.

After the fibrogranular bodies undergo the transitions first noted, the pores and AL differentiate within the fibrogranular bodies (Kessel and Beams, 1969) (Figs. 35–38). Discovery of this transformation demonstrated for the first time that morphogenesis of pores (and membrane) could be influenced to occur in association with a specific product produced earlier in time. The studies also demonstrated that pores are somehow necessary for the subsequent alteration of the fibrogranular subunits. This is true because, with time, more and more AL form within the fibrogranular bodies. Furthermore, apparent stages in subunit organization into polyribosomes have been noted (Figs. 43 and 44) (Kessel, 1983c). Moreover, eventually, the subunits of the fibrogranular bodies decrease drastically in number (Fig. 45), the amount of RER attached to the ends of the AL increases markedly in both number and length (Figs. 46 and 47), and the polyribosome compliment also increases. Late in oogenesis, the fibrogranular bodies are no longer recognizable, and stacks of AL are difficult to locate in the ooplasm, but stacks of RER are much more frequently encountered.

Based on the apparent origin of the fibrogranular bodies, their staining reactions, the size and nature of subunits, the differentiation of AL and their associations with polyribosomes and RER, these features collectively have led to the view (Kessel, 1983a,c) that AL in the dragonfly oocyte somehow function to release stored developmental information in the form of functioning polyribosomes (and/or RER), that could synthesize various unknown products at the appropriate developmental period. In fact, polyribosomes and microtubules are occasionally observed within the fibrogranular bodies (Figs. 41 and 42) in which the AL are also differentiating. The microtubules are separated from the fibrogranular body material (see also Fig. 48) by a thin, clear zone that has been reported recently (Stebbings and Hunt, 1982) as possibly being related to electrostatic repulsion between microtubules and ribosomes. Conceivably, the microtubule protein could be synthesized on these polysomes.

The sequence just postulated has been strengthened by somewhat similar observations recently (Kessel, 1981b) for the spermatocytes and spermatids of *Drosophila*. It has been shown that protein synthesis, but not RNA synthesis, occurs in spermatids (Brink, 1968, Hennig, 1967; Olivieri and Olivieri, 1965). In addition, it has been shown in organ culture (Gould-Somero and Holland, 1975) that protein synthesis in spermatids continues even in the presence of 3′-deoxyadenosine. This has led to the conclusion that at least a considerable part of the RNA necessary for spermiogenesis is synthesized in spermatocytes. As an interesting corollary to these data, it was recently demonstrated (Kessel, 1981b) that fibrogranular bodies appear in the cytoplasm of *Drosophila* spermatocytes. Then, in the spermatid stage, AL differentiate within the fibrogranular bodies (Fig. 49), after which they then become intimately associated with or connected to RER and polyribosomes. Thus, the fibrogranular material appears at a time when RNA synthesis is intense, and AL differentiation within the fibrogranular bodies occurs at a later developmental period characterized by intense protein synthesis. The extensive sequence of events illustrated in the *Drosophila* spermatocyte/spermatid system and the cytochemical information about the composition of the fibrogranular bodies in the dragonfly oocyte are consistent with the postulated functional role (Kessel, 1981a,b, 1982, 1983a,b,c) of AL in the release of stored developmental information and polyribosomal processing and assembly.

## 5.8. Composition of Fibrogranular Bodies (or Dense Cytoplasmic Masses) in Other Oocytes

In oocytes of *Nereis pelagica*, extrusion of nuclear material occurs, and the extruded material condenses into cytoplasmic clumps or bodies of considerable electron density. These cytoplasmic bodies exhibit basophilia and stain with the Palekar-Bernhard's technique, both of which indicate a high RNA content. The formation of these cytoplasmic masses is described to continue into the first stages of vitellogenesis, at which time some of the aggregates or bodies appear to transform into a diffuse material containing granules (smaller than ribosomes) and a fibrillar material. With high-resolution autoradiography using [$^{3}$H]uridine, the radioactivity was first observed in the nucleolus and later in the nucleoplasm, but after 3 days, label was observed in the cytoplasmic aggregates (fivefold higher than in surrounding cytoplasm). The studies by Dhainaut (1970a,b,c) indicate that the masses of cytoplasmic material that has a fibrogranular appearance are derived from the nucleus and have an RNA component.

Fibrogranular bodies initially appear during prophase of the first meiotic division in ovarian oocytes of the golden hamster. The fibrogranular bodies reach a peak incidence during the diplotene or early dictyate stage of meiotic prophase. The fibrogranular bodies then appear to break down into smaller masses that are dispersed throughout the cytoplasm. Cytochemical and autoradiographic studies on ovarian oocytes of the golden hamster indicate that granular–fibrillar cytoplasmic bodies and intermitochondrial substance contain both basic protein and ribonucleic acid (Weakley, 1971). Weakley (1971) proposed that the granular–fibrillar bodies represent stored inactive maternal messenger RNA that is synthesized during oogenesis for subsequent use in the developing embryo.

A specialized dense material has been noted in the ooplasm of previtellogenic squid oocytes. This material was hypothesized to have a nuclear origin, since it seemed to resemble nuclear extrusions and appeared to be related to the formation of AL in these oocytes (Dhainaut and Richard, 1972). In more recent times, AL have been shown to be closely associated with or to be present within fibrogranular type bodies (or comparable cytoplasmic masses) in myriapod eggs (Herbaut, 1972), beetle oocytes (Chu *et al.*, 1982), in oocytes of a crustacean, *Oniscus ocellus* (Fig. 53), in full-grown oocytes of the newt *Cynops pyrrhogaster* (Imoh, 1982), and in early spermatocytes, oocytes, and developing vitelline cells (Stranoch, 1979) of a fish gill parasite, *Diplozoon paradoxum* (Trematoda, Monogenea). In some of the chimpanzee oocytes examined by Hertig and Adams (1967), they found some electron-dense material between the nuclear envelope and a stack of AL. Cytoplasmic bodies in maturing *Xenopus* oocytes characterized as fibrillar bodies have mitochondria associated with their periphery. Numerous grains are located over the fibrillar body (GMA section) after hybridization with $^{125}$I-labeled rRNA (Angelier *et al.*, 1976).

Another term used to characterize cytoplasmic masses of material in various spermatocytes of different animals is the "chromatoid body." The chromatoid body has been described to contain considerable amounts of protein (Eddy and Ito, 1971), but chromatoid bodies have been shown to be labeled with [$^{3}$H]uridine in rat spermatids (Söderström and Parvinen, 1976). The chromatoid body appears to have a nucleolar origin in some male germ cells (Schjeide *et al.*, 1972). Both AL (Söderström, 1981; Schjeide *et al.*, 1972) and mitochondria (Schjeide *et al.*, 1972) have been noted to be closely associated with the chromatoid bodies in some species. Large fibrillar bodies in *Xenopus* oocytes surrounded by mitochondria appear to contain DNA complementary to ribosomal RNA (rDNA) (Steinert *et al.*, 1976).

Homogeneous material surrounded or indented by mitochondria have been observed with considerable frequency in the cytoplasm of many germ cells, both female and male. This so called intermitochondrial cement may be traversed by an occasional annulate lamella in teleost spermatogenic stages, for example (Clérot, 1976). Clérot *et al.* (1977) have reported a method for the isolation of intermitochondrial cement from tench (*Tina tinca*) spermatocytes. When this cement was analyzed (Toury *et al.*, 1977), it was described to contain 28, 18, and 5 S RNA, transfer RNA, proteins, cytochromes, and lipid. RNA and protein were the major constituents, while the lipid was a contaminant from the mitochondrial membranes. It was concluded (Toury *et al.*, 1977) that the intermitochondrial cement was associated with the origin, formation, and biogenesis of mitochondria.

The polar granules present in *Drosophila* oocytes and segregated into pole cells during development bear some structural resemblances to chromatoid bodies, nuage, and fibrogranular bodies. Based on a variety of studies by Mahowald and colleagues (Mahowald, 1962, 1963, 1968; Mahowald and Tiefert, 1970; Illmensee and Mahowald, 1976), it has been shown that AL differentiate in association with polar granules at a certain period of development. The polar granules contain RNA and protein (Mahowald, 1971a,b; Waring *et al.*, 1978) and it has been suggested that the basic protein in the polar granules contains maternal messenger RNA and is the site at which coding occurs for those proteins necessary for the formation of primordial germ cells (see Chapter 13 for a discussion of polar granules).

## 6. Morphogenesis of Pores

The means by which pores of AL are formed is uncertain. The event of pore formation appears to be quite rapid; in fact, Scheer (1972, 1973) has calculated that some 500 pores per minute are formed in the germinal vesicle of lampbrush-stage *Xenopus laevis* oocytes. Furthermore, it is difficult to observe pore formation in annulate lamellae under regulated conditions at precise periods. The formation of pores in cytoplasmic AL was earlier related to an event involving fusion of vesicles (Kessel, 1963). This sequence was proposed on the basis of primary fixation of the cells in osmium tetroxide (Kessel, 1963). In addition, intranuclear AL have also been described to result from the fusion of intranuclear vesicles derived from a blebbing of the *inner* membrane of the nuclear envelope in this case (Kessel, 1965, 1983b). However, results of fixation of *Necturus* oocytes with glutaraldehyde indicate that long membranous lamellae are released from the outer layer of the nuclear envelope (Figs. 23 and 24). Then, pores form within some of the membranous lamellae in regions far removed from the nucleus (Fig. 25). The precise mechanisms involved in the formation of pores in these membranes have yet to be elucidated.

There are several cell types, not including oocytes, in which single pore complexes have been observed in a lamella of endoplasmic reticulum (Hoage and Kessel, 1968; Maul, 1977b; Kessel, 1983a). In this case, it has been considered that a pore might form by a localized fusion process. In this process, the two membranes would make focal contact, followed by fusion and release of a small segment so as to form a pore. In such a proposed step it could be expected that a membranous vesicle might result as membrane is released when the pore is formed. There are, however, only rare instances in which membranous vesicles are observed in areas of forming AL and which could be hypothesized to have resulted from pore formation (cf. Fig. 22 and review by Kessel, 1983a). But, the event may be a rapid one, and the membranous vesicle could disappear rapidly (see review by Kessel, 1983a).

In those cases in which AL arise within specific cytoplasmic masses (fibrogranular bodies), the mechanisms of pore morphogenesis are equally unclear. In these cases, it appears that the pores can form *de novo* within and perhaps under the influence of the fibrogranular material. It is not yet clear what effect ribonucleoproteins may exert on membrane alteration (see Reyser *et al.*, 1970). A very early stage in pore formation within a fibrogranular body of the dragonfly egg is illustrated in Fig. 36. In one area denoted by the arrow in this illustration, a single pore is evident, and it is attached to only a very small amount of membrane. This suggests that it is a nascent pore.

## 7. Experimental Alteration

The limited number of studies dealing with the experimental alteration of annulate lamellae have generally involved cells other than oocytes. Those studies that indicate some correlation between a specific event and the presence or absence (or increased or decreased amount) of AL are summarized in the following outline:

1. Some studies have reported AL in cells exposed to various viruses, including Simian adenovirus 30 (Merkow *et al.*, 1968a,b), rubella virus (Kim and Boatman, 1967), porcine polioencephalomyelitis virus (Koestner *et al.*, 1968), and Rous sarcoma virus (Courington and Vogt, 1967).
2. Stacks of AL were noted in epithelial cells in some parts of the ductus epididymidis from rats that had been bilaterally castrated for about 3 months. Van De Velde and Van De Velde (1968) indicate that the presence of AL in the epithelium of the ductus epididymidis of fetal rats is related to the absence of androgen, for AL were observed to undergo a breakdown when androgen therapy was provided to long-term castrated rats.
3. Dhainaut (1973) indicated that in the absence of a brain in nereid

worms (in absence of a cerebral hormone and brain activity), the oocytes seem to have large numbers of AL.

4. When certain cultured cells, such as KB cells, are treated with macromomycin (a protein with antibiotic and antitumor activity), one of the major effects produced after prolonged treatment is a proliferation of cellular organelles, especially AL (Vandré *et al.*, 1979).
5. Chick embryos reared at a reduced temperature of 90°F (compared with the normal 100°F) were found to have enhanced numbers of AL in myocardial cells (Merkow and Leighton, 1966).
6. Lymphocytes exposed to phytohemagglutinin show the resultant basophilic lymphoblasts to contain annulate lamellae not observed in control cells (Procicchiani *et al.*, 1968).
7. Progesterone treatment of full-grown *Rana pipiens* oocytes eventually results in the breakdown of numerous stacks of AL (Kessel and Subtelny, 1981). In these studies, manually defolliculated, full-grown oocytes were cultured in progesterone for varying lengths of time.
   a. Kessel and Subtelny (1981) noted that the stacks of AL were present in such oocytes up to 9 hr after addition of progesterone (Fig. 26). However, stacks of annulate lamellae were not observed in those oocytes that had been cultured in progesterone for 12–22 hrs at 18°C. Rather, in comparable places in the animal pole ooplasm where AL had been observed, Kessel and Subtelny (1981) observed large areas consisting of closely packed membranous vesicles (Fig. 27). These results led Kessel and Subteiny (1981) to conclude that after induction of maturation in *Rana pipiens* eggs with progesterone, the AL undergo a vesiculation such that they can no longer be recognized as a distinct organelle. Thus, progesterone appears to cause the breakdown of AL just as it affects the germinal vesicle nuclear envelope (see Chapter 1); however, the breakdown of AL occurs a few hours before breakdown of the NE under the experimental conditions used.
   b. Structural alteration of AL in the *Rana pipiens* oocyte appears to be the first morphological change that occurs during the induction of meiotic maturation. The functional significance of this alteration is not clear; nor is it known what fate might befall the vesicles resulting from the alteration or whether AL reappear during embryonic development. It would be of interest to determine if AL breakdown is in any way related to the reported abrupt rise in protein synthesis, in which newly synthesized RNA is not required, that occurs in hormone-stimulated amphibian oocytes (reviewed in Smith, 1975). It has also been found (Masui and Markert, 1971) that the maturing cytoplasm contains a factor that becomes effective in inducing maturation (MPF) durng the interval approximately 6–20 hr after progester-

one administration (see Chapter 6). It is of interest here that the reported increase in cytoplasmic maturation promoting factor occurs when AL become structurally altered.

c. In a related study, Steinert *et al.* (1974) studied the ultrastructure of *Xenopus laevis* oocytes after they were injected with cytoplasm from oocytes that had undergone the progesterone-induced maturation. In such experimentally treated *Xenopus laevis* oocytes, they found that AL seemed to have undergone a breakdown into vesicles and ER cisternae. Thus, the alteration of AL may be a common feature of amphibian oocyte maturation.

8. In several different cell types maintained in culture, it has been reported that prolonged exposure of these cells (e.g., about 48 hr) to sublethal doses of various antitubulins (e.g., colchicine, vinblastine sulfate, and others) results in the presence of increased amounts of AL compared to untreated (control) cells. In such experimentally treated cells, other organelles have been reported to increase in number as well (Krishan *et al.*, 1968; Kessel and Katow, 1984a; reviewed in Kessel, 1983a). DeBrabander and Borgers (1975) in a study of prolonged exposure to antitubulins on an MO cell line in culture indicated that the enhanced development of AL was induced by the disintegration of microtubules. In a study of prolonged antitubulin treatment on cultured $\alpha$ L929 cells and chick limb bud mesenchyme cells, Kessel and Katow (1984a) observed different results. While culture of the $\alpha$ L929 cells for 48 hr in colchicine or vinblastine sulfate resulted in multinucleated cells and enhanced development of several organelles, including AL, this was not observed in the chick limb bud cells in culture. Treatment of chick limb bud mesenchyme cells with the same antitubulins for the same time period did not result in either multinucleate cells or enhanced development of AL. The observed effect of enhanced AL after prolonged antitubulin culture in certain cells under some conditions may be more complex than perhaps was formerly realized. Thus, the previously proposed suggestion that the pores in the nuclear envelope and AL might be sites of tubulin synthesis or microtubule polymerization as suggested by DeBrabander and Borgers (1975) might not be valid.

Kallenback (1982) has studied the origin of centrioles in sea urchin eggs subjected to hypertonic stress. Kallenbach reports, "Those centrioles which appear away from the nuclear surface originate exclusively at sites associated with annulate lamellae."

## 8. Proposed Functions

Postulated functions for annulate lamellae abound and largely reflect variations in morphological relationships and structure. Annulate lamellae have neither been isolated nor biochemically analyzed. Some of the proposed functions for AL are better substantiated than others and have been reported based on the study of a variety of cell types—especially oocytes. Among the proposed functions are (1) formation of nuclear envelope (Hu, 1971; Spindler and Hemleben, 1982); (2) biogenesis of mitochondria (Boquest, 1980; Babbage and King 1970); (3) steroid production (Nagano, 1966); (4) synthesis of viral material (Kim and Boatman, 1967); (5) synthesis of specific enzymes (Benzo, 1972); (6) tubulin synthesis or polymerization (DeBrabander and Borgers, 1975); (7) release, packaging, or assembly of stored developmental information (Kessel, 1968d, 1981a,b, 1982, 1983a,b,c); and (8) processing and assembly of functioning polyribosomes (Kessel, 1981a,b, 1982, 1983a,b,c).

The latter two functions are clearly related and could help explain the generally observed condition that AL are particularly well developed in the female gamete and to a lesser extent in the male gamete. These interrelated functions might also help explain the fact that AL are frequently connected to RER and that AL are noted to be closely related to a variety of cell products. For example, many of the lipid droplets present in vitellogenic *Acmaea* oocytes are closely surrounded by concentric layers of RER and interconnected AL (Figs. 50–52). It has been suggested that AL may be involved in some manner with the processing and assembling of ribosomal subunits or precursors into functioning polyribosomes that, in turn, could be associated with the synthesis of the multitude of lipid droplets that eventually fill the ooplasm of the full-grown oocyte (Kessel, 1982).

## 9. Intranuclear Annulate Lamellae

Porous membranes, called intranuclear annulate lamellae (IAL), can be found inside the nuclei of some cells, primarily in oocytes, zygotes (morulae), spermatogenic cells, and embryonic cells. They have also been described in a few plant cells and in both normal and abnormal cells in culture (reviewed by Kessel, 1983a). The IAL are single and may be linear, branched, curved, or circular (Fig. 54). The IAL are characteristically located in the peripheral nucleoplasm in close proximity to the nuclear envelope. The pores in IAL are similar or identical to those in the NE and in cytoplasmic AL. On the basis of those studies that are available regarding the origin of IAL, it appears that membrane is elaborated or budded from the inner nuclear membrane and that the resulting intranuclear vesicles then transform (perhaps involving a fusion process) into progressively longer intranuclear lamellae that become annulate (Kessel, 1964b, 1966a, 1968d, 1983b). Some studies have pointed out that IAL may contact or bear a close association with either nucleolar material or chromatin, depending on the cell type. Nucleolar fragments are in contact with IAL in, for example, *Nereis* oocytes (Dhainaut, 1970a,b), and the ascidian *Styela* (Kessel,

1983b) as well as a number of other cells (reviewed by Kessel, 1983a). Hillman and Tasca (1969) noted IAL in contact with nucleolar fragments in early cleavage-stage mouse embryos and indicated that the IAL decrease in number as the nucleoli also decrease in number. Chromatin has been described as being closely associated with IAL in, for example, spermatogonia (Folliot, 1968), mouse embryos (Hillman and Tasca, 1969), and rabbit zygotes (Maul and Cross, 1974). Maul (1970a) concluded that IAL in somatic cells were accidentally incorporated as pieces of the NE in the reforming nucleus after mitosis. Later, however, Maul and Cross (1974) indicated that this contention might have to be reevaluated.

Whereas it is clear that the IAL in cells containing them represent an amplification of the NE, their functional significance is entirely unknown. Kessel (1983b) has speculated that the presence of IAL in certain ascidian oocytes and their close relationship with apparent nucleolar fragments or bodies may suggest a role in processing or assembly of the ribosomal RNA and perhaps other forms of RNA into functioning polyribosomes. Such a proposal implies that the pores of IAL are functioning in a manner similar to, but an amplification of, that postulated for the pores in the NE and cytoplasmic AL as well. These suggestions are highly speculative, however, and much work is required to understand what role(s) the IAL play in those cells that contain these interesting constituents.

## 10. Concluding Remarks

From the background information detailed about annulate lamellae in this chapter, it is apparent that these membranes comprise a significant organelle in a variety of cells, especially the female gamete. The porous membranes comprising AL are the least understood of all the cellular organelles. Yet AL are important constituents in the developmental process of cells. Whereas critical information about the functional significance of AL is lacking, a number of points deserve emphasis.

The similarity between the NE and AL has been established especially in terms of the structurally complicated pore complex. The subunits that may be associated with the pores include small fibrils and small granules, both of which have been related to RNP structures. The formation of AL (or pores) does not appear to be uniform but may range from a derivation from membrane elaborated by the NE or other organelles to a *de novo* origin. The AL clearly are closely related in function to the rough-surfaced endoplasmic reticulum. The direct attachment frequently observed between the two organelles, as well as the association of ribosomes or polyribosomes with both membrane systems, further attests to a close functional relationship.

It has also been pointed out that the origin of AL occurs within a distinct cellular product in a number of cell types, especially oocytes and spermatocytes. There is some evidence that this product is derived from the nucleus (nucleolus) and that it includes RNA and proteins of an unidentified nature. The dispersal of this fibrogranular material during subsequent development, as AL differentiate within the material, is associated with the development of extensive RER and many polyribosomes. These relationships have resulted in an hypothesis that considers pores in AL important in the release of stored informational material and consequently involved in gene expression. The hypothesis (Kessel, 1981a,b, 1983a) that AL are somehow involved with the mobilization of stored gene products or with the processing of long-lived developmental information seems amenable to more rigorous testing. In the not-too-distant future, greater insight should be available into the significance of this interesting eukaryotic organelle.

ACKNOWLEDGMENTS. Some of the investigations described in this chapter were supported by research allocations from the National Science Foundation, U.S. Public Health Service, and the Whitehall Foundation.

## Note Added in Proof

Stafstrom and Staehelin (1984) recently described annulate lamellae in the syncytial *Drosophila* blastoderm embryo as it cleaves to form the cellular blastoderm. These workers indicate an inverse relationship between the number of nuclear pores and the number of pores in AL and conclude that AL represent stored nuclear pores.

## References

Aaronson, R., Coruzzi, L., and Schmidt, K., 1982, Attempted isolation of the nuclear pore complex: Evidence for a polydisperse complex of polypeptides, in: *The Nuclear Envelope and the Nuclear Matrix* (G. Maul, ed.), pp. 13–29, Alan R. Liss, New York.

Abelson, H. T., and Smith, G. H., 1970, Nuclear pores: The pore–annulus relationship in thin section, *J. Ultrastruct. Res.* **30**:558–588.

Aeschlimann, A., and Hecker, H., 1970, Présence de membranes annelées dans les ovocytes d'*Ornithodorus moubata*, Murray (Ixodoidea; Argasidae), *Acta Tropi.* **27**:268–270.

Afzelius, B., 1955, The ultrastructure of the nuclear membrane of the sea urchin oocyte as studied with the electron microscope, *Exp. Cell Res.* **8**:147–158.

Afzelius, B., 1957, Electron microscopy on the basophilic structures of the sea urchin eggs, *Z. Zellforsch. Mikrosk. Anat.* **45**:660–675.

Aizenshtadt, T., and Detlaf, T., 1972, Ultrastructure of stellate sturgeon oocytes during maturation. Report 1. Annulate lamellae and Golgi complex, *Sov. J. Dev. Biol.* **3**:220–229.

Anderson, E., 1969, Ooctye–follicle cell differentiation in two species of amphineurans (Mollusca), *Mopalia mucosa* and *Chaetopleura apliculata*, *J. Morphol.* **129**:89–126.

Anderson, E., 1970, A cytological study of the centrifuged whole, half, and quarter eggs of the sea urchin, *Arbacia punctulata*, *J. Cell Biol.* **47**:711–733.

Angelier, N., Bouteille, M., Charret, R., Curgy, J., Delain, E., Fakan, S., Geusken, S. M., Guelin, M., Lacroix, J., Laval, M., Steinert, G., and Van Assel, S., 1976, Detection of nucleic acids by means of ultrastructural autoradiography, *J. Microsc. Biol. Cell.* **27**:215–230.

Anteunis, A., 1964, Les membranes annelées dans l'oeuf d'*Artemia salina*. Étude de microscopie électronique, *Bull Assoc. Anat.* 49th Reunion, Madrid, 1964, 168.

Atkinson, K., and Bryam, J., 1976, The structure of the ovarian ball and oogenesis in *Moniliformis dubius* (Acanthocephala), *J. Morphol.* **148**:391–426.

Avrameas, S, 1970, Immunoenzyme techniques: Enzymes as markers for the localization of antigen and antibodies, *Int. Rev. Cytol.* **27**:349–385.

Avrameas, S., and Bouteille, M., 1968, Ultrastructural localization of antibody by antigen label with peroxidase, *Exp. Cell Res.* **53**:166–176.

Babbage, P. C., and King, P. E., 1970, Post fertilization functions of annulate lamellae in the periphery of the egg of *Spirorbus borealis* (Daudin) (Serpulidae = Annelida), *Z. Zellforsch. Mikrosk. Anat.* **107**:15–22.

Baca, M., and Zamboni, L., 1967, The fine structure of human follicular oocytes, *J. Ultrastruct. Res.* **19**:354–381.

Baca, M. K., and Johnson, H. G., 1966, Polysomes on a nuclear membrane fraction as intermediates in the transfer of ribonucleic acid from the nucleus to the cytoplasm, *Nature* **209**:893–895.

Baker, T. G., and Franchi, L. L., 1967, The fine structure of oogonia and oocytes in human ovaries, *J. Cell Sci.* **2**:213–224.

Bal, A. K., Jubinville, F., Cousineau, G. H., and Inoué, S., 1968, Origin and fate of annulate lamellae in *Arbacia punctulata* eggs, *J. Ultrastruct. Res.* **25**:15–28.

Balinsky, B. I., and Devis, R. J., 1963, Origin and differentiation of cytoplasmic structures in the oocytes of *Xenopus laevis*, *Acta Embryol. Morphol. Exp.* **6**:55–108.

Barton, B., and Hertig, A. T., 1972, Ultrastructure of annulate lamellae in primary oocytes of chimpanzees *(Pan troglodytes)*, *Biol. Reprod.* **6**:98–108.

Beams, H., and Kessel, R., 1980, Ultrastructure and vitellogenesis in the oocyte of the crustacean, *Oniscus asellus*, *J. Submicrosc. Cytol.* **12**:17–27.

Beaulaton, J., 1968a, Modifications ultrastructurales des cellules sécrétrices de la glande prothoracique de vers a soie, au cours des deux dernier ages larvaires. III. Les lamelles annelées et leur dégradation, *J. Microsc.* **7**:895–906.

Beaulaton, J., 1968b, Sur l'action d'enzymes au niveau des pores nucléaires et d'autres structures de cellules sécrétrices prothoraciques incluses en épon, *Z. Zellforsch. Microskop. Anat.* **89**:453–461.

Bellairs, R., 1967, Aspects of the development of yolk spheres in the hen's oocyte, studied by electron microscopy, *J. Embryol. Exp. Morphol.* **17(2)**:267–281.

Benzo, C. A., 1972, The annulate lamellae of chick embryo liver cells in organ culture, *Anat. Rec.* **174**:399–405.

Bernhard, W., 1969, A new staining procedure for electron microscopical cytology, *J. Ultrastruct. Res.* **27**:250–265.

Blanchette, E. J., 1961, A study of the fine structure of the rabbit primary ooctyte, *J. Ultrastruct. Res.* **5**:349–363.

Boquist, L., 1980, On the relationship between annulate lamellae and mitochondria in human parathyroid adenomas, *Z. Zellforsch. Anat. Forsch. (Leipzig)* **94**:241–249.

Bottke, W., 1973, Zur Ultrastruktur des Ovars von *Viviparus contectus* (Millet 1813) (Gastropoda, Prosobranchia). II. Die Occyten, *Z. Zellforsch. Mikrosk. Anat.* **138**:239–259.

Brink, N. G., 1968, Protein synthesis during spermatogenesis in *Drosophila melanogaster*, *Mutat. Res.* **5**:192–194.

Brown, D. D., 1967, The genes for ribosomal RNA and their transcription during amphibian development, *Curr. Topics Dev. Biol.* **2**:47–73.

Chu, H., Norris, D., and Rao, K., 1982, Sorbic acid-induced differences in the ultrastructural development of oocytes in the microbially ectosymbiotic female of *Xyleborus ferrugineus* (Fabr.) (Coleoptera, Scolytidae), *J. Morphol.* **173**:313–324.

Clawson, G., and Smuckler, E., 1982, A model for nucleocytoplasmic transport of ribonucleoprotein particles, in: *The Nuclear Envelope and the Nuclear Matrix*, (G. Maul, ed.), pp. 271–278, Alan R. Liss, New York.

Clérot, J. C., 1968, Mise en évidence par cytochimie ultrastructurale de l'émission de protéines par le noyau d'auxocytes de Batraciens, *J. Microsc.* **7**:973–992.

Clérot, J. C., 1976, Les groupements mitochondriaux des cellules germinales des poissons Teléostéens Cyprinidés. I. Étude ultrastructurale, *J. Ultrastruct. Res.* **54**:461–475.

Clérot, J. C., Toury, R., and André, J., 1977. Les groupements mitochondriaux des cellules germinales des poissons Téléostéens Cyprinides. III. Méthode d'isolement du "ciment" intermitochondrial, *Biol. Cell.* **30**:217–224.

Conway, C., 1971, Evidence for RNA in the heavy bodies of sea urchin eggs, *J. Cell Biol.* **47**:40A.

Conway, C. M., and Metz, C. B., 1974, *In vitro* maturation of *Arbacia punctulata* oocytes and intiation of heavy body formation. An ultrastructural study, *Cell Tissue Res.* **150**:271–279.

Courington, D., and Vogt, P. K., 1967, Electron microscopy of chick fibroblasts infected by defective Rous sarcoma virus and its helper, *J. Virol.* **1**:400–414.

Cruickshank, W. J., 1972, The formation of "accessory nuclei" and annulate lamellae in the oocytes of the flour moth *Anagasta Kühniella*, *Z. Zellforsch. Mikrosk. Anat.* **130**:181–192.

Cutler, L., Chaudhry, A., and Montes, M., 1974, Alkaline phosphatase activity associated with the nuclear pore in normal and neoplastic salivary gland tissue, *J. Histochem. Cytochem.* **22**:1113–1117.

Damarasinghe, D., Petterborg, L., Zeagler, J., Tiang, K., and Reiter, R., 1983, On the occurrence of a myeloid body in pinealocytes of the white-footed mouse, *Peromyscus leucopus*, *Cell Tissue Res.* **228**:649–659.

Danieli, G., and Burighel, P., 1974, Nuclear pore–polyribosome associations in the midgut cells of the ascidian *Botryllus schlosseri*, *J. Submicrosc. Cytol.* **6**:95–101.

DeBrabander, M., and Borgers, M., 1975. The formation of annulated lamellae induced by the disintegration of microtubules, *J. Cell Sci.* **19**:331–340.

Dhainaut, A., 1966, Presence de membranes annelées extra- et intranucléaires au cours de l'ovogenése naturelle et expérimentelle chez *Nereis pelagica* L., *C. R. Soc. Biol.* **160**:729–754.

Dhainaut, A., 1970a. Étude en microscopie électronique et par autoradiographie á haute résolution des extrusions nucléaires au course de l'ovogenése de *Nereis pelagica* L., *J. Microsc.* **9**:99–118.

Dhainaut, A., 1970b, Étude l'évolution ovocytaire en absence d'hormone cérébrale chez les Néréidiens, *Bull. Soc. Zool.* Fr. **95**:99–118.

Dhainaut, A., 1970c, Étude cytochimique et ultrastructurale de l'évolution ovocytaire de *Nereis pelagica* L. (Annélide Polychéte). I. Ovogenèse naturelle, *Z. Zellforsch. Mikroskop. Anat.* **104**:375–389.

Dhainaut, A., 1973, Mode de formation des lamelles annelées resultant de l'évolution, en condition anhormonale, des ovocytes de *Nereis diversicolor* (Annélide Polychéte), *Z. Zellforsch. Mikrosk. Anat.* **137**:481–492.

Dhainaut, A., and Richard, A., 1972, Processus d'elaboration des lamelles annelées dans les ovocytes de *Sepia officinalis* Céphalopode), *C. R. Acad. Sci. Paris* **275**:417–420.

Dumont, J., and Anderson, E., 1967, Vitellogenesis in the horseshoe crab, *Limulus polyphemus*, *J. Microsc.* **6**:791–806.

Dunne, M., 1965, Light and electron microscopic studies on the oocyte of the lizard, *Scelophorus undulatus*, M.S. thesis, University of Iowa, Iowa City, Iowa.

DuPraw, E. J., 1965, The organization of nuclei and chromosomes in honeybee embryonic cells, *Proc. Natl. Acad. Sci. U.S.A.* **53**:161–168.

Eckelbarger, L., and Grassle, J., 1982, Ultrastructure of the ovary and oogenesis in the polychaete *Capitella jonesi* (Hartman, 1959), *J. Morphol.* **171**:305–320.

Eckert, W., Franke, W., and Scheer, U., 1972, Actinomycin D and the central granules in the nuclear pore complex: Thin sectioning versus negative staining, *Z. Zellforsch. Mikroskop. Anat.* **127**:230–239.

Eddy, E., and Ito, S., 1971, Fine structural and autoradiographic observations on dense perinuclear cytoplasmic material in tadpole oocytes, *J. Cell Biol.* **49**:90–108.

Elliott, R., and Arhelger, R., 1966, Fine structure of parathyroid adenomas, *Arch. Pathol. Lab. Med.* **81**:200–212.

Everingham, J. W., 1968. Attachment of intranuclear annulate lamellae to the nuclear envelope, *J. Cell Biol.* **37:540–550.**

Fabergé, A., 1973, Direct demonstration of eight-fold symmetry in nuclear pores, *Z. Zellforsch. Mikroskop. Anat.* **136:**183–190.

Fabergé, A., 1974, The nuclear pore complex: Its free existence and a hypothesis as to its origin, *Cell Tissue Res.* **151:**403–415.

Feldherr, C., 1972, Structure and function of the nuclear envelope, *Adv. Cell Mol. Biol.* **2**:273–305.

Feldherr, C., 1974, The binding characteristics of the nuclear annuli, *Exp. Cell Res.* **85**:271–277.

Fisher, H., and Cooper, T., 1967, Electron microscope observations on the nuclear pores of HeLa cells, *Exp. Cell Res.* **48**:620–622.

Folliot, R., 1968, Les lamelles annelées intranucléaires des cellules du tissu germinal mâle avant le méiose chez *Philaenus spumarius* L. (Insecte, Homoptére), *Z. Zellforsch. Mikroskop. Anat.* **92:**115–129.

Fonzo, S., and Esponda, P., 1974, Relationship between annulate lamellae and ribosomal structures during meiosis in *Chorthippus apicalis*, *Cytobiologie* **9:**66–71.

Franke, W., 1970, On the universality of nuclear pore complex structure, *Z. Zellforsch. Mikroskop. Anat.* **105:**405–429.

Franke, W. W., 1974, Structure, biochemistry, and functions of the nuclear envelope, *Int. Rev. Cytol. (Suppl.)* **71:**235.

Franke, W. W., and Falk, H., 1970, Appearance of nuclear pore complexes after Bernhard's staining procedure, *Histochemie* **24:**266–278.

Franke, W., and Scheer, U., 1970a, The ultrastructure of the nuclear envelope of amphibian oocytes: A reinvestigation. I. The mature oocyte, *J. Ultrastruct. Res.* **30:**288–316.

Franke, W. W., and Scheer, U., 1970b, The ultrastructure of the nuclear envelope of amphibian oocytes: A reinvestigation. II. The immature oocyte and dynamic aspects, *J. Ultrastruct. Res.* **30:**317–327.

Franke, W. W., and Scheer, U., 1971, Some structural differentiations in the HeLa cell: Heavy bodies, annulate lamellae, and catte de maillet endoplasmic reticulum, *Cytobiologie* **4:**317–329.

Franke, W. W., and Scheer, U., 1974, Structures and functions of the nuclear envelope, in: *The Cell Nucleus*, Vol. 1 (H. Busch, ed.), pp. 220–347, Academic Press, New York.

Franke, W. W., and Scheer, U., 1975, Biochemical and structural aspects of nucleocytoplasmic transfer of ribonucleoproteins at the nuclear envelope level: Facts and theses, in: *Biochemistry of the Cell Nucleus. Mechanism and Regulation of Gene Expression*, Vol. 33 (E. J. Hidvegi, J. Sumegi, and P. Venetianer, eds.), pp. 205–211, North-Holland/American Elsevier.

Franke, W., Deumling, B., Ermen, B., Tarasch, E., and Kleinig, H., 1970, Nuclear membranes from mammalian liver. I. Isolation and general characterization, *J. Cell Biol.* **46:**379–395.

Gall, J. G., 1967, Octagonal nuclear pores, *J. Cell Biol.* **32:**391–400.

Gould-Somero, M., and Holland, L., 1975, Oocyte differentiation in *Urechis caupo* (Echinura): A fine structural study, *J. Morphol.* **147:**475–506.

Gouranton, J., 1969, L'Enveloppe nucléaire, *Ann. Biol.* **8:**385–409.

Green, N., 1982, The nuclear pore—a biological grommet? *Nature* **297:**287–288.

Gross, P. R., Philpott, D. E., and Nass, S., 1960, Electron microscopy of the centrifuged sea urchin egg with a note on the structure of the ground cytoplasm, *J. Biophys. Biochem. Cytol.* **7:**135–142.

Gulyas, B. J., 1971, The rabbit zygote: Formation of annulate lamellae, *J. Ultrastruct. Res.* **35**:112–126.

Gulyas, B. J., 1972, The rabbit zygote. II. The fate of annulate lamellae during first cleavage, *Z. Zellforsch. Mikrosk. Anat.* **133**:187–200.

Gwynn, J., Barton, R., and Jones, P., 1971, Formation and function of cytoplasmic annulate lamellae in the eggs of *Pomatoceros triqueter* L., *Z. Zellforsch. Mikrosk. Anat.* **112**:309–395.

Hadek, R., and Swift, H., 1961, Electron microscopic study on the oocyte and blastocyst in the rabbit, *Anat. Rec.* **139**:234 (abstr.).

Halkka, L., and Halkka, O., 1975, Accumulation of gene products in the oocytes of the dragonfly *Cordula aenea* L. I. The nematosomes, *J. Cell Sci.* **19**:103–115.

Halkka, L., and Halkka, O., 1977, Accumulation of gene products in the oocytes of the dragonfly *Cordulia aenea*. II. Induction of annulate lamellae within dense masses during diapause, *J. Cell Sci.* **26**:217–228.

Hanzely, L. and Olah, L., 1973, Fine structure and distribution of nuclear pores in root tip cells of *Allium sativum*, *Trans. Am. Microsc. Soc.* **92**:35–43.

Harris, P., 1967a, Nucleolus-like bodies in sea urchin eggs, *Am. Zool.* **7**:753–754.

Harris, P., 1967b, Structural changes following fertilization in the sea urchin egg, *Exp. Cell Res.* **48**:569–581.

Harris, J., 1978, The biochemistry and ultrastructure of the nuclear envelope, *Biochem. Biophys. Acta* **515**:55–104.

Harrison, G. A., 1966, Some observations on the presence of annulate lamellae in alligator and sea gull adrenal cortical cells, *J. Ultrastruct. Res.* **14**:158–166.

Harvey, E. B., 1956, *The American Arbacia and Other Sea Urchins*, Princeton University Press, Princeton, New Jersey.

Hennig, W., 1967, Untersuchungen zu Struktur und Funktion des Lampenbürsten-Y-chromosomes in der Spermatogenese von *Drosophila*, *Chromosoma (Berl.)* **22**:294–357.

Henshaw, E. E., Revel, M., and Hiatt, H. H., 1965, A cytoplasmic particle bearing messenger ribonucleic acid in rat liver, *J. Mol. Biol.* **14**:241–256.

Herbaut, C., 1972, Étude cytochimique et ultrastructurale de l'ovogenèse chez *Lithobius forficatus* L. (Myriapode Chilopode). Evolution des constituants cellulairs, *Wilhelm roux Arch.* **170**:115–134.

Hertig, A. T., 1968, The primary human oocyte: Some observations on the fine structure of Balbiani's vitelline body and the origin of annulate lamellae, *Am. J. Anat.* **122**:107–138.

Hertig, A. T., and Adams, E. C., 1967, Studies on the human oocyte and its follicle. I. Ultrastructural and histochemical observations on the primordial follicle stage, *J. Cell Biol.* **34**:647–675.

Hillman, N. W., and Tasca, R. J., 1969, Ultrastructural and autoradiographic studies of mouse cleavage stages, *Am. J. Anat.* **126**:151–174.

Hoage, T., and Kessel, R., 1968, An electron microscope study of the process of differentiation and growth during spermatogenesis in the drone honey bee (*Apis mellifera* L.) with special reference to centriole replication and elimination, *J. Ultrastruct. Res.* **24**:6–23.

Holland, N. D., 1971, The fine structure of the ovary of the feather star *Nemaster rubiginosa*. (Echinodermata, Crinoidea), *Tissue Cell* **3**:161–175.

Hsu, W. S., 1967, The origin of annulate lamellae in the oocyte of the Ascidian, *Boltenia villosa* Stimpson, *Z. Zellforsch. Mikrosk. Anat.* **82**:376–390.

Hu, F., 1971, Ultrastructural changes in the cell cycle of cultured melanoma cells, *Anat. Rec.* **170**:41–56.

Heubner, E., Tobe, S. S., and Davey, K. G., 1975, Structural and functional dynamics of oogenesis in *Glossina austeni*: Vitellogenesis with special reference to the follicular epithelium, *Tissue Cell* **7**:535–558.

Humeau, C., and Temple, D., 1969, Origine des lamelles annelées dans les ovocytes d'Amphibiens, *C. R. Soc. Biol.* **163**:1898–1902.

Illmensee, K., and Mahowald, A. P., 1976, The autonomous function of germ plasm in a somatic region of the *Drosophila* eggs, *Exp. Cell Res.* **97**:127–140.

Imoh, H., 1982, Behaviour of annulate lamellae during the maturation of occytes in the newt, *Cynops pyrrhogaster*, *J. Embryol. Exp. Morphol.* **70:**153–169.

Infante, A. A., and Nemer, M., 1968, Heterogeneous ribonucleoprotein particles in the cytoplasm of sea urchin embryos, *J. Mol. Biol.* **32:**543–565.

Jacob, J., and Danieli, G. A., 1972, Electron microscope observations on nuclear pore–polysome associations, *Cell Differ.* **1:**119–125.

Jasper, D. K., 1976, ER–Annulate lamellar associations of mucosal epithelial cells of the rat jejunum, *Cell Tissue Res.* **175:**417–420.

Joklik, W. K., and Becker, Y., 1965, Studies on the genesis of polyribosomes. II. The association of nascent messenger RNA with the 40S subribosomal particle, *J. Mol. Biol.* **13:**511–520.

Kallenbach, R. J., 1982, 'De novo' centrioles originate at sites associated with annulate lamellae in sea-urchin eggs, *Biosci. Rep.* **2:**959–966.

Kartenbeck, J., Zentgraf, H., Scheer, U., and Franke, W. W., 1971, The nuclear envelope in freeze-etching, *Adv. Anat. Embryol. Cell Biol.* ***45:**1–55.*

Kaye, G. I., and Johnston, I., 1973, The nuclear envelope: Current problems of structure and function, *Sub-Cell. Biochem.* **2:**127–166.

Kessel, R. G., 1963, Electron microscope studies on the origin of annulate lamellae in oocytes of *Necturus*, *J. Cell Biol.* **19:**391–414.

Kessel, R. G., 1964a, Electron microscope studies on oocytes of an echinoderm, *Thyone briareus*, with special reference to the origin and structure of the annulate lamellae, *J. Ultrastruct. Res.* **10:**498–514.

Kessel, R. G., 1964b, Intranuclear annulate lamellae in oocytes of the tunicate, *Styela partita*, *Z. Zellforsch. Mikrosk. Anat.* **63:**37–51.

Kessel, R. G., 1965, Intranuclear and cytoplasmic annulate lamellae in tunicte oocytes, *J. Cell Biol.* **24:**471–487.

Kessel, R. G., 1966a, Ultrastructure and relationships of ooplasmic components in tunicates, *Acta Embryol. Morphol. Exp.* **9:**1–24.

Kessel, R. G., 1966b, Some observations on the ultrastructure of the oocyte of *Thyone briareus* with special reference to the relationship of the Golgi complex and endoplasmic reticulum in the formation of yolk, *J. Ultrastruct. Res.* **16:**305–319.

Kessel, R. G., 1968a, Fine structure of annulate lamellae, *J. Cell Biol.* **36:**658–664.

Kessel, R. G., 1968b, An electron microscope study on differentiation and growth in oocytes of *Ophioderma panamensis*, *J. Ultrastruct. Res.* **22:**63–89.

Kessel, R. G., 1968c, Electron microscope studies on developing oocytes of a coelenterate medusa with special reference to vitellogenesis, *J. Morphol.* **126:**211–248.

Kessel, R. G., 1968d, Annulate lamellae, *J. Ultrastruct. Res. (Suppl.)* **10:**1–82.

Kessel, R. G., 1969a, Fine structure of pore–annulus complex in the nuclear envelope and annulate lamellae in germ cells, *Z. Zellforsch. Mikrosk. Anat.* **94:**411–453.

Kessel, R. G., 1969b, The effect of glutaraldehyde fixation of the elucidation of the morphogenesis of annulate lamellae in oocytes of *Rana pipiens*, *Z. Zellforsch. Mikrosk. Anat.* **94:**454–461.

Kessel, R. G., 1973, Structure and function of the nuclear envelope and related cytomembranes, in: *Progress in Surface and Membrane Science*, Vol. 6 (J. F. Danielli, M. D. Rosenberg, and D. A. Cadenhead, eds.), pp. 243–329, Academic Press, New York.

Kessel, R. G., 1981a, Annulate lamellae and polyribosomes in young oocytes of the rainbow trout, *Salmo gairdneri*, *J. Submicrosc. Cytol.* **13(2):**231–252.

Kessel, R. G., 1981b, Origin, differentiation, distribution and possible functional role of annulate lamellae during spermatogenesis in *Drosophila melanogaster*, *J. Ultrastruct. Res.* **75:**72–96.

Kessel, R. G., 1982, Differentiation of *Acmaea digitalis* oocytes with special reference to lipid–endoplasmic reticulum–annulate lamellae–polyribosome relationships, *J. Morphol.* **171:**225–243.

Kessel, R. G., 1983a, Structure and function of annulate lamellae, *Int. Rev. Cytol.* **82:**181–303.

Kessel, R. G., 1983b, Intranuclear membranes (vesicles, lamellae, annulate lamellae) in oocytes of the ascidian, *Styela*, *J. Submicrosc. Cytol.* **15:**773–785.

Kessel, R. G., 1983c, Fibrogranular bodies, annulate lamellae and polyribosomes in the dragonfly oocyte, *J. Morphol.* **176**:171–180.
Kessel, R. G., and Beams, H. W., 1969, Annulate lamellae and "yolk nuclei" in oocytes of the dragonfly, *Libellula pulchella*, *J. Cell Biol.* **42**:185–201.
Kessel, R., and Katow, H., 1983a, Effect of prolonged antibulin culture on annulate lamallae in mouse alpha L929 fibroblasts, *J. Morphol.* **179**:291–304.
Kessel, R., and Katow, H., 1983b, Effect of prolonged antibulin culture on the ultrastructure of anterior limb bud cells of the chick embryo, *J. Morphol.* **179**:263–271.
Kessel, R. G., and Subtelny, S. S., 1981, Alteration of annulate lamellae in the in vitro, progesterone-treated, full-grown *Rana pipiens* oocyte, *J. Exp. Zool.* **217**:119–135.
Kim, K. S., and Boatman, E. S., 1967, Electron microscopy of monkey kidney cell cultures infected with rubella virus, *J. Virol.* **1**:205–212.
King, P. E., and Fordy, M. R., 1970, The formation of "accessory nuclei"in the developing oocytes of the parasitoid hymenopterans *Ophion leteus* (L.) and *Aphanteles glomeratus* (L.), *Z. Zellforsch. Mikrosk. Anat.* **109**:158–170.
King, P. E., and Richards, J. G., 1968, Accessory nuclei and annulate lamellae in hymenopteran oocytes, *Nature* **218**:488–489.
Koestner, A., Lasza, L., and Lindig, O., 1966, Electron microscopy of tissue cultures infected with porcine polioencephalomyelitis virus, *Am. J. Pathol.* **48**:129–147.
Koshiba, K., Smetana, K., and Busch, H., 1970, Hypertrophy of granular endoplasmic reticulum and annulate lamellae in Earle's L cells exposed to vinblastine sulphate, *J. Cell Biol.* **39**:211–216.
Koulish, S., 1965, Ultrastructure of differentiating oocytes in the trematode *Georgodernia attenuata*. I. The "nucleolus-like" cytoplasmic body and some lamellar membrane systems, *Dev. Biol.* **12**:248–250.
Krauskopf, C., 1968a, Elktronenmikroskopische Untersuchungen über die Struktur der Oozyte und des 2-Zellenstadiums beim Kaninchen. I. Oozyte, *Z. Zellforsch. Mikrosk. Anat.* **92**:275–295.
Krauskopf, C., 1968b, Elktronenmikroskopische Untersuchungen über die Struktur der Oozyte und des 2-Zellenstadiums beim Kaninchen. II. Blastomeren, *Z. Zellforsch. Mikrosk. Anat.* **92**:297–312.
Kress, A., 1982, Ultrastructural indications for autosynthetic proteinaceous yolk formation in amphibian oocytes, *Experientia* **38**:761–771.
Krishan, A., Hsu, D., and Hutchins, P., 1968, Hypertrophy of granular endoplasmic reticulum and annulate lamellae in Earle's L cells exposed to vinblastine sulfate, *J. Cell Biol.* **39**:211–216.
LaCour, L. F., and Wells, B., 1972, The nuclear pores of early meiotic prophase nuclei of plants, *Z. Zellforsch. Milrosk. Anat.* **123**:178–194.
Latham, H., and Darnell, J., 1965, Entrance of mRNA into HeLa cell cytoplasm in puromycin-treated cells, *J. Mol. Biol.* **14**:13–22.
Leduc, E. H., Avrameas, S., and Bouteille, M., 1968, Ultrastructural localization of antibody in differentiating plasma cells, *J. Exp. Med.* **127**:109–118.
Longo, F., and Anderson, E., 1969, Cytological events leading to the formation of the two-cell stage in the rabbit: Association of the maternally and paternally derived genomes, *J. Ultrastruct. Res.* **29**:86–118.
Lukanidin, E. M., Zalmanzon, E. S., Komaromi, L., Samarina, O. P., and Georgiev, G. P., 1972, Structure and function of informofers, *Nature New Biol.* **238**:193–197.
Lumsden, R. D., 1965, Macromolecular structure of glycogen in some cyclophyllidean and trypanorhynch cestodes, *J. Parasit.* **51**:501–515.
Mahowald, A. P., 1962, Fine structure of pole cells and polar granules in *Dropophila melanogaster*, *J. Exp. Zool.* **151**:201–204.
Mahowald, A. P., 1963, Ultrastructural differentiations during formation of the blastoderm in the *Drosophila* embryo, *Dev. Biol.* **8**:186–204.
Mahowald, A. P., 1968, Polar granules of *Drosophila*. II. Ultrastructural changes during early embryogenesis, J. Exp. Zool. **167**:237–262.

Mahowald, A. P., 1971a, Polar granules of *Drosophila*. IV. Cytochemical studies showing loss of RNA from polar granules during early stages of embryogenesis, *J. Exp. Zool.* **176**:345–352.

Mahowald, A. P., 1971b, Polar granules of *Drosophila*. III. The continuity of polar granules during the life cycle of *Drosophila*, *J. Exp. Zool.* **176**:329–344.

Mahowald, A. P., and Tiefert, M., 1970, Fine structural changes in the *Drosophila* oocyte nucleus during a short period of RNA synthesis, *Wilhelm Roux Arch.* **165**:8–25.

Mancuso, V., 1964, Ultrastructural changes in the cytoplasm of *Ciona intestinalis*, *Acta Embryol. Morphol. Exp. (Palermo)* **7**:269–295.

Masui, Y., and Markert, C., 1971, Cytoplasmic control of nuclear behavior during meiotic maturation of frog oocytes, *J. Exp. Zool.* **177**:129–146.

Matsuura, S., Morimoto, T., Tashiro, Y., Higashinakagawa, T., and Muramatsu, M., 1974, Ultrastructural and biochemical studies on the precursor ribosomal particles isolated from rat liver nucleoli, *J. Cell Biol.* **63**:629–640.

Maul, G., 1970a, On the relationship between the Golgi apparatus and annulate lamellae, *J. Ultrastruct. Res.* **30**:368–384.

Maul, G., 1970b, The presence of intranuclear annulate lamellae shortly after mitosis in human melanoma cells in vitro, *J. Ultrastruct. Res.* **31**:375–380.

Maul, G., 1971, On the octagonality of the nuclear pore complex, *J. Cell Biol.* **51**:558–563.

Maul, G., 1977a, The nuclear and the cytoplasmic pore complex: Structure, dynamics, distribution and evolution, *Int. Rev. Cytol. (Suppl.)* **6**:75–186.

Maul, G., 1977b, Annulate lamellae and single pore complexes in normal, SV 40-transformed and tumor cells in vitro, *Exp. Cell Res.* **104**:233–245.

Maul, G. G., and Cross, P. C., 1974, Thymidine incorporation by annulate lamellae of rabbit zygotes, *J. Ultrastruct. Res.* **47**:115–124.

Maul, G., Price, J., and Lieberman, M., 1971, Formation and distribution of nuclear pore complexes in interphase, *J. Cell Biol.* **51**:405–418.

Maul, G., Maul, H., Scogna, J., Lieberman, M., Stein, G., Hsu, B., and Borun, T., 1972, Time sequence of nuclear pore formation in phytohemagglutinin-stimulated lymphocytes and in HeLa cells during the cell cycle, *J. Cell Biol.* **55**:433–437.

McConkey, E. H., and Hopkins, J. W., 1965, Subribosomal particles and transport of messenger RNA in HeLa cells, *J. Mol. Biol.* **14**:257–270.

McCullough, D., 1952, Fibrous structures in the ground cytoplasm of the *Arbacia* egg, *J. Exp. Zool.* **119**:47–64.

Mentré, P., 1969, Présence d'acide ribonucléique dans l'anneau osmiophile et le granule central des pores nucléaires, *J. Microsc.* **8**:51–68.

Mepham, R., and Lane, G., 1969, Nucleopores and polyribosome formation, *Nature* **221**:288–289.

Merchant, H., 1970, Ultrastructural changes in preimplantation rabbit embryos, *Cytologia* **35**:319–334.

Merkow, L., and Leighton, J., 1966, Increased numbers of annulate lamellae in myocardium of chick embryos incubated at abnormal temperatures, *J. Cell Biol.* **28**:127–137.

Merkow, L., Epstein, S., Caito, B., and Bartus, B., 1967, The cellular analysis of liver carcinogenesis: Ultrastructural alterations within hyperplastic liver nodules induced by 2-fluorenylacetamide, *Cancer Res.* **27**:1712–1721.

Merkow, L. P., Slifkin, M., Pardo, M., and Rapoza, N., 1968a, The histopathology and ultrastructure of tumors induced by simian adenovirus 30, *Cancer Res.* **28**:1180–1190.

Merkow, L. P., Slifkin, M., Pardo, M., and Rapoza, N., 1968b, Studies on the pathogenesis of simian adenovirus-induced tumors. III. The histopathology and ultrastructure of intracranial neoplasms induced by SV20, *Cancer Res.* **41**:1051–1070.

Merriam, R. W., 1959, The origin and fate of annulate lamellae in maturing sand dollar eggs, *J. Biophys. Biochem. Cytol.* **5**:117–121.

Merriam, R. W., 1961, On the fine structure and composition of the nuclear envelope, *J. Biophys. Biochem. Cytol.* **11**:559–570.

Meusy, J-J., 1968, Ultrastructure de la zone germinative et des gonies du testicule et de l'ovaire d'*Orchestia gammarella* P. (Crustacé Amphipode), *Ann. Sci. Nat. Zool. Biol. Animale* **10**:101–116.

Monneron, A., and Bernhard, W., 1969, Fine structural organization of the interphase nucleus in some mammalian cells, *J. Ultrastruct. Res.* **27**:266–288.

Nagano, T., 1966, Some observations on the fine structure of the Sertoli cells in the human testis, *Z. Zellforsch. Mikrosk. Anat.* **73**:89–106.

Okada, E., and Waddington, C. H., 1959, The submicroscopic structure of the *Drosophila* egg, *J. Embryol. Exp. Morphol.* **7**:583–597.

Olivieri, G., and Olivieri A., 1965, Autoradiographic study of nucleic acid synthesis during spermatogenesis in *Drosophila melanogaster*, *Mutat. Res.* **2**:366–380.

Palade, G. E., 1955, Studies on the endoplasmic reticulum. II. Simple dispositions in cells in situ, *J. Biophys. biochem. Cytol.* **1**:567–582.

Pasteels, J. J., Castinaux, P., and Vandermurssche, G., 1958, Cytochemical localizations and ultrastructure in the fertilized unsegmented egg of *Paracentrotus lividus*, *J. Biophys. Biochem. Cytol.* **4**:575–578.

Petit, J., 1973, Étude morphologique et cytochimique de deux types de groupements mitochondriaux dans les jeunes ovocytes de *Polydesmus angustus* Latz. (Myriapode, Chilopode), *J. Microsc.* **17**:41–54.

Procicchiani, G., Miggliano, V., and Arancia, G., 1968, A peculiar structure of membranes in PHA-stimulated lymphocytes, *J. Ultrastruct. Res.* **22**:195–205.

Rebhun, L. I., 1956, Electron microscopy of basophilic structures of some invertebrate oocytes. I. Periodic lamellae and the nuclear membrane, *J. Biophys. biochem. Cytol.* **2**:*93–104*.

Rebhun, L. I., 1961, Some electron microscope observations on membranous basophilic elements of invertebrate eggs, *J. Ultrastruct. Res.* **5**:208–225.

Reverberi, G., and DeLeo, G., 1972, The oocyte of *Amphioxus* examined by the electron microscope, *Acta Embryol. Exp.* **1972**:65–84.

Rogers, M. E., 1968, Ribonucleoprotein particles in the amphibian oocyte nucleus. Possible intermediates in ribosome synthesis, *J. Cell Biol.* **36**:421–432.

Ruthmann, A., 1958, Basophilic lamellar systems in the crayfish spermatocyte, *J. Biophys. Biochem. Cytol.* **4**:267–274.

Ryser, U., 1970, Die Ultrastruktur der Mitosekerne in den Plasmodien von *Physarum polycephalum*, *Z. Zellforsch. Mikrosk. Anat.* **110**:108–130.

Sachs, M., and Anderson, E., 1970, A cytological study of artificial parthenogenesis in the sea urchin *Arbacia punctulata*, *J. Cell Biol.* **47**:140–158.

Samarina, O. P., Molnar, J., Lukanidin, E. M., Bruskov, V. I., Krichevskaya, A. A., and Georgiev, G. P., 1967, Reversible dissociation of nuclear ribonucleoprotein particles containing mRNA into RNA and protein, *J. Mol. Biol.* **27**:187–191.

Samarina, O. P., Lukanidin, E. M., Molnar, J., and Georgiev, G. P., 1968, Structural organization of nuclear complexes containing DNA-like RNA, *J. Mol. Biol.* **33**:251–256.

Sanchez, S., 1968, Effets de l'actinomycine D sur les constituants cellulaires et le metabolise de l'ARN de L'ovocyte d'oursin (*Paracentrotus lividus*). Étude autoradiographique, *Exp. Cell Res.* **50**:19–31.

Sanchez, S., and Lanet, A., 1966, Distribution des protéines basiques nucléaires et cytoplasmiques dans l'ovocyte encroissane de *Paracentrotus lividus*, *C. R. Soc. Biol.* **160**:145–147.

Scanbissi, F. S., and Trentini, M., 1979, Ultrastructural observations on the oogenesis of *Triops cancriformis* (Crustacea, Notostraca), *Cell Tissue Res.* **201**:361–368.

Scheer, U., 1970, The ultrastructure of the nuclear envelope of amphibian oocytes: A reinvestigation. III. Actinomycin-induced decrease in central granules within the pores, *J. Cell Biol.* **45**:445–449.

Scheer, U., 1972, The ultrastructure of the nuclear envelope of amphibian oocytes. IV. On the chemical nature of the nuclear pore complex, *Z. Zellforsch. Mikrosk. Anat.* **127**:127–148.

Scheer, U., 1973, Nuclear pore flow rate of ribosomal RNA and chain growth rate of its precursor during oogenesis of *Xenopus laevis*, *Dev. Biol.* **30**:13–28.

Scheer, U., and Franke W., 1969, Negative staining and adenosine triphosphatase activity of annulate lamellae of newt oocytes, *J. Cell Biol.* **42:**519–533.

Scheer, U., and Franke, W. W., 1972, Annulate lamellae in plant cells: Formation during microsporogenesis and pollen development in *Canna generalis* Bailey, *Planta* **107:**145–152.

Scherrer, K., and Marcaud, L., 1968, Messenger RNA in avian erythroblasts at the transcriptional and translational levels and the problem of regulation in animal cells, *J. Cell Physiol. (Suppl.)* **72:**181–212.

Scherrer, K., Marcaud, L., Zajdela, F., London, I. M., and Gros, F., 1966, Patterns of RNA metabolism in a differentiated cell: A rapidly labeled, unstable 60S RNA with messenger properties in duck erythroblasts, *Proc. Natl. Acad. Sci. USA* **56:**1571–1578.

Schjeide, O. A., Nicholls, T., and Graham, G., 1972, Annulate lamellae and chromatoid bodies in the testes of a cyprinid fish *(Pimephales notatus)*, *Z. Zellforsch. Mikrosk. Anat.* **129:**1–10.

Schulz, H., 1957, Elektronenmikroskopische Untersuchungen eines Mammakarzinoms der Ratte, *Oncologia (Basel)* **10:**307–319.

Shaklai, M., and Tavassoli, M., 1982, Preferential localization of lanthanum to nuclear-pore complexes, *J. Ultrastruct. Res.* **81:**139–144.

Smith, F., and Berlin, J. D., 1977, Cytoplasmic annulate lamellae in human spermatogenesis, *Cell Tissue Res.* **176:**235–242.

Smith, L. D., 1975, Molecular events during oocyte maturation in: *The Biochemistry of Animal Development*, Vol. 3 (R, Weber, ed.), pp. 1–46, Academic Press, New York.

Smith, S. J., Adams, H. R., Smetana, K., and Busch, H., 1969, Isolation of the outer layer of the nuclear envelope. Composition of the RNA, *Exp. Cell Res.* **55:**185–197.

Söderström, K.-O., and Parvinen, M., 1976, Incorporation of ($^3$H)-uridine by the chromatoid body during rat spermatogenesis, *J. Cell Biol.* **70:**239–246.

Söderström, K.-O., 1981, Annulate lamellae in spermatids of the rat, *Z. Mikrosk. Anat. Forsch.* Leipzig **95:**845–854.

Spindler, M., and Hemleben, Ch., 1982, Formation and possible function of annulate lamellae in planktic foraminifer, *J. Ultrastruct. Res.* **81:**341–350.

Stafstrom, J. P., and Staehelin, L. A., 1984, Are annulate lamellae in the *Drosophila* embryo the result of overproduction of nuclear pore components?, *J. Cell Biol.* **98:**699–708.

Stebbings, H., and Hunt, C., 1982, The nature of the clear zone around microtubules, *Cell Tissue Res.* **227:**609–617.

Steinert, G., Baltus, E., Hanocq-Quertier, J., and Brachet, J., 1974, Ultrastructure of *Xenopus laevis* oocytes after injection of an extract from progesterone treated oocytes, *J. Ultrastruct. Res.* **49:**188–210.

Steinert, G., Thomas, C., and Brachet, J., 1976, Localization by in situ hybridization of amplified ribosomal DNA during *Xenopus laevis* oocyte maturation (a light and electron microscopy study), *Proc. Natl. Acad. Sc. USA* **73:**833–836.

Stick, R., and Krohne, G., 1982, Immunological localization of the major architectural protein associated with the nuclear envelope of the *Xenopus laevis* oocyte, *Exp. Cell Res.* **138:**319–330.

Stranock, S. D., 1979, Annulate lamellae in the reproductive tissues of a fish parasite, *J. Anat.* **129:**885 (abst.).

Swift, H., 1956, The fine structure of annulate lamellae, *J. biophys. Biochem. Cytol.* **2:**415–419.

Tardini, A., Vitali-Massz, L., and Mansani, F., 1960, Ultrastructuttura dell'ovocita umano maturo. I. Rapporti fra cellule della corona radiate, pellucida ed ovoplasma, *Arch. Vecchi Anat. Pat.* **33:**281–305.

Taylor, G., and Anderson, E., 1969, Cytochemical and fine structural analysis of oogenesis in the gastropod, *Ilyanassa obsoleta*, *J. Morphol.* **129:**211–238.

Thomopoulos, G., and Kastritsis, C., 1983, Activities of the nuclear envelope in the salivary glands of *Drosophila*, *Cell Tissue Res.* **228:**497–509.

Toury, R., Clérot, J. C., and André, J., 1977, Les groupements mitochondriaux des cellules germinales des poissons Téléostéens Cyprindés. IV. Analyse biochemique des constituants du "ciment" intermitochondrial isolé, *Biol. Cell.* **30:**225–232.
Van De Velde, R., and Van De Velde, S., 1968, Annulate lamellae in the ductus epididymidis of fetal and castrated adult rats, *Anat. Rec.* **161:**427–440.
Vandré, D., Shephard, V., and Montgomery, R., 1979, Effects of macromomycin on the ultrastructure and biological properties of cultured mammalian cells, *Cancer Res.* **39:**4091–4097.
Verhey, C. A., and Moyer, F. H., 1967, Fine structural changes during sea urchin oogenesis, *J. Exp. Zool.* **164:**195–226.
Wang, J. -J., and Hsü, C.-Y., 1974, Ultrastructural changes in oocytes and follicle cells during tadpole development, *Bull. Inst. Zool. Acad. Sini,* **13:**75–86.
Wang, J., Hsü, C., and Liang, H., 1980, Fine structural changes in tadpole oocytes after heat treatment, *Acta Embryol. Morphol. Exp.* **1:**43–57.
Ward, R. T., and Ward, E., 1975, Evidence for diaphragms in the pores of annulate lamellae and the nuclear envelope, *J. Morphol.* **146:**81–96.
Waring, G. L., Allis, C. D., and Mahowald, A. P., 1978, Isolation of polar granules and the identification of polar granule-specific protein, *Dev. Biol.* **66:**199–206.
Wartenberg, H., and Stegner, H. E., 1960, Über die electronenmikroskopische Feinstruktur des Menschlichen ovarialeies, *Z. Zellforsch. Mikrosk. Anat.* **52:**450–474.
Weakley, B., 1969, Annulate lamellae in the oocyte of the golden hamster, *Z. Zellforsch. Mikrosk. Anat.* **96:**229–236.
Weakley, B., 1971, Basic protein and ribonucleic acid in the cytoplasm of the ovarian oocyte in the golden hamster, *Z. Zellforsch. Mikrosk. Anat.* **112:**69–84.
Wetherbee, R., West, J., and Wynne, M., 1974, Annulate lamellae in postfertilization development in *Polysiphonia novae-angliae* (Rhodophyta), *J. Ultrastruct. Res.* **49:**401–414.
Wischnitzer, S., 1970a, The annulate lamellae, *Int. Rev. Cytol.* **27:**65–100.
Wischnitzer, S., 1970b, The annulate lamellae of salamander oocytes: Morphological and functional aspects, *Wilhelm Roux Archiv.* **164:**279–292.
Yamamoto, K., and Onozato, H., 1965, Electron microscope study on the growing oocyte of the goldfish during the first growth phase, *Mem. Fac. Fisheries Hokkaido Univ.* **13:**79–106.
Yasuzumi, G., and Tsubo, J., 1966, The fine structure of nuclei as revealed by electron microscopy. III. Adenosine triphosphatase activity in the pores of the nuclear envelopes of mouse choroid plexus epithelial cells, *Exp. Cell Res.* **43:**281–292.
Yoo, B., and Bayley, S., 1967, The structure of pores in isolated pea nuclei, *J. Ultrastruct. Res.* **18:**651–660.
Zamboni, L., 1971, *Fine Morphology of Mammalian Fertilization,* pp. 70–73, Harper and Row, New York.
Zamboni, L., and Mastroianni, L., Jr., 1966a, Electron microscopic studies on rabbit ova. I. The follicular oocyte, *J. Ultrastruct. Res.* **14:**95–117.
Zamboni, L., and Mastroianni, L., Jr., 1966b, Electron microscopic studies on rabbit ova. II. The penetrated tubal ovum, *J. Ultrastruct. Res.* **14:**118–132.
Zamboni, L., Thompson, R., and Smith, D., 1972, Fine morphology of human oocyte maturation in vitro, *Biol. Reprod.* **7:**425–457.
Zbarsky, I., 1978, An enzyme profile of the nuclear envelope, *Int. Rev. Cytol.* **54:**295–360.

Chapter 5

# Egg Envelopes in Vertebrates

JAMES N. DUMONT and ANNA RUTH BRUMMETT

## 1. Introduction

We have attempted to summarize in this chapter the available information about the origin, structure, composition, and function of the acellular envelopes that enclose vertebrate eggs. The information presented here is a collation of data from a variety of sources in a form that facilitates comparisons of the characteristics of the egg envelopes. Although little attention has been given to the envelopes of the invertebrates, it should be clear that, if this group were included, the similarities would exceed the differences. In the course of reviewing the literature on the envelopes of vertebrate eggs we have been impressed by the similarities in origin, structure, composition, and function. It is our hope that these similarities will be obvious to the reader and, further, that this chapter will encourage a broader, more unified perspective of the apparent evolutionary homologies of animal egg envelopes.

## 2. Teleosts

The teleost egg is surrounded by a single envelope, generally termed the **chorion,** which is complex in both origin and structure. The envelope appears to have primary (of oocyte origin), secondary (of follicle cell origin), and probably tertiary (of oviduct origin) components. It is proteinaceous and, in general, forms a tough protective envelope for the egg and developing embryo.

A wide ranging terminology has been used to designate the egg envelope of the fish. Most commonly encountered are: zona pellucida, pellucid membrane, corona radiata, zona radiata (interna and externa), vitelline

JAMES N. DUMONT • Biology Division, Oak Ridge National Laboratory, Oak Ridge, Tennessee 37830. ANNA RUTH BRUMMETT • Biology Department, Oberlin College, Oberlin, Ohio 44074 membrane, vitelline envelope, and chorion. Historical aspects of pertinent nomenclature were recently reviewed by Laale (1980) and are not dealt with here. The term **vitelline envelope** seems most appropriate, since it relates the envelope directly to what appears to be the homologous structure in a broad spectrum of organisms. However, since the term chorion is very widely used, we shall continue to so designate the mature envelope that surrounds the ovulated egg. It is the chorion that (1) becomes separated by the perivitelline space from the egg surface at the time of activation, (2) hardens to serve as a protective covering for the developing embryo, and (3) is finally softened and thinned from the inside by the action of hatching enzymes released by the contained larva.

## 2.1. Origin and Structure of the Vitelline Envelope

In very young oocytes, the beginnings of chorion formation can be observed as a thin layer of electron-dense material between the short microvilli that adorn the surface of the oocytes (Azevedo, 1974; Corporiccio and Connes, 1977; Riehl, 1978; E. Anderson, 1967, 1974; Tesoriero, 1977a; Shackley and King, 1977; Flegler, 1977; Jollie and Jollie, 1964; Ulrich, 1969; Wourms, 1976; Lonning, 1972; Dumont and Brummett, 1980; Stehr and Hawkes, 1983). As oogenesis continues, additional material is synthesized and added proximal to the oocyte. The envelope grows in thickness and complexity, has been shown to be rich in polysaccharides (Tesoriero, 1977b; Tiberi-Marquis, 1973), and is frequently referred to as the **zona radiata externa.** Whereas no truly definitive experiment to establish the provenance of the teleost vitelline envelope has been accomplished, existing circumstnatial evidence indicates that the principal, and in some cases the exclusive, source of the envelope is the oocyte itself (Tesoriero, 1977a). However, certain specializations that later adorn the envelope are most likely of follicle cell origin. In those species whose envelopes are adorned with **chorionic fibrils,** the zona externa of the envelope very early exhibits periodic postrusions, or hillocks, that constitute the point of attachment of the chorionic fibrils (Fig. 1) (Tsukahara, 1971; Dumont and Brummett, 1980; Hart and Donovan, 1983). The fibrils have their beginning as very dense material associated with the hillocks and extend into the spaces between neighboring follicle cells. The components of the fibrils, which are considered by most investigators to be derived from the follicle cells, exhibit a regular pattern of densely-packed filaments of tubules (E. Anderson, 1967; Wourms, 1976; Wourms and Sheldon, 1976; Tsukahara, 1971; Flugel, 1967; Busson-Mabillot, 1973; Dumont and Brummett, 1980; Hart *et al.*, 1984).

As the oocyte continues to grow and develop, the vitelline envelope (i.e., developing chorion) increases in thickness and complexity. Additional material **(zona interna)** is apparently elaborated by the oocyte and deposited below the zona externa described above. The material is clearly

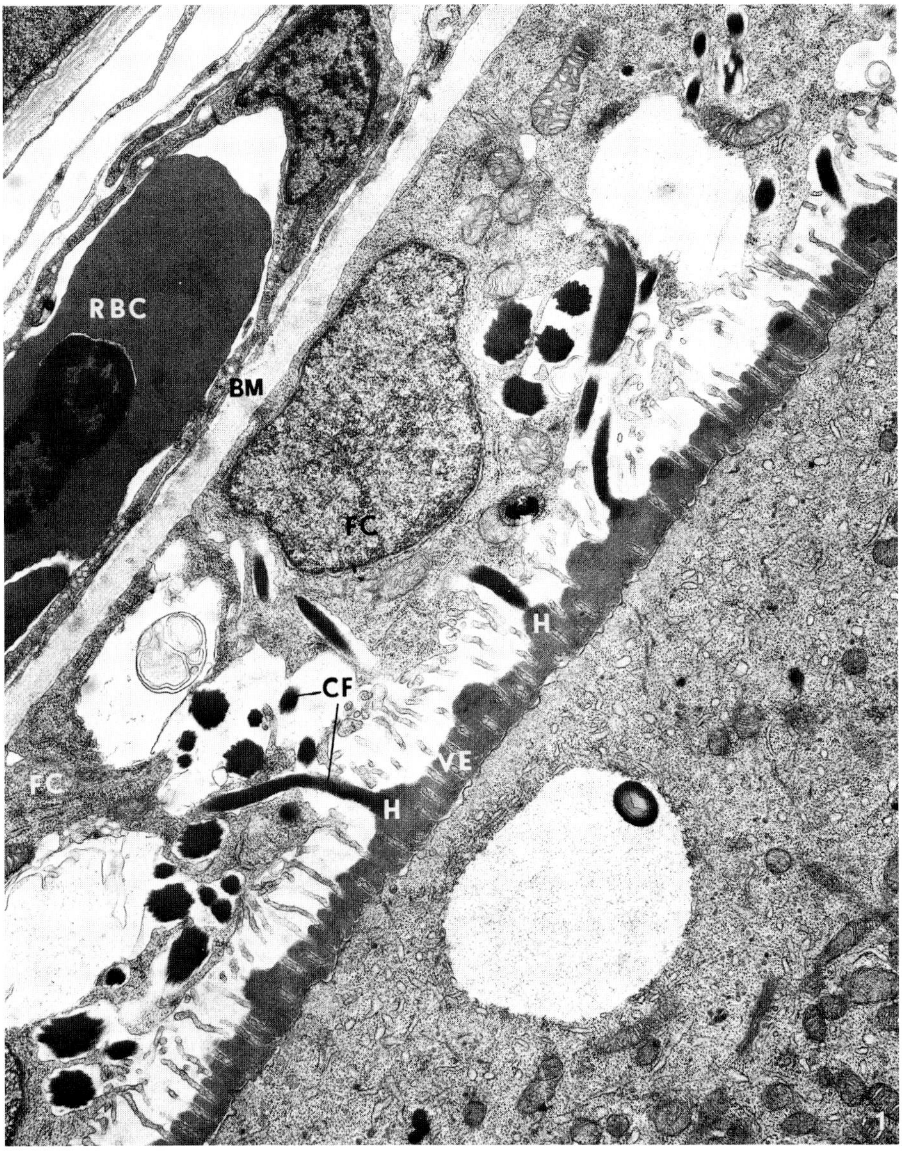


**Figure 1.** The surface of an oocyte and surrounding follicular components from a medium-size follicle of *Fundulus*. Long chorionic fibrils (CF) are attached to small protrusions (hillocks, H) on the developing vitelline envelope (VE) and extend into the spaces between the follicle cells (FC). Distal to the (BM) is a capillary containing red blood cells (RBC). ×10,300. (From Dumont and Brummett, 1980.)

distinguishable from that of the zona externa both in terms of its electron density and its composition. Whereas the external zone is granular or particulate, the underlying layers tend to be fibrillar or filamentous with the components exhibiting architecturally complex relationships (Fig. 2) (Busson-Mabillot, 1973; Riehl and Schulte, 1972; Corporiccio and Connes, 1977; Flegler, 1977; Wourms, 1976; Dumont and Brummett, 1980). Those layers that make up the interna are composed essentially of protein, whereas the externa is rich in polysaccharides (Tiberi-Marquis, 1973; Busson-Mabillot, 1973). Microvilli of increasing length occupy very narrow canals (pore

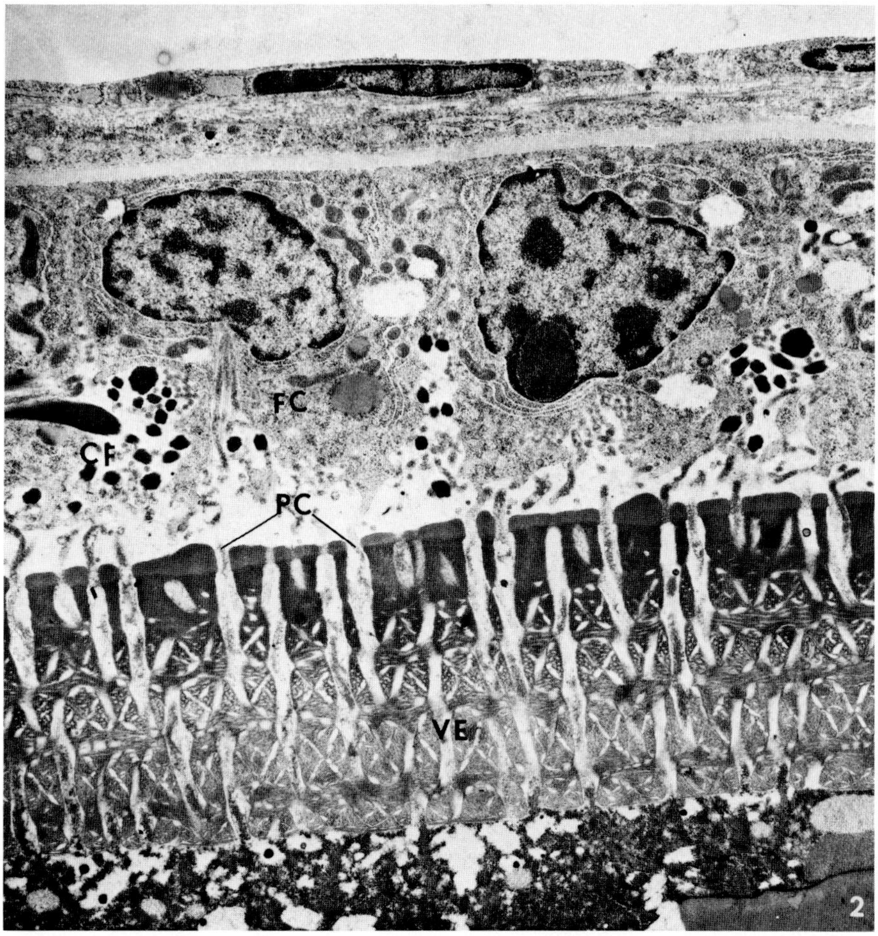


**Figure 2.** The mature vitelline envelope (VE) of *Fundulus* is stratified and consists of nine separate strata. Chorionic fibrils (CF) appear in cross and tangential section in spaces between the follicle cells (FC). Pore canals (PC) containing microvilli from the oocyte and from the follicle cells penetrate the envelope. ×7000. (Dumont and Brummett, unpublished.)

canals) in the thickening envelope. These produce the visible radiations characteristic of the envelope, which led to the terms **zona radiata** and **corona radiata.** Most evidence indicates that the vitelline envelope of the teleosts is derived from products synthesized by the oocyte. Thus, in accordance with accepted terminology, it should be considered a primary envelope. The chorionic fibrils, however, probably represent components of secondary (follicle cell) origin.

### 2.2. Structure of the Chorion

Near the end of oogenesis the vitelline envelope begins to exhibit changes in its ultrastructural characteristics (Dumont and Brummett, 1980). The architectural complexity of the inner portion gradually becomes less obvious as the envelope becomes more compact. Microvilli occupying the patent canals that traverse the envelope are apparently withdrawn or degenerate, and the patency of individual canals is lost (Figs. 3 and 6). Except for the micropyle (see Section 2.3) the mature vitelline envelope or chorion of the *Fundulus* egg appears to have no canals extending through it. The envelope of the carp egg, however, is reported to retain patent canals that are believed to permit the transport of cortical vesicle material released at the time of fertilization (Kudo, 1982). It is possible that, in some species at least, swelling of the egg may account for these changes by stretching and compacting the envelope (Stehr and Hawkes, 1983). Unique hexagonal patterning of the chorion of the sole has also recently been reported by Stehr and Hawkes (1983). These investigators suggest that the components forming the pattern are synthesized by the follicle cells and deposited around the circumference of these hexagonally packed cells. Thus, the shape of the follicle cell determines the pattern.

### 2.3. Origin and Structure of the Micropyle

The early literature on micropyle formation was reviewed by Laale (1980). Recent studies include those of Riehl (1977); Ohta and Teranishi (1982); Ohta and Takano (1982); Takano and Ohta (1982). Early in oogenesis, a single, large, conical, modified follicle cell, the **micropylar cell**—or the *Zapfenzellen* of Riehl (1977)—occupies an indented region of the developing vitelline envelope. As the oocyte grows and its vitelline envelope differentiates the micropylar cell maintains contact with the oocyte surface by means of a broad protoplasmic extension; thus the micropyle is formed (Fig. 4). In the studies of Takano and Ohta (1982) on the pond smelt, Ohta and Takano (1982) on the Pacific herring, and Ohta and Teranishi (1982) on the loach, a conspicuous feature of the broad protoplasmic process of the micropylar cell is a large bundle of microtubules oriented perpendicular to

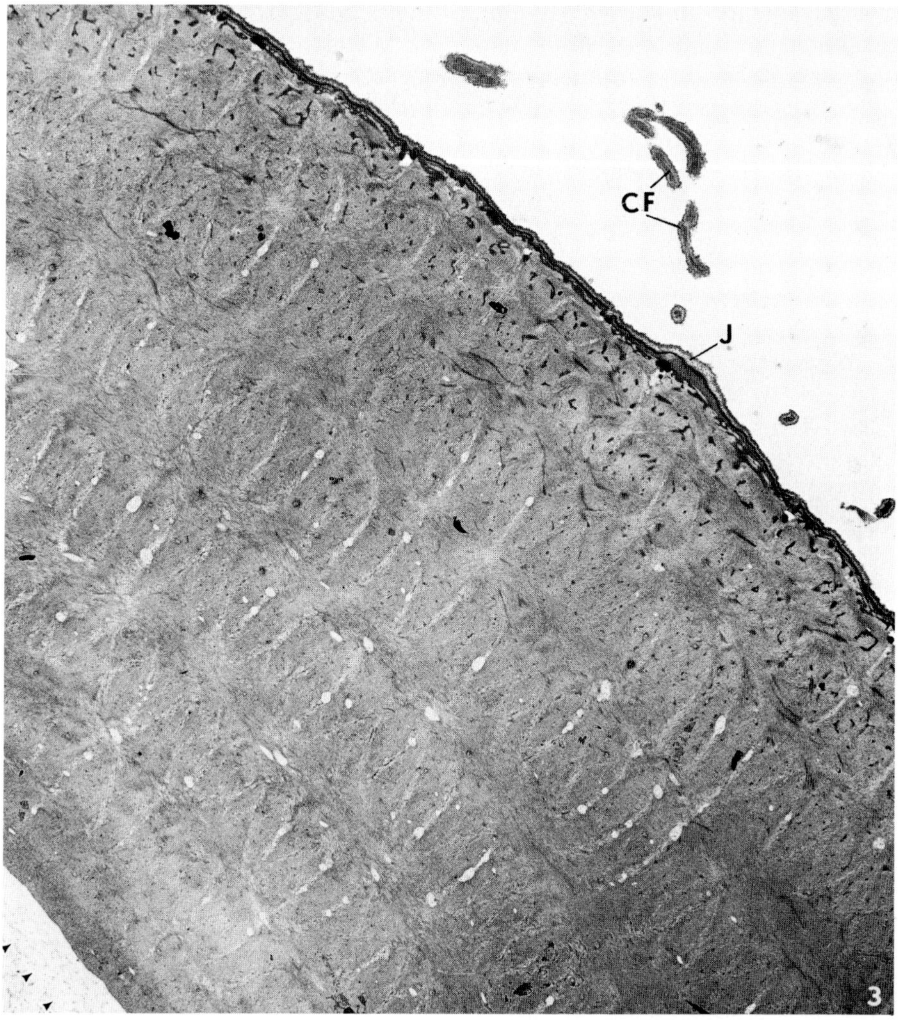


**Figure 3.** A nearly perpendicular section through the chorion of an ovulated *Fundulus* egg. The elaborate architecture characteristic of the vitelline envelope has been replaced, although there is still some indication of alternating strata. The pore canals are obliterated; the entire envelope is much more compact and appears finely filamentous. Internally, the envelope has become covered by a thin layer of material (arrows) and externally by a thin layer of jelly (J). CF: chorionic fibrils. ×3700. (From Dumont and Brummett, 1980.)

the oocyte surface. Similar features have been observed by Brummett and Dumont (unpublished) in developing ooctyes of *Fundulus*.

The fundamental pattern of individual micropyles appears to be the same in all species, but there are species-specific variations that frequently permit identification of the eggs on the basis of micropylar characteristics

(Iwamatsu and Ohta, 1981; Brummett and Dumont, 1979; Kuchnow and Scott, 1977; Kobayashi and Yamamoto, 1981; Riehl, 1977, 1980a,b). Basically, micropyle structure consists of a more or less conical pit (vestibule) leading to a patent canal that terminates at the surface of the ooctye (Fig. 5A). The canal is slightly tapered, and its opening at the surface of the ooctye measures only slightly larger than the diameter of the head of the sperm of that species. On the inner side of the chorion, the micropylar region appears as a conical hillock (reflecting the conical depression on the outer side) (Fig. 5B).

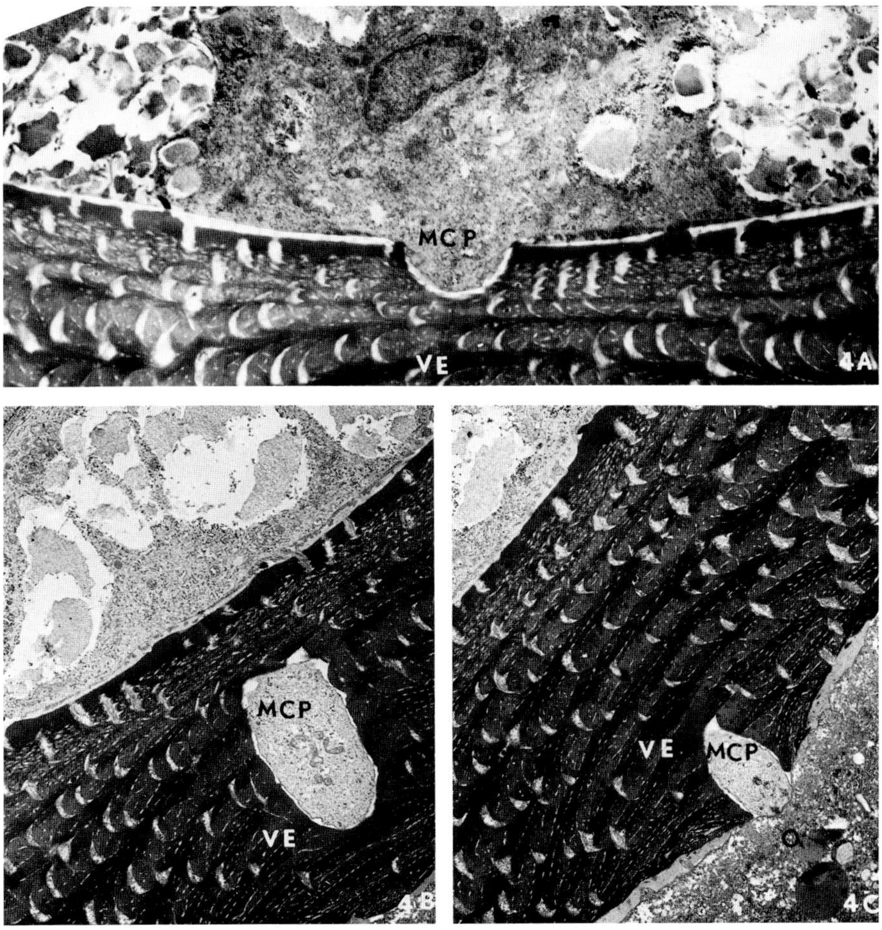


**Figure 4.** A series of three serial oblique sections through the developing micropyle of a *Fundulus* oocyte. The three micrographs show the progression of the broad micropyle cell process (MCP) through the vitelline envelope (VE). The cytoplasm is unique in that it contains large numbers of ribosomes and microtubules oriented along the long axis of the process. O: oocyte. ×5470. (Brummett and Dumont, unpublished observations.)

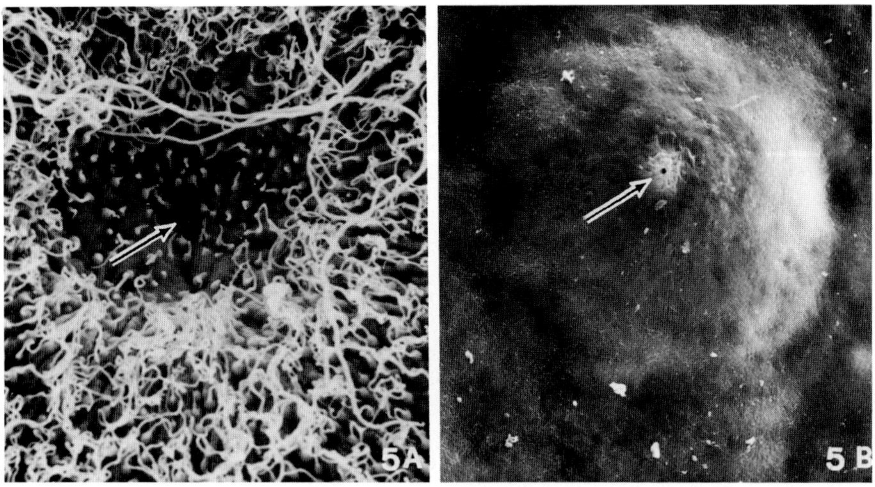


**Figure 5.** *Fundulus* micropyle as observed with scanning electron microscopy. Micrograph of the external opening (arrow) of the *Fundulus* micropyle, which is set into a funnel-shaped depression. Numerous chorionic fibrils surround the micropyle. ×850. (B) Micrograph of the micropylar cone on the internal surface of the chorion. The cone fits into a depression of similar size on the oocyte surface. The apex of the cone bears the opening of the micropylar canal (arrow). ×590. (From Dumont and Brummett, 1980.)

## 2.4 Alterations at Ovulation

At the time of (or subsequent to) ovulation, a final layer (the **jelly coat**) may be deposited on the chorionic surface. This has been described and pictured in *Fundulus* (Dumont and Brummett, 1980; Brummett and Dumont, 1981) (Figs. 3 and 6A,B) and in the medaka (Hart *et al.*, 1984). The source of this material may be the follicle cells or perhaps specialized cells of the ovarian lumen (Brummett *et al.*, 1982). Thus, this final covering—when it is present—constitutes either a secondary or a tertiary component of the fully mature egg envelope.

## 2.5. Specializations and Species Variations

The eggs of teleosts vary considerably in size and to some extent in shape (most are round, but some are ovoid or pear shaped). Size appears to be correlated directly with the amount of yolk present in the ovum, the time required for development from fertilization to hatching (though the latter can be modified considerably by environmental temperature), and with the number of eggs produced (fecundity). Although the egg envelopes of different species are built around fundamentally similar characteristics, there is considerable variation. Of particular importance is whether the

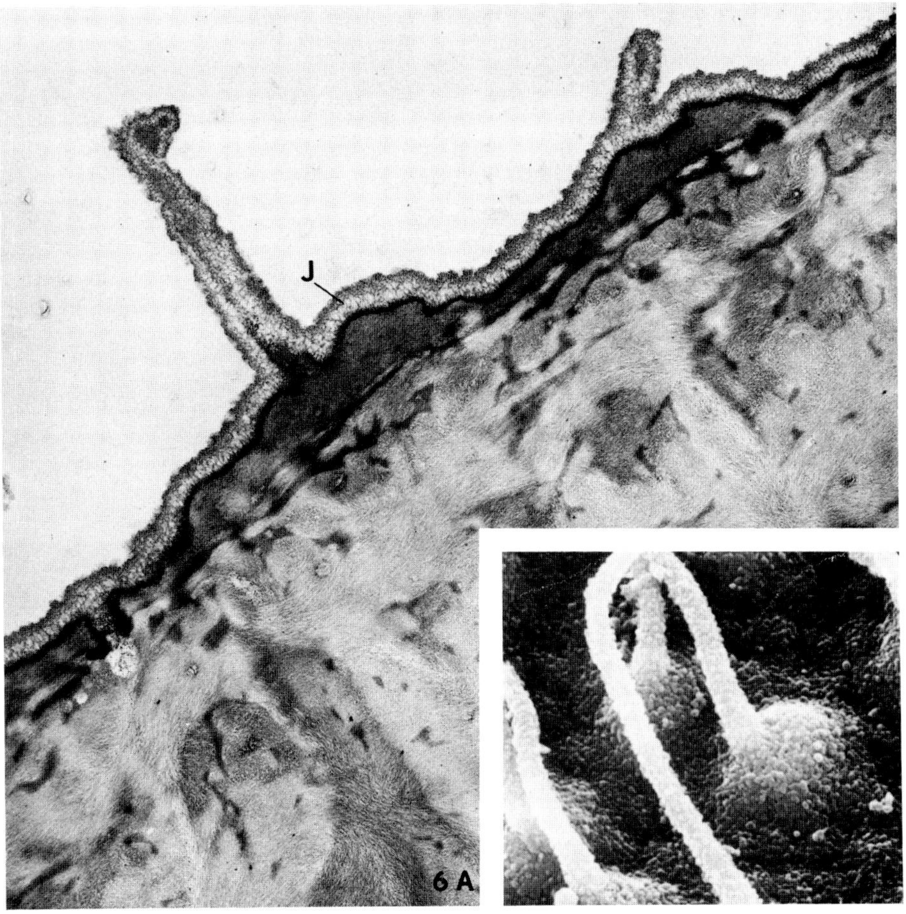


**Figure 6.** Chorion of the ovulated *Fundulus* egg. High magnification micrograph of the external surface of the chorion of an ovulated *Fundulus* egg. Note the spongy appearance of the jelly layer (J) on its surface and on the surface of the chorionic fibril. The external layers of the chorion are more compact than those of the vitelline envelope. ×14,500. (B) Scanning electron micrograph of chorionic fibrils of an ovulated *Fundulus* egg. Note the nodular appearance of the jelly, which covers the chorion and fibrils. ×6,300. (From Dumont and Brummett, 1980.)

species is oviparous, ovoviviparous, or viviparous and whether the egg is pelagic or demersal (Stehr and Hawkes, 1979; Flegler, 1977).

Among the more striking morphological differences of the eggs of teleosts are chorionic fibrils. These adornments vary dramatically both in length and in architecture. The elaborate fibrillar structures that ornament the surface of the annual fishes *Cynolebias melanotaenia* and *C. ladigesi* have been described in considerable detail by Wourms and Sheldon (1976).

It has generally been assumed that the function of the chorionic fibrils is to attach the oviposited egg to vegetation or some other environmental substrate (E. Anderson, 1974; Brummett, 1966; Brummett and Dumont, 1982) and to facilitate the retention of moisture at the egg surface when it is exposed at low tide (Kuchnow and Scott, 1977). In the case of *Oryzias*, Tsukahara (1971) and Hart *et al.* (1984), have described two topographically distinct types of chorionic fibrils which they refer to as nonattaching and attaching. The nonattaching fibrils show a regular spatial distribution over the chorion surface and are thought to function in maintaining the integrity of the egg cluster. The second type consists of a group of thin fibrils 25–30 $\mu$m in length that are confined to a single area of the chorion and intertwine with each other and with similarly intertwined groups arising from other eggs in the cluster. Together they form an umbilical-like cord, which suspends the egg cluster from the gonoduct of the female for several hours after laying. Brummett and Dumont (1981) also described chorionic fibrils of two different thicknesses on the eggs of the southern race of *Fundulus heteroclitus;* those of the northern race of the same species, however, have only one type of fibril.

## 2.6. Alterations at Fertilization

When the teleost egg is activated by a penetrating sperm, the egg responds by cortical vesicle breakdown and the appearance of an obvious perivitelline space. The previously soft chorion, which before this time was closely opposed to the egg plasmalemma, separates from the latter, hardens, and becomes taut (Kao and Chambers, 1954). Factors that have been implicated in the hardening of the chorion include calcium ions, components of the colloid released from the cortical vesicles, and a hardening enzyme or a phospholipid derived from the egg cortex (see review by Yamamoto, 1961). Ohtsuka (1960) concluded that hardening results from an oxidation of sulfhydryl groups of protein and glycol groups of polysaccharide components of the soft chorion (see review by Blaxter, 1969). It is noteworthy that a comparable set of reactions involving similar reaction components is involved in alterations of the amphibian vitelline envelope at the time of fertilization (see Section 6.4).

On the basis of ultrastructure and cytochemistry, Kudo (1982) believes that a distinction between the vitelline envelope (e.g., the chorion before egg activation) and the **fertilization envelope** (the chorion after egg activation) is justified. He interprets his data to indicate that cortical alveolar material released at the time of fertilization reaches the outer surface of the chorion through patent chorionic canals and effects a morphological change in the outermost layer of the chorion.

The jelly, which covers the egg at the time of oviposition, also appears to undergo a physicochemical change as the chorion hardens after egg activation. Initially, the change may involve increased adhesiveness—presumably enhancing the adherence of the eggs to environmental substrates or to each other. This adhesiveness subsequently disappears in most species (see review by Laale, 1980).

## 2.7. Functions of the Envelope

### 2.7.1. Role in Fertilization

The egg and the sperm of the teleost have, of course, evolved together and are well adapted to effect rapid, monospermic, species-specific fertilization. The thick chorion prevents the entry of the acrosomeless sperm except at the micropyle, the inner diameter of which can generally accommodate only one sperm. In addition, there is some evidence that fertilization is further facilitated by the presence of a sperm attractant in the micropylar region of the chorion (Suzuki, 1958). Thus, the structure of the chorion ensures that a single sperm will contact the egg at the preferred fertilization site; polyspermy is prevented, and effective monospermic fertilization is facilitated. It is possible, then, that the micropyle not only serves as an access route to the egg, but that it also facilitates fertilization by attracting (binding?) sperm to the appropriate region of the chorion. It is conceivable that this could also be a function of the jelly. Kudo (1982) describes ultrastructural changes in the chorion of the egg of *Cyprinus carpio* resulting from the addition of cortical vesicle material to the outer layer of the vitelline envelope. There the material serves as a physical barrier and may also serve both protective and attachment functions for the developing embryo.

### 2.7.2. Protection

The primary function of the chorion, once fertilization has occurred, is the protection of the developing embryo from environmental hazards. The chorionic fibrils and the outer jelly coat cause the egg to adhere to substrates in the spawning area. For those species whose spawn are subject to tidal fluctuations, the chorion with its jelly coat provides protection against desiccation and rapid salinity changes. The embryo is also protected against the deleterious effects of bacteria, fungi, and small invertebrates in the environment. In summary, the chorion and its enclosed perivitelline fluid provide a sterile and stable environment for the developing embryo. The importance of this environment is easily demonstrated when one attempts to raise dechorionated embryos in the laboratory: the fragility of the developing embryo in the absence of its chorion becomes readily apparent.

#### 2.7.3. Permeability

The intact chorion of the teleost egg is freely permeable to $H_2O$, $CO_2$, $O_2$, salts, dyes, and amino acids. The colloidal content of the cortical vesicles, released to the perivitelline space at the time of fertilization causes a rapid imbibition of water from the environment (and possibly from the egg itself). Eggs with elastic chorions (e.g., *Oryzias*) swell; those with rigid chorions (e.g., *Fundulus*) remain the same size but become more turgid (Lonning and Davenport, 1980; Kao and Chambers, 1954). Eggs with elastic chorions are the most fragile and typically develop in quieter, more protected, waters. In the case of the annual fishes that undergo a diapause buried in mud during the dry season, it has been suggested that the elaborate ornamentation of the chorion (the secondary component) may function as a chorionic respiratory system comparable to the chorionic plastron of some aquatic insect eggs (Wourms and Sheldon, 1976).

### 2.8. Alterations at Hatching

The chorion remains intact around the developing embryo until the latter reaches a relatively late larval stage. At that time, hatching enzyme(s) secreted by the larva cause it to become noticeably softer and thinner; muscular movements of the enclosed larva result in the tearing of the envelope and the release of the fry (Schoots *et al.*, 1982). While the hatching "chorionase" can be concentrated simply by permitting a number of fry to hatch in a small volume of medium (Milkman, 1954), attempts to use it for the chemical dechorionation of embryos at earlier stages of development have met with little success. That this should be the case is not too surprising: (1) The chorions may change with time, and those of earlier embryos may not have reached a stage of maturation susceptible to the enzyme; and (2) only the outer surfaces of the young intact chorions are exposed to the enzyme, whereas normal exposure to effect hatching is from the inside. When *pieces of chorion* are exposed to "chorionase", the dissolution of the innermost layer begins to occur within a few minutes, whereas the outer layer is not dissolved even after prolonged incubations with the enzyme (Kaighn, 1964; Yamamoto and Yamagami, 1975).

## 3. Lampreys

The basic structure of the vitelline envelope of lamprey oocytes is similar to that of the teleost (Busson-Mabillot, 1967). Whether there is a preferred sperm entry site (i.e., whether the sperm's access to the egg is limited to a particular region) is not known. There is no micropyle, however, and the sperm possesses an acrosome that presumably releases enzymes to digest a path through the envelope. Afzelius *et al.* (1968) report that the long acrosomal filament and the head of the sperm appear to reach the egg surface through pore canals in the envelope. The pore canals are believed to be formed by pseudopodial connections between the oocyte and follicle cells as the envelope is formed during oogenesis.

## 4. Chondrichthyes

Little is known about the details of the formation of the vitelline envelope in the chondrichthyans. Ginsberg (1972) reports that a delicate homogeneous layer first appears on the surface of the young oocyte and that a second layer with distinct radial striations subsequently appears underneath the first. Both layers are considered to be formed by the oocyte and together they constitute the zona radiata. As in the case of the birds and reptiles, fertilization must occur before egg case formation is completed. The latter is a tertiary envelope synthesized and secreted by the shell gland. Biochemical and biophysical studies have revealed that collagen is the primary component of the shell (see review by Wourms, 1977). The morphology of the egg case varies considerably from species to species, but basically two types are produced. Oviparous species produce a permanent egg case within which embryonic development is completed. The developmental picture in viviparous species is more variable (see Wourms, 1977). In some instances (e.g., sting ray), no egg case is formed; in others a temporary case is formed from which the embryo emerges and completes development *in utero;* in still others the egg case is retained throughout the *in utero* developmental period, sometimes being incorporated into the placenta (TeWinkle, 1963).

The egg envelopes of chimaeras show close similarities with those of selachians. The eggs are large and yolky and covered by a firm capsule.

## 5. Primitive Osteichthyes

The eggs of chondrostean fishes are enclosed in a well-developed two- or three-layered zona radiata and an outer jelly coat that gives the egg a highly adhesive quality (Cherr and Clark, 1982). Trypsin affects the integrity of the envelopes, and they are mucopolysaccharide positive. Several micropyles (2 to 40) are distributed over the animal pole of the egg. Ultrastructurally, the micropyles are similar to those of teleost eggs (see Fig. 5). The cortical cytoplasm extends up into the innermost region of the micropyle (Ginsberg, 1972).

Only scant data have been accumulated on oogenesis in the lung fishes. A typical zona radiata develops and is gradually reduced in thickness. Insemination is external, and spawned and fertilized eggs are covered with a firm envelope that resembles a horny capsule. It is surrounded by a jelly coat resembling that of amphibians (see Section 3.2.).

# 6. Amphibians

## 6.1. Origin and Structure of the Vitelline Envelope

The amphibian ovarian follicle is a comparatively simple association of cellular and acellular components (Dumont and Brummett, 1978). In *Xenopus* the vitelline envelope begins to form between the oocyte surface and the overlying follicle cells early in oocyte development (stage II) (Dumont, 1972). At first it appears as isolated patches of fibrillar material that later become continuous over the surface of the oocyte (Fig. 7). At its

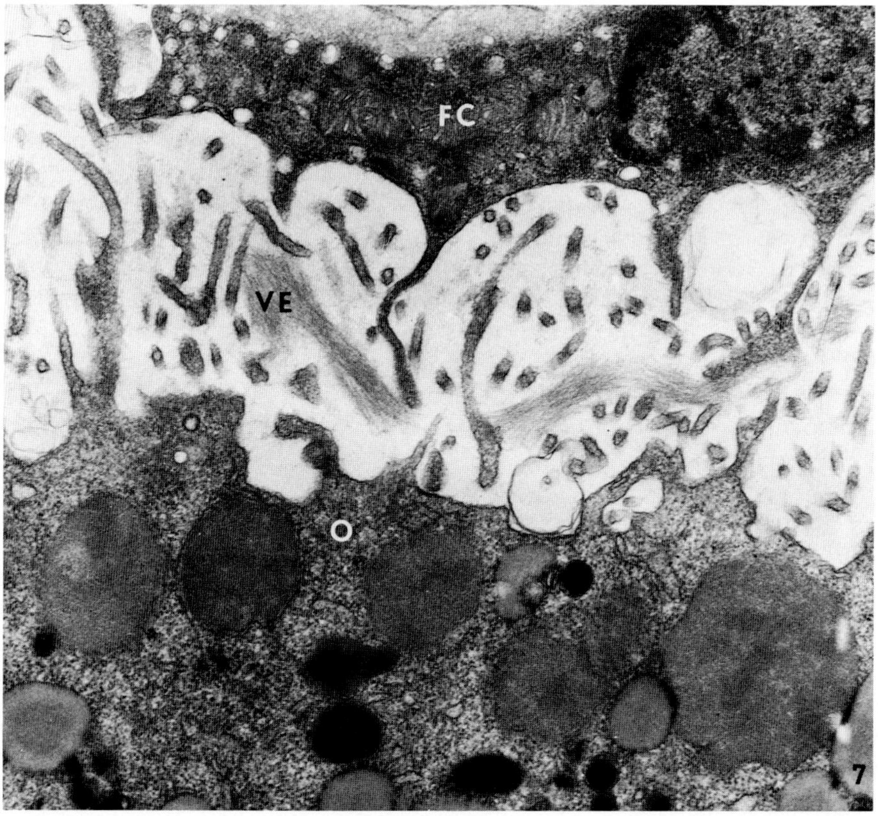


**Figure 7.** Early stage II oocyte of *Xenopus* with wisps of finely filamentous material—the developing vitelline envelope (VE)—in the space between the follicle cells (FC) and the oocyte (O). ×6180. (From J. N. Dumont, unpublished observations.).

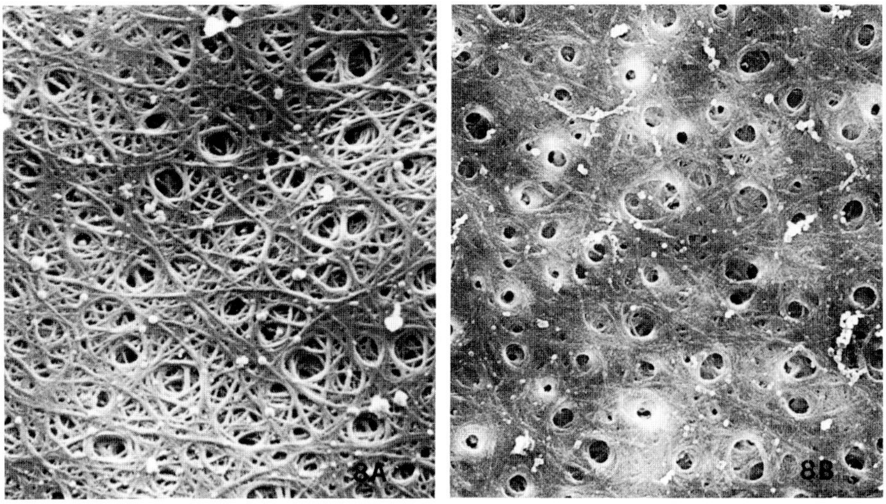


**Figure 8.** The *Xenopus* vitelline envelope of preovulatory and postovulatory oocytes. (A) Scanning electron micrograph of the surface of the vitelline envelope of a mature Stage VI *Xenopus* oocyte. The surface of the envelope is textured and appears as a mat of fibrils. Openings or pores allow the penetration of microvilli and account for the permeability of the envelope. ×10,000. (B) Surface of a coelomic egg of *Xenopus*. The surface is smoother than that of the ovarian oocyte, although the pores and tunnels remain. Smoothing of the surface may be the result of factors that cause detachment of the follicle cells before ovulation. ×5900. (From Dumont and Brummett, 1978.)

maximum thickness, it is composed of loosely packed 4–7-nm filaments woven into an intricate mat (see Fig. 9 and 10) (Grey *et al.*, 1974; Dumont and Brummett, 1978). Scanning electron microscopy reveals that the envelope is punctuated by relatively large pores or channels through which long macrovilli from the follicle cells pass to make contact with the oocyte (Fig. 8A). Thus, the vitelline envelope is a porous structure and permits the passage of relatively large molecules (including ferritin, dextran, horseradish peroxidase, thorium dioxide, and trypan blue, as well as yolk precursors) from the circulatory system of the follicle to the oocyte surface (Dumont, 1978; Brummett and Dumont, 1977).

In the salamanders the vitelline envelope (zona pellucida, Wischnitzer, 1963, 1964) consists of two layers: a proximal zona radiata (a classic term now infrequently used in the amphibian literature) and a peripheral (outer) homogeneous layer. More recent studies of the salamander, as well as of bufonids and ranids, indicate close structural similarities among the vitelline envelopes of all amphibian groups.

The origin of the vitelline envelope in amphibians remains problematic. Suffice it to say that for the homologous mammalian structure (the zona pellucida) good evidence has now been presented to confirm its origin

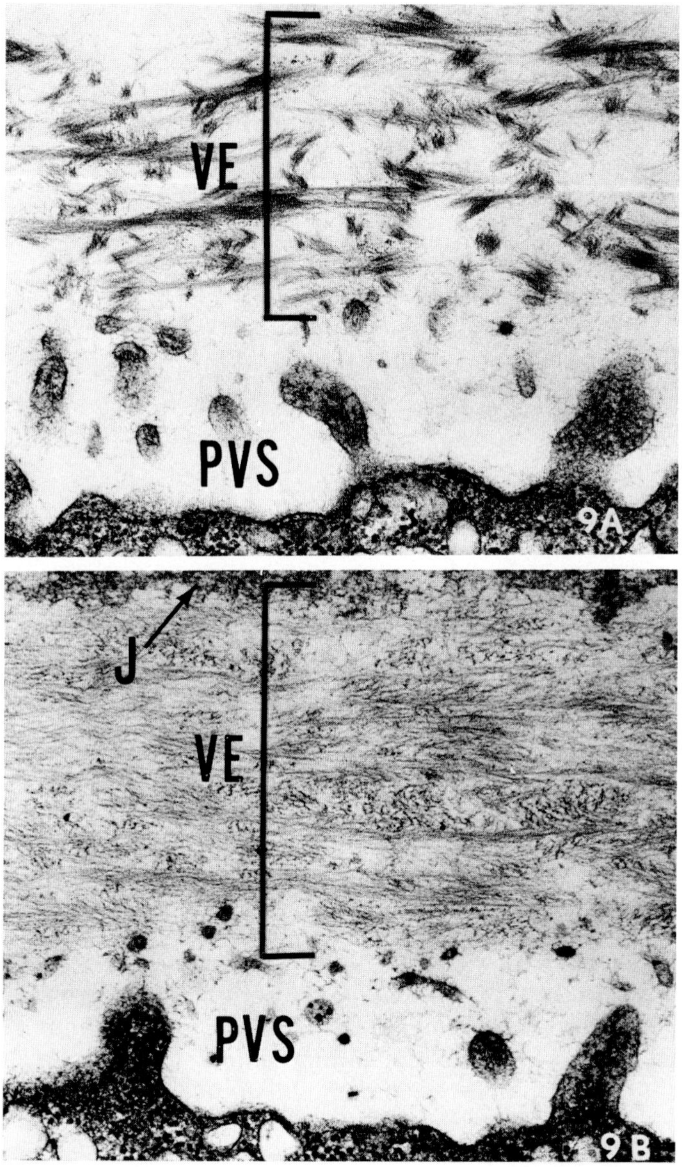


**Figure 9.** Changes in the structure of the vitelline envelope during passage through the oviduct. (A) Vitelline envelope (VE) of a coelomic *Xenopus* egg. The envelope is composed of bundles of fibrils that lie parallel to the egg. The envelope is composed of bundles of fibrils that lie parallel to the egg surface. The bundles are more dispersed than in the ovarian oocyte. PVS, perivitelline space. ×39,000. (B) Vitelline envelope (VE) of an oviposited *Xenopus* egg. The filaments are now more dispersed, for example, discrete bundles are less obvious. The inner portion of the J1 jelly coat (J) is seen on the surface of the envelope. PVS, perivitelline space. ×39,000. (From Grey *et al.*, 1977.)

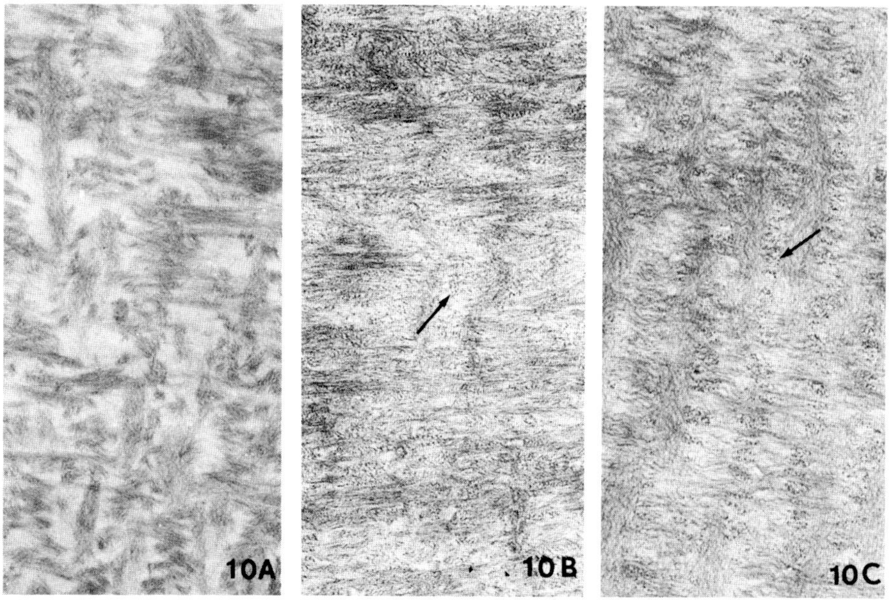


**Figure 10.** Transformation of the *Bufo* vitelline envelope by pars recta material. (A) Vitelline envelope of Bufo is similar to that of *Xenopus* but with bundles of filaments oriented perpendicular as well as parallel to the oocyte surface. (B) *Bufo* vitelline envelope from an egg that had entered the oviduct where particulate material (arrow) is added. At this time, the envelope fibrils disperse, and the channels disappear. (C) *Bufo* envelope treated with pars recta extract. The extract causes the same changes as observed *in vivo*, illustrating that the alterations are caused by influences from the oviduct. ×31,370. (From Katagiri *et al.*, 1982.)

as being solely from the oocyte (see Section 11.1). Thus, considering the similarities it seems reasonable to consider comparable origins for both.

## 6.2. The Envelope of Coelomic Eggs

Scanning electron micrographs of vitelline envelopes of ovulated coelomic eggs show a much smoother appearance than those of ovarian oocytes. (Fig. 8B). The follicle cells become detached from the envelope just before ovulation, and it has been suggested that the smoothing of the surface may be the result of the same factor or factors that bring about the dissolution of the relationship between the follicle cells, vitelline envelope, and/or oocyte (Dumont and Brummett, 1978). In contrast to the loosely packed filaments of ovarian eggs, the vitelline envelope filaments of coelomic eggs become tightly packed into larger bundles that are more randomly arranged (see review, Schmell *et al.*, 1983) (Fig. 9). The coelomic egg envelope prevents the penetration of sperm and thus fertilization.

However, if the envelope is torn, ruptured or altered chemically, fertilization of coelomic eggs is possible (Elinson, 1973; Katagiri, 1974).

## 6.3. The Envelope of Oviducal Eggs

Striking and discrete structural changes occur in the vitelline envelope as the egg passes into the initial segment of the oviduct. Using *Xenopus*, Grey *et al.* (1977) have shown that after the passage of the egg through the oviduct the fibril bundles disperse, and the channel spaces disappear despite the fact that the fibril diameter and envelope thickness remain unchanged (Fig. 9B). Dispersal of the fibrils occurs in the anterior region of the oviduct, or *in vitro* by raising or lowering the pH of the medium. Katagiri *et al.* (1982) report similar alterations of *Bufo* vitelline envelopes. In *Bufo*, particulate material from the pars recta is also added to the envelope as the egg passes into the oviduct. That this material is involved in the structural transformation of the envelope has been shown by Yoshizaki and Katagiri (1981) and Katagiri *et al.* (1982) (Fig. 10A–C). The alterations induced by this material from the oviduct enhance the susceptibility of the vitelline envelope to the lytic activity of the sperm-derived lysins and thus increase fertilizability (Yoshizaki and Katagiri, 1982; Miceli *et al.*, 1978a,b). In *Xenopus* the added component has a molecular weight of 57,000, and — at the time of its addition—other components of the envelope are modified (Schmell *et al.*, 1983). It is not clear whether the increased ability of sperm to penetrate the envelope is due directly to alterations of the envelope itself or whether the role of the secretions is simply to facilitate or induce the acrosome reaction. Finally, it is interesting that pars recta extracts from hibernating animals are ineffective in increasing fertilization.

Schmell *et al.* (1983) propose the following hypothesis regarding the change in form and function of the vitelline envelope after exposure to oviducal secretions: An oviducal protease selectively hydrolyzes the 43,000-$M_r$ envelope component, causing a conformational molecular change that ultimately alters the interaction of the envelope with sperm—either sperm binding or envelope susceptibility to sperm lysins (acrosin). The result, of course, is an increased probability of successful fertilization.

## 6.4. Alterations at Fertilization: The Fertilization Envelope

Significant structural changes also occur in the vitelline envelope at the time of fertilization that are directly related to the cortical reaction. During this event cortical granule lectins, expelled into the perivitelline space, pass through the vitelline envelope and interact with the jelly to alter the envelope both chemically and physically. After this interaction the physically and chemically altered vitelline envelope is referred to as the fertilization envelope (Grey *et al.*, 1974). The fertilization envelope is characterized by the appearance of a superficial layer (the **fertilization** or **F layer**) with two distinct morphological regions. One, which is immediately adjacent to the outer margin of the envelope, is a thin condensed band that completely covers the surface. Peripheral to this band, and lying in contact with the innermost jelly layer, is a more irregular component that appears dispersed and flocculent (Fig. 11A–C). The layer is formed by the interaction of J1 jelly, cortical granule exudate, and divalent ions. The reaction most likely involves galactoside residues of sulfated jelly macromolecules, for in the absence of jelly a discrete F layer is not formed. Schmell *et al.* (1983) have summarized the details of these envelope alterations and have formulated an hypothesis that incorporates the possible interactions of cortical granule lectins and enzymes and jelly ligands.

Concomitant with fertilization and the cortical reaction the fertilization envelope lifts from the egg's surface leaving an intervening space, the perivitelline space. The mechanism by which this space is formed has been attributed to osmotic changes brought about by alterations in the permea-

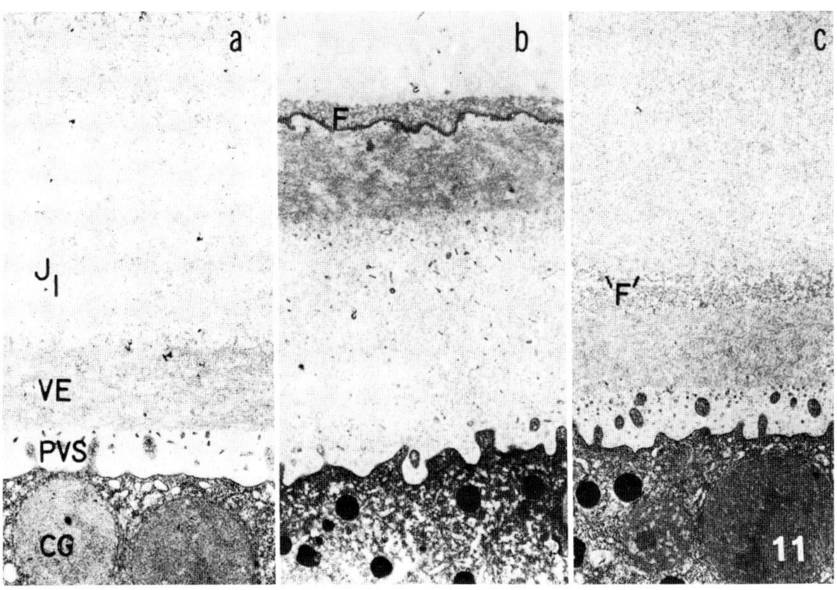


**Figure 11.** Envelopes of *Xenopus* eggs. (A) Unfertilized. (B) Fertilized. (C) Treated with cortical granule exudate. In the unfertilized egg, cortical granules (CG) are present in the ooplasm. The perivitelline space (PVS), vitelline envelope (VE), and J1 jelly (J1) are depicted. In the fertilized egg, the cortical granule contents have been expelled into the perivitelline space, the envelope (now the fertilization envelope) has lifted from the egg surface, and the F layer (F) has formed between the envelope and the jelly. The formation of the F layer can also be induced by treatment of unfertilized eggs with isolated cortical granule exudate. ×9000. (From Wyrick *et al.*, 1974.)

bility of the egg envelope. Indeed, permeability differences between the vitelline envelope and fertilization envelope have now been demonstrated that can account directly for the osmotic elevation of the envelope and formation of the perivitelline space (Nishihara and Hedrick, 1977; Greve and Hedrick, 1978).

### 6.5. Composition and Character of the Vitelline and Fertilization Envelopes

Immunological studies using anti-vitelline envelope and anti-fertilization envelope sera show that the vitelline envelope contains at least three of the same antigenic molecules as the fertilization envelope. The fertilization envelope, however, contains two additional antigens—one derived from the jelly layer (J1) and the other derived from the cortical granule components (Wolf *et al.*, 1976; Wyrick *et al.*, 1974). Both envelopes contain about 16% carbohydrate and 84% protein. Fucose, glucosamine, galactosamine, mannose, galactose, and sialic acid are present in both but in quantitatively different amounts. Amino acid analysis does not exhibit unique types or amounts, although the relative content of the envelopes differs significantly. For example, the aromatic amino acid content of the fertilization envelope is higher than that of the vitelline envelope. Lipids have not been detected in significant amounts (Wolf *et al.*, 1976).

Polyacrylamide gel electrophoresis identifies a variety of glycoproteins ranging from 16,000 to 125,000 $M_r$. Those with molecular weights of 40,000 and 33,000 represent two-thirds of the mass of the envelope. Electrophoretic patterns from the fertilization envelope display the same components as the vitelline envelope except that a 77,000-$M_r$ material is absent and a 67,500-$M_r$ component appears, presumably from the proteolysis of the 77,000-$M_r$ component. If fertilization envelopes are prepared in the presence of calcium, additional components are found that are apparently derived from the cortical granule material that becomes deposited in the F layer.

Distinct changes in the physical properties of the envelope also occur and are coincident with the chemical changes at the time of fertilization. For example, the fertilization envelope is much less soluble, more resistant to enzymatic digestion, and has more structural rigidity than the vitelline envelope (Miceli *et al.*, 1971; Wolf, 1974; Wolf *et al.*, 1976). In fact, in *Xenopus*, decreased solubility of the envelope can be correlated with the progression of the cortical reaction, implying that cortical granule material reacts with components of the envelope in such a way as to reduce solubility. The physicochemical changes are clearly capable of forming a block to polyspermy (Wyrick *et al.*, 1974; Grey *et al.*, 1976). For example, sperm in the presence of jelly extract are capable of binding to and penetrating isolated vitelline envelopes, but neither bind to nor penetrate fertilization envelopes. The block is not only at the superficial F layer; sperm cannot penetrate from the inner surface either, indicating that the entire fertilization envelope is significantly different from the vitelline envelope.

It is interesting to recall that urodele eggs are polyspermic. These eggs lack cortical granules (see Wischnitzer, 1966) so that the formation of a *Xenopus*-type fertilization layer involving contents of the cortical granules is not possible.

## 6.6. Origin and Composition of the Jelly Envelope

Historically, many functions have been attributed to the jelly layers. For example, it has been postulated that they are important in altering the physiology of the ovum (Humphries, 1961), that they serve as an energy (nutrient) source for the developing embryo (Nace *et al.*, 1960; Humphries, 1961), and that they possess immunological characteristics that prevent (or at least regulate) interspecies fertilization and prevent polyspermy (Shaver and Barch, 1960; Shivers, 1962; Shivers and Metz, 1962; Shaver *et al.*, 1970; Shaver, 1966). That at least some jelly components are essential for normal fertilization has been unequivocally demonstrated.

The jelly is synthesized and secreted by the oviduct under the influence and control of hormones. Histological and histochemical studies of the oviducts of a variety of amphibians have been published but will not be discussed in detail here (Humphries and Hughes, 1959; Boisseau, 1973a,b, 1975; Xavier, 1973; McLaughlin and Humphries, 1978; Lee, 1967b, 1969). Suffice it to note that as the egg passes through the oviduct it acquires sequential and unique layers of jelly, the properties of which can be correlated with specific segments of the oviduct.

Morphologically, the jelly coat of amphibian eggs consists of three or more layers, depending on the species and the technique used to study it (Nace *et al.*, 1960; Salthe, 1963; Steinke and Benson, 1970; Shivers and James, 1970a; Freeman, 1968). From proximal to distal the layers are designated J1, J2, and J3. Each layer shows characteristic histochemical staining reactions, but, in general, glycoproteins (acid and neutral mucopolysaccharides) are the major components (Monroy, 1965; Humphries, 1966; Humphries *et al.*, 1968; Yurewicz *et al.*, 1975; Lee, 1967a). Analyses of the jelly of *Rana* by cytochemical and biochemical procedures show that the significant polysaccharides are hexoses and hexosamines (galactosamine, glucosamine, galactose and fucose). In *Rana* the relative proportions of the components are protein, 28–48%, hexosamines, 28–45%; hexoses, 20–25%; and sialic acid 15–18% (Lee, 1967a). These data are similar to those obtained for other species (Minganti, 1955; Yurewicz *et al.*, 1975; Viarengo and Falcone, 1977). Each jelly layer has its own unique chemical composition. Clearly, carbohydrate is the major component of the amphibian eggs.

Sulfate in the jelly, especially in J1, has been documented by many investigators, including Humphries (1966), Freeman (1968), Steinke and Benson (1970), and Kelly *et al.* (1970). Radioactive sulfur incorporation into the jelly coats was demonstrated by Humphries (1970), Pereda (1970), and Hedrick *et al.* (1974). Little if any of the sulfate is incorporated into the developing embryo, which tends to discount the hypothesis that the jelly is used by the embryo as a food source (see Hedrick *et al.*, 1974).

## 6.7. Functions of the Envelopes

### 6.7.1. Role in Fertilization

Typically, fertilization of coelomic eggs, which have not yet acquired their complement of jelly, or from eggs that have had the jelly removed is possible only under very specific and stringent conditions (cf. Aplington, 1957; Elinson, 1971; Katagiri, 1963, 1965, 1966; Wolf and Hedrick, 1971). Although jellyless eggs are not usually fertilizable, pretreatment of the sperm by contact with jelly or with jelly extract, makes them capable of activating jellyless eggs. This conditioning of the sperm has been homologized with capacitation of mammalian sperm (Shivers and James, 1970b, 1971). That the "capacitating factor(s)" are located in the jelly is demonstrated by the fact that exposure of sperm to extracts of other tissues fails to make sperm capable of activating oocytes. Available evidence suggests that the acrosome reaction does not occur during the process of capacitation.

The addition of water extract from jellied oocytes (egg–water or jelly–water) to sperm and to dejellied eggs improves fertilization, although comparatively high sperm concentrations may be required for fertilization in some cases. Such observations clearly demonstrate the requirement for jelly. When eggs possess only the J1 layer, fertilization rates are very low: sperm do not appear to be capable of entering or penetrating this layer (Barbieri and d'Atenor, 1973; Katagiri, 1965). Ooctyes that possess J1 and J2 become fertilized at a rate less than normal, and oocytes with J3 only are fertilizable but only at very high sperm concentrations (Elinson, 1971; Raisman and Pisano, 1970). It should be noted that fertilization of dejellied eggs is possible if the vitelline envelope is altered (Elinson, 1973).

The mature egg of the salamander *Notophthalmus viridescens* becomes surrounded by five jelly layers as it passes down the oviduct. Fertilization is internal, and polyspermy is normal. McLaughlin and Humphries (1978) removed oocytes from various positions along the oviduct and tested their fertilizability. As is the case with *Rana* and *Xenopus* the success of fertilization increases as the number of jelly layers increases (see also Nadamitsu, 1957). In the case of the salamander and other amphibians, hydration of the jelly is an important factor for sperm penetration (McLaughlin and

Humphries, 1978; del Pino, 1973). Hydration may deplete ions that block sperm entry, since the addition of saline solutions reverses the hydration process and the blocks to sperm penetration.

The integrity of the jelly envelopes has been attributed to the presence of disulfide bonds by Gusseck and Hedrick (1971), who have shown that in both *Xenopus* and *Rana* [and in *Bufo* by Katagiri (1963)] the jelly coats can be removed (dissolved) by a variety of agents known to reduce disulfide bonds. Furthermore, these workers postulate that the sperm may contain free sulfhydryl groups, which allow it to penetrate the jelly layers by dissolving a hole in it that is later closed by the reoxidation of the sulfhydryl groups. While an attractive hypothesis, it does oppose the earlier view that sperm contain trypsinlike enzymes (lysins) that function to aid jelly penetration (see Monroy, 1965). However, more current findings suggest that lysins are involved only in penetration of the vitelline envelope.

### 6.7.2. Antigenic Properties

Immunological properties of the jelly coats are important in the initiation of sperm–egg interactions. Shaver and Barch (1960) and Shaver *et al.* (1970) examined antigenic properties of *Rana* jelly coats, and—as might be expected—there are several antigenic components. Antigenic components vary in the different jelly layers, and some are shared with other species, whereas others are unique to a given species (Shaver, 1966; Barch and Shaver, 1963). Frequently, unique species components are in the outer and middle jelly layers, whereas shared, cross-species components are in the inner layers. Shaver *et al.* (1970) show that *Rana* eggs treated with antisera to *Rana* oviduct (thus *Rana* jelly) are not fertilized, but if they are treated with antisera to oviducts from other amphibians, fertilization rates actually improve.

Although cross fertilization, especially among the ranids, is possible (Elinson, 1975), characteristics of jelly coats are responsible for prevention of cross fertilization in certain species. For example, *Hyla arborea japonica* eggs cannot be fertilized by *Rana chensinensis* sperm (Katagiri, 1966), fertilization rates are low between *Xenopus mulleri* eggs and *Xenopus laevis* sperm (Blackler and Gecking, 1972), and cross-fertilization between *Rana clamitans* eggs and *Rana pipiens* sperm is not possible despite the fact that sperm migration through the jelly is normal. The reciprocal cross, however, is successful (Elinson, 1974). When *Rana clamitans* coelomic eggs are allowed to pass through the oviducts of *Rana pipiens* and acquire *Rana pipiens* jelly, they become fertilizable with *Rana pipiens* sperm. In the reciprocal jelly transfer *Rana pipiens* eggs can be fertilized with *Rana clamitans* sperm, but *Rana pipiens* sperm are ineffective. Brun and Kobel (1977) found that *Xenopus borealis* sperm can fertilize *Xenopus laevis* eggs, but the recriprocal cross is not successful. Thus, the jelly appears to be involved in blocking or at least reducing the possibility of interspecies fertilization. However, not to be discounted is the possibility that the vitelline envelope is also a factor in blocking cross-fertilizations.

### 6.7.3. Effects on Metabolic Activity

Oviducal secretions are postulated to alter metabolic activity in the egg. Legname et al. (1972) have shown that in certain respects the metabolism of coelomic eggs resembles that of adult tissues and later stage embryos. On the other hand oviduct eggs and early embryo segmentation stages differ in several respects from adult tissues. For example, respiratory activity is increased in eggs that have passed through the oviduct. In addition, fumarate and pyruvate oxidizing capacity increases, while the capacity to oxidize citrate decreases. Legname *et al.* (1972) suggest that the pars recta secretion might influence metabolism by regulating the relative velocities of enzyme reactions and thus their relative participation in the specific pathways. Smith and Ecker (1970) report a reduction in protein synthesis that may be related to exposure to secretions of the oviduct. Finally, Glick and Shaver (1963) have shown that *Rana pipiens* embryos derived from eggs from the mid- and lower portion of the oviduct have faster cleavage rates than embryos derived from eggs from the terminal ovisac.

## 6.8. Alterations at Hatching

Just before hatching, the embryo secretes a hatching enzyme that is capable of dissolving both the fertilization envelope and the jelly coats. There is a gradual loss in the average dry weight of the fertilization envelope, but the relative amounts of protein and carbohydrate remain unchanged, although chemical breakdown is shown by the disappearance of high-molecular-weight material during the later stages of hatching. There is a gradual disorganization and loosening of the bundles of filaments that make up the fertilization envelope as well as a reduction in their number as hatching nears. However, dense material thought to represent F-layer components remains essentially intact (Yoshizaki, 1978).

The jelly is also modified during the hatching process. Specifically, the proximal layer (J1) displays an increased sensitivity to UV radiation. Susceptibility of jelly to hatching enzyme may be species specific, since *R. clamitans* embryos developing inside *R. pipiens* jelly fail to hatch (Elinson, 1974). Katagiri (1975) reports that J1 and the fertilization envelope are only degraded by hatching enzyme from the same species but the outer jelly layers are digested by hatching enzymes from a variety of sources.

## 7. Reptiles

Except for the egg shell, information about the egg envelopes of reptilian eggs is minimal. Follicle morphology has been examined by Munson (1904), Neaves (1971), Taddei (1972), and Betz (1963); others have concentrated on changes in morphology as the follicles and oocytes grow and differentiate (Aitken *et al.*, 1976; Rahil and Narbaitz, 1973).

### 7.1. Origin and Structure of the Vitelline Envelope

Munson (1904) reported that the "egg membrane" (i.e., vitelline envelope) of the tortoise, *Clemmys marmorata*, consists of an outer homogeneous layer that, when torn, has a fibrous appearance. Beneath this is another layer with radial striations. Thing (1918) reported that it begins as a thin one-layered cuticle that, in tangential sections, shows the beginnings of a reticular network. A second layer, radially striated, soon appears. The final envelope was described as consisting of a homogeneous substance with numerous canals enclosing filaments or microvilli. Rahil and Narbaitz (1973) in a study of developing oocytes of the turtle *Pseudemys scripta elegans* confirm and extend these earlier observations. At the beginning of vitelline envelope formation glycogen-like granules, which appear to originate in the follicle cells, reach the intermicrovillous spaces apparently by apocrine secretion. The microvilli increase in length, and filamentous material appears among them. This layer increases in thickness and finally results in a vitelline envelope consisting of (1) an outer mainly filamentous zone, (2) a darker intermediate zone containing numerous channels occupied by oocyte microvilli, and (3) a clear zone near the oocyte.

In the turtle *Chelonia mydas* L., the vitelline envelope makes its initial appearance as a layer of periodic acid-Schiff (PAS)-positive granules between the single layer of follicle cells and the oocyte surface (Aitken et al., 1976). As it continues to develop it comes to consist of two layers, both of which have radial striations. Aitkin *et al.* (1976) report some very puzzling aspects for follicles $> 10$ mm in diameter in this species; namely, that in follicles in this size range, the vitelline envelope disappears, and later the granulosa also disappears with the result that the theca lies next to the oocyte. (However, in follicles which are interpreted to be either atretic or corpora lutea there are cells which appear to be granulosa cells.)

In lizards, the follicle cells investing the oocyte may be single or multilayered. There are three types of cells: small, intermediate, and large pyriform cells. Studies by Neaves (1971) and Taddei (1972) have demonstrated that the pyriform cells establish protoplasmic confluence with the oocyte cytoplasm, and these intercellular bridges result in channels in the vitelline envelope as it develops. These reports, unlike the case in most other vertebrates, have convincingly confirmed early reports by light microscopists that cytoplasmic bridges exist between oocyte and follicle cells. Betz (1963) has also reported the presence of protoplasmic bridges between the pyriform follicle cells and ooplasm in the snake, *Natrix rhombifera*, producing striations in the vitelline envelope. The vitelline envelope is at first a homogeneous layer but later differentiates into an outer homogeneous layer and an inner striated layer in the lizard, *Calotes versicolor* (Varma, 1970).

### 7.2. The Albumen

The formation of albumen in reptiles appears to be a mystery. Badham (1971) has published one of the few studies of its formation. In newly laid eggs of the agamid lizard, *Amphibolurus barbatus barbatus*, albumen is absent, but when the eggs are incubated in moist sand a clear viscous liquid develops to surround the yolk and embryo. The absorbed water remains external to the developing embryo and receives a colloidal component from the yolk. If this account is accurate, accepted terminology would make the albumen layer in this egg a primary envelope, whereas in most species it appears to be a tertiary envelope. Badham (1971) suggests that albumen prevents the embryo and yolk from making direct contact with the permeable parchment shell—thereby buffering it against changes in ambient moisture. Clark (1953) also reported that in the egg of the black snake *Coluber c. constrictor* the volume of albumen increases with absorption of environmental water and acts as a reservoir for nitrogenous excretory products (urea) liberated into it from the allantois. Ewert (1979) and Tracy (1980) report that the albumen in turtle eggs appears to have both fungicidal and bactericidal properties. Tracy (1980) suggests that two evolutionary strategies have been employed in reptiles. First, in the case of hard calcareous eggs (many chelonians and all crocodilians), the egg is invested with substantial albumen at the time of oviposition. Second, in the case of parchment-shelled eggs (e.g., lizards), the egg has little (or no?) albumen at the time of oviposition but is capable of rapid and extensive absorption of water from the environment.

### 7.3. Shell Membrane and Eggshell

In the eggs of many reptiles the egg shell consists of two parts: a fibrous shell membrane (sometimes two distinct layers) adjacent to the albumen and a calcareous layer attached to the outer surface of the shell membrane. In other species there is no calcareous component, and the egg shell is composed only of the shell membrane (Packard et al., 1982).

With respect to shells, the reptiles fall into three groups: (1) parchment-shelled eggs (mostly squamates) that indent easily and expand through water imbibition; (2) brittle-shelled eggs (crocodilians, some chelonians, a few squamates) that do not indent or expand significantly (late in development air pockets may develop interior to the fibrous layer or between the latter and the mineral layer); and (3) hard- but expansible-shelled eggs (some chelonians) (Ewert, 1979; Packard *et al.*, 1982). The structure of the egg shell places certain constraints on the capacity of the egg to respond to variations in ambient water, and these constraints transcend taxonomic boundaries (Packard *et al.*, 1982).

Insofar as is known, the calcareous component of the reptilian egg shell occurs as calcite in shells of squamates and crocodilian eggs and primary aragonite in the shells of turtle eggs (Packard, 1980; Packard *et al.*, 1982, 1984b). In *Cheloniamydas* L., calcium deposition begins simultaneously along the entire length of the shell-producing region of the oviduct. The calcified regions of oviposited eggs consist mainly of aragonite crystals and a small amount of calcite (<5%) (see also Solomon and Baird, 1976, 1977; Packard and Packard, 1979; Packard, 1980). The shell membrane fibers act as nucleation sites for mineralization. There is no apparent pore structure in the shell and no evidence of an outer cuticular layer. Protection against bacteria and fungi, therefore, must be due to the soft shell membrane and the albumen layer (Solomon and Baird, 1976). In the shells of eggs of kinosternid turtles, however, a few scattered pores are found in the calcareous layer, and a discontinuous organic layer may cover part of the outer surface (Packard *et al.*, 1984b). Evidence is also provided that in this group of turtles organic cores on the outer surface of the shell membrane may serve as nucleation sites for individual shell units (Packard *et al.*, 1984c).

Calcium is apparently removed from the shell by developing turtle embryos. This activity, in addition to supplying necessary calcium to the developing embryo, may also improve gas exchange as the embryo develops and may also facilitate hatching by making it easier for the would-be hatchling to pip the shell (Bustard *et al.*, 1969; Packard *et al.*, 1984a,b).

## 8. Birds

The avian egg has been described as having five envelopes. the innermost is the very thin bilayered vitelline envelope, which is of dual origin. The next enveloping layer is the **albumen** (egg white), which consists of a mixture of proteins (mostly albumins) and about 85% water. Egg white of greater density forms two rope-like **chalazae,** which are believed to function to keep the yolk at the center of the albumen. On the surface of the albumen is a bilayered **shell membrane,** which consists of matted layers of what is generally considered to be keratin. The two layers are closely apposed except at the blunt end of the egg where an air space is interposed between them. The outermost envelope is the **shell,** which consists primarily of calcium carbonate and is pierced by numerous fine pores. A thin **cuticle** covers the shell.

## 8.1. Origin and Structure of the Vitelline Envelope

Studies of the vitelline envelope by Bellaris *et al.* (1963) show that it consists of two principal layers separated by a thin continuous layer. According to these authors the ovarian egg of the chicken is not surrounded by a vitelline envelope until the final growth phase 7–10 days before ovulation. During this phase only the inner layer is laid down. The fibrillar material of the inner layer of the nascent vitelline envelope appears to be moulded around protoplasmic projections from the follicle cells and follows the contours of the underlying oocyte (Press, 1964). In larger follicles, the ovum and the granulosa (follicle) cells are separated by a thick homogeneous acellular layer within which are elongate electron-dense aggregates oriented parallel to the oocyte surface. Wyburn *et al.* (1965) believe that this layer corresponds to the zona pellucida of mammals and prefer to call it the "perivitelline layer" rather than vitelline envelope.

The layer termed the **zona radiata** (Wyburn *et al.*, 1965) is the layer consisting primarily of microvilli extending from the surface of the ovum and a few processes extending from granulosa cells through the so-called "perivitelline layer" to the surface of the oocyte. The question of whether the "perivitelline layer" is derived from the oocyte or from the follicle cells has not been answered in any definitive way. To refer to the envelope as the "perivitelline layer," however, is in our view unfortunate, for it certainly appears to be comparable, whatever its origin, to the vitelline envelope described in fishes and amphibians. The papers of Bakst and Howarth (1977a,b) are also somewhat misleading with regard to egg envelope nomenclature. Like Wyburn *et al.* (1965), Bakst and Howarth call the acellular layer developed in the follicle the "perivitelline layer" and refer to the *plasma membrane* of the oocyte as the *vitelline membrane.* We suggest that in the interest of both consistency and clarity both of these usages should be set aside. The plasmalemma of the ooctye is just that—the oolemma, i.e., the plasma membrane of the ovum. The proximal proteinaceous acellular layer, which is laid down in the follicle and which invests the oocytes, is the inner layer of the vitelline envelope. Hamilton (1972) believes there is sufficient evidence of a primary vitelline envelope secreted by the oocyte itself even though it might not represent the entire zona radiata of mature oocytes.

It seems clear that before ovulation only the inner layer of the vitelline envelope is present around the avian oocyte. All oviducal eggs, however, whether taken from the magnum (where the albumen is added) or from the isthmus of the uterus (where the shell membrane is formed) possess the inner, the outer, and the intermediate "continuous layer" of the complete vitelline envelope. In this, the oviducal eggs are indistinguishable from laid eggs.

The innermost layer of the vitelline envelope that overlies the ovum at the time of ovulation has been described as a network of anastomosing fibers (Bellairs *et al.*, 1963) (Fig. 12). A slightly porous ground substance extends from its upper surface to nearly half its thickness but is absent from the underside (Bakst and Howarth, 1977a). Bellairs *et al.* (1963) describe this innermost layer as being "unlike collagen but like the proteins that accompany collagen in connective tissue." Scanning electron microscope studies have shown that the inner face of the mature vitelline envelope consists of a branched meshwork of large fibers (Chernoff and Overton, 1977). The intermediate continuous sheetlike layer interposed between the inner and outer layers of the vitelline envelope has been described as an amorphous electron dense layer (Anderson *et al*, 1974; Bakst and Howarth, 1977a).

The outer layer of the vitelline envelope is laid down in the first few centimeters of the oviduct (B. Anderson *et al.*, 1974) and is filamentous

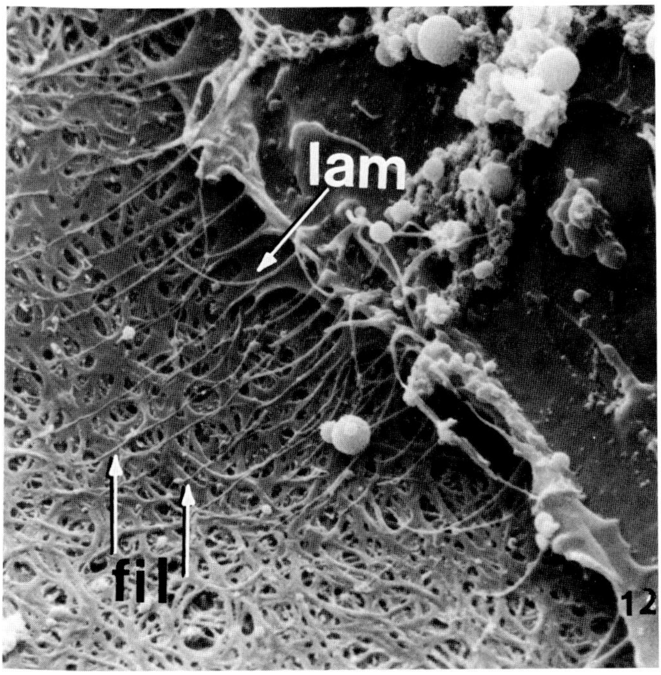


**Figure 12.** Inner surface of the chicken vitelline envelope and the expanding extraembryonic epiblast (upper right). The envelope (lower left) has an appearance and texture similar to that of the amphibian and mammalian envelopes. Filopodial (fil) and lamellipodial (lam) extensions from the epiblast are shown. ×5380. (From Chernoff and Overton, 1977.)

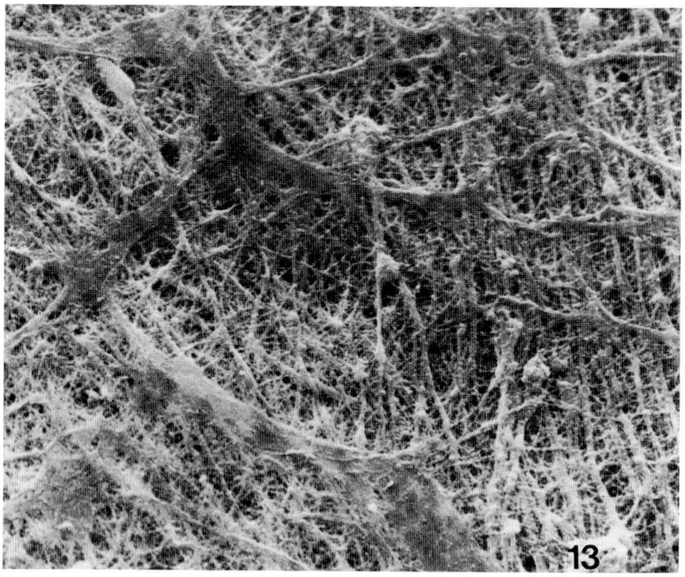


**Figure 13.** Outer surface of the vitelline envelope of the chicken egg showing a randomly-oriented meshwork of fibrils. ×5300. (From Chernoff and Overton, 1977.)

with interspersed granular material. The studies of Chernoff and Overton (1977) reveal the outer face to be composed of a dense meshwork of fibrils (Fig. 13). In amino acid composition this layer most closely resembles lysozyme, conalbumen, and avidin (Bellairs *et al.*, 1963). Cunningham and Ylander (1980) offer evidence that both the continuous layer and the outer layer of the envelope are composed of mucin.

## 8.2. Albumen

The albumen begins to be deposited around the vitelline membrane-invested oocyte as soon as it passes through the infundibulum into the magnum of the oviduct. Albumen is secreted by goblet cells as strands of mucinous gel that, deposited as a sievelike sheet, form the inner part of the chalaziferous layer (Romanoff and Romanoff, 1949). As the egg rotates slowly down the spiral folds of the oviduct, additional mucin as well as a more liquid albumen are secreted. It has been shown experimentally that the rotation of the egg as it descends the oviduct causes the mucinous fibers to become twisted and thus form the two chalazae (Conrad and Phillips, 1938).

Only about 50% of the albumen layer is formed by the time the egg enters the isthmus of the oviduct where the shell membrane is formed. In the uterus a very liquid mineral solution is secreted that readily penetrates the porous shell membrane, diluting the albumen and doubling its volume.

The chief solid constituents of the albumen are simple proteins (75% ovalbumin, 3% ovoconalbumin, 2% ovoglobulin) and glycoproteins (13% ovomucoid, 7% ovomucin). The most abundant minerals are sulfur, potassium, sodium, and chloride. Lesser amounts of phosphorus, calcium, and magnesium are present along with a trace of iron (Hamilton, 1952; Romanoff and Romanoff, 1949; Romanoff, 1967).

## 8.3. Shell Membrane

When the egg enters the isthmus of the oviduct its presence immediately causes the tubular glands to secrete a granular, keratinlike material (Romanoff and Romanoff, 1949). The granules absorb water and coalesce into sticky fibers that become felted into a lacelike membrane applied to the surface of the albumen. By the time the egg is fully into the isthmus it is covered with a tightly fitting finely fibrous membrane. An outer layer of coarser, more loosely woven fibers is deposited before the egg moves out of the isthmus. The membranes subsequently expand to form a much looser covering, and a small amount of pigment (ooporphyrin) is added to give the shell membrane a slightly pink color. Thus, the shell membrane consists of two layers: an inner layer and a thicker outer layer (Fig. 14). The inner layer is bounded by a limiting membrane where it is in contact with the albumen (Bellairs and Boyde, 1969). Except for size, there appears to be no difference in the structure or arrangement of the fibers of the two layers. Wong *et al.* (1984) have provided evidence that both types of fibers are collagenous. Each fiber consists of a core with a covering matrix. Histochemical studies (Robinson and King, 1968) and TEM studies of ruthenium red-stained membranes (Wong *et al.*, 1984) indicate that the covering matrix consists of mucopolysaccharides (glycosaminoglycans). At certain points, the fibers of the outer layer become embedded in the calcite of the **mammillary knobs** of the egg shell (Bellairs and Boyde, 1969; Becking, 1975). This union serves to strengthen the shell, which would otherwise be brittle or fragile (Hamilton, 1952).

## 8.4. Calcareous Shell

As the egg, invested with its loose shell membrane, enters the uterus, the uterine epithelium secretes an aqueous substance that diffuses into the egg causing it to swell. Thus, the egg's investing layers become snug, providing a firm surface for the deposition of the multilayered **calcareous shell**. Tiny granules of calcium salts deposited on the surface of the shell

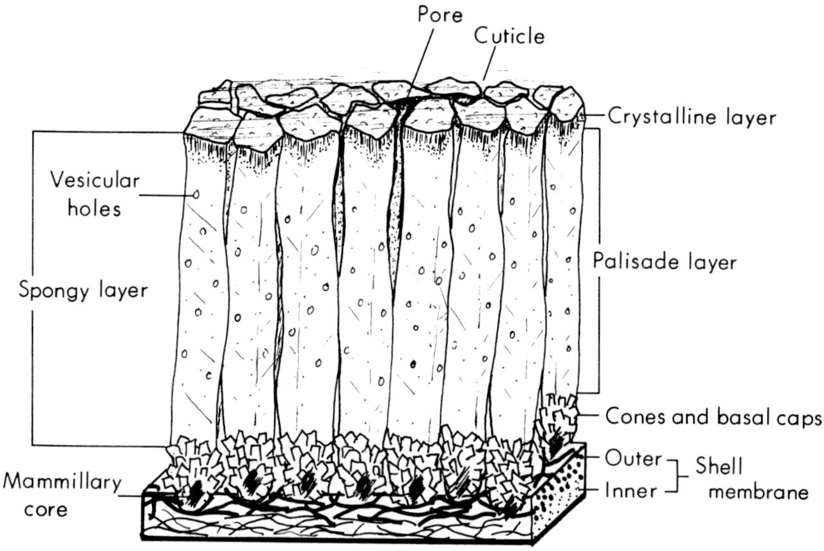


**Figure 14.** Diagrammatic representation of the complicated structure of the avian egg shell and shell membranes. The bilayered shell membrane is composed of large and small filaments. On its outer surface are embedded proteinaceous mammillary cores, upon which crystallization begins forming cones and the basal caps. Calcite crystallization continues, and the thicker spongy layer is formed. The crystals are perforated by small vesicular holes. Pores are seen between the crystals. The surface is covered by a thin organic cuticle. (Redrawn and modified from Becking, 1975.)

membrane, grow by accretion. At the center of each mineral concretion is a small amount of a protein matrix material (the mammillary core). Each becomes a knoblike structure and together they constitute the **mammillary layer** (Romanoff and Romanoff, 1949; Becking, 1975). From each mammillary core randomly oriented calcite crystals grow, forming the basal caps (Fig. 14).

Superimposed on the mammillary layer is the so-called **spongy layer,** which is composed of an inner palisade layer and a peripheral or surface compact crystalline layer (Fig. 14). The spongy layer is actually very compact and becomes spongy only after decalcification of the shell. Its mineral content is largely calcite and provides the shell with its rigidity and strength. Covering the external surface of the shell is a very thin, continuous layer of protein called the **cuticle,** which is permeable to gases (Becking, 1975) (Fig. 14). Shell pigments (porphyrins) derived from red blood cells may be absorbed by both the ground substance of the shell or its more superficial layers. The shell has numerous unevenly distributed pores that facilitate respiration of the embryo. The outer openings of the **pores** are covered with the cuticle, and perpendicular unbranched pore canals penetrate the spongy layer, forming connecting passages that provide for gas exchange between the external surface of the surface of the shell and air spaces in the mammillary layer and, ultimately, the developing embryo (Romanoff and Romanoff, 1949; Hamilton, 1952).

## 8.5. Functions of the Envelopes

### 8.5.1. Role in Fertilization

Our knowledge of the details of fertilization in the bird is meager. Fertilization occurs very soon after ovulation and normally before the egg reaches the magnum of the oviduct (Olsen and Neher, 1948).

The mature sperm possesses an acrosome and has a long and very slender nucleus (McIntosh and Porter, 1967; Bakst and Howarth, 1977b). It is reasonable to assume, on the basis of studies involving other species, that the release of acrosomal enzymes facilitates sperm penetration of the vitelline envelope. Since only the innermost layer of the vitelline envelope surrounds the egg at the time of ovulation, sperm penetration involves only the inner layer of the vitelline envelope (Bellairs *et al.*, 1963). Bakst and Howarth (1977b) present evidence that the outer layers of the vitelline envelope prevent sperm penetration and serve as a block to pathological polyspermy.

Acrosin extracted from cock sperm has been shown to preferentially hydrolyze the inner layer of the vitelline envelope, which covers the blastodisc (Ho and Meizel, 1975; Bakst and Howarth, 1977b; Howarth and Digby, 1973). Such studies indicate that the inner vitelline envelope is different from other areas. Bakst and Howarth (1977b) suggest the possibility of a concentrated array of sperm receptor sites in this region.

### 8.5.2. Functions During Embryonic Development

The vitelline envelope initially covers the entire zygote, including the blastodisc located at the animal pole. As the blastodisc begins to cleave, the vitelline envelope undergoes changes apparently brought about by the embryonic cells. It is suggested that these alterations weaken the envelope and bring about its dissolution over the surface of the developing embryo (Jensen, 1969). As the blastodisc undergoes expansion over the yolk, the extraembryonic portion of the epiblast precedes the hypoblast, and uses the underside of the vitelline envelope as its physical substrate while migrating; i.e., the ring of marginal cells clings to the lower surface of the vitelline envelope (Bellairs *et al.*, 1969; Chernoff and Overton, 1977).

A special transitory component, called the *subembyronic liquid*, is derived from the yolk and albumen. It exists only from the 2nd to the 15th day of embryonic development in the chick and provides a watery solution of electrolytes for the formation of fluids in the amniotic and allantoic sacs (Romanoff, 1967).

The egg albumen is the primary source of water for the developing embryo. It also provides proteins (equal to the amount provided by the yolk), minerals, and water-soluble vitamins. During the first few days of incubation, the wet weight of the yolk increases as water diffuses into it from the albumen. Concomitantly, the wet weight of the albumen decreases at an exceptionally fast rate. It continues to diminish at a slower rate until the 13th day, when it again loses weight rapidly and finally disappears (Romanoff, 1967). The albumen also provides both mechanical and chemical protection. Bactericidal action is provided by **avidin,** a biotin-binding protein (and a bacterial-growth inhibitor), and by lytic action on some kinds of bacteria (Hamilton, 1952).

The shell membranes provide a substrate for shell formation, enclose the albumen and prevent it from clogging the pores of the shell, separate the chorioallantoic membrane from the shell during incubation, and—at the same time—permit the ready diffusion of carbon dioxide from the embryo to the outside and oxygen from the outside to the embryo. The shell membranes also appear to serve as barriers to bacteria (Bellairs and Boyde, 1969).

The egg shell, of course, plays a very important protective role for the developing embryo, but—in addition—it supplies (from its matrix and calcareous structures) chemical substances essential for embryonic development. Organic matter and mineral, in approximately equal amounts, are transported from the shell to the egg content (Romanoff, 1967). The egg shell is the principal source of calcium from the chick embryo (Tuan, 1983).

Thus, the yolk, the albumen, and the shell provide essential nutrients and/or minerals to the embryo. Together with the shell membrane, they also provide both physical and microbial protection. "Through an efficient, selective osmotic intercommunication, these parts maintain a biologically suitable environment in the highly complex physio-chemical system within the shell" (Romanoff, 1969, p. 179).

## 8.6. Alterations at Hatching

Early in development the vitelline envelope ruptures over the animal pole and gradually slips down over the yolk to become incorporated into the albumen sac, which forms a plug at the base of the yolk sac. By the 16th day of incubation (chicken), the contents of the albumen sac are practically all absorbed, and by the 20th day, the yolk sac has been completely taken into the body of the embryo and practically all of the amniotic fluid has been absorbed. The head of the chick normally lies next to the air chamber between the two layers of the egg membrane, and on the 20th day the beak pierces the inner layer of the membrane and slowly breathes the air contained within the chamber. The beak subsequently pierces the shell, which is somewhat weakened by loss of calcium, and the chick begins to breathe normally.

## 9. Monotremes

There have been very few investigations of the egg envelopes of monotremes using modern techniques. Much of the descriptive information dates back into the last century (e.g., Caldwell, 1887). Griffiths (1978) has presented a summary of information on the subject. Caldwell's report (1887) is based on extensive studies of both *Echidna* and the platypus. He reported that in very young oocytes a thin vitelline envelope can be detected between the surface of the oocyte and the single layer of follicle cells. The vitelline envelope thickens with oocyte growth but then thins out as the oocyte accumulates yolk. The envelope is perforated by protoplasmic processes (microvilli) from the oocyte and follicle cells. When the oocyte reaches its maximum size the cells of the follicular epithelium also increase in size, divide rapidly, and secrete a dense homogeneous substance next to the vitelline envelope. Caldwell (1887) refers to this as the "pro-albumen." Flynn and Hill (1939) regard this secretion as a homologue of the liquor folliculi of the graafian follicle of eutherian mammals. At ovulation, the mature oocyte is received into the fallopian tube, in which it is fertilized. The ovum at the time of fertilization is surrounded by a thin vitelline envelope and a layer of "pro-albumen."

When the egg reaches the uterus, it has its full complement of envelopes: the **albuminous layer**, reported by Hughes *et al.* (1975) to be a mucoid coat of acidic glycoprotein, plus the shell membrane (or shell). The egg increases in size in the uterus (presumably by absorption and synthesis) and as it grows the albumen is absorbed and the vitelline envelope comes to lie next to the two-layered shell membrane. The shell membrane and the vitelline envelope have, in the meantime, both increased in thickness. As the blastocyst grows, a third layer of the shell appears (Griffiths, 1978). Evidence indicates that the substance of this layer is secreted by uterine glands. The thick shell of the newly laid egg is comprised of the basal layer, a rodlet layer, and a third layer consisting of an inner region of coarse granules with open spaces and an outer layer of larger and more irregular granules. The shell is porous (Hughes, 1977), but the pores are small and not numerous. The shell appears to consist largely of ovokeratin (Hughes *et al.*, 1975).

When the eggs are laid they are somewhat elliptical and covered with a sticky substance of unknown origin that causes them to adhere to one another (one to three eggs are laid) and presumably to the hairs of the pouch in the case of the *Echidna* (Burrell, 1927). At the time of hatching, the young are presumed to rip the shell by the combined use of caruncle and egg tooth (Griffiths, 1978).

## 10. Marsupials

Because of the unusual reproductive cycles among the marsupials, research efforts have been placed on (a) the uniqueness of the oocyte organelles and nucleus, (b) the polyovular follicles, (c) oocyte development and the control of folliculogenesis, and (d) embryonic membranes and implantation (cf. Harrison and Weir, 1977; Hughes, 1977). The structure of the ovarian follicle follows the same basic pattern as that of other mammals. Generally the mature follicles are quite large although oocyte size does not increase concomitantly (Lyne and Hollis, 1983). As in other mammals, oocyte and cumulus (follicle) cell microvilli penetrate the envelope, but in some forms there are areas that are bare of cells (Lyne and Hollis, 1983; Ullmann, 1979).

According to Hughes (1974) the protein- and carbohydrate-rich vitelline envelope (zona pellucida) is of ovarian cell origin and fully formed at the time of ovulation. At the time of fertilization the oocyte is surrounded only by the zona (cf., Talbot and DiCarlantonio, 1984). Typically, the envelope of marsupials is thinner than that of other mammals and disappears as the blastocyst begins to expand. (Hughes, 1974; Sharman, 1961). Talbot and DiCarlantonio (1984) report that until fertilization, opossum oocytes are invested by two layers, the zona pellucida and a previously undescribed extracellular component present in the perivitelline space. This extracellular matrix component consists of hyaluronic acid-rich filaments and protein-rich granules. Sperm hyaluronidase digests the matrix material at the time of fertilization.

As the ovum passes through the Fallopian tube, it acquires a **mucoid coat** (thin layer of albumen) which has been characterized as a highly sulfated glycoprotein. This coat varies considerably in thickness among the marsupials and disappears by the time the embryo has become bilaminar (Hughes, 1974).

The shell membrane is present in all marsupials and begins to overlay the mucoid coat in tubal eggs. However, deposition apparently continues even after the eggs enter the uterus. Its origin is uncertain, but Hughes (1974) believes it is derived from uterine secretions produced by the endometrial glands. The shell membrane is nonelastic but can be deformed, is a resistant proteinaceous material containing disfulfide bonds, and resembles keratin (Hughes, 1974; Tyndale-Biscoe, 1975). Krause and Cutts (1983) have shown that the shell membrane of the opossum consists of a mat of interwoven fibrils, and their scanning micrographs show it to have a striking resemblance to the vitelline envelope of lower forms and of placental mammals as well. With this resemblance in structure it is not surprising that it is permeable to a variety of large molecules, including nutrients derived from uterine secretions (Hughes and Shorey, 1973; Renfree, 1973). The shell membrane is retained longer than the vitelline envelope or the mucoid coat, being present in some species for two-thirds of the gestation period (Hughes, 1974; Sharman, 1961). In most marsupial embryos, it is retained until the formation of fetal rudiments. Its presence delays implantation (since it intervenes between the embryo and the uterine wall), and fragments of the membrane can reportedly still be collected from the uterus when the embryos are at full term.

## 11. Placental Mammals

### 11.1. Origin and Structure of the Vitelline Envelope (Zona Pellucida)

The origin of the mammalian **zona pellucida** (vitelline envelope), like that of lower vertebrates, has been debated for over a century. Some proposed its origin from the oocyte, others from the follicle cells, and still others from both sources. The question now seems to be resolved by the studies of Bliel and Wassarman (1980a), who used cultured denuded oocytes (i.e. oocytes without associated cellular material) in various stages of development and radioisotope procedures, to show that oocytes synthesize and secrete zona pellucida proteins. Developing zonae become labeled with both methionine and fucose, whereas zonae on fully-grown oocytes do not—indicating that by the final growth stages of the follicle the zona is complete. Their data provide the most direct evidence available that the zona pellucida is the product of synthetic activity of the oocyte alone and that the follicle cells play little, if any, role in the production of materials for this envelope. Corroborating evidence comes from studies of Bousquet *et al.* (1981). These investigators, who used immunofluorescent techniques to visualize zona proteins, conclude that in humans and hamsters the oocyte is the major source of zona pellucida material. More recently Dietl and Czuppon (1984) have also suggested, partly on the basis of studies of the details of porcine zona structure, that the zona components must originate from the oocyte.

During early stages of zona formation, patches of finely filamentous material appear between the oocyte and follicle cells (i.e., in the perivitelline space) and eventually form a continuous envelope around the ovum (Kang, 1974; Odor, 1965; Chiquoine, 1960) (Fig. 15). The mature zona varies in thickness depending on the species (Austin, 1961) and is composed of a meshwork of fine filaments with numerous fenestrations (Anderson *et al.*, 1978; Baranska *et al.*, 1975; Jackowski and Dumont, 1979; Phillips and Shalgi, 1980s) and in this respect is similar to the vitelline envelope of amphibians and birds. The zona is highly porous and permits a variety of

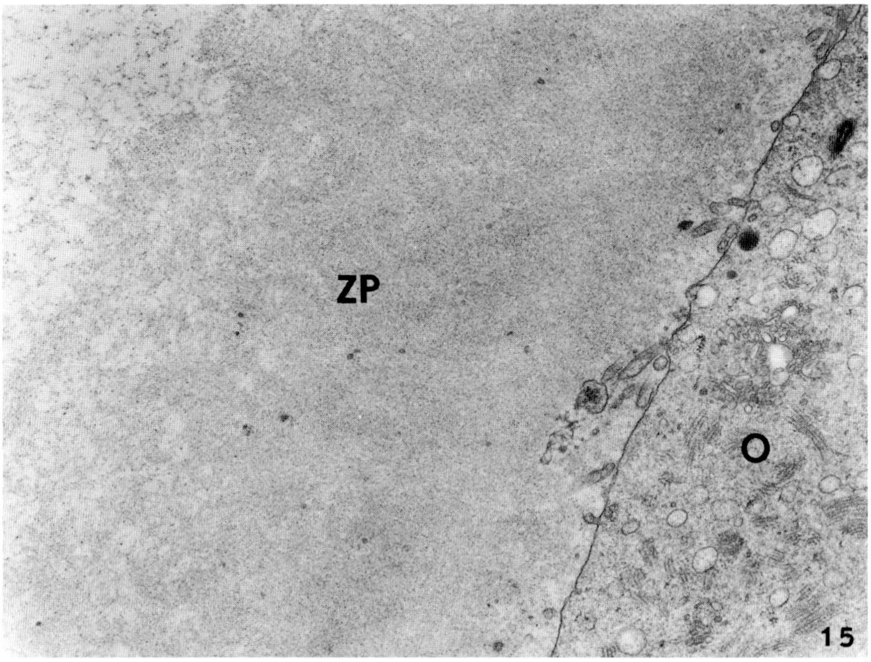


**Figure 15.** Zona pellucida (ZP) of a mouse egg. The zona is architecturally less complex than the vitelline envelopes of lower vertebrates. It appears as a very finely filamentous envelope surrounding the egg. Being less structured than other vitelline envelopes, its margins are less distinctly defined. O: oocyte. ×15,200. (Courtesy of S. Downs, Jackson Laboratory, Bar Harbor, Maine.)

large molecules to penetrate to the oocyte surface (Glass, 1963; Hastings *et al.*, 1972; Sellins and Jenkinson, 1975; Gwatkin, 1967; Anderson *et al.*, 1978). Wartenberg and Stegner (1960) and Stegner and Wartenberg (1961) depict, on the basis of histochemical staining properties, a bilayered zona around the human ovum; the inner layer is rich in neutral polysaccharides, whereas the thin outer layer stains for acid mucopolysaccharides. It has been suggested that this layer might be analogous to the jelly layers in other forms (Piko, 1969). In the rat, Odor (1960) describes a granular layer on the outer surface of the zona while Hope (1965) describes a flocculant layer on the outer surface of the rhesus zona. The internal surface of the hamster zona is particulate and lacks fenestrations (Phillips and Shalgi, 1980b).

## 11.2. Composition of the Zona Pellucida

A rather extensive literature exists regarding the composition of the zona pellucida. Early studies have shown the presence of sulfated mucopolysaccharides (Gothie, 1958; Moricand and Gothie, 1955), hyaluronic and sialic acid (Soupart and Noyes, 1964; Silva Sasso, 1959), acid and neutral polysaccharides (Wartenberg and Stegner, 1960; Oakberg and Tyrell, 1975), and glycoproteins (Kang, 1974; Gwatkin, 1976, 1977).

Comprehensive studies of the biochemical components of the mouse zona have been presented by Bliel and Wassarman (1980b), who report that 80% of the weight of the fully mature mouse zona is protein represented by three distinct species designated ZP1, ZP2, and ZP3 with molecular weights of 200,000, 120,000, and 83,000, respectively. SDS-polyacrylamide electrophoresis identifies them as glycoproteins, and amino acid analysis illustrates that each peptide moiety is unique. They represent respectively 36%, 47%, and 17% of the total zona protein. The major component, ZP2, is synthesized by the oocyte (Greve *et al.*, 1982). The zona of the pig appears to be somewhat more complex than that of the mouse (Dunbar *et al.*, 1980). For example, there appears to be a larger variety of macromolecular components that noncovalently interact to provide structural integrity. Mammalian zonae are similar to the vitelline envelopes of lower vetebrates and some invertebrates, not only in terms of their general chemical constituents but also in terms of function.

## 11.3. Alterations at Fertilization

Alterations in the appearance of the mouse zona after fertilization have been reported (Jackowski and Dumont, 1979; Baranska *et al.*, 1975). The observations suggest that very soon after sperm entry, the zona becomes smoother. Baranska *et al.*, (1975) examined zonae from early mouse embryos as well as morulae and blastocysts and report that four distinct layers (instead of the two that can be identified before fertilization) are visible by electron microscopy after ruthenium red staining. The innermost layer is coarse and granular and is thought to result from the interaction with cortical granule material expelled from the oocyte at the time of fertilization. The peripheral layer may be an accretion of material derived from the oviduct or uterus (Daniel, 1970; Kirschner, 1972). On the other hand, Burgos *et al.*, (1976) report that the zona of the rat does not change from the one-cell to the blastocyst stage, and Phillips and Shalgi (1980a) report no changes in the mouse or hamster zonae after fertilization.

Physicochemical changes also occur in the zona pellucida after fertilization. For example, zonae from unfertilized eggs of a variety of mammalian species are more readily solubilized than are zonae from zygotes and embryos (Inoue and Wolf, 1974, 1975a; Smithberg, 1953; Chang and Hunt, 1956; Gwatkin, 1964; Dunbar *et al.*, 1980; Gould *et al.*, 1971; Cholewa-Stewart and Massaro, 1972). Observations such as these imply that structural and/or chemical alterations after fertilization arise from the interaction of cortical granule material with the zona—the zona reaction

(Braden *et al.*, 1954; Barros and Yanagimachi, 1971). It has been suggested that these changes involve a cross-linking mechanism utilizing ovoperoxidase (Schmell and Gulyas, 1980). There is little evidence that alterations induced by interactions with oviduct or uterine secretions produce these physicochemical changes.

Modification of the zona pellucida proteins after the zona reaction has been shown by Bliel and Wassarman (1980c, 1983) and by Bliel *et al.* (1981). The zona proteins, ZP1, ZP2, and ZP3 are present in zonae of both eggs and embryos. However, 60 to 70% of ZP2 is modified after fertilization, apparently by proteolytic cleavage, into a 90,000-$M_r$ molecule designated ZP2f. Neither sperm nor oviduct components are required; the cortical granule exudate alone is sufficient to trigger the conversion of ZP2 to ZP2f, suggesting that cortical granule proteases may be responsible for the chemical modification. Functionally, the result is thought to be the formation of the slow block to polyspermy.

## 11.4. Functions of the Zona Pellucida

### 11.4.1. Sperm Receptors

It is clear that the zona provides receptors for sperm and—after fertilization—a barrier to polyspermy (Gwatkin, 1976, 1977; Hartmann *et al.*, 1972; Sato, 1979; Bliel and Wassarman, 1980c). Fertilization induces changes that prevent both additional sperm binding and penetration (Braden *et al.*, 1954; Piko, 1969; Barros, and Yanagimachi, 1972; Inoue and Wolf, 1975a,b). The association of the sperm with the zona pellucida takes place in two distinct steps. The first is a loose attachment (generally not species specific) such that sperm can be readily dislodged by gentle washing. The second step involves a firmer binding and is species specific (Hanada and Chang, 1972; Hartman *et al.*, 1972).

Certain plant lectins that bind specific saccharide moieties also bind to the zona. Dunbar *et al.*, (1980) have shown that lectin binding to the pig zona is assymetrical, indicating that the distribution of carbohydrates is not uniform. Similar observations in a variety of mammals have also been made by Nicolson *et al.*, (1975). Some lectins (wheat germ agglutinins) prevent sperm binding and penetration, whereas other (concanavalin A and *Ricinus* and *Dolichos* agglutinins) prevent penetration but not binding (Oikawa *et al.*, 1973). Binding may also be prevented by antisera against zonae in a variety of species (for reviews, see Shivers, 1974; and Shivers and Dudkiewicz, 1974; Dunbar and Shivers, 1976).

The sperm receptor component of the zona pellucida has been identified by Bliel and Wassarman (1980c) as the ZP3 glycoprotein. They have shown that purified ZP3 from unfertilized eggs is capable of competitively reducing sperm binding to 10% of control levels. However, ZP3 isolated from embryos as early as the 2-cell stage is ineffective in reducing binding, indicating that the receptor (ZP3) is altered (ZP3f) at the time of fertilization or very shortly afterward, possibly by a modification of the carbohydrate moieties. In any case, it is rendered ineffective as a sperm receptor or as an inducer of the acrosome reaction (Bliel and Wassarman, 1980c, 1983).

### 11.4.2. Acrosomal Reaction

Sperm binding to the zona is a prerequisite for the acrosomal reaction—a process that involves the fusion of the sperm plasma membrane and the acrosomal membrane, thus exposing the contents of the acrosomal vesicle. The latter contains proteolytic enzymes (acrosin) responsible for digesting the zona to allow passage of the sperm to the surface of the ovum (Gwatkin, 1976, 1977; McRorie and Williams, 1974; Florman and Storey, 1982). The ZP3 zona protein has been shown to be responsible for the induction of the acrosomal reaction. Its ability to induce the reaction, like its ability to bind sperm, is lost after fertilization (Bliel and Wassarman, 1983). The relationships and functions of the zona in sperm-egg interactions have been reviewed by Wassarman (1982) and Wassarman and Bliel (1982).

### 11.4.3. Zygote/Blastocyst Protection

The zona also provides protection for the zygote and developing embryo as they pass through the oviduct (Modlinski, 1970; Gwatkin, 1963) and assists in the maintenance of a specific biochemical environment around the embryo—a milieu known to differ in significant ways from that of the uterine lumen. Some embryos (mouse and sheep, for example) develop and grow normally in culture with or without the zona (Rottmann and Lampeter, 1981; Mintz, 1962; Trounson and Moore, 1974). In general, early cleavage embryos from a variety of species require protection from the uterine environment. This has led to the suggestion that some metabolic changes occur in the blastocyst and/or uterus that make it possible for the blastocyst to survive after shedding the zona.

### 11.4.4. Antigenicity

The antigenicity of the zona has generated a great deal of interest as a means of fertility control (Dunbar and Shivers, 1976; Aitkin *et al.*, 1981). Antibodies against the zona from a variety of species are effective in preventing sperm binding (and thus fertilization). Such antibodies induce the formation of a precipitate on the zona surface, which apparently excludes sperm binding (Dudkiewicz *et al.*, 1976). However, Wood *et al.*, (1981) have shown that, in the rabbit, immunization against zonae inhibits ovulation. This suggests that functions other than sperm binding to and/or penetration of zonae may be affected by zona antibodies. Long-term active immunization has the potential for creating an effective means of fertility control, especially in view of the crossreactivity of zona antigens among divergent mammalian species (Bavister, 1980; Swenson and Dunbar, 1982; Aitken and Richardson, 1980).

### 11.5. Alterations at Hatching

In preparation for implantation in the uterus, the blastocyst rids itself of the zona through a hatching process. Interest surrounding this phenomenon basically involves the questions: Does the uterus contribute to the hatching process? Is the hatching process mechanical, enzymatic, or both as it is in the fish and amphibians? Many studies involving the transplantation of the blastocyst to ectopic sites or to *in vitro* conditions show that hatching can be accomplished outside the uterus. Whereas no evidence exists to suggest that the blastocyst itself secretes lytic enzymes, there is evidence to support the view that such enzymes in the uterus do indeed exist and function to weaken the zona. On the other hand, since blastocysts hatch in ectopic sites and *in vitro* in the absence of such lytic enzymes, mechanical rupture of the zona is also presumed. Bergström (1972) reports that thinning of the zona in mouse blastocysts is related to blastocyst expansion and not to lysis. He supports this conclusion with microscopic observations that reveal cracked and fragmented zonae and trophoblasts protruding from openings in an otherwise intact zona—a picture suggestive of an active hatching, as opposed to a generalized lytic (dissolution) process. It is probable that both mechanical and enzymatic uterine factors are in fact involved in the normal hatching process (McLaren, 1970; for a review, see Sherman and Wudl, 1976).

## 12. Summary

As the material presented in this chapter was being collated, our existing perceptions about the basic similarities of vertebrate (and indeed most, if not all, invertebrate) egg envelopes became increasingly strengthened. Perhaps without exception, all vertebrate and invertebrate eggs acquire a "vitelline" envelope. Interestingly, its filamentous ultrastructure and chemical composition—basically protein and carbohydrate—is similar in all species as is its permeability to large molecules. Furthermore, many (if not all) of its functions are shared among the animal phyla as is its potential to become altered at the time of fertilization and, in its altered state, to provide a new set of *modi operandi*. It provides sperm receptors that are generally species specific and helps prevent polyspermy; it protects the developing embryo yet yields at the time of hatching.

In most vertebrate eggs (including some mammals), a jelly or albumen coat is added to the vitelline envelope. These components may vary immensely in thickness, but again their basic chemical composition is common to all. The functions of these envelopes, while perhaps somewhat less clear than those of the vitelline envelope, are related to species-specific fertilization and to embryonic protection. Albumen serves a nutritional role—most clearly shown in the birds. Finally, the shell membrane and shell present in diverse groups contribute additional adaptations for embryo protection.

Vertebrate egg envelopes, then, are basically similar; the modifications, including the addition of shell membranes and shells in some groups, reflect adaptations to differing reproductive strategies and to the environmental exigencies with which the egg must cope.

With the growth of our understanding about the structure, chemistry, function, and evolution of egg envelopes new questions will continually be formulated. Many will be the same as those asked years ago but they will be answered with newer techniques and with greater insight.

ACKNOWLEDGMENTS. We thank Ms. Rhonda Epler for her assistance with the preparation of photographs and for her help with literature searches. Oak Ridge National Laboratory is operated by Union Carbide Corporation under contract W-7405-eng-26 with the United States Department of Energy.

## References

Afzelius, B. A., Nicander, L., and Sjoden, I, 1968, Fine structure of egg envelopes and the activation changes of cortical alveoli in the river lamprey, *Lampetra fluviatilis*, *J. Embryol. Exp. Morphol.* **19**:311–318.

Aitken, R. J., and Richardson, D. W., 1980, Immunization against zona pellucida antigens, in: *Immunological Aspects of Reproduciton and Fertility Control* (J. P. Hearn, ed.) pp. 173–201, MPT Press Ltd., Lancaster.

Aitken, R. J., Rudak, E. A., Richardson, D. W., Dor, J., Djahanbahkch, O., and Templeton, A. A., 1981, The influence of anti-zona and anti-sperm antibodies on sperm–egg interations, *J. Reprod. Fertil.* **62**:597–606.

Aitken, R. N. C., Solomon, S. E., and Amaroso, E. C., 1976, Observations on the histology of the ovary of the Costa Rican green turtle, *Chelonia mydas* L., *J. Exp. Mar. Biol. Ecol.* **24**:198–204.

Anderson, B., Kim, N. B., Rhea, R. P., and Rosenberg, M. D., 1974, Nucleoside triphosphate hydrolase activities of avian vitelline membrane and oviductal lumen, *Dev. Biol.* **37**:306–316.

Anderson, E., 1967, The formation of the primary envelope during oocyte differentiation in teleosts, *J. Cell Biol.* **35**:193–212.

Anderson, E., 1974, Comparative aspects of the ultrastructure of the female gamete, *Int. Rev. Cytol.* **4**:1–70.

Anderson, E., Wilkinson, R. F., Lee, G., and Meller, S., 1978, A correlative microscopical analysis of differentiating ovarian follicles of mammals, *J. Morphol.* **156**:339–366.

Aplington, H. W., Jr., 1957, The insemination of body cavity and oviducal eggs of amphibia, *Ohio J. Sci.* **57**:91–99.

Austin, C. R., 1961, *The Mammalian Egg*, Charles C. Thomas, Springfield, Ill.

Azevedo, C., 1974, Evolution des envelopes ovocytaires, au cours de l'ovogenese, chez un Teleosteen vivipare, *Xiphophorus helleri*, *J. Microsc. (Paris)* **21**:43–54.

Badham, J., 1971, Albumen formation in eggs of the agamid *Amphibolurus barbatus barbatus*, *Copeia* **1791**:543–545.

Bakst, M. R., and Howarth, B., Jr., 1977a, The fine structure of the hen's ovum at ovulation, *Biol. Reprod.* **17**:361–369.

Bakst, M. R., and Howarth, B., Jr., 1977b, Hydrolysis of the hen's perivitelline layer by cock sperm, *Biol. Reprod.* **17**:370–379.

Baranska, W., Konwinski, M., and Kujawa, M., 1975, Fine structure of the zona pellucida of unfertilized cells and embryos, *J. Exp. Zool.* **192**:193–202.

Barbieri, F. D., and Budeguer de Atenor, M. S., 1973, Role of oviducal secretions in the fertilization of *Bufo arenarum* oocytes, *Arch. Biol. (Bruxelles)* **84**:501–511.

Barch, S. H., and Shaver, J. R., 1963, Regional antigenic differences in frog oviduct in relation to fertilization. *Am. Zool.* **3**:157–165.

Barros, C., and Yanagimachi, R., 1971, Introduction of the zona reaction in golden hamster eggs by cortical granule material, *Nature* **233**:268–269.

Barros, C., and Yanagimachi, R., 1972, Polyspermy-prevention mechanisms in the golden hamster egg, *J. Exp. Zool.* **180**:251–266.

Bavister, B. D., 1980, Recent progress in the study of early events in mammalian fertilization, *Dev. Growth Diff.* **22**:385–402.

Becking, J. H., 1975, The ultrastructure of the avian eggshell, *Ibis* **117**:143–151.

Bellairs, R., and Boyde, A., 1969, Scanning electron microscopy of the shell membranes of the hen's egg. *Z. Zellforsch.* **96**:237–249.

Bellairs, R., Boyde, A., and Heaysman, J. E., 1969, The relationship between the edge of the chick blastoderm and the avian vitelline membrane, *Arch. Entwichlungsmech. Org.* **163**:113–121.

Bellairs, R., Harkness, M., and Harkness, R. D., 1963, The vitelline membrane of the hen's egg: A chemical and electron microscopical study, *J. Ultrastruct. Res.* **8**:339–359.

Bergström, S., 1972, Shedding of the zona pellucida of the mouse blastocyst in normal pregnancy, *J. Reprod. Fertil.* **31**:275–277.

Betz, T. N., 1963, The ovarian histology of the diamond-backed water snake, *Natrix rhombifera* during the reproductive cycle, *J. Morphol.* **113**:245–260.

Blackler, A. W., and Gecking, C. A., 1972, Transmission of sex cells of one species through the body of a second species of the genus *Xenopus*. II. Interspecific mating, *Dev. Biol.* **27**:385–394.

Blaxter, J. H. S., 1969, Development: Eggs and larvae, in: *Fish Physiology: Reproduction and Growth, Bioluminescence, Pigments, and Poisons*, Vol. III (W. S. Hoar and D. J. Randall, eds.), pp. 177–252, Academic Press, New York.

Bliel, J. D., and Wassarman, P. M., 1980a, Synthesis of zona pellucida proteins by denuded and follicle-enclosed mouse oocytes during culture in vitro, *Proc. Natl. Acad. Sci. USA* **77**:1029–1033.

Bliel, J. D., and Wassarman, P. M., 1980b, Structure and function of the zona pellucida: Identification and characterization of the proteins of the mouse oocyte's zona pellucida, *Dev. Biol.* **76**:185–202.

Bliel, J. D., and Wassarman, P. M., 1980c, Mammalian sperm–egg interaction: Identification of a glycoprotein in mouse egg zonae pellucidae possessing receptor activity for sperm, *Cell* **20**:873–882.

Bliel, J. D., and Wassarman, P. M., 1983, Sperm–egg interactions in the mouse: Sequence of events and inducation of the acrosome reaction by a zona pellucida glycoprotein, *Dev. Biol.* **95**:317–324.

Bliel, J. D., Bealle, C. F., and Wassarman, P. M., 1981, Mammalian sperm–egg interaction: Fertilization of mouse eggs triggers modification of the major zona pellucida glycoprotein, ZP2, *Dev. Biol.* **86**:189–197.

Boisseau, C., 1973a, Etude ultrastructurale de l'oviducte du triton *Pleurodeles waltlii* Michah. I. Ultrastructure des cellules epitheliales de l'oviducte moyen differencié *J. Micros* **18**:341–358.

Boisseau, C., 1973b, Etude ultrastructurale de l'oviducte du triton *Pleurodeles waltlii* Michah. II. Morphogenèse des glandes et differenciation des cellules epitheliales de l'oviducte moyen, *J. Microsc.* **18**:359–382.

Boisseau, C., 1975, Etude ultrastructurale de l'oviducte du triton *Pleurodeles waltlii* Michah. III. Action de l'oestradiol-17 sur la differenciation des cellules secretrices de l'oviducte moyen de la femelle immature, *Gen. Comp. Endocrinol.* **27**:430–449.

Bousquet, D., Léveillé, M. C., Roberts, K. D., Chapdelaine, A., and Bleau, G., 1981, The cellular origin of the zona pellucida antigen in human and hamster, *J. Exp. Zool.* **215**:215–218.

Braden, A. W. H., Austin, C. R., and David, H. A., 1954, The reaction of the zona pellucida to sperm penetration, *Aust. J. Biol. Sci.* **7**:391–409.

Brummett, A. R., 1966, Observations of the eggs and breeding season of *Fundulus heteroclitus* at Beaufort, North Carolina, *Copeia* **1966**:616–620.

Brummett, A. R., and Dumont, J. N., 1977, Intracellular transport of vitellogenin in *Xenopus* oocytes: An autoradiographic study. *Dev. Biol.* ***60**:482–486.*

Brummett, A. R., and Dumont, J. N., 1979, Initial stages of sperm penetration into the egg of *Fundulus heteroclitus, J. Exp. Zool.* **210**:417–434.

Brummett, A. R., and Dumont, J. N., 1981, A comparison of chorions from eggs of northern and southern populations of *Fundulus heteroclitus, Copeia* **1981**:607–614.

Brummett, A. R., Dumont, J. N., and Larkin, J. R., 1982, The ovary of *Fundulus heteroclitus, J. Morphol.* **173**:1–16.

Brun, R., and Kobel, H. R., 1977, Observations on the fertilization block between *Xenopus borealis* and *Xenopus laevis laevis, J. Exp. Zool.* **201**:135–138.

Burgos, M. H., Segal, S. J., and Passantino, T., 1976, Surface changes of the rat embryo before implantation, *Fertil. Steril.* **27**:1085–1094.

Burrell, H., 1927, *The Platypus: Its Discovery, Zoological Position, Form and Characteristics, Habits, Life History, etc.,* Angus and Robertson Limited, Australia.

Busson-Mabillot, S., 1967, Structure ovarienne chez la lamproie adulte *Lampetra planeri* (Bloch). 1. La zone pellucide. Morphogenèse et constitution histochemique, *J. Micros.* **6**:577–598.

Busson-Mabillot, S., 1973, Evolution des enveloppes de l'ovocyte et de l'oeuf chez un poisson teleostéen, *J. Micros.* **18**:23–55.

Bustard, H. R., Simkiss, K., and Jenkins, N. K., 1969, Some analyses of artificially incubated eggs and hatchlings of green and loggerhead sea turtles, *J. Zool.* **158**:311–315.

Caldwell, W. H., 1887, XVI. *The Embryology of Monotremata and Marsupialia*—Part I, *Philos. Trans. R. Soc. Lond.* Ser. B. **178**:463–486.

Chang, M. C., and Hunt, D. M., 1956, Effects of proteolytic enzymes on the zona pellucida of fertilized and unfertilized mammalian eggs, *Exp. Cell Res.* **11**:479–499,

Chernoff, E. A. G., and Overton, J., 1977, Scanning electron microscopy of chick epiblast expansion on the vitelline membrane, *Dev. Biol.* **57**:33–46.

Cherr, G. N., and Clark, W. H., Jr., 1982, Fine structure of the envelope and micropyles in the eggs of the white sturgeon, *Acipenser Transmontanus Richardson, Dev. Growth Diff.* **24(4)**:341–352.

Chiquoine, A. D., 1960, The development of the zona pellucida of the mammalian ovum, *Am. J. Anat.* **106**:149–170.

Cholewa-Stewart, J., and Massaro, E. J., 1972, Thermally induced dissolution of the murine zona pellucida, *Biol. Reprod.* **7**:166–169.

Clark, H., 1953, Metabolism of the black snake embryo. I. Nitrogen excretion, *J. Exp. Biol.* **30**:492–501.

Conrad, R. M., and Phillips, R. E., 1938, The formation of the chalazae and inner thin egg white in the hen's egg, *Poult. Sci.* **17**:143–146.

Corporiccio, B., and Connes, R., 1977, Etude ultrastructurale des enveloppes periovocytaires et periovulaires de *Dicentrarchus labrax* L. (Poisson teleostéen), *Ann. Sci. Nat. Zool. (Paris)* **19**:351–368.

Cunningham, F. E., and Ylander, D. M., 1980, Ultrastructure of vitelline membranes from normal and mottled egg yolks, *Poult. Sci.* **59**:2449–2455.

Daniel, J. C., 1970, Uterine proteins and mammalian development, in: *Advances in Biosciences, 6th Shering Symposium on Intrinsic and Extrinsic Factors in Early Mammalian Development* (G. Raspe, ed.), pp. 191–203, Pergamon Press, Oxford.

Dietl, J., and Czuppon, A. B., 1984, Ultrastructural studies of porcine zona pellucida during the solubilization process by Li-3,5-diiodosalicylate, *Gamete Res.* **9**:45–54.

Dudkiewicz, A. B., Shivers, C. A., and Williams, W. C., 1976, Ultrastructure of the hamster zona pellucida treated with zona precipitating antibody, *Biol. Reprod.* **14**:175–185.

Dumont, J. N., 1972, Oogenesis in *Xenopus laevis* (Daudin). I. Stages of oocyte development in laboratory maintained animals, *J. Morphol.* **136**:153–180.

Dumont, J. N., 1978, Oogenesis in *Xenopus laevis* (Daudin). VI. The route of injected tracer transport in the follicle and developing oocyte, *J. Exp. Zool.* **204**:193–218.

Dumont, J. N., and Brummett, A. R., 1978, Oogenesis in *Xenopus laevis* (Daudin). V. Relationships between developing oocytes and their investing follicular tissues, *J. Morphol.* **155**:73–98.

Dumont, J. N., and Brummett, A. R., 1980, The vitelline envelope, chorion, and micropyle of *Fundulus heteroclitus* eggs, *Gamete Res.* **3**:25–44.

Dunbar, B. S., and Shivers, C. A., 1976, Immunological aspects of sperm receptors on the zona pellucida of mammalian eggs, *Immunol. Comm.* **5**:375–385.

Dunbar, B. S., Wardrip, N. J., and Hedrick, J. L., 1980, Isolation, physiochemical properties, and macromolecular composition of zona pellucida from porcine oocytes, *Biochemistry* **19**:356–365.

Elinson, R. P., 1971, Fertilization of partially jellied and jellyless oocytes of the frog *Rana pipiens*, *J Exp. Zool.* ***176**:415–428.*

Elinson, R. P., 1973, Fertilization of frog body cavity eggs enhanced by treatments affecting the vitelline coat, *J. Exp. Zool.* **183**:291–302.

Elinson, R. P., 1974, A block to cross-fertilization located in the egg jelly of the frog *Rana clamitans*, *J. Embryol. Exp. Morphol.* **32**:325–335.

Elinson, R. P., 1975, Fertilization of green frog *(Rana clamitans)* eggs in their native jelly by bullfrog *(Rana catesbeiana)* sperm, *J. Exp. Zool.* 193:419–423.

Ewert, M. A., 1979, The embryo and its egg: Development and natural history, in: *Turtles: Perspectives and Research* (M. Harless and H. Morlock, eds.), pp. 333–413, Wiley, New York.

Flegler, C., 1977, Electron microscope studies on the development of the chorion of the viviparous teleost *Dermogenys pusillus* (Hemirhamphidae), *Cell Tissue Res.* **179**:255–270.

Florman, H. M., and Storey, B. T., 1982, Mouse gamete interactions: The zona pellucida is the site of the acrosome reaction leading to fertilization in vitro, *Dev. Biol.* **91**:121–130.

Flugel, H., 1967, Licht- und elektronenmikroskopische untersuchungen an Oozyten und Eiern einiger Knochenfische, *Z. Zellforsch.* **83**:82–116.

Flynn, T. T. and Hill, J. P., 1939, The development of the Monotremata. Part IV. Growth of the ovarian ovum, maturation, fertilization, and early cleavage. *Trans. Zool. Soc. London* **24**:445–622.

Freeman, S. B., 1968, A study of the jelly envelopes surrounding the egg of the amphibian, *Xenopus laevis*, *Biol. Bull.* **135**:501–513.

Ginsberg. A. S., 1972, *Fertilization in Fishes and the Problem of Polyspermy* (T. A. Detlaf, ed.)

Academy of Science of the USSR–Institute of Developmental Biology (translated from the Russian), Israel Programs for Scientific Translations, Jerusalem.

Glass, L. E., 1963, Transfer of native and foreign serum antigens to oviductal mouse egg, *Am. Zool.* **3**:135–156.

Glick, R. N., and Shaver, J. R., 1963, The fertilizability of frog eggs from different levels of the female reproductive tract, *Exp. Cell Res.* **32**:615–618,.

Gothie, S., 1958, Contribution à l'étude de la membrane pellucide de l'oeuf de lapine a l'aide du $^{35}$S, *J. Physiol. (Paris)* **50**:293–294.

Gould, K., Zaneveld, L. J. D., Srivastava, P. N., and Williams, W. L., 1971, Biochemical changes in the zona pellucida of the rabbit ova induced by fertilization and sperm enzymes, *Proc. Soc. Exp. Biol. Med.* **136**:6–10.

Greve, L. C., and Hedrick, J. L., 1978, An immunocytochemical localization of the cortical granule lectin in fertilized and unfertilized eggs of *Xenopus laevis*, *Gamete Res.* **1**:13–18.

Greve, J., Salzmann, G. S., Roller, R. J., and Wassarman, P. M., 1982, Biosynthesis of the major zona pellucida glyco-protein secreted by oocytes during mammalian oogenesis, *Cell* **31**:749–759.

Grey, R. D., Wolf, D. P., and Hedrick, J. L., 1974, Formation and structure of the fertilization envelope in *Xenopus laevis*, *Dev. Biol.* **36**:44–61.

Grey, R. D., Working, P. K., and Hedrick, J. L., 1976, Evidence that the fertilization envelope blocks sperm entry in eggs of *Xenopus laevis*: Interaction of sperm with isolated envelopes, *Dev. Biol.* **54**:52–60.

Grey, R. D., Working, P. K., and Hedrick, J. L., 1977, Alteration of structure and penetrability of the vitelline envelope after passage of eggs from coelom to oviduct in *Xenopus laevis*, *J. Exp. Zool.* **201**:73–84.

Griffiths, M., 1978, *The Biology of the Monotremes*, Academic Press, New York.

Gusseck, D. J., and Hedrick, J. L., 1971, A molecular approach to fertilization. I. Disulfide bonds in *Xenopus laevis* jelly coat and a molecular hypothesis for fertilization, *Dev. Biol.* **25**:337–347.

Gwatkin, R. B. L., 1963, Studies on the zona pellucida of the mouse egg, *J. Reprod. Fertil.* **6**:325 (abstr.).

Gwatkin, R. B. L., 1964, Effect of enzymes and acidity on the zona pellucida of the mouse egg before and after fertilization, *J. Reprod. Fertil.* **7**:99–105.

Gwatkin, R. B. L., 1967, Passage of mengovirus through the zona pellucida of the mouse morala, *J. Reprod. Fertil.* **13**:577–578.

Gwatkin, R. B. L., 1976, Fertilization, in: *The Cell Surface in Animal Embryogenesis and Development* (G. Poste and G. L. Nicolson, eds.), pp. 1–54, Elsevier/North-Holland, New York.

Gwatkin, R. B. L., 1977, *Fertilization Mechanisms in Man and Mammals*, Plenum Press, New York.

Hamilton, H. L., 1952, *Lillie's Development of the Chick*, 3rd ed., Holt, New York.

Hanada, A., and Chang, M C., 1972, Penetration of zona-free eggs by spermatozoa of different species, *Biol. Reprod.* **6**:300–309.

Harrison, R. J., and Weir, B. J., 1977, Structure of the mammalian ovary, in: *The Ovary*, 2nd Ed., Vol. 1 (Lord S. Zuckerman and B. J. Weir, eds.), pp. 113–217, Academic Press, New York.

Hart, N. H., and Donovan, M., 1983, Fine structure of the chorion and site of sperm entry in the egg of *Brachydanio*, *J. Exp. Zool.* **227**:277–296.

Hart, N. H., Pietri, R., and Donovan, M., 1984, The structure of the chorion and associated surface filaments in *Oryzias*—evidence for the presence of extracellular tubules, *J. Exp. Zool.* **230**:273–296.

Hartmann, J. F., Gwatkin, R. B. L., and Hutchinson, C. F., 1972, Early contact interactions between mammalian gametes in vitro: Evidence that the vitellus influences adherence between sperm and zona pellucida, *Proc. Natl. Acad. Sci. USA* **9**:2767–2769.

Hastings, R. A., Enders, A. C., and Schlofke, S., 1972, Permeability of the zona pellucida to protein tracers, *Biol. Reprod.* **7**:288–296.

Hedrick, J. L., Smith, A. J., Yurewicz, E. C., Oliphant, G., and Wolf, D. P., 1974, The incorporation and fate of [$^{35}$S]-sulfate in the jelly coat of *Xenopus laevis* eggs, *Biol. Reprod.* **11**:534–542.

Ho, J. J. L., and Meizel, S., 1975, Hydrolysis of the hen egg vitelline membrane by cock sperm acrosin and other enzymes, *J. Exp. Zool.* **194**:429–438.

Hope, J., 1965, The fine structure of the developing follicle of the Rhesus ovary, *J. Ultrastruct. Res.* **12**:592–619.

Howarth, B., Jr., and Digby, S. T., 1973, Evidence for the penetration of the vitelline membrane of the hen's ovum by a trypsin-like acrosomal enzyme, *J. Reprod. Fertil.* **33**:123–125.

Hughes, R. L., 1977, Egg membranes and ovarian function during pregnancy in the monotremes and marsupials, in: *Reproduction and Evolution* (J. H. Calaby and C. H. Tyndale-Biscoe, eds.), pp. 281–291, Australian Academy of Sciences, Canberra.

Hughes, R. L., and Shorey, C. D., 1973, Observations on the permeability properties of the egg membranes of the marsupial, *Trichosurus vulpecula J. Reprod. Fertil.* **32**:25–32.

Hughes, R. L., Thomson, J. A., and Shorey, C. D., 1975, Reproduction in the platypus, *Ornithorhynchus anatinus*, with particular reference to the evolution of viviparity, *J. Reprod. Fertil.* **43**:374–375.

Humphries, A. A., 1961, Experiments on the normal meiotic block in the ovum of *Triturus viridescens*, *Biol. Bull.* **120**:29–37.

Humphries, A. A., Jr., 1966, Observations on the deposition, structure, and cytochemistry of the jelly envelopes of the egg of the newt, *Triturus viridescens*, *Dev. Biol.* **13**:214–230.

Humphries, A. A. Jr., 1970, Incorporation of $^{35}$S sulfate into the oviducts and egg jelly of the newt, *Notophthalmus viridescens*, Exp. Cell Res. **59**:157–161.

Humphries, A. A., Jr., Freeman, S. B., and Workman, W. M., 1968, Sialic acid in egg capsules and oviducts of the newt, *Notophthalmus viridescens*, *Biol. Bull.* **134**:266–271.

Humphries, A. A., and Hughes, W. N., 1959, A study of the polysaccharide histochemistry of the oviduct of the newt, *Triturus viridescens*, *Biol. Bull.* **116**:446–451.

Inoue, M., and Wolf, D. P., 1974, Comparative solubility properties of the zonal pellucidae of unfertilized and fertilized mouse ova, *Biol. Reprod.* **11**:558–565.

Inoue, M., and Wolf, D. P., 1975a, Fertilization-associated changes in the murine zona pellucida: A time sequence study, *Biol. Reprod.* **13**:546–551.

Inoue, M., and Wolf, D. P., 1975b, Sperm binding characteristics of the murine zona pellucida, *Biol. Reprod.* **13**:340–346.

Iwamatsu, T., and Ohta, T., 1981, Scanning electron microscope observations on sperm penetration in teleostean fish, *J. Exp. Zool.* **218**:261–277.

Jackowski, S., and Dumont, J. N., 1979, Surface alterations of the mouse zona pellucida and ovum following in vivo fertilization: Correlation wtih the cell cycle, *Biol. Reprod.* **20**:150–161.

Jensen, C., 1969, Ultrastructural changes in the avian vitelline membrane during embryonic development, *J. Embryol. Exp. Morphol.* **21**:467–484.

Jollie, W. P., and Jollie, L. G., 1964, The fine structure of the ovarian follicle of the ovoviviparous Poecilid fish, *Lebistes reticulatus*, *J. Morphol.* **114**:479–502.

Kaighn, M. E., 1964, A biochemical study of the hatching process in *Fundulus heteroclitus*, *Dev. Biol.* **9**:56–80.

Kang, Y., 1974, Development of the zona pellucida in the rat oocyte, *Am. J. Anat.* **139**:535–566.

Kao, C.-Y., and Chambers, R., 1954, Internal hydrostatic pressure of the *Fundulus* egg. I. The activated egg, *J. Exp. Biol.* **31**:139–149.

Katagiri, C., 1963, On the fertilizability of the frog egg. III. Removal of egg envelopes from unfertilized egg, *J. Fac. Sci. Hokkaido Univ. Ser. VI (Zool.)* **15**:202–211.

Katagiri, C., 1965, The fertilizability of coelomic and oviductal eggs of toad *Bufo bufo formosus*, *J. Fac. Sci. Hokkaido Univ., Ser. 6* **15**:633–643.

Katagiri, C., 1966, Fertilization of dejellied uterine toad eggs in various experimental conditions, *Embryologia* **9**:159–169.

Katagiri, C., 1974, A high frequency of fertilization in premature and mature coelomic toad eggs after enzymic removal of the vitelline membrane, *J. Embryol. Exp. Morphol.* **31**:573–581.

Katagiri, C., 1975, Properties of the hatching enzyme from frog embryos, *J. Exp. Zool.* **193**:109–118.

Katagiri, C., Iwao, Y., and Yoshizaki, N., 1982, Participation of oviducal pars recta secretions in inducing the acrosome reaction and release of vitelline coat lysin in fertilizing toad sperm, *Dev. Biol.* **94**:1–10.

Kelly, J. W., Berger, E. R., and Schuel, H., 1970, Sulfate in egg cortical granules, egg jelly, and oviduct glands, *Fed. Proc.* **29**:706.

Kirschner, C., 1972, Immuno-histologic studies on the synthesis of a uterine specific protein in the rabbit and its passage through the blastocyst covering, *Fertil. Steril.* **23**:131–136.

Kobayashi, W., and Yamamoto, T. S., 1981, Fine structure of the micropylar apparatus of the chum salmon egg, with a discussion of the mechanism for blocking polyspermy, *J. Exp. Zool.* **217**:265–275.

Krause, W. J., and Cutts, J. H., 1983, Ultrastructural observations on the shell membrane of the North American oppossum (*Didelphis virginiana*), *Anat. Rec.* **207**:335–338.

Kuchnow, K. P., and Scott, J. R., 1977, Ultrastructure of the chorion and its micropyle apparatus in the mature *Fundulus heteroclitus* (Walbaum) ovum, *J. Fish. Biol.* **10**:197–201.

Kudo, S., 1982, Ultrastructure and ultracytochemistry of fertilization envelope formation in the carp egg, *Dev. Growth Diff.* **24(4)**:327–339.

Laale, H. W., 1980, The perivitelline space and egg envelopes of bony fishes: A review, *Copeia* **1980**:210–226.

Lee, P. A., 1967a, Studies of frog oviducal jelly secretion. I. Chemical analysis of secretory product, *J. Exp. Zool.* **166**:99–106.

Lee, P. A., 1967b, Studies of frog oviducal jelly secretion. II. Cytology of secretory cycle, *J. Exp. Zool.* **166**:107–120.

Lee, P. A., 1969, Histology of the oviduct of the leopard frog. *Rana pipiens*, *Anat. Rec.* **165**:493–502.

Legname, A. H., Legname, H. S., Sanchez, S. S., Sanchez Riera, A. N., and Fernandez, S. N., 1972, Metabolic changes in *Bufo arenarum* oocytes induced by oviducal secretions, *Dev. Biol.* **29**:283–292.

Lonning, S., 1972, Comparative electron microscope studies of teleostean eggs with special reference to the chorion, *Sarsia* **49**:41–48.

Lonning, S., and Davenport, J., 1980, The swelling egg of the long rough dab, *Hippoglossides platessoides limandoides* (Bloch), *J. Fish. Biol.* **17**:359–378.

Lyne, A. G., and Hollis, D. E., 1983, Observations in Graafian follicles and their oocytes during lactation and after removal of pouch young in the marsupials *Isoodon macrourus* and *Perameles nasuta*, *Am. J. Anat.* **166**:41

McIntosh, J. R., and Porter, K. R., 1967, Microtubules in the spermatids of the domestic fowl, *J. Cell Biol.* **35**:153–173.

McLaren, A., 1970, The fate of the zona pellucida in mice, *J. Embryol. Exp. Morphol.* **23**:1–9.

McLaughlin, E. W., and Humphries, A. A., Jr., 1978, The jelly envelopes and fertilization of eggs of the newt, *Notophthalmus viridescens*, *J. Morphol.* **158**:73–90.

McRorie, R. A., and Williams, W. L., 1974, Biochemistry of mammalian fertilization, *Annu. Rev. Biochem.* **43**:777–803.

Miceli, D. C., del Pino, E. J., Barbieri, F. D., Mariano, M. I., and Raisman, J. S., 1977, The vitelline envelope-to-fertilization envelope transformation in the toad *Bufo arenarum*, *Dev. Biol.* **59**:101–110.

Miceli, D. C., Fernandez, S. N., Raisman, J. S., and Barbieri, F. S., 1978a, A trypsin-like oviducal proteinase involved in *Bufo arenarum* fertilization, *J. Embryol. Exp. Morphol.* **48**:79–91.

Miceli, D. C., Frenandez, S. N., and del Pino, E. J., 1978b, An oviductal enzyme isolated by affinity chromatography which acts upon the vitelline envelope of *Bufo arenarum* coelomic oocytes, *Biochem. Biophys. Acta.* **526**:289–292.

Milkman, R., 1954, Controlled observation of hatching in *F. heteroclitus, Biol. Bull.* **107**:300.

Minganti, A., 1955, Chemical investigations on amphibian egg jellies, *Exp. Cell Res. (Suppl.)* **3**:248–251.

Mintz, B., 1962, Experimental study of the developing mammalian egg: Removal of the zona pellucida, *Science* **138**:594–595.

Modlinski, J. A., 1970, The role of the zona pellucida in the development of the mouse egg in vivo, *J. Embryol. Morphol.* **23**:539–547.

Monroy, A., 1965, *Chemistry and Physiology of Fertilization*, Holt, Rinehart, New York.

Moricard, R., and Gothie, S., 1955, Etude de la repartition en $S^{35}$ dans les cellules folliculaires periovocytaires au cours de l'ovogenèse et de la terminaison de la première mitose de maturation chez la lapin adulte, *C. R. Soc. Biol.* **149**:1918–1922.

Munson, J. P., 1904, Researches on the oogenesis of the tortoise, *Clemmys marmorata, Am. J. Anat.* **3**:311–347.

Nace, G. W., Suyama, T., and Smith, N., 1960, Early development of special proteins, in: *Symposium on Germ Cells and Development*, pp. 564–603, Instituto Internazionale di Embriologia Fondazione A. Baselli, Milano.

Nadamitsu, S., 1957, Fertilization of coelomic and oviductal eggs of *Triturus pyrrogaster* (Boie.), *J. Sci. Hiroshima Univ. Ser. B* **17**:51–53.

Neaves, W. B., 1971, Intercellular bridges between follicle cells and oocyte in the lizard, *Anolis carolinensis, Anat. Rec.* **170**:285–301.

Nicolson, G. L., Yanagimachi, R., and Yanagimachi, H., 1975, Ultrastructural localization of lectin-binding sites on the zonae pellucidae and plasma membranes of mamalian eggs, *J. Cell Biol.* **66**:263–274.

Nishihara, T., and Hedrick, J. L., 1977, A molecular mechanism for envelope elevation at fertilization, *Fed. Proc.* **36**:811.

Oakberg, E. F., and Tyrell, P. D., 1975, Labeling the zona pellucida of the mouse oocyte, *Biol. Reprod.* **12**:477–482.

Odor, D. L., 1960, Electron microscopic studies on ovarian oocytes and unfertilized tubal ova in the rat, *J. Biophys. Biochem. Cytol.* **7**:567–574.

Odor, D. L., 1965, The ultrastructure of unilaminar follicles of the hamster ovary, *Am. J. Anat.* **116**:493–521.

Ohta, H., and Takano, K., 1982, Ultrastructure of micropylar cells in the pre-ovulatory follicles of Pacific herring, *Clupea pallasi Valenciennes, Bull. Fac. Fish. Hokkaido Univ.* **33(2)**:57–64.

Ohta, H., and Teranishi, T., 1982, Ultrastructure and histochemistry of granulosa and micropylar cells in the ovary of the loach, *Misgurnus anguillicaudatus (Cantor), Bull. Fac. Fish. Hokkaido Univ.* **33(1)**:1–8.

Ohtsuka, E., 1960, On the hardening of the chorion of the fish egg after fertilization. III. The mechanism of chorion hardening in *Oryzias latipes, Biol. Bull.* **118**:120–128.

Oikawa, T., Yanagimachi, R., and Nicolson, G. L., 1973, Wheat germ agglutin blocks mammalian fertilization, *Nature* **241**:256–259.

Olsen, M. W., and Neher, B. H., 1948, The site of fertilization in the domestic fowl, *J. Exp. Zool.* **109**:355–366.

Packard, M. J., 1980, Ultrastructural morphology of the shell and shell membrane of eggs of common snapping turtles *(Chelydra serpentina), J. Morphol.* **165**:187–204.

Packard, M. J., and Packard, G. C., 1979, Structure of the shell and tertiary membranes of eggs of softshell turtles *(Trionyx spiniferus), J. Morphol.* **159**:131–144.

Packard, M. J., Packard, G. C., and Boardman, T. J., 1982, Structure of egg shells and water relations of reptilian eggs, *Herpetologica* **38**:136–155.

Packard, M. J., Short, T. M., Packard, G. C., and Gorell, T. A., 1984a, Sources of calcium for embryonic development in eggs of the snapping turtle *Chelydra serpentina*, *J. Exp. Zool.* **230**:81–87.

Packard, M. J., Hirsch, K. F., and Iverson, J. B., 1984b, Structure of shells from eggs of kinosternid turtles, *J. Morphol.* **181**:9–20.

Packard, M. J., Iverson, J. B., and Packard, G. C., 1984c, Morphology of shell formation in eggs of the turtle *Kinesternon flavescens*, *J. Morphol.* **181**:21–28.

Pereda, J., 1970, Etude histochimique des mucopolysaccharides de l'oviducte et des gangues muqueuses des ovocytes de *Rana pipiens*: Incorporation du $^{35}SO_4^{2-}$, *Dev. Biol.* **21**:318–330.

Phillips, D. M., and Shalgi, R. M., 1980a, Surface architecture of the mouse and hamster zona pellucida and oocyte, *J. Ultrastruct. Res.* **72**:1–12.

Phillips, D. M., and Shalgi, R. M., 1980b, Surface properties of the zona pellucida, *J. Exp. Zool.* **213**:1–8.

Piko, L., 1969, Gamete structure and sperm entry in mammals, in: *Fertilization Comparative Morphology, Biochemistry and Immunology* (C. Metz and A. Monroy, eds.), pp. 325–404, Academic Press, New York.

del Pino, E. M., 1973, Interactions between gametes and environment in the toad *Xenopus laevis* (Daudin) and their relationship to fertilization, *J. Exp. Zool.* **185**:121–132.

Press, N., 1964, An unusual organelle in avian ovaries, *J. Ultrastruct. Res.* **10**:528–546.

Rahil, K. S., and Narbaitz, R., 1973, Ultrastructural studies on the relationship between follicular cells and growing oocytes in the turtle *Pseudemys scripta elegans*, *J. Anat.* **115**:175–186.

Raisman, J., and Pisano, A., 1970, Fertilization of jelly-less *Bufo arenarum* oocytes in the presence of high sperm concentrations, *Acta Embryol. Exp.* **1970**:3–11.

Renfree, M. N., 1973, The composition of fetal fluids of the marsupial *Macropus eugenii*, *Dev. Biol.* **33**:62–79.

Riehl, R., 1977, Licht- und elektronenmikroskopisch Untersuchungen zu Bau und Entwicklung der Mikropylen von *Noemacheilus barbatulus* (L.) und *Gobio gobio* (L.) (Pisces, Teleostei), *Zool. Anz.* **198**:313–327.

Riehl, R., 1978, Licht- und elektronenmikroskopische Untersuchungen an den Oocyten der Susswasser-Teleosteer *Nocmacheilus barbatulus* (L.) und *Gobio gobio* (L.) (Pisces, Teleostei), *Zool. Ans.* **201**:199–219.

Riehl, R., 1980a, Feinbau, Entwicklung und Bedeutung der Eihullen bei Knochenfischen, *Z. Binnenfisch.* **30(5)**:33–37.

Riehl, R., 1980b, Micropyle of some salmonius and coregonius, *Env. Biol. Fish.* **5**:59–66.

Riehl, R., and Schulte, E., 1972, Licht- und elektronenmikroskopische Untersuchungen au den Eihullen der Elritze *Phoxinus phoxinus* (L.), Teleostei, Cyprinidae, *Protoplasma* **92**:147–162.

Robinson, D. S., and King, N. R., 1968, Mucopolysacchairides of an avian shell membrane. *J. R. Microsc. Soc.* **88**:13–22.

Romanoff, A. L., 1967, *Biochemistry of the Avian Embryo—A Quantitative Analysis of Prenatal Development*, Wiley New York.

Romanoff, A. L., and Romanoff, A. J., 1949, *The Avian Egg*, Wiley, New York.

Rottmann, O. J., and Lampeter, W. W., 1981, Development of early mouse and rabbit embryos without zona pellucida, *J. Reprod. Fertil.* **61**:303–306.

Salthe, S. N. 1963, The egg capsules in the amphibia, *J. Morpholo.* **118**:161–171.

Sato, K., 1979, Polyspermy prevention mechanisms in mouse eggs fertilized in vitro, *J. Exp. Zool.* **210**:353–359.

Schmell, E. D., and Gulyas, B. J., 1980, Ovaperoxidase activity in ionophore treated mouse eggs. II. Evidence for the enzymes role in the hardening of the zona pellucida, *Gamete Res.* **3**:279–290.

Schmell, E. D., Gulyas, B. J., and Hedrick, J. L., 1983, Egg surface changes during fertilization and the molecular mechanism of the block to polyspermy, in: *Mechanisms and Control of Animal Fertilization* (J. F. Hartmann, ed.), pp. 365–413, Academic Press, New York.

Schoots, A. F., Stikkelbroek, J. M., Bekhuis, J. F., and Denuce, J. M., 1982, Hatching in teleostean fishes: Fine structural changes in the egg envelope during enzymatic breakdown *in vivo* and *in vitro*, *J. Ultrastruct. Res.* **80:**185–196.

Sellins, M. H., and Jenkinson, E. J., 1975, Permeability of the mouse zona pellucida to immunoglobulins, *J. Reprod.* Biol. **42:**153–157.

Shackley, S. E., and King, P. E., 1977, Oogenesis in a marine teleost, *Blennius pholis L.*, *Cell Tissue Res.* **181:**105–128.

Sharman, G. B., 1961, The embryonic membranes and placentation in five genera of diprotodont marsupials, *Proc. Zool. Soc. Lond.* **137:**197–220.

Shaver, J. R., 1966, Immunobiological studies of the jelly-coats of anuran eggs, *Am. Zool.* **6:**75–87.

Shaver, J. R., and Barch, S. H., 1960, Experimental studies on the role of jelly coat material in fertilization in the frog, *Acta Embryol. Morphol. Exp.* **3:**180–189.

Shaver, J. R., Barch, S. H., and Umpierre, C. C., 1970, Interspecific relationships of oviducal materials as related to fertilization in amphibia, *J. Embryol. Exp. Morphol.* **24:**209–225.

Sherman, M. I., and Wudl, L. R., 1976, The implanting mouse blastocyst, in: *The Cell Surface in Animal Embryogenesis and Development* (G. Poste and G. L. Nicolson, eds.), pp. 81–126, Elsevier/North-Holland, New York.

Shivers, C. A., 1962, Localization of the inhibitory effect of anti-jelly serum on fertilization in frog eggs by fluorescein-togged antibodies, *Am. Zool.* **2:**448.

Shivers, C. A., 1974, Immunological interference with fertilization, in: *Immunological Approaches to Fertility Control* (E. Diczfalusy, ed.), pp. 59–74. Karolinska Institute, Stockholm, Sweden.

Shivers, C. A., and Dudkiewicz, A. G., 1974, Inhibition of fertilization with specific antibodies, in: *Physiology and Genetics of Reproduction.* Part B (E. M., Coutinho and F. Fuchs, eds.), pp. 81–96, Plenum Press, New York.

Shivers, C. A., and James, J. M., 1970a, Morphology and histochemistry of the oviduct and egg-jelly layers in the frog, *Rana pipiens*, *Anat. Rec.* **166:**541–556.

Shivers, C. A., and James J. M., 1970b, Capacitation of frog sperm, *Nature* **227:**183–184.

Shivers, C. A., and James, J. M., 1971, Fertilization of antiserum-inhibited frog eggs with "capacitated" sperm, *Biol. Reprod.* **5:**229–235.

Shivers, C. A., and Metz, C. B., 1962, Inhibition of fertilization in frog eggs by univalent fragments of rabbit antibody, *Proc. Soc. Exp. Biol. Med.* **110:**385–387.

Silva Sasso, D. W., 1959, Existence of hyaluronic acid in the zona pellucida of the rabbit's ovum, *Acta Anat.* **36:**352–357.

Smith, L. D., and Ecker, R. E., 1970, Uterine suppression of biochemical and morphogenetic events in *Rana pipiens*, *Dev. Biol.* **22:**622–637.

Smithberg, M., 1953, The effect of different proteolytic enzymes on the zona pellucida of mouse ova, *Anat. Rec.* **117:**554.

Solomon, S. E., and Baird, T., 1976, Studies on the egg shell (oviducal and ovoposited) of *Chelonia mydas* L., *J. Exp. Mar. Biol. Ecol.* **22:**145–160.

Solomon, S. E., and Baird, T., 1977, Studies on the soft-shell membranes of the egg shell of *Chelonia mydas.* L., *J. Exp. Mar. Biol. Ecol.* **27:**83–92.

Soupart, P., and Noyes, R. W., 1964, Sialic acid as a component of the zona pellucida of the mammalian ovum, *J. Reprod. Fertil.* **8:**251–253.

Stegner, H. E., and Wartenberg, H., 1961, Elektronenmikroskopische und histotopochemische Untersuchungen uber Struktur und Bildung der Zona pellucida menschlicher eizelleen, *Z. Zellforsch.* **53:**702–713.

Stehr, C. M., and Hawkes, J. W., 1979, The comparative ultrastructure of the egg membranes and associated pore structures in the starry flounder, *Platichthys stellata (Pallas)*, and pink salmon, *Oncorhynchus gorbuscha (Walbaum)*, *Cell Tissue Res.* **202:**347–356.

Stehr, C. M., and Hawkes, J. W., 1983, The development of the hexagonally structured egg envelope of C-O sole *(Pleuronichthys coenosus)*, *J. Morphol.* **178:**267–284.

Steinke, J. H., and Benson, D. G., Jr., 1970, The structure and polysaccharide cytochemistry of the jelly envelopes of the egg of the frog. *Rana pipiens*, *J. Morphol.* **130**:57–66.

Suzuki, R., 1958, Sperm activation and aggregation during fertilization in some fishes, *Embryologia* **4**:93–102.

Swenson, C. E., and Dunbar, B. S., 1982, Specificity of sperm–zona interaction, *J. Exp. Zool.* **219**:97–104.

Taddie, C., 1972, Significance of pyriform cells in ovarian follicle of *Lacerta sicula*, *Exp. Cell Res.* **72**:562–566.

Takano, K., and Ohta, H., 1982, Ultrastructure of micropylar cells in the ovarian follicles of the pond smelt, *Hypomesus transpacificus nipponensis*, *Bull, Fac. Fish. Hokkaido Univ.* **33(2)**:65–78.

Talbot, P., and DiCarlantonio, G., 1984, Ultrastructure of opossum oocyte investing coats and their sensitivity to trypsin and hyaluronidase, *Dev. Biol.* **103**:159–167.

Tesoriero, J. V., 1977a, Formation of the chorion (zona pellucida) in the teleost *Oryzias latipes*, *J. Ultrastruct. Res.* **59**:282–291.

Tesoriero, J. V., 1977b, Formation of the chorion (zona pellucida) in the teleost *Oryzias latipes*. II. Polysaccharide cytochemistry of early oogenesis, unpublished manuscript; see *J. Histochem. Cytochem.* **25**:1376–1380.

TeWinkle, L. E., 1963, Notes on the smooth dogfish, *Mustelus canis*, during the first three months of gestation. II. Structural modifications of yolk-sacs and yolk-stalks correlated with increasing absorptive function, *J. Exp. Zool.* **152**:123–137.

Thing, A., 1918, The formation and structure of the zona pellucida in the ovarian eggs of turtles, *Am. J. Anat.* **23**:237–258.

Tiberi-Marquis, J., 1973, Quelques aspects histochimiques de l'ovogenese chez *Betta splendens* R. (Teleosteen Anabantide), *Histochemie* **33**:139–158.

Tracy, C. R., 1980, Water relations of parchment-shelled lizard (*Sceloparus undulatus*) eggs, *Copeia* **1980**:478–482.

Trounson, A. O., and Moore, N. W., 1974, The survival and development of sheep eggs following complete or partial removal of the zona pellucida, *J. Reprod. Fertil.* **41**:97–105.

Tsukahara, J., 1971, Ultrastructural study on the attaching filaments and villi of the oocyte of *Oryzias latipes* during oogenesis, *Dev. Growth Diff.* **13**:173–180.

Tuan, R. S., 1983, Supplemental eggshell restores calcium transport in chorioallantoic membrane of cultured shell-less chick embryos, *J. Embryol. Exp. Morphol.* **74**:119–131.

Tyndale-Biscoe, H., 1975, *Life of Marsupials*, American Elsevier, New York.

Ullmann, S. L., 1979, Observations on the primoridal oocyte of the bandicoot *Isoodon macrourus* (Peramelidae, marsupialia), *J. Anat.* **128**:619–631.

Ulrich, E., 1969, Etude des ultrastructures au cours de l'ovogenese d'un poisson Teleosteen, la danio, *Brachydanio rerio* (Hamilton-Buchanan), *J. Microsc. (Paris)* **8**:447–478.

Varma, S. K., 1970, Morphology of ovarian changes in the garden lizard, *Calotes versicolor*, *J. Morphol.* **131**:195–210.

Viarengo, A., and Falcone, R., 1977, A chemical analysis of *Rana esculenta* egg jelly envelope, *Acta Embryol. Exp.* **3**:357–362,.

Wartenberg. H., and Stegner, H. E., 1960, Uber die elektronenmikroskopische Feinstructur des menschlichen Ovarioleies, *Z. Zellforsch.* **52**:450–474.

Wassarman, P. M., 1982, Fertilization, in: *Cell Interactions and Development: Molecular Mechanisms* (K. M. Yamada, ed.), pp. 1–27, Wiley, New York.

Wassarman, P. M., and Bliel, J. D., 1982, The role of the zona pellucida glycoproteins as regulators of sperm–egg interactions in the mouse, in: *Cellular Recognition* (W. Frazier, L. Glaser, and D. Gottlieb, eds), pp. 845–863, Alan R. Liss, New York.

Wischnitzer, S., 1963, The ultrastructure of the layers enveloping yolk-forming oocytes from *Triturus viridescens*, *Z. Zellforsch.* **60**:452–462.

Wischnitzer, S., 1964, An electron microscope study of the formation of the zona pellucida in oocytes from *Triturus viridescens*, *Z. Zellforsch.* **64**:196–209.

Wischnitzer, S., 1966, The ultrastructure of the cytoplasm of the developing amphibian egg, *Adv. Morph.* **5**:131–179.

Wolf, D. P., 1974, On the contents of the cortical granules from *Xenopus laevis* eggs, *Dev. Biol.* **38**:14–29.

Wolf, D. P., and Hedrick, J. L., 1971, A molecular approach to fertilization. II. Viability and artifical fertilization of *Xenopus laevis* gametes, *Dev. Biol.* **25**:348–395.

Wolf, D. P., Nishihara, T., West, D. M., Wyrick, R. E., and Hedrick, J. L., 1976, Isolation, physicochemical properties, and the macromolecular composition of the vitelline and fertilization envelopes from *Xenopus laevis* eggs, *Biochemistry* **15**:3671–3678.

Wong, M., Hendrix, M. J. C., von der Mark, K., Little, C., and Stern, R., 1984, Collagen in the egg shell membranes of the hen, *Dev. Biol.* **104**:28–36.

Wood, D. M., Liv, C., and Dunbar, B. S., 1981, Effect of alloimmunization and heteroimmunization with zonae pellucidae on fertility in rabbits, *Biol. Reprod.* **25**:439–450.

Wourms, J. B., 1976, Annual fish oogenesis. I. Differentiation of the mature oocyte and formation of the primary envelope, *Dev. Biol.* **50**:338–354.

Wourms, J. P., 1977, Reproduction and development in Chondrichthyan fishes, *Am. Zool.* **17**:379–410.

Wourms, J. P., and Sheldon, H. 1976, Annual fish oogenesis. II. Formation of the secondary egg envelope, *Dev. Biol.* **50**:355–366.

Wyburn, G. M., Aitken, R. N. C., and Johnston, H. S., 1965, The ultrastructure of the zona radiata of the ovarian follicle of the domestic fowl, *J. Anat.* **99**:469–484.

Wyrick, R. E., Nishihara, T., and Hedrick, J. L., 1974, Agglutination of jelly coat and cortical granule components and the block to polyspermy in the amphibian *Xenopus laevis*, *Proc. Natl. Acad. Sci. USA* **71**:2067–2071.

Xavier, F., 1973, Le cycle des voies genitales femelles de *Nectophrynoides occidentalis* angel, amphibien anoure vivipare, *Z. Zellforsch.* **140**:509–534.

Yamamoto, T., 1961, Physiology of fertilization in fishes, *Int. Rev. Cytol.* **12**:361–405.

Yamamoto, M., and Yamagami, K., 1975, Electron microscopic studies on choriolysis by the hatching enzyme of the teleost. *Oryzias latipes*, *J. Fac. Sci. Univ. Tokyo* **410**:123–127.

Yoshizaki, N., 1978, Disintegration of the vitelline coat during the hatching process in the frog, *J. Exp. Zool.* **203**:127–134.

Yoshizaki, N., and Katagiri, Ch., 1981, Oviducal contribution to alteration of the vitelline coat in the frog, *Rana japonica*. An electron microscopic study, *Dev. Growth Dif.* **23**:495–506.

Yoshizaki, N., and Katagiri, C., 1982, Acrosome reaction in sperm of the toad, *Bufo bufo japonicus*, *Gamete Res.* **6**:343–352.

Yurewicz, E. C., Oliphant, G., and Hedrick, J. L., 1975, The macromolecular composition of *Xenopus laevis* egg jelly coat, *Biochemistry* **14**:3101–3107.

Chapter 6

# Oocyte Maturation in Amphibians

JAMES L. MALLER

## 1. Introduction

Although the amphibian oocyte possesses a vast store of structural and regulatory components, the oocyte itself is extremely repressed in terms of its rate of protein synthesis. It can remain in this repressed state for a period of months or years, as it grows in the ovary by virtue of deposition of yolk platelets. The quiescent state of oocyte metabolism is broken when the process of oocyte maturation is initiated by hormones. Studies over many years have demonstrated that *in vivo* the epithelial layer of follicle cells surrounding each oocyte is stimulated by pituitary-derived luteinizing hormone (LH) to synthesize and secrete progesterone, which acts directly on the oocyte to induce maturation of the oocyte into an unfertilized egg (Masui, 1967; Maller and Krebs, 1980). The interaction of progesterone with the isolated oocyte appears to be a sufficient trigger for a normal oocyte maturation response, as judged by cytological analysis and by the ability of isolated oocytes responding to exogenously applied progesterone to undergo normal embryonic development (Smith *et al.*, 1968; Drury and Schorderet-Slatkine, 1975).

Treatment of the oocyte with progesterone activates a number of biosynthetic processes. Thus, oocyte maturation can be viewed as the initial step in the activation of metabolism for development (see Section 4). From this perspective, it is reasonable to assume that information about the mechanisms involved in utilizing stored components during oocyte maturation will have general significance relative to the control of early development after fertilization. The analysis of regulatory mechanisms during oocyte maturation is enhanced by the fact that the oocyte can be microinjected with known quantities of purified substances while undergoing oocyte maturation. Since the oocyte is a single cell, injected materials have access to all regions of the egg; this procedure provides a powerful tool for

JAMES L. MALLER • Department of Pharmacology, University of Colorado School of Medicine, Denver, Colorado 80262.

studying the effects of purified substances on the intact living cell. This chapter discusses the use of this technique, which has contributed heavily to the current picture of the regulation of oocyte maturation.

### 1.1. Cyclic Nucleotides and the Initiation of Maturation

The experimental analysis of oocyte maturation has identified early events (i.e., those that occur rapidly after exposure to progesterone *in vitro*) and late events that occur several hours after progesterone, culminating in the breakdown of the germinal vesicle (**GVBD**). GVBD is a convenient morphological end point for many studies of oocyte maturation. Because the oocytes from different females vary widely in the absolute time transpiring from progesterone to GVBD in 50% of responding oocytes, the specific order of events can be better judged by using a relative time scale; 1.0 $GVBD_{50}$ refers to the time required after progesterone administration for 50% of the oocytes to undergo GVBD (Wasserman and Masui, 1975). After progesterone administration, within 0.1 $GVBD_{50}$, a rapid drop occurs in the level of cyclic adenosine monophosphate (cAMP) to about 40–60% of basal (Maller *et al.*, 1979; Morrill *et al.*, 1977; Schorderet-Slatkine *et al.*, 1982; Speaker and Butcher, 1977). Compared with the regulation of cAMP by hormones in somatic cells, a drop of this magnitude is small; thus, the question of its significance has been raised. In eukaryotic cells, cAMP is believed to exert its effects exclusively through the activation of a protein kinase. The mechanism of activation involves the dissociation of an inactive complex of regulatory and catalytic subunits, as indicated in equation (1):

$$R_2C_2 + 4\text{cAMP} \rightleftharpoons (R(\text{cAMP})_2)_2 + 2C \qquad (1)$$

The regulatory (or R) subunit is thus an inhibitor of the catalytic subunit when bound to it; cAMP relieves the inhibition by binding to R, causing dissociation of R and C to yield a free catalytic subunit that expresses all the protein kinase activity. The molar ratio of regulatory to catalytic subunits in cells approximates 1.0; thus, changes in the concentration of either R or C should produce a change in the expression of protein kinase activity as well as a change in protein phosphorylation.

The catalytic and regulatory subunits from different tissues or different species are highly conserved and can interact with each other. In order to assess the significance of the drop in cAMP with progesterone, homogeneous preparations of either R or C from muscle tissue were microinjected into oocytes, and the effects on subsequent progesterone-induced GVBD determined (Maller and Krebs, 1977). As shown in Figure 1, microinjection of the catalytic subunit totally blocks progesterone action at an effective concentration of $10^{-7}$ M. This result indicates that the decrease in

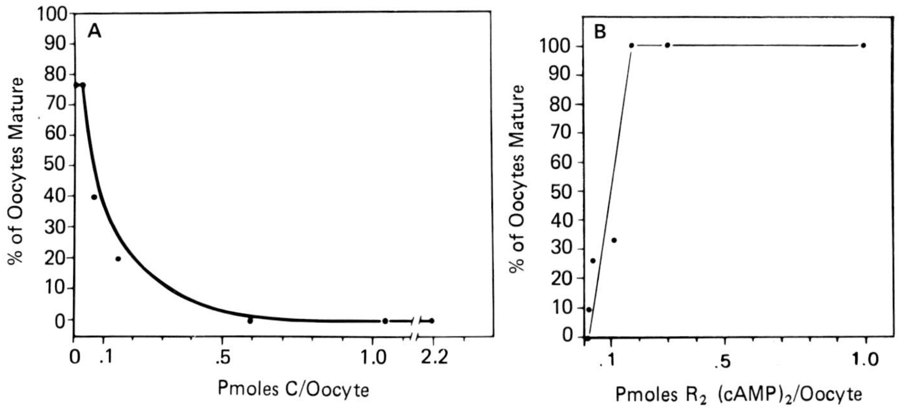


**Figure 1.** Effect of microinjected protein kinase subunits on oocyte maturation. (A) Dose–response curve for catalytic subunit. Oocytes were injected with varying concentrations of homogeneous catalytic subunit of cAMP-dependent protein kinase, subsequently induced to mature with progesterone, and the frequency of germinal vesicle breakdown (GVBD) was determined. Oocytes contain approximately 0.001 ml cell water; 0.1 pmole oocyte thus represents an internal concentration of $1 \times 10^{-7}$ M. (B) Dose–response curve for regulatory subunit. The homogeneous type II regulatory subunit–cAMP complex was injected in increasing concentrations into oocytes, and the oocytes were observed to undergo GVBD in the absence of progesterone. (From Maller and Krebs, 1977.)

cAMP is necessary for oocyte maturation, since increasing the effect of cAMP blocks the hormonal response. When the regulatory subunit was microinjected into oocytes, GVBD was found to occur in the absence of progesterone, at an effective intracellular concentration of R of about $10^{-7}$ M. The ability of R to induce oocyte maturation indicates that the decrease in cAMP is sufficient to trigger oocyte maturation, since merely decreasing the effect of cAMP via protein kinase triggers the biological response. Control experiments demonstrated that the injected isolated subunits could interact with the endogenous oocyte protein kinase components. Subsequent work has shown that this inhibition of catalytic subunit leads to a significant decrease in oocyte protein phosphorylation after microinjection (Boyer *et al.*, 1980). These findings suggest a model in which the prophase arrest of the resting oocyte is maintained by a phosphoprotein substrate of cAMP-dependent protein kinase (Maller and Krebs, 1977). Removal of this protein is both necessary and sufficient to cause oocyte maturation. It also appears to be rate limiting, since the time required for GVBD is inversely proportional to the amount of protein kinase inhibitor microinjected (Huchon *et al.*, 1981).

These results demonstrate the power of microinjection techniques in that a modest decrease in cAMP actually represents the crucial event in stimulating the conversion of an oocyte into an unfertilized egg. It seems likely that cAMP is also stored in the oocyte for use in early development, since the major portion of total cAMP is in the vegetal pole. Thus, the observed modest decrease in cAMP with progesterone may represent a much greater decrease in the level of metabolically active cAMP in the oocyte.

## 1.2. Role of Protein Phosphatase Activity

The results discussed above provide convincing evidence that the level of cAMP-dependent protein kinase activity decreases with progesterone. This is predicted to cause a decrease in the steady-state level of phosphorylation of substrate proteins. Of course, the steady-state level of phosphorylation of any protein is a balance between the protein kinase and protein phosphatase activities. This raises the question of whether progesterone might, in addition to decreasing kinase activity, increase protein phosphatase activity as a means of decreasing the level of phosphorylation of the phosphoprotein(s) hypothesized to maintain the prophase block of the oocyte. The possibility of phosphatase regulation is suggested by the recent findings in mammalian systems that cAMP can also regulate phosphatase activity via the phosphorylation of phosphatase inhibitor proteins (Nimmo and Cohen, 1978).

Two types of inhibitor proteins have been described: (1) **phosphoinhibitor-1,** which inhibits phosphatase only when phosphorylated by cAMP-dependent protein kinase (Huang and Glinsmann, 1976); and (2) **dephosphoinhibitor-2,** which is only inhibitory when dephosphorylated (Hemmings *et al.*, 1982). These inhibitors act only on a specific class of phosphatase termed **protein phosphatase-1.** With inhibitor-1, a change in cAMP thus leads to a reciprocal change in protein phosphatase activity, having the effect of potentiating the effect of cAMP on the phsophorylation state of a protein.

This idea is schematically represented in Figure 2. A prediction of this scheme is that changes in the level of active phosphatase inhibitor should alter the level of phosphorylation of proteins that are substrates for cAMP-dependent protein kinase and thus affect the initiation of oocyte maturation by progesterone. As shown in Figure 3, this is indeed the case. Microinjection of dephosphoinhibitor-2 delays GVBD induced by progesterone; control experiments demonstrate that the injected dephosphoinhibitor-2 slows the rate of dephosphorylation of phosphorylated proteins (Foulkes and Maller, 1982). Similar results are obtained with phosphoinhibitor-1 injections (Huchon *et al.*, 1981b). These results demonstrate that the activity of protein phosphatase-1 can have an effect on oocyte maturation. However, it remains to be demonstrated in the intact oocyte that the activity of protein phosphatase-1 changes in response to progesterone, possibly via inhibitor-1 dephosphorylation.

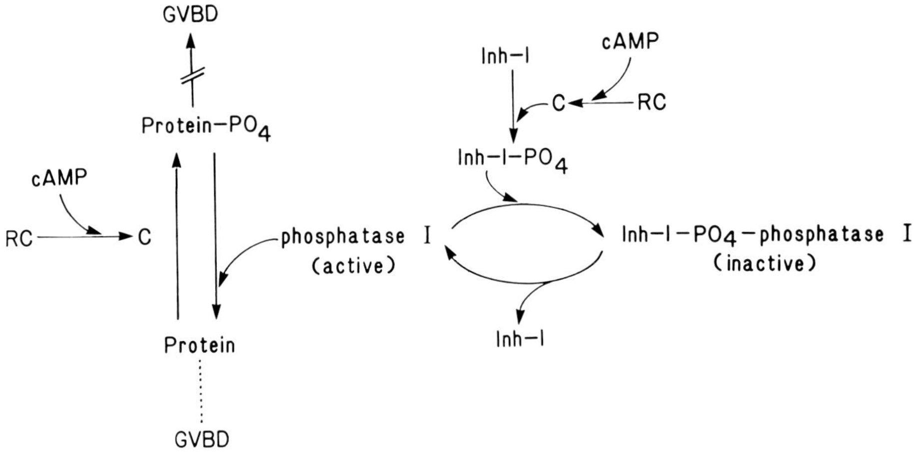


**Figure 2.** Model proposed for concerted control of the phosphorylation state of the putative maturation-inhibiting phosphoprotein. The model postulates that a high steady-state level of the maturation-inhibiting phosphoprotein is necessary and sufficient to maintain the prophase block of the oocyte and that the level of this protein is determined by the rate of phosphorylation by cAMP-dependent protein kinase and the rate of dephosphorylation by protein phosphatase(s).

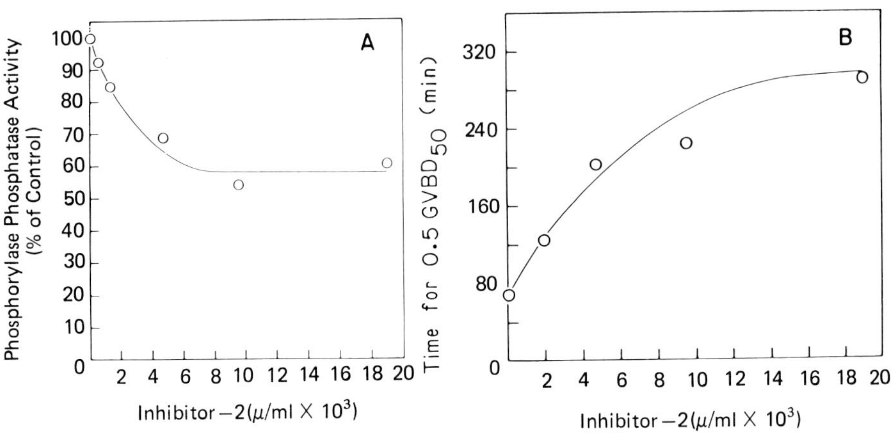


**Figure 3.** Effect of microinjected phosphatase inhibitor. (A) Dose–response curves for inhibition of [$^{32}$P]phosphorylase *a* dephosphorylation. [$^{32}$P]phosphorylase *a* was mixed with increasing concentrations of inhibitor-2 and injected in a constant volume (60 nl; >2000 cpm) into 15–30 oocytes. After 3 min, the reaction was terminated by dropping oocytes into ice cold 10% trichloroacetic acid, and the extent of dephosphorylation was determined by liquid scintillation counting of soluble and insoluble fractions. (B) Dose-dependent delay of GVBD by inhibitor-2. Inhibitor-2 was injected at varying concentrations into each of 25–30 oocytes. Controls were injected with buffer. After 15–30 min, oocytes were treated with progesterone (10 μM), and the time course of GVBD was determined. (From Foulkes and Maller, 1982.)

### 1.3. Mechanism of Cyclic Nucleotide Decrease

An important question concerns the mechanism by which the crucial decrease in cAMP is achieved after progesterone stimulation of the oocyte. The interest in this question transcends the problem of oocyte maturation, because it has been generally assumed that steroids bind cytosolic receptors and act exclusively at the level of transcription, independently of cAMP fluctuations. The direct evidence described above, which shows that progesterone rapidly lowers the level of cAMP, raises the question of how this occurs and its general significance. Formally, the decrease in cAMP could occur due to excretion of cAMP from the cell, inhibition of adenylate cyclase, activation of phosphodiesterase, or some combination thereof. Several laboratories have shown that no cAMP appears in the medium of oocytes incubated with progesterone, indicating that excretion is not a plausible mechanism. The possibility that adenylate cyclase activity could be regulated by progesterone was suggested by the demonstration that cholera toxin pretreatment of oocytes totally blocks the ability of progesterone to cause GVBD (Maller *et al.*, 1979). Cholera toxin is a potent irreversible activator of adenylate cyclase in eukaryotic cells (for a recent review, see Johnson, 1982). The activation is achieved via ADP-ribosylation of one of the adenylate cyclase subunits, leading to a persistent state of stimulation. The ability of cholera toxin to block oocyte maturation is consistent with the effects of microinjected protein kinase subunits on oocyte maturation; further study has demonstrated that oocyte adenylate cyclase is activated by cholera toxin with the same dose–response curve observed for the inhibition of GVBD by cholera toxin (Maller *et al.*, 1979; Sadler and Maller, 1981). These studies demonstrate that the activity of adenylate cyclase is capable of affecting the oocyte maturation process. The first direct evidence that progesterone regulates adenylate cyclase activity came from Ozon's laboratory with reports that progesterone decreases the rate of conversion of microinjected ($\alpha$-$^{32}$P)-ATP into cAMP in the intact oocyte (Mulner *et al.*, 1979). This finding was extended significantly first by Sadler and Maller (1981) and then by others with the demonstration of progesterone-dependent inhibition of cyclase in purified plasma membranes (Jordana *et al.*, 1981; Finidori-Lepicard *et al.*, 1981; Sadler and Maller, 1981). The concentration of progesterone needed for half-maximal inhibition ($2 \times 10^{-7}$ M) is similar to the concentration required for 50% GVBD, supporting the physiological relevance of the inhibition.

These results raise the question of the mechanism by which progesterone inhibits the cyclase, particularly whether it is similar to the mechanism of inhibition used by other hormones with somatic cells. Addressing this question requires understanding the cyclic stimulation cycle of adenylate cyclase activity in somatic cells. This general scheme, illustrated diagrammatically in Figure 4, incorporates findings from many laboratories utilizing adenylate cyclase from many sources, including mutant cells that

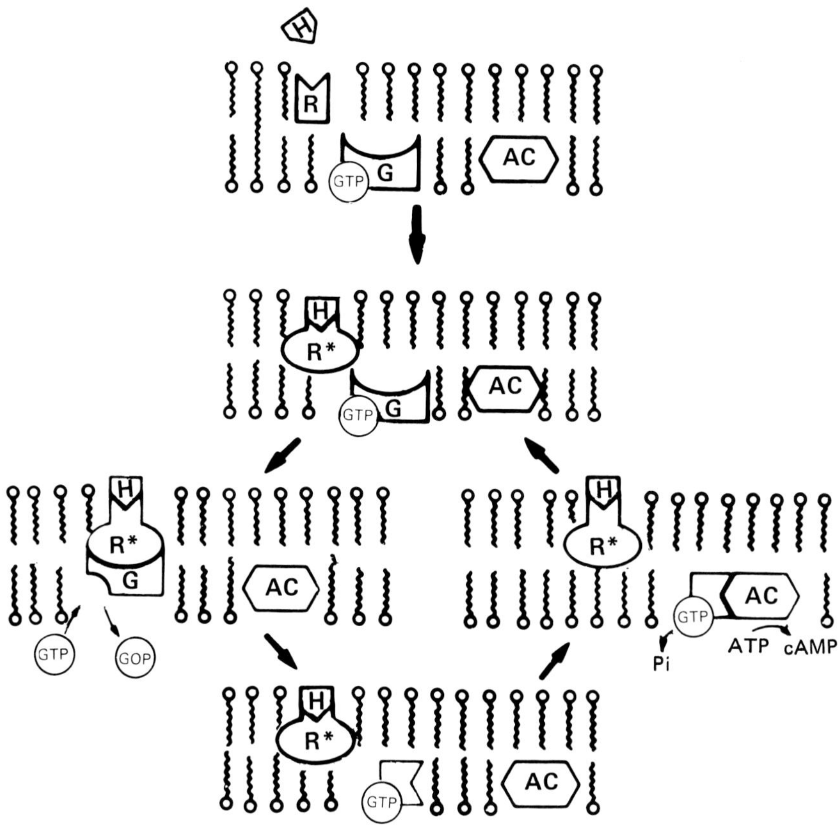


**Figure 4.** General scheme of the regulatory GTPase cycle of adenylate cyclase. The enzyme is activated by receptor-mediated exchange of GDP for GTP on the G subunit, which then interacts with the catalytic subunit (AC) to increase activity. Activity is terminated by hydrolysis of GTP on the G protein, which promotes dissociation of G and C. R, receptor; H, hormone.

are defective in one or more components of the adenylate cyclase system. The cyclase is composed of at least three subunits: a hormone receptor, a guanine nucleotide-binding protein (G), and a catalytic subunit. Enzyme activity cycles between an active state in which GTP is bound to G, thereby promoting G/C interaction and leading to increased cAMP synthesis. The binding of hormone to receptor promotes the exchange of GDP on the G protein for GTP, giving rise to the activated state, and the active cycle is terminated by GTP hydrolysis to GDP, generating an inactive state (Cassel *et al.*, 1977). Several predictions follow from this model, which is generally accepted at present. One is that treatment of cyclase with nonhydrolyzable analogs of GTP such as GTP$\gamma$S or Gpp(NH)p should cause a persistent activation of the system; this has been widely observed in many systems, including the oocyte. Indeed, the ability of cholera toxin to activate adenylate cyclase irreversibly is due to inhibition of GTPase activity following ADP ribosylation, effectively preventing the turn-off reaction for cyclase (Cassel and Selinger, 1977, Johnson, 1982) (Fig. 4).

In avian systems, the rate-limiting step in the activation of cyclase has been demonstrated to be the exchange of GDP for GTP (Cassel *et al.*, 1977). Thus, hormonal inhibition of cyclase could occur either by decreasing the rate of guanine nucleotide exchange or by increasing the rate of GTP hydrolysis, or both. If the inhibition occurred by increasing GTP hydrolysis, the cyclase activated by cholera toxin or by nonhydrolyzable analogues of GTP would not be subject to hormonal inhibition. However, as shown in Figure 5, progesterone was able to inhibit oocyte adenylate cyclase activity under these conditions, indicating that GTPase activity was not essential for inhibition by progesterone. This finding places progesterone in a different class than most other hormones that inhibit adenylate cyclase, since they are not active with Gpp(NH)p and have been shown in some cases to simulate GTPase activity (Koski and Klee, 1981).

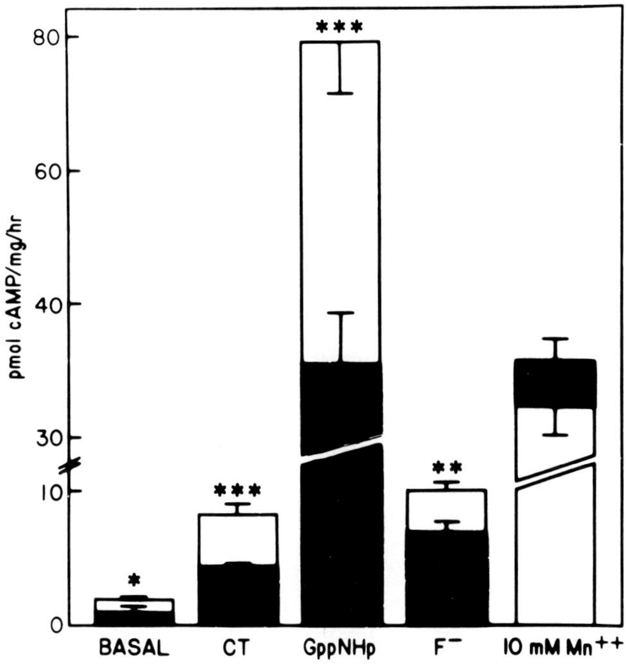


**Figure 5.** Inhibition of plasma membrane adenylate cyclase by progesterone. Plasma membranes were manually dissected from oocytes and assayed for adenylate cyclase activity in the presence of various stimulators (150 $\mu$M Gpp (NH) p, 4 mM NaF, or 10 mM $MnSO_4$. In the case of cholera toxin (CT), oocytes were preincubated with 5 nM toxin before membrane isolation. Open bars: absence of progesterone. Closed bars: presence of progesterone (30 $\mu$M). *, $P < 0.075$; **, $P < 0.025$; ***, $P < 0.005$. (From Sadler and Maller, 1981.)

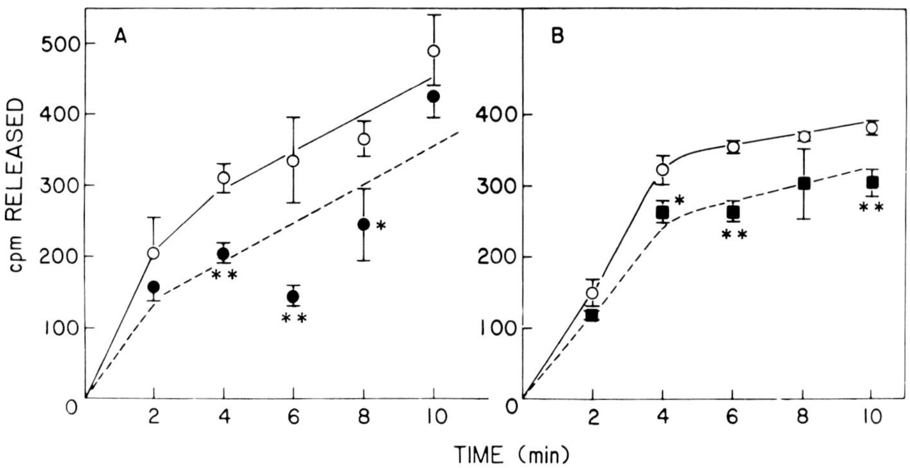


**Figure 6.** Effect of 2′,5′-dideoxyadenosine and progesterone on guanine nucleotide release from oocyte membranes. Aliquots (0.02 ml) of membrane suspensions that had been preloaded with [$^3$H]-GTP were added to 0.13 ml volumes of release buffer containing GTPγS and incubated for various times at room temperature. The samples were then microfuged, and 0.1 ml aliquots of the supernatants were analyzed by liquid scintillation spectrometry for the level of guanine nucleotide release. (A), control (— ○ —); + 100 μM 2′,5′-DDA) (- - ● - -). (B), control (— ○ —); +30 μM progesterone) (- - ■ - -). Each point represents the mean of triplicate determination ± SE. The experiment in A was carried out with oocyte membranes from a different frog than that in B, resulting in the different absolute amounts of guanine nucleotide. (From Sadler and Maller, 1983.)

The only class of hormone other than progesterone that is capable of inhibiting adenylate cyclase activity stimulated by Gpp(NH)p is a class of adenosine compounds known as "P-site" agonists, so named because activity at this site requires an unsubstituted purine ring. Interesting features of the P site for adenosine include the evidence that it is located on the inner surface of the plasma membrane and that the concentration of specific P-site agonists is much higher in amphibian than in mammalian cells (Sayhoun *et al.*, 1976; Londos *et al.*, 1981). Because of the similarities in the inhibition of cyclase by progesterone and the P-site agonists, a detailed comparative study was undertaken by Sadler and Maller (1983). Inhibition of oocyte adenylate cyclase by progesterone and a potent P-site agonist, 2′ :5′-dideoxyadenosine, is nonadditive, suggesting a common mechanism of action. These studies culminated in the demonstration (Fig. 6) that both types of hormones decrease the rate of guanine nucleotide exchange on the membrane cyclase, suggesting that inhibition is associated with decreasing the turn on for cyclase rather than stimulating the turn-off reaction (Fig. 4). These studies provide the first evidence that hormonal regulation of cyclase in any system occurs at this step in the cyclic activation process (Fig. 4) and suggest that the properties of P-site inhibition may have physiological relevance. Because the P site may be on the inner surface of the plasma membrane, it is possible that the activated state of cyclase resulting from external hormone–receptor interaction is subject to modification intracellularly by compounds that act at the P site.

## 2. Interrelations of Calcium and Cyclic AMP

The data demonstrating that steroids inhibit adenylate cyclase activity are consistent with the ability of microinjected protein kinase subunits to control maturation and the observed drop in cAMP with progesterone in the intact oocyte. Further evidence in this regard lies in the finding that the order of potency of various agonists for inducing GVBD parallels their potency *in vitro* for inhibiting adenylate cyclase. An area that requires further work is the analysis of whether the inhibition of cyclase is sufficient to account for the observed drop in cAMP, or whether the decrease is concerted, involving a concurrent activation of phosphodiesterase. This possibility is discussed in greater detail in Section 2.2.

Adenylate cyclase is the classic transducer of information from the outside of the cell to the inside. The regulation of this enzyme reaches its most sophisticated level in the mammalian central nervous system, which is composed of highly differentiated nerve cells that regulate adenylate cyclase by a multitude of hormones and neurotransmitters. It is impressive that the beginning of development with oocyte maturation uses as the initiating signal the transduction system associated with highly differentiated somatic cells. This situation is a reflection of the fact that although the oocyte is often considered an undifferentiated cell, it is in fact a cell that has accumulated most of the information needed for a new individual and thus already possesses the sophisticated transduction systems associated with the development of specific organ systems during embryogenesis. It seems likely that most of the control systems described in somatic cells will turn out to be present in the oocyte.

### 2.1. Role of Surface Receptor

One of the subunits of the adenylate cyclase complex is a receptor (see Fig. 4). The concept that a steroid receptor is linked to adenylate cyclase is rather novel based on the dogma that steroid receptors are cytosolic proteins that migrate to the nucleus upon steroid binding. However, before the study of steroid-regulated membrane adenylate cyclase, there were several indications of initial steroid action at the oocyte surface. Smith and Ecker (1971) and Masui and Markert (1971) showed that microinjection of progesterone did not induce oocyte maturation, although induction would have been expected on the basis of a cytosolic receptor model. Subsequent work (Godeau *et al.*, 1978; Ishikawa *et al.*, 1977) showed that steroid bound to an extracellular polymer that is not taken up by oocytes is capable of inducing GVBD.

Although all these studies suggested that a steroid receptor exists on the oocyte surface, it was important to examine this aspect directly, since a totally new type of hormonal regulation was being postulated. With peptide or catecholamine ligands, the most straightforward approach to the question of a surface receptor would be direct binding studies. However, lipophilic compounds such as steroids dissolve in membranes to very high levels nonspecifically, precluding direct binding assays. An alternative approach is photoaffinity labeling of membranes with an activatable steroid ligand. Upon excitation at an appropriate wavelength, the steroid undergoes a molecular rearrangement, generating a free radical that binds covalently to adjacent proteins. The greater the specificity or affinity of a protein for the steroid, the more covalent labeling should occur. Since the products of excitation are charged, they can be separated from covalently labeled protein by SDS gel electrophoresis. Figure 7 summarizes an experiment carried out with this approach using the photoactivatable steroid, R5020, which is capable of causing GVBD in *Xenopus* oocytes at a concentration of $2 \times 10^{-6}$ M. A single labeled protein appears at a molecular weight of 110,000; most of this labeling is displaced if a 100-fold excess of nonradioactive R5020 is present during the preequilibration in the dark, indicating that the binding is saturable (Fig. 7A). A kinetic analysis of the initial rate of photolabeling gives an apparent $K_d$ of 1.2 $\mu$M for R5020 (Fig. 7B). In parallel experiments, adenylate cyclase activity was measured in photolabeled membranes, and inhibition was observed that is correlated with the amount of R5020 covalently bound. Half-maximal inhibition is seen at about 1 $\mu$M R5020 (Fig. 7C). This result is very important, since the concept of a receptor requires the demonstration that the binding of hormone is transduced into a biologically relevant response. Both the $K_d$ for binding of R5020 and the dose required for inhibition of adenylate cyclase are equivalent to the concentration of R5020 required for GVBD, strongly supporting the physiological relevance of the photolabeled protein. These results provide direct evidence that a steroid receptor exists on the oocyte membrane, and it appears that oocyte maturation represents a novel mechanism of steroid hormone action in which membrane steroid receptors are linked to the adenylate cyclase system. An important line of investigation that remains to be fully documented is the structure–activity relationships of the membrane receptor relative to steroid binding and induction of GVBD.

### 2.2. Phosphodiesterase

It is possible that progesterone could activate phosphodiesterase (PDE) at the same time as it inhibits adenylate cyclase, resulting in a concerted

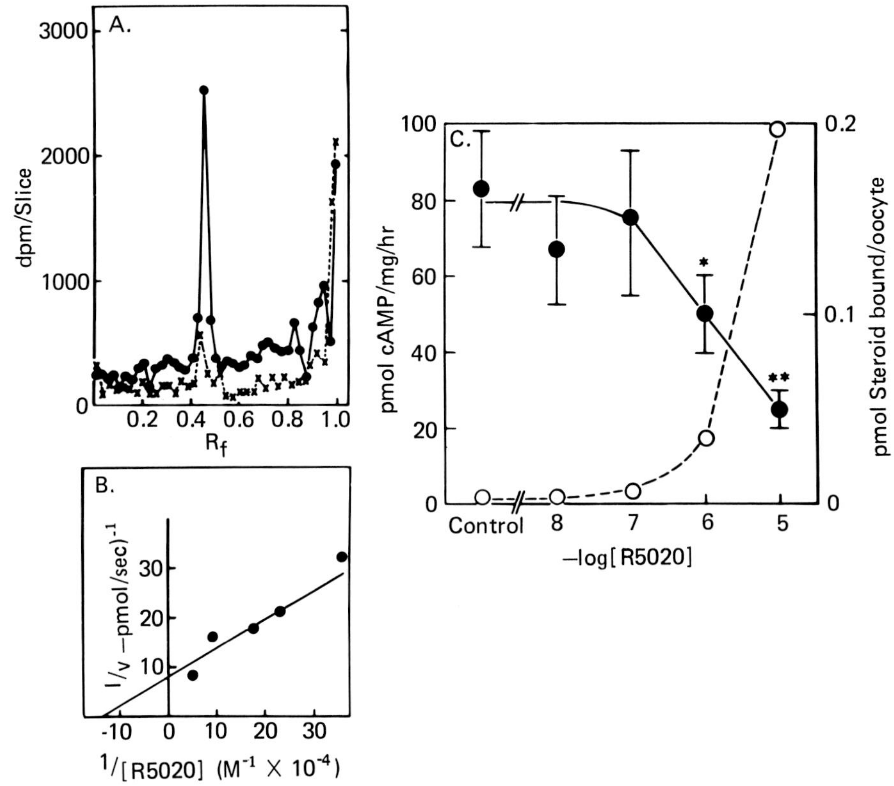


**Figure 7.** Receptor-mediated inhibition of adenylate cyclase by progesterone. Plasma membranes were manually dissected from oocytes and incubated in the dark with [$^3$H]-R5020. After photoactivation, membranes were electrophoresed on 8.5% SDS gels, sliced, and counted. (A) Gel electrophoretic profile of labeled membrane protein at $2 \times 10^{-8}$ M R5020 (circles); 100-fold excess of unlabeled R5020 (x). (B) Kinetic analysis of binding to the 110,000-dalton protein. The initial rate of photolabeling was determined at different concentrations of R5020, and the results were expressed as a double reciprocal plot. Data indicate an apparent $K_d$ of 1.2 $\mu$M, close to the concentration of R5020 required for a half-maximal effect. (C) Effect of covalent labeling on adenylate cyclase activity. The amount of R5020 covalently bound to the 110,000-dalton protein was measured (— o —) by gel electrophoresis at different concentrations of R5020, and in a parallel set of membranes, the adenylate cyclase activity (— • —) was determined after of photolabeling. (From Sadler and Maller, 1982.)

decrease in the level of cAMP. An attractive mechanism suggested by this hypothesis is that the calmodulin-activated PDE could be involved. Calmodulin is an acidic low-molecular-weight heat-stable protein that activates many enzymes in target cells and is believed to represent the basis for expression of responses that follow an increase in free cytosolic calcium during hormonal stimulation. One of the best-characterized calmodulin-dependent enzymes is phosphodiesterase (Lin and Cheung, 1980). The activation of PDE by calmodulin occurs according to equation (2):

$$\begin{aligned} Ca^{2+} + CaM &\rightleftharpoons Ca\ CaM \\ Ca^{2+}CaM + PDE &\rightleftharpoons Ca\ CaM\ PDE \end{aligned} \tag{2}$$

The increased activity with calmodulin observed in enzyme from brain is due both to an increase in $V_{max}$ and to a decrease in the $K_m$. Both kinetic changes are believed to be important, since the PDE has a relatively high $K_m$ for cAMP (Lin and Cheung, 1980).

The hypothesis that calmodulin-dependent PDE participates in the decline in cAMP (Maller and Krebs, 1977) is supported by the finding from aequorin-injected albino oocytes that an increase in free calcium parallels in time the decrease in cAMP (Wasserman *et al.*, 1980). Moreover, addition of phosphodiesterase inhibitors to oocytes blocks GVBD as well as the degradation of microinjected cAMP (O'Connor and Smith, 1976; Allende *et al.*, 1977). The possibility that CaM PDE is involved is also suggested by the ability of microinjected calmodulin to cause GVBD in oocytes from some females (Maller and Krebs, 1980; Wasserman and Smith, 1981), an effect that is blocked in oocytes preinjected with catalytic subunit of cAMP-dependent protein kinase.

The putative importance of PDE in oocyte maturation has led to preliminary characterization of the enzyme in *Xenopus* oocytes. As reported for mammalian tissues, *Xenopus* oocytes contain both a form with a low $K_m$ (1 $\mu$M) for cAMP and a form with a high $K_m$ (100 $\mu$M) for cAMP (O'Connor and Smith, 1976; Allende *et al.*, 1977; Mulner *et al.*, 1980). As in mammalian systems, only the high-$K_m$ form in oocytes is activated by calmodulin in oocytes, and the enzyme is easily activated irreversibly by proteolysis, resulting in a loss of calmodulin stimulation (Echeverria *et al.*, 1981; Orellana *et al.*, 1981).

Several investigators have attempted to determine whether the degradation of microinjected cAMP is increased upon addition of progesterone. No increase in the rate of degradation has been observed, but very large amounts of cAMP have been injected, indicating that only the high-$K_m$ form would be measured. The actual concentration of free cAMP in the oocyte after injection is difficult to establish because of the unknown magnitude of the stored cAMP reservoir. The injections also assume that the increase in free calcium with progesterone would mix with calmodulin and the injected cAMP to permit activation of PDE, which is difficult to establish. An interesting series of experiments that have not yet been carried out would involve determining whether the rate of degradation of microinjected cAMP is inhibited by co-injected specific calmodulin antagonists to a greater extent after progesterone administration.

### 2.3. Calmodulin Systems

A large number of calmodulin-activated systems have been described in various somatic cells (Cheung, 1980). It has not been established how many calmodulin systems coexist in the same cell, but it is generally believed that the pleiotropic response that occurs with calcium-mobilizing hormones is in part due to diverse calmodulin systems being activated in much the same way that the pleiotropic effects of cAMP are attributed to the broad substrate specificity of cAMP-dependent protein kinase. An important question, then, is how many calmodulin systems there are in oocytes and which ones are activated in response to progesterone. Presumably, an answer to this question would provide an explanation for the ability of injected calmodulin to cause GVBD in some oocytes. In addition to phosphodiesterase, three other calmodulin systems have been identified in oocytes. One system, reported by Wasserman and Smith (1981), is a soluble protein kinase activity that phosphorylates several unidentified substrates in the oocyte. Maller (1983) reported a particulate protein kinase activity stimulated by calmodulin that phosphorylates three membrane proteins. A similar particulate kinase activity has been described in many tissues (Schulman *et al.*, 1980), and in brain the phosphorylation of some membrane proteins by the enzyme is correlated with effects on release of neurotransmitters. A third calmodulin system that has been identified in oocytes is protein phosphatase IIB (Foulkes and Maller, 1982). This enzyme prefers the $\alpha$-subunit of phosphorylase kinase as substrate relative to the $\beta$-subunit, and it has been proposed to be an active phosphatase against protein phosphatase inhibitor-1 (Cohen, 1982). As inhibitor-1 is inhibitory only when phosphorylated by cAMP-dependent protein kinase (Fig. 2), the decrease in cAMP with progesterone would be expected to reduce the level of phosphorylation of inhibitor-1 by decreasing protein kinase activity, but in addition, the activation of phosphatase IIB by calcium/calmodulin could further reduce the phosphorylation level of inhibitor-1. The mechanism could be thought of as a protein phosphatase cascade. Whether this type of mechanism operates with progesterone remains to be established.

It is likely that there are other calmodulin-dependent enzymes in the oocyte; clearly, much additional work is needed to establish which of these are important in the process of oocyte maturation. An approach that may prove useful in pursuing this question is affinity chromatography of oocyte extracts on calmodulin-Sepharose columns.

## 3. Maturation-Promoting Factor

A late event in oocyte maturation is the appearance of the maturation-promoting factor (MPF). This factor was named on the basis of the observation that cytoplasm from progesterone-treated oocytes causes GVBD rapidly upon microinjection into recipient oocytes. Several features of this activity are of interest. Time-course studies show that the activity first appears at about 0.4 $GVBD_{50}$ and half-maximal GVBD in recipient-injected oocytes occurs with cytoplasm taken from progesterone-treated oocytes at

0.6 $GVBD_{50}$ (Wasserman and Masui, 1975). Thus, MPF appears to be formed in response to progesterone treatment, and once formed, it is capable of stimulating GVBD directly. Further evidence in this regard comes from the finding that MPF can cause GVBD in the presence of cycloheximide, whereas GVBD induced by progesterone is totally blocked if cycloheximide is administered before the appearance of MPF (Wasserman and Masui, 1975; Maller *et al.*, 1977). The MPF activity in the crude preparations is trypsin sensitive and heat labile, consistent with MPF being a protein.

Another fundamental feature of MPF came from studies demonstrating that the cytoplasm of injected *recipient* oocytes subsequently could serve as a source of active MPF for injection into second recipient oocytes (Smith and Ecker, 1971; Masui and Markert, 1971). This serial injection procedure has been carried through 10 recipient oocytes (Reynhout and Smith, 1974); since only about 1% of the volume of an oocyte is injected into each recipient, the tenth recipient cytoplasm contains at most an approximately a $10^{12}$ dilution of the original hormone-treated cytoplasm. This tremendous dilution with no loss of activity makes it unlikely that an active component of the original hormone-treated oocyte is being transferred in an unmodified form. Instead, this phenomenon indicates that MPF activity is amplified in recipient oocytes. This feature suggests that MPF (or the information needed to make MPF) is stored in an inactive form in the oocyte.

A question that is of great importance concerns the significance of MPF. Is the factor involved in cell division regulation in other species during meiosis and mitosis, or is it restricted to the process of oocyte maturation in amphibians? Subsequent studies showed that MPF is not species-specific among different amphibians and a similar activity assayed in *Xenopus* oocytes could be detected in somatic cells during mitosis (Sunkara *et al.*, 1979). MPF activity was also observed in starfish and sand dollar eggs (Kishimoto and Kanatani, 1976). These results have provided convincing evidence that MPF is of general significance in the control of cell division in eukaryotic cells. The amphibian oocyte is still the model system of choice for study of MPF, because the assay relies on microinjection and the oocyte is very sensitive to small quantities of microinjected MPF.

### 3.1. Role of Protein Phosphorylation

The increase in MPF activity in microinjected recipients has the features of an autoamplification process, i.e., active MPF generates more active MPF inside the oocyte. Several types of autoamplification reactions could theoretically account for this type of a process. One conceivable mechanism is **proteolysis.** If MPF were activated by proteolysis, injections of small amount of protease could lead to MPF activity. MPF could also generate more protease activity. In starfish, it has been reported that protease inhibitors block MPF appearance. However, this could be due to prevention of yolk breakdown required to supply amino acids for protein synthesis. (Protein synthesis is required for MPF appearance.)

Another type of reaction is **protein phosphorylation.** Virtually all protein kinases autophosphorylate themselves; in cases in which this has been studied extensively, as, for example in the case of phosphorylase kinase, this phosphorylation is associated with increased activity against exogenous substrates (Wang *et al.*, 1976). Autophosphorylation of protein kinases can occur either by intermolecular or intramolecular reactions. The amplification of MPF in recipient oocytes could theoretically be accounted for by transfer of an activated protein kinase, which then phosphorylates an inactive store of this kinase, leading ultimately to GVBD as a consequence of increased phosphorylation of exogenous substrates. This idea was supported by the findings of Maller *et al.* (1977), which demonstrate an immediate three- to fivefold increase in total cell protein phosphorylation upon injection of MPF. With progesterone treatment, a similar increase in phosphorylation occurs, but after a lag, coincident with the appearance of MPF as determined by the microinjection assay. The increased phosphorylation is not a result of a change in the specific activity of the ATP pool. An individual oocyte analysis of progesterone-treated oocytes demonstrates that the increased phosphorylation invariably precedes GBVD. This was important to establish, because the asynchrony of a population responding to hormone makes it difficult to determine the order of events if only data generated by a large population of cells are examined. The possibility that phosphorylation of specific substrates accounts for the action and/or amplification of MPF is further supported by the demonstration that the phosphorylation burst occurs when MPF is injected into cycloheximide-treated recipients but does not occur when cycloheximide blocks the action of progesterone (Maller *et al.*, 1977).

The mechanism by which MPF preparations increase protein phosphorylation is not yet established. MPF could be a protein kinase that phosphorylates new substrates, a protein phosphatase that removes phosphate from oocyte proteins, permitting exchange with ATP via protein kinases, or MPF could increase the availability for phosphorylation of stored substrates in the oocyte. There are insufficient data to distinguish among these possibilities, although it is clear that crude MPF preparations contain both protein kinase and protein phosphatase activity (Maller *et al.*, 1977; Wu and Gerhart, 1980). The direct way to prove that phosphorylation underlies MPF activity would be to purify MPF activity from crude preparations to homogeneity based on the microinjection assay for GVBD and to demonstrate protein kinase activity in the purified protein. Although a direct approach, the attempts to purify MPF on the basis of the microinjection assay have not been very successful to date. The best purification of about 50-fold over crude cytosol has been reported by Gerhart's laboratory (Wu and Gerhart, 1980). Their studies show that the activity of MPF preparations in inducing GVBD has a very strict (seventh-order) dependence on the volume injected. Thus, a small dilution of MPF to give a lower concentration of activity requires that much more total activity be injected in order to give an equivalent response (Wu and Gerhart, 1980; Maller, 1983). The basis for this phenomenon is not understood, but it causes great difficulty in purification procedures that result in dilution of MPF activity.

An alternative approach to MPF purification based on the hypothesis that phosphorylation underlies the activity would be to identify specific substrates that are phosphorylated in MPF preparations. Isolation of such substrates would facilitate purification to homogeneity of the relevant protein kinase using an *in vitro* kinase assay with the identified substrate. The activity of the purified kinase for GVBD could then be assessed by microinjection. Progress with this approach has been reported in several laboratories after the identification (Maller and Sadler, 1981) of ribosomal protein S6 as one of the substrates for MPF-induced phosphorylation (Hanocq-Quertier and Baltus, 1981; Nielsen *et al.*, 1982; Kaltoff *et al.*, 1982). The phosphorylation of S6 is changed from 0–1 moles phosphate/mole S6 to 4–5 moles phosphate/mole S6. The phosphorylation of S6 has been reported in a large number of systems in response to diverse growth-promoting stimuli (Thomas *et al.*, 1982). The precise function in protein synthesis that is affected by S6 phosphorylation has not been identified, but it has been proposed that S6 phosphorylation plays a role in mRNA recruitment (Duncan and McConkey, 1982; Thomas *et al.*, 1982). Further study by Thomas and Maller demonstrates that MPF preparations purified 50-fold over crude cytosol contain an S6 protein kinase activity that can phosphorylate oocyte 40 S ribosome subunits at the same sites *in vitro* as are phosphorylated *in vivo* in response to progesterone treatment or MPF injection (Nielsen *et al.*, 1982; Martin-Perez *et al.*, 1984). These results suggest that purification of S6 kinase activity from progesterone-treated oocytes might lead to further purification of MPF. However, it has not yet been established that S6 phosphorylation is necessary for MPF action. In any case the study of S6 phosphorylation seems warranted on the basis of the strong correlation of the event with proliferative stimuli.

The fact that the only identified substrate for MPF-associated protein phosphorylation is a ribosomal protein has focused attention on the effect of MPF on protein synthesis. Since cycloheximide blocks the appearance of MPF in response to progesterone, it is clear that protein synthesis is required for MPF appearance. However, it is not required for the action of MPF to induce GVBD. Recently, intensive effort has been given to the question of whether the amplification of MPF in recipient oocytes requires protein synthesis. Studies by Smith and collaborators showed that MPF causes an immediate twofold change in the absolute rate of protein synthesis in injected oocytes (Wasserman *et al.*, 1982). The mechanism of this increase is primarily mRNA recruitment, although an increase in translational efficiency for some proteins could not be excluded (Richter *et al.*, 1982). Gerhart *et al.* (1984) investigated the course of MPF amplification in MPF-injected oocytes in the presence and absence of cycloheximide. They observed that cycloheximide-treated oocytes contain less MPF than controls, although still enough to cause GVBD, but MPF activity declines near the time of meiosis I and does not reappear later. In control oocytes, MPF activity reaches higher levels and also declines at meiosis I, but the activity of MPF returns to maximal levels by 1 h after meiosis I (Gerhart *et al.*, 1984). This oscillation of MPF activity is similar to that seen in somatic cells at mitosis, suggesting that protein synthesis is required for the reappearance of activity after cytokinesis but not necessarily for generating enough additional MPF after injection to cause GVBD. Since the total MPF level in cycloheximide-treated oocytes is lower than in controls, the results imply that amplification occurs by both protein synthesis-dependent and protein synthesis-independent mechanisms. In terms of MPF-induced phosphorylation of S6, it is tempting to speculate that ribosomes containing maximally phosphorylated S6 preferentially select MPF mRNA or mRNA for proteins involved in MPF activation from the oocyte store of mRNA. Whether this type of mechanism operates remains to be determined, but it would provide an explanation for a partial dependence of MPF amplification on protein synthesis.

### 3.2. Role of Increased Intracellular pH

In other systems of growth stimulation, the phosphorylation of S6 is correlated with an increase in intracellular pH ($pH_i$). It has not been established whether either event is causal for the other, but Pouysségur *et al.* (1982) have proposed that increased $pH_i$ is coupled to S6 phosphorylation. These ideas have prompted investigation of possible $pH_i$ changes in amphibian oocyte maturation. The question of $pH_i$ changes is also of interest because increased $pH_i$ is associated with the activation of development at fertilization (Epel, 1978) in some species of sea urchins, and there are a number of parallels in developmental events between sea urchin egg fertilization and amphibian oocyte maturation (Smith, 1981). Lee and Steinhardt (1981) were the first to report an increase in intracellular pH of oocytes treated with progesterone using pH microelectrodes. Subsequent studies in several laboratories have confirmed an increase in $pH_i$ using the uptake of a radiolabeled weak acid as a measure of $pH_i$ (Houle and Wasserman, 1983; Stith and Maller, 1983; Cicirelli *et al.*, 1983). With both techniques, there is an increase of 0.3 to 0.4 pH units (pH 7.3–7.7), beginning at about 0.2 $GVBD_{50}$ (Houle and Wasserman, 1983; Stith and Maller, 1983). When MPF is injected, $pH_i$ increases immediately to about the same level as with progesterone (Lee and Steinhardt, 1981).

The importance of the increase in $pH_i$ is difficult to assess. Studies with sodium-free medium show that the $pH_i$ increase with progesterone is totally blocked, but the ability of progesterone to induce GVBD is not inhibited (Stith and Maller, 1984). The same experiments show the phosphorylation of S6 to be totally unaffected by amiloride, casting doubt about the hypothesis of coupling of these events (Pouysségur *et al.*, 1982). However, since both events are correlated in many growth systems and both are rapidly induced by MPF in oocytes, it seems clear that continued investigation of these processes may provide new insights into growth regulation during development.

## 4. Summary

This chapter presents a complex picture of the regulation of oocyte maturation. The events that have been identified can be expressed temporally on a relative time scale, as shown in Figure 8. A major question that lies ahead concerns which of these events is causal for subsequent events and for GVBD. Clearly, the first event at 0.1 $GVBD_{50}$, a decrease in cAMP, is both necessary and sufficient to induce GVBD. The increase in protein synthesis at 0.3 $GVBD_{50}$ is apparently necessary, and the activity of MPF appearing at 0.4 $GVBD_{50}$ is apparently sufficient, as judged by the microinjection assay. It is important to identify further steps after MPF appearance in order to establish the mechanism of action of MPF. The fact that events highly correlated with cell growth (i.e., $pH_i$ increase, S6 phosphorylation, and MPF appearance) all occur during amphibian oocyte mat-

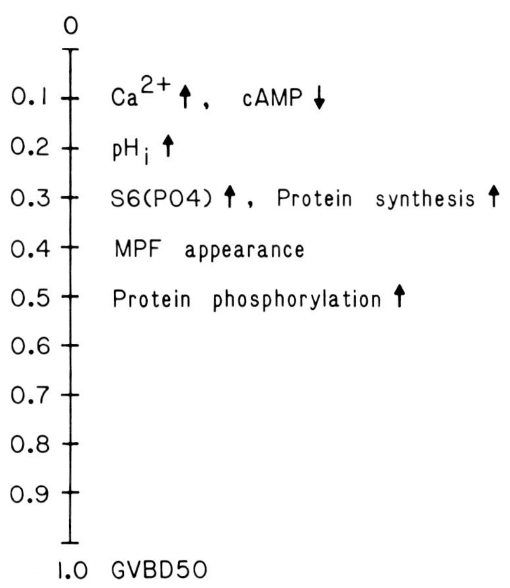


**Figure 8.** Relative timing of biochemical events during progesterone-induced maturation of *Xenopus* oocytes. The period from hormone addition to 50% germinal vesicle breakdown (GVBD) has been normalized to a scale of 0–1.0 to facilitate comparison of data from many females, which differ widely in the absolute time required for detection of specific events. What is indicated is the timing of the initial change in the measured parameter becoming statistically significant. It is not known whether the sequence in time of the changes represents a sequential dependence of each event on the preceding one or the activation of a series of independent pathways by progesterone.

uration provides strong support for continued use of the oocyte in studying the general control of cell proliferation, particularly with the technique of microinjection.

Investigation of the process of amphibian oocyte maturation has also led to the discovery of a totally new type of steroid hormone mechanism in which a plasma membrane steroid receptor is linked to the adenylate cyclase system. The oocyte system is the first example of a steroid hormone effect mediated by a direct effect on a target enzyme without a requirement for protein or RNA synthesis. The generality of this type of steroid hormone mechanism remains to be established, but its similarity to the P site on adenylate cyclase is suggestive evidence of a general mechanism, since the P site is present in all eukaryotic plasma membranes. Most steroid hormone systems have been investigated with respect to the mechanism of increased protein synthesis after steroid addition, with little attention to any cell division occurring. In the oocyte, the cell division effect of steroids is the predominant one; it is possible that this type of effect is mediated by a different receptor system than the one associated with changes in gene expression in somatic cells.

The information described in this chapter hints at the complexity underlying the activation of development during the process of oocyte maturation. The finding that so many of the specific events and general mechanisms operating in oocyte maturation are also used in differentiated tissues strongly indicates that further study of amphibian oocyte maturation will lead to findings of general significance in terms of how information stored in the oocyte is used during development to result in a new member of the species.

## References

Allende, C. C., Bravo, R., and Allende, J. E., 1977, Comparison of *in vivo* and *in vitro* properties of cyclic adenosine 3′:5′-monophosphate phosphodiesterase of amphibian oocytes, *J. Biol. Chem.* **252**:4662–4666.

Boyer, J., Belle, R., Huchon, D., and Ozon, R., 1980, *In ovo* protein kinase activity during progesterone-induced maturation of *Xenopus laevis* oocytes, in: *Steroids and Their Mechanism of Action in Non-mammalian Vertebrates* (G. Delrio and J. Brachet, eds.), pp. 85–92, Raven Press, New York.

Cassel, D., and Selinger, Z., 1977, Mechanism of adenylate cyclase activation by cholera toxin: Inhibition of GTP hydrolysis at the regulatory site, *Proc. Natl. Acad. Sci. USA* **74**:3307–3311.

Cassel, D., Levkovitz, H., and Selinger, Z., 1977, The regulatory GTPase cycle of turkey erythrocyte adenylate cyclase, *J. Cyclic Nucleotide Res.* **3**:393–406.

Cheung, W. Y., 1980, Calmodulin plays a pivotal role in cellular regulation, *Science* **207**:19–27.

Cicirelli, M. F., Robinson, K. R., and Smith, L. D., 1983, Internal pH of *Xenopus* oocytes: A study of the mechanism and role of pH changes during meiotic maturation, *Dev. Biol.* **100**:133–146.

Cohen, P., 1982, The role of protein phosphorylation in neural and hormonal control of cellular activity, *Nature* **296:**613–620.

DeRobertis, E. M., Partington, G. A., Longthorne, R. F., and Gurdon, J. B., 1977, Somatic nuclei in amphibian oocytes: Evidence for selective gene expression, *J. Embryol. Exp. Morphol.* **40:**199–214.

Drury, K. C., and Schorderet-Slatkine, S., 1975, Effects of cycloheximide on the "autocatalytic" nature of the maturation promoting factor (MPF) in oocytes of *Xenopus laevis*, *Cell* **4:**269–274.

Duncan, R., and McConkey, E. H., 1982, Rapid alterations in initiation rate and recruitment of inactive RNA are temporally correlated with S6 phosphorylation, *Eur. J. Biochem.* **123:**539–544.

Echeverria, M., Orellana, O., Jedlicki, E., Plaza, M., Allende, C., and Allende, J., 1981, Regulation of a cyclic nucleotide phosphodiesterase from *Xenopus laevis* ovary by calmodulin and calcium, *Biochem. Int.* **2:**539–545.

Epel, D., 1978, Mechanisms of activation of sperm and egg during fertilization of sea urchin gametes, *Curr. Top. Dev. Biol.* **12:**185–246.

Finidori-Lepicard, J., Schorderet-Slatkine, S., Hanoune, J., and Baulieu, E. E., 1981, Progesterone inhibits membrane-bound adendylate cyclase in *Xenopus laevis* oocytes, *Nature* **292:**255–257.

Foulkes, J. G., and Maller, J. L., 1982, *In vivo* action of protein phosphatase inhibitor-2 in *Xenopus* oocytes, *FEBS Lett.* **158:**155–160.

Gerhart, J., Wu, M., and Kirschner, M. W., 1984, Cell cycle dynamics of an M-phase–specific cytoplasmic factor in *Xenopus laevis* oocytes and eggs, *J. Cell Biol.* **98:**1247–1255.

Godeau, J. F., Schorderet-Slatkine, S. Hubert, P., and Baulieu, E. E., 1978, Induction of maturation in *Xenopus laevis* oocytes by a steroid linked to a polymer, *Proc. Natl. Acad. Sci. USA* **75:**2353–2357.

Hanocq-Quertier, J., and Baltus, E., 1981, Phosphorylation of ribosomal proteins during maturation of *Xenopus laevis* oocytes, *Eur. J. Biochem.* **120:**351–355.

Hemmings, B. A., Resink, T., and Cohen, P., 1982, Activation of the MgATP-dependent protein phosphatase by glycogen synthase kinase 3 does involve a phosphorylation reaction, *FEBS Lett.* **150:**319–324.

Huang, F. L., and Glinsmann, W. H., 1976, Separation and characterization of two phosphorylation phosphatase inhibitors from rabbit skeletal muscle, *Eur. J. Biochem.* **70:**419–426.

Huchon, D., Ozon, R., Fischer, E. H., and Demaille, J. G., 1981a, The pure inhibitor of cyclic AMP-dependent protein kinase initiates *Xenopus laevis* meiotic maturation: A 4-step scheme for meiotic maturation, *Mol. Cell. Endocrinol.* **22:**211–222.

Huchon, D., Ozon, R., and Demaille, J. G., 1981b, Protein phosphatase-1 is involved in *Xenopus* oocyte maturation, *Nature* **294:**358–359.

Ishikawa, K., Hanaoka, Y., Kondo, Y., and Imai, K., 1977, Primary action of steroid hormone at the surface of amphibian oocyte in the induction of germinal vesicle breakdown, *Mol. Cell. Endocrinol.* **9:**91–100.

Johnson, G. L., 1982, Cholera toxin action and the regulation of hormone-sensitive adenylate cyclase, in: *Molecular Action of Toxins and Viruses* (P. Cohen and P. Van Heyningen, eds.), pp. 33–49, Elsevier Biomedical Press, New York.

Jordana, X., Allende, C. C., and Allende, J. E., 1981, Guanine nucleotides are required for progesterone inhibition of amphibian oocyte adenylate cyclase, *Biochem. Int.* **3:**527–532.

Kaltoff, H., Darmer, D., Towbin, H., Gordon, J., Amons, R., Moller, W., and Richter, D., 1982, Ribosomal protein S6 from *Xenopus laevis* ovaries, *Eur. J. Biochem.* **122:**439–443.

Kishimoto, T., and Kanatani, H., 1976, Cytoplasmic factor responsible for germinal vesicle breakdown and meiotic maturation in starfish oocyte, *Nature* **260:**321–322.

Koski, K., and Klee, W. A., 1981, Opiates inhibit adenylate cyclase by stimulating GTP hydrolysis, *Proc. Natl. Acad. Sci. USA* **78:**4185–4189.

Lee, S. C., and Steinhardt, R. A., 1981, pH changes associated with meiotic maturation in oocytes of *Xenopus laevis*, *Dev. Biol.* **85:**358–369.

Lin, Y. M., and Cheung, W. Y., 1980, $Ca^{+}$-dependent cyclic nucleotide phosphodiesterase, in: *Calcium and Cell Function: Calmodulin* (Y. Cheung, ed.), pp. 79–107, Academic Press, New York.

Londos, C., Wolff, J., and Cooper, D. M. F., 1981, Adenosine as a regulator of adenylate cyclase, in: *Purinergic Receptors* (G. Burnstock, ed.), pp. 287–323, Chapman and Hall, New York.

Maller, J. L., 1983, Interaction of steroids with the cyclic nucleotide system in amphibian oocytes, *Adv. Cyclic Nucleotide Res.* **15**:295–336.

Maller, J. L., and Krebs, E. G., 1977, Progesterone-stimulated meiotic cell division in *Xenopus* oocytes: Induction by regulatory subunit and inhibition by catalytic subunit of adenosine 3′:5′-monophosphate-dependent protein kinase, *J. Biol. Chem.* **252**:1712–1718.

Maller, J. L., and Krebs, E. G., 1980, Regulation of oocyte maturation, *Curr. Top. Cell. Regul.* **16**:217–311.

Maller, J. L., and Sadler, S. E., 1981, Regulation of steroid induced cell division in amphibian oocytes by protein phosphorylation, *Proc. Cold Spring Harbor Conf. Cell Prolif.* **8**:1127–1141.

Maller, J. L., Wu, M., and Gerhart, J. C., 1977, Changes in protein phosphorylation accompanying maturation of *Xenopus laevis* oocytes, *Dev. Biol.* **58**:295–312.

Maller, J. L., Butcher, F. R., and Krebs, E. G., 1979, Early effect of progesterone on levels of cyclic adenosine 3′:5′-monophosphate in *Xenopus* oocytes, *J. Biol. Chem.* **254**:579–582.

Martin-Perez, J., Maller, J. L., and Thomas, G., 1984, Progesterone- and MPF-induced S6 phosphorylation during meiotic maturation of *Xenopus laevis* oocytes: in vivo and in vitro sites of phosphorylation, *J. Biol. Chem.* (submitted).

Masui, Y., 1967, Relative roles of the pituitary, follicle cells and progesterone in the induction of oocyte maturation in *Rana pipiens*, *J. Exp. Zool.* **166**:365–376.

Masui, Y., and Markert, C. L., 1971, Cytoplasmic control of nuclear behavior during meiotic maturation of frog oocytes, *J. Exp. Zool.* **177**:129–146.

Morrill, G. A., Schatz, F., Kostellow, A. B., and Poupko, J. M., 1977, Changes in cyclic AMP levels in the amphibian ovarian follicle following progesterone induction of meiotic maturation, *Differentiation*, **8**:97–104.

Mulner, O., Huchon, D., Thibier, C., and Ozon, R., 1979, Cyclic AMP synthesis in *Xenopus laevis* oocytes: Inhibition by progesterone, *Biochim. Biophys. Acta* **582**:179–184.

Mulner, O., Cartaud, A., and Ozon, R., 1980, Cyclic AMP phosphodiesterase activities in *Xenopus laevis* oocytes, *Differentiation* **16**:31–39.

Nielsen, P. J., Thomas, G., and Maller, J. L., 1982, Increased phosphorylation of ribosomal protein S6 during meiotic maturation of *Xenopus* oocytes, *Proc. Natl. Acad. Sci. USA* **79**:2937–2941.

Nimmo, G. A., and Cohen, P., 1978, The regulation of glycogen metabolism: Purification and characterization of protein phosphatase inhibitor-1 from rabbit skeletal muscle, *Eur. J. Biochem.* **87**:341–351.

O'Connor, C. M., and Smith, L. D., 1976, Inhibition of oocyte maturation by theophylline: Possible mechanism of action, *Dev. Biol.* **52**:318–322.

Orellana, O., Allende, C. C., and Allende, J. E., 1981. Trypsin activates the calmodulin sensitive cyclic nucleotide phosphodiesterase of amphibian oocytes, *Biochem Int.* **3**:663–668.

Pouysségur, J., Chambard, J. C., Franchi, A., Paris, S., and van Obberghen-Schilling, E., 1982, Growth factor activation of an amiloride-sensitive $Na^{+}/H^{+}$ exchange system in quiescent fibroblasts. Coupling to ribosomal protein S6 phosphorylation, *Proc. Natl. Acad. Sci. USA* **79**:3935–3939.

Reynhout, J. K., and Smith, L. D., 1974, Studies on the appearance and nature of a maturation-inducing factor in the cytoplasm of amphibian oocytes exposed to progesterone, *Dev. Biol.* **38**:394–400.

Richter, J. D., Wasserman, W. J., and Smith, L. D., 1982, The mechanism for increased protein synthesis during *Xenopus* oocyte maturation, *Dev. Biol.* **89**:159–167.

Sadler, S. E., and Maller, J. L., 1981, Progesterone inhibits adenylate cyclase in *Xenopus* oocytes: Action on the guanine nucleotide regulatory protein, *J. Biol. Chem.* **256**:6368–6373.

Sadler, S. E., and Maller, J. L., 1982, Identification of a steroid receptor on the surface of *Xenopus* oocytes by photoaffinity labeling, *J. Biol. Chem.* **257**:355–361.

Sadler, S. E., and Maller, J. L., 1983, Inhibition of *Xenopus* oocyte adenylate cyclase by progesterone and 2′,5′-dideoxyadenosine is associated with slowing of guanine nucleotide exchange, *J. Biol. Chem.* **258**:7935–7941.

Sayhoun, N., Schmitges, C. J., Siegel, M. I., and Cuatrecasas, P., 1976, 2′-Deoxyadenosine-3′-monophosphate: A naturally occurring inhibitor of adenylate cyclase in amphibian and mammalian cells, *Life Sci.* **19**:1961–1970.

Schorderet-Slatkine, S., Schorderet, M., and Baulieu, E. E., 1982, Cyclic AMP-mediated control of meiosis: Effects of progesterone, cholera toxin and membrane-active drugs in *Xenopus laevis* oocytes, *Proc. Natl. Acad. Sci. USA*, **79**:850–854.

Schulman, H., Huttner, W., and Greengard, P., 1980, Calcium-dependent protein phosphorylation in mammalian brain and other tissues, in: *Calcium and Cell Function, Vol. 1: Calmodulin*, (W. Y. Cheung, ed.), pp. 219–252, Academic Press, New York.

Smith, L. D., 1981, Frog oocytes and sea urchin eggs: Steroid hormones and sperm induce similar events, in: *Progress in Developmental Biology* (H. W. Sauer, ed.), pp. 35–48, Gustav Fischer Verlag, Stuttgart and New York.

Smith, L. D., and Ecker, R. E., 1971, The interaction of steroids with *Rana pipiens* oocytes in the induction of maturation, *Dev. Biol.* **25**:233–247.

Smith, L. D., Ecker, R. E., and Subtelny, S., 1968, *In vitro* induction of physiological maturation in *Rana pipiens* oocytes removed from the ovarian follicles, *Dev. Biol.* **17**:627–643.

Speaker, M. G., and Butcher, F. R., 1977, Cyclic nucleotide fluctuations during steroid-induced meiotic maturation of frog oocytes, *Nature (London)* **267**:848–849.

Stith, B. J., and Maller, J. L., 1984, The effect of insulin on intracellular pH and ribosomal protein S6 phosphorylation in oocytes of *Xenopus laevis*, *Dev. Biol.* **102**:79–89.

Sunkara, P. S., Wright, D. A., and Rao, P. N., 1979, Mitotic factors from mammalian cells induce germinal vesicle breakdown and chromosome condensation in amphibian oocytes, *Proc. Natl. Acad. Sci. USA* **76**:2799–2802.

Thomas, G., Martin-Perez, J., Seigmann, M., and Otto, A. M., 1982, The effect of serum, EGF, $PGF_{2\alpha}$ and insulin on S6 phosphorylation and the initiation of protein and DNA synthesis, *Cell* **30**:235–242.

Wang, J. H., Stull, J. T., Huang, T. S., and Krebs, E. G., 1976, A study of the autoactivation of rabbit muscle phosphorylase kinase, *J. Biol. Chem.* **251**:4521–4527.

Wasserman, W. J., and Masui, Y., 1975, Effects of cycloheximide on a cytoplasmic factor initiation meiotic maturation in *Xenopus* oocytes, *Exp. Cell. Res.* **91**:381–388.

Wasserman, W. J., and Smith, L. D., 1981, Calmodulin triggers the resumption of meiosis in amphibian oocytes, *J. Cell. Biol.* **89**:389–394.

Wasserman, W. J., Pinto, L. H., O'Connor, C. M., and Smith, L. D., 1980, Progesterone induces a rapid increase in $(Ca^{++})_{in}$ of *Xenopus laevis* oocytes, *Proc. Natl. Acad. Sci. USA* **77**:1534–1536.

Wasserman, W. J., Richter, J. D., and Smith, L. D., 1982, Protein synthesis during maturation-promoting factor and progesterone-induced maturation in *Xenopus* oocytes, *Dev. Biol.* **89**:152–158.

Wu, M., and Gerhart, J. C., 1980, Partial purification and characterization of the maturation-promoting factor from eggs of *Xenopus laevis*, *Dev. Biol.* **79**:465–477.

Chapter 7

# Oocyte–Somatic Cell Interactions During Oocyte Growth and Maturation in the Mammal

JOHN J. EPPIG

## 1. Introduction

Compared with most somatic cells, the oocyte is a very large cell in which the area of plasma membrane—the oolemma—is relatively small in relationship to the large volume of cytoplasm. Consequently, various systems of communication have evolved between the oocyte and somatic cells to supply the oocyte with sufficient nutrients for oogenesis and early embryogenesis. In some species, oocyte support cells have direct cytoplasmic continuity with the ooplasm (see Raven, 1961, for review). In contrast, oocytes of other species, including mammals, do not have a direct cytoplasmic bridge to the supporting cells. Rather, a specialization of juxtaposed plasma membranes of the oocyte and adjacent somatic cells provides the avenue of communication between these cells. This specialization is called a **gap junction.** Through this coupling pathway, nutrients are passed to the oocyte; also through this pathway pass molecules that regulate oocyte development. The nature and function of this cell-to-cell communication system during oocyte growth and meiotic maturation are the subject of this chapter.

## 2. Metabolic Cooperativity

**Metabolic cooperativity** occurs when a metabolite is moved from one cell to another, via gap junctions, to participate in a metabolic pathway. When thin-sectioned cells are examined by electron microscopy, the juxtaposed plasma membranes comprising the gap junction are separated by

JOHN J. EPPIG • The Jackson Laboratory, Bar Harbor, Maine 04609.

a 2–4-$\mu$m extracellular space, or gap, that can be penetrated and visualized by electron-opaque molecules such as lanthanum. The gap-junctional area contains numerous 8–9-nm particles in electron microscopic freeze-fracture images (Fig. 1). These particles apparently contain the channels that can be traversed by molecules of $<2000M_r$.

That gap junctions provide the pathway for metabolic cooperativity was demonstrated in classic experiments wherein tissue culture cells unable to convert hypoxanthine to the nucleotide inosine monophosphate (IMP-negative cells) were able to incorporate labeled exogenous hypoxanthine into nucleic acid, but only when the mutant cells were grown in coculture with wild-type (IMP-positive) cells (Gilula *et al.*, 1972; Pitts and Simms, 1977). Contact of the two cell types and the resultant formation of gap junctions between apposing plasmalemmas were required for the IMP-negative cells to incorporate hypoxanthine into nucleic acid. When mutant cells were unable to form gap junctions with wild-type cells, no labeled hypoxanthine was incorporated into nucleic acid in the mutant cells, even though the cells were in contact. Incorporation was unaffected in the wild-type cells. These results suggest that the wild-type cells converted labeled hypoxanthine to IMP, which was subsequently transferred to the IMP-negative cells. They further suggested that the pathway for the transfer was the gap junction (Gilula *et al.*, 1972; Pitts and Simms, 1977).

Gap junctions may also provide the pathway whereby one cell type can regulate the function of another cell type (Lawrence *et al.*, 1978). An example of this process is seen in the following experiment. Ovarian granulosa cells produce plasminogen activator when stimulated with follicle-stimulating hormone (FSH), probably via a cyclic adenosine monophosphate (cAMP)-dependent mechanism. Likewise, the increase of myocardial cell beat frequency in culture, promoted by norepinephrine, is probably mediated by cAMP. Cocultured granulosa cells and myocardial cells form both homologous (same cell type) and heterologous (different cell type) gap junctions. When heterologous gap junctions were established in cocultures of myocardial and granulosa cells, the beat frequency of myocardial cells was increased with FSH; granulosa cells secreted plasminogen activator when norepinephrine was added to the medium (Lawrence *et al.*, 1978). Myocardial cells, not coupled to granulosa cells, do not increase beat fre-

→

**Figure 1.** Electron micrographs of freeze-fracture preparations of ovarian gap junctions. (A) Homologous gap junctions between two granulosa cells (From Anderson *et al.*, 1978.) (B) Heterologous gap junctions between a granulosa cell and an oocyte. Note that the oocyte–granulosa cell gap junctions are very much smaller than the granulosa cell–granulosa cell gap junctions. Although the individual particles in both types of junction are about the same size (7–9 nm), there are fewer of them in the heterologous junction. GJ, gap junction; mv, microvilli. Scale bars: 0.1 $\mu$m. (From Anderson and Albertini, 1976. Reproduced from *The Journal of Cell Biology*, 1976, Vol. 71, pp. 680–686 by copyright permission of The Rockefeller University Press.)

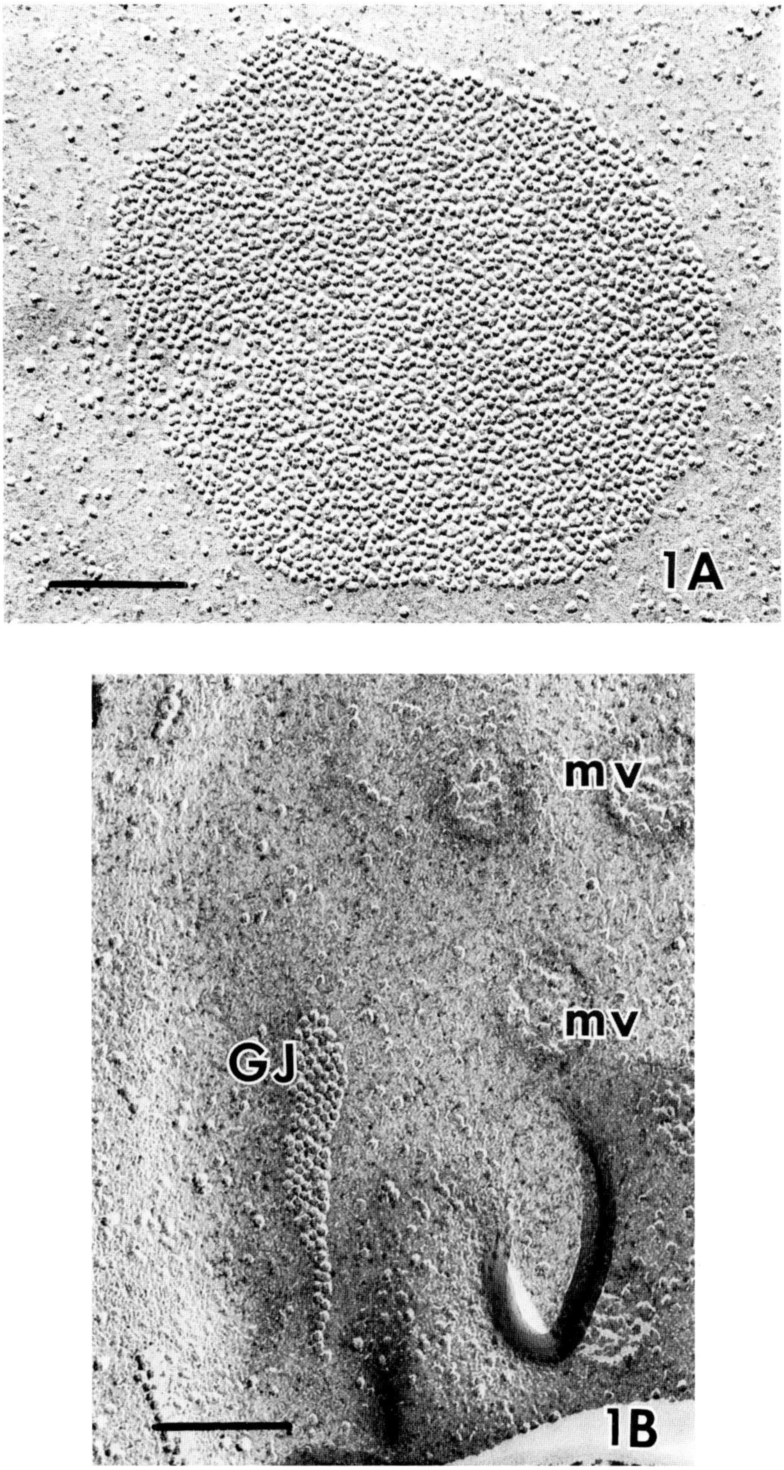
1A
mv
mv
GJ
1B

quency in response to FSH, nor do granulosa cells not coupled to myocardial cells secrete plasminogen activator in response to norepinephrine at the concentrations used. Therefore, specific hormonal stimulation of one cell type resulted in a change in the activity of the other cell type when the cells were coupled by heterologous gap junctions. Since the hormonal responses of both cell types are probably mediated by cAMP, it was suggested that cAMP may have been the molecule transmitted via the heterologous gap junctions that affected the activity of the cell type not directly responsive to the specific hormonal stimulus (Lawrence *et al.*, 1978).

## 3. Occurrence of Gap Junctions During Oogenesis

Heterologous gap junctions between granulosa cells and oocytes are present throughout the oocyte growth phase in mammals (Anderson and Albertini, 1976). It is not clear as to when the gap junctions are first established, but Anderson and Albertini (1976) report that they are present before the deposition of the zona pellucida. During oocyte growth, cytoplasmic processes from the granulosa cells traverse the zona pellucida, and small gap junctions are seen at points of juxtaposition of the oolemma and the granulosa cell processes (Fig. 1B) (Anderson and Albertini, 1976). These cytoplasmic processes retract, and the heterologous junctions uncouple by the time of ovulation (Zamboni, 1974; Dekel *et al.*, 1978; Gilula *et al.*, 1978).

## 4. Role of Cell-to-Cell Communication During Oocyte Growth

During the growth phase of mammalian oocytes, their diameters increase from 15–20 μm to 70–85 μm, while they are arrested in meiotic prophase. This is not only a phase of increasing size, but also of differentiation. Changes occur in the ultrastructural appearance of nuclear and cytoplasmic components, and cortical granules are elaborated. Several proteins are synthesized by growing oocytes no longer produced by full-grown oocytes (Schultz *et al.*, 1979; Kaplan *et al.*, 1982) (see also Chapter 11, this volume). In addition, when the oocytes achieve a diameter of 60–65 μm, they acquire competence to resume the first meiotic division, reducing the number of chromosomes from the diploid to the haploid number, and producing the first polar body (meiotic maturation).

The period of growth and differentiation depends on the communication of the oocytes with granulosa cells. When mid-growth phase oocytes are cultured either without companion granulosa cells or resting on a monolayer of granulosa cells (which may form only a limited number of junctional contacts with oocytes), the growth of the oocytes is severely retarded. On the other hand, when the normal associations between the granulosa cells and oocytes are intact, the oocytes grow at a rate similar to that of oocytes growing *in vivo*. In addition, during oocyte growth *in vitro* many oocytes acquire competence to resume maturation (Eppig, 1977, 1979a). These results suggest that cell-to-cell communication between the granulosa cells and oocytes is required for oocyte growth and differentiation. However, oocyte growth and differentiation may be separable processes. Mouse oocytes cultured on a monolayer of fibroblasts did not grow but acquired meiotic competence at a time corresponding to oocytes developing *in vivo* (Canipari *et al.*, 1984). Nevertheless, the very low rate at which granulosa cell-denuded oocytes grow when cultured on a monolayer of somatic cells implies that the influence of the granulosa cells is mediated by the normal heterologous junctional associations rather than by a diffusible factor.

Such studies imply, but do not prove, that oocyte growth is promoted by metabolic coupling via gap junctions between the granulosa cells and oocytes. Addressing this question, it was found that mid-growth phase oocytes denuded of granulosa cells do not incorporate as much labeled uridine, choline, leucine, or 2-deoxyglucose as do granulosa cell-enclosed oocytes (Eppig, 1979a; Heller *et al.*, 1981).

Various lines of evidence have suggested that the additional amount of radiolabeled metabolites in the granulosa cell-enclosed oocytes gain access to the oocyte via the gap junctions that couple them to the granulosa cells. First, the patterns of phosphorylated derivatives of uridine in granulosa cell-enclosed oocytes more closely resemble the pattern in granulosa cells than in granulosa cell-denuded oocytes. Second, treating granulosa cell-enclosed oocytes with agents known to interfere with the integrity of gap junctions in other cell types disrupts metabolic cooperativity between oocytes and granulosa cells. Third, when granulosa cell-enclosed oocytes are removed from medium containing the labeled metabolite, label continues to accumulate in the oocyte (Heller *et al.*, 1981). Subsequent experiments showed that the rate of oocyte growth is positively correlated with the number of adherent granulosa cells and the extent of metabolic cooperativity between granulosa cells and the oocyte (Brower and Schultz, 1982). These data provide strong evidence for a nutritional role for granulosa cells in oocyte growth that is mediated by gap junctions.

## 5. Oocyte Maturation in Mammals

### 5.1. General Concepts of Meiotic Regulation in Mammals

Mammalian oocytes are arrested in prophase I of meiosis from about the time of birth until the oocytes have nearly completed their growth phase. Oocyte maturation *in vivo* occurs under two conditions: (1) in suitably developed antral follicles, after the endogenous preovulatory gonadotropin surge or the administration of an appropriate dose of exogenous gonadotropins, or (2) in atretic follicles. In most of the commonly studied mammalian species, only matured ova are normally ovulated. These ova are usually enclosed within a mass of mucified cumulus cells. The second meiotic division occurs after the ovum is penetrated by the sperm. At this time, one chromatid from each chromosome is segregated into the second polar body, but the chromosome number (ploidy) remains unchanged.

In 1935, Pincus and Enzmann reported that germinal vesicle-stage oocytes removed from antral follicles undergo spontaneous germinal vesicle breakdown (GVBD) in culture. This observation led to the hypothesis that antral follicles exert an inhibitory influence on meiotic maturation, and that removal of the oocyte from this inhibitory action results in oocyte maturation. This hypothesis implies that maturation *in vivo* is the result of gonadotropins terminating or overcoming the inhibitory action of the follicle. This concept is different from that presented for starfish and amphibians, wherein maturation is thought to be the result of the direct stimulatory action of the hormones 1-methyladenine and progesterone, respectively (see Chapters 1 and 6).

Alternative explanations to the follicular inhibition of maturation hypothesis have been presented (Biggers and Powers, 1979). It was suggested that the spontaneous maturation of oocytes might simply be an artifactual response of the oocyte to its removal from the follicular environment and its culture in an artificial environment. This argument appeared to be supported by experiments showing that oocytes spontaneously maturing in culture have a low frequency of fertilization and development to term. Thus, the maturation seemed to be abnormal or incomplete. However, it was recently demonstrated that mouse oocytes matured and fertilized *in vitro* have the same developmental capacity shown by oocytes matured *in vivo* (Schroeder and Eppig, 1984).

Some other arguments can be directed against the "artificial environment" hypothesis for spontaneous maturation. When oocytes, metabolically coupled to their companion granulosa cells, are placed in culture before the stage at which they are competent to mature, they grow in culture to the stage in which maturational competence has been acquired (Eppig, 1977). Yet many of the oocytes developing under these artificial conditions *in vitro* do not mature until their coupling to the granulosa cells is disrupted. In addition, when porcine cumulus cell-enclosed oocytes are placed on hemisections of follicle wall in culture, oocyte maturation occurs much less frequently than when cultured in the absence of the hemisections (Leibfried and First, 1980a). These data indicate that it is not the artificial environment of the culture system itself that promotes spontaneous maturation, but rather the association of the oocytes with the granulosa cells that maintains meiotic arrest.

Another argument against the artificial environment hypothesis is the common observation that oocytes in atretic antral follicles undergo maturation. One of the first morphological manifestations of follicular atresia is the degeneration of the membrana granulosa; pyknosis is observed among the granulosa cells. This is followed by oocyte maturation. The matured oocyte in an atretic follicle is often divested of its cumulus cells. In this case, the oocyte is certainly not being placed in an artificial environment of a culture tube or dish, nor is the follicle being mechanically disrupted. Rather, the integrity of the granulosa cell environment of the oocyte degenerates. These changes are evident even when oocyte and follicular development occur in the absence of gonadotropins. Thus, it can be argued once again that GVBD results from the loss of normal granulosa cell–oocyte associations, rather than the oocyte being placed in an artificial environment.

Considerable research effort has been devoted to exploring the mechanisms of oocyte maturation inhibition in mammalian follicles. However, one cannot eliminate the possibility that positive actions are taken by the follicle as well, in response to the preovulatory gonadotropin surge, in order to promote normal oocyte maturation.

Most studies on the mechanism of oocyte maturation inhibition have been based on the premise that there is a maturation-inhibiting factor in the follicle. However, these studies have been fraught with controversy due to conflicting results from different laboratories, even when oocytes from the same species were investigated. Nevertheless, a common thread that has emerged is that cell-to-cell communication between the oocyte and its companion cumulus/granulosa cells is crucial to meiotic control mechanisms.

## 5.2. Role of cAMP in the Maintenance of Meiotic Arrest

The concept that cAMP may play a role in the regulation of meiosis in mammals is rooted in evidence using amphibian oocytes (see Chapter 6) and in studies with mouse and rat oocytes showing that derivatives of cAMP and phosphodiesterase inhibitors reversibly prevent the spontaneous maturation of oocytes *in vitro* (Cho *et al.*, 1974; Wassarman *et al.*, 1976; Magnusson and Hillensjo, 1977; Dekel and Beers, 1978). GVBD will occur, however, when oocytes are cultured for about 45 min in the absence of 3-isobutyl-1-methyladenine (IBMX), a phosphodiesterase inhibitor, and are then transferred to medium containing IBMX, which would have blocked GVBD if the oocytes had been incubated in it continuously. This finding implies that a drop in the cAMP levels in the oocyte is somehow related to an irreversible commitment to GVBD. Consistent with this hypothesis is the observation that a decrease in oocyte cAMP levels precedes GVBD and during the commitment period in the mouse oocyte (Fig. 2). Preventing this decrease with phosphodiesterase inhibitors prevents GVBD (Schultz *et al.*, 1983a). Results from these studies suggest a role for cAMP in regulating the resumption of meiosis.

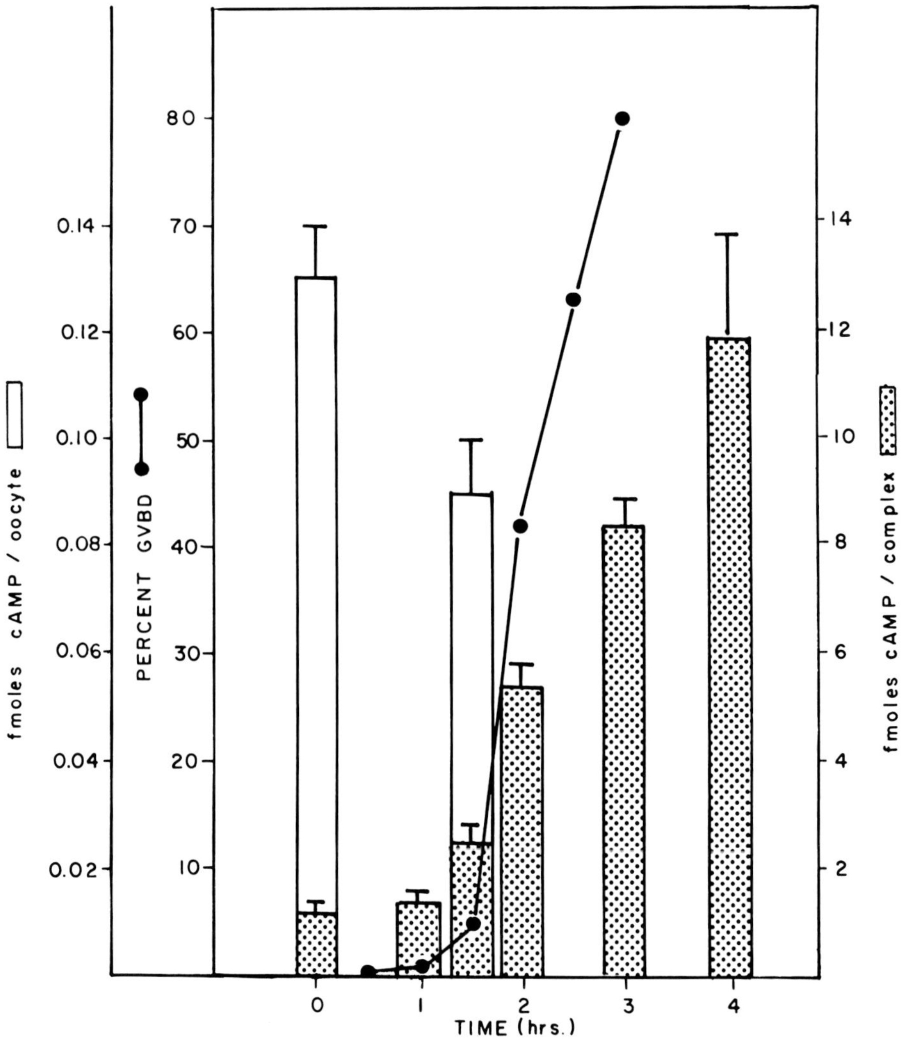


**Figure 2.** Correlation of cAMP levels in mouse oocytes (□) and oocyte–cumulus cell complexes (▩) with the time of GVBD (—●—) in oocytes *in vivo*. Oocyte maturation was induced by injecting PMSG primed mice with hCG. Time is indicated as hours after injection of hCG. The cAMP levels in the oocytes decrease before the time of maturation. cAMP increases in the cumulus cells while it is decreasing in the oocytes. cAMP, cyclic adenosine monophosphate; GVBD, germinal vesicle breakdown; hCG, human chorionic gonadotropin; PMSG, pregnant mare serum gonadotropin. (Data redrawn from Schultz *et al.*, 1983a, with permission of the authors.)

It has been hypothesized that cAMP may gain access to the oocyte from cumulus cells via the gap junctions that metabolically couple the cumulus cells and oocyte. In addition, it has been suggested that disruption of the communication between cumulus cells and oocytes before ovulation may reduce the flow of cAMP to the oocytes and thus terminate meiotic arrest (Dekel and Beers, 1978, 1980; Gilula *et al.*, 1978). Two critical aspects of this attractive hypothesis have been amenable to experimental testing. First, does a reduction of cumulus cell–oocyte coupling occur before or after GVBD? Second, does cAMP equilibrate between cumulus cells and the oocyte?

Measurement of ionic coupling between rat cumulus cells and the oocyte at various times after LH treatment of intact follicles showed that GVBD occurs before uncoupling (Dekel *et al.*, 1981). However, measurement of ionic coupling is essentially an all-or-nothing measurement. That is, some degree of uncoupling may occur before GVBD, but this may have been undetected. As discussed previously, another way of measuring coupling, on a more quantitative basis, is by measuring metabolic cooperativity by the metabolic transfer method. When this was done using cumulus cell–oocyte complexes from mice, no reduction of metabolic cooperativity was detected before GVBD (Eppig, 1982a). In the mouse, GVBD occurred by 3 hr after the injection of an ovulatory dose of human chorionic gonadotropin (hCG), but a decrease in metabolic coupling was not seen until 9 hr after hCG (Fig. 3) (Eppig, 1982a). Similarly, LH promoted GVBD in cultured intact sheep follicles, but junctional communication between the cumulus cells and the oocyte remained intact (Moor *et al.*, 1981). It was concluded from these experiments that GVBD is not initiated as a result of the termination of coupling between cumulus cells and the oocyte. However, it remains a possibility that GVBD is initiated by a qualitative or quantitative change in a factor transferred to the oocyte via the intact coupling pathway.

Does cumulus cell cAMP equilibrate with the oocyte? To test this possibility, levels of cAMP in mouse cumulus cells were elevated more than 20-fold with either FSH or cholera toxin in the presence of IBMX. Cholera toxin is an activator of adenylate cyclase. Under these conditions, no difference in cAMP levels in cumulus cell-enclosed oocytes versus denuded oocytes was detected (Schultz *et al.*, 1983b). Thus, it appears that cAMP remained compartmentalized within the cumulus cells. However, phosphorylated derivatives of uridine and 2-deoxyglucose did equilibrate between the cumulus cells and the oocyte, demonstrating that the coupling pathway remained intact and functional throughout the time course of this experiment. The possibility that oocyte cAMP was secreted or that undetected quantities of cumulus cell cAMP was transferred to the oocyte could not be eliminated. On the other hand, cAMP hydrolysis in the oocyte was unlikely, since IBMX was present at all times. Loss of oocyte cAMP because

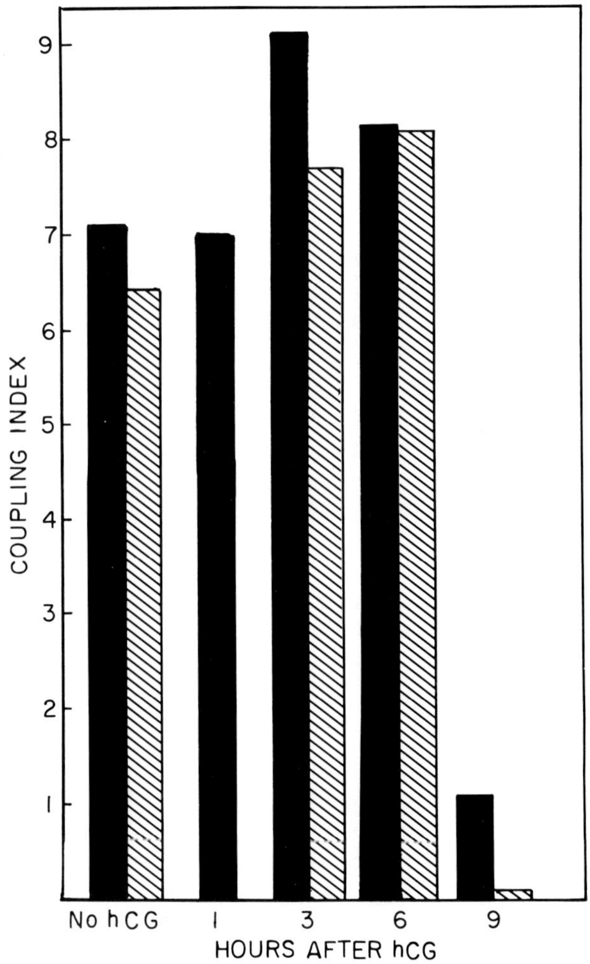


**Figure 3.** Metabolic cooperativity between cumulus cells and oocytes in the mouse after the induction of maturation with hCG. At the various times indicated, oocyte–cumulus cell complexes were isolated from the largest graafian follicles and metabolic cooperativity measured by the metabolite transfer method. Intact oocyte–cumulus cell complexes and oocytes denuded of their cumulus cells were incubated together in medium containing either [$^3$H]uridine (■) or [$^3$H]choline (▧).formula *A-B/B*, where *A* was the radioactivity in the oocytes labeled in intact complexes and *B* was the radioactivity measured in the denuded oocytes (Heller *et al.*, 1981). Thus, oocytes coupled to cumulus cells acquire more radiolabel than do uncoupled or denuded oocytes. More than 90% of the complexes isolated 3 hr after hCG had undergone GVBD. Since the coupling indexes demonstrated no reduction in metabolic cooperativity until after 6 hr, uncoupling cannot be the initiator of GVBD. GVBD, germinal vesicle breakdown; hCG, human chorionic gonadotropin. (Data from Eppig, 1982a).

of potential damage incurred during the removal of cumulus cells was also unlikely, since derivatives of uridine that accumulated in the oocyte as a result of metabolic cooperativity were not lost. It is not clear how cAMP is compartmentalized in the cumulus cells, whereas molecules of similar size and charge, such as phosphorylated derivatives of uridine, are not. Perhaps cAMP-binding proteins are present in sufficient amounts, either constitutively or after stimulation, to bind the very high levels of cAMP generated by FSH in the presence of IBMX.

Even though no evidence of cAMP transfer to oocytes is detected after stimulation of cumulus cells with FSH, there is a transient inhibition of GVBD (Fig. 4) that was positively correlated with the elevation of cAMP in the cumulus cells (Schultz *et al.*, 1983b; Eppig *et al.*, 1983). This result suggests that some factor(s) of cumulus-cell origin other than cAMP is active in the inhibition of GVBD. This GVBD-inhibitory factor seems to be dependent on cAMP for its activity and/or generation, because its activity is correlated with elevation of cAMP in the cumulus cells (Fig. 5).

This hypothesis has been tested using concentrations of dibutyryl cAMP (dbcAMP) that are suboptimal for the inhibition of GVBD *in vitro*. Since this analogue of cAMP readily penetrates plasma membranes, it probably has equal access to both the cumulus cell-enclosed and denuded oocytes. It was found that the suboptimal concentrations of dbcAMP had a greater inhibitory effect on GVBD in cumulus cell-enclosed oocytes than in denuded oocytes (Fig. 6) (Eppig *et al.*, 1983). Such experiments show that dbcAMP has an action on cumulus cells that results in an oocyte maturation-inhibiting effect that is greater than the inhibitory action of the analogue alone. These results support the hypothesis that a factor of cumulus-cell origin inhibits oocyte maturation by a cAMP-dependent process. If cAMP itself is not transferred to the oocyte, it is likely that cAMP promotes the activation and/or generation of an inhibitory factor by the cumulus cells. The action of FSH, cholera toxin, and suboptimal concentrations of dbcAMP in inhibiting GVBD depends on an intact coupling pathway between the cumulus cells and oocyte (Schultz *et al.*, 1983a; Eppig *et al.*, 1983). Therefore, a putative maturation-inhibiting factor must have the physical properties that would permit passage through the heterologous gap junctions.

Oocyte maturation-inhibiting (OMI) substances have been reported to be present in pig, hamster, and human follicular fluid (Tsafriri and Channing, 1975; Gwatkin and Andersen, 1976; Chari *et al.*, 1983). Furthermore, studies indicate that OMI is produced by porcine granulosa cells in culture (Tsafriri and Channing, 1975; Centola *et al.*, 1981). The inhibitory action of porcine OMI is reversed by treatment with LH (Tsafriri *et al.*, 1976, 1977). It also appears that cumulus cells are required for OMI function (Hillensjo *et al.*, 1979). Preliminary biochemical characterization of OMI indicate a molecular weight of $<1000$, thus enabling it to traverse gap junctions (Tsafriri *et al.*, 1982). The existence of OMI has been controversial, since

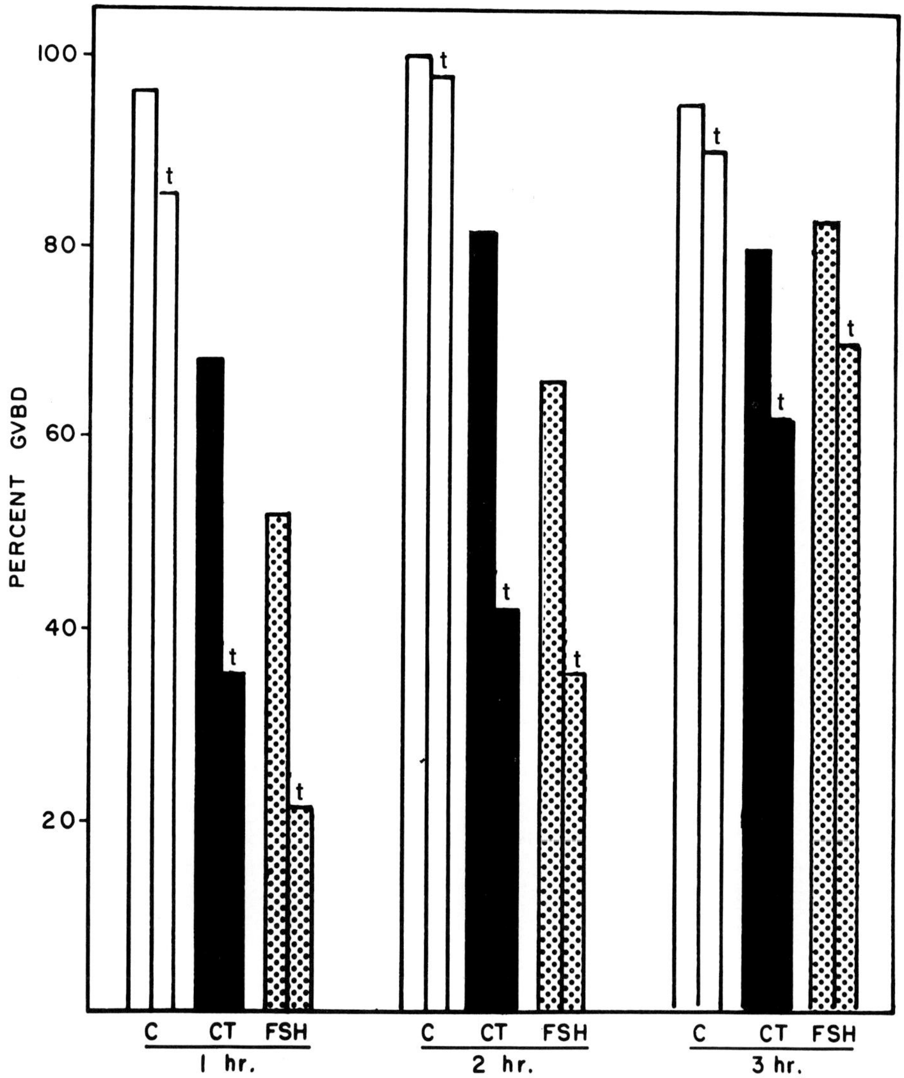


**Figure 4.** Effect of FSH and cholera toxin on oocyte maturation. Oocyte–cumulus cell complexes isolated from mice were incubated in either control medium (□) or media containing 2.5 μg/ml FSH (⁙), or 2 μg/ml cholera toxin (■). The percentage of GVBD was determined after 1,2, and 3 hr of culture. A *t* over the bar indicates that 2 μg/ml testosterone was added to the medium. Both FSH and CT transiently inhibited maturation, and testosterone acted synergistically with either FSH or CT to inhibit GVBD. C, control; CT, cholera toxin; FSH, follicle-stimulating hormone; GVBD, germinal vesicle breakdown. (Data from Schultz *et al.*, 1983b.)

no effect of follicular fluid on maturation has been reported by some investigators (Jagiello *et al.*, 1977; Leibfried and First, 1980b; Racowsky and McGaughey, 1982a; Fleming *et al.*, 1983). The reasons for these discrepancies may reside in several areas (e.g., the physiological state of the follicles that were the source of the follicular fluid, the state of the cumulus cell–oocyte complexes used in the assays, and the experimental protocols).

The relationship, if any, between OMI and the putative inhibiting factor proposed by Schultz *et al.* (1983b) and Eppig *et al.* (1983) is not clear,

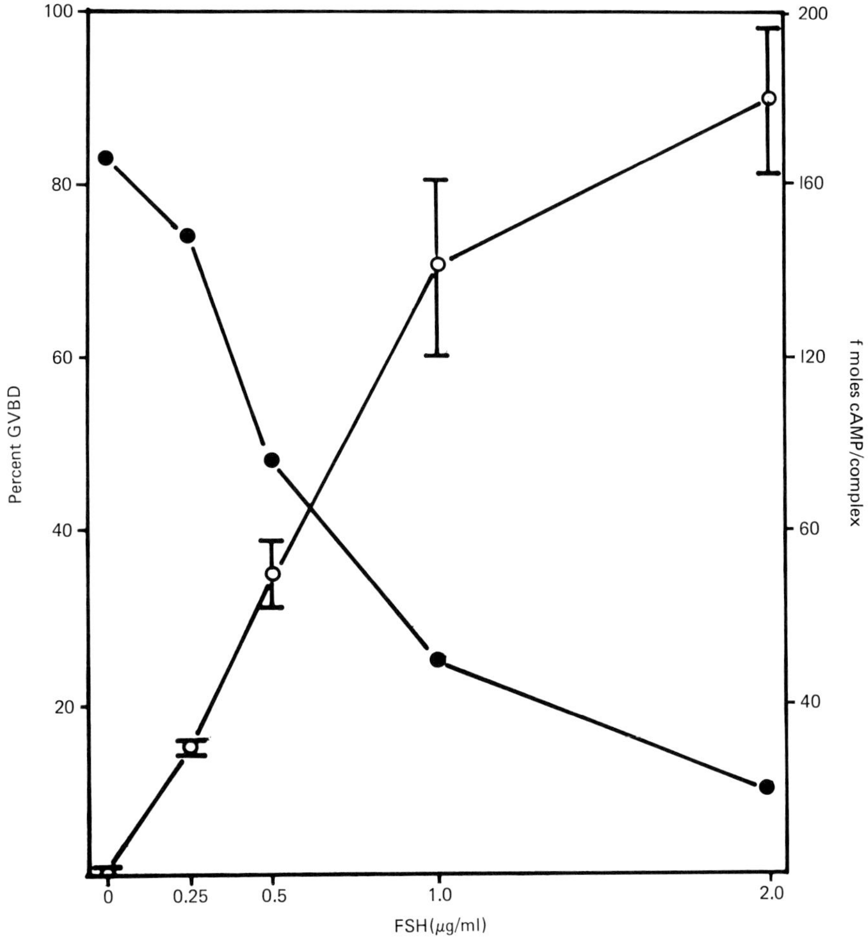


**Figure 5.** Effect of FSH on cAMP levels in isolated cumuli oophori and on oocyte maturation. Increasing the concentration of FSH results in a dose-dependent increase in the cAMP levels (— ○ —) in mouse cumuli oophori and a concomitant inhibition of maturation (— ● —). cAMP, cyclic adenosine monophosphate; FSH, follicle-stimulating hormone. (Data from Eppig *et al.*, 1983.)

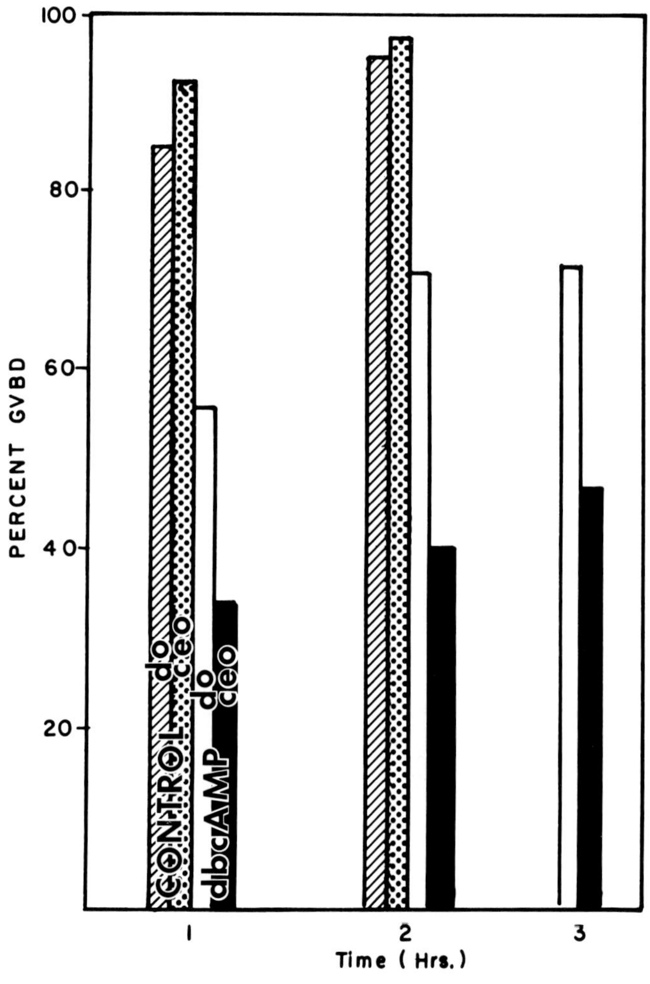


**Figure 6.** Effect of 100 $\mu$M dbcAMP on the maturation of cumulus cell-enclosed and denuded mouse oocytes. The percentage of cumulus cell-enclosed (CEO) (⁙) and denuded oocytes (do) (▨) that had matured by 1 and 2 hr culture was the same in control medium (no dbcAMP). By contrast, when 100 $\mu$M dbcAMP was added to the medium, the percentage of denuded oocytes (□) that matured was consistently greater than the percentage of cumulus cell-enclosed oocytes (■) that matured. Thus, dbcAMP has a greater inhibitory effect on cumulus cell-enclosed oocytes than on denuded oocytes. (Data from Eppig *et al.*, 1983.)

and the establishment of a relationship must await the identification of the factor(s). However, if a relationship does exist, then the activity of OMI should be augmented as a result of elevating cAMP levels in cumulus cells or oocytes. A low-molecular-weight fraction (PM10; Tsafriri *et al.*, 1976) of porcine follicular fluid has a significant, but rapidly transient, inhibitory effect on the maturation of cumulus cell-enclosed mouse oocytes (Downs and Eppig, 1984). Raising the level of cAMP in the cumulus cells with FSH in the presence of the PM10 fraction has a much greater inhibitory effect than the PM10 or FSH alone (Fig. 7). In addition, suboptimal concentrations of dbcAMP also act synergistically with the PM10 fraction to reversibly suppress maturation. The active component of the PM10 fraction does not appear to be a lipid or a peptide, but has a molecular weight less than 1000 (Downs and Eppig, 1984). Thus, the action of the PM10 fraction of porcine follicular fluid in inhibiting GVBD in mouse oocytes apparently depends on the levels of cAMP in the cumulus cell–oocyte complexes. Perhaps the conflicting results obtained with follicular fluid preparations is related to the cAMP levels in the various cumulus cell–oocyte assay systems.

Transient elevation of cumulus cell cAMP with FSH or cholera toxin resulted in a transient inhibition of GVBD. Obviously, GVBD-inhibitory mechanisms *in vivo* must be more prolonged. It seems likely that the entire mass of granulosa cells participates in the maintenance of oocyte meiotic arrest. Granulosa cells are coupled with one another by homologous gap junctions (Albertini and Anderson, 1974), thereby rendering the mass of granulosa cells into a syncytium. It is possible that the mass of granulosa cells may pass active inhibitor through its syncytial network to the oocyte, with cumulus cells acting as a conduit. In addition, inhibitor may be

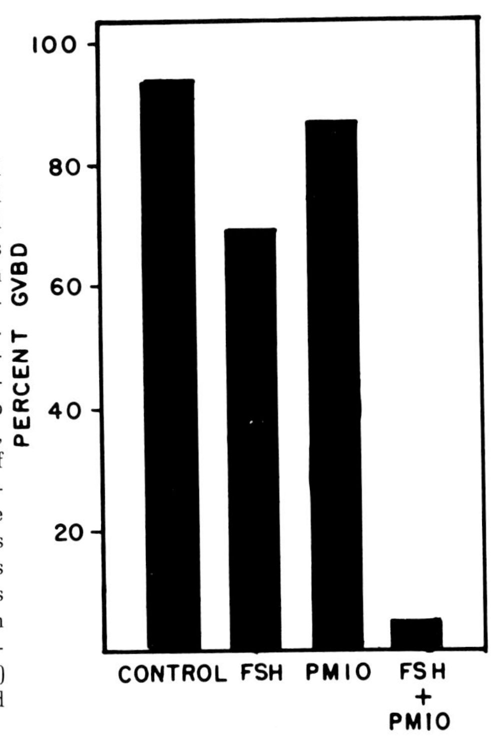


**Figure 7.** Effect of porcine follicular fluid on the maturation of cumulus cell-enclosed mouse oocytes. A fraction of follicular fluid having molecular weight of <1000 was obtained by passing the fluid through an Amicon PM10 filter. Hence, groups containing this material are designed PM10. Cumulus cell-enclosed oocytes were incubated in control medium or medium containing 2.5 $\mu$g/ml FSH, PM10 (equivalent to 67% of its concentration in follicular fluid), or FSH plus PM10, and the frequency of GVBD was determined after 1.5 hr incubation. At that time, no effect of PM10 alone was seen, but when the oocyte–cumulus cell complexes were incubated in FSH plus PM10, the inhibitory action of FSH was greatly enhanced. (The PM10 preparation was obtained from Dr. S. Pomerantz, University of Maryland School of Medicine.) (From S.M. Downs and Eppig, unpublished data.)

secreted into the follicular fluid and may then be incorporated into the cumulus cell–oocyte complex.

## 5.3. Role of Steroid Hormones in the Regulation of Maturation

Steroid hormones in follicular fluid undergo quantitative and qualitative changes after the preovulatory gonadotropin surge. Of particular note is the change in the ratios of estrogens and progestins, with the latter predominating before ovulation. Since the follicular fluid bathes the cumulus cell–oocyte complex, and since progesterone is well known to promote maturation in amphibians, it was obvious that steroid hormones should be tested for their effects on mammalian oocyte maturation.

Luteinizing hormone induces both the maturation of oocytes and production of progesterone in intact follicles *in vitro* (Tsafriri, 1978). However, when the follicles are incubated in cyanoketone or aminoglutethimide—two inhibitors of progesterone production—there is no effect on LH-induced oocyte maturation (Tsafriri, 1978). Furthermore, incubation of follicles in media containing progesterone does not stimulate maturation (Tsafriri, 1978). These results suggest that LH does not promote oocyte maturation via the synthesis of progesterone. Nevertheless, it seems possible that exposure to steroids before the preovulatory gonadotropin surge may be required to promote competency to undergo maturation. However, competency for spontaneous GVBD is acquired during oocyte growth *in vitro* in the absence of hormone additions to the culture media (Eppig, 1977). Although this result suggests that steroid hormones are not required for the acquisition of competence to undergo GVBD, it does not eliminate a role for basal levels of progesterone that may be produced by the somatic cells in the culture system. Nor do these experiments address the potential role for steroid hormones in the acquisition of competency for the complete maturation of the oocyte that would result in the production of a gamete capable of fertilization and development.

Studies on the response of isolated oocytes to steroid hormones are difficult to interpret. Although the concentrations of steroids in follicular fluid are high, it is difficult to mimic those concentrations for studies *in vitro* because the form of the steroids in follicular fluid has not been sufficiently evaluated. For example, how much steroid is free or bound to other components in follicular fluid? Are the free or the bound forms physiologically significant? When steroids are added to culture media, what proportion of the amount added is in a physiologically relevant form? With these problems as a backdrop, it is not surprising that studies reporting the effects of steroid hormones on oocyte maturation *in vitro* are often conflicting and inconsistent.

Since estrogens predominate in the follicular fluid of follicles with meiotically arrested oocytes, 17β-estradiol has been tested extensively for inhibitory effects on spontaneous maturation. At concentrations above 10 $\mu$g/ml, 17$\beta$-estradiol has been reported to have some inhibitory effect on GVBD in porcine oocytes (McGaughey, 1977). However, others have reported using mouse oocytes that high concentrations of progesterone or testosterone can also have an inhibitory action on GVBD (Eppig and Koide, 1978; Smith and Tenny, 1980). At lower concentrations, various steroids seem to suppress later stages of maturation (McGaughey, 1977; Smith and Tenny, 1980). Other investigators have reported that progesterone can promote spontaneous maturation (Bae and Foote, 1975; Gould and Graham, 1976). The culture systems used for such experiments can affect the response to steroid hormones as demonstrated by the report that the presence or absence of protein in the medium determine whether 17$\beta$-estradiol suppresses porcine oocyte maturation (Racowsky and McGaughey, 1982b). That study found 17$\beta$-estradiol ineffective in inhibiting maturation when protein was present. Since the environment of the oocyte–cumulus cell complex contains ample protein, it was suggested that 17$\beta$-estradiol may not have a direct physiological role in inhibiting maturation (Racowsky and McGaughey, 1982b).

It was recently shown that steroid hormones may act synergistically with a cAMP-dependent maturation-inhibiting factor within the oocyte. As described above, when either FSH or cholera toxin is used to raise the level of cumulus cell cAMP, oocyte maturation is inhibited (Schultz *et al.*, 1983b). The inhibition is directly correlated with the levels of cAMP in the cumulus cells (Schultz *et al.*, 1983b; Eppig *et al.*, 1983). It was also reported that testosterone, used at concentrations that do not affect maturation when used alone, acts synergistically with FSH or cholera toxin to inhibit maturation (Fig. 4) (Schultz *et al.*, 1983b). Likewise, testosterone and dbcAMP act synergistically to inhibit mouse oocyte maturation to a significantly greater extent than does dbcAMP alone (Eppig *et al.*, 1983). A similar synergistic effect of dbcAMP and testosterone was reported using porcine oocytes (Rice and McGaughey, 1981). When testosterone is added to cultures containing FSH, the level of cAMP in the mouse cumulus cells is not raised above the level stimulated by FSH alone. Testosterone does not promote the transfer of cAMP from cumulus cells to the oocytes, nor does it affect the kinetics of cAMP degradation in the cumulus cells or the extent of metabolic coupling between the cumulus cells and oocyte (Schultz *et al.*, 1983b; Eppig *et al.*, 1983). Thus, testosterone apparently acts synergistically with a cAMP-dependent process to inhibit maturation. Forskolin, an activator of adenylate cyclase, which raises the cAMP levels in denuded oocytes (Schultz *et al.*, 1983a), and dbcAMP both act synergistically with testosterone to inhibit maturation of denuded oocytes (Eppig *et al.*, 1983). Therefore, the oocyte itself can be a site for the synergistic action of testosterone with the cAMP-dependent process. Furthermore, since the elevated levels of cAMP generated in cumulus cells may not be transferred to the oocyte, and since it is probable that a cAMP-dependent factor is

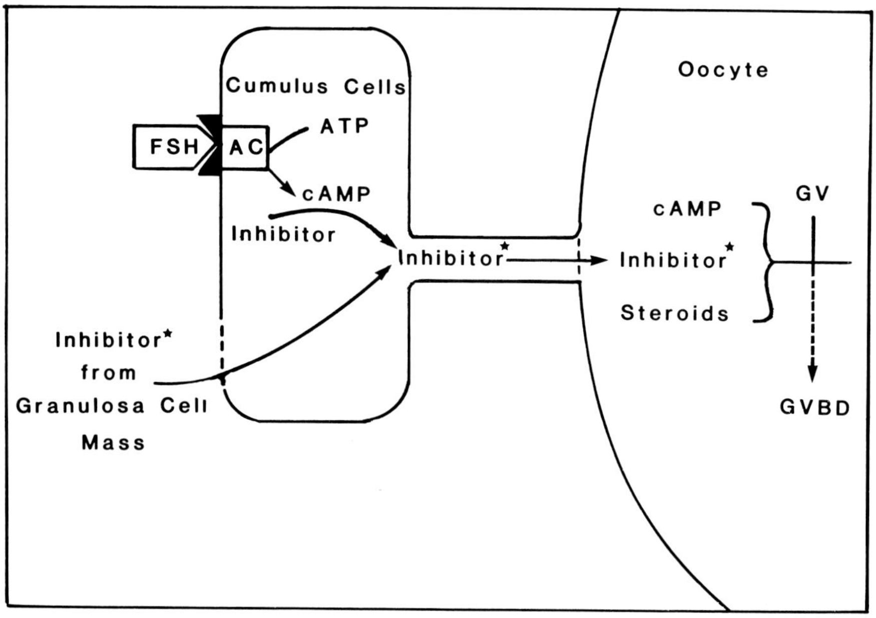


**Figure 8.** Working model of inhibition of maturation in mouse oocytes. Inhibiting factor in the cumulus cells is activated as a result of a cAMP-dependent process. For example, FSH, interacting with its receptor, activates cumulus cell adenylate cyclase to elevate cAMP levels, which results in the production and/or activation of inhibitor. Active inhibitor (*) may also enter the cumulus cells via gap junctions from the mass of mural granulosa cells. The active inhibitor is transferred to the oocyte by the heterologous gap junctions. Inhibiting factor, cAMP, and steroids function within the oocyte to inhibit maturation. Inhibiting factor may also be taken up from the follicular fluid into the cumulus cells and/or oocyte and interact with a cAMP-dependent process. AC, adenylate cyclase; ATP, adenosine triphosphate; cAMP, cyclic adenosine monophosphate; FSH, follicle-stimulating hormone. (From Eppig *et al.*, 1983.)

transferred from the cumulus cells to the oocyte, it is reasonable to hypothesize that testosterone acts synergistically with the cAMP-dependent factor in the oocyte to inhibit maturation (see working model depicted in Fig. 8). The synergism with the cAMP-dependent inhibiting factor is not specific to testosterone: Progesterone, 17$\beta$-estradiol, and dihydrotestosterone are also reported to act synergistically with the cAMP-dependent factor under various conditions (Eppig *et al.*, 1983). The mechanisms and physiological relevance of the steroid hormone synergisms in meiotic regulation are unknown.

## 5.4. Cytoplasmic Maturation of Mammalian Oocytes

In addition to the nuclear events of oocyte maturation, changes occur in the cytoplasm of the oocyte as well. A developmental program of alterations in the patterns of protein synthesis and phosphorylation is specific to the stages of maturation (Warnes *et al.*, 1977; Schultz and Wassarman, 1977; Van Blerkom and McGaughey, 1978a; Schultz *et al.*, 1983a). The patterns of protein synthesis are similar, but not identical, during maturation *in vivo* and *in vitro* (Warnes, 1977; Van Blerkom and McGaughey, 1978a). The differentiation of the ooplasm during maturation is also manifest in its ability to promote the swelling of the sperm head after its entry into the ooplasm. The sperm head does not swell to form a normal male pronucleus after penetration of the sperm into the ooplasm of an immature oocyte, and embryonic development is impossible. The ability of the ooplasm to promote the formation of a normal male pronucleus increases with time after the induction of oocyte maturation *in vivo* (Iwamatsu and Chang, 1972; Niwa and Chang, 1975). The component(s) of the ooplasm that promotes the development of the male pronucleus has been referred to as **male pronuclear growth factor (MPGF)** (Thibault, 1977).

Some studies seemed to indicate that spontaneous oocyte maturation *in vitro* does not constitute normal or complete maturation, since male pronuclear growth failed to occur, or because the frequency of embryonic development was very low (Thibault, 1977; Moor and Trounson, 1977; Trounson *et al.*, 1977). This argument seemed to be supported by the observed differences in the protein synthetic patterns during maturation *in vivo* and during spontaneous maturation (Warnes *et al.*, 1977; Van Blerkom and McGaughey, 1978a). However, Van Blerkom and McGaughey (1978b) reported that spontaneously matured rabbit oocytes could be fertilized and undergo cleavage and preimplantation development in spite of the minor differences that they had observed in protein synthesis in oocytes matured *in vivo* and *in vitro*. Shalgi *et al.* (1979) found that treatment of isolated cumulus cell–oocyte complexes with biological grade LH resulted in a greatly increased frequency of fertilization and development to term when the matured rat ova were transferred to the oviducts of sterile inseminated females. It is not certain, however, whether the acquisition of developmental competence is due to the induction of a qualitative change in the maturing oocyte (e.g., the induction of MPGF activity) or to the mucification of the cumulus oophorus (see Section 6.1), which may facilitate access of the sperm to the ovum.

These studies indicate that the culture conditions that permit spontaneous nuclear maturation were not always sufficient to promote the acquisition of the capacity for embryonic development. In view of the promising results achieved in some studies (Cross and Brinster, 1970; Mukherjee, 1972; Van Blerkom and McGaughey, 1978b; Shalgi *et al.*, 1979; Motlik and Fulka, 1981; Nishimoto *et al.*, 1982; Fulka *et al.*, 1982), it seemed likely that developmental failures resulted from deficiencies in experimental systems. Accordingly, it has now been shown that the developmental capacity of mouse oocytes that mature spontaneously *in vitro* is normal (Schroeder and Eppig, 1984). This conclusion is based on experiments showing equal frequencies of fertilization and development to live fertile offspring from oocytes matured *in vivo* or spontaneously matured *in vitro* (Schroeder and Eppig, 1984). It seems likely that similar results will be obtained using oocytes from other species after the culture systems for the maturation of oocytes, fertilization, and embryo culture are refined for each species.

## 6. Hormonal Control of Cumulus Cell Function

The regulation of cumulus cell function has developed into an active area of research because of the metabolic cooperativity between the cumulus cells and the oocyte, and because of emerging evidence that the hormonal control of oocyte function is mediated by the cumulus cells.

### 6.1. Cumulus Expansion

Before the preovulatory surge of gonadotropins, the cumulus cells are tightly packed around the oocyte (Fig. 9A). However, dramatic changes occur in the cumulus oophorus after the surge: The cumulus cells secrete hyaluronic acid, a nonsulfated glycosaminoglycan (GAG), which expands the spaces between the cumulus cells until the cumulus oophorus finally consists of dispersed cells embedded in a matrix of hyaluronic acid (Fig. 9B). Cumulus expansion, or **mucification,** appears to be initiated after GVBD has occurred in the oocyte (Dekel *et al.*, 1979; Eppig, 1982a).

FSH stimulates the expansion of cumuli oophori isolated from the ovaries of all the mammalian species that have been studied (Fig. 10A–C) (Thibault, 1972; Dekel and Kraicer, 1978; Eppig, 1979b; Hillensjo and Channing, 1980; Ball *et al.*, 1982). Isolated cumulus cell–oocyte complexes of the mouse do not expand in response to highly purified preparations of human chorionic gonodotropin (hCG) or LH (Eppig, 1979b,c). However, isolated pig cumuli oophori do mucify in response to highly purified LH (Hillensjo and Channing, 1980). The action of gonadotropins in stimulating the expansion of isolated cumuli oophori appears to be mediated by cAMP. This conclusion is based on the findings that elevating cumulus cell cAMP levels with

**Figure 9.** Photomicrographs of sections of mouse ovaries (A) 48 hr after priming with PMSG, and (B) 10 hr after inducing oocyte maturation in the primed mice with hCG. (A) The germinal vesicle (arrow)-stage oocyte is surrounded by two to three layers of tightly packed cumulus cells. The oocyte–cumulus cell complex appears to be continguous with the mural granulosa cells. (B) The germinal vesicle has broken down, and the cumulus cells have been dispersed and embedded in a matrix containing hyaluronic acid. The oocyte, with its expanded cumulus oophorus, has become detached from the mass of mural granulosa cells in preparation for ovulation. cc, cumulus cells; hCG, human chorionic gonadotropin; mgc, mural granulosa cells. Scale bars: 100 $\mu$m.

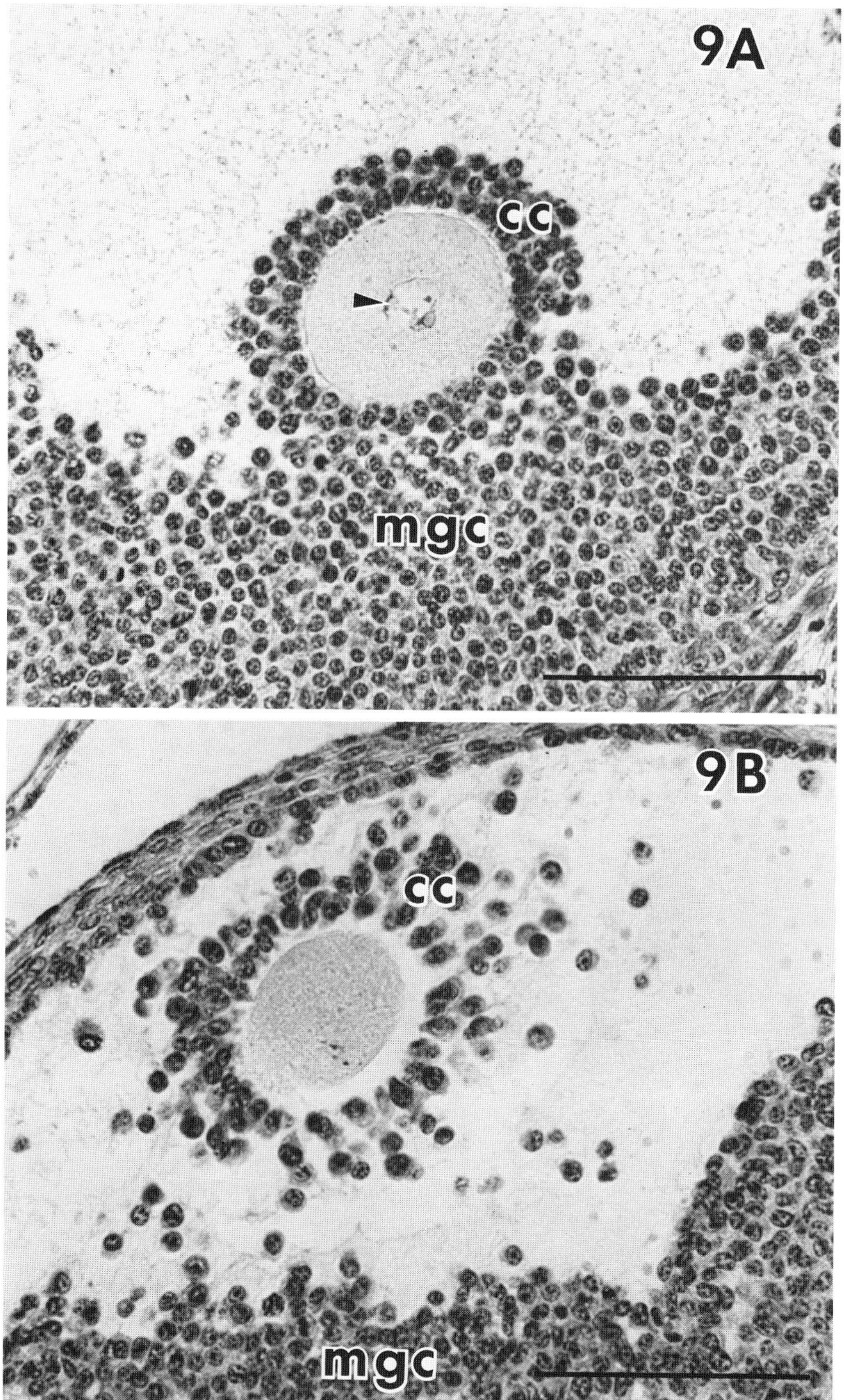
9A
cc
mgc
9B
cc
mgc

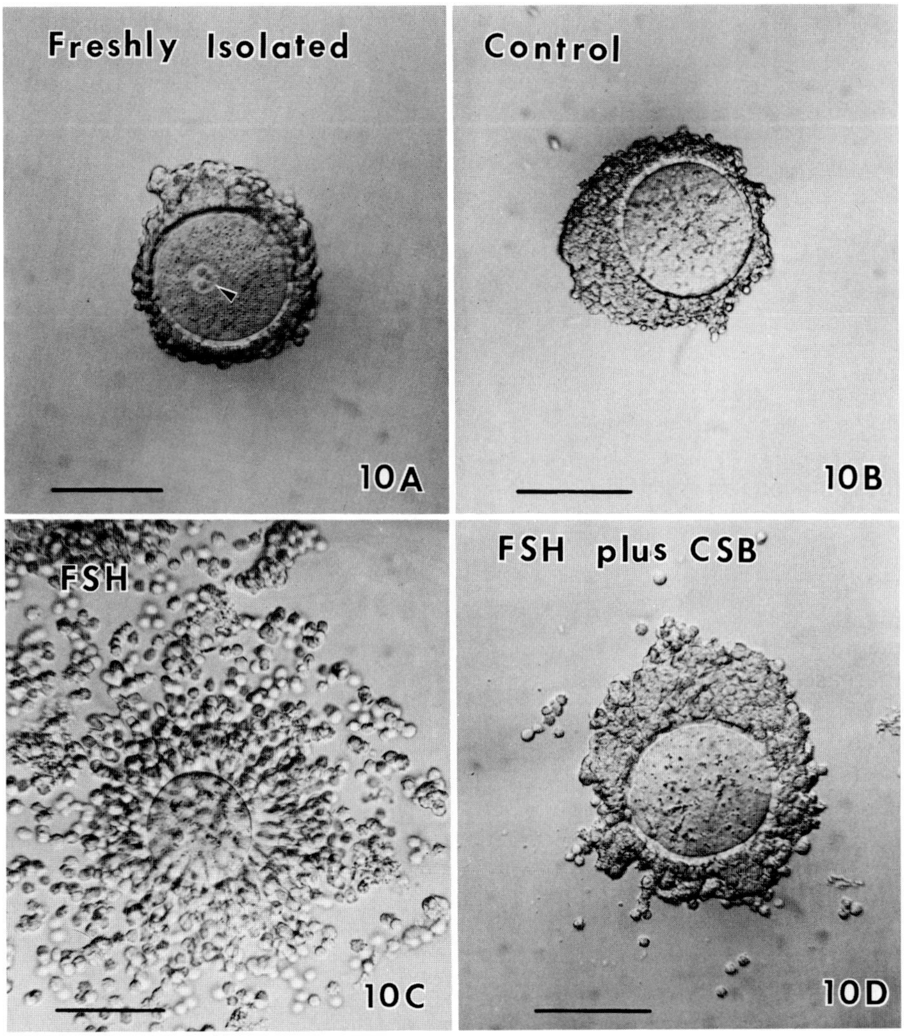


**Figure 10.** Effect of FSH and chondroitin sulfate B on the expansion of cumuli oophori isolated from mice. Living oocyte–cumulus cell complexes photographed using a microscope equipped with Nomarski interference optics. (A) Freshly isolated oocyte–cumulus cell complex showing intact germinal vesicle (arrow) and compact cumulus oophorus. (B) Oocyte–cumulus cell complex after incubation in control medium for 15 hr. The oocyte has undergone spontaneous maturation, but the cumulus cells remain tightly packed around the oocyte. (C) Oocyte–cumulus cell complex after incubation in medium containing 2.5 $\mu$g/ml FSH for 15 hr. The cumulus cells have become dispersed and embedded in a matrix containing hyaluronic acid. (D) Oocyte–cumulus cell complex after incubation in medium containing FSH plus 2.0 mg/ml chondroitin sulfate B for 15 hr. CSB suppressed FSH-stimulated cumulus expansion but did not affect spontaneous oocyte maturation. CSB, chondroitin sulfate B; FSH, follicle stimulating hormone. Scale bars: 100 $\mu$m.

cAMP analogues, phosphodiesterase inhibitors, or cholera toxin also stimulates cumulus expansion (Dekel and Kraicer, 1978; Eppig, 1979b,c).

Although purified LH or hCG does not stimulate the expansion of isolated cumuli oophori from mice, injection of highly purified hCG into mice does stimulate cumulus expansion *in vivo* (Eppig, 1980). In addition, highly purified hCG stimulates cumulus expansion in organ cultures of intact follicles (Eppig, 1980). This result suggests that the preovulatory LH surge probably initiates cumulus expansion *in vivo* even though mouse cumuli oophori may not respond directly to LH. Therefore, LH stimulation of expansion is probably mediated by other follicular components. One possibility would be that LH promotes the production of prostaglandins that, in turn, stimulate cumulus expansion. In support of this hypothesis, it has been reported that prostaglandin $E_2$ stimulates the expansion of isolated mouse cumuli oophori (Eppig, 1981a) and cumuli oophori of intact rat follicles (Phillips and Dekel, 1982). In addition, indomethacin, a prostaglandin synthetase inhibitor, suppresses LH-induced cumulus expansion in mice. Injection of prostaglandin $E_2$ reverses the inhibition with indomethacin (Downs and Longo, 1983).

Mural granulosa cells of the rat and pig have 10–15-fold more LH receptors than do a corresponding number of cumulus cells (Lawrence *et al.*, 1980; Channing *et al.*, 1981). In addition, increasing proximity of the mural granulosa cells to the follicle-limiting basal lamina is correlated with increased numbers of LH receptors (Zeleznik *et al.*, 1974; Amsterdam *et al.*, 1975). Mural granulosa cells constitute a metabolic syncytium as a result of the gap junctions that conjoin adjacent cells. It seems likely that at least some of the cumulus cells are coupled to mural granulosa cells at the point of attachment of the cumulus cell–oocyte complex to the follicle wall. Thus, an LH-induced cumulus expansion-stimulating signal, originating in the mural granulosa cells, may be transmitted to the cumulus cells through the metabolic syncytium mediated by gap junctions.

The highly sensitive expansion response of cumulus cells to FSH presents in interesting paradox. It has been known for years that FSH is required for follicular development from at least the early antrum stage. FSH has an important role in estrogen production, cell proliferation, and the development of LH receptors by granulosa cells. In addition, immature mice are often primed with pregnant mare serum gonadotropin (PMSG), containing both FSH and LH activity, to promote follicular development. The PMSG, however, does not stimulate cumulus expansion *in vivo*, even though the fragile complexes isolated from the unprimed mice expand in response to PMSG in culture (Eppig, 1980). Hence, the paradox: Why do cumuli oophori that respond to FSH or PMSG in culture not expand in response to FSH or PMSG during follicular development? It has been proposed that some follicular component inhibits the expansion response of cumuli oophori until that inhibition is terminated by the preovulatory surge of LH (Eppig, 1980).

Sulfated GAG, such as chondroitin sulfate B (dermatan sulfate) and heparan sulfate, have been reported in follicular fluid at concentrations ranging from 1 mg/ml (Yanagishita *et al.*, 1979) to as high as 5 mg/ml (Ax and Ryan, 1979). It is likely that most of the sulfated GAG are covalently bound to a core protein to form proteoglycans. Sulfated GAG were found to inhibit FSH-stimulated expansion of isolated cumuli oophori from mice (Fig. 10D) (Eppig, 1981b). The potency of the sulfated GAG roughly correlates with the degree of sulfation of the GAG: heparin > heparan sulfate ≧ chondroitin sulfate B > chondroitin sulfate C > chondroitin sulfate A. Although these properties are correlative, a direct relationship has not been demonstrated. The sulfated GAG inhibits the induction of hyaluronic acid-synthesizing activity in cumulus cells (Fig. 11) but does not inhibit the activity in lysates of FSH-induced cells (Eppig and Ward-Bailey, 1984). Therefore, the sulfated GAG act on a cellular function whereby FSH induces hyaluronic acid-synthesizing activity, but the sulfated GAG are not direct enzyme inhibitors. The site of action of the sulfated GAG is distal to FSH-stimulated elevation of cumulus cell cAMP: The GAG blocks

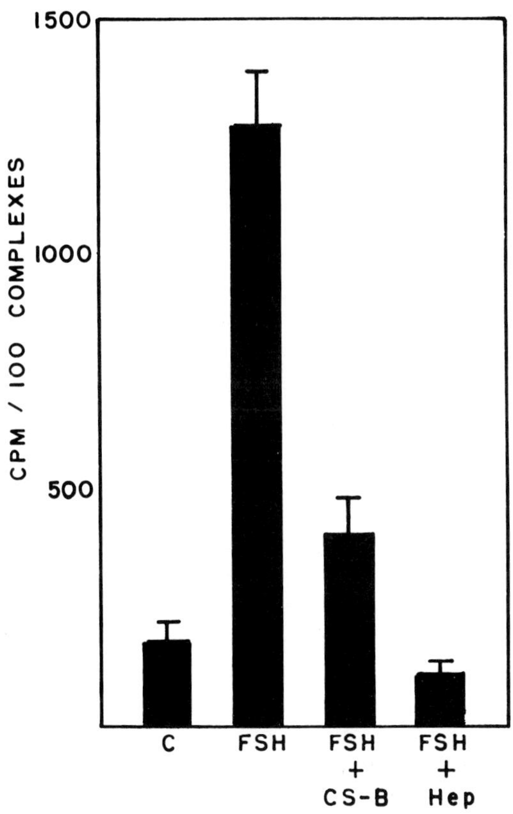


**Figure 11.** Effect of sulfated glycosaminoglycans on FSH-induced hyaluronic acid synthesizing activity of cumulus cells. Activity was measured in cell lysates after the intact cumuli oophori were cultured for 6 hr in control medium (C), or medium containing 2.5 $\mu$g/ml follicle-stimulating hormone (FSH), FSH plus 5 mg/ml chondroitin sulfate B (CS-B), or FSH plus 100 $\mu$g/ml heparin (Hep). The assay measures the incorporation of $^{14}$C-labeled UDP-glucuronic acid into *Streptomyces* hyaluronidase-sensitive, cetylpyridinium chloride-precipitable counts.

dbcAMP-stimulated cumulus expansion (Eppig, 1981c) but not FSH-stimulated elevation of cumulus cell cAMP (Eppig *et al.*, 1982). The action of sulfated GAG on FSH-stimulated cumulus cell functions is differential; i.e., whereas they inhibit FSH-stimulated cumulus expansion, they do not inhibit FSH-stimulated phosphatidylcholine biosynthesis or progesterone secretion (Eppig *et al.*, 1982; Schultz *et al.*, 1983c).

The physiological context for an action of sulfated GAG in the regulation of cumulus cell function remains to be resolved. It has been hypothesized that sulfated GAG may inhibit the precocious expansion of the cumulus oophorus in response to FSH indigenous to the antral follicle before the preovulatory gonadotropin surge (Eppig, 1981b; Eppig *et al.*, 1982). The LH surge may reduce the concentration of sulfated GAG in follicular fluid to ineffective levels by reducing their synthetic rates and/or increasing the volume of the follicular fluid. Clearly, further experimentation is required to test these hypotheses. In addition, it remains to be demonstrated that native ovarian proteoglycans affect the expansion of cumuli oophori in a manner similar to the sulfated GAG moiety alone.

### 6.2. Cumulus Cell–Oocyte Uncoupling

The metabolic coupling of cumulus cells and the oocyte terminates before ovulation, but after GVBD (Zamboni, 1974; Dekel *et al.*, 1978; Gilula *et al.*, 1978; Moor *et al.*, 1981; Eppig, 1982a). The timing of cumulus cell–oocyte uncoupling and cumulus expansion has been determined after inducing ovulation in mice with biological grade hCG (Eppig, 1982a). The expansion of the cumulus oophorus is first seen at the outermost, peripheral cumulus cells and progresses, with time, inward to include the cumulus cells immediately adjacent to the oocyte. This sequence is also seen in the expansion of rat cumuli oophori (Dekel *et al.*, 1979). The mucification of the innermost cumulus cells does not occur until at least 9 hr after the injection of hCG (Eppig, 1982a). There is no reduction in the cumulus cell–oocyte coupling index until 9 hr after hCG injection (Eppig, 1982a). This coincidence suggests that the expansion of the cumulus cells adjacent to the oocyte may be the mechanism whereby coupling is terminated.

Follicle-stimulating hormone not only stimulates the expansion of isolated cumuli oophori, but promotes the uncoupling of the cumulus cells from the oocytes as well (Eppig, 1982b; Eppig and Ward-Bailey, 1982). However, it was found that sulfated GAG, which inhibits FSH-stimulated cumulus expansion (Eppig, 1981a), does not inhibit FSH-stimulated cumulus cell–oocyte uncoupling (Fig. 12) (Eppig and Ward-Bailey, 1982; Salustri and Siracusa, 1983). These results show that cumulus expansion is not the cause of cumulus cell-oocyte uncoupling. FSH stimulates changes in gran-

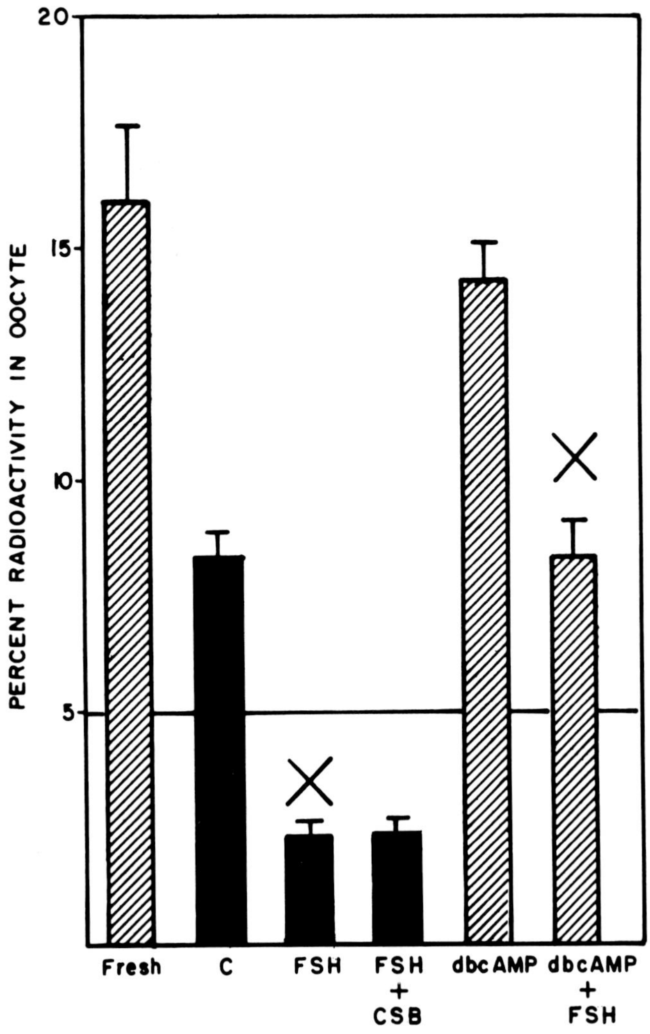


**Figure 12.** Factors affecting the metabolic cooperativity between cumulus cells and oocytes. In these experiments, metabolic cooperativity is expressed as the percentage of radioactive choline (or its metabolites) found in the oocyte relative to the radioactivity measured in the entire oocyte–cumulus cell complex. This method internally controls for any variations in the rate of uptake of the labeled choline into the cumulus cells such as occur in response to FSH or dbcAMP. When the percentage of radiolabel in the oocyte exceeds 5% of the label in the entire complex, the oocytes are considered to be coupled to the cumulus cells. The oocytes are considered to be uncoupled from the cumulus cells if less than 5% of the radiolabel is in the oocyte, since the same amount of label is found in denuded oocytes incubated in the same medium. (▨) Oocytes in the germinal vesicle stage at the time that metabolic cooperativity was measured. (■) Oocytes that had spontaneously matured. An X over the bar indicates that cumulus expansion had occurred in that group of complexes. Complexes were incubated for 18 hr in control medium (C), medium containing 2.5 $\mu$g/ml follicle-stimulating hormone (FSH), FSH plus 5 mg/ml chondroitin sulfate B (CSB), 0.25 $\mu$M dibutyryl cyclic adenosine monophosphate (dbcAMP), or dbcAMP plus FSH before the measurement of metabolic cooperativity. (Data from Eppig and Ward-Bailey, 1982.)

ulosa cell shape, apparently via a cAMP-mediated action on the cell's cytoskeletal system (Lawrence *et al.*, 1979). Perhaps cumulus cell–oocyte uncoupling is brought about by similar FSH-induced actions on the cumulus cell cytoskeleton.

Partial uncoupling of cumulus cells and oocytes occurs during spontaneous maturation in culture (Eppig and Ward-Bailey, 1982). When spontaneous maturation is prevented by addition of dbcAMP or IBMX, at concentrations below those necessary to promote cumulus expansion, partial uncoupling is prevented (Fig. 12). FSH induces complete uncoupling. However, when FSH is added to the cultures containing dbcAMP or IBMX, only partial uncoupling occurs (Eppig and Ward-Bailey, 1982). In the experiments, FSH did not reverse the inhibitory action of dbcAMP or IBMX on GVBD. Therefore, these results suggest that the GV-stage oocyte plays a role in the maintenance of cumulus cell–oocyte metabolic cooperativity, since complete uncoupling is observed only in oocytes that have undergone GVBD.

## 7. Role of Somatic Cell–Oocyte Interactions in Ovarian Teratocarcinogenesis and Parthenogenetic Development

About one-half the females of mouse strain LT/Sv acquire spontaneous ovarian teratomas (Stevens and Varnum, 1974). Striking evidence has been presented that these teratomas are derived from oocytes that develop parthenogenetically within the ovaries. LT/Sv ovaries contain normal-appearing cleaved eggs and embryos up to the blastocyst stage. Later embryonic stages become disorganized and subsequently develop into teratomas. In addition, 10–20% of the ovulated oocytes from strain LT/Sv mice undergo spontaneous activation, and implanted embryos are found in virgin females. Occasionally these implanted parthenotes develop to the primitive streak stage before disorganization and finally death ensue.

The parthenogenetic embryos that give rise to ovarian teratomas in strain LT/Sv mice develop in atypical follicles deficient in the number of granulosa cells that surround the developing oocyte (Fig. 13) (Eppig, 1978). Parthenogenesis occurs, therefore, both after ovulation from normal graafian follicles and in the atypical granulosa cell-deficient follicles. Junctional complexes are found between the apposing membranes of the granulosa cells and oocytes of granulosa cell-deficient folicles, indicating the potential for cell-to-cell communication is similar to that described for oocytes in normal follicles.

Parthenogenetic activation and development to blastocyst have been demonstrated after the spontaneous maturation of LT/Sv oocytes *in vitro*

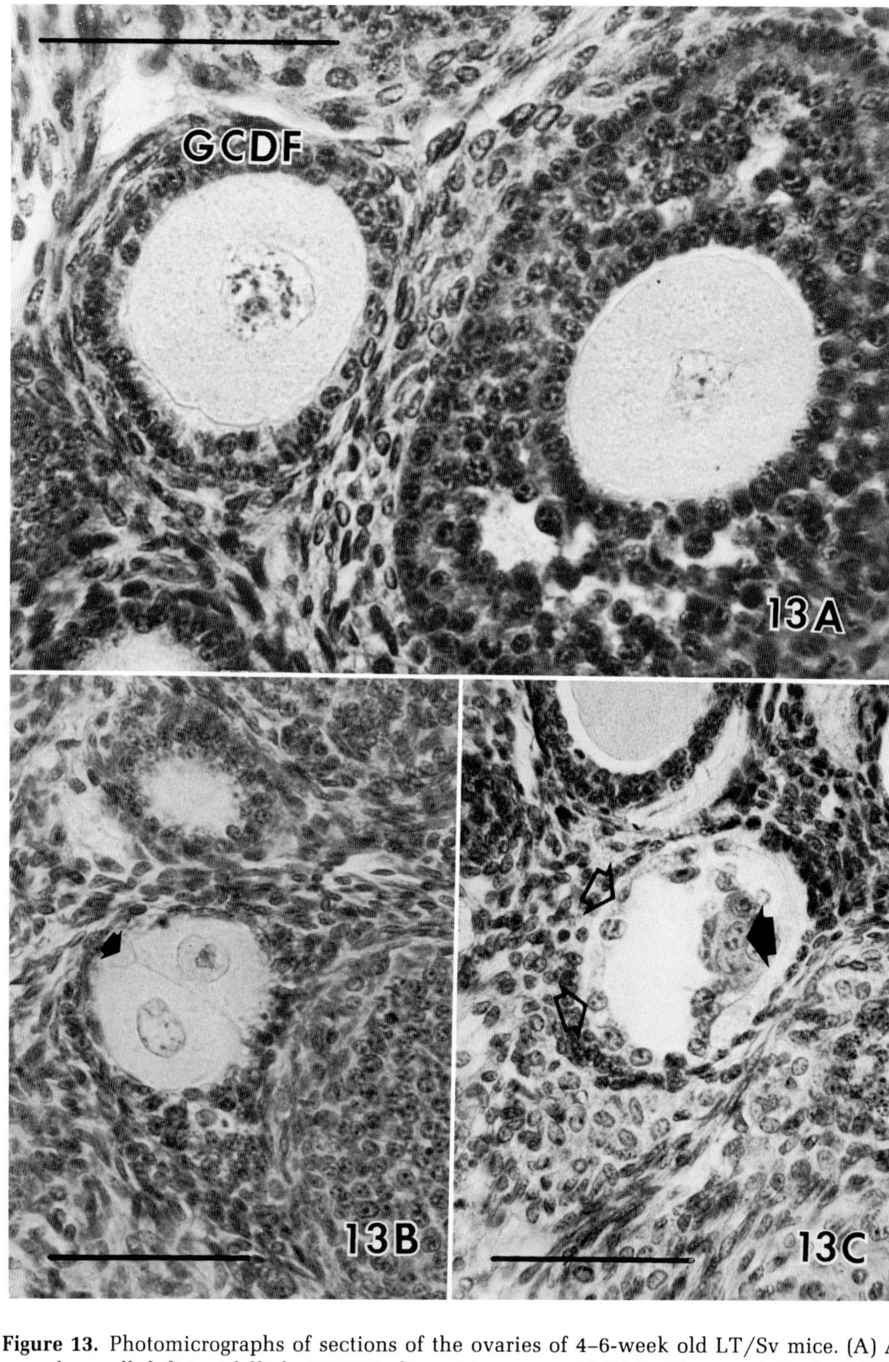


**Figure 13.** Photomicrographs of sections of the ovaries of 4–6-week old LT/Sv mice. (A) A granulosa cell-deficient follicle (GCDF) adjacent to a normal follicle. The oocyte in the GCDF is the same size as the full-grown oocyte in the normal follicle. While the oocyte in the normal

(Eppig, 1981d). In addition, $F_1$ hybrids of LT/Sv with several other inbred strains produce ova that undergo parthenogenetic development after maturation *in vitro*. The frequency of parthenogenetic development is directly related to the duration of metabolic cooperativity between the cumulus cells and the maturing oocyte (Fig. 14) (Eppig, 1982b). Therefore, parthenogenetic activation and preimplantation embryonic development appear to be dependent on the interaction of the maturing oocyte with its companion somatic cells. It is not known whether the potential for parthenogenetic activation is intrinsic to the LT/Sv oocyte or whether the somatic cells provide the stimulus for activation. The somatic cells may also provide the maturing oocytes with metabolites needed to support subsequent embryogenesis.

The model derived from the previously described studies on the control of meiotic maturation (see Fig. 8) may also provide an explanation for the mechanism for parthenogenesis in the granulosa cell-deficient follicles of strain LT/Sv mice. In that model, the large number of granulosa cells in the wall of normal antral follicles constitutes the principal source of the putative oocyte maturation inhibiting factor. Hence, the prolonged maintenance of meiotic arrest probably requires the participation of this large mass of mural granulosa cells. In strain LT/Sv, the granulosa cell-deficient follicles probably do not provide a sufficient amount of inhibiting factor to maintain meiotic arrest. Consequently, when the oocytes reach the stage of development in which they are competent to resume meiosis, they do so because there is an insufficient mass of granulosa cells to provide them with inhibitor. Some fraction of these oocytes maturing in the granulosa cell-deficient follicles subsequently becomes parthenogenetically activated and has the potential for developing into teratomas.

ACKNOWLEDGMENTS The research conducted in the author's laboratory was supported by grants from the National Science Foundation and the American Cancer Society. A special thanks to Dr. Everett Anderson and Dr. Richard Schultz for the use of their micrographs and data and to Dr. Richard Schultz, Dr. Steven Downs, Dr. Patricia Ward-Bailey, Dr. Janan Eppig, and Dr. Joe Nadeau for their helpful comments on this chapter.

---

follicle is surrounded by many layers of granulosa cells, the oocyte in the GDCF is enclosed by only one to three layers of granulosa cells. (From Eppig, 1978.) (B) Parthenogenetically cleaved egg in the ovary of a 4-week-old LT/Sv mouse. The first polar body can be seen (arrow) in addition to the two-blastomeres, indicating that the first meiotic division had occurred. Note that the parthenote is in a granulosa cell-deficient follicle. (C) A blastocyst developed from a parthenogenetically activated egg within the ovary of a 6-week-old LT/Sv mouse. The blastocyst has a normal-appearing inner cell mass (closed arrow) and trophectoderm (open arrow). Blastocysts such as this one sometimes give rise to ovarian tetratomas. Scale bars: 100 $\mu$m.

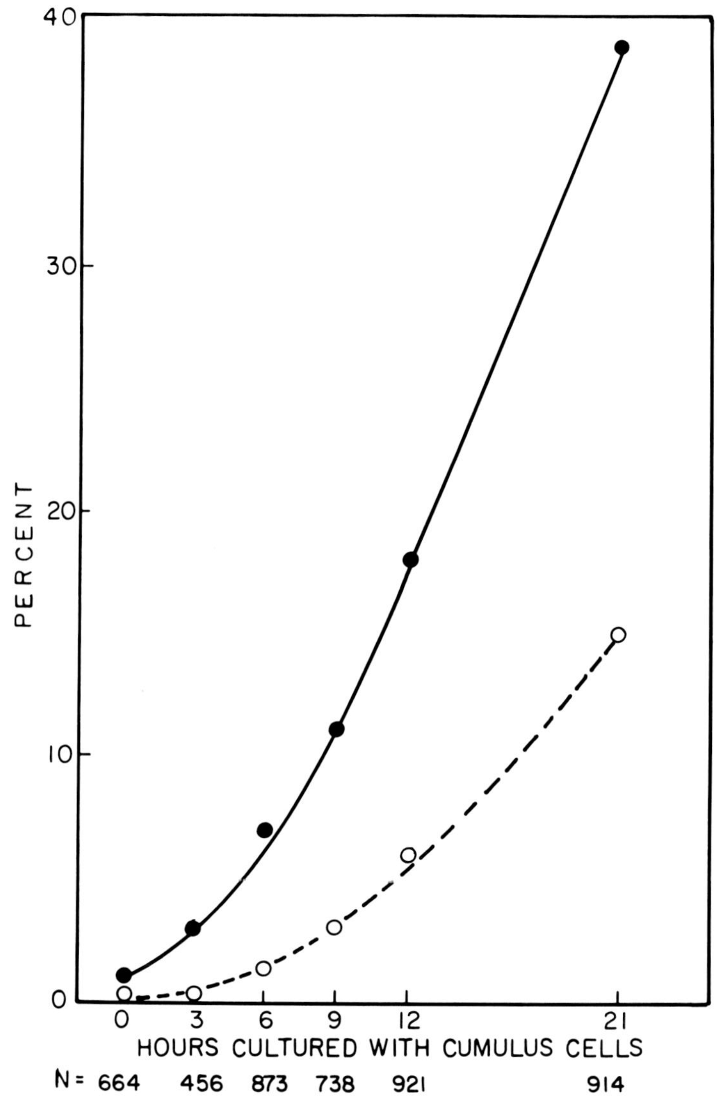


**Figure 14.** Effect of cumulus cells on the acquisition of parthenogenetic developmental competence by maturing (C57BL/6J × LT/Sv)$F_1$ oocytes. Cumulus cells were removed from oocytes maturing *in vitro* at the times indicated, and the percentage of ova that subsequently cleaved and developed parthenogenetically was determined. Solid circles (●) indicate percent of ova that cleaved to the 2-cell stage, while open circles (○) show percent of ova that developed into blastocysts. Clearly, the frequency of parthenogenetic development is related to the interactions of cumulus cells with the maturing oocyte. (From Eppig, 1982b.)

## References

Albertini, D. F., and Anderson, E., 1974, The appearance and structure of the intercellular connections during the ontogeny of the rabbit ovarian follicle with special reference to gap junctions, *J. Cell Biol.* **63**:234–250.

Amsterdam, A., Koch, Y., Lieberman, M. E., and Linder, H. R., 1975, Distribution of binding sites for human chorionic gonadotropin in the preovulatory follicle of the rat, *J. Cell Biol.* **67**:894–900.

Anderson, E., and Albertini, D. F., 1976, Gap junctions between the oocyte and companion follicle cells in the mammalian ovary, *J. Cell Biol.* **71**:680–686.

Anderson, E., Wilkinson, R. F., Lee, G., and Meller, S., 1978, A correlative microscopical analysis of differentiating ovarian follicles of mammals, *J. Morphol.* **156**:339–366.

Ax, R. L., and Ryan, R. J., 1979, The porcine ovarian follicle. IV. Mucopolysaccharides at different stages of development, *Biol. Reprod.* **20**:1123–1132.

Bae, I. H., and Foote, R. H., 1975, Effects of hormones on the maturation of rabbit oocytes recovered from follicles of various sizes, *J. Reprod. Fertil.* **42**:357–360.

Ball, G. D., Bellin, M. E., Ax, R. L., and First, N. L., 1982, Glycosaminoglycans in bovine cumulus-oocyte complexes: Morphology and chemistry, *Mol. Cell. Endocrinol.* **28**:113–122.

Biggers, J. D., and Powers, R. D., 1979, Comments on the control of meiotic maturation in mammals, in: *Ovarian Follicular Development and Function* (A. R. Midgely and W. A. Sadler, eds.), pp. 365–373, Raven Press, New York.

Brower, P. T., and Schultz, R. M., 1982, Intercellular communication between granulosa cells and mouse oocytes: Existence and possible nutritional role during oocyte growth, *Dev. Biol.* **90**:144–153.

Canipari, R., Palombi, F., Riminucci, M., and Mangia, F., 1984, Early programming of maturation competence in mouse oogenesis, *Dev. Biol.* **102**:519–524.

Centola, G. M., Anderson, L. D., and Channing, C. P., 1981, Oocyte maturation inhibition activity in porcine granulosa cells, *Gamete Res.* **4**:451–462.

Channing, C. P., Bae, I. H., Stone, S. L., Anderson, L. D., Edelson, S., and Fowler, S. C., 1981, Porcine granulosa and cumulus cell properties. LH/hCG receptors, ability to secrete progesterone and ability to respond to LH, *Mol. Cell. Endocrinol.* **22**:359–370.

Chari, S., Hillensjo, T., Magnusson, C., Sturm, G., and Daume, E., 1983, In vitro inhibition of rat oocyte meiosis by human follicular fluid fractions, *Arch. Gynecol.* **233**:155–164.

Cho, W. K., Stern, S., and Biggers, J. D., 1974, Inhibitory effect of dibutyryl cAMP on mouse maturation in vitro, *J. Exp. Zool.* **187**:383–386.

Cross, P. C., and Brinster, R. L., 1970, In vitro development of mouse oocytes, *Biol. Reprod.* **3**:298–307.

Dekel, N., and Beers, W. H., 1978, Rat oocyte maturation in vitro: relief of cyclic AMP inhibition with gonadotropins, *Proc. Natl. Acad. Sci. USA* **75**:4369–4373.

Dekel, N., and Beers, W. H., 1980, Development of the rat oocyte in vitro: inhibition and induction of maturation in the presence or absence of the cumulus oophorus, *Dev. Biol.* **75**:247–254.

Dekel, N., and Kraicer, P. F., 1978, Induction in vitro of mucification of rat cumulus oophorus by gonadotropins and adenosine 3′,5′-monophosphate, *Endocrinology* **102**:1797–1802.

Dekel, N., Kraicer, P. F., Phillips, D. M., Sanchez, R. S., and Segal, S. J., 1978, Cellular associations in the rat oocyte–cumulus cell complex: Morphology and ovulatory changes, *Gamete Res.* **1**:47–57.

Dekel, N., Hillensjo, T., and Kraicer, P. F., 1979, Maturational effects of gonadotropins on the cumulus–oocyte complex of the rat, *Biol. Reprod.* **20**:191–197.

Dekel, N., Lawrence, T. S., Gilula, N. B., and Beers, W. H., 1981, Modulation of cell-to-cell communication in the cumulus–oocyte complex and the regulation of oocyte maturation by LH, *Dev. Biol.* **80**:356–362.

Downs, S. M., and Eppig, J. J., 1984, Cyclic adenosine monophosphate and ovarian follicular fluid act synergistically to inhibit mouse oocyte maturation, *Endocrinology* **114:**418–427.

Downs, S. D., and Longo, F. J., 1983, Prostaglandins and preovulatory maturation in mice, *J. Exp. Zool.* **288:**99–108.

Eppig, J. J., 1977, Mouse oocyte development in vitro with various culture systems, *Dev. Biol.* **60:**371–388.

Eppig, J. J., 1978, Granulosa cell deficient follicles. Occurrence, structure, and relationship to ovarian teratocarcinogenesis in strain LT/Sv mice, *Differentiation* **12:**111–120.

Eppig, J. J., 1979a, A comparison between oocyte growth in coculture with granulosa cells and oocytes with granulosa cell–oocyte junctional contact maintained in vitro, *J. Exp. Zool.* **209:**345–353.

Eppig, J. J., 1979b, Gonadotropin stimulation of the expansion of cumulus oophori isolated from mice: General conditions for expansion in vitro, *J. Exp. Zool.* **208:**111–120.

Eppig, J. J., 1979c, FSH stimulates hyaluronic acid synthesis by oocyte–cumulus cell complexes from mouse preovulatory follicles, *Nature* **281:**483–484.

Eppig, J. J., 1980, Regulation of cumulus oophorus expansion by gonadotropins in vivo and in vitro, *Biol. Reprod.* **23:**545–552.

Eppig, J. J., 1981a, Prostaglandin $E_2$ stimulates cumulus expansion and hyaluronic acid synthesis by cumuli oophori isolated from mice, *Biol. Reprod.* **25:**191–195.

Eppig, J. J., 1981b, Ovarian glycosaminoglycans: Evidence for a role in regulating the response of the oocyte–cumulus cell complex to FSH, *Endocrinology* **108:**1992–1994.

Eppig, J. J., 1981c, Regulation by sulfated glycosaminoglycans of the expansion of cumuli oophori isolated from mice, *Biol. Reprod.* **25:**599–608.

Eppig, J. J., 1981d, Preimplantation embryonic development of spontaneous mouse parthenotes after oocyte meiotic maturation in vitro, *Gamete Res.* **4:**3–13.

Eppig, J. J., 1982a, The relationship between cumulus cell–oocyte coupling, oocyte meiotic maturation, and cumulus expansion, *Dev. Biol.* **89:**268–272.

Eppig, J. J., 1982b, The relationship between parthenogenetic embryonic development and cumulus cell–oocyte intercellular coupling during oocyte meiotic maturation, *Gamete Res.* **5:**229–237.

Eppig, J. J., and Koide, S. L., 1978, Effects of progesterone and oestradiol-17$\beta$ on the spontaneous meiotic maturation of mouse oocytes, *J. Reprod. Fertil.* **53:**99–101.

Eppig, J. J., and Ward-Bailey, P. F., 1982, The mechanism of cumulus cell–oocyte uncoupling: Evidence for the participation of both cumulus cells and oocytes, *Gamete Res.* **6:**145–154.

Eppig, J. J., and Ward-Bailey, P. F., 1984, Sulfated glycosaminoglycans inhibit hyaluronic acid synthesizing activity in mouse cumuli oophori, *Exp. Cell Res.* **50:**459–465.

Eppig, J. J., Ward-Bailey, P. F., Potter, J. E. R., and Schultz, R. M., 1982, Differential action of sulfated glycosaminoglycans on follicle-stimulating hormone-induced functions of cumuli oophori isolated from mice, *Biol. Reprod.* **27:**399–406.

Eppig, J. J., Freter, R. R., Ward-Bailey, P. F., and Schultz, R. M., 1983, Inhibition of oocyte maturation in the mouse: Participation of cAMP, steroids, and a putative maturation inhibitory factor, *Dev. Biol.* **100:**39–49.

Fleming, A. D., Khalil, W., and Armstrong, D. T., 1983, Porcine follicular fluid does not inhibit maturation of rat oocytes *in vitro*, *J. Reprod. Fertil.* **69:**665–670.

Fulka, J., Jr., Pavlok, A., and Fulka, J., 1982, In vitro fertilization of zona-free bovine oocytes matured in culture, *J. Reprod. Fertil.* **64:**495–499.

Gilula, N. B., Reeves, O. R., and Steinbach, A., 1972, Metabolic coupling, ionic coupling and cell contacts, *Nature* **235:**262–265.

Gilula, N. B., Epstein, M. L., and Beers, W. H., 1978, Cell-to-cell communication and ovulation: A study of the cumulus cell-oocyte complex, *J. Cell Biol.* **78:**58–75.

Gould, K. G., and Graham, C. E., 1976, Maturation in vitro of oocytes recovered from prepubertal rhesus monkeys, *J. Reprod. Fertil.* **46:**269–270.

Gwatkin, R. B. L., and Andersen, O. F., 1976, Hamster oocyte maturation in vitro: Inhibition by follicular components, *Life Sci.* **19**:527–536.
Heller, D. T., Cahill, D. M., and Schultz, R. M., 1981, Biochemical studies of mammalian oogenesis: Metabolic cooperativity between granulosa cells and growing mouse oocytes, *Dev. Biol.* **84**:455–464.
Hillensjo, T., Channing, C. P., Pomerantz, S. H., and Schwartz-Kripmer, A., 1979, Intrafollicular control of oocyte maturation in the pig, *In Vitro* **15**:32–39.
Hillensjo, T., and Channing, C. P., 1980, Gonadotropin stimulation of steroidogenesis and cellular dispersion in cultured porcine cumuli oophori, *Gamete Res.* **3**:223–240.
Iwamatsu, T., and Chang, M. C., 1972, Sperm penetration in vitro of mouse oocytes at various times during maturation, *J. Reprod. Fertil.* **31**:237–247.
Jagiello, G., Graffeo, J., Ducayen, M., and Prosser, R., 1977, Further studies of inhibitors of in vitro, mammalian oocyte maturation, *Fertil. Steril.* **28**:476–481.
Kaplan, G., Abreu, S. L., and Bachvarova, R., 1982, rRNA accumulation and protein synthetic patterns in growing mouse oocytes, *J. Exp. Zool.* **220**:361–380.
Lawrence, T. S., Beers, W. H., and Gilula, N. B., 1978, Transmission of hormonal stimulation by cell-to-cell communication, *Nature* **272**:501–506.
Lawrence, T. S., Ginzberg, R. D., Gilula, N. B., and Beers, W. H., 1979, Hormonally induced cell shape changes in cultured rat ovarian granulosa cells, *J. Cell Biol.* **80**:21–36.
Lawrence, T. S., Dekel, N., and Beers, N. B., 1980, Binding of human chorionic gonadotropin by rat cumuli oophori and granulosa cells: A comparative study, *Endocrinology* **106**:1114–1118.
Leibfried, L., and First, N. L., 1980a, Follicular control of meiosis in the porcine oocyte, *Biol. Reprod.* **23**:705–709.
Leibfried, L., and First, N. L., 1980b, Effect of bovine and porcine follicular fluid and granulosa cells on maturation of oocytes in vitro, *Biol. Reprod.* **23**:699–704.
Magnusson, C., and Hillensjo, T., 1977, Inhibition of maturation and metabolism of rat oocytes by cyclic AMP, *J. Exp. Zool.* **201**:138–147.
McGaughey, R. W., 1977, The culture of pig oocytes in minimal medium, and the influence of progesterone and estradiol-17 on meiotic maturation, *Endocrinology* **100**:39–45.
Moor, R. M., and Trounson, A. O., 1977, Hormonal and follicular factors affecting maturation of sheep oocytes in vitro and their subsequent developmental capacity, *J. Reprod. Fertil.* **49**:101–109.
Moor, R. M., Osborn, J. C., Cran, D. G., and Walters, D. E., 1981, Selective effect of gonadotrophins on cell coupling, nuclear maturation and protein synthesis in mammalian oocytes, *J. Embryol. Exp. Morphol.* **61**:347–365.
Motlik, J., and Fulka, J., 1981, Fertilization of rabbit oocytes co-cultured with granulosa cells, *J. Reprod. Fertil.* **63**:425–429.
Mukherjee, A. B., 1972, Normal progeny from fertilization in vitro of mouse oocytes matured in culture and spermatozoa capacitated in vitro, *Nature* **237**:397–398.
Nishimoto, T., Yamada, I., Niwa, K., Mori, T., Nishimura, T., and Iritani, A., 1982, Sperm penetration in vitro of human oocytes matured in a chemically defined medium, *J. Reprod. Fertil.* **64**:115–119.
Niwa, K., and Chang, M. C., 1975, Fertilization of rat eggs in vitro at various times before and after ovulation with special reference to fertilization of ovarian oocytes matured in culture, *J. Reprod. Fertil.* **43**:435–451.
Phillips, D. M., and Dekel, N., 1982, Effect of gonadotropins and prostaglandin on cumulus mucification in cultures of intact follicles, *J. Exp. Zool.* **221**:275–282.
Pincus, G., and Enzmann, E. V., 1935, The comparative behavior of mammalian eggs in vivo and in vitro. I. The activation of ovarian eggs, *J. Exp. Med.* **62**:665–675.
Pitts, J. D., and Simms, J. W., 1977, Permeability of junctions between animal cells: Intercellular transfer of nucleotides but not macromolecules, *Exp. Cell Res.* **104**:153–163.
Racowsky, C., and McGaughey, R. W., 1982a, Further studies of the effects of follicular fluid

and membrana granulosa cells on the spontaneous maturation of pig oocytes, *J. Reprod. Fertil.* **66**:505–512.

Racowsky, C., and McGaughey, R. W., 1982b, In the absence of protein, estradiol suppresses meiosis of porcine oocytes in vitro, *J. Exp. Zool.* **224**:103–110.

Raven, C. P., 1961, *Oogenesis*, Pergamon Press, New York.

Rice, C., and McGaughey, R. W., 1981, Effect of testosterone and dibutyryl cAMP on the spontaneous maturation of pig oocytes, *J. Reprod. Fertil.* **62**:245–256.

Salustri, A., and Siracusa, G., 1983, Metabolic coupling cumulus expansion and meotic resumption in mouse cumuli oophori cultured *in vitro* in the presence of FSH or dbcAMP, or stimulated *in vivo* by hCG, *J. Reprod. Fert.* **68**:335–341.

Schroeder, A. C., and Eppig, J. J., 1984, The developmental capacity of mouse oocytes that matured spontaneously *in vitro* is normal, *Dev. Biol.* **102**:493–497.

Schultz, R. M., and Wassarman, P. M., 1977, Specific changes in the pattern of protein synthesis during meiotic maturation of mammalian oocytes in vitro, *Proc. Nat. Acad. Sci. USA* **74**:538–541.

Schultz, R. M., Letourneau, G. E., and Wassarman, P. M., 1979, Program of early development in the mammal: Changes in the patterns and absolute rates of tubulin and total protein synthesis during oocyte growth in the mouse, *Dev. Biol.* **73**:120–133.

Schultz, R. M., Montgomery, R., and Belanoff, J., 1983a, Regulation of mouse oocyte meiotic maturation: Implication of a decrease in oocyte cAMP and protein dephosphorylation in commitment to resume meiosis, *Dev. Biol.* **97**:264–273.

Schultz, R. M., Montgomery, R., Ward-Bailey, P. F., and Eppig, J. J., 1983b, Regulation of oocyte maturation in the mouse: Possible roles of intercellular communication, cAMP and testosterone, *Dev. Biol.* **95**:294–304.

Schultz, R. M., Ward-Bailey, P. F., and Eppig, J. J., 1983c, Site of action for the suppression of follicle-stimulating hormone-induced cumulus expansion by sulfated glycosaminoglycans, in: *Factors Regulating Ovarian Function* (G. S. Greenwald and P. E. Terranova, eds), pp. 71–74, Raven Press, New York.

Shalgi, R., Dekel, N., and Kraicer, P. F., 1979, The effect of LH on the fertilizability and developmental capacity of rat oocytes matured in vitro, *J. Reprod. Fertil.* **55**:429–435.

Smith, D. M., and Tenney, D. Y., 1980, Effect of steroids on mouse oocyte maturation in vitro, *J. Reprod. Fertil.* **60**:331–338.

Stevens, L. C., and Varnum, D. S., 1974, The development of teratomas from parthenogenetically activated ovarian eggs, *Dev. Biol.* **37**:369–380.

Thibault, C., 1972, Final stages of mammalian oocyte maturation, in: *Oogenesis* (J. D. Diggers and A. W. Schuetz, eds.), pp. 397–411, University Park Press, Baltimore.

Thibault, C., 1977, Are follicular maturation and oocyte maturation independent processes?, *J. Reprod. Fertil.* **51**:1–15.

Trounson, A. O., Willadsen, S. M., and Rowson, L. E. A., 1977, Fertilization and developmental capacity of bovine follicular oocytes matured in vitro and in vivo and transferred to the oviducts of rabbits and cows, *J. Reprod. Fertil.* **51**:321–327.

Tsafriri, A., 1978, Oocyte maturation in mammals, in: *The Vertebrate Ovary, Comparative Biology and Evolution* (R. E. Jones, ed.), pp. 409–442, Plenum Press, New York.

Tsafriri, A., and Channing, C. P., 1975, An inhibitory influence of granulosa cells and follicular fluid upon porcine oocyte meiosis in vitro, *Endocrinology* **96**:922–927.

Tsafriri, A., Pomerantz, S., and Channing, C. P., 1976, Inhibition of oocyte maturation by porcine follicular fluid: Partial characterization of the inhibitor, *Biol. Reprod.* **14**:511–516.

Tsafriri, A., Channing, C. P., Pomerantz, S. H., and Lindner, H. R., 1977, Inhibition of maturation of isolated rat oocytes by porcine follicular fluid, *J. Endocrinol.* **75**:285–291.

Tsafriri, A., Dekel, N., and Bar-Ami, S., 1982, The role of oocyte maturation inhibitor in follicular regulation of oocyte maturation, *J. Reprod. Fertil.* **64**:541–551.

Van Blerkom, J., and McGaughey, R. W., 1978a, Molecular differentiation of the rabbit ovum. I. During oocyte maturation in vivo and in vitro, *Dev. Biol.* **63**:139–150.

Van Blerkom, J., and McGaughey, R. W., 1978b, Molecular differentiation of the rabbit ovum. II. During the preimplantation development of in vivo and in vitro matured oocytes, *Dev. Biol.* **63**:151–164.

Warnes, G. M., Moor, R. M., and Johnson, M. H., 1977, Changes in protein synthesis during maturation of sheep oocytes in vivo and in vitro, *J. Reprod. Fertil.* **49**:331–335.

Wassarman, P. M., Josefowicz, W. J., and Letourneau, G. E., 1976, Meiotic maturation of mouse oocytes in vitro: Inhibition of maturation at specific stages of nuclear progression, *J. Cell Sci.* **22**:531–545.

Yanagishita, M. M., Rodbard, D., and Hascall, V. C., 1979, Isolation and characterization of proteoglycans from porcine ovarian follicular fluid, *J. Biol. Chem.* **254**:911–920.

Zamboni, L., 1974, Fine morphology of the follicle wall and follicle cell–oocyte association, *Biol. Reprod.* **10**:125–149.

Zeleznik, A. J., Midgley, A. R., Jr., and Reichert, L. E., Jr., 1974, Granulosa cell maturation in the rat: Increased binding of human chorionic gonadotropin following treatment with follicle stimulating hormone in vivo, *Endocrinology* **95**:818–825.

# II

# Gene Expression: Regulation and Consequences for Oogenesis and Early Development

Chapter 8

# RNA Synthesis and Storage During Insect Oogenesis

SPENCER J. BERRY

## 1. Introduction

At fertilization, insect embryos contain large amounts of RNA in the form of transfer RNA and ribosomes to support protein synthesis, in addition to the mRNA necessary to program this activity (see review by Berry, 1982, for specific examples). The exclusive transcription of these RNAs on the maternal—rather than the zygote—genome is evident from several diverse types of experiments. Inhibitors of nucleic acid synthesis have very little effect on pregastrulation development. The necessary role of protein synthesis during these same stages is apparent from the extreme sensitivity of embryos to inhibitors of protein synthesis (Lockshin, 1966). Incorporation of amino acids into proteins during cleavage can be demonstrated to be extensive, while little incorporation of nucleotides into RNA can be observed (Fig. 1).

All animal embryos examined to date rely predominantly on stored RNA synthesized during oogenesis for the synthesis of new proteins required throughout the cleavage stages up to a point shortly before gastrulation (Davidson, 1976). Gerhart (1980) described a period termed **mid-blastula transition** (MBT) in amphibians during which the developing embryo begins extensive transcription of its own genome.

The actual information content of maternal mRNA is very poorly understood. Hough-Evans *et al.* (1980) measured the sequence complexity for stored oocyte mRNA of the housefly *Musca domestica* and reported a value of $2.4 \times 10^7$. If all this mRNA represented translatable messages, it would code for 12,000 different polypeptides of average size. The number of transcripts for each unique sequence is estimated to be 760,000, so this represents a substantial pool of information both in volume and in diver-

SPENCER J. BERRY • Department of Biology, Hall-Atwater and Shanklin Laboratories, Wesleyan University, Middletown, Connecticut 06457.

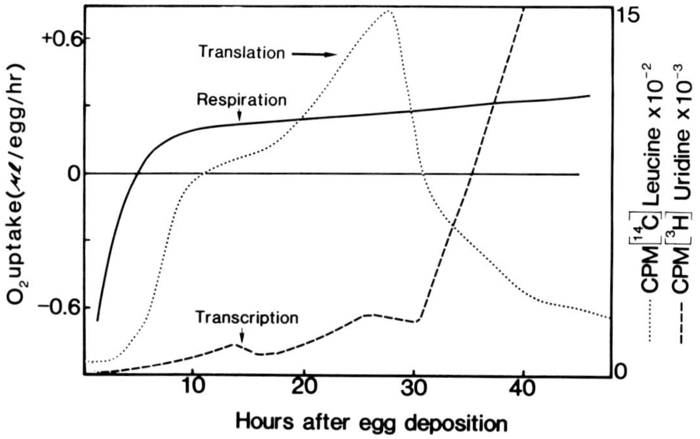


**Figure 1.** Some metabolic parameters of early development of the embryos of the tobacco hornworm, *Manduca sexta*. Since fertilization is internal, the initial rates of translation and transcription were determined on unfertilized eggs excised from the ovary. Transcription and translation were measured by direct injection of labeled precursors into the developing embryos. Respiration was measured in a Warburg respirometer. The syncytial blastoderm stage occurs at 20–24 hr, gastrulation at 30–35 hr.

sity. Some maternal messengers may be of general, rather than specifically developmental, significance. For example, Jäckle (1980b) identified actin as one of the polypeptides synthesized on maternal mRNA in *Smittia*. However, certain maternal gene products play specific roles in development, as demonstrated by the influence of maternal-effect mutant genes (see Chapter 13).

Insect embryos are excellent material in which to study the role of maternal RNAs for a number of compelling reasons. Relatively well-synchronized embryos of a number of insects can be obtained in large quantities throughout the year. The unparalleled number of maternal-effect mutants in *Drosophila*, and to some extent the commercial silkworm, *Bombyx mori*, present the potential for identifying both normal and mutant gene products and for investigating their mechanism of action during development. Many insect eggs are relatively large, facilitating such manipulations as the injection of exogenous nucleic acids, labeled precursors, or inhibitors (e.g., actinomycin D and puromycin). Smaller eggs can often be permeabilized to permit the introduction of similar compounds when injection would be difficult (Limbourg and Zalokar, 1973). Such manipulated embryos can be expected to survive to definable end points, such as the larva or the adult. Recently, it was found that some insect embryos can be disrupted and that the cell-free preparations from disrupted embryos can faithfully reproduce certain biochemical events that occur in the intact embryos. Finally, in a number of insect ovaries, oocyte

RNA is synthesized in trophic cells rather than in the oocyte, which may permit study of the transcription process separated from the context of other aspects of oogenesis, such as the accumulation of yolk.

## 2. Organization of Ovarian Follicles

Insect ovaries are divided into three general classes that coincidentally reflect the source of oocyte RNA. These three types of ovary are illustrated in Fig. 2 (see also Chapter 2). The somewhat indirect rationale for the nomenclature is summarized by Telfer (1975) in his indispensable review of polytrophic follicles. **Panoistic follicles** represent the least evolved mechanism in which the RNA is transcribed on the chromosomes of the egg nucleus and sometimes on extrachromosomal DNA found transiently in the oocyte nucleus. This mode is directly comparable to the mechanisms observed in amphibian and echinoderm oocytes. Often lampbrush chromosomes and supernumerary nucleoli are associated with panoistic oocytes. Panoistic follicles are associated with more "primitive" insects, such as dragonflies and cockroaches. In **meroistic follicles,** the site of RNA and ribonucleoprotein (RNP) synthesis is in nuclei other than the oocyte nucleus (germinal vesicle). Two types of meroistic ovaries are generally recognized: **polytrophic** and **telotrophic.** Polytrophic follicles contain a variable number of nurse cells within the follicle, whereas telotrophic oocytes remain attached to a syncytium of trophic cells by cytoplasmic extensions, the trophic cords. The trophic syncytium is outside the follicle proper, at the anterior end of the ovary. In polytrophic ovaries, the genome is reiterated enormously (1000 or more times) in each nurse cell nucleus; similar amplification is accomplished by the trophic syncytium in telotrophic ovaries.

To support this extensive transcriptional activity in nurse cells, the DNA is synthesized during the development of the follicle; incorporation of tritiated thymidine can be demonstrated by autoradiography. The specific patterns of incorporation vary in different species (see Telfer, 1975), but incorporation can be detected up to the point described by Pollak and Telfer (1969), when the nurse cell cytoplasm is injected into the ooplasm, and nurse cell nuclei become pyknotic.

In addition to the oocyte and trophic cells, when present, the ovarian follicles are surrounded by a palisade of mesodermally derived follicle cells. Follicle cells facilitate the entry of bloodborne proteins into the oocyte and may synthesize some of the proteins of the oocyte (see Chapter 2). At the end of the period of yolk accumulation (vitellogenesis), the follicle cells synthesize the chorion proteins of the egg shell. There is no compelling evidence that follicle cells contribute to the nucleic acid content of the oocyte, however. The chromosomes of the germinal vesicle of panoistic

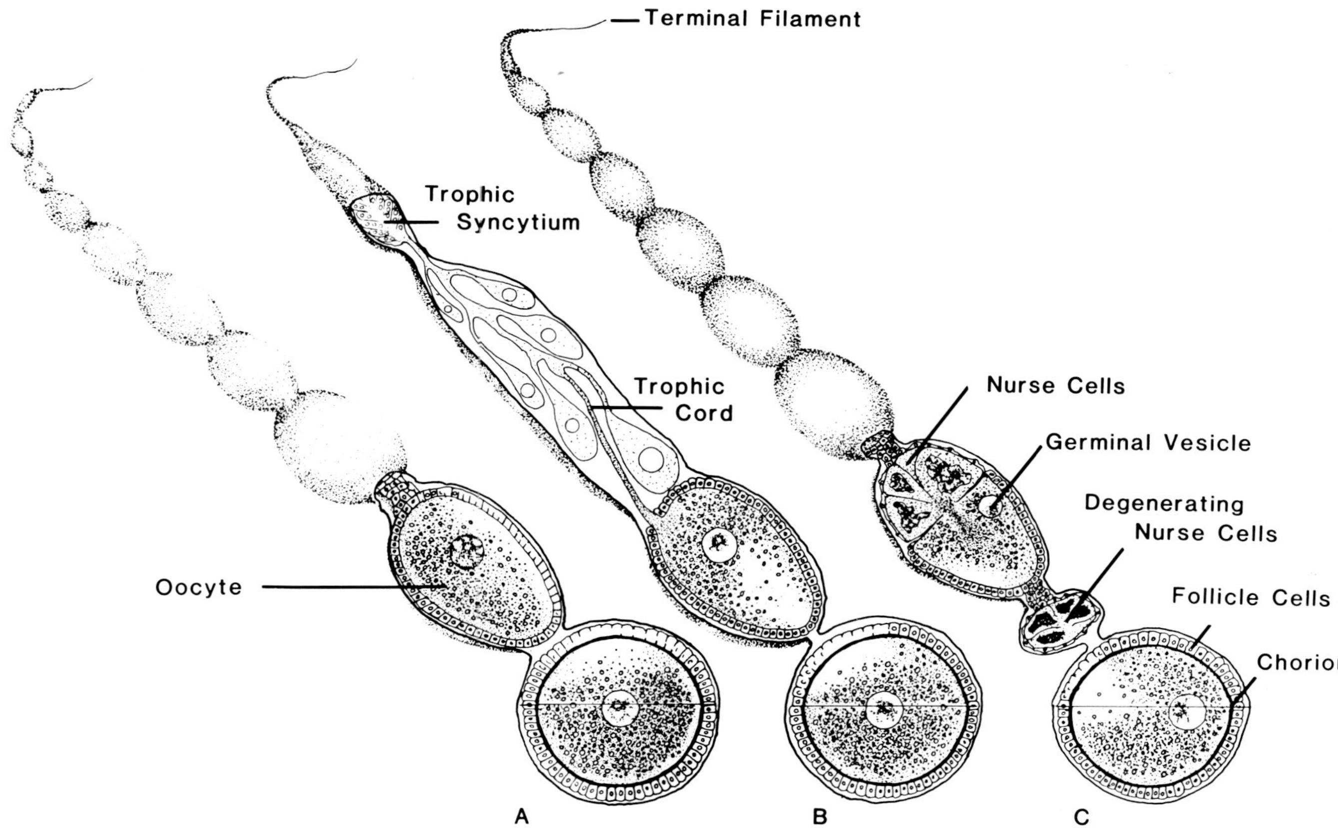


**Figure 2.** The three major types of insect ovary. (A) Panoistic. There are no trophic cells. (B) Meroistic-telotrophic. A trophic syncytium is connected to developing oocytes by means of trophic cords. (C) Meroistic-polytrophic. Trophic nurse cells are included in each ovarian follicle.

oocytes, the trophic cell chromosomes, and the extrachromosomal bodies are the major, probably the sole, template for nonmitochondrial RNA stored in oocytes.

The polytrophic ovaries of Diptera and Lepidoptera are the most extensively studied of the three types. This account focuses most extensively on follicles of this type; details concerning panoistic and telotrophic follicles are discussed where data are available. In addition to the review of polytrophic follicles by Telfer (1975), more extensive recent reviews have also been published (e.g., King *et al.*, 1982, on insect gamete formation; King and Büning, 1983, on oocyte and nurse cell formation; and Huebner, 1983, on development and ultrastructure of the telotrophic ovary).

## 2.1. Formation of Follicles

The formation of follicles in the dipteran *Drosophila melanogaster* and the lepidopteran *Hyalophora cecropia* have been described in detail by King and associates (Brown and King, 1964; King and Aggarwal, 1965; King *et al.*, 1968; King, 1970). Oogonia, which are located in the terminal tip of the ovariole, undergo divisions to produce stem oogonia and cystocytes. The oogonia continue the mitotic cycle, while the cystocytes undergo divisions with incomplete cytokinesis, leading to clusters of cells with cytoplasmic continuity through **ring canals** or **fusomes** and with nuclei containing the 4C DNA complements. In the fruitfly, *Drosophila*, four rounds of cystocyte division produce 16-cell clusters, whereas in the moth *Hyalophora cecropia* three rounds of division produce clusters of 8 cells. These clusters of cystocytes are then surrounded by mesodermal elements that will become the follicle cells. The follicle cell–cystocyte clusters enter the ovarioles and remain attached to other clusters by contacts among the mesodermal cells. The result is a series of developing follicles, with the oldest (and most mature) follicle nearest the oviduct and the youngest nearer the terminal filament. In many insects, a graded sequence of follicles is produced that recapitulates the development of the egg. In most insects, the mature, chorionated egg passes the spermathecal gland and is fertilized by stored sperm just before oviposition.

The geometry of the divisions that produce the cystocytes dictates that the first two daughter cells will have the largest number of cytoplasmic connections (three in the case of *H. cecropia* and four in the case of *D. melanogaster*). One of these two cells will become the oocyte, and the rest will become nurse cells. The molecular basis for determining which cell will become the oocyte is unknown, but a number of structural features have been observed to characterize the potential oocyte. In the case of the water beetle, *Dytiscus*, a large, extrachromosomal DNA mass, the **Giardina body,** forms in the prospective germinal vesicle at the two-cell stage (Giardina, 1901). Early formation of synaptonemal complexes by the chromosomes of

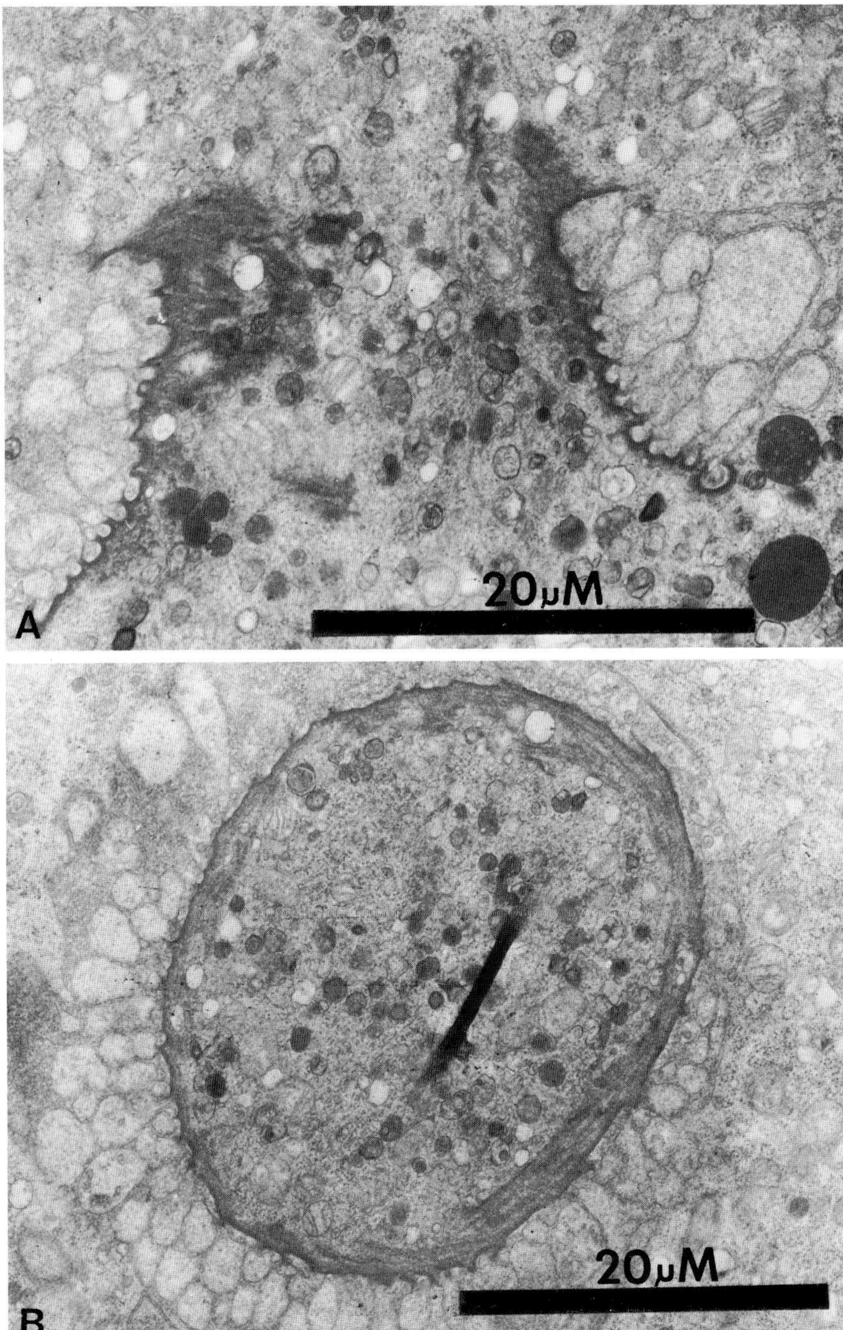


**Figure 3.** Intercytoplasmic connections between nurse cells and oocytes in the follicles in *Antheraea polyphemus*. (A) Longitudinal section. (B) Cross section. These sections are from early vitellogenic follicles.

the two prospective oocytes in *Drosophila* was described by Carpenter (1975). Mahowald and Strassheim (1970) have found that most of the centrioles in the *Drosophila* cystocyte cluster migrate to the cell that will become the oocyte.

Differences have also been detected in the structure of the ring canals formed at the different cystocyte divisions (Koch and King, 1969), but the general structure consists of interconnected cytoplasmic connections surrounded by, and sometimes containing, various types of fibrillar material (Fig. 3). It seems clear that the function of these cytoplasmic connections is to provide channels for the shipment of transcription products and perhaps other organelles such as mitochondria from the synthetically active nurse cells into the quiescent oocyte. The trophic syncytium in telotrophic ovaries serves the same function as nurse cells, and the trophic cords provide longer channels having functions similar to those of the ring canals.

## 2.2. Cytology of Follicles

Histochemical examination of the ovarian follicles reveals substantial cytoplasmic basophilia in the nurse cells and in the cells of the trophic syncytium, suggesting the presence of large amounts of nucleic acid. Conventional basophilic dyes stain the cytoplasm very darkly, and the basophilia is lost after ribonuclease treatment. This result is also consistent with the presence of large amounts of cytoplasmic RNA. Ultraviolet excitation of sections of *H. cecropia* nurse cells stained with acridine orange produces the intense orange fluorescence characteristic of RNA uniformly distributed in the cytoplasm. Electron micrographs show extraordinary amounts of free ribosomal material, relatively little membrane, modest numbers of mitochondria, and occasional stacks of annulate lamellae (Fig. 4). This distribution of cytoplasmic elements would be entirely consistent with a modest amount of constituent protein synthesis and the stockpiling of ribonucleoprotein particles. Fibrillar structures such as microtubules seem restricted to the regions around the ring canals.

The volume of nurse cell nuclei is large compared with the volume of diploid nuclei and reflects the spectacular increases in ploidy in these cells. Estimates of the ploidy of the nurse cells varies from slightly more than 1000 ploid for *Drosophila* to more than 30,000 for the moth, *Antheraea polyphemus*. Large methyl green-positive ring nucleoli that also show the green fluorescence typical of DNA after acridine orange staining are found in lepidopteran nurse cell nuclei (Fig. 5). Often large accumulations of RNA are closely associated with these DNA-containing structures. The nature of the complexes has not been thoroughly examined, but it seems unlikely that they represent specifically amplified ribosomal DNA, because amplification of rDNA has not been detected in the nurse cells of Lepidoptera (Cave and Sixby, 1976; Cave, 1978). However, rDNA amplification is detectable

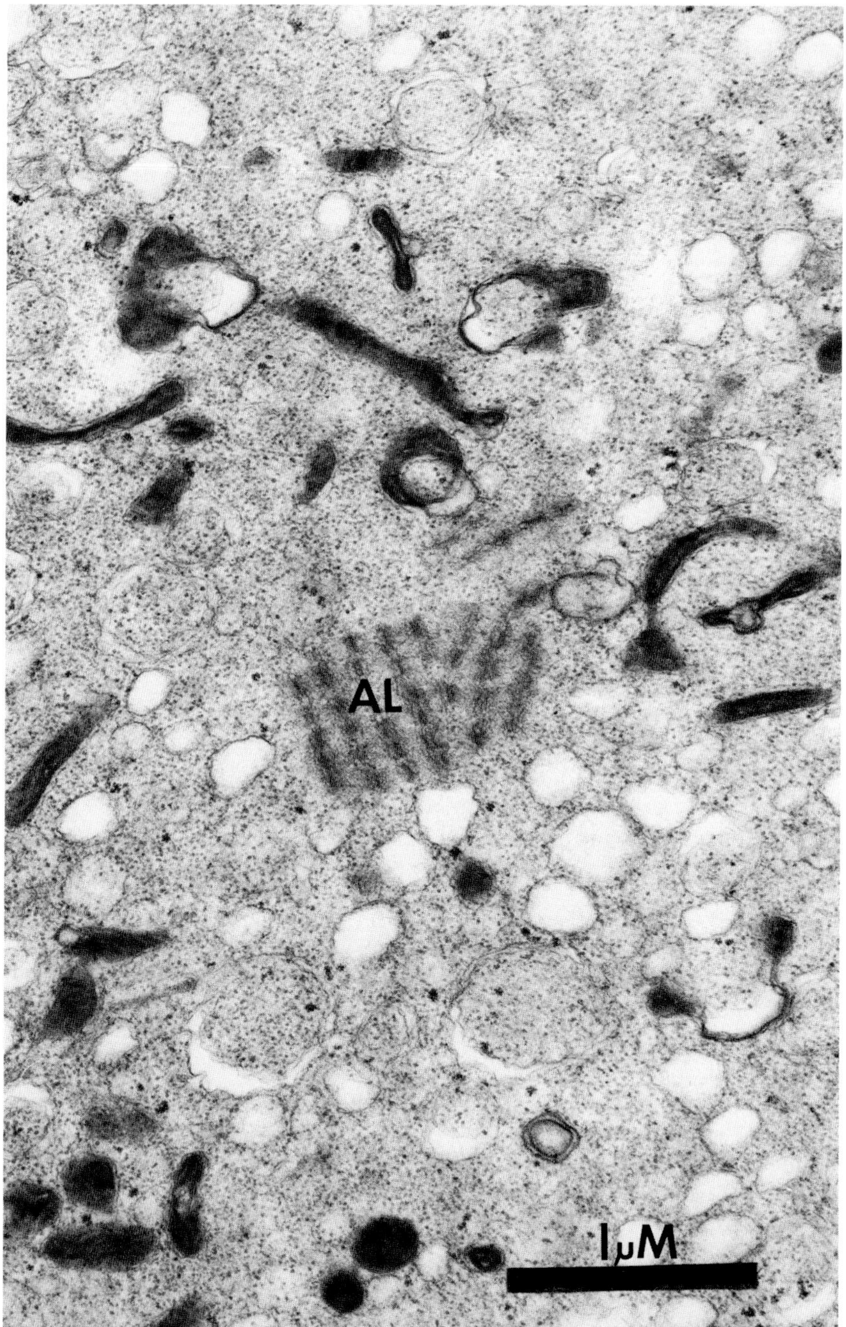


**Figure 4.** Cytoplasm of the nurse cells of *Hyalophora cecropia*. Note the absence of rough endoplasmic reticulum and the presence of a dense population of "free" ribosomes. Dense mitochondria surround a stack of annulate lamellae (AL).

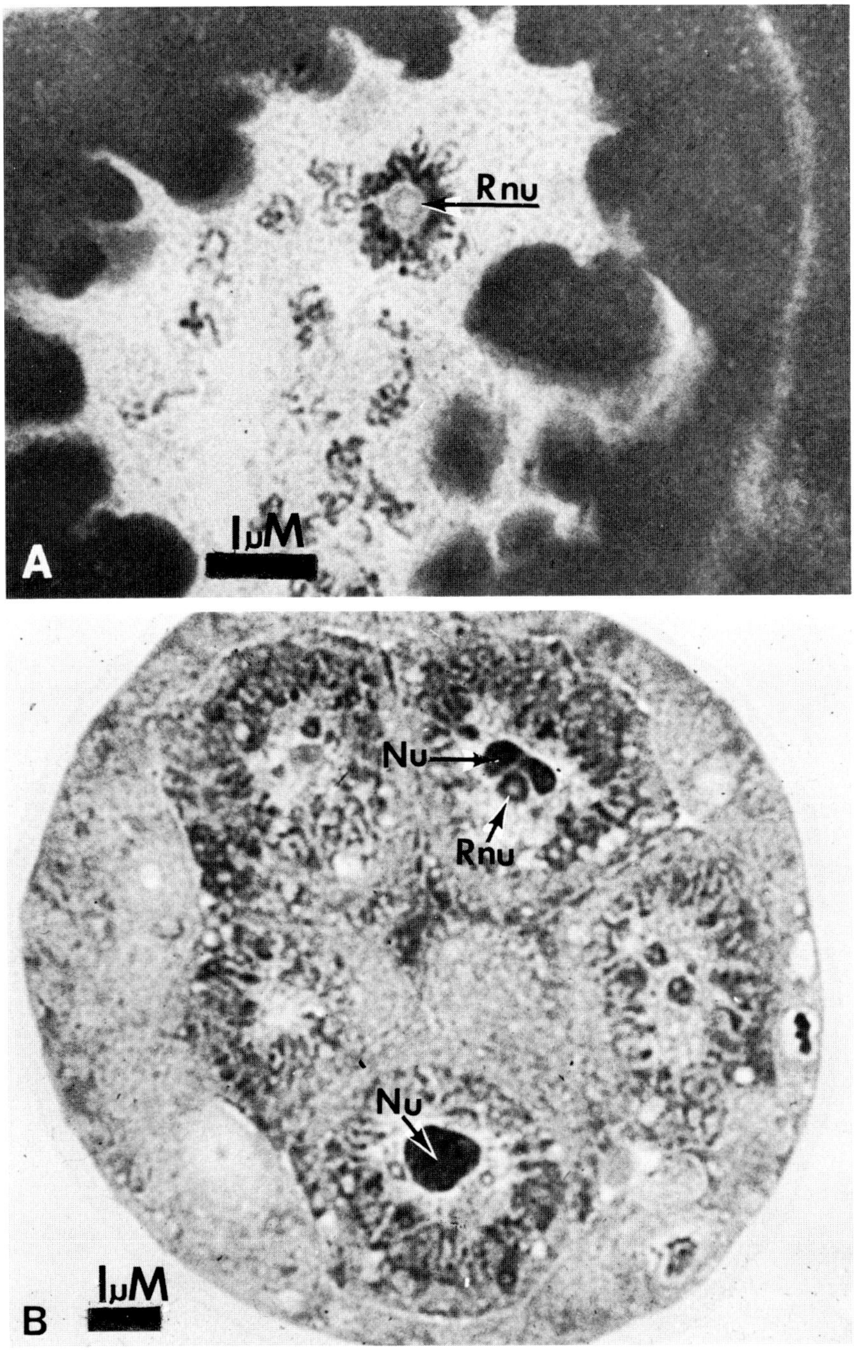


**Figure 5.** Sections of nurse cells of the moth *Actias luna* stained with methyl green–thionine to distinguish DNA from nucleoprotein complexes. (A) Nurse cell from a mid-vitellogenic follicle showing a large ring nucleolus (Rnu) surrounded by chromosomes. The ring nucleolus stains bright green, indicating a high concentration of DNA. (B) Previtellogenic region of the ovary. Rnu, intensely green ring nucleolus. Nu, intense purple-stained nucleolus, indicating large amounts of ribonucleoprotein.

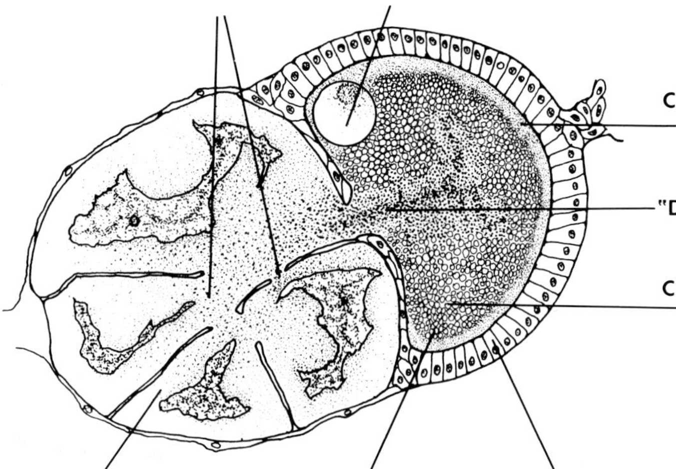


**Figure 6.** Typical mid-vitellogenic follicle from the ovary of the moth *Hyalophora cecropia.*

and extensive in panoistic ovaries (see review by Cave, 1982). The extensive reiteration of the entire trophic cell genome in insects with meroistic ovaries would seem to obviate the necessity for specific amplification. Insects with panoistic ovaries face the same need for selective amplification of the rDNA locus as amphibians and other organisms with panoistic ovaries. By contrast, the giant ring nucleoli in the nurse cells of moths may represent some alternative mechanism, such as the association of the ribosomal RNA loci of several chromosomes in some form of limited polyteny.

The developing oocyte is an object of enormous complexity, and its structure is poorly understood. Much of the available information on the cytoarchitecture of insect eggs is summarized by Zissler and Sander (1982). The developing oocyte is divided into regions that may be of developmental significance. An outer cortical region is the site of pinocytotic activity during vitellogenesis, but it is relatively free of yolk granules, particularly during the later stages of oocyte maturation. More centrally located is a region rich in yolk vesicles of various types, at the center of which is another yolk-free delta, where material transported from the nurse cells arrives. These regions are shown diagrammatically in Fig. 6. The germinal vesicle is considerably larger than the nuclei of the adjacent follicle cells but seems to contain very vague and undefined structures (Fig. 7). In spite

---

**Figure 7.** Germinal vesicles in *Actias luna* oocytes. (A) Standard histological preparation stained with methyl green-thionine. Most of the nucleoplasm appears to be empty, except for the denser body at the upper end. (B) Phase-contrast micrograph of an autoradiogram after

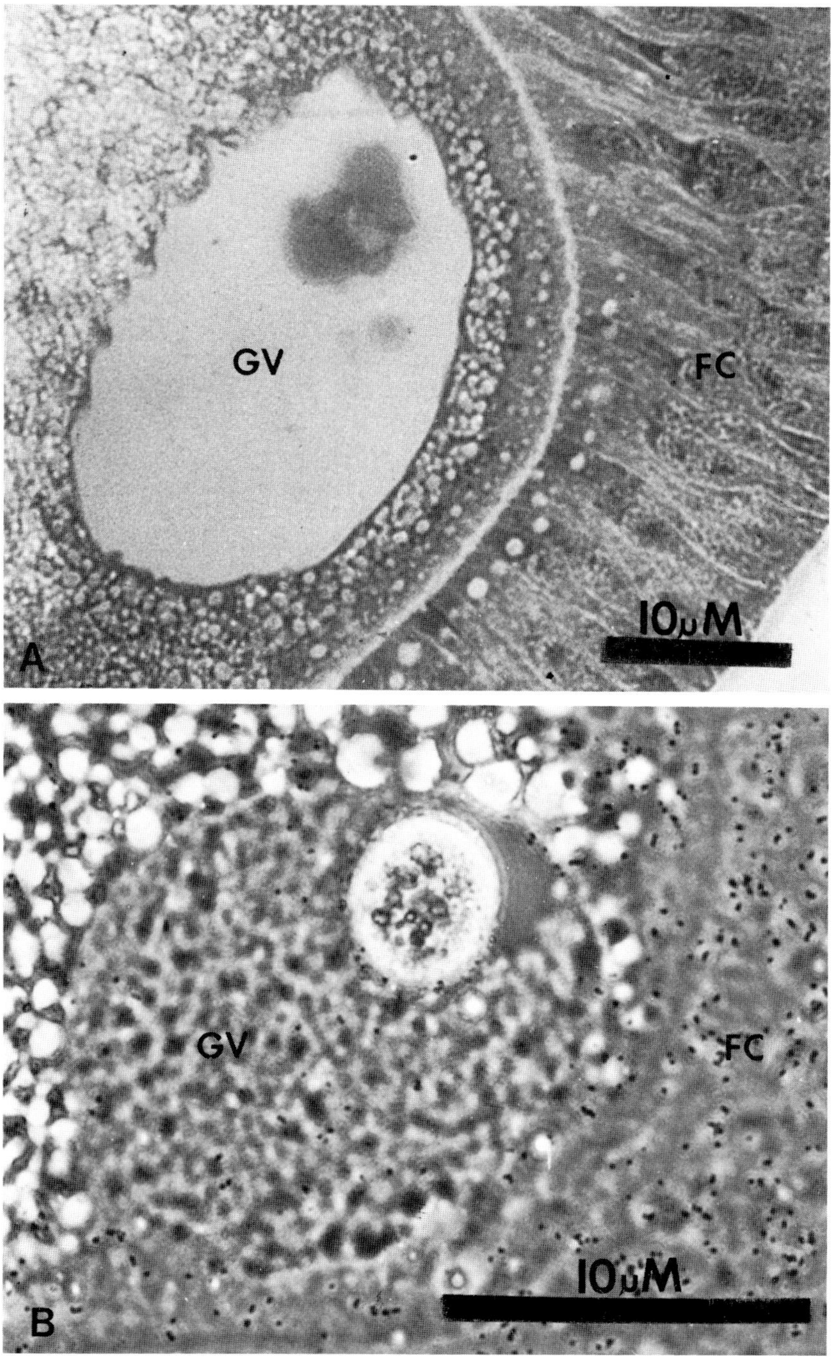


treatment with [$^3$H]-poly (U). Phase contrast reveals some structure in the nucleoplasm. Black dots are silver grains caused by radioactive disintegrations from poly (U), presumably bound to the poly(A) "tails" of mRNA. Most of the grains are over follicle cells on the right-hand side of the micrograph, but some are located over the germinal vesicle, possibly indicating the presence of some mRNA. GV, germinal vesicle; FC, follicle cells.

of its size, the germinal vesicle retains the 4C amounts of DNA, and the structure that can be discerned is thought to represent inactive chromosomes and perhaps an inactivating apparatus involved in the supression of transcription.

## 3. Autoradiographic Studies of Ovarian RNA Synthesis

The first timed studies of ovarian RNA synthesis were accomplished using autoradiographic techniques to examine the pattern of incorporation of various labeled ribonucleosides in sectioned ovarian material. Bier (1963) examined autoradiographs of sections of developing follicles of the housefly, *Musca domestica*, made at different times after administration of labeled uridine. The pattern of incorporation involved heavy initial labeling of the nurse cell nuclei, followed by labeling of the nurse cell cytoplasm—apparently at the expense of the nuclear label. The germinal vesicle was unlabeled during the entire oogenic period, but in later stages the oocyte cytoplasm was found to accumulate label. A similar pattern of labeling of the oocytes of the kelp fly, *Coelopa frigida*, was described by Muhlach and Schwalm (1977) (see Fig. 8). The most plausible explanation of the labeling pattern is that RNA was initially synthesized in the nurse cell nuclei, transferred to nurse cell cytoplasm, and finally transported to the ooplasm. The intense labeling of nurse cell nuclei, as opposed to the relative "silence" of the germinal vesicle, certainly suggests that the enormous stores of RNA in insect oocytes are synthesized in the nurse cells. Similar results were obtained for the giant moth, *Hyalophora cecropia*, by Pollak and Telfer (1969), who measured the amounts of RNA in the nurse cell clusters and the developing oocytes. They found that each nurse cell cluster contains approximately 1 $\mu$g of RNA, which remains constant during the period of vitellogenesis. The oocyte RNA content increases gradually, until it reaches a value of 3 $\mu$g in the mature oocyte. Harris and Forrest (1967) found that the smaller, newly fertilized egg of *Oncopeltus* contains approximately 0.18 $\mu$g of ribosomal RNA. These values are many times the content of diploid somatic cells and represent a general pattern observed during insect oogenesis.

In hemimetabolous insects, which do not undergo a dramatic metamorphosis, ovarian development begins in the nymphal stages. In holometabolous insects, in which the familiar larval–pupal–adult metamorphosis occurs, differentiation of eggs and sperm is initiated during the pupal–adult transformation. Choi and Nagl (1977) followed the synthesis of nucleic acid during the development of the telotrophic ovary of the hemimetabolus bug *Gerris najas*. They found extensive uridine and thymidine incorporation in the trophocytes as early as the third instar, and RNA synthesis and transfer of RNA from trophocytes to oocyte seem most extensive during the fifth instar.

One of the disadvantages of the autoradiographic techniques is that

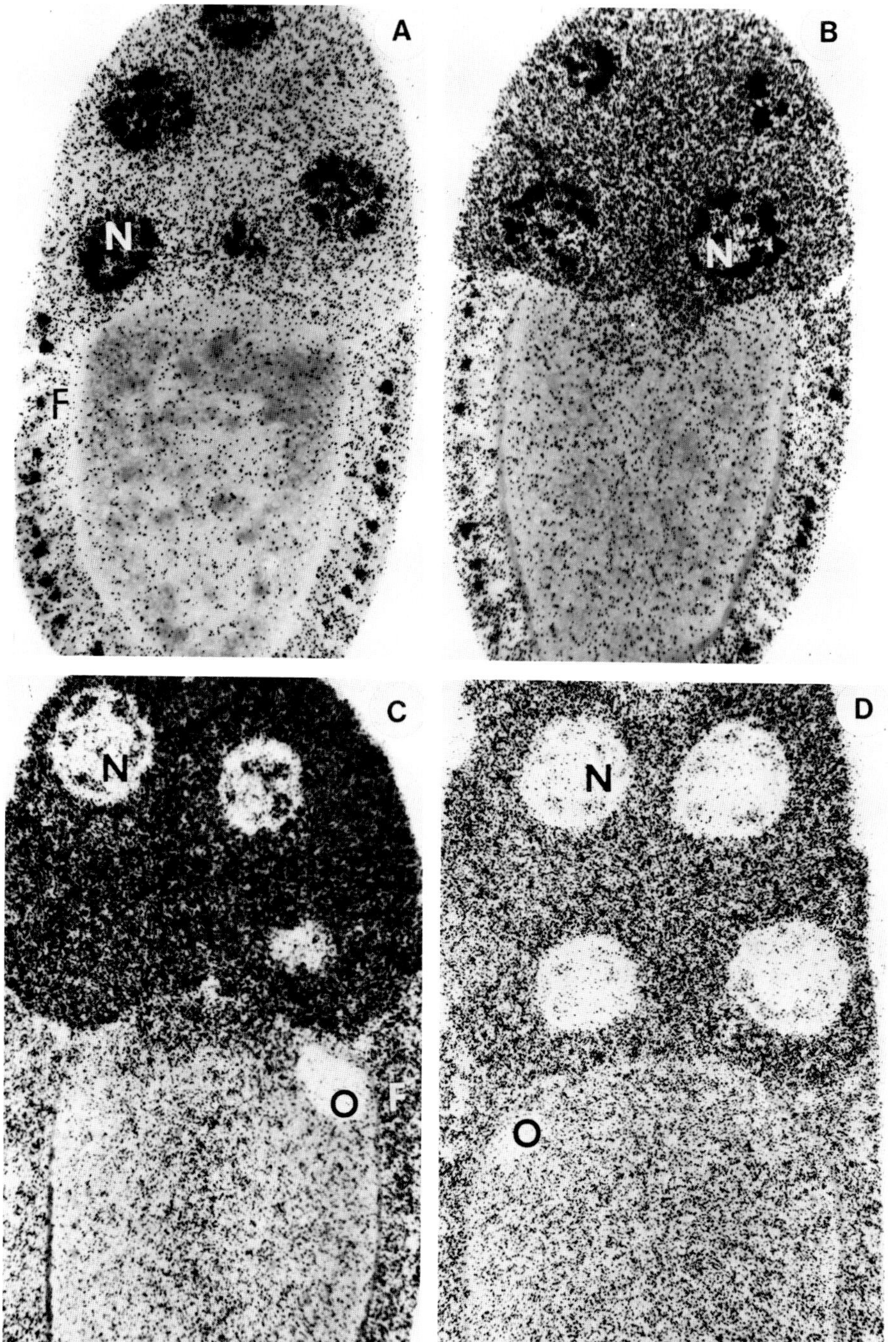


**Figure 8.** Autoradiograms of sections of the ovarian follicles of the kelp fly *Coelopa frigida* after administration of [$^3$H]uridine at (A) 35-min incubation, (B) 1-hr incubation, (C) 4-hr incubation, and (D) 8-hr incubation. F, follicle cells; N, nurse cell nucleus; O, germinal vesicle. (From Muhlach and Schwalm, 1977, with permission.)

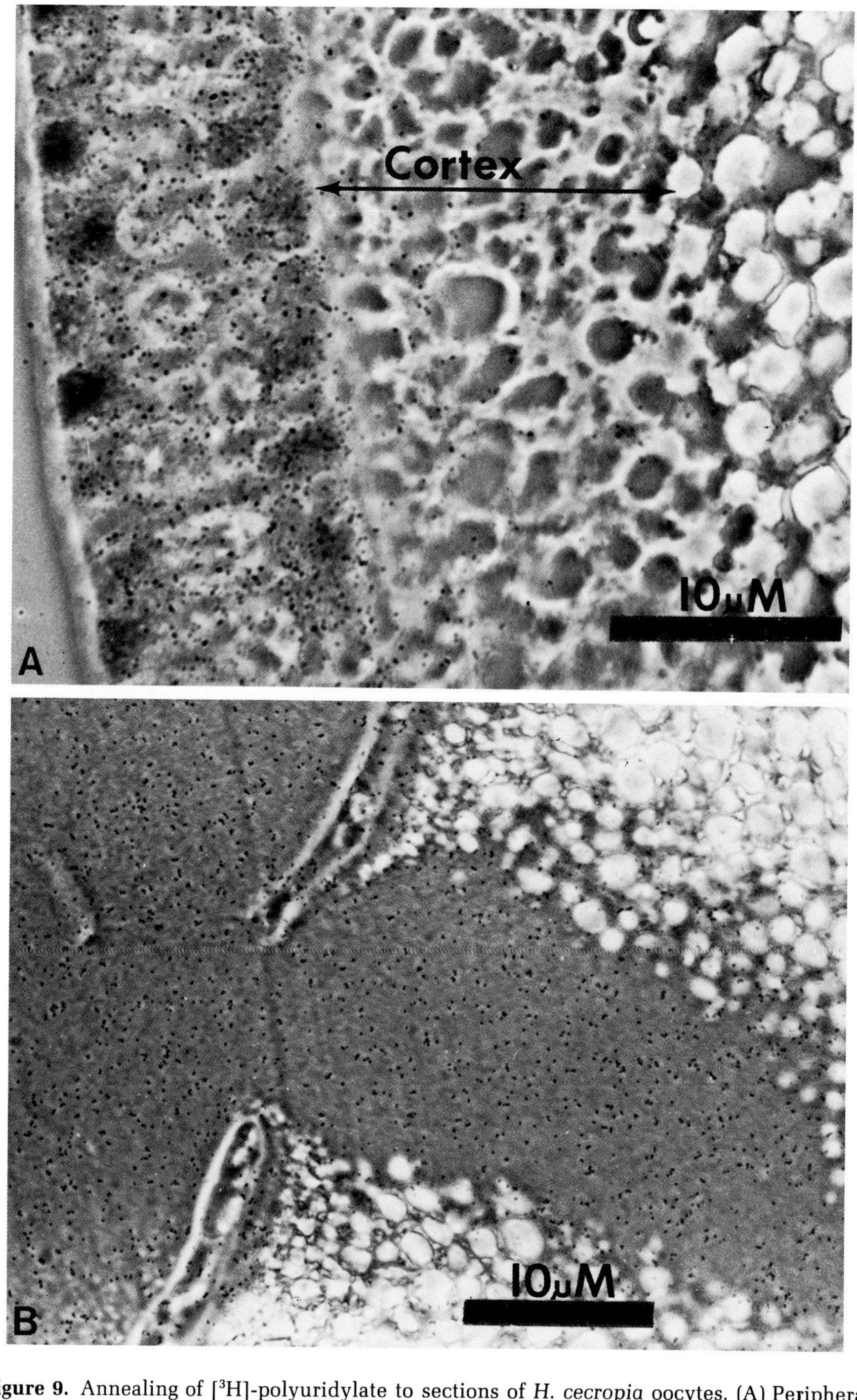


**Figure 9.** Annealing of [$^3$H]-polyuridylate to sections of *H. cecropia* oocytes. (A) Peripheral ooplasm and follicle cells. The follicle cells, on the left, which are active in protein synthesis, show extensive annealing, as indicated by the dense population of silver grains. A significant number of grains appear over the central region (cortex), whereas there is no label above background over the central ooplasm on the right. (B) Nurse cell–oocyte junction. Heavy labeling

they do not identify the type of RNA synthesized, except under special circumstances. Cave (1973) showed by *in situ* hybridization that labeled satellite rDNA was bound to the extra chromosomal body in the oocytes of the cricket *Achaeta domesticus* (see Section 4.1.1.), indicating the presence of rRNA genes in this body.

Capco and Jeffrey (1979) made use of the ability of radioactive polyuridine to anneal to the 3′ polyadenylic acid tracts in mRNA, in order to locate the sites of accumulation of message in the developing follicles in the telotrophic ovary of the milkweed bug, *Oncopeltus fasciatus*. Poly(A)$^+$ RNA was found to accumulate in the trophocyte cytoplasm and was detected in the trophic cords and in the oocytes. The poly(U)-binding sites were found to be uniformly distributed over the ooplasm during most of oogenesis. A preferential distribution was noted at the anterior and posterior periplasm in the final stages of vitellogenesis, when contact with the trophic syncytium is lost. This differential distribution disappears after chorion formation. These investigators suggest that these results may indicate a source of mRNA other than the trophic syncytium but also point out that masking of the poly(A)$^+$ regions of the message or other effects might also account for the variation in apparent localization.

Nichols and Duhl conducted a similar study of the distribution of poly(U)-binding in the polytrophic follicles of moths (L. Nichols and G. Duhl, personal communication). These workers found extensive binding in the nurse cell nuclei and cytoplasm and particularly in the ring canals connecting oocyte and nurse cell cytoplasm (see Fig. 9B). In the developing oocytes, binding was extensive in the central yolk-free endoplasm and in the cortical periplasm. No detectable binding of poly(U) was detected in the oocyte nuclei in either study of developing oocytes. These results are entirely consistent with the synthesis of the bulk of oocyte messenger RNA in the trophic cells and the subsequent transport to the ooplasm. The data cannot take into account possible masking or processing of the poly(A) tracts, and thus "cryptic" messages could be overlooked. Another possible factor that might distort the pattern of binding could be the engagement of message in translationally active polysomes. In both studies cited here, the follicle cells show extensive binding of poly(U) at stages in which they are known to be translationally active, suggesting that translation does not mask the poly(A) sites.

## 4. Biochemical Analysis of Oocyte RNA

Clearly, examination of the molecules of RNA synthesized in the follicles would be required to verify the models suggested by cytochemical and autoradiographic studies.

---

of the nurse cell cytoplasm on the left is confluent with the "delta" region of the oocyte. (See Fig. 6 for the locations of the regions indicated.)

## 4.1. Ribosomal RNA

More than 90% of the cytoplasmic RNA is ribosomal. To support the synthesis of large numbers of ribosomal RNAs, the template consists of 200–500 tandemly repeated ribosomal genes per diploid somatic genome for most insects examined. The primary transcript from these ribosomal genes contains both 18 S and 28 S ribosomal RNA, which is then processed to yield the two separate RNA chains. The ribosomal genes of *Drosophila* are located in two clusters, one on each of the sex chromosomes. Long and Dawid (1980) have tabulated the available data on the ribosomal loci in eukaryotes. For more detailed descriptions of the details of ribosomal RNA synthesis and processing in insects, reviews by King and Büning (1985) and Berry (1985) may be helpful.

### 4.1.1. Quantitative Aspects of Ovarian Ribosomal RNA

Gall *et al.*, (1969) made use of the observation that among the undifferentiated oogonia or stem cells in the tip of the polytrophic ovary of dytiscid water beetles may be found a modest number of oocytes having a very conspicuous extrachromosomal Giardina body. The DNA extracted from this portion of the ovary was found to be enriched for a heavy satellite with buoyant density of 1.721 g/cm$^3$ as opposed to 1.693 for the "main-band" DNA. Hybridization with ribosomal RNA indicated that ribosomal DNA is located in this satellite peak. Thus, the amplified DNA of the Giardina body contains the ribosomal template that presumably could be transcribed to produce rRNA for the oocyte ribosomes. Lima-de-Faria *et al.* (1969) found amplified rDNA by similar methods in the DNA bodies of the panoistic ovaries of the cricket *Acheta*. In both examples, the satellite did not consist exclusively of rDNA but apparently contained other DNA as well. In his excellent review of amplification of rDNA in insect oocytes, Cave (1962) gives a detailed list of the various DNA-containing bodies in insect oocyte nuclei, particularly those of crickets. In the cricket, these DNAs are circular and have a molecular weight of $\sim 37 \times 10^6$, about 12% of which codes for the large and small ribosomal subunit. For further detailed information on amplified ribosomal DNAs, the review of King and Büning (1985) as well as that of Cave should be consulted.

The presence of highly polyploid nurse cells should eliminate the necessity for specific amplification of rDNA in both telotrophic and polytrophic follicles. With the exception of the DNA-containing structures such as the Giardina body, which are found in the oocyte of some polytrophic ovaries, this seems to be the general rule. Cave (1975) was unable to detect amplification of the rDNA above the level of ploidy in the trophic tissues of the telotrophic ovary of the milkweed bug, *Oncopeltus*. In the diptera, Renkawitz and Kunz (1975) found rDNA to be modestly overrepresented (135%) in the ovaries of *Calliphora erythrocephala*, but underrepresented in *Drosophila hydei* (51%) and *Sarcophaga barbata* (47%). In the polytene chromosomes of the salivary glands of *Drosophila melanogaster*, rDNA is considerably underreplicated (Spear and Gall, 1974). Jacobs-Lorena *et al.* (1980) have examined the amplification of the rDNA, as compared with the 5 S RNA genes in the developing ovary of *D. melanogaster*. He found approximately 30% underreplication of the rDNA in the ovary, as compared with 80% in the salivary gland. The 5 S RNA is associated with active ribosomes but is transcribed on separate genes and thus is not a component of the primary rRNA transcript. The 5 S RNA genes are amplified in parallel with the increase in overall ploidy. Evidence for the active transcription of genes for 18 S and 28 S ribosomal RNA can be gained from the observation that the increase in ribosomal RNA in the oocyte is 15 times as great as the increase in rDNA. In their work on the lepidopteran ovary, Cave and Sixby (1976) found no amplification of rDNA in the ovary of either *Bombyx mori* or *Antheraea pernyi* (Cave, 1978).

#### 4.1.2. Synthesis and Transport of Oocyte rRNA

Hughes and Berry (1970) dissected the clusters of nurse cells from the follicles of *Antheraea pernyi* and thus were able to study the synthesis of rRNA in a context uncomplicated by the contribution of follicle cell somatic ribosomal RNA synthesis. These workers were able to identify a 39 S primary transcript that was cleaved to 30 S and 21 S fragments and finally processed to 28 S and 17 S RNA. In a more extensive study of the synthesis of rRNA in the ovary of the housefly, *Musca*, by Hall and Cummings (1975), a 38 S primary transcript was detected. They report an initial cleavage into 32 S and 20 S molecules. The 32 S fragment was then converted to 30 S and finally 28 S RNA, representing approximately 50% of the original transcript. The smaller fragment was converted to 18 S RNA, which contained 24% of the primary transcript. The half-life of the 38 S molecule was calculated to be 12 min, which is relatively long.

Evidence for the transport of newly synthesized ribosomes from trophic cells into the developing oocyte comes from the studies of Hughes and Berry (1970) on the polytrophic ovaries of *A. polyphemus* and Davenport (1976) on the telotrophic ovaries of *Oncopeltus*. Both cases show a distinct lag between the appearance of labeled ribosomes in the trophic cell cytoplasm and in the oocyte. When the nurse cell cluster was dissected from *A. polyphemus* follicles, no labeled ribosomes were detected in the oocyte. When the trophic cords in the *Oncopeltus* ovary were ligatured, labeled ribosomes failed to accumulate in the oocytes. These experimental results indicate ribosomes are synthesized in the trophic cells and migrate via intercellular connectives to the oocytes.

Chooi (1976) prepared chromosomes from *Drosophila* nurse cells and embryos using the technique of O. Miller (Miller and Beatty, 1969). This method consists of lysing the nuclei gently with detergent and collecting the chromosomes on an electron microscope grid. If the chromosomes are sufficiently spread, they appear as a single fiber with the partially synthesized RNAs appearing as branches on alternating sides of the chromosome. Since several RNA polymerase molecules can transcribe the same rRNA gene simultaneously and the RNA branch elongates as transcription progresses, the complex assumes a configuration resembling an inverted, two-dimensional Christmas tree (Fig. 10). The longest branches represent the most nearly complete transcripts preceded by a graded series of shorter, less complete transcripts. Since the ribosome locus consists of a series of tandemly repeated RNA genes, the distance between the "bottom" of one Christmas tree and the top of the next represents nontranscribed spacer DNA. In the nurse cells, the primary ribosomal transcripts are separated by longer nontranscribed spacers. This difference suggests that a different ribosomal locus is transcribed in the nurse cells. Genetic analysis involving deletion of the sex chromosomes that carry the ribosomal genes indicates that the long spacer rDNA is associated with the X chromosomes, but not the Y. It is not known whether there are sequence differences in the ribosomes transcribed from these two sites and whether differences in maternal versus embryo ribosomes affect translation during early embryogenesis.

## 4.2. Messenger RNA

Eukaryotic messenger RNAs have characteristic structures at both ends. The 5′ end is characterized by a terminal "cap" structure consisting of a methyl-7-guanosine residue linked to the penultimate nucleotide by a 5′ to 5′ triphosphate. The cap structure is thought to be important in the formation of a complex with the ribosomal small subunit. This binding is the first step in the formation of the initiation complex, which is essential for the translation of message into protein (Filipowicz, 1978). The 3′ terminus is characterized by a series of up to 200 AMP residues linked by the

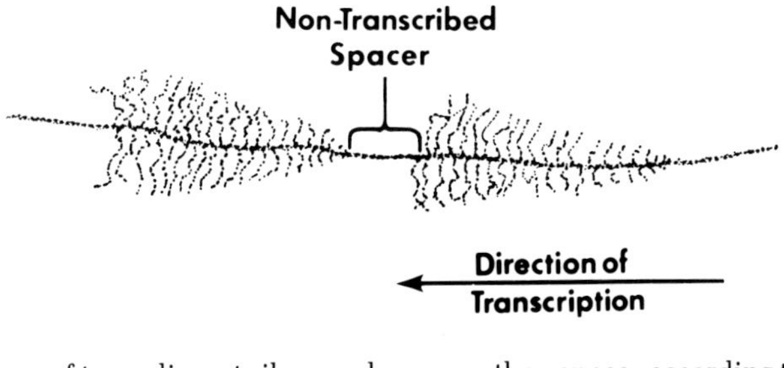


**Figure 10.** Diagram of two adjacent ribosomal genes as they appear according to the spreading technique of Miller and Beatty (1969). The "branches" of the trees are the growing rRNA chains.

conventional 3′-5′ phosphodiester bond. Some messages, particularly histone messages, have truncated or missing poly(A) tails. Several functions have been suggested for the poly(A) tail, but none have been definitively established. The ability of the poly(A) tail to attach to either redundant sequences of UMP or dTMP residues has been exploited to purify messenger RNA or to detect the presence of poly(A)$^+$ messages (see Section 3).

In bacterial systems, messenger RNAs are characterized by their short half-lives, which are estimated to be on the order of 2 hr. In insect somatic cells, including some showing active protein synthesis, the half-life may depend on the particular messenger species and may be as long as 6 days. In the translationally inactive oocytes of insects, the messenger RNA may be stored for periods of several weeks but retain the ability to initiate protein synthesis after fertilization. The transcription of this mRNA on nurse cell or premeiotic oocyte chromosomes and its storage without activity until fertilization led to the earlier concept of a "masked maternal message." Recently, the masking hypothesis was questioned in sea urchins (e.g., Moon *et al.*, 1982); alternative mechanisms involving sequestration of the message, absence of initiation factors, or inactivation of ribosomes have been proposed to explain the absence of prefertilization translational activity. Fertilization may not always be a necessary signal for the activation of translation of maternal mRNA, particularly in dipterans. Two examples of prefertilization translational activity are the reports of Pietruschka and Bier (1972) in *Musca* and Zalokar (1976) in *Drosophila*.

#### 4.2.1. Synthesis and Transport of mRNA

Telfer, in his review of the literature before 1974 (Telfer, 1975) found no compelling evidence for the synthesis of mRNA by either nurse cells or the trophic syncytium. There were references to "unstable" RNA fractions in oocytes of insects, but since maternal mRNAs are stored rather than translated or degraded, instability is not an expected characteristic of oocyte mRNA. Two factors are probably responsible for the failure to detect mRNA in follicle preparations: (1) mRNA makes up a small (2–3%) fraction of the total cellular RNA, and (2) it is contained in a ribonucleoprotein particle with a buoyant density close to that of the small ribosomal subunit, which masks it in sucrose-gradient profiles.

The development of poly(U) Sepharose and oligo(dT) cellulose columns, which trap RNA by hybridization to the poly(A) tails has made possible the detection and purification of mRNA and mRNP. Winter (1974) detected a fraction of ribonucleoprotein particles of lower buoyant density than monosomes in the ovary of the bug *Dysdercus intermedius*. This "slow" fraction contained 7–9 S RNA molecules with associated poly(A) that would stimulate translation in a cell-free system prepared from ribosomes and other elements obtained from other tissues. This appears to have been the first demonstration of mRNA in insect oocytes.

Paglia *et al.* (1976a) detected a poly(A)-containing RNP fraction that appeared as a shoulder on the heavier side of the transfer RNA peak in sucrose density gradients. This peak contained more than 95% of the poly(A)$^+$ RNA of mature, unfertilized oocytes of a number of moths, including *Antheraea polyphemus, Hyalophora cecropia, Actias luna,* and *Manduca sexta.* The putative mRNA was contained in particles with a buoyant density of 1.46 g/cm$^3$ in cesium chloride gradients. The buoyant density value is taken to indicate a higher ratio of protein to nucleic acid than that of ribosomes ($p$ = 1.69 for oocyte monosomes). When the particles were partially purified on oligo(dT) columns and dried on electron microscope grids, they were found to be 180–250 nm in diameter, considerably smaller than that found for monosomes treated in the same way (Paglia *et al.*, 1976b). When the RNA was extracted from these mRNP particles, it was found to band in a region of >20 S to about 12 S, with the main peak at about 15 S (Kastern *et al.*, 1982). The 15S value compares well with 17 S values reported for *Drosophila* embryo mRNA by Lamb and Laird (1976) and Anderson and Lengyel (1981).

One discrepancy in the moth oocyte data is the very low estimate for the buoyant density of mRNP particles (approximately 7 S in sucrose-density gradients). Close inspection of the gradients revealed that a layer of blue pigment was associated with the region containing the transfer RNA. Directly under the blue layer was a thin band of yellow pigment, which contained the mRNP. When 0.5% Triton X-100 detergent was added to the sucrose gradients, the yellow pigment appeared at the very top of the gradients, and the mRNP was associated with the monosome fraction. It is not clear whether the association with the yellow pigment represents simple adventitious binding of mRNP or whether there may be some functional significance to the association. The results demonstrate that the mRNP particles are in fact large enough to contain the RNA and protein ascribed to them.

Paglia *et al.* (1976a) also studied the transport of ribosomal, transfer, and particularly messenger RNA, into the oocyte from the nurse cells. Excised follicles were incubated in medium containing [$^3$H]adenosine to label newly synthesized RNA. Actinomycin D was then added to the medium to halt further RNA synthesis, and the appearance of labeled poly(A)$^+$ RNA fractions in the oocytes was followed by poly(U) Sepharose column chromatography. The results indicate that labeled RNAs (including mRNA) continued to accumulate in the ooplasm after all new RNA synthesis had been halted by the inhibitor. The only source of the labeled ooplasmic RNA was transport from the nurse cell cytoplasm. Figure 11 shows the pattern of accumulation of labeled RNA in the ooplasm when follicles were incubated in medium containing labeled adenosine for increasing lengths of time. After a 2-hr incubation, small amounts of tRNA

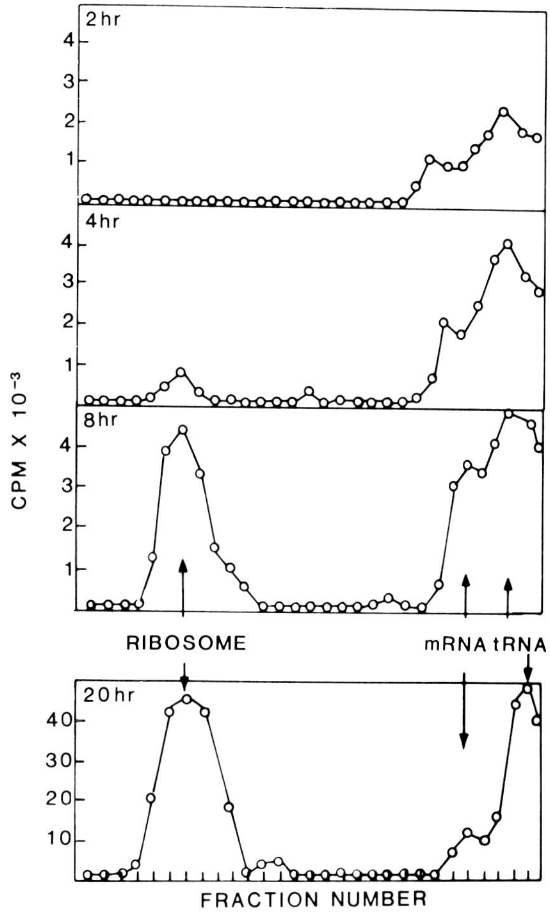


**Figure 11.** Accumulation of radioactive RNA in *Hyalophora cecropia* oocytes. Follicles were incubated in medium containing [$^3$H]adenosine for the times indicated. At the end of the incubation period, nurse and follicle cells were removed, and the ooplasm was examined on sucrose-density gradients. Note the change in the scale of the y axis in the lowest panel.

and mRNA were detectable. Labeled ribosomes were first detected after 4 hr, and label in all three peaks increased until the experiment was terminated after 20-hr incubation.

Chooi (1976) also observed the formation of non-rRNA transcripts in Miller spreads of *Drosophila* nurse cell chromosomes. She found evidence for "giant" transcripts, some averaging 2.2 $\mu$m in length, including one transcript 5.3 $\mu$m in length. This is several times the length of blastoderm-stage embryo transcripts. A similar giant transcript from the developing oocytes of the newt *Notophthalmus* has been extensively studied in Gall's laboratory (Diaz *et al.*, 1981). Both groups suggest that the RNA polymerase may ignore "stop" signals in the gene and continue to transcribe through regions downstream from the normal end of the gene. The evidence available for *Drosophila* oocytes (Lamb and Laird, 1976; Anderson and Lengyel, 1981) does not indicate that oocyte messages are unusually long, however. The explanation for the discrepancy in the length of hn and cytoplasmic mRNAs may be that messenger processing is somewhat different during oogenesis than during normal somatic mRNA production.

### 4.2.2. Controls of Translation of Maternal mRNA

Kastern and Berry (1976) reported that oocyte mRNA of *Manduca* lacks the 5′-terminal cap structure. This naturally occurring uncapped mRNA translates with low efficiency when used to challenge a cell-free translation system prepared from wheatgerm (Kastern *et al.*, 1982). In the presence of S-adenosyl-homocysteine, an inhibitor of methylation, efficiency of translation of the oocyte message is even further reduced. Message extracted from early embryos proved to be capped and to translate efficiently, suggesting that the lack of a functional cap may prevent translation. A recent report (Caldwell and Emerson, 1982) indicates that sea urchin maternal message may also lack the methylated cap structure before fertilization but that it acquires the cap during the period of early embryogenesis when translation is stimulated.

The precise mechanisms whereby the fertilized egg activates translation are still obscure. What evidence there is suggests that dissociation of an inhibitor may be involved. Winter (1974) treated mRNP particles obtained from the oocytes of *Dysdercus* with elevated concentrations of KCl and was able to show that this made the normally untranslatable particles active in her cell-free preparations. Jäckle (1980a) was able to demonstrate the ability of "masked mRNP" from *Smittia* oocytes, supporting the translation of actin message after salt treatment. Similar salt treatment has been shown to activate the mRNPs from sea urchin oocytes (see review by Raff, 1980). Jäckle (1980b) has also found evidence for stage-specific activation of actin message in *Smittia* embryos. Mermod *et al.* (1980) found three DNA-binding proteins to be synthesized on message associated with polysomes of early *Drosophila* embryos but not in the "free" mRNP particles. This finding suggests that the activation of stored mRNAs may not be a simple one but that it may involve a mechanism for discriminating between specific messages.

One major problem with the investigations of translation control in insect embryos is that the assays have been conducted in cell-free preparations made from heterologous systems such as wheatgerm or rabbit reticulocytes or, in the case of Winter (1974), elements prepared from other nonovarian tissues excised from adult insects. The amount of information that can be obtained concerning control mechanisms from these systems is limited because oocyte-specific control molecules could not be expected to operate in preparations derived from nonoocyte sources. A source of concern is the observation that a distinct rRNA locus is transcribed in *Drosophila* nurse cells; also, the possibility that different 5 S genes are transcribed, as has been observed in oocytes of the frog, *Xenopus* (Ford and Southern, 1973), may indicate fundamental differences in somatic—as opposed to early embryo—translation mechanisms. Even the mRNA transcripts from nurse cells are much larger than from somatic cells. All these observed differences may represent trivial variations in oocyte transcription mechanisms as opposed to those of somatic cells, but the possibility that translation mechanisms in early embryogenesis may differ in significant details cannot be dismissed.

A possible approach to this problem would be the use of homologous cell-free preparations of oocytes and embryos in the search for control mechanisms. Recently it was demonstrated that homogenates of oocytes and early embryos can be prepared that display translational and transcriptional behavior similar to that observed *in vivo* (J. Hooper, personal communication). Figure 12 shows data obtained from homogenates of *Manduca* embryos of different ages. Incorporation of amino acids is sensitive to inhibitors of eukaryotic—but not prokaryotic—protein synthesis. That the preparations are able to initiate translation is shown by the stimulation caused by addition of exogenous message and sensitivity to edeine, an inhibitor of initiation. A comparison of the rates of translation *in vitro* shown in Fig. 12 with *in vivo* rates in Fig. 1 shows good agreement for the two methods and suggests that the *in vivo* control mechanisms are preserved in the cell-free preparations.

The significance of the maternal mRNAs in the determination of the structures of the embryo is still debatable. The crucial organizing role of the *esc* gene product referred to in Chapter 13 is undeniable, but the mech-

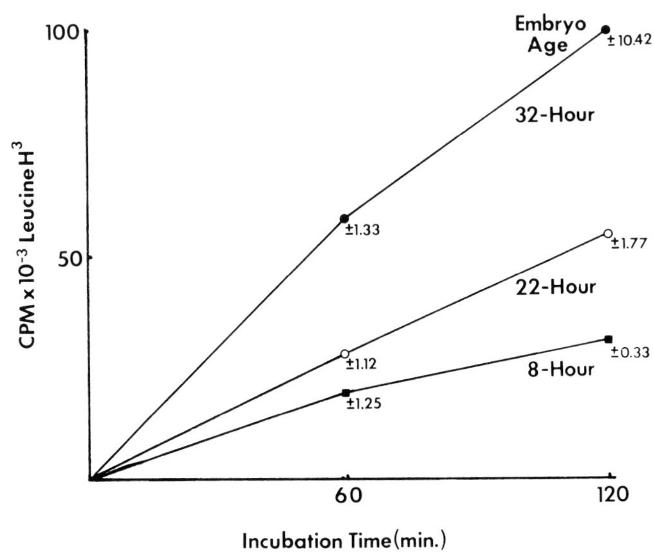


**Figure 12.** Rates of incorporation of [$^3$H]leucine by cell-free preparations of *Manduca sexta* embryos at different stages of development. Compare with the *in vivo* results in Fig. 1. (Courtesy of J. Hooper.)

anism is obscure. A series of papers from the laboratory of Kalthoff (e.g., Kalthoff *et al.*, 1975; Kandler-Singer and Kalthoff, 1976) have shown that a peculiar developmental anomaly—double abdomen—can be induced by a number of methods. The defect, which consists of mirror-image duplication of the abdominal segments and loss of the anterior segments, can be produced by centrifugation of the eggs, by UV irradiation of the anterior half, and by treatment of the anterior half with low doses of ribonuclease. In the last case, the egg is submerged in a solution of ribonuclease, and a small puncture is made at the prospective anterior end. The resulting embryos show formation of the mirror-image double abdomen. A number of explanations of these results are possible, but the simplest is that destruction of the mRNA that is responsible for formation of anterior elements causes a reorganization of the remaining developmental information. The loss of ribosomal and transfer RNA to nuclease activity could be compensated for by the undigested molecules of tRNA and RNA, but the lost information in specific mRNAs could not be retrieved. The UV induction of double abdomen could be explained by a similar mechanism, and centrifugation could be viewed as causing a reorganization of the ooplasm containing the specific anterior mRNAs. Centrifugation is the only one of these methods that can produce the reciprocal result—double anteriors (Kalthoff *et al.*, 1982).

A most unexpected mechanism, the repair of maternal mRNA, was also uncovered by Kalthoff (see Kalthoff and Jäckle, 1982, for review). Repair of DNA has been extensively studied, but this appears to be the first demonstration of repair of RNA. RNA repair in *Smittia* embryos is photoreversible and involves the removal of pyrimidine dimers induced by UV irradiation (Kalthoff *et al.*, 1978; Jäckle and Kalthoff, 1978, 1980). These investigators propose that the eggs would normally be exposed to UV irradiation by the sun and would accumulate pyrimidine dimers. At the same time, the visible light from the sun would stimulate the repair mechanism to correct these potentially lethal alterations of the mRNA. The uncovering of this unusual mechanism may be further proof of the central role of maternal mRNA, the loss of which would be irretrievable in the case of the early embryo.

An alternate model of determination in insects involves the organizing activity of gradients of diffusible morphogens. One of the clearest arguments for this point of view is the double gradient model of Meinhardt (1977). According to this model, the relative concentrations of an activator and an inhibitor would determine the structures formed along the anterior-posterior axis of the embryo. Both the gradient model and the particulate determinant model have inherent strengths and weaknesses. Gradient models best explain the general organization of the embryo along axes but leave open the nature of the diffusing molecules and place a heavy burden of eliciting specific genetic programs on simple differences in the amount of morphogen. Particulate determinant models provide sufficient potential in terms of information content but do not address problems of distribution. It is possible that some combination of the two models may provide a more satisfactory working hypothesis. This question is addressed again in section 7.

## 5. Transport of RNA and RNP from Trophic Cells to Ooplasm

In the follicles of the panoistic ovary, the transport of RNA from nucleus to cytoplasm presents the general and nontrivial problem faced by all cells. Mechanisms for the directed movement of large elements such as ribosomes to the nuclear envelope and into the cytoplasm are largely unknown, but the transport from the cytoplasm of one cell to another over considerable distances adds a new dimension to the problem.

Perhaps the most convincing demonstration of migration of material from trophic cells to oocytes is the observation that injection of a dye such as trypan blue into the nurse cells of an excised panoistic follicle results in the accumulation of the dye in the oocytes after a few hours' incubation. If small amounts of dye are introduced, none remains in the nurse cells. When trypan blue is injected into the oocyte, it remains there, and none migrates into the nurse cells. More elegantly designed experiments of this sort were performed by Telfer and associates. Fluorescein-conjugated lysozyme (FLy) carries a net basic charge at neutral pH, whereas methylcarboxylation of the conjugate (McFLy) carries a net negative charge. The fluorescein moiety makes it possible to detect minute amounts of these substances. When the basic FLy was injected into single nurse cells of *H. cecropia*, it remained there, but when the acidic McFLy was substituted, it migrated into the oocyte (Woodruff and Telfer, 1980). A similar result was obtained with the telotrophic ovary of *Rhodnius*. McFLy was transported from the trophic syncytium among the trophic cords to the oocyte, while FLy remained close to the site of injection and did not diffuse much even in the trophic cells (Telfer *et al.*, 1981). These results suggest that charged complexes might move along an electrical gradient whose negative pole would be situated in the trophic tissues and the positive pole in the oocyte. A further interesting observation, (Huebner, 1981) was the migration of dyes such as procion yellow and unconjugated fluorescein from oocytes to follicle cells via gap junctions. The larger conjugates such as McFLy were excluded from migrating into the follicle cells, however. Thus, there is communication between the oocyte and follicle cells that could permit the exchange of small molecules but not of complexes the size of RNP particles.

The possible role of electrical fields in controlling early development was suggested by Lionel Jaffe on the basis of his studies of the developing egg of the seaweed *Fucus* (Jaffe, 1969). Woodruff and Telfer (1973) placed microelectrodes in the follicles of *H. cecropia* and were able to demonstrate the electrical polarization of the follicle. With the development of the vibrating probe (Jaffe and Nuccitelli, 1974), it became possible to measure current flow in a noninvasive manner. Jaffe and Woodruff (1979) used the vibrating probe to show that a current of up to 20 $\mu A/cm^2$ passes from the anterior end of the oocyte to the nurse cells. A current of about 100 nA transverses the entire follicle. A smaller, but similar electrical current has been shown to pass from the trophic syncytium to the oocytes in the telotrophic ovaries of *Rhodnius* (Telfer *et al.*, 1981) and *Dysdercus* (Dittmann *et al.*, 1981). In the case of the *Rhodnius* current, a substantial increase was observed when juvenile hormone was added to the surrounding medium. Juvenile hormone is known to be involved in oocyte maturation in many species, and one role it may play is in maintaining the electrical gradient (Telfer *et al.*, 1981). In the panoistic ovary of *Locusta*, Wollberg *et al.* (1976) reported electrical currents traversing various elements of the ovariole. Observations of this sort have led workers (e.g., DeLoof, 1983) to suggest that the insect ovary might function as a microelectrophoresis chamber for the transport of maternal RNA species in the oocytes.

## 6. Activation of Transcription of the Embryo Genome

The activation of transcription from the embryo genome has not been extensively examined in insects. Kastern *et al.* (1981) eliminated the absence of RNA polymerases as a suppression mechanism. These workers found evidence of large amounts of all three DNA-dependent RNA polymerases in the cytoplasm of unfertilized eggs of *Manduca*. RNA polymerase II, which is responsible for the synthesis of mRNA, was examined in detail because of its role in supplying "new" information and because of its extreme sensitivity to inhibition by the mushroom toxin, $\alpha$-amanitin. In a partially purified state, oocyte RNA polymerase II was found to be present at a concentration of $10^{-2}$ pmoles, or $6 \times 10^9$ molecules per cell, and was able to transcribe calf thymus DNA in a cell-free assay. On the basis of these results, it would seem likely that the DNA of early embryo chromosomes is unavailable for transcription, perhaps because of the presence of an inhibitor.

Newport and Kirschner (1982a,b) proposed that a transcription suppressor prevents RNA synthesis in *Xenopus* until the mid-blastula transition. The suppressor is thought to be present in the ooplasm until a critical threshold amount of DNA (24 ng) accumulates at cleavage number 12. Exogenous DNA injected into the embryo before cleavage 12 can induce precocious transcription, presumably by binding the remaining free suppressor molecules. The presence of a similar transcription suppressor has not been demonstrated in insect embryos, but a considerable body of evidence suggests the presence of some similar mechanism.

## 7. RNA in Insect Morphogenesis: An Alternate Model of "Determination"

Is there a resolution of the apparent contradiction in the particulate as opposed to the gradient determination model? There is a tradition in science, best demonstrated by the combination of the wave and quantum theories of the nature of light, of combining two models with complementary inadequacies into one. The strength of gradient models lies in their overall organizational properties, whereas particulate models provide for the requirements for diverse, yet specific types of informational particles. Present gradient models propose that differences in the concentration of morphogens impose a latent "body image" on the developing embryo. Asymmetries in the distribution of activator and inhibitor molecules along the embryonic axes would specify specific programs of gene activation that, in turn, would lead to different programs of transcription in the nuclei. It is difficult to imagine how these concentration differences would provide the information to cause the activation of a large number of different programs. Particulate determination models would involve the strategic placement of information in the form of mRNA or proteins that could program nuclei, but they leave open the mechanism for distribution.

The nature of insect development invites speculation of this sort, because the early syncytial nature of the embryo poses no problems of barriers to diffusion of morphogens. The apparently critical role of the peripheral cytoplasm in determining the fate of the totipotent nuclei has been classically viewed as suggesting a role for particulate determinants. In many different sorts of embryos, the cortical cytoplasm has been implicated in determination processes, because centrifugation of oocytes often leads to the relocation of most of the observable organelles, but this relocation sometimes has had little effect in changing the determination of various regions of the egg. The cortex is often cited as the element least affected by centrifugation.

The preceding section summarized some of the growing body of evidence that transport of RNP from trophic cells to oocytes is accomplished by an electrophoretic mechanism. Electrophoresis could be invoked as the organizing gradient responsible for distributing mRNPs in a specific array, if two conditions could be met. First, there must be a method for discriminating between particular mRNPs on the basis of net charge, and second, there would have to be some mechanism to trap and hold the mRNPs at the electrically assigned locations. Reorganizations such as those observed in the induction of double abdomens would further require either an ability to reestablish the electrical field, or more probably, to retain the polarity during the postoogenetic period until well after fertilization. In the case of the experiments of Kandler-Singer and Kalthoff (1976), in which double abdomens were induced by puncturing the anterior pole in the presence of ribonuclease, the RNase could be expected to destroy the anterior deter-

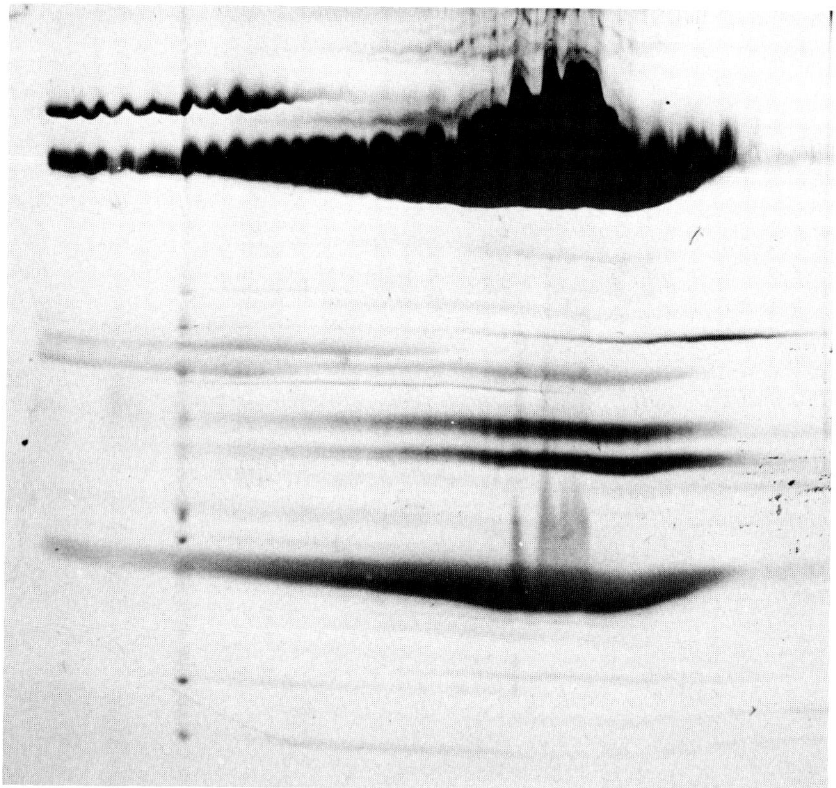

**Figure 13.** Two-dimensional separation of the mRNP proteins of *Antheraea polyphemus* oocytes. mRNP was removed from sucrose gradients and separated in the first dimension on 0.7% agarose gels. The agarose gels were then treated with urea; separation in the second dimension was carried out in 10% acrylamide gels in the presence of sodium dodecyl sulfate. Note the nonhomogeneous distribution of the minor bands. (Courtesy of B. Jarnot.)

minants. If there were an electrical field in the untreated embryo with the positive pole in the anterior region, puncturing might be expected to interrupt current flow. Once the embryo had resealed the damaged surface, the field could be reorganized at the margin of the damaged region of the cortex near the center of the anterior-posterior axis. A reconstruction of the electrical gradient would lead to displacement of the positive pole to the center and to negative poles at the posterior and the new anterior end. The new gradient with its axis of symmetry in the center could reorient the undamaged posterior determinants to form the double abdomen.

There is little objective evidence to support a model such as this, but some of the results obtained recently in my laboratory utilizing the eggs and embryos of Lepidoptera are consistent with it. When the mRNP fraction from oocytes is separated by electrophoresis in very dilute agarose gels (0.7% agarose), a series of broad bands is resolved. When these gels are

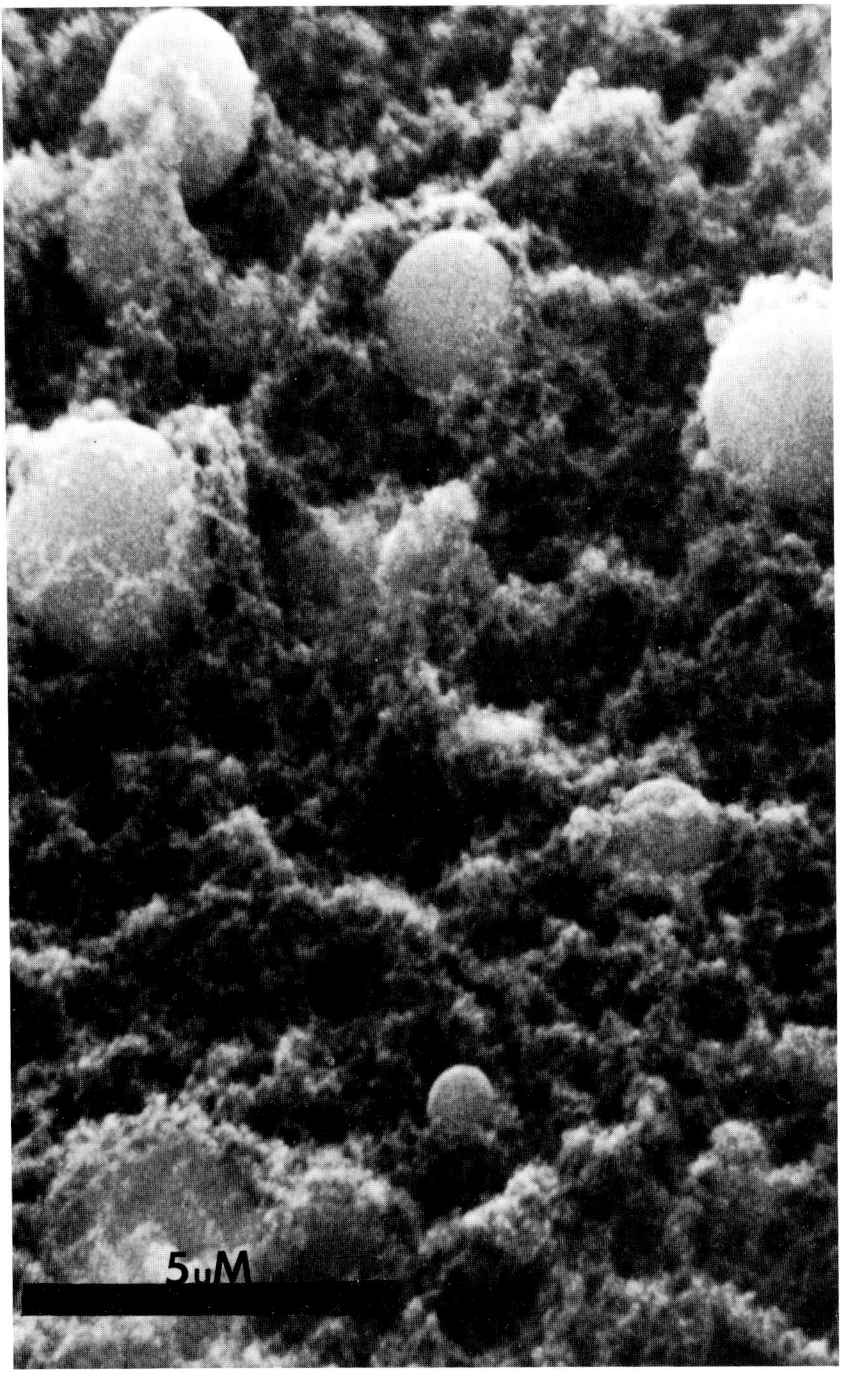


**Figure 14.** Cytoskeleton of the cortical cytoplasm of the unfertilized oocyte of *Antheraea polyphemus* viewed from the inner edge of the cortex. The spherical objects are yolk platelets. Prepared by detergent extraction of fragments of mature chorionated oocytes.

treated with dissociating agents such as urea and the released proteins separated by electrophoresis in a second dimension, up to 30 separate bands of proteins can be resolved. One such gel is shown in Fig. 13. Of particular interest is the nonuniform distribution of the low-abundance proteins. Such results indicate that mRNP particles can be separated by electrophoresis and that the different proteins associated with the particles could determine the electrophoretic behavior of the particles.

The requirement for a mechanism to immobilize the mRNAs could be satisfied by binding to cytoskeletal elements in the egg cortex. The nature of the cortex of insect eggs has not been extensively examined, but an organized cytoskeleton can be observed in this region. Figure 14 shows the structure of the cortex of an unfertilized egg of *A. polyphemus* after detergent extraction. Moon *et al.* (1983) report that mRNA is associated with the cytoskeleton of sea urchin eggs. *In situ* hybridization with poly(U) (Fig. 9A) indicates that mRNA is associated with the cortex of lepidopteran eggs. This cytoskeleton could bind the mRNA and hold it in place. Finally, in the preparation of cell-free homogenates of early embryos, it became clear that if the pieces of chorion with their attached cortex were not removed from the preparation, more extensive and uniform rates of chloramphenicol-insensitive translation were obtained. This result indicates that some portion of the protein synthesis apparatus is located in the cortex.

None of the evidence cited above is more than circumstantial, but it is consistent with a model of determination in which mRNPs are distributed by an electrical gradient along the anterior-posterior axis of the embryo and suggests the direction that more critical experiments to test the model might take.

Acknowledgments. I am grateful to Professors Erwin Heubner, Jurgen Büning, and Robert King for providing me with copies of their manuscripts before publication. The expert editorial advice of Bruce Jarnot and Judith Hooper and their willingness to share unpublished data have been invaluable in supporting the intellectual framework of the chapter. Finally, I wish to thank Peggy Everett for her skill and patience in preparing the manuscript.

## References

Anderson, K. V., and Lengyel, J. A., 1981, Changing rates of DNA and RNA synthesis in *Drosophila* embryos, *Dev. Biol.* **82**:127–138.

Berry, S. J., 1982, Maternal direction of oogenesis and early embryogenesis in insects, *Annu. Rev. Entomol.* **267**:205–227.

Berry, S. J., 1985, Insect nucleic acids, in: *Comprehensive Insect Physiology, Biochemistry and Pharmacology*, Vol. 10 (G. A. Kerkut and L. I. Gilbert, eds.), Pergamon Press, Oxford (in press).

Bier, K., 1963, Synthese, interzellularer transport, und abbau von ribonucleinsaure im ovar der stubenfliege *Musca domestica*, *J. Cell Biol.* **16**:436–440.

Brown, E. H., and King, R. C., 1964, Studies on the events resulting in the formation of an egg chamber in *Drosophila melanogaster*, *Growth* **28**:41–81.

Caldwell, D. G., and Emerson, C. F., 1982, Methylation of 5′ terminal caps is activated following fertilization of sea urchin eggs, *Forty-first Symposium of the Society for Developmental Biology* (abst).

Capco, D., and Jeffery, W., 1979, Origin and spatial distribution of maternal messenger RNA during oogenesis of an insect, *Oncopeltus fasciatus*, *J. Cell Sci.* **39**:63–76.

Carpenter, A. T. C., 1975, Electron microscopy of meiosis in *Drosophila melanogaster* females. I. Structure, arrangement, and temporal change of the synaptonemal complex in wild-type, *Chromosoma* **51**:157–182.

Cave, M. D., 1973, Synthesis and characterization of amplified DNA oocytes of the house cricket, *Acheta domesticus* (Orthoptera: Gyrillaidae), *Chromosoma* **42**:1–22.

Cave, M. D., 1975, Absence of ribosomal DNA amplification in the meroistic (telotrophic) ovary of the large milkweed bug *Oncopeltus fasciatus* (Dallas) (Hemiptera: Lygaeidae), *J. Cell Biol.* **66**:401–469.

Cave, M. D., 1978, Absence of amplification of ribosomal DNA in the polytrophic meroistic ovary of the giant silkmoth, *Antheraea pernyi*, *Wilhelm Roux Arch. Dev. Biol.* **184**:135–142.

Cave, M. D., 1982, Morphological manifestations of ribosomal DNA amplification during insect oogenesis, in: *Insect Ultrastructure*, Vol. 1 (R. C. King and H. Akai, eds.), pp. 86–117, Plenum Press, New York.

Cave, M. D., and Sixby, J., 1976, Absence of ribosomal DNA amplification in a meroistic polytrophic ovary, *Exp. Cell Res.* **101**:23–30.

Choi, W. C., and Nagl, W., 1977, Patterns of DNA and RNA synthesis during the development of ovarian nurse cells in *Gerris najas* (heteroptera), *Dev. Biol.* **61**:262–272.

Chooi, W. Y., 1976, RNA transcription and ribosomal protein assembly in *Drosophila melanogaster*, in: *Handbook of Genetics*, Vol. 5: *Molecular Genetics* (R. C. King, ed.), pp. 219–265, Plenum Press, New York.

Davenport, D., 1976, Transport of ribosomal RNA into the oocytes of the milkweed bug, *Oncopeltus fasciatus*, *J. Insect Physiol.* **22**:925–926.

Davidson, E. H., 1976, *Gene Action in Early Development*, 2nd ed., Academic Press, New York.

DeLoof, A., 1983, The meroistic insect ovary as a miniature electrophoresis chamber, *Comp. Biochem. Physiol.* **74A**:3–9.

Diaz, M. O., Barsacchi-Pilone, G., Mahon, K. A., and Gall, J. G., 1981, Transcripts from both strands of a satellite DNA occur on lampbrush chromosome loops of the Newt Notophthalmus, *Cell* **24**:649–659.

Dittmann, F., Ehni, R., and Engels, W., 1981, Bioelectric aspects of the hemipteran telotrophic ovariole. (Dysdercus intermedius), *Wilhelm Roux Arch.* **190**:221–225.

Filipowicz, W., 1978, Functions of the 5′-terminal $M^7G$ cap in eukaryotic mRNA, *FEBS Lett.* **96**:1–11.

Ford, P. J., and Southern E. M., 1973, Different sequences for 5S RNA in kidney cells and ovaries of *Xenopus laevis*, *Nature New Biol.* **241**:7–12.

Gall, J. G., MacGregor, H. C., and Kidston, M. E., 1969, Gene amplification in the oocytes of Dytiscid water beetles, *Chromosoma* **26**:169–187.

Gerhart, J. G., 1980, Mechanisms regulating pattern formation in the amphibian egg and early embryo, in: *Biological Regulation and Development*, Vol. 2 (R. F. Goldberger, ed.), pp. 133–315, Plenum Press, New York.

Giardina, A., 1901, Origine dell'oöcite e delle cellule nutrici nei *Dytiscus*, *Int. Mschr. Anat. Physiol.* **18**:417–484.

Hall, T. J., and Cummings, M. R., 1975, *In vitro* synthesis and processing of ribosomal RNA in the housefly ovary, *Dev. Biol.* **46**:233–242.

Harris, S. E., and Forrest, H. S., 1967, RNA and DNA synthesis in developing eggs of the milkweed bug *Oncopeltus fasciatus* (Dallas), *Science* **156:**1613–1615.

Hough-Evans, B. R., Jacobs-Lorena, M., Cummings, M. R., Britten, R. J., and Davidson, E. H., 1980, Complexity of RNA in eggs of *Drosophila melanogaster* and *Musca domestica*, *Genetics* **95:**81–94.

Huebner, E., 1981, Oocyte–follicle cell interaction during normal oogenesis and atresia in an insect, *J. Ultrastruct. Res.* **74:**95–104.

Huebner, E., 1984, The ultrastructure and development of the telotrophic ovary, in: *Insect Ultrastructure*, Vol. 2 (R. C. King and H. Akai, eds.), Plenum Press, New York (in press).

Hughes, M., and Berry, S. J., 1970, The synthesis and secretion of ribosomes by nurse cells of *Antheraea polyphemus*, *Dev. Biol.* **23:**651–664.

Jäckle, H., 1980a, Two-dimensional gel analysis after removal of major proteins reveals stage-dependent proteins in early insect development, *FEBS Lett.* **118:**191–194.

Jäckle, H., 1980b, Actin messenger in maternal RNP particles from an insect embryo (*Smittia* spec., Chironomidae, Diptera), *Wilhelm Roux Arch. EntwMech.* **188:**225–233.

Jäckle, H., and Kalthoff, K., 1978, Photoreactivation of RNA in UV-irradiated insect eggs (*Smittia* sp., Chironomidae, Diptera). I. Photosensitized production and light-dependent disappearance of pyrimidine dimers, *Photochem. Photobiol.* **27:**309–315.

Jäckle, H., and Kalthoff, K., 1980, Photoreversible UV-inactivation of messenger RNA in an insect embryo (*Smittia* spec. Chironomidae, Diptera), *Photochem. Photobiol.* **32:**749–761.

Jacobs-Lorena, M., Hough-Evans, B. R., Britten, R. J., and Davidson, E. H., 1980, Complexity of RNA in developing oocytes of *Drosophila melanogaster*, *Dev. Biol.* **76:**509–513.

Jaffe, L. F., 1969, On the centripetal course of development, the *Fucus* egg, and self-electrophoresis, *Dev. Biol. Suppl.* **3:**83–111.

Jaffe, L. F., and Nuccitelli, R., 1974, An ultrasensitive vibrating probe for measuring steady extracellular currents, *J. Cell Biol.* **63:**614–628.

Jaffe, L. F., and Woodruff, R. I., 1979, Large electrical currents traverse developing cecropia follicles, *Proc. Natl. Acad. Sci. USA* **76:**1328–1332.

Kalthoff, K., and Jäckle, H., 1982, Photoreactivation of pyrimidine dimers generated by a photosensitized reaction in RNA in insect embryos (*Smittia* spec), in: *Trends in Photobiology* (C. Helene, M. Chariler, T. H. Montenay-Garestier, and G. Laustriat, eds.), pp. 173–188, Plenum, New York.

Kalthoff, K., Kandler-Singer, I., Schmidt, O., and Zissler, D., 1975, Mitochondria and polarity in the egg of *Smittia* spec. (Diptera, Chironomidae): UV irradiation, respiration measurements ATP determinations and application of inhibitors, *Wilhelm Roux Arch. EntwMech.* **178:**99–121.

Kalthoff, K., Urban, K., and Jäckle, H., 1978, Photoreactivation of RNA in UV-irradiated insect eggs (*Smittia* sp., Chironomidae, Diptera). II. Evidence for heterogeneous light-dependent repair activities, *Photochem. Photobiol.* **27:**317–322.

Kalthoff, K., Rau, K., and Edmond, J. C., 1982, Modifying effects of ultraviolet irradiation of the development of abnormal body patterns in centrifuged insect embryos (*Smittia* sp. Chironomidae, diptera), *Dev. Biol.* **91:**413–422.

Kandler-Singer, I., and Kalthoff, K., 1976, RNase sensitivity of an anterior morphogenetic determinant in an insect egg (Smittia spec., Chironomidae, Diptera), *Proc. Natl. Acad. Sci. USA* **73:**3739–3743.

Kastern, W. H., and Berry, S. J., 1976, Non-methylated guanosine as the 5′ terminus of capped mRNA from insect oocytes, *Biochem. Biophys. Res. Commun.* **71:**37–44.

Kastern, W. H., Underberg, D. A., and Berry, S. J., 1981, DNA-dependent RNA polymerases I and II in insect oocytes, *Dev. Biol.* **87:**383–389.

Kastern, W. H., Swindlehurst, M., Aaron, C., Hooper, J., and Berry, S. J., 1982, Control of mRNA translation in oocytes and developing embryos of giant moths. 1. Function of the 5′ terminal "cap" in the tobacco hornworm, *Manduca sexta*, *Dev. Biol.* **89:**437–449.

King, R. C., 1970, *Ovarian Development in Drosophila melanogaster*, 227, Academic Press, New York and London.

King, R. C., and Aggarwal S. K., 1965, Oogenesis in *Hyalophora cecropia*, *Growth* **29:** 17–83.

King, R. C., and Büning, J., 1985, The origin and functioning of insects oocytes and nurse cells, in: *Comprehensive Insect Physiology, Biochemistry and Pharmacology*, Vol. 1 (G. A. Kerkut and L. I. Gilbert, eds.), Pergamon Press, Oxford (in press).

King, R. C., Aggarwal, S. K., and Aggarwal, U., 1968, The development of the female *Drosophila* reproductive system, *J. Morphol.* **124:**143–166.

King, R. C., Cassidy, J. D., and Rousset, A., 1982, The formation of clones of interconnected cells during gametogenesis in insects, in: *Insect Ultrastructure*, Vol. 1 (R. C. King and H. Akai, eds.), pp. 3–31, Plenum Press, New York.

Koch, E. A., and King, R. C., 1969, Further studies on the ring canal system of the ovarian cystocytes of *Drosophila melanogaster*, *Z. Zellforsch.* **102:**129–152.

Lamb, M. M., and Laird, C. D., 1976, The size of Poly(A)-containing RNAs in *Drosophila melanogaster* embryos, *Biochem. Genet.* **14:**357–371.

Lima-de-Faria, A., Birnstiel, M., and Jaworska, H., 1969, Amplication of ribosomal cistrons in the heterochromatin of *Acheta*, *Genetics* Suppl 1 **61:**145–159.

Limbourg, B., and Zalokar, M., 1973, Permeabilization of *Drosophila* eggs, *Dev. Biol.* **35:**382–387.

Lockshin, R. A., 1966, Insect embryogenesis: Macromolecular synthesis during early development, *Science* **154:**775–776.

Long, E. O., and Dawid, I. B., 1980, Repeated genes in eukaryotes, *Annu. Rev. Biochem.* **49:**727–764.

Mahowald, A. P., and Strassheim, J. M., 1970, Intracellular migration of centrioles in the germarium of *Drosophila melanogaster*, *J. Cell Biol.* **45:**306–320.

Meinhardt, H., 1977, A model for pattern formation in insect embryogenesis, *J. Cell Sci.* **23:**117–139.

Mermod, J. J., Schatz, G., and Crippa, M., 1980, Specific control of messenger translation in *Drosophila* oocytes and embryos, *Dev. Biol.* **75:**177–186.

Miller, O. L., and Beatty, B. R., 1969, Visualization of nucleolar genes, *Science* **164:**955–957.

Moon, R. T., Danilchik, M. V., and Hille, M. B., 1982, An assessment of the masked message hypothesis: Sea urchin messenger ribonucleoprotein complexes are efficient templates for *in vitro* protein synthesis, *Dev. Biol.* **93:**389–403.

Moon, R. T., Nicosia, R. F., Olsen, C., Hille, M. B., and Jeffery, W. R., 1983, The cytoskeletal framework of sea urchin eggs and embryos: Developmental changes in the association of messenger RNA, *Dev. Biol.* **95:**447–458.

Muhlach, W. L., and Schwalm, F. E., 1977, Utilization of uridine in developing ovarioles of the kelp fly, *Coelopa*, *J. Insect Physiol.* **23:**931–937.

Newport, J., and Kirschner, M., 1982a, A major developmental transition in early *Xenopus* embryos: I. Characterization and timing of cellular changes at the midblastula stage, *Cell* **30:**675–686.

Newport, J., and Kirschner, M., 1982b, A major developmental transition in early *Xenopus* embryos: II. Control of the onset of transcription, *Cell* **30:**687–696.

Paglia, L. M., Berry, S. J., and Kastern, W. H., 1976a, Messenger RNA synthesis, transport, and storage in silkmoth ovarian follicles, *Dev. Biol.* **51:**173–181.

Paglia, L. M., Kastern, W. H., and Berry, S. J., 1976b, Messenger ribonucleoprotein particles in silkmoth oogenesis, *Dev. Biol.* **51:**182–189.

Pietruschka, F., and Bier, K., 1972, Autoradiographische Untersuchungen zur RNS-und Protein-Synthese in der fruhen Embryogenese von *Musca domestica*, *Wilhelm Roux Arch. EntwMech.* **169:**56–69.

Pollak, S. B., and Telfer, W. H., 1969, RNA in cecropia moth ovaries: Sites of synthesis, transport, and storage, *J. Exp. Zool.* **170:**1–24.

Raff, R., 1980, Masked messenger RNA and the regulation of protein synthesis in eggs and embryos, in: *Cell Biology*, Vol. 4 (D. M. Prescott, ed.), pp. 107–136, Academic Press, New York.

Renkawitz, R., and Kunz, W., 1975, Independent replication of the ribosomal RNA genes in the polytrophic–meroistic ovaries of *Calliphora erythrocephala, Drosophila hydei,* and *Sarcophaga barbata, Chromosoma* **53:**131–140.

Spear, B. B., and Gall, J. G., 1974, Independent control of ribosomal gene replication in polytene chromosomes of *D. melanogaster, Proc Natl. Acad. Sci. USA* **70:**1359–1363.

Struhl, G., 1981, A gene product required for correct initiation of segmental determination in *Drosophila, Nature* **293:**36–41.

Telfer, W. H., 1975, Development and physiology of the oocyte–nurse cell synctium, *Adv. Insect Physiol.* **11:**223–320.

Telfer, W. H., Woodruff, R. I., and Huebner E., 1981, Electrical polarity and cellular differentiation in meroistic ovaries, *Am. Zool.* **21:**675–686.

Winter, H., 1974, Ribonucleoprotein-partikel aus dem telotroph-meroistischen Ovar von *Dysderrcus intermedius* Dis. (Heteroptera, Pyrrhoc.) und ihr Verhalten im zellfreien Proteinsynthesesystem, *Wilhelm Roux Arch. EntwMech.* **175:**103–127.

Wollberg, Z., Cohen, E., and Kalina, M., 1976, Electrical properties of developing oocytes of the migratory locust, *Locusta migratoria, J. Cell Physiol.* **88:**145–158.

Woodruff, R. I., and Telfer, W. H., 1973, Polarized intercellular bridges in the ovarian follicles of the Cecropia moth, *J. Cell Biol.* **58:**172–188.

Woodruff, R. I., and Telfer, W. H., 1980, Electrophoresis of proteins in intercellular bridges, *Nature* **286:**84–86.

Zalokar, M., 1976, Autoradiographic study of protein and RNA formation during early development of Drosophila eggs, *Dev. Biol.* **49:**425–437.

Zissler, D., and Sander, K., 1982, The cytoplasmic architecture of the insect egg cell, in: *Insect Ultrastructure,* Vol. 1 (R. C. King and H. Akai, eds.), pp. 189–221, Plenum Press, New York.

Chapter 9

# Functional Organization of the Amphibian Oocyte Nucleus

ULRICH SCHEER and MARIE-CHRISTINE DABAUVALLE

## 1. Introduction

Nuclei of amphibian oocytes ("germinal vesicles") can grow to considerable size. For example, the nucleus of a mature *Xenopus laevis* oocyte has a diameter of about 0.5 mm and, after isolation, can be seen with the naked eye. This exceptional size (for comparison, the diameter of somatic cell nuclei usually ranges from 5 to 30 μm), the ease and rapidity with which they can be manually isolated, and the occurrence of lampbrush chromosomes and amplified nucleoli makes amphibian nuclei choice material not only for studies of chromosome structure and genetic activity at different levels of resolution, but also for analysis of the biochemical composition of individual nuclei and nuclear components as well. A single manually isolated nucleus from an amphibian oocyte is sufficient to analyze its major protein constituents by gel electrophoresis, and a few nuclei provide sufficient RNA for analysis by gel electrophoresis or electron microscopic spreading methods. In addition, the size of amphibian oocyte nuclei greatly facilitates the introduction of certain substances by microinjection. In fact, *Xenopus* oocytes are presently used in numerous laboratories as "living test tubes" to study the expression of cloned DNA sequences after microinjection into their nuclei with the aim of identifying those DNA sequences necessary for transcriptional initiation and termination events, as well as the factors involved in gene regulation. Another important approach is the microinjection of proteins into amphibian oocytes in order to analyze their resulting nucleocytoplasmic distribution and to classify them as "karyophilic," "karyophobic," or "amphiphilic" proteins. This approach requires a very rapid isolation of nuclei in order to prevent redistribution

ULRICH SCHEER and MARIE-CHRISTINE DABAUVALLE • Institute of Cell and Tumor Biology, German Cancer Research Center, D-6900 Heidelberg 1, Federal Republic of Germany.

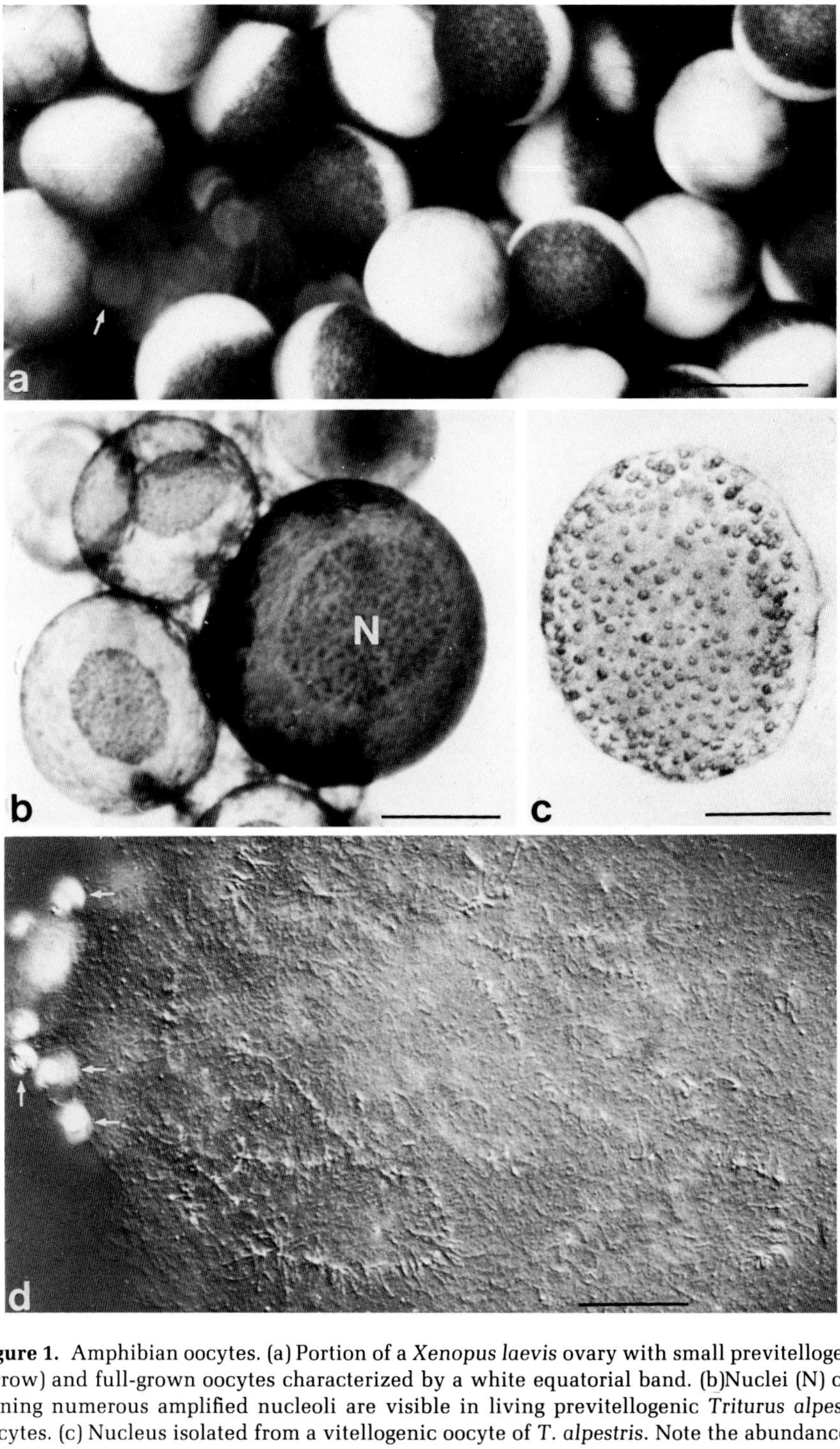


**Figure 1.** Amphibian oocytes. (a) Portion of a *Xenopus laevis* ovary with small previtellogenic (arrow) and full-grown oocytes characterized by a white equatorial band. (b)Nuclei (N) containing numerous amplified nucleoli are visible in living previtellogenic *Triturus alpestris* oocytes. (c) Nucleus isolated from a vitellogenic oocyte of *T. alpestris*. Note the abundance of

or leakage of proteins and therefore excludes in most cases the use of conventionally isolated nuclei from somatic cells.

Thus, nuclei of amphibian oocytes are not only of general interest for studies that attempt to correlate the structure and function of nuclear components, but also provide a very useful experimental system for micromanipulation.

For the purpose of this chapter it is sufficient to divide amphibian oogenesis into three main stages: (1) previtellogenesis, (2) vitellogenesis, and (3) postvitellogenesis or maturity. A portion of an ovary from *Xenopus laevis* is shown in Fig. 1a to illustrate the difference in size between translucent previtellogenic and full-grown oocytes (the commonly used staging system is that of Dumont, 1972.)

In the first part of this chapter we discuss the main components of the amphibian oocyte nucleus, i.e., **lampbrush chromosomes, amplified nucleoli,** and the **nuclear envelope,** which can be recognized in the living oocyte (Fig. 1b) or in freshly isolated nuclei (Fig. 1c,d). It is worth recalling that these structures were identified and described in detail almost a century ago using stained paraffin sections of ovaries (Born, 1894; Carnoy and Lebrun, 1897, 1898). Next, we consider the soluble phase of oocyte nuclei—the **nucleoplasm** or nuclear sap—which contains the majority of nuclear proteins. Of special interest are histone and nonhistone proteins, which are stored in a soluble form and are most likely used to package DNA into chromatin during the extremely rapid cell divisions of early embryonic development. Finally, we discuss the possibility of interfering with gene expression in the living oocyte by microinjection of specific antibodies into its nucleus.

## 2. Lampbrush Chromosomes

### 2.1. General Morphology

Lampbrush chromosomes are the most conspicuous and are certainly the most attractive constituents of amphibian oocyte nuclei. As chromosomes of the diplotene stage of meiotic prophase, they occur in the form of homologous pairs, being joined by one or more chiasmata (Fig. 2a). Because of their size, lampbrush chromosomes can be readily observed under the light microscope (Fig. 2a,b) (for methods of preparation, see Callan and Lloyd, 1960; Gall, 1966). The axis of each chromosome appears as a linear

amplified nucleoli. (d) In optical cross sections of the *T. alpestris* nucleus (the nuclear envelope has been removed for clarity), the tight packaging of the lampbrush chromosomes is evident (arrows denote some amplified nucleoli). Scale bars: 1 mm (a), 0.2 mm (b,c), and 50 $\mu$m (d).

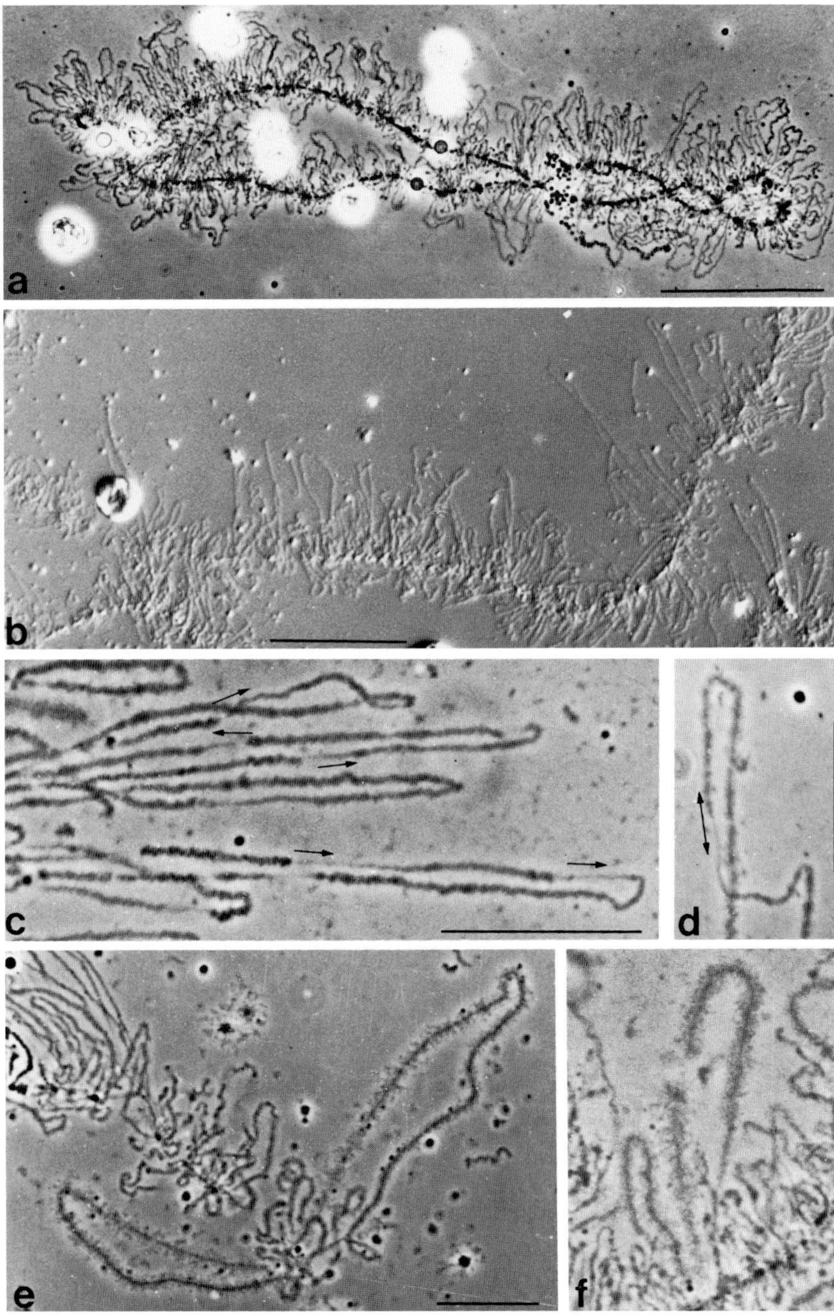


**Figure 2.** Isolated lampbrush chromosomes from *Pleurodeles waltlii* (a), *Triturus alpestris* (b–e) and *T. cristatus* (f), photographed with phase contrast (a, c–f) and Nomarski interference contrast optics (b). The matrix material of the giant loops (e,f) contains numerous fine fibrils of increasing lengths, so that a thin and thick loop end can be distinguished. Other loops contain two or more thin–thick gradients of identical or opposite polarities (arrows in c and d). Scale bars: 50 $\mu$m (a,b) and 20 $\mu$m (c–f).

aggregate of dark globules, the **chromomeres,** from which numerous **loops** extend laterally. In *Triturus* species, most of the loops are 30–50 μm long, but some can well exceed 100 μm in axial length (Fig. 2e). Although the chromosomal axis normally appears single, each chromosome consists of two identical chromatids, so that the loops extend as symmetrical sister pairs (Fig. 2e). Specific loops, which are recognizable by size and/or unique morphology, occur at genetically fixed positions and can serve as reference structures to identify the chromosomes. "Working maps" of lampbrush chromosomes have been constructed for numerous amphibian species (Callan and Lloyd, 1975).

The precise stage at which lampbrush chromosomes form is difficult to determine, because the unfolding of the lateral loops is a gradual process. It is clear, however, that the lampbrush form develops very early in diplotene shortly after the pachytene stage, lasting until the oocyte approaches maturity.

The lateral loops are the manifestation of ongoing transcription (see also Section 8). When lampbrush chromosomes are spread for electron microscopy according to the procedure introduced by O. Miller (Miller and Beatty, 1969; Miller and Hamkalo, 1972), numerous individual lateral fibrils are seen attached to the loop axes (Fig. 3a,b). These lateral fibrils have been identified as nascent ribonucleoprotein (RNP) transcripts, each containing a growing RNA chain complexed with proteins. They are anchored to the chromatin axis by virtue of a 12–15-nm-thick particle containing an RNA polymerase molecule (for details, see Miller *et al.*, 1972; Scheer *et al.*, 1979a; Sommerville, 1981). When viewed under the light microscope, numerous loops appear asymmetrical in the sense that they reveal a continuously increasing thickness of the RNP matrix between their two insertion sites on the chromomere (Fig. 2e,f). This polarized RNP distribution reflects the increasing transcript length as the polymerases move along the loop axis from the thin end containing the initiation site to the thick end containing the termination site for transcription. These loops therefore contain one transcriptional unit. On the other hand, loops with multiple thin–thick gradients of identical, opposite, or varying polarities can also be found reflecting the occurrence of several transcription units within a single loop (Fig. 2c,d) (Scheer *et al.*, 1979a). This interpretation is confirmed by electron microscopic spread preparations (Fig. 3c). A situation such as that shown in Fig. 3c, in which the origins of two transcription units of opposite polarity are separated by an extremely short spacer region, clearly excludes the possibility that in the living cell the spacer has formed a part of the chromomere and was pulled out artificially due to mechanical stress during the spreading procedure.

Whereas the progressive size increase of the nascent RNP fibrils can be clearly seen in electron microscopic spreads near the origin of transcription, i.e., in the proximal portions of transcriptional units (Fig. 3a–c), the lateral fibrils of more distal regions are difficult to trace because of their

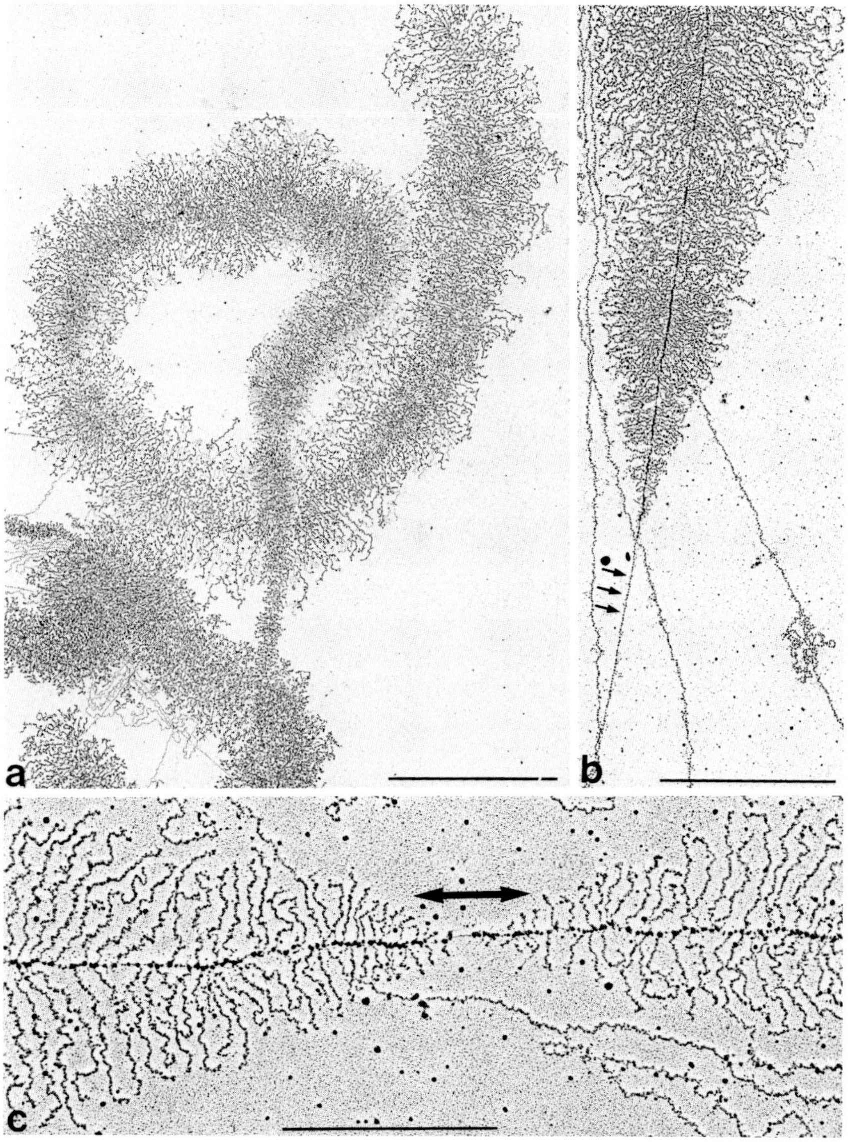

**Figure 3.** Electron microscopic spread preparation of lampbrush chromosomes from *Pleurodeles waltlii.* Note the high packing density of transcribing RNA polymerase particles. The nontranscribed chromatin colinear with the origin of the transcription unit shown in (b) has a beaded appearance (arrows). Two closely adjacent transcription units of opposite polarities (denoted by the double-headed arrow) are shown in c. Scale bars: 5 $\mu$m (a), 2 $\mu$m (b) and 1 $\mu$m (c).

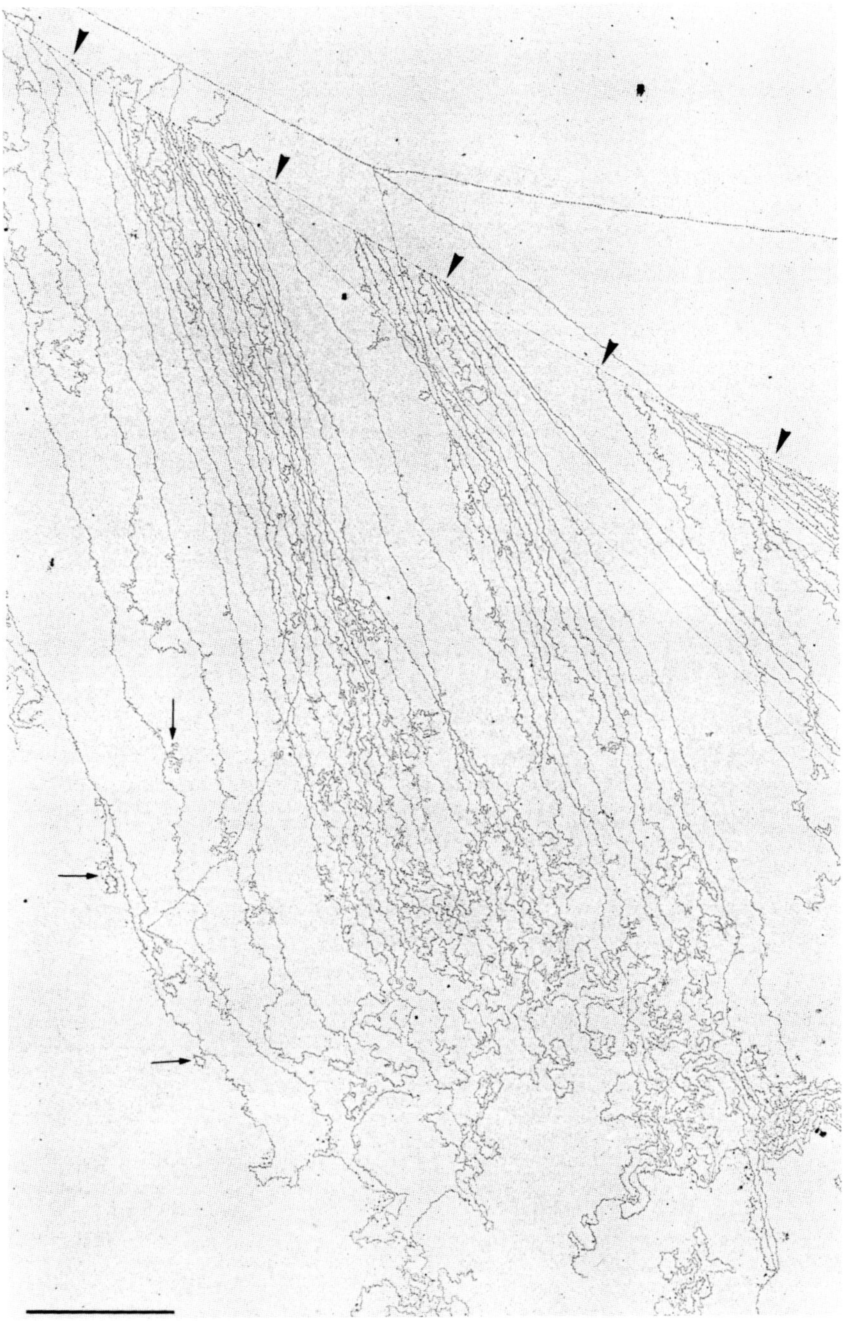

**Figure 4.** Terminal region of a transcription unit from a lampbrush chromosome of *Triturus helveticus*. The chromatin axis is denoted by arrowheads. The extremely long nascent ribonucleoprotein (RNP) fibrils are folded at numerous sites into complex ring- and bushlike structures (e.g., at the arrows). Scale bar: 2$\mu$m.

tendency to form extensive aggregates. In favorable spreads, however, nascent RNP fibrils can be seen to reach lengths of $\geq 20$ $\mu$m (Fig. 4).

## 2.2. Structural Organization of Nascent RNP Transcripts

A comparison of the highly compact state of the RNP matrix as seen at the light microscopic level (Fig. 2a–d) with the extended configuration of individual lateral RNP fibrils in electron microscopic spread preparations (Fig. 4) demonstrates that *in vivo* the RNP transcripts are tightly packed but can be transformed experimentally into extended fibrils. This transformation is brought about by exposure of the lampbrush chromosomes to mildly alkaline buffers of very low ionic strength (0.1–1 mM borate buffer, pH 8–9; for details, see Miller *et al.*, 1972; Miller and Hamkalo, 1972; Scheer *et al.*, 1979a). Electron micrographs of sections through flat embedded lampbrush chromosomes isolated in a solution containing 0.1 M saline probably reflect the natural packaging state of nascent RNP transcripts (Fig. 5a) (Mott and Callan, 1975; Spring and Franke, 1981). Under such conditions, the RNP material consists of periodically arranged globular particles with a size ranging from about 25 to 40 nm. These globular units of compacted RNP are unraveled by lowering the ionic strength, as seen in spread preparations (Fig. 5b–d). An intermediate stage in this unfolding process is presented in Fig. 5b. Maximally extended RNP transcripts appear as thin fibrillar structures 5–12 nm thick (Figs. 4 and 5c,d). Occasionally, however, they display a particulate morphology due to the presence of closely spaced 20-nm particles. Such regions can alternate with regions of smooth morphology along a given fibril (Fig. 5c). Thus, under the specific spreading conditions, nascent RNP material apparently can assume two alternative morphological aspects but is—at the next higher level of organization—compacted into linear arrays of globular structures with diameters of 25–40 nm.

Lateral fibrils commonly display other, unique morphological features such as loop- and ringlike structures distributed along their entire lengths (Fig. 5d) (for further examples, see Sommerville, 1981). The most commonly found structure is an RNP loop of variable circumference connected to the main axis of the lateral fibril via a very thin stem, which most likely represents a double-stranded RNA region with few, if any, associated proteins (Fig. 5d) ( Sommerville, 1981). It is important to point out that similar or more complex branching or bushlike RNP structures are also found in somatic cells of amphibian (Fig. 5e) and other species (e.g., Beyer *et al.*, 1981).

Loop-stem structures do not occur at random but rather at corresponding positions in the lateral fibrils of a transcriptional unit, shown schematically in Figure 6. Such a situation indicates that stable RNA duplex structures form at defined positions by a foldback mechanism of the nascent RNA chain shortly after synthesis of inverted repeat sequences.

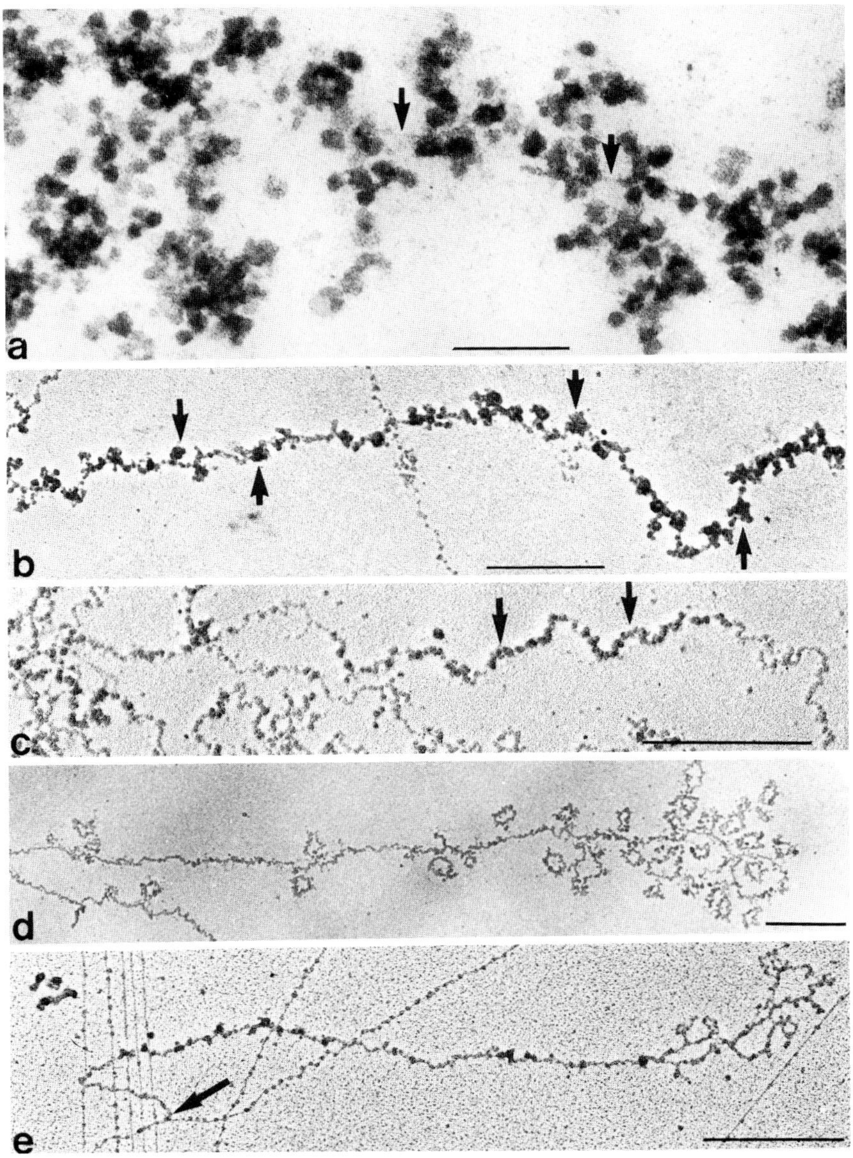

**Figure 5.** Morphology of nascent ribonucleoprotein (RNP) as seen in an ultrathin section through flat-embedded lampbrush chromosomes of *Pleurodeles waltlii* (a) and in spread preparations of lampbrush chromosomes from *P. waltlii* (b), *Triturus cristatus* (c), and *Necturus maculosus* (d). For comparison, spread chromatin from cultured *Xenopus* kidney cells with an attached RNP fibril is shown (e). (a) The RNP material of lampbrush chromosomes isolated in 0.1 M saline appears as strings of 25–40-nm granules around the loop axis. The loop axis is denoted by arrows. (b–d) Low salt treatment unfolds these linear aggregates of particles into fibrillar strands. (b) An intermediate state of the relaxation process is depicted, where the arrows denote some residual large globular units of condensation. (c) The extended RNP fibrils occasionally show a beaded morphology (arrows). (d) Complex folding of the RNP fibril into numerous circular structures. (e) Corresponding structures are also found in somatic cell nuclei (arrow points to the attachment site of the RNP fibril to the beaded chromatin axis). Scale bars: 0.5 $\mu$m (b–e) and 0.2 $\mu$m (a).

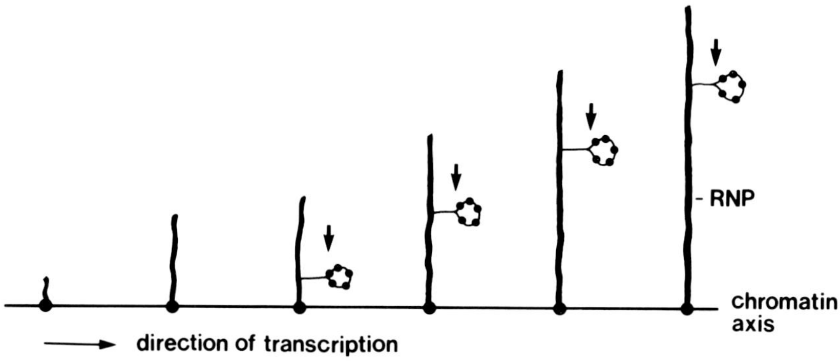


**Figure 6.** Analysis of a transcription unit from a lampbrush chromosome of *Triturus helveticus*. The scheme illustrates that loop formation is not a random process but occurs at corresponding sites along the nascent ribonucleoprotein (RNP) fibrils (arrows). Compare the length reduction of the third RNP fibril with that of the second RNP fibril, caused by loop formation.

## 2.3. Proteins Bound to Nascent RNA

Morphological evidence from spread preparations suggests that proteins bind to nascent RNA with the possible exception of double-stranded regions (see Section 2.2). Whereas the protein composition of RNP particles containing heterogeneous nuclear RNA (hnRNP) isolated from *Triturus* oocytes has been studied in detail by sodium dodecyl sulfate polyacrylamide gel electrophoresis (SDS-PAGE) (Sommerville, 1981), it is as yet not clear whether the same set of proteins is also associated with nascent RNA. Immunofluorescence microscopy has been used to show that the ubiquitous core proteins of hnRNP particles with molecular weights ranging from 34,000 to 40,000 are constitutive elements of the RNP material of all loops of lampbrush chromosomes (Martin and Okamura, 1981). This observation indicates that proteins immunologically related to mammalian RNP core proteins bind to nascent RNA and are probably involved in folding the growing chain. In addition, there is considerable evidence for the occurrence of loop-specific or RNA sequence-specific proteins in addition to these ubiquitous packaging proteins (Sommerville, 1981).

## 2.4. Sizes of Chromosome Loops, Transcription Units, and hnRNA

The relative amount of DNA engaged in transcription at any one time can be calculated directly from light microscopic preparations of lampbrush chromosomes by determining the axial lengths of all lateral loops per chromosome set. Irrespective of large variations of genome sizes, especially of tailed amphibia, the percentage of genomic DNA contained in the lateral loops is relatively constant at 5–10% (Scheer and Sommerville, 1982). A similar figure has been obtained from molecular hybridization experiments (see Davidson, 1976).

In *Triturus* species, most of the lateral loops have axial lengths of 30–50 μm (Fig. 2b–d). The dimensions of transcriptional units as measured in electron microscopic spread preparations are similar (3–30 μm) (Fig. 3a), although this technique tends to underestimate the true lengths of transcriptional units due to the difficulty in identifying their termini. Primary transcripts of such transcriptional units—provided that they represent contiguous RNA chains—are therefore expected to have molecular weights ranging between $3 \times 10^6$ and at least $30 \times 10^6$.

An electron micrograph showing hnRNA molecules isolated from nuclei of previtellogenic *Triturus* oocytes by a method that minimizes chain breakage is presented in Fig. 7. It is worth mentioning that a substantial proportion of the RNA molecules form stable intramolecular duplex structures under the partially denaturing conditions of surface spreading. These are visible as hairpin or loop-stem structures, which indicate the extensive occurrence of inverted repeat sequences (for details, see Sommerville and Scheer, 1982). It is tempting to correlate such RNA loops with RNP loops seen in chromatin spreads (see Section 2.2). The results of contour-length measurements of RNA molecules are summarized in Fig. 7b. It is apparent that most of the nuclear hnRNA molecules are relatively short, comparable to the sizes of poly(A)$^+$ RNA isolated from total oocyte RNA (Fig. 7c), and they do not exceed 10 μm in length (corresponding to a molecular weight of $10 \times 10^6$). Thus, nascent RNA molecules contained in apparently continuous lateral RNP fibrils such as those shown in Fig. 4 must be nicked at several sites, with the integrity of the RNP being maintained through protein–protein interactions or through double-stranded RNA regions (see Sommerville, 1981; Scheer and Sommerville, 1982).

## 2.5. Model of Lampbrush Transcription Units

On the basis of molecular hybridization of hnRNA and on *in situ* hybridization of various DNA probes to nascent RNA transcripts attached to chromosomal loops (DNA/RNA-transcript hybridization), the following general picture of a lampbrush transcription unit has emerged (Sommerville and Malcolm, 1976; Varley *et al.*, 1980a; Diaz *et al.*, 1981; Callan, 1982; Gall *et al.*, 1983). The origin of a transcription unit (usually the thin end of the loop), reflects initiation of transcription at a promoter sequence adjacent to the protein-coding region. The RNA polymerases continue transcription beyond the coding region without interruption all along the loop axis. According to this model, a primary transcript would contain gene sequences including introns as well as flanking noncoding sequences with various types of repetitive sequences. In fact, repetitive sequences are tran-

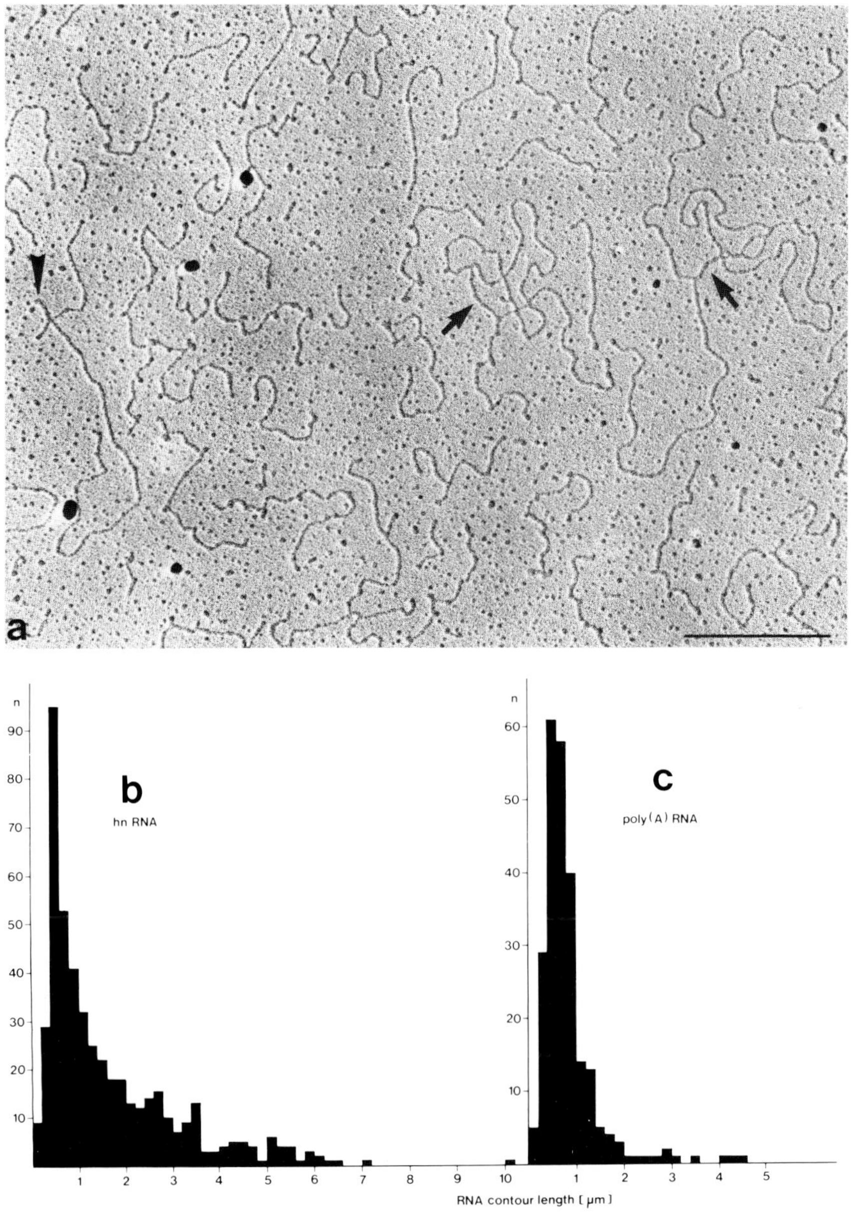


**Figure 7.** Analysis of hnRNA (a,b) and total poly $(A)^+$ RNA (c) from oocytes of *Triturus cristatus*. (a) Electron micrograph of spread hnRNA molecules isolated from nuclei of previtellogenic oocytes. Intrastrand duplex formation results in loop-stem structures (double-stranded "stem" regions denoted by arrows) or in hairpin structures (arrowhead). Scale bar: 0.5 $\mu$m. The histograms illustrate the contour length distributions of hnRNA (b) and polyadenylated RNA from total ovaries (c).

scribed on lateral loops, as shown by DNA/RNA transcript hybridization (Varley *et al.*, 1980a,b; Diaz *et al.*, 1981; Sommerville and Scheer, 1982; Jamrich *et al.*, 1983) and are also present in isolated hnRNA molecules, as shown by their ability to form rapidly intermolecular base-paired structures (Sommerville and Scheer, 1982). In addition, a sizeable amount of poly(A)$^+$ RNA isolated from the cytoplasm of *Xenopus* oocytes also contains interspersed repetitive sequences, in contrast to the poly(A)$^+$ RNA isolated from tadpoles (Anderson *et al.*, 1982) (see Chapter 12).

## 2.6. Visualization of Tandemly Repeated, Very Short Transcription Units That Do Not Form Typical Lateral Loops

Besides the huge transcription units, which are readily visualized at the light microscopic level as lateral loops, small genes such as those coding for 5 S rRNA, tRNA, small nuclear RNA (especially U1; Forbes *et al.*, 1983) and the recently discovered OAX-RNA (Wakefield *et al.*, 1983; Ackerman, 1983) are also transcribed in lampbrush chromosomes. By incubating isolated oocyte nuclei in the presence of different concentrations of $\alpha$-amanitin, Schultz *et al.* (1981) were able to demonstrate convincingly that RNA polymerase II is responsible for all RNA synthesis on lateral loops, whereas polymerase III mediates RNA synthesis at a few sites scattered on the chromomeric axes of the chromosomes. This observation clearly excludes the possibility that genes transcribed by RNA polymerase III (such as the highly reiterated 5 S rRNA genes; see Chapter 10) form a typical loop.

The special chromatin organization shown in Fig. 8a,b probably represents tandemly arranged transcription units of 5 S rRNA genes. This type of chromatin organization is characterized by a periodic alternating pattern of thick and thin regions—the thick regions containing two or three closely spaced RNA polymerase particles, which are separated from each other by a beaded chromatin spacer. Nascent RNA-containing fibrils are not visible because they are below the critical size of resolution (for details, see Franke *et al.*, 1976a). The size and the pattern of arrangement of the thickened regions along a chromatin strand are compatible with the known sequence arrangement of DNA containing 5 S rRNA genes (for details, see Scheer, 1982).

Another class of slightly larger transcription units containing approximately 940 base pairs of DNA is shown in Fig. 8c. They are also tandemly repeated, separated by chromatin spacers with a nucleosomal configuration, and occur in clusters of 100 or more copies (Scheer, 1981). At the moment, they cannot be correlated with repetitious genes of defined genetic content known thus far.

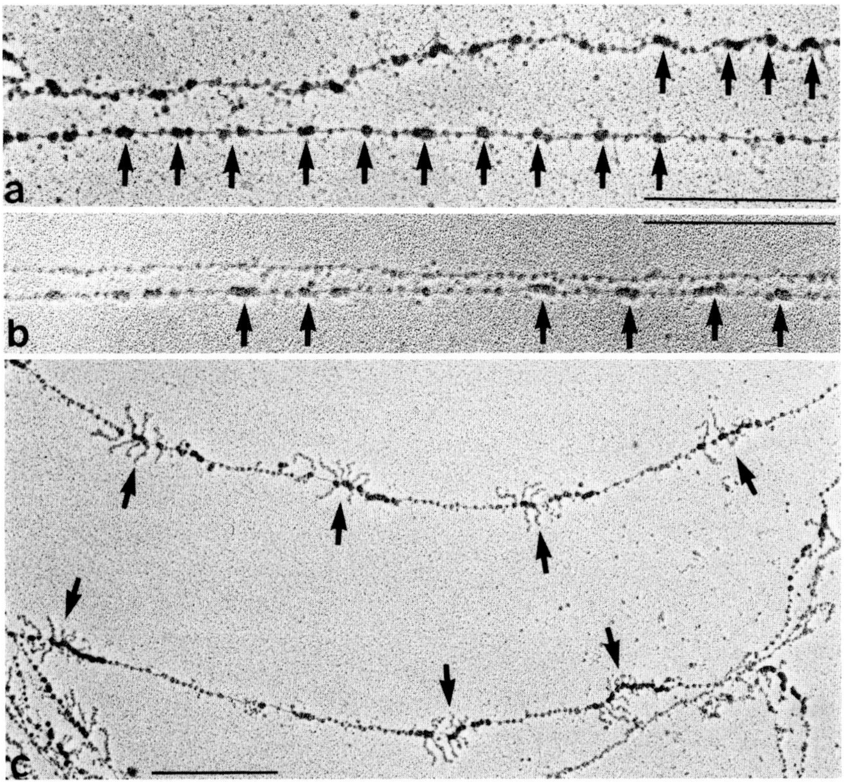


**Figure 8.** Spread lampbrush chromatin from *Pleurodeles waltlii* (a,c) and *Xenopus laevis* (b) oocytes. The periodic alternating thick–thin pattern seen in (a) and (b) probably reflects tandemly arranged, transcriptionally active 5 S rRNA genes (arrows) separated by beaded spacer chromatin. Another class of repeated genes is shown in c (arrows). Here, the transcripts are long enough to be visualized as lateral fibrils. Scale bars: 0.5 μm.

## 2.7. Chromatin Organization of Chromomeres and Lateral Loops

The transcriptionally inactive chromatin contained in the chromomeres shows a beaded, nucleosomal organization when spread for electron microscopy (Fig. 9a). This organization, which is characteristic of inactive chromatin (Olins and Olins, 1974; Oudet *et al.*, 1975), extends next to the origin of transcription units (Figs. 3b and 9a). By using immunofluorescence techniques, the presence of histones in the transcriptionally inactive chromomeric regions of lampbrush chromosomes can be demonstrated directly (Fig. 9b,c) (see also Sommerville *et al.*, 1978; Scheer *et al.*, 1979b).

Maximally transcribing loop axes are saturated with RNA polymerases with no detectable interspersed nucleosomal particles and are thought to occur in an extended state equivalent to the B conformation of DNA (Fig.

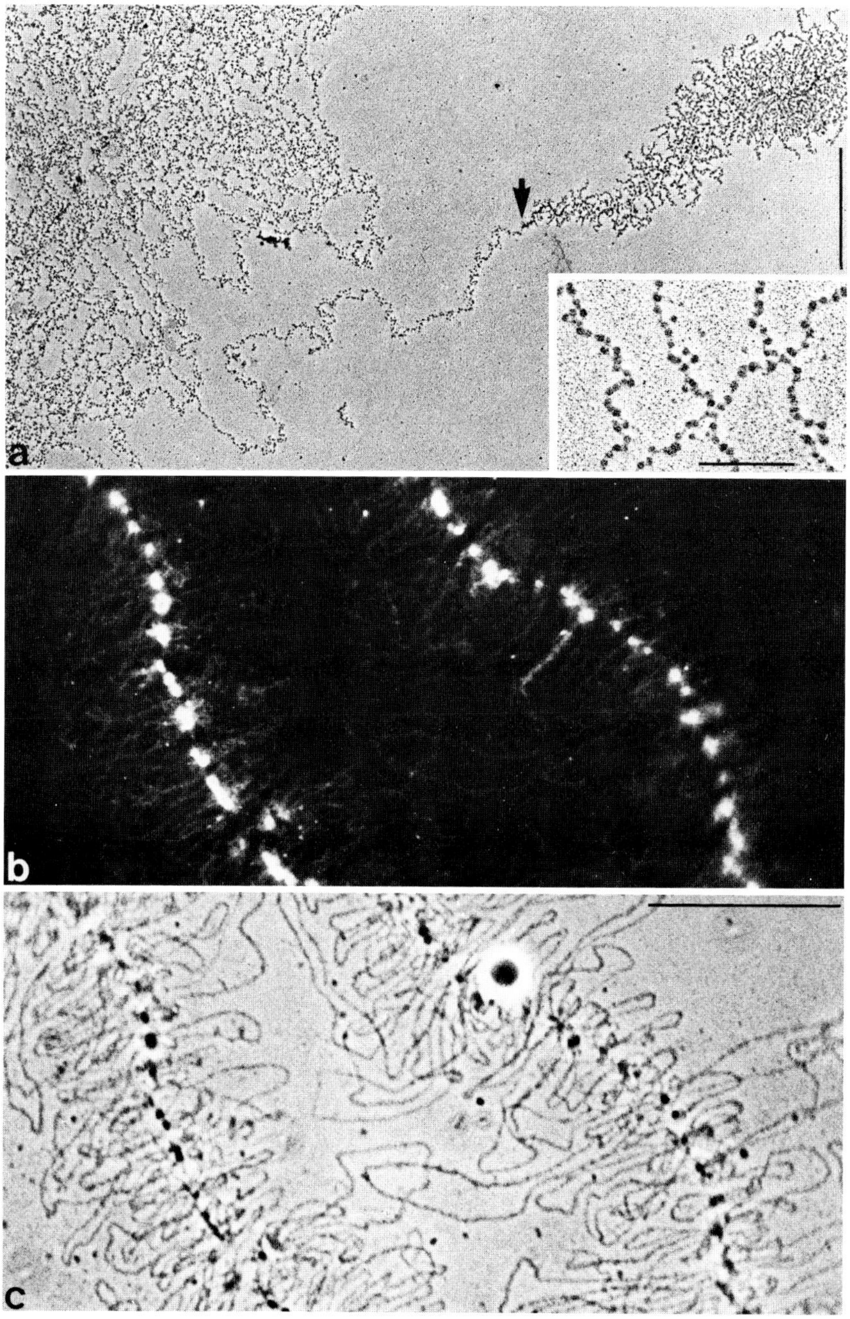


**Figure 9.** Chromomere organization. (a) Nucleosomal organization of transcriptionally inactive chromatin contained in the chromomere of a lampbrush chromosome of *Triturus helveticus* (arrow indicates the origin of a transcription unit). Inset shows, at higher magnification, the "beads-on-a-string" appearance of chromatin. (b) Histones are located in the chromomeres of lampbrush chromosomes (in this case from *T. cristatus*), as demonstrated by immunofluorescence microscopy using antibodies to histone H2B. (c) Corresponding phase-contrast photograph. Scale bars: 1 $\mu$m (a), 0.2 $\mu$m (inset in a) and 20 $\mu$m (c).

3a) (see Franke *et al.*, 1976b, 1978; Scheer *et al.*, 1979a). Indirect evidence suggests that histones are still associated with the transcriptionally active loop chromatin, although in an apparently nonnucleosomal mode (see Section 8). In states of reduced RNA synthesis (e.g., in maturing oocytes), the RNA polymerases are more distantly spaced, and the chromatin of such "gaps" assumes a beaded nucleosomal organization (Fig. 10a,b). Situations

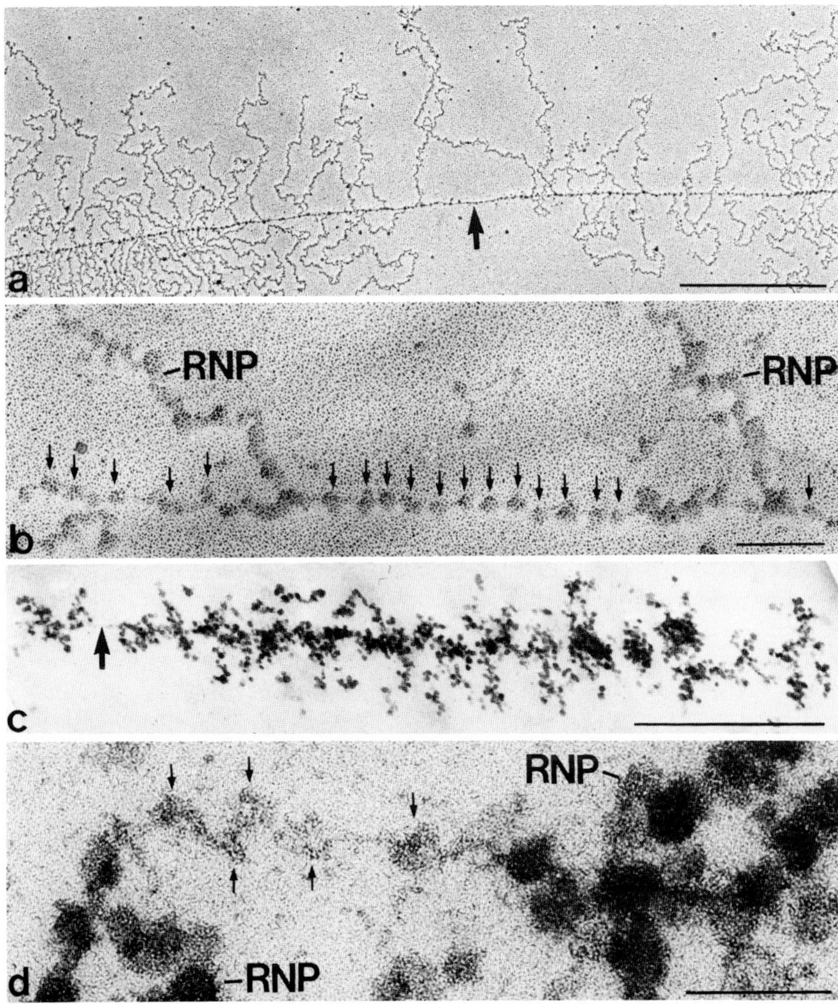


**Figure 10.** Morphology of transcript-free regions of loop chromatin as seen in electron microscopic spread preparations of lampbrush chromosomes of *Triturus cristatus* (a,b) and in ultrathin section through flat-embedded lampbrush chromosomes of *Pleurodeles waltlii* (c,d). Gaps within transcriptional units are recognized at low magnification (a,c) (arrows). Higher magnification displays the presence of nucleosomal beads between the distantly spaced transcripts (b,d, arrows). Scale bars: 1 $\mu$m (a,c) and 0.1 $\mu$m (b,d).

such as that shown in Fig. 10b, in which transcribing polymerases are surrounded on both sides by nucleosomes, suggest that nucleosomes unfold directly in front of a polymerase and reform rapidly following transcription (Scheer, 1978; McKnight *et al.*, 1978).

The same structural features are found when chromosomes are prepared under conditions that largely preserve their native state (Fig. 10c,d). Intragenic chromatin stretches free of transcripts invariably reveal the presence of nucleosomal beads (Fig. 10d). Thus, it appears that nucleosomes will form transiently within transcription units whenever transcribing RNA polymerases are sufficiently far apart.

## 2.8. Function of Loop Transcription

The conventional view assumes that lampbrush chromosomes generate a complex set of maternal as well as of polysomal mRNAs (see Sommerville, 1977; MacGregor, 1980; Davidson, 1976). However, this concept is difficult to reconcile with the finding that not only the total poly(A) RNA but also individual poly(A) RNA sequences reach their final levels early in oogenesis and remain thereafter at an approximately constant level throughout oogenesis (Rosbash and Ford, 1974; Golden *et al.*, 1980; van Dongen *et al.*, 1981, 1983). This apparent paradox was explained recently by Anderson *et al.* (1982), who showed that nuclei of *Xenopus* lampbrush stage oocytes do indeed synthesize poly(A) RNA of a complexity similar to that already stored in the ooplasm but that this RNA production is just sufficient to compensate for the continuous degradation due to turnover of the very large pool of maternal mRNA.

In order to prove that transcription of lampbrush chromosomes leads in fact to functional mRNAs, we have used an experimental approach originally developed by DeRobertis and Gurdon (1977). This approach involves injecting somatic cell nuclei into amphibian oocytes and analyzing changes in the synthetic pattern of proteins by two-dimensional gel electrophoresis. However, instead of injecting somatic cell nuclei, we have transferred single, manually isolated nuclei of early vitellogenic oocytes of *Pleurodeles waltlii* into full-grown oocytes of *Xenopus laevis*. The rationale of this experiment was to identify the synthesis of *Pleurodeles*-specific proteins in the *Xenopus* oocyte. A comparison of the [$^{35}$S]methionine-labeled total proteins of mature *Xenopus* (Fig. 11a) and early vitellogenic *Pleurodeles* (Fig. 11b) oocytes by two-dimensional gel electrophoresis reveals (besides some common spots such as actin) marked differences in the overall pattern of the polypeptides. This is shown more clearly when only a selected area of the gels is analyzed with actin as a reference spot (Fig. 12a,b). Figure 12d shows the result of a nuclear transfer experiment. *Xenopus* oocytes containing *Pleurodeles* oocyte nuclei were incubated for a few days, then labeled for 24 hr with L-[$^{35}$S]methionine, and the proteins were analyzed

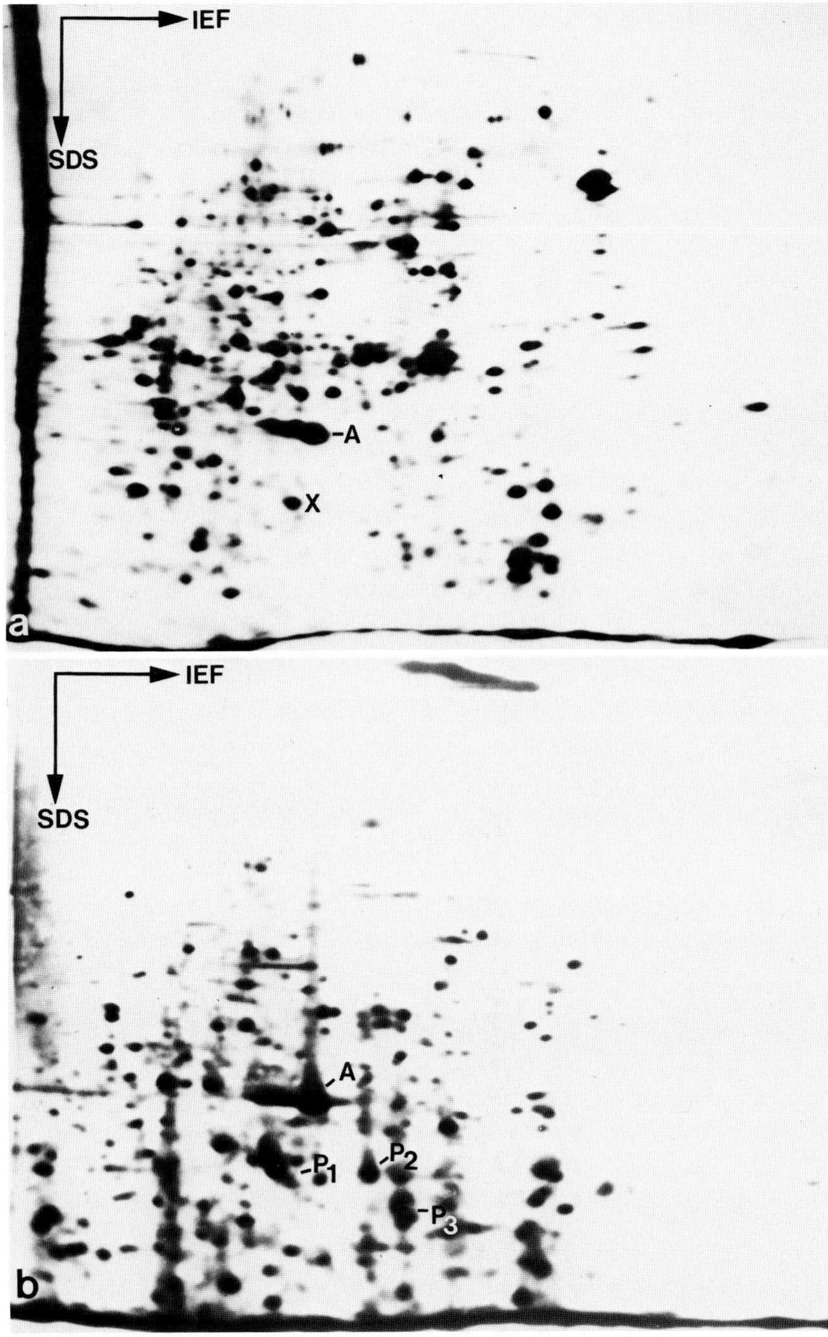


**Figure 11.** Radioactive proteins from full-grown oocytes of *Xenopus laevis* (a) and from previtellogenic oocytes of *Pleurodeles waltlii* (b) analyzed by two-dimensional gel electrophoresis (first dimension: isoelectric focusing, IEF; second dimension: 12% polyacrylamide in the presence of SDS). Oocytes were incubated for 24 hr in Barth medium containing [$^{35}$S]methionine and were then homogenized; the proteins of the low-speed supernatant were analyzed. Gels were processed for autoradiography. A, actin; X, *Xenopus*-specific protein; $P_{1\text{-}3}$, *Pleurodeles*-specific proteins.

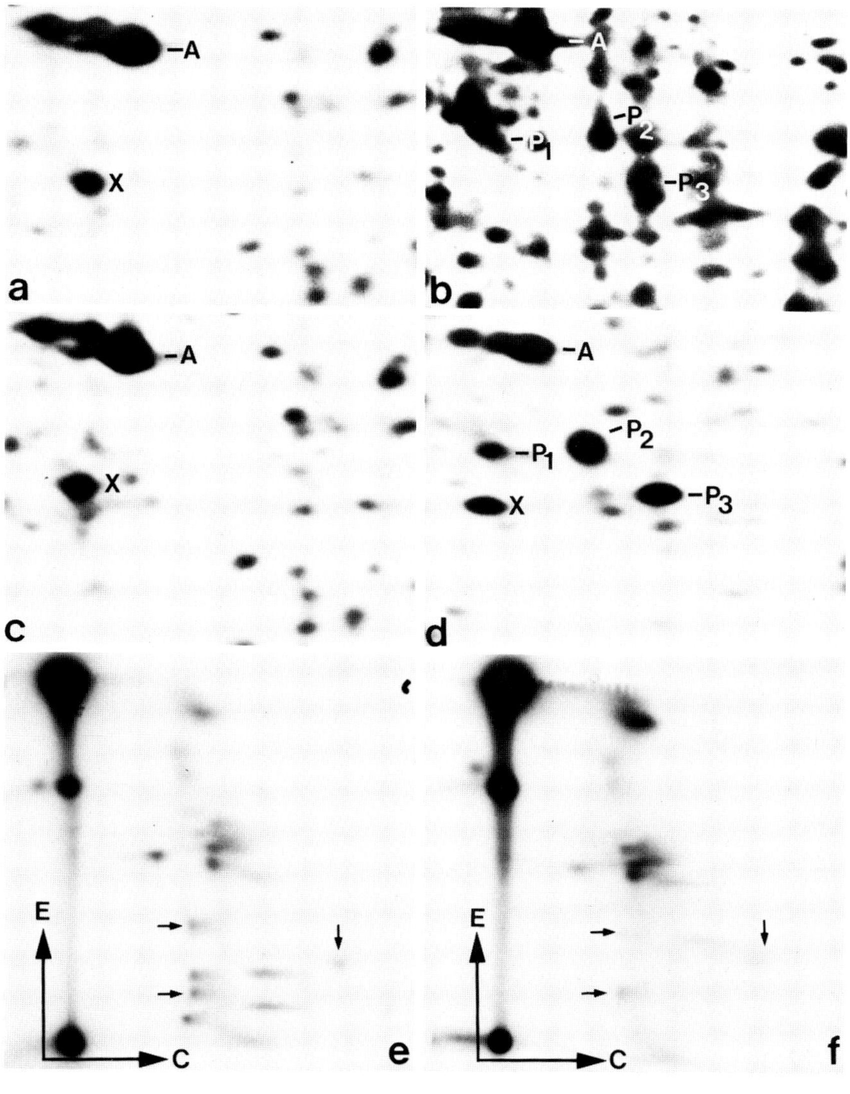


**Figure 12.** (a,b) The regions below actin were selected from the gels of the previous figure and are shown at higher magnification. Actin (A) serves as reference; X, *Xenopus*-specific protein, $P_{1\text{-}3}$, *Pleurodeles*-specific proteins. After transfer of the nucleus from a previtellogenic *P. waltlii* oocyte into a mature *X. laevis* oocyte, successfully healed oocytes were incubated for 5 days, followed by labeling with [$^{35}$S]methionine for 24 hr. *Xenopus*- and *Pleurodeles*-specific proteins are synthesized (d). As a control, *Xenopus* oocytes were injected with $\alpha$-amanitin (final concentration, 1mg/ml shortly after the nuclear transfer, followed by labeling. As seen in (c) *Pleurodeles*-specific proteins are absent. (e,f) Tryptic peptide maps of *in vivo*-labeled $P_2$ (e, authentic $P_2$; f, $P_2$ synthesized in *Xenopus* oocytes after nuclear transfer). The pattern of cleavage products is essentially identical, with some minor spots denoted by arrows (e, f). E, electrophoresis; C, chromatography.

by two-dimensional gel electrophoresis. It is evident that *Pleurodeles*-specific proteins (designated $P_1$–$P_3$ in Fig. 12d) are synthesized. The occurrence of $P_1$–$P_3$ depends on transcriptional activity, since injection of $\alpha$-amanitin at concentrations sufficient to inhibit lampbrush chromosome transcription prevents their synthesis (Fig. 12c). That $P_1$–$P_3$ are actually proteins encoded by the genome of *Pleurodeles* is suggested by the similarity of the peptide maps between authentic $P_2$ (Fig. 12e) and $P_2$ synthesized in *Xenopus* oocytes after transplantation of *Pleurodeles* oocyte nuclei (Fig. 12f).

Taken together, these results provide direct evidence to support the view that lampbrush chromosomes are synthesis sites for functional mRNAs.

## 3. Amplified Nucleoli

### 3.1. General Features

Ribosomal RNA genes are amplified in amphibian oocytes (Brown and Dawid, 1968). As a consequence, a huge number of extrachromosomal rRNA genes occur (in addition to the few hundred rRNA genes clustered at the chromosomal nucleolus organizer regions) in numerous nucleoli that are not associated with the lampbrush chromosomes (Fig. 13a,b). Although the amplification process begins in premeiotic oogonia, the main period for the selective replication of rDNA is pachytene of meiotic prophase, followed by completion in early diplotene. In *Xenopus laevis*, the amount of extrachromosomal rDNA per oocyte nucleus is about 30 pg, corresponding to approximately 2 million copies of rRNA genes, which are distributed in about 1000 amplified nucleoli (reviewed in MacGregor, 1972).

It is obvious that amplification of the rDNA enables amphibian oocytes to support unusually high rates of rRNA synthesis. Thus, a single *Xenopus* oocyte synthesizes in stages of maximal growth about 300,000 ribosomes per second (Scheer, 1973), whereas a normal somatic cell produces in the same time span only 10–100 ribosomes (Hadjiolov, 1980). The majority of the oocyte ribosomes are stored in the cytoplasm as monosomal particles for use in future embryonic growth.

In growing oocytes, the amplified nucleoli are distributed toward the periphery of the nucleus and are often firmly attached to the inner nuclear membrane by a network of fibrillar strands (Franke and Scheer, 1970). Amplified nucleoli appear either as compact spheroidal bodies 2–10 $\mu$m in diameter (Figs. 1c; 13a, and 17a) or as ringlike structures (Fig. 13b). In several amphibian species, transitions between both nucleolar forms have been described in specific stages of oogenesis (see MacGregor, 1972).

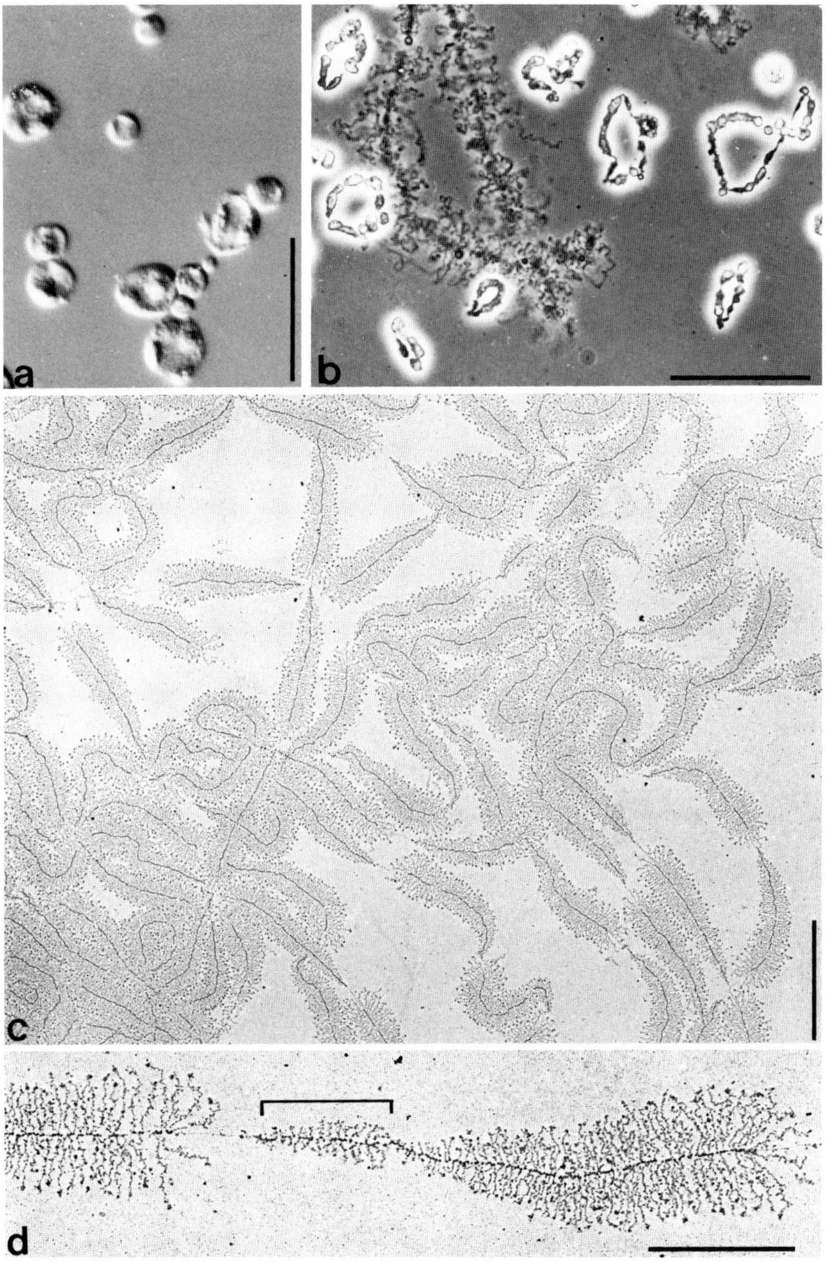


**Figure 13.** Amplified nucleoli (a,b) and nucleolar chromatin (c,d). In light microscopic preparations, amplified nucleoli appear as compact spheroidal entities (a, from *Triturus alpestris*, Nomarski interference contrast) or as ringlike structures (b, from *Pleurodeles waltlii*, phase contrast). When spread for electron microscopic study, nucleolar chromatin displays the characteristic tandem arrangement of transcriptionally active rRNA genes (c, from *P. waltlii*; d, from *X. laevis*). Note the small transcriptional unit in the spacer region between two adjacent rRNA genes (bracket, d). Scale bars: 50 $\mu$m (a,b), 2 $\mu$m (c), and 1 $\mu$m (d). (Figure 13d courtesy of Dr. M. Trendelenburg.)

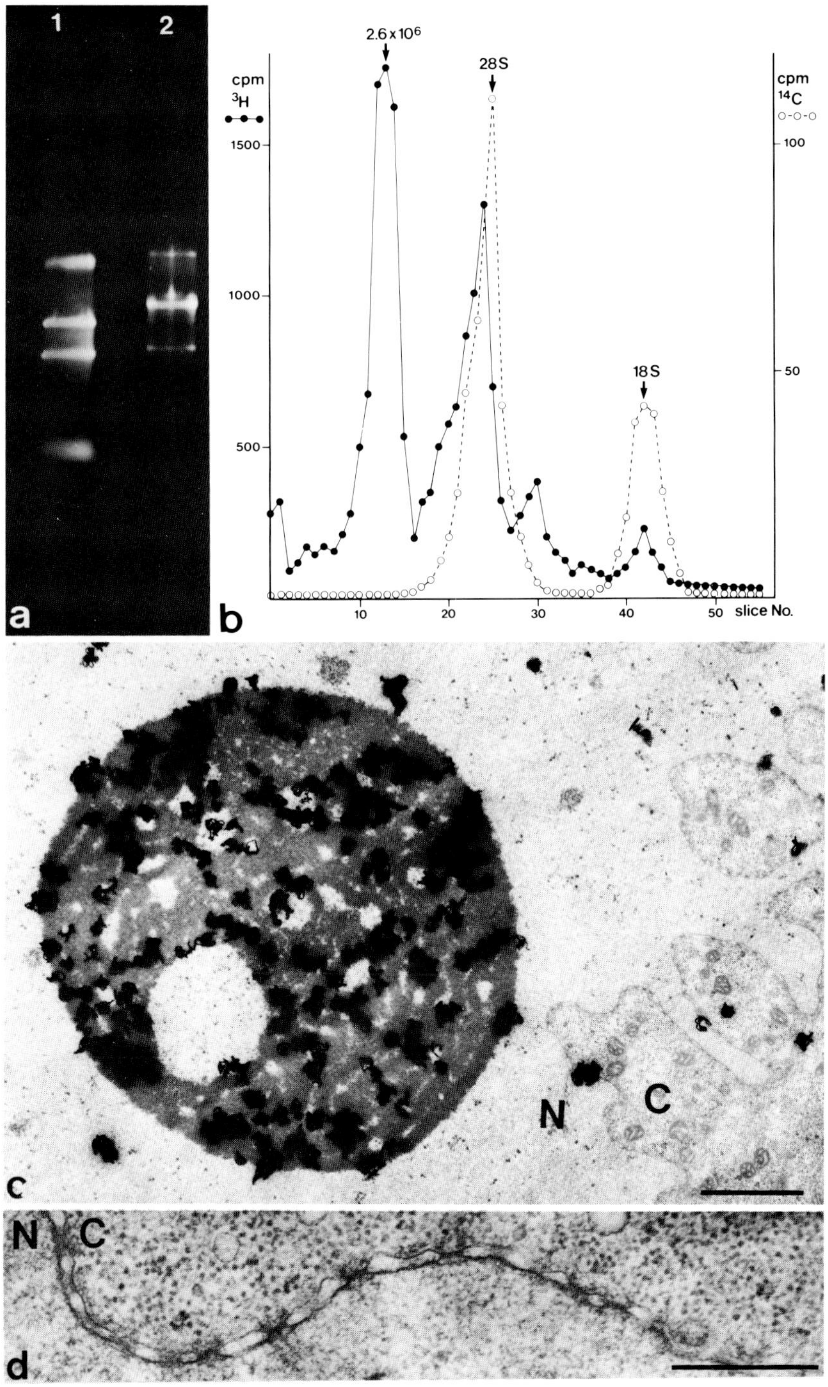


**Figure 14.** (a,b) Gel electrophoretic analysis of RNA from manually isolated nuclei. Total RNA from 150 *Xenopus laevis* oocyte nuclei was separated on a 1.5% agarose gel and stained with ethidium bromide (a, lane 2). The three bands correspond to 40 S pre-rRNA, 30 S, and 18 S rRNA. Reference RNAs are, from top to bottom, tobacco mosaic virus RNA, *Escherichia*

### 3.2. Visualization of Transcribing rRNA Genes

By introducing a new spreading technique to prepare chromatin for electron microscopic examination, Miller and Beatty (1969) were able to visualize directly the process of transcription of amplified rRNA genes from amphibian oocytes. An example of spread nucleolar chromatin from a *Pleurodeles* oocyte is shown in Fig. 13c. Numerous maximally active rRNA genes are seen in linear arrays separated from each other by so-called nontranscribed spacers. Each gene contains about 100 nascent RNP transcripts arranged in a length gradient, thereby giving rise to the characteristic Christmas tree appearance. Each transcript is anchored to the chromatin axis by a 12–15 nm particle that contains a RNA polymerase I complex (for details, see Franke *et al.*, 1979; Scheer and Zentgraf, 1982). It should be emphasized that the "nontranscribed" spacer regions not infrequently contain small transcription units ("prelude complexes") (Fig. 13d), which (in *Xenopus laevis*) seem to initiate at positions corresponding to reduplicated promoterlike sequences (Trendelenburg, 1981). It has been hypothesized that spacer transcription might serve to concentrate RNA polymerases around the ribosomal gene promoter (Moss, 1983).

### 3.3. Synthesis of rRNA and Formation of Preribosomes

The amplified rRNA genes are transcribed as 40 S pre-rRNA molecules with an apparent molecular weight of $2.6 \times 10^6$ (Fig. 14a,b) (for review, see Sollner-Webb *et al.*, 1982). When total RNA extracted from manually isolated vitellogenic *Xenopus* oocytes is analyzed by gel electrophoresis, it is evident that rRNA species predominate (Fig. 14a). The prevalence of rRNA synthesis can also be demonstrated by autoradiography after labeling the oocytes with [$^3$H]uridine. Most of the silver grains are located over the peripheral, amplified nucleoli (Fig. 14c).

Specific ribosomal proteins bind to pre-rRNA molecules while they are still attached to the template. This has been shown convincingly in *Drosophila* by combining Miller spreads with immunolocalization techniques (Chooi and Leiby, 1981). By contrast, other ribosomal proteins such as S 1 attach to preribosomes only at later stages of their maturation pathway, i.e.,

---

*coli* rRNAs, and tRNA (a, lane 1). After incubation of *Triturus alpestris* oocytes in a medium containing [$^3$H]uridine, nuclei were isolated and the RNA was analyzed on a composite gel (2.25% acrylamide, 0.5% agarose) together with added $^{14}$C-labeled rRNA from *Xenopus* as internal reference (b). Note that 28 S rRNA migrates ahead of the nuclear radioactivity peak. (c) In cytological preparations, silver grains are preferentially localized over the amplified nucleoli (electron microscopic autoradiography after labeling the oocytes of *T. alpestris* with [$^3$H]uridine). (d) Typical ribosomal particles are absent from the nucleoplasm, as demonstrated in the ultrathin section through the nuclear periphery of a *T. alpestris* oocyte. N, nucleus; C, cytoplasm. Scale bars: 2 $\mu$m (c) and 0.5 $\mu$m (d).

after translocation of the newly synthesized rRNA into the granular component of the nucleolus (Hügle *et al.*, 1985). The coordinate association with ribosomal and nonribosomal proteins and processing of the pre-rRNAs leads to the formation of different kinds of nucleolar RNP particles (Rogers, 1968), which finally mature into ribosomal subunits (reviewed in Hadjiolov, 1980). The final maturation steps occur in amphibian oocytes immediately before, during, or shortly after the passage through the nuclear pores, since not only morphologically identifiable ribosomal particles are absent from the nucleoplasm (Fig. 14d), but the 28 S rRNA component of the large subunit of cytoplasmic ribosomes is also missing (Fig. 14b). Instead, an RNA precursor sedimenting at about 30 S is detected in isolated nuclei by gel electrophoresis (Fig. 14b).

The protein complement of preribosomal particles of amphibian oocyte nuclei is so far unknown. Recent experiments based on microinjection of ribosomal proteins into oocytes of *Xenopus laevis* suggest that certain ribosomal proteins migrate into the nucleus and take part in the formation of preribosomal particles (Kalthoff and Richter, 1982).

### 3.4. Regulation of rRNA Synthesis

Although rDNA amplification takes place very early in oogenesis (the pachytene stage), synthesis of rRNA reaches its peak value only after the onset of yolk deposition (Ford, 1972; Sommerville, 1977; Scheer *et al.*, 1976a). On the basis of biochemical data, it has been calculated that previtellogenic and mature oocytes of *Triturus alpestris* synthesize rRNA at only about 0.01% and 13%, respectively, of the rate of vitellogenic oocytes (Scheer *et al.*, 1976a). In states of reduced synthetic activity the rRNA genes reveal a reduced number of RNP transcripts (Fig. 15a–c). This observation indicates that regulation of rRNA synthesis occurs at the level of transcription by modulation of the frequency of initiation events. Each gene represents an independent unit of regulation, since genes fully loaded with RNA polymerases can alternate with sparsely loaded genes or with totally inactive ones (Fig. 15a) (Scheer *et al.*, 1976a).

### 3.5. Structure of Nucleolar Chromatin at Different States of Activity

It is now generally accepted that transcriptionally active nucleolar chromatin—including gene and spacer regions—is organized differently from inactive chromatin, which is packed into nucleosomes (Franke *et al.*, 1976b; Scheer, 1978; Labhart and Koller, 1982; for reviews, see Franke *et al.*, 1979; Scheer and Zentgraf, 1982). The absence of nucleosomes is especially evident in transcription units with a reduced number of transcripts

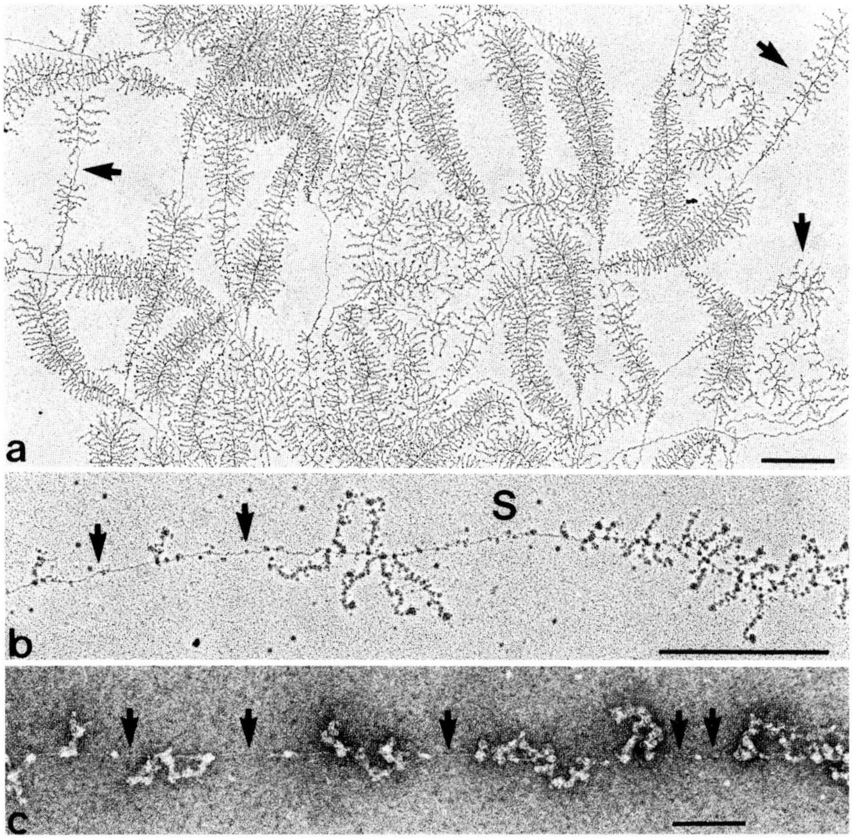

**Figure 15.** Spread preparations of nucleolar chromatin from full-grown oocytes of *Pleurodeles waltlii* (a,b) and *Triturus cristatus* (c) after positive staining and additional metal shadowing (a,b) or negative staining (c). Several rRNA genes are only sparsely covered by transcripts (arrows, in a). The transcript-free intragenic regions (arrows in b and c) and the spacer (S, b) exhibit a nonbeaded organization. Scale bars: 1 $\mu$m (a,b) and 0.1 $\mu$m (c).

(Fig. 15b,c): The chromatin intervals between adjacent RNA polymerase particles are thin and nonbeaded. The same holds true for the spacer regions, although—depending on the specific spreading conditions—irregularly sized and distributed particles can be present, which are, however, of a nonnucleosomal nature (Fig. 15b) (Scheer, 1980). When the contour lengths of transcription units, spacer regions, and the total repeat units are compared with the corresponding lengths of deproteinized rDNA, it is obvious that nucleolar chromatin occurs in electron microscopic spread preparations in a fully extended state equivalent to the length of the B-form DNA (Scheer *et al.*, 1977; Reeder *et al.*, 1978). The important question as to whether histones are associated with transcriptionally active rDNA is as yet not settled. Biochemical data indicate that histones with the exception of H1 are present in the rDNA-containing nucleolar "cores" isolated from amplified nucleoli of *Xenopus* oocytes (Higashinakagawa, 1982). However, unless the histones:rDNA ratio is determined, it cannot be excluded that active nucleolar chromatin is deficient in histones. Thus, at the moment it is not clear whether the specific organization of active nucleolar chromatin reflects a deficiency in histones or a nonnucleosomal arrangement of the full complement of histones.

Completely inactive nucleolar chromatin assumes a nucleosomal organization indistinguishable from that of other portions of inactive chromatin (Scheer, 1978). Thus, it appears that the extended conformation reflects a state of transcribability but that the actual transcriptional intensities of individual rRNA genes are controlled by other factors.

## 3.6. Protein Filaments

Recent evidence suggests that protein filaments occur within the nucleoli. These filaments might be involved in maintaining nucleolar structural integrity and/or in storage, maturation, and intranucleolar translocation of the preribosomal particles.

### 3.6.1. Medusoid Fibril Bodies

In spread preparations of *Xenopus laevis* oocyte nuclei, transcribed rRNA genes are often seen in conspicuously close vicinity to roundish aggregates of a distinct filament type (Fig. 16a). These filamentous aggregates, which have been termed **medusoid fibril bodies** (Moreno Diaz de la Espina *et al.*, 1982), contain two structurally different components: the filaments proper with a diameter of 8–12 nm and a distinctly beaded appearance and densely stained 18–30-nm particles, which are attached at numerous sites to these filaments (Fig. 16b). It has been suggested that the medusoid fibrils form part of the nucleolar cortex and provide structural support for the attachment of nucleolar RNP particles (Moreno Diaz de la Espina *et al.*, 1982).

### 3.6.2. Residual Skeleton Structures

Amplified nucleoli can be isolated in a highly pure state by fluorescence-activated particle sorting using an argon ion laser (Fig. 17a) (Franke *et al.*, 1981a). Extraction of the nucleoli in a buffer of high ionic strength (1 M KCl) containing 1% of the nonionic detergent Triton X-100 leaves a residual fibrillar meshwork of about 3–5-nm-thick filaments (Fig. 17b,c). Digestion of the nucleoli with DNAse and RNAse before the salt extraction does not alter the morphological appearance of these fibrillar aggregates; trypsin, on the other hand, effectively destroys them (Franke *et al.*, 1981a).

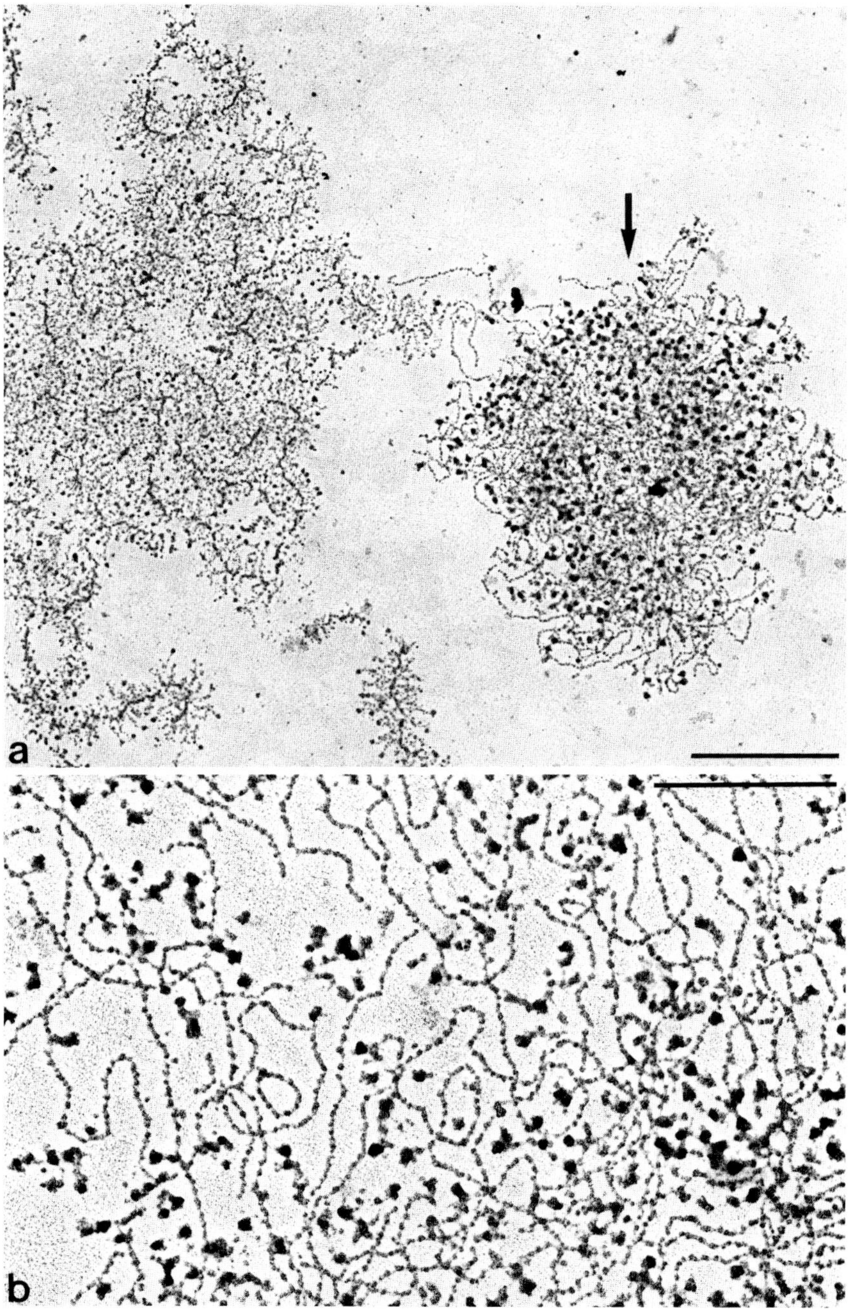


**Figure 16.** Spread preparation of nucleolus-enriched material from *Xenopus laevis* oocytes. (a) The "medusoid body" composed of a network of fibrils (arrow) lies next to a group of rRNA genes. (b) Numerous large particles are attached to the filaments, which exhibit a periodic substructure. Scale bars: 1 $\mu$m (a) and 0.5 $\mu$m (b).

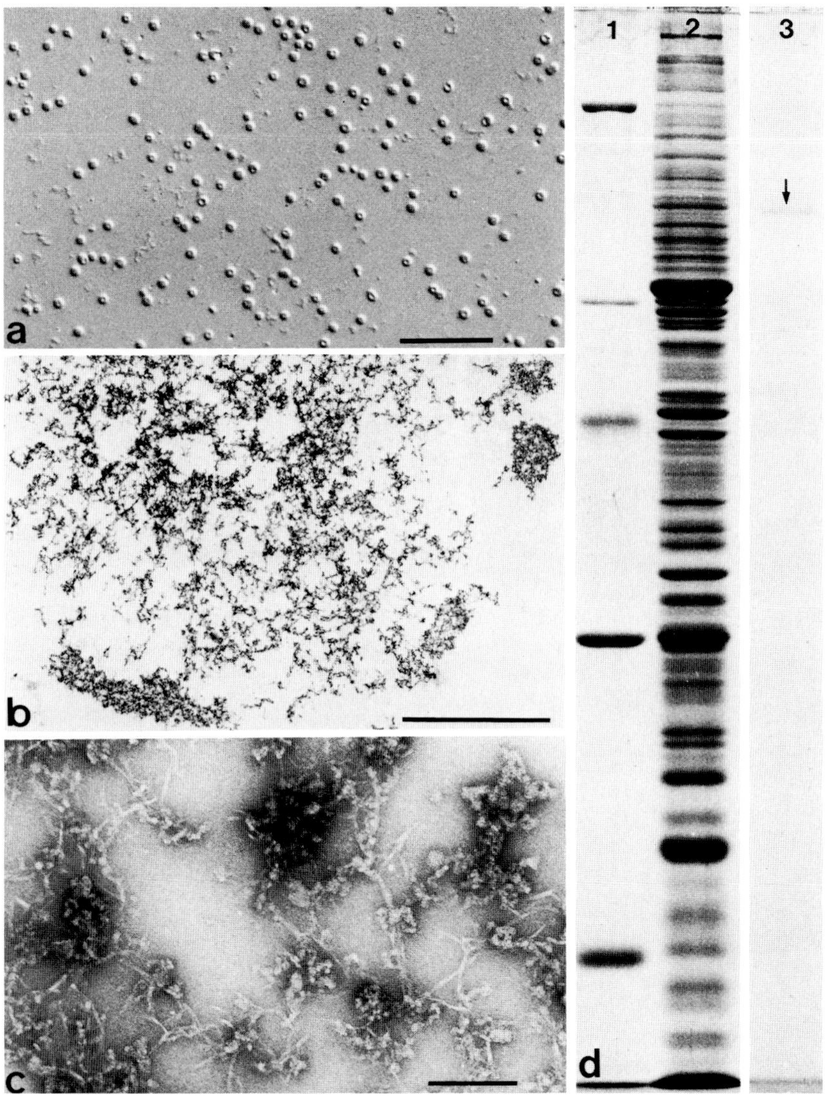


**Figure 17.** Insoluble protein component of amplified nucleoli from *Xenopus laevis* oocytes. Nucleoli isolated by fluorescence-activated particle sorting are shown in a (Nomarski interference contrast). After extraction with 1 M KCl and 1% Triton X-100, residual fibrillar structures remain as seen in ultrathin sections (b) and in negatively stained preparations (c). Upon one-dimensional gel electrophoresis, only one predominant polypeptide with a molecular weight of 145,000 is recognized in such residual fractions (d, lane 3, arrow). For comparison, the complex protein pattern of total nuclei is also shown (d, lane 2). Reference proteins are, from top to bottom, myosin heavy chain, phosphorylase a, bovine serum albumin (BSA), actin, and chymotrypsinogen (d, lane 1). Scale bars: 50 $\mu$m (a), 1 $\mu$m (b), and 0.1 $\mu$m (c).

Fractions enriched in high-salt-resistant nucleolar fibrils contain only one major protein with a molecular weight of approximately 145,000 (Fig. 17d, lane 3) and an isoelectric pH value of about 6.15 (Franke *et al.*, 1981a). Antibodies against this protein were used to demonstrate that the nucleolar skeletal fibrils are not uniformly distributed throughout the amplified nucleoli but are more concentrated toward their peripheries (Krohne *et al.*, 1982; Benavente *et al.*, 1984a,b). Furthermore, these workers showed that immunologically related proteins occur also in nucleoli of somatic *Xenopus* cells.

# 4. The Nuclear Envelope

## 4.1. Morphology and Distribution of Pore Complexes

The nuclear envelope of amphibian oocyte nuclei is perforated by numerous pore complexes, which—in freeze-fractured (Fig. 18a) or negatively stained (Fig. 18b) preparations—appear as ringlike structures with diameters of 70–80 nm. In transverse section, it is obvious that the pores are sites of local fusion of the inner and outer nuclear membrane (Fig. 19d). Pore complexes are ubiquitous structures found in all eukaryotic cells and have a very characteristic and highly regular architecture, the most prominent nonmembranous components being eight symmetrically arranged annulus subunits on either pore margin (Figs. 18b and 19d) (for reviews, see Franke and Scheer, 1974; Franke *et al.*, 1981b). The number of pore complexes per unit area is remarkably high in amphibian oocytes (50–60 pores/$\mu m^2$) (Fig. 18a,b), so that about 25% of the surface area of the nucleus is occupied by pores. This high density—together with the enormous size of the germinal vesicle—results in a total number of about 38 million pore complexes per nucleus of a mature *Xenopus* oocyte. They are formed during oogenesis at a rate of about 8 pores/sec (Scheer, 1973).

## 4.2. Architectural Components of the Pore Complexes: High-Salt-Resistant Proteins

Isolated oocyte nuclei can be manually subfractionated into nuclear envelopes and gelled nuclear contents (Fig. 19a–c). The nuclear envelopes thus obtained are extremely pure and represent a fraction that is highly enriched in pore complexes (Fig. 19d) (Scheer, 1972). When exposed to mechanical stress (such as during surface spreading on a drop of water), the pore complexes remain intact. With progressive disintegration of the interporous membrane, a network of resistant fibrils appear, which seems to interconnect all pore complexes (Fig. 19e; Scheer *et al.*, 1976b). After extraction with 1.5 M KCl and 1% Triton X-100, the basic architecture of

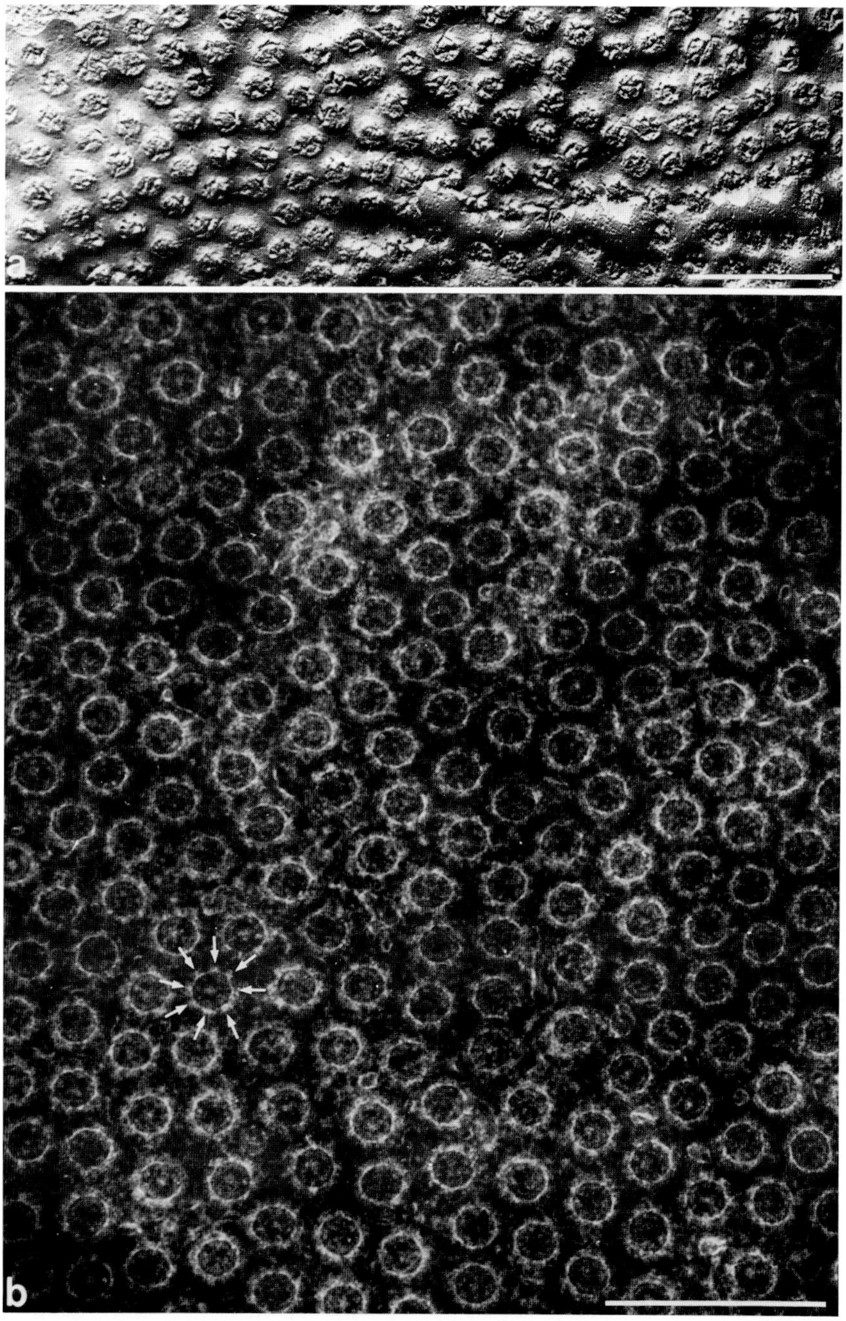


**Figure 18.** Nuclear envelope from *Xenopus laevis* oocytes as seen in freeze-etched (a) and negatively stained (b) preparations. Note the eight symmetrically arranged annulus subunits lying on the pore rims (e.g., at the arrows in b) and the central granules. Scale bars: 0.5 $\mu$m.

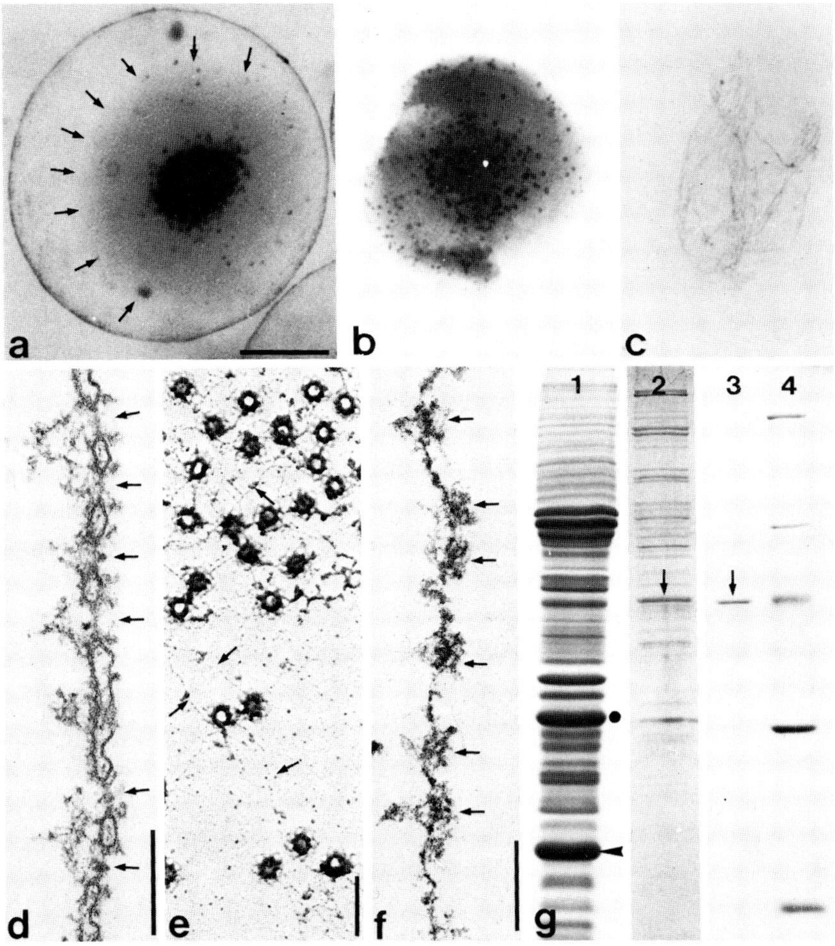


**Figure 19.** Isolation and high salt extraction of the nuclear envelope. (a–c) Isolation procedure. An isolated nucleus (in this case from *Triturus alpestris*) is transferred into a medium containing 5–10 mM $MgCl_2$, in which the nucleoplasm forms a gel (denoted by arrows, a). The nuclear envelope is then punctured and separated from the gelled nucleoplasm (b,c). Note the purity of the nuclear envelope fraction as demonstrated by electron microscopy of ultrathin sections (d, port complexes denoted by arrows). Mechanical stress (e) or extraction with 1 M KCl plus 1% Triton X-100 (f) destroys the interporous membrane and exhibits a network of fibrils interconnecting the very stable pore complexes as shown by surface spreading (e) or ultrathin sectioning (f) (interporous fibrils denoted by arrows, e; high salt-stable pore complexes denoted by arrows, f). (g) One-dimensional gel electrophoresis of proteins from total nuclei (lane 1), nuclear envelopes (lane 2), and extracted nuclear envelopes (lane 3) shows the enrichment of one polypeptide band with a molecular weight of 68,000 (arrows). The dot denotes actin and the arrowhead indicates nucleoplasmin (lane 1). Reference proteins are, from top to bottom, myosin heavy chain, phosphorylase a, bovine serum albumin (BSA), actin, and chymotrypsinogen (lane 4). (Figures 19 f,g courtesy of Dr. G. Krohne.) Scale bars: 0.2 mm (a; a–c are magnified to the same scale) and 0.2 $\mu$m (d–f).

the pore complexes is still retained; in ultrathin sections, it is evident that they are interconnected by a thin layer ("lamina") at the level of the inner nuclear membrane (for further details, see Franke *et al.*, 1981b; Benavente *et al.*, 1984b). The protein composition of the extracted nuclear envelopes is very simple with only one major component having an apparent molecular weight of 68,000 (Fig. 19g, lane 3) (Krohne *et al.*, 1978, 1981). Immunological data indicate that the high-salt-resistant protein with a molecular weight of 68,000 is not only a structural component of the lamina proper, but contributes to the architecture of the pore complexes as well (Stick and Krohne, 1982; Benavente *et al.*, 1984b).

### 4.3. Permeability of the Nuclear Envelope and Nucleocytoplasmic Flow Rate of Ribosomal RNP through the Pore Complexes

Various substances—including proteins and colloidal gold particles of known sizes—have been microinjected into living oocytes in order to study the permeability characteristics of the nuclear envelope (reviewed in Bonner, 1978; Paine and Horowitz; 1980; Feldherr and Ogburn, 1982). The results are consistent with the view that macromolecules enter the nucleus by passing through the pore complexes and that the passage is restricted to a central channel with a patent opening much smaller than the actual pore diameter. The upper size limit for passage through the pore complexes is about 10–15 nm; this is also the size to which large RNP particles constrict during their nucleocytoplasmic translocation (Franke and Scheer, 1970). Experiments involving mechanical disruption of the nuclear envelope in intact oocytes have suggested that the nuclear envelope does not play a major role for the intracellular compartmentation of endogenous proteins (Feldherr and Ogburn, 1982). However, the model view of free diffusion through the pores and selective retention in the nucleoplasm to explain the unequal nucleocytoplasmic distribution of numerous proteins (Bonner, 1978) is not consistent with recent experiments favoring instead a selective entry mechanism for nucleoplasmin (Dingwall *et al.*, 1982) (see also Section 5.1). Furthermore, it has been shown that the cytoplasm of *Xenopus laevis* oocytes contains an abundant class of relatively small, soluble proteins, which are excluded from the nucleus ("karyophobic" proteins) (Dabauvalle and Franke, 1984). Thus, at the moment it is not clear to what extent the pore complexes are involved in the regulation of nucleocytoplasmic exchange and the maintenance of the differential distribution of proteins in both compartments (see also DeRobertis, 1983).

In growing oocytes of *Xenopus laevis*, two to three rRNA molecules are transferred every minute through a pore complex to the cytoplasm (Scheer, 1973). It is interesting to note that this rate does not represent the maximal transport capacity of a pore complex for ribosomal RNP material. In macronuclei of exponentially growing *Tetrahymena pyriformis*, for instance, the flow rate is higher by a factor of at least ten (Franke and Scheer, 1974). Thus, it is unlikely that the main function of the millions of pore complexes on the surface of oocyte nuclei is simply to provide enough export sites for ribosomal RNP material.

## 5. The Nucleoplasm

### 5.1. Identification of Some Major Stored Proteins: Nucleoplasmin, N1/N2, Histones, HMG Proteins, and Actin

The term nucleoplasm was originally used to describe the nonstructured ground substance of cell nuclei. In the present context the **nucleoplasmic fraction** is defined preparatively as the supernatant fraction obtained after centrifugation of a nuclear homogenate at 100,000 × g for 1 hr.

Total proteins extracted from manually isolated oocyte nuclei reveal a complex pattern upon gel electrophoresis (Fig. 20a). A major protein is **nucleoplasmin** (Laskey and Earnshaw, 1980), with a molecular weight of approximately 30,000. Nucleoplasmin is quantitatively recovered in the high speed supernatant fraction of nuclei and can be readily identified by two-dimensional gel electrophoresis (Fig. 20b). The amount of nucleoplasmin in oocyte nuclei can be determined by quantitative scanning densitometry of Coomassie blue-stained gels and determination of the total protein content of individual nuclei. Nuclei from mature *Xenopus laevis* oocytes contain about 0.25 $\mu$g of nucleoplasmin, which is equivalent to a concentration of 4.2 mg/ml (Krohne and Franke, 1980; Mills *et al.*, 1980), whereas nuclei from *Pleurodeles* oocytes contain 0.4 $\mu$g, which is equivalent to 6.1 mg/ml; i.e., about $7.5 \times 10^{12}$ molecules per nucleus. Nucleoplasmin is located exclusively in the nucleus and is not detectable in the cytoplasmic fraction (Fig. 20c–e).

Other major nucleoplasmic proteins are the very acidic, high-molecular-weight proteins N1 and N2, which—depending on the species—occur as a pair of closely related proteins (Fig. 20a, lane 2) or as a single protein termed N1/N2 (Fig. 20a, lane 3) (cf. Bonner, 1978; Dabauvalle and Franke, 1982). In *Xenopus* oocytes, N1 and N2 (100,000 and 110,000 $M_r$, respectively) represent approximately 12% of the total nuclear protein; i.e., 0.34 $\mu$g per nucleus, which is equivalent to a concentration of 5.2 mg/ml.

N1/N2 and nucleoplasmin are typical "karyophilic" proteins, since they are located with a remarkable specificity in the oocyte nucleus (Fig. 20c–e) and, upon microinjection into the cytoplasm, are rapidly sequestered into the nucleus (Bonner, 1978; DeRobertis *et al.*, 1978; Dabauvalle and Franke, 1982; Dingwall *et al.*, 1982; DeRobertis, 1983) (see also Section 4.3).

Analysis of the nucleoplasmic proteins from *Xenopus laevis* oocytes by

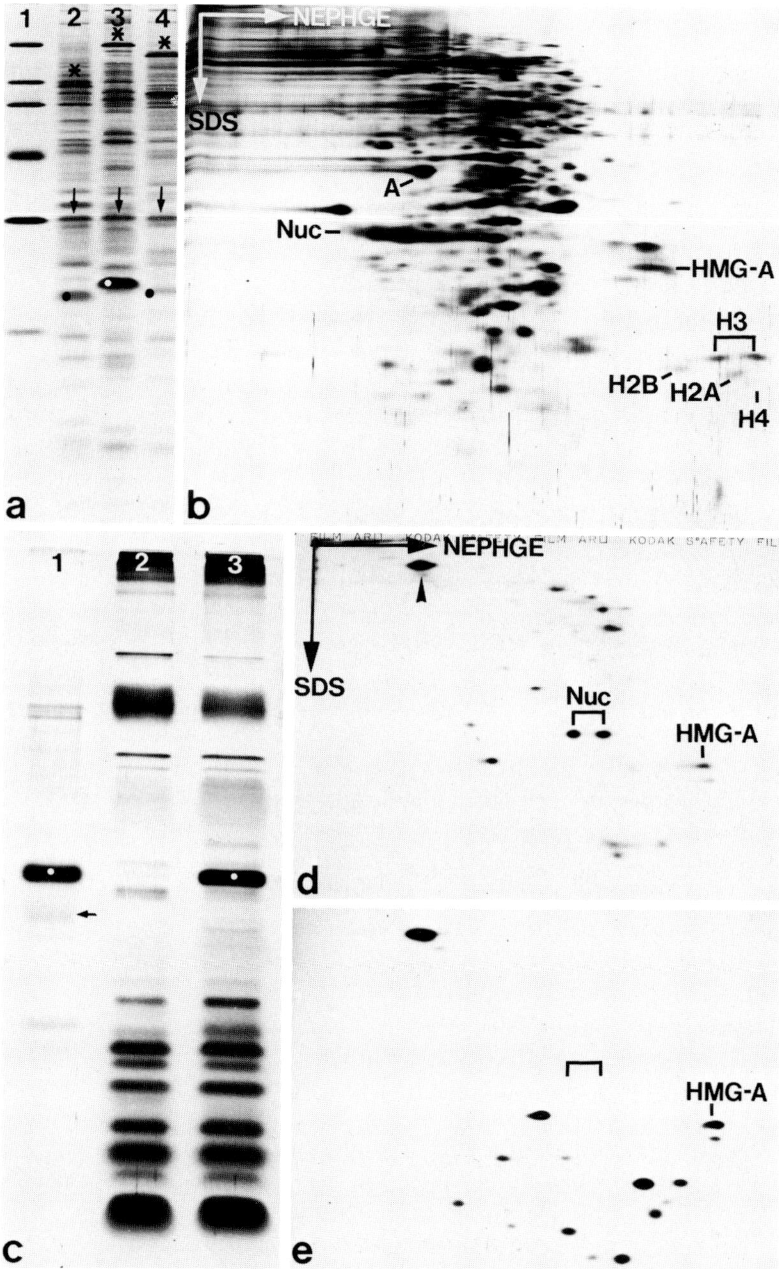


**Figure 20.** Analysis of nuclear proteins from amphibian oocytes by one- and two-dimensional gel electrophoresis. (a) Total proteins from oocyte nuclei of *Xenopus laevis* (lane 2), *Pleurodeles waltlii* (lane 3), and *Rana temporaria* (lane 4). Coomassie blue staining. Dots denote nucleoplasmin, arrows actin, and asterisks N1/N2. Reference proteins are (from top to bottom) myosin heavy chain, $\beta$-galactosidase, phosphorylase a, bovine serum albumin (BSA), actin, and chymotrypsinogen (lane 1). (b) Two-dimensional gel electrophoresis (first dimension: nonequilibrium pH gradient electrophoresis, NEPHGE; second dimension: 18% polyacrylamide containing SDS) of the high-speed supernatant of 10 *X. laevis* nuclei. Silver stain-

two-dimensional gel electrophoresis with nonequilibrium pH gradient electrophoresis as the first dimension (Fig. 20b) indicates the presence of a large pool of stored histones (Woodland, 1980; Kleinschmidt and Franke, 1982) and of the HMG-1 like protein HMG-A (A for amphibia) (Kleinschmidt *et al.*, 1983). It should be pointed out that these proteins, which are normally bound to DNA, occur in a free form in large excess over the DNA present in oocyte nuclei. The stored histones H3 and H4 are associated with the highly acidic proteins N1/N2, thus forming defined, negatively charged complexes (Kleinschmidt and Franke, 1982). Recently, a second type of a soluble acidic histone complex was described in the nuclei of *Xenopus laevis* oocytes, which contains all four core histones in association with nucleoplasmin (Kleinschmidt *et al.*, 1985). By contrast, HMG-A occurs in free monomeric form and is apparently not bound to other proteins (Kleinschmidt *et al.*, 1983). *Xenopus* oocyte nuclei contain about 14 ng HMG-A, which is equivalent to a concentration of 0.2 mg/ml. Approximately equal amounts of HMG-A are also stored in the cytoplasm (Fig. 20d,e) (Kleinschmidt *et al.*, 1983).

Actin is present in relatively large amounts in isolated nuclei (Fig. 20a) and in the high-speed supernatant (Fig. 20b). *Xenopus* oocyte nuclei contain about 0.15 $\mu$g of actin, which is equivalent to a concentration of $\backsim$4.5 mg/ml (Clark and Rosenbaum, 1979). Similar values have been reported for *Pleurodeles* oocyte nuclei (Gounon and Karsenti, 1981). In both species, the amount of globular (G) actin exceeds that of filamentous (F) actin. It has been reported that 37% of the total actin of freshly isolated *Xenopus* oocyte nuclei occurs in form of F actin (Clark and Rosenbaum, 1979), whereas the corresponding value is less than 10% in *Pleurodeles* nuclei (Gounon and Karsenti, 1981). These differences in the relative amounts of F actin are apparently reflected by differences of the consistency of the nuclear sap, which is very stiff in *Xenopus* and almost fluid in *Pleurodeles*. The soluble G actin behaves differently from muscle actin, since it does not assemble spontaneously into filaments *in vitro*.

Although the role of nuclear actin is not yet clear, recent evidence suggests that this protein may be involved in transcriptional events. Injection of antibodies to actin into amphibian oocyte nuclei interferes with RNA polymerase II-mediated transcription and causes retraction of the lateral

---

ing. Note the presence of the core histones and HMG-A. (c) Proteins extracted with 5% PCA from total *X. laevis* oocytes (lane 3) and manually separated nuclei (lane 1) and cytoplasm (lane 2). Nucleoplasmin (denoted by white dots) is not detectable in the cytoplasm. Arrow denotes HMG-A. Silver staining. (d,e) Two-dimensional gel electrophoresis of radioactively labeled proteins extracted with 5% PCA from nuclei (d) and cytoplasmic portions (e) of *P. waltlii* oocytes. Nucleoplasmin is recovered from the nuclear fraction (d) and is not detectable in the cytoplasm (e; the position of nucleoplasmin is indicated by bracket). N1/N2 is also confined to the nucleus (arrowhead, d), whereas HMG-A is present in both compartments. The high-molecular-weight acidic polypeptide of the cytoplasm (e) is not related to N1/N2. Nuc, nucleoplasmin; A; actin.

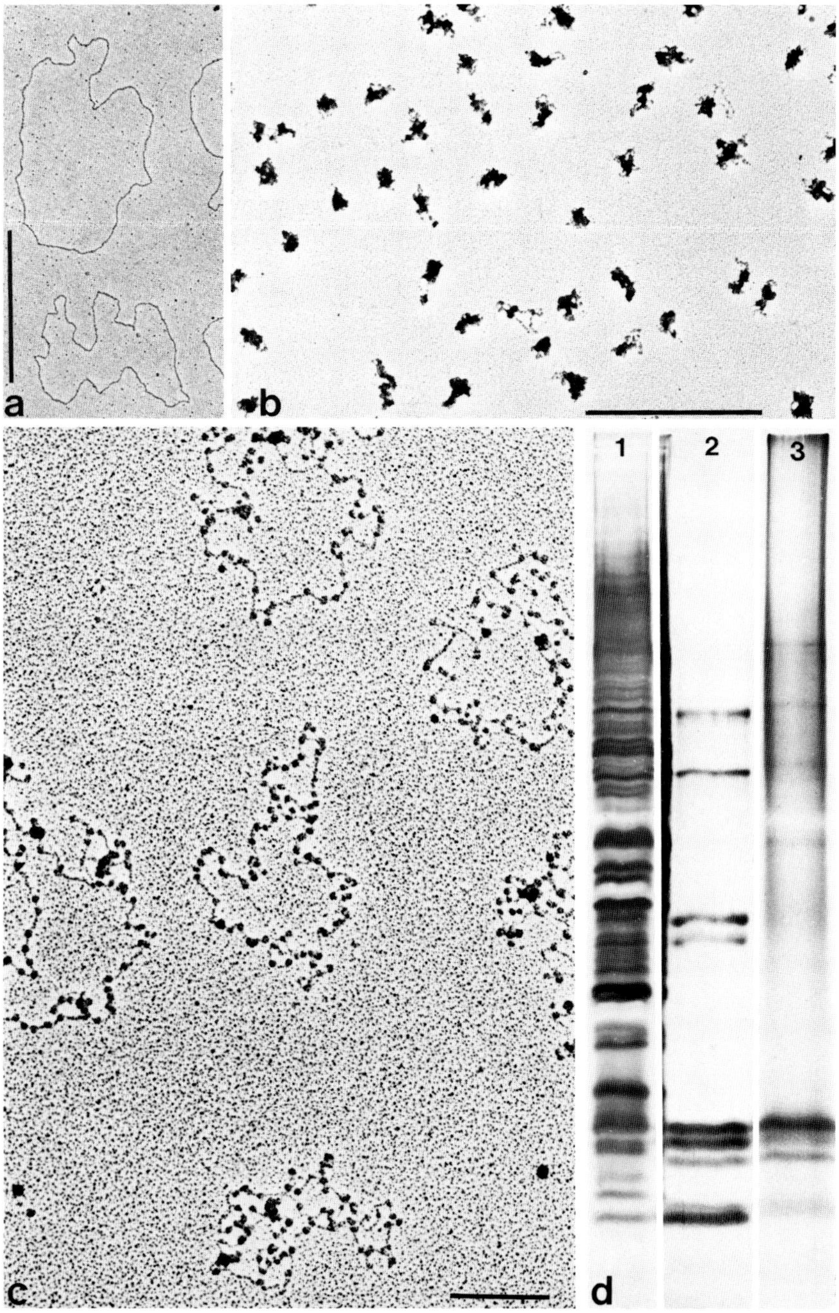


**Figure 21.** *In vitro* chromatin assembly. (a) The recombinant plasmid shown is converted into nucleoprotein complexes by incubation with the 100,000 × g supernatant fraction from isolated *X. laevis* oocyte nuclei. The newly formed nucleoprotein complexes appear—depending on the spreading conditions—as compact particles (b, 0.1 M salt) or as relaxed beaded chromatin circles (c, low salt). (d) Major proteins bound to the plasmid DNA are the four core histones as shown by one-dimensional gel electrophoresis and silver staining (d, lane 3). Sev-

loops of the lampbrush chromosomes. The same inhibitory effect is also observed upon injection of actin binding proteins from different sources, such as fragmin from the slime mould *Physarum polycephalum* and the actin modulator from pig stomach smooth muscle (Scheer *et al.*, 1984). Although this experimental approach is unable to distinguish whether the actin involved in transcriptional events includes actin polymers or monomers, we favor the concept that monomeric or oligomeric nuclear actin interacts directly with the transcriptional machinery. This possibility is supported by the recent finding that a soluble protein factor required *in vitro* for accurate transcription by RNA polymerase II contains actin (Egly *et al.*, 1984). In this connection, it is worth mentioning that, when transcription of the lampbrush chromosomes from *Triturus* or *Pleurodeles* species is inhibited by drugs such as actinomycin D or $\alpha$-amanitin or by injection of antibodies into the cell nucleus (see Section 8), the formation of an extensive fibrillar network containing actin is induced after dispersing the nuclear contents in 0.1 M saline containing 0.1 mM $CaCl_2$ (Fig. 23d). The same phenomenon, though to a lesser degree, is also seen in stages of natural inactivation: i.e., in fully mature oocytes. This observation indicates that the equilibrium between G and F actin in the oocyte nucleus is somehow dependent on transcriptional activity of the lampbrush chromosomes and thus might imply that actin is involved, directly or indirectly, in transcriptional processes or post-transcriptional maturation events of the hnRNP.

## 5.2. Chromatin Assembly

It has been shown recently that double-stranded DNA is converted into chromatinlike structures by incubation in a supernatant fraction obtained from homogenized *Xenopus laevis* eggs (reviewed in Laskey and Earnshaw, 1980). The components involved in this assembly process include the large pool of stored core histones and nucleoplasmin, which is believed to catalyze the ordered interaction between histones and DNA. Chromatin assembly is also achieved by using the nucleoplasmic fraction of *Xenopus* oocyte nuclei. The newly formed nucleoprotein complexes can be purified by sucrose gradient centrifugation and either visualized in the electron microscope (Fig. 21b,c) or analyzed by SDS-gel electrophoresis (Fig. 21d) (for details, see Scheer *et al.*, 1980).

The morphology of these nucleoprotein complexes is dependent on the ionic strength of the spreading medium. In the presence of 0.1 M salt con-

---

eral minor nonhistone proteins are also detected. Note the absence of H1. The complex protein pattern of the incubation mixture is shown in lane 1. Reference proteins are (from top to bottom) glutamate dehydrogenase, actin, calf thymus H1, and core histones (lane 2). Scale bars: 1 $\mu$m (a,b) and 0.2 $\mu$m (c).

centrations, they appear as very compact particles (Fig. 21b), whereas low salt treatment relaxes them into chromatin rings with a regularly beaded aspect (Fig. 21c). That the beads represent nucleosomes is indicated by an approximate twofold compaction of the original DNA in the chromatin (the contour length of the chromatin circle is about one-half that of the naked DNA; compare Fig. 21a,c). The major protein components of the newly formed nucleoprotein complexes are the four core histones, demonstrating that they are indeed chromatinous structures, which, however, lack histone H1 (Fig. 21d, lane 3). How closely their nonhistone protein composition resembles that of the *in vivo*-formed embryonic chromatin is an interesting question that awaits further analysis.

## 6. Storage at Two Different Levels: Proteins and Maternal mRNA

As shown in Section 5 of this chapter, *Xenopus* oocyte nuclei contain sufficient amounts of protein components for future assembly of chromatin during early embryogenesis, when extremely rapid nuclear divisions take place (Woodland, 1980). It is interesting to note that amphibian oocytes store not only the proteins needed for the assembly of embryonic chromatin but the corresponding mRNAs as well. When total *Xenopus* ovary RNA is translated in a rabbit reticulocyte lysate and the proteins are analyzed by two-dimensional gel electrophoresis, the four core histones and HMG-A are among the major translation products (Fig. 22a,b) (see also Rud-

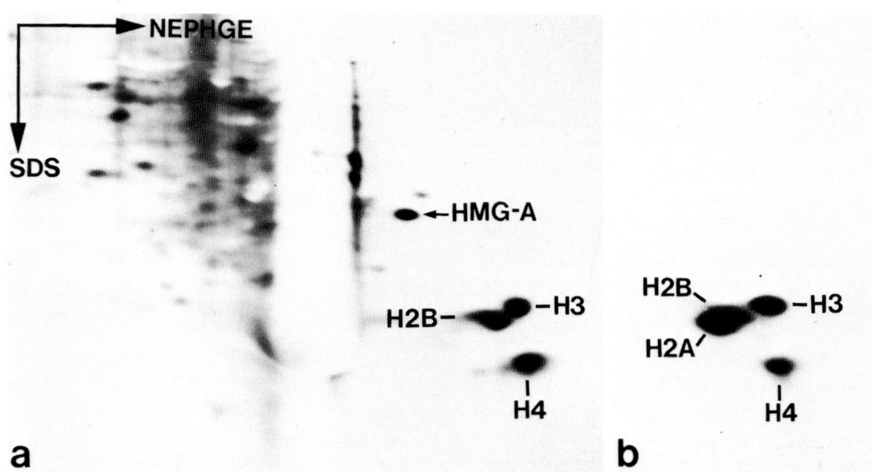


**Figure 22.** Two-dimensional gels of the translation products obtained by incubating total *Xenopus laevis* ovary mRNA in a rabbit reticulocyte lysate containing [$^{35}$S]methionine (a) or $^{14}$C-labeled amino acids (b). Major translation products are the core histones and HMG-A.

erman *et al.*, 1979). The large pool of histone mRNAs stored in *Xenopus* oocytes (for quantitative determinations see Van Dongen *et al.*, 1981) is mobilized during maturation, resulting in an about 50-fold rise in histone synthesis (Woodland, 1980). As a consequence, the histones (and probably also HMG-A) contained in newly assembled embryonic chromatin are a mosaic of molecules made months before and stored in the oocyte and of molecules newly synthesized by translation of the preformed maternal mRNAs (see also Woodland, 1980, 1982).

## 7. Microinjection of DNA into Oocyte Nuclei

The purpose of injecting purified gene sequences into oocyte nuclei is generally to examine the function of specific nucleotide sequences of the introduced DNA for regulation of transcription. This aspect is beyond the scope of this chapter. However, some information about the functional status of the germinal vesicle can be obtained from such studies.

Most of the injected DNA molecules are converted into chromatin structures similar in appearance to the *in vitro*-assembled chromatin (see Section 5.2), whereas only a relatively small proportion is transcribed. Trendelenburg and Gurdon (1978) injected a recombinant plasmid containing a full repeat unit of *Xenopus* rDNA into *Xenopus* oocyte nuclei and examined its fate by electron microscopic spread preparations. They observed that—in contrast to nontranscribed chromatin circles, which showed a regularly beaded morphology—the transcribed circles were free of nucleosomes except for a region containing the plasmid vector. Thus, the conformation of the transcribed chromatin containing the exogenous rDNA is indistinguishable from that of the endogeneous nucleolar chromatin. The occurrence of a few maximally active injected genes next to a majority of transcriptionally inactive ones packaged into nucleosomes implies that some component present in limiting amounts in the oocyte nucleus is necessary to induce and maintain their extended nonnucleosomal state.

The *Xenopus* oocyte nucleus is not only capable of transcribing the injected genes but also of processing the transcripts and producing functional mRNAs (Gurdon and Melton, 1981; Etkin, 1982; Green *et al.*, 1983). Electron microscopic analysis indicates that proteins bind in an apparent specific mode to the nascent RNA chains, since the lateral fibrils of transcription units formed on injected genes show a characteristic ultrastructural organization and foreshortening of the RNA (Trendelenburg, 1983). It is remarkable that in the *Xenopus* oocyte nucleus heterologous RNA molecules synthesized on injected genes are converted into RNP structures by association with oocyte-specific RNA-binding proteins and are correctly processed and spliced. This indicates that the maturation steps for the production of functional mRNAs follow a general principle.

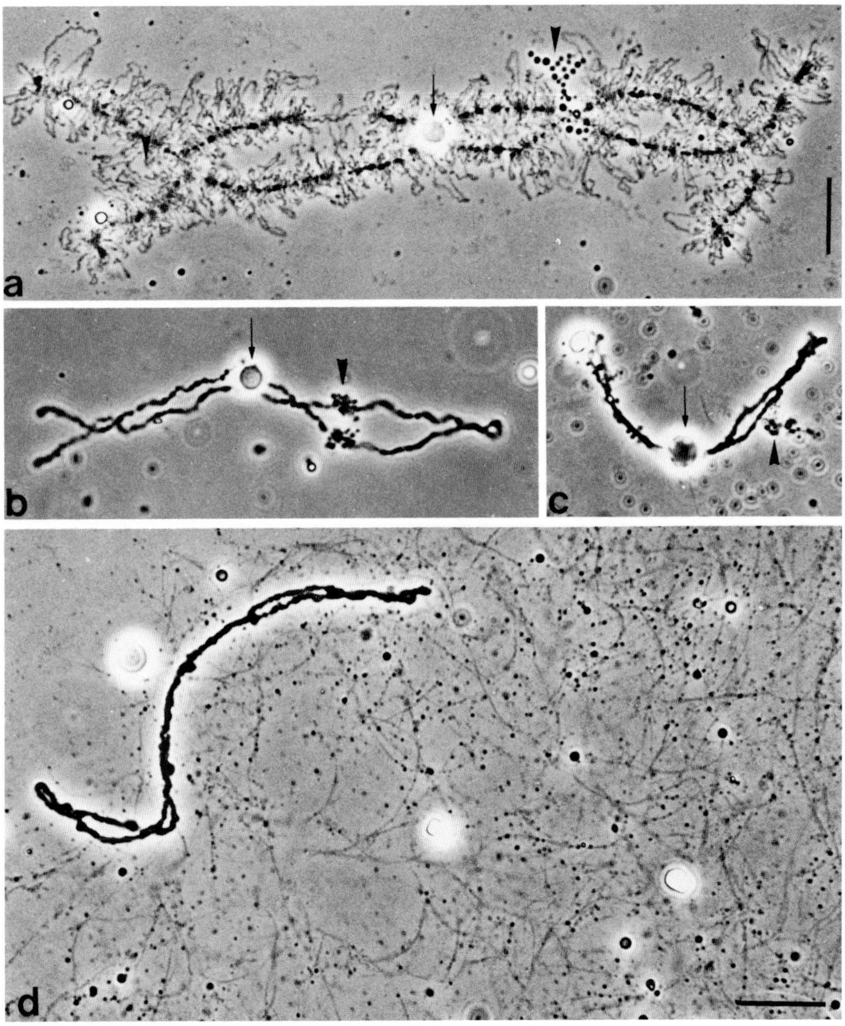


**Figure 23.** (a–c) Retraction of the lateral loops of lampbrush chromosomes after injection of antibodies to histone H2B into nuclei of *Pleurodeles waltlii* oocytes. Shown here is chromosome bivalent No. XI, identified by the sphere (arrows) and the granular loop (arrowheads). (a) Control preparation after injection of nonimmune IgG. (b,c) Chromosome preparations 1 hr and 2 hr, respectively, after injection of the antibodies to H2B. Note that the granular loop does not retract into the chromosome axis (arrowheads, b,c). The condensed chromosomes are often embedded in a finely filamentous network. This is especially clearly seen after complete inhibition of transcription by incubation of the oocytes for several hours in the presence of 50 $\mu$g/ml actinomycin D (d). Scale bars: 20 $\mu$m (a–c are magnified to the same scale).

## 8. Injection of Antibodies into the Oocyte Nucleus: A Means of Interfering with Gene Expression in the Living Cell

When antibodies directed against RNA polymerase II are injected into the nucleus of *Pleurodeles* oocytes, transcription of the lateral loops is immediately and effectively inhibited (Bona *et al.*, 1981). Inhibition of transcription of the lampbrush chromosomes results in dramatic structural rearrangements, which are visible in the light microscope as retraction of the lateral loops. It is expected that this test system will provide a valuable means to decipher the highly complex process of gene expression, since it offers the possibility of interfering selectively with the whole cascade of transcription and post-transcription events involved in the formation of functional mRNAs. In fact, this experimental approach permitted the proposal of a novel function of nuclear actin in transcription of protein-coding genes (Scheer *et al.*, 1984) (see Section 5.1). Antibodies to chromosomal components such as histone H2B and HMG-1 also induce loop retraction (Fig. 23b,c) (Scheer *et al.*, 1979b; Kleinschmidt *et al.*, 1983). A possible explanation for this finding is that, by binding of the antibodies to their DNA-associated antigens, a steric hindrance of the further progression of the RNA polymerase II is imposed, which in turn results in the premature release of the transcripts followed by the collapse of the loop axis. The observed loop retraction is therefore indicative of the presence of histones and HMG-1 on the heavily transcribed loop chromatin.

Antibodies directed against oocyte nuclear RNP and hnRNP core proteins do not inhibit transcription. However, it is an intriguing idea that injection of antibodies that bind to the nascent RNA itself or to defined RNA associated proteins might be used in the future to alter normal processing pathways.

ACKNOWLEDGMENTS. We thank Dr. W. W. Franke for many helpful suggestions during the preparation of the manuscript, our colleagues Dr. H. Zentgraf, Dr. M. Trendelenburg, and Dr. G. Krohne for providing us with illustrative material, and Dr. J. Sommerville, of the University of St. Andrews, Scotland, for reading and correcting the manuscript. This work received financial support from the Deutsche Forschungsgemeinschaft (grant Sche 157/5-4).

## References

Ackerman, E. J., 1983, Molecular cloning and sequencing of OAX DNA: An abundant gene family transcribed and activated in Xenopus oocytes, *EMBO J.* **2**:1417–1422.

Anderson, D. M., Richter, J. D., Chamberlin, M. E., Price, D. H., Britten, R. J., Smith, L. D., and Davidson, E. H., 1982, Sequence organization of the poly(A) RNA synthesized and accumulated in lampbrush chromosome stage *Xenopus laevis* oocytes, *J. Mol. Biol.* **155**:281–309.

Benavente, R., Krohne, G., Stick, R., and Franke, W. W., 1984a, Electron microscopic immunolocalization of a karyoskeletal protein of molecular weight 145,000 in nucleoli and perinucleolar bodies of *Xenopus laevis*, *Exp. Cell. Res.* **151**:224–235.

Benavente, R., Krohne, G., Schmidt-Zachmann, M. S., Hügle, B., and Franke, W. W., 1984b, Karyoskeletal proteins and the organization of the amphibian oocyte nucleus, *J. Cell Sci.* (in press).

Beyer, A. L., Bouton, A. H., and Miller, O. L., 1981, Correlation of hnRNP structure and nascent transcript cleavage, *Cell* **26**:155–165.

Bona, M., Scheer, U., and Bautz, E. K. F., 1981, Antibodies to RNA polymerase II (B) inhibit transcription in lampbrush chromosomes after microinjection into living amphibian oocytes, *J. Mol. Biol.* **151**:81–99.

Bonner, W. M., 1978, Protein migration and accumulation in nuclei, in: *The Cell Nucleus*, Vol. 6 (H. Busch, ed.), pp. 97–148, Academic Press, New York.

Born, G., 1894, Die Struktur des Keimbläschens im Ovarialei von *Triton taeniatus*, *Arch. Mikrosk. Anat.* **43**:1–79.

Brown, D. D., and Dawid, I. B. 1968, Specific gene amplification in oocytes, *Science* **160**:272–280.

Callan, H. G., 1982, Lampbrush chromosomes, *Proc. R. Soc. Lond. B* **214**:417–448.

Callan, H. G., and Lloyd, L., 1960, Lampbrush chromosomes of crested newts *Triturus cristatus* (Laurenti), *Proc. R. Soc. Lond. B* **243**:135–212.

Callan, H. G., and Lloyd, L., 1975, Working maps of the lampbrush chromosomes of amphibia, in: *Handbook of Genetics*, Vol. 4 (R. C. King, ed.), pp. 57–77, Plenum Press, New York.

Carnoy, J. B., and Lebrun, H., 1897, La vésicule germinative et les globules polaires, *La Cellule* **12**:191–295.

Carnoy, J. B., and Lebrun, H., 1898, La vésicule germinative et les globules polaires, *La Cellule* **14**:113–200.

Chooi, W. Y., and Leiby, K. R., 1981, An electron microscopic method for localization of ribosomal proteins during transcription of ribosomal DNA: A method for studying protein assembly, *Proc. Natl. Acad. Sci. USA* **78**:4823–4827.

Clark, T. G., and Rosenbaum, J. L., 1979, An actin filament matrix in hand-isolated nuclei of *Xenopus laevis* oocytes, *Cell* **18**:1101–1108.

Dabauvalle, M.-C., and Franke, W. W., 1982, Karyophilic proteins: Polypeptides synthesized in vitro accumulate in the nucleus on microinjection into the cytoplasm of amphibian oocytes, *Proc. Natl. Acad. Sci. USA* **79**:5302–5306.

Dabauvalle, M.-C., and Franke, W. W., 1984, Karyophobic proteins. A category of abundant soluble proteins which accumulate in the cytoplasm, *Exp. Cell Res.* **153**:308–326.

Davidson, E. H., 1976, *Gene Activity in Early Development*, 2nd ed., Academic Press, New York.

DeRobertis, E. M., 1983, Nucleocytoplasmic segregation of proteins and RNAs, *Cell* **32**:1021–1025.

DeRobertis, E. M., and Gurdon, J. B., 1977, Gene activation in somatic nuclei after injection into amphibian oocytes, *Proc. Natl. Acad. Sci. USA* **74**:2470–2474.

DeRobertis, E. M., Longthorne, R. F., and Gurdon, J. B., 1978, Intracellular migration of nuclear proteins in *Xenopus* oocytes, *Nature* **272**:254–256.

Diaz, M. O., Barsacchi-Pilone, G., Mahon, K. A., and Gall, J. G., 1981, Transcripts from both strands of a satellite DNA occur on lampbrush chromosome loops of the newt *Notophthalmus*, *Cell* **24**:649–659.

Dingwall, C., Sharnick, S. V., and Laskey, R. A., 1982, A polypeptide domain that specifies migration of nucleoplasmin into the nucleus, *Cell* **30**:449–458.

Dumont, J. N., 1972, Oogenesis in *Xenopus laevis* (Daudin). I. stages of oocyte development in laboratory maintained animals, *J. Morphol.* **136**:153–180.

Egly, J. M., Miyamoto, N., Moncollin, V., and Chambon, P., 1984, Is actin a transcription factor for RNA polymerase B?, *EMBO J.* **3**:2363–2371.

Etkin, L. D., 1982, Analysis of the mechanisms involved in gene regulation and cell differentiation by microinjection of purified genes and somatic cell nuclei into amphibian oocytes and eggs, *Differentiation* **21**:149–159.

Feldherr, C. M., and Ogburn, J. A. 1982, The uptake of nuclear proteins in oocytes, in: *The Nuclear Envelope and the Nuclear Matrix* (G. G. Maul, ed.), pp. 63–74, Alan R. Liss, New York.

Forbes, D. J., Kornberg, T. B., and Kirschner, M. W., 1983, Small nuclear RNA transcription and ribonucleoprotein assembly in early Xenopus development, *J. Cell Biol.* **97**:62–72.

Ford, P. J., 1972, Ribonucleic acid synthesis during oogenesis in *Xenopus laevis*, in: *Oogenesis* (J. D. Biggers and A. W. Schuetz, eds.), pp. 167–191, University Park Press, Baltimore.

Franke, W. W., and Scheer, U., 1970, The ultrastructure of the nuclear envelope of amphibian oocytes: A reinvestigation, *J. Ultrastruct. Res.* **30**:317–327.

Franke, W. W., and Scheer, U., 1974, Structures and functions of the nuclear envelope, in: *The Cell Nucleus*, Vol. 1 (H. Busch, ed.), pp. 219–347, Academic Press, New York.

Franke, W. W., Scheer, U., Spring, H., Trendelenburg, M. F., and Krohne, G., 1976a, Morphology of transcriptional units of rDNA, *Exp. Cell. Res.* **100**:233–244.

Franke, W. W., Scheer, U., Trendelenburg, M. F., Spring, H., and Zentgraf, H., 1976b, Absence of nucleosomes in transcriptionally active chromatin, *Cytobiologie* **13**:401–434.

Franke, W. W., Scheer, U., Trendelenburg, M. F., Zentgraf, H., and Spring, H., 1978, Morphology of transcriptionally active chromatin, *Cold Spring Harb. Symp. Quant. Biol.* **42**: 755–772.

Franke, W. W., Scheer, U., Spring, H., Trendelenburg, M. F., and Zentgraf, H., 1979, Organization of nucleolar chromatin, in: *The Cell Nucleus*, Vol. 7 (H. Busch, ed.), pp. 49–95, Academic Press, New York.

Franke, W. W., Kleinschmidt, J. A., Spring, H., Krohne, G., Grund, C., Trendelenburg, M. F., Stoehr, M., and Scheer, U., 1981a, A nucleolar skeleton of protein filaments demonstrated in amplified nucleoli of *Xenopus laevis*, *J. Cell Biol.* **90**:289–299.

Franke, W. W., Scheer, U., Krohne, G., and Jarasch, E. D., 1981b, The nuclear envelope and the architecture of the nuclear periphery, *J. Cell Biol.* **91**:39–50.

Gall, J. G., 1966, Techniques for the study of lampbrush chromosomes, in: *Methods in Cell Physiology*, Vol. 2 (D. M. Prescott, ed.), pp. 37–60, Academic Press, New York.

Gall, J. G., Diaz, M. O., Stephenson, E. C., and Mahon, K. A., 1983, The transcription unit of lampbrush chromosomes, in: *Gene Structure and Regulation in Development* (S. Subtelny and F. C. Kafatos, eds.), pp. 137–146. Alan R. Liss, New York.

Golden, L., Schafer, U., and Rosbash, M., 1980, Accumulation of individual $pA^+$ RNAs during oogenesis in *Xenopus laevis*, *Cell* **22**:835–844.

Gounon, P., and Karsenti, E., 1981, Involvement of contractile proteins in the changes in consistency of oocyte nucleoplasm of the newt *Pleurodeles waltlii*, *J. Cell Biol.* **88**:410–421.

Green, M. R., Maniatis, T., and Melton, D. A., 1983, Human $\beta$-globin pre-mRNA synthesized in vitro is accurately spliced in *Xenopus* oocyte nuclei, *Cell* **32**:681–694.

Gurdon, J. B., and Melton, D. A., 1981, Gene transfer in amphibian eggs and oocytes, *Annu. Rev. Genet.* **15**:189–218.

Hadjiolov, A. A., 1980, Biogenesis of ribosomes in eukaryotes, in: *Subcellular Biochemistry*, Vol. 7 (D. B. Roodyn, ed.), pp. 1–80, Plenum Press, New York.

Higashinakagawa, T., 1982, Organization of *Xenopus* ribosomal genes in chromatin, in: *The Cell Nucleus*, Vol. 11 (H. Busch and L. Rothblum, eds.), pp. 255–291, Academic Press, New York.

Hügle, B., Hazan, R., Scheer, U., and Franke, W. W., 1985, Localization of ribosomal protein S1 in the granular component of the interphase nucleolus and its distribution during mitosis *J. Cell Biol.* (in press).

Jamrich, M., Warrior, R., Steele, R., and Gall, J. G., 1983, Transcription of repetitive sequences on Xenopus lampbrush chromosomes, *Proc. Natl. Acad. Sci. USA* **80**:3364–3367.

Kalthoff, H., and Richter, D., 1982, Subcellular transport and ribosomal incorporation of microinjected protein S6 in oocytes from *Xenopus laevis*, *Biochemistry* **21**:741–745.

Kleinschmidt, J. A., and Franke, W. W., 1982, Soluble acidic complexes containing histones H3 and H4 in nuclei of *Xenopus laevis* oocytes, *Cell* **29:**799–809.

Kleinschmidt, J. A., Scheer, U., Dabauvalle, M. -C., Bustin, M., and Franke, W. W., 1983, High mobility group proteins of amphibian oocytes: A large storage pool of a soluble high mobility group-1-like protein and involvement in transcriptional events, *J. Cell Biol.* **97:**838–848.

Kleinschmidt, J. A., Fortkamp, E., Krohne, G., Zentgraf, H., and Franke, W. W., 1985, Co-existence of two different types of soluble histone complexes in nuclei of *Xenopus laevis* oocytes, *J. Biol. Chem.* (in press).

Krohne, G., and Franke, W. W., 1980, Immunological identification and localization of the predominant nuclear protein of the amphibian oocyte nucleus, *Proc. Natl. Acad. Sci. USA* **77:**1034–1038.

Krohne, G., Franke, W. W., and Scheer, U., 1978, The major polypeptides of the nuclear pore complex, *Exp. Cell Res.* **116:**85–102.

Krohne, G., Dabauvalle, M. -C., and Franke, W. W., 1981, Cell type-specific differences in protein composition of nuclear pore complex-lamina structures in oocytes and erythrocytes of *Xenopus laevis*, *J. Mol. Biol.* **151:**121–141.

Krohne, G., Stick, R., Kleinschmidt, J. A., Moll, R., Franke, W. W., and Hausen, P., 1982, Immunological localization of a major karyoskeletal protein in nucleoli of oocytes and somatic cells of *Xenopus laevis*, *J. Cell Biol.* **94:**749–754.

Labhart, P., and Koller, T., 1982, Structure of the active nucleolar chromatin of *Xenopus laevis* oocytes, *Cell* **28:**279–292.

Laskey, R. A., and Earnshaw, W. C., 1980, Nucleosome assembly, *Nature* **286:**763–767.

MacGregor, H. C., 1972, The nucleolus and its genes in amphibian oogenesis, *Biol. Rev.* **47:**177–210.

MacGregor, H. C., 1980, Recent developments in the study of lampbrush chromosomes, *Heredity* **44:**3–35.

Martin, T. E., and Okamura, C. S., 1981, Immunocytochemistry of nuclear hnRNP complexes, in: *The Cell Nucleus*, Vol. 9 (H. Busch, ed.), pp. 119–144, Academic Press, New York.

McKnight, S. L., Bustin, M., and Miller, O. L., 1978, Electron microscopic analysis of chromosome metabolism in the *Drosophila melanogaster* embryo, *Cold Spring Harb. Symp. Quant. Biol.* **42:**741–754.

Miller, O. L., and Beatty, B. R., 1969, Visualization of nucleolar genes, *Science* **164:**955–957.

Miller, O. L., and Hamkalo, B. A., 1972, Visualization of RNA synthesis on chromosomes, *Int. Rev. Cytol.* **33:**1–25.

Miller, O. L., Beatty, B. R., and Hamkalo, B. A., 1972, Nuclear structure and function during amphibian oogenesis, in: *Oogenesis* (J. D. Biggers and A. W. Schuetz, eds.), pp. 119–128, University Park Press, Baltimore.

Mills, A. D., Laskey, R. A., Black, P., and DeRobertis, E. M., 1980, An acidic protein which assembles nucleosomes in vitro is the most abundant protein in *Xenopus* oocyte nuclei, *J. Mol. Biol.* **139:**561–568.

Moreno Diaz de la Espina, S., Franke, W. W., Krohne, G., Trendelenburg, M. F., Grund, C., and Scheer, U., 1982, Medusoid fibril bodies: A novel type of nuclear filament of diameter 8 to 12 nm with periodic ultrastructure demonstrated in oocytes of *Xenopus laevis*, *Eur. J. Cell Biol.* **27:**141–150.

Moss, T., 1983, A transcriptional function for the repetitive ribosomal spacer in *Xenopus laevis*, *Nature* **302:**223–228.

Mott, M. R., and Callan, H. G., 1975, An electron-microscope study of the lampbrush chromosomes of the newt *Triturus cristatus*, *J. Cell. Sci.* **17:**241–261.

Olins, A. L., and Olins, D. E., 1974, Spheroid chromatin units ($\nu$ bodies), *Science* **183:**330–332.

Oudet, P., Gross-Bellard, M., and Chambon, P., 1975, Electron microscopic and biochemical evidence that chromatin structure is a repeating unit, *Cell* **4:**281–300.

Paine, P. L., and Horowitz, S. B., 1980, The movement of material between nucleus and cytoplasm, in: *Cell Biology: A Comprehensive Treatise*, Vol. 4 (D. M. Prescott and L. Goldstein, eds.), pp. 299–338, Academic Press, New York.

Reeder, R. H., Wahn, H. L., Botchan, P., Hipskind, R., and Sollner-Webb, B., 1978, Ribosomal genes and their proteins from *Xenopus*, *Cold Spring Harb. Symp. Quant. Biol.* **42**:1167–1177.

Rogers, M. E., 1968, Ribonucleoprotein particles in the amphibian oocyte nucleus, *J. Cell Biol.* **36**:421–432.

Rosbash, M., and Ford, P. J., 1974, Polyadenylic acid-containing RNA in *Xenopus laevis* oocytes, *J. Mol. Biol.* **85**:87–101.

Ruderman, J. V., Woodland, H. R., and Sturgess, E. A., 1979, Modulations of histone messenger RNA during the early development of *Xenopus laevis*, *Dev. Biol.* **71**:71–82.

Scheer, U., 1972, The ultrastructure of the nuclear envelope of amphibian oocytes: IV. On the chemical nature of the nuclear pore complex material, *Z. Zellforsch.* **127**:127–148.

Scheer, U., 1973, Nuclear pore flow rate of ribosomal RNA and chain growth rate of its precursor during oogenesis of *Xenopus laevis*, *Dev. Biol.* **30**:13–28.

Scheer, U., 1978, Changes of nucleosome frequency in nucleolar and non-nucleolar chromatin as a function of transcription: An electron microscopic study, *Cell* **13**:535–549.

Scheer, U., 1980, Structural organization of spacer chromatin between transcribed ribosomal RNA genes in amphibian oocytes, *Eur. J. Cell Biol.* **23**:189–196.

Scheer, U., 1981, Identification of a novel class of tandemly repeated genes transcribed on lampbrush chromosomes of *Pleurodoles waltlii*, *J. Cell Biol.* **88**:599–603.

Scheer, U., 1982, A novel type of chromatin organization in lampbrush chromosomes of *Pleurodeles waltlii*: Visualization of clusters of tandemly repeated, very short transcriptional units, *Biol. Cell.* **44**:213–220.

Scheer, U., and Sommerville, J., 1982, Sizes of chromosome loops and hnRNA molecules in oocytes of amphibia of different genome sizes, *Exp. Cell Res.* **139**:410–416.

Scheer, U., and Zentgraf, H., 1982, Morphology of nucleolar chromatin in electron microscopic spread preparations, in: *The Cell Nucleus*, Vol. 11 (H. Busch and L. Rothblum, eds.), pp. 143–176, Academic Press, New York.

Scheer, U., Trendelenburg, M. F., and Franke, W. W., 1976a, Regulation of transcription of genes of ribosomal RNA during amphibian oogenesis: A biochemical and morphological study, *J. Cell Biol.* **69**:465–489.

Scheer, U., Kartenbeck, J., Trendelenburg, M. F., Stadler, J., and Franke, W. W., 1976b, Experimental disintegration of the nuclear envelope. Evidence for pore-connecting fibrils, *J. Cell Biol.* **69**:1–18.

Scheer, U., Trendelenburg, M. F., Krohne, G., and Franke, W. W., 1977, Lengths and patterns of transcriptional units in the amplified nucleoli of oocytes of *Xenopus laevis*, *Chromosoma* **60**:147–167.

Scheer, U., Spring, H., and Trendelenburg, M. F., 1979a, Organization of transcriptionally active chromatin in lampbrush chromosome loops, in: *The Cell Nucleus*, Vol. 7 (H. Busch, ed.), pp. 3–47, Academic Press, New York.

Scheer, U., Sommerville, J., and Bustin, M., 1979b, Injected histone antibodies interfere with transcription of lampbrush chromosome loops in oocytes of *Pleurodeles*, *J. Cell Sci.* **40**:1–20.

Scheer, U., Sommerville, J., and Müller, U., 1980, DNA is assembled into globular supranucleosomal chromatin structures by nuclear contents of amphibian oocytes, *Exp. Cell Res.* **129**:115–126.

Scheer, U., Hinssen, H., Franke, W. W., and Jockusch, B. M., 1984, Microinjection of actin-binding proteins and actin antibodies demonstrates involvement of nuclear actin in transcription of lampbrush chromosomes. *Cell* **39**:111–122.

Schultz, L. D., Kay, B. K., and Gall, J. G., 1981, In vitro RNA synthesis in oocyte nuclei of the newt *Notophthalmus*, *Chromosoma* **82**:171–187.

Sollner-Webb, B., Wilkinson, J. K., and Miller, K. G., 1982, Transcription of *Xenopus* ribosomal

RNA genes, in: *The Cell Nucleus*, Vol. 12 (H. Busch and L. Rothblum, eds.), pp. 31–67, Academic Press, New York.

Sommerville, J., 1977, Gene activity in the lampbrush chromosomes of amphibian oocytes, *Int. Rev. Biochem.* **15:**79–156.

Sommerville, J., 1981, Immunolocalization and structural organization of nascent RNP, in: *The Cell Nucleus*, Vol. 8 (H. Busch, ed.) pp. 1–57, Academic Press, New York.

Sommerville, J., and Malcolm, D. B., 1976, Transcription of genetic information in amphibian oocytes, *Chromosoma* **55:**183–208.

Sommerville, J., and Scheer, U., 1982, Transcription of complementary repeat sequences in amphibian oocytes, *Chromosoma* **86:**95–113.

Sommerville, J., Crichton, C., and Malcolm, D., 1978, Immunofluorescent localization of transcriptional activity on lampbrush chromosomes, *Chromosoma* **66:**99–114.

Spring, H., and Franke, W. W., 1981, Transcriptionally active chromatin in loops of lampbrush chromosomes at physiological salt concentrations as revealed by electron microscopy of sections, *Eur. J. Cell Biol.* **24:**298–308.

Stick, R., and Krohne, G., 1982, Immunological localization of the major architectural protein associated with the nuclear envelope of the *Xenopus laevis* oocyte, *Exp. Cell Res.* **138:**319–330.

Trendelenburg, M. F., 1981, Initiations of transcription at distinct promoter sites in spacer regions between pre-rRNA genes in oocytes of *Xenopus laevis:* An electron microscopic analysis, *Biol. Cell* **42:**1–12.

Trendelenburg, M. F., 1983, Progress in visualization of eukaryotic gene transcription, *Hum. Genet.* **63:**197–215.

Trendelenburg, M. F., and Gurdon, J. B., 1978, Transcription of cloned *Xenopus* ribosomal genes visualized after injection into oocyte nuclei, *Nature* **276:**292–294.

Van Dongen, W., Zaal, R., Moorman, A., and Destrée, O., 1981, Quantitation of the accumulation of histone messenger RNA during oogenesis in *Xenopus laevis*, *Dev. Biol.* **86:**303–314.

Van Dongen, W., Moorman, A., and Destrée, O., 1983, The accumulation of the maternal pool of histone H1A during oogenesis in Xenopus laevis, *Cell Diff.* **12:**257–264.

Varley, J. M., MacGregor, H. C., Nardi, I., Andrews, C., and Erba, H. P., 1980a, Cytological evidence of transcription of highly repeated DNA sequences during the lampbrush stage in *Triturus cristatus carnifex*, *Chromosoma* **80:**289–307.

Varley, J. M., MacGregor, H. C., and Erba, H. P. 1980b, Satellite DNA is transcribed on lampbrush chromosomes, *Nature* **283:**686–688.

Wakefield, L., Ackerman, E., and Gurdon, J. B., 1983, The activation of RNA synthesis by somatic nuclei injected into amphibian oocytes, *Dev. Biol.* **95:**468–475.

Woodland, H. R., 1980, Histone synthesis during the development of *Xenopus*, *FEBS Lett.* **121:**1–7.

Woodland, H. R., 1982, The translational control phase of early development, *Biosci. Rep.* **2:**471–491.

Chapter 10

# 5 S Ribosomal Gene Transcription During *Xenopus* Oogenesis

ANGELA KRÄMER

## 1. Introduction

In contrast to somatic cells, *Xenopus* oocytes accumulate large numbers of ribosomes for their subsequent use in early embryogenesis (Davidson, 1976). To accommodate the need for ribosomal RNAs, the genes encoding the large ribosomal RNAs (i.e., 18 S, 28 S, and 5.8 S rRNA) are amplified in the oocyte to yield about $2 \times 10^6$ extrachromosomal copies per nucleus (Brown and Weber, 1968b; Gall, 1968). Since the genes coding for 5 S rRNA are not part of the repeating unit that encodes the large rRNAs (Brown and Weber, 1968a; Brown and Dawid, 1968), initial investigations on 5 S rRNA focused on elucidating the organization and structure of these genes.

The presence of two *Xenopus* 5 S rRNA gene families, which are developmentally regulated, has generated further interest in this system. The somatic gene family is expressed in both oocytes and somatic cells, whereas the oocyte-type genes are only expressed during oogenesis and early embryogenesis (Wegnez *et al.*, 1972; Ford and Southern, 1973; Brown *et al.*, 1981). Thus, the introduction of a second set of 5 S rRNA genes, which are only transcribed when large amounts of the RNA are needed, provides extra templates comparable to the amplified 18 S and 28 S rRNA genes.

The question of how the expression of the related 5 S gene families is regulated has led to an investigation of the basic mechanisms of 5 S rRNA synthesis. The use of *Xenopus* oocyte microinjection and the development of soluble cell-free systems, both of which direct the accurate transcription of cloned genes by RNA polymerase III, has facilitated the search for signals encoded in the DNA that are essential to transcription initiation and

ANGELA KRÄMER • Division of Molecular Biology and Biochemistry, Institute of Cell and Tumor Biology, German Cancer Research Center, D-6900 Heidelberg 1, Federal Republic of Germany.

termination. Moreover, these reconstituted systems have contributed to the evaluation of the proteins involved in 5 S rRNA synthesis.

This chapter attempts to provide an overview of our present understanding of 5 S rRNA gene transcription in *Xenopus* oocytes. The differential expression of oocyte versus somatic genes is covered briefly. Further information on this topic can be found in several recent review articles (Pelham *et al.*, 1981a; Roeder *et al.*, 1981; Korn and Bogenhagen, 1982; Wormington *et al.*, 1983; Brown, 1984).

## 2. Isolation and Structural Organization of 5 S rRNA Genes in *Xenopus*

The purification of 5 S DNA from *Xenopus* was possible before the development of recombinant DNA technology, because—in this organism—the 5 S genes represent a highly reiterated gene family (about 0.7% of the total nuclear DNA). A tandem arrangement of the 5 S genes and a lower content of guanosine (G) and cytidine (C) residues (33%) than in bulk DNA permitted its purification by a series of centrifugations (Brown *et al.*, 1971). Equilibrium centrifugation of DNA complexed with $Ag^+$ shifts the 5 S DNA to the light side of bulk DNA. Since 5 S DNA—due to its high content of adenosine (A) and thymidine (T) residues—binds less actinomycin D than any other DNA component, the final purification was achieved after complexing the DNA with the antibiotic and subjecting the complex to equilibrium centrifugation. Renaturation experiments and electron microscopic examination of partially denatured 5 S DNA demonstrated that each repeating unit consists of a region of high GC content, containing the sequence coding for 5 S RNA, alternating with a spacer element of low GC content.

Information on the primary structure of the 5 S genes was derived from sequencing either 5 S rRNA, isolated 5 S DNA, or cloned 5 S DNA sequences. Fingerprint analysis of $^{32}P$-labeled 5 S rRNA isolated from oocytes of *X. laevis* indicates that this RNA is somewhat heterogeneous and that it contains sequences not present in somatic 5 S rRNA. Although the gene products in oocytes and in somatic cells are of the same size, their nucleotide sequences differ in several positions (Brownlee *et al.*, 1972; Wegnez *et al.*, 1972; Ford and Southern, 1973). To date, three different families of *X. laevis* 5 S genes have been identified and cloned. In addition, two related families in *X. borealis* have been isolated.

The major class of 5 S genes in *X. laevis*, the oocyte-type genes, are present in 24,000 copies per haploid genome and are tandemly reiterated. They are located at the ends of most chromosomes (Pardue *et al.*, 1973; Harper *et al.*, 1983). One repeating unit consists of about 750 base pairs (bp) and can be divided into an AT-rich region and a GC-rich part; the latter contains the 120 base pairs of the coding sequence. The length heterogeneity observed within this gene family can be attributed to a variable number of repeats of a 15-nucleotide sequence in part of the AT-rich region (Carroll and Brown, 1976a; Fedoroff and Brown, 1978). On average, five to six tandem copies of this oligonucleotide are present. Repeats of different lengths are found adjacent to each other (Carroll and Brown, 1976b). The GC-rich part comprises a short sequence of nontranscribed spacer and the 5 S rRNA coding sequence, followed by a duplication, which includes 73 base pairs of the 5′ sequences flanking the coding region and 101 base pairs of the coding region itself. This duplication, or pseudogene sequence, differs from the 5 S gene coding portion by 10 nucleotides, whereas the remaining sequence of the duplication has diverged by 13 nucleotides from the corresponding portion of the 5′ flanking sequences (Jacq *et al.*, 1977; Miller *et al.*, 1978). The function of the pseudogenes is not clear. Although they are transcribed in cell-free extracts (Korn and Brown, 1978; Bogenhagen and Brown, 1981) and after injection into oocytes of *Xenopus* (Miller and Melton, 1981), transcripts have not been detected *in vivo* (Wegnez *et al.*, 1972; Ford and Southern, 1973; Jacq *et al.*, 1977). These genes may not be transcribed *in vivo;* alternatively, the gene products could be unstable, or termination could be inefficient, leading to transcripts of sizes different from 5 S RNA, which might not appear as a distinct species when examined by gel electrophoresis (Miller and Melton, 1981).

A second minor type of oocyte 5 S genes, the trace oocyte-type genes, are present in 1300 copies per haploid genome. Like the major oocyte-type 5 S genes, they are arranged in tandem (Brown *et al.*, 1977; Peterson *et al.*, 1980). This class of genes is located on the distal end of the long arm on chromosome 13 of *X. laevis* (Harper *et al.*, 1983). The trace oocyte genes resemble the major class of 5 S genes in that they contain an AT-rich untranscribed region and a GC-rich sequence containing the coding region followed by a duplication of almost all the spacer sequence. The duplicated sequence has diverged by about 40% from the spacer sequence preceding the coding region and does not include any parts of the sequences coding for 5 S RNA. In contrast to the major oocyte-type 5 S genes, the trace oocyte-type genes do not contain a sequence related to the repeated 15 nucleotides. Therefore, the repeats are homogeneous in length, with 310 base pairs per repeat (Peterson *et al.*, 1980).

A third class of 5 S genes is represented by the somatic-type 5 S gene family. These genes are present in about 400 copies per haploid complement and differ from the oocyte-type families in having GC-rich spacer sequences. The coding sequence differs from that of the oocyte-type 5 S gene by 6 nucleotides. As in the oocyte-type 5 S genes, the gene sequences are arranged in tandem arrays of 880 base pairs and do not exhibit any length heterogeneity (Peterson *et al.*, 1980). *In situ* hybridization studies have localized these genes to the end of the long arm of chromosome 9 (Harper *et al.*, 1983).

Genes cloned from *X. borealis* corresponding to the major oocyte and somatic-type 5 S genes of *X. laevis* exhibit similar characteristics. The oocyte-type genes are present in 9000 copies per haploid genome. The repeats contain AT-rich spacers and are heterogeneous in length (1000–2000 base pairs) due to a variable number of a repeated 21-nucleotide sequence. In addition, the number of genes and pseudogenes and their relative positions in each cluster vary (Korn and Brown, 1978). The 5 S genes of the somatic type are present in about 700 copies per haploid genome. They have a homogeneous length of 850 base pairs and a GC-rich spacer and contain no pseudogene within them.

## 3. 5 S rRNA Synthesis During Oogenesis

### 3.1. Levels of 5 S rRNA

The genes for the large ribosomal RNAs are amplified in amphibian oocytes, whereas the genes for 5 S rRNA are reiterated and present in about 100,000 copies in the tetraploid oocyte (see above). Synthesis of 18 S and 28 S rRNA is extremely low in the young oocyte. Accumulation of these RNA species begins in early vitellogenesis (Ford, 1971). In previtellogenic oocytes, 5 S rRNA and tRNA each represent about 40% of the total cellular RNA (Mairy and Denis, 1971). At this stage in oogenesis, the molar ratio of 5 S rRNA to 18 S and 28 S rRNA was calculated to be 110:1 (Ford, 1971). Therefore, an amount of 5 S rRNA has accumulated, allowing for the large increase in the ribosomal content of the oocyte during vitellogenesis. In early vitellogenesis, the ratio of 5 S rRNA to 18 S and 28 S rRNA decreases drastically to about 10:1 (Ford, 1971). At this stage, more than 80% of the total RNA is represented by large RNAs, which are composed of DNA-like RNA (hnRNA, mRNA-precursors) and the large ribosomal RNAs. 5 S rRNA now accounts for only 10% of the RNA content of the oocyte (Mairy and Denis, 1971). The low percentage of 5 S rRNA, however, is not attributable to a degradation of the small rRNA but to the activation of the genes coding for 18 S and 28S rRNA (Mairy and Denis, 1971; Ford, 1971). By labeling isolated oocytes from different developmental stages, Mairy and Denis (1971) demonstrated that 5 S rRNA genes are still active in the mature oocyte. The amount of 5 S rRNA increases from about 50 ng per oocyte in early vitellogenesis to 80 ng in the mature oocyte, concomitant with the synthesis of 18 and 28 S rRNA. The steady decrease in the proportions of 5 S rRNA relative to the large ribosomal RNAs is therefore caused by the rapid accumulation of 18 S and 28 S rRNA. These results are compatible with a noncoordinated expression of the genes for 5 S rRNA and 18 S and 28 S rRNA. This view is supported by the fact that the genes that code for 5 S rRNA are not interspersed with those that code for the large ribosomal rRNAs (Brown and Weber, 1968b) and that the two classes of RNA are transcribed by different RNA polymerases (Roeder, 1976).

### 3.2. Storage of 5 S RNA During Oogenesis

During the early studies of 5 S rRNA synthesis, it was proposed that the small RNAs (5 S rRNA as well as tRNA) do not occur free in the cell sap, but are complexed with proteins to form ribonucleoprotein (RNP) particles. The first such particles described were 42 S RNP particles (Ford, 1971) containing 5 S rRNA and tRNA in addition to protein.

These particles are distinct from the small ribosomal subunit (40 S), in that they have a different sedimentation behavior in the presence of EDTA and high salt concentrations. They contain about 50% of the 5 S rRNA in young oocytes and 90% of the tRNA (Denis and Mairy, 1972). The particles are found in large amounts in previtellogenic oocytes, when only a few ribosomes are present; they are no longer detectable at late developmental stages (Ford, 1971; Denis and Mairy, 1972; Dixon and Ford, 1982). Once vitellogenesis begins, stored 5 S rRNA is transferred to the accumulating ribosomes; tRNA is then released into the cytoplasm (Mairy and Denis, 1972).

Biochemical examination of the 42 S particle (Picard *et al.*, 1980) has demonstrated that it contains 5 S rRNA and tRNA in a ratio of 1:3 as well as two proteins of 50,000 and 40,000 molecular weight (in a ratio of 2:1). Peptide mapping of ribosomal proteins and the proteins present in the 42 S particle do not show any similarities, suggesting that the proteins in the storage particle are distinct from ribosomal proteins (Delaunay *et al.*, 1975).

The remaining 5 S rRNA that is not associated with 42 S RNP particles (50%) in the young oocyte is sequestered in 7 S RNP particles (Picard and Wegnez, 1979). Each of these small particles contains one molecule of 5 S rRNA bound to a protein of 40,000 $M_r$. This protein is absent both from ribosomes and 42 S RNP particles (Picard and Wegnez, 1979; Honda and Roeder, 1980).

Therefore, it is evident that 5 S rRNA is associated with different proteins (7 S/42 S RNP proteins and a ribosomal protein) during the course of oogenesis. The transfer probably takes place once oocytes enter vitellogenesis and ribosomes appear. The functions of the 7 S and 42 S RNP particles are not completely understood, but they most likely function as storage particles, thereby protecting the RNA from endoribonucleases (Wegnez and Denis, 1973). Recently, an additional function of the 7 S RNP protein in the regulation of transcription became apparent. It was observed that this protein acts as a positive transcription factor, which is necessary for the accurate initiation of transcription of 5 S genes by RNA polymerase III. Results leading to this proposal are discussed in detail below.

## 4. Transcription of 5 S rRNA Genes

### 4.1. Test Systems

Our present knowledge of the mechanism of 5 S rRNA synthesis in *Xenopus*—both on the level of the proteins involved and on the level of signals encoded in the DNA template—was gained by transcription of 5 S genes in either injected oocytes or reconstituted systems. *Xenopus* oocyte nuclei, when injected with purified genomic DNA or cloned 5 S rRNA genes, accurately synthesize the expected gene products (Brown and Gurdon, 1977, 1978; Gurdon and Brown, 1978). These experiments not only indicate that these DNAs are functional in the oocyte, but they also show that a single repeating unit of *Xenopus* 5 S DNA contains all the information necessary for correct initiation and termination of transcription. However, when purified *Xenopus* RNA polymerase III was used to transcribe 5 S DNA *in vitro*, RNA synthesis was nonspecific and random (Parker and Roeder, 1977). In contrast, a chromatin template isolated from *Xenopus* ovaries was transcribed in an asymmetrical fashion by the purified enzyme, yielding predominantly 5 S rRNA. This study demonstrated for the first time the selective transcription of a particular gene in a reconstituted cell-free system. The requirement of a chromatin template to achieve selective transcription of 5 S genes by RNA polymerase III implies that components not present in the purified enzyme but stably associated with chromatin are required to yield specific gene products.

Subsequent to these studies, soluble cell-free systems were developed that were either prepared from isolated germinal vesicles (Birkenmeier *et al.*, 1978) or derived from whole ovaries or somatic cells of *Xenopus* (Ng *et al.*, 1979; Weil *et al.*, 1979). These cell-free systems contain all the components required for accurate 5 S rRNA synthesis when programmed with cloned 5 S genes. Maximal 5 S rRNA synthesis, up to 90% of the total RNA synthesized during the reaction period (B. S. Shastry, P. L. Martin, and R. G. Roeder, personal communication), occurs at 5 mM $MgCl_2$ in ovarian extracts (8 mM in germinal vesicle extracts), requires low salt concentrations (60–80 mM KCl), and is resistant to low, but sensitive to high, concentrations of $\alpha$-amanitin, confirming that transcription is mediated by RNA polymerase III. Transcription is linear for several hours, but a lag period of about 15 min is observed in both the cytoplasmic and the germinal vesicle systems. Preincubation of an extract with the template before the addition of the labeled ribonucleoside-triphosphate eliminates this lag period (Birkenmeier *et al.*, 1978). In contrast to total RNA synthesis by RNA polymerase III, specific transcription of 5 S genes is inhibited by high concentrations of template DNA. Plasmid DNAs that do not contain 5 S genes inhibit 5 S rRNA synthesis as well (see Ng *et al.*, 1979). Therefore, it appears that certain transcription components become sequestered by nonspecific interaction with excess DNA.

An additional feature of these systems is that the specificity of 5 S gene transcription observed *in vivo* cannot be reproduced *in vitro*. For example, plasmids containing oocyte-type 5 S genes are efficiently transcribed in extracts derived from cultured *Xenopus* kidney cells, which contain—but do not express—these genes *in vivo* (Weil *et al.*, 1979). Components involved in the *in vivo* control might not be retained or might be nonfunctional in the cell-free extracts. Also, in intact cells, the DNA is present in a chromatin template, which could be modified such that transcription of certain genes is repressed. The purified DNAs used for *in vitro* transcription, however, may be viewed as derepressed templates that can be transcribed by the extracts. Despite the lack of elements that control 5 S gene expression *in vivo*, the use of these extracts has contributed a great deal to our understanding of the events leading to the synthesis, *in vitro*, of gene products that are indistinguishable from their *in vivo* counterparts.

## 4.2. DNA Sequences Required for Specific Transcription of 5 S rRNA Genes

### 4.2.1. Initiation Sequences

Promoter elements in prokaryotes are confined to the 5′ flanking regions of the genes. Two regions of conserved nucleotides that are functionally important are located about 6–12 nucleotides upstream from the start site (Pribnow box) and around residue 35 (Pribnow, 1975; Gilbert, 1976). In contrast to the situation in prokaryotes, indications that most of the spacer sequences of 5 S genes are dispensable for initiation of transcription came from two studies. First, cloned *X. borealis* 5 S genes, separated by only about 80 nucleotides, are accurately transcribed *in vitro* (Korn and Brown, 1978). Second, a cloned 5 S gene from *X. laevis*, in which all but 49 nucleotides preceding the coding region had been deleted, was a fully active template after injection into oocytes (Fedoroff, 1979).

To define a promoter for 5 S genes more precisely, D. Brown and colleagues constructed a collection of deletion mutants of a somatic gene from *X. borealis* (Sakonju *et al.*, 1980; Bogenhagen *et al.*, 1980). The deletions extended into the coding region to varying degrees, either from the 5′ or 3′ flanking regions. 5 S DNA sequences were thereby progressively replaced by vector DNA. The ability of the mutant templates to direct 5 S rRNA synthesis was tested in a germinal vesicle extract from *X. laevis*. The resulting transcripts were characterized by their migration in high-resolution urea/polyacrylamide gels, by RNAase $T_1$ oligonucleotide fingerprinting, and by determination of the 5′ terminal nucleotide of the primary transcript. Moreover, an assay was developed to detect correct transcription initiation in 3′ deletion mutants, which lack a functional termination site and are therefore not transcribed into a single RNA species. This assay employed the addition of cordycepin-triphosphate (3′ dATP) to the *in vitro* system, causing premature termination of transcription and thereby generating an array of shortened RNA molecules, which could be examined in polyacrylamide gels.

The results of these studies indicate that not only all the sequences flanking the 5 S coding region to the 5′ and 3′ ends—but also a portion of the sequences within the coding region itself—can be deleted without loss of ability to initiate close to the usual start site or in vector sequences corresponding to the position at which initiation usually occurs. Only nucleotides 50–83 of the coding sequence are required to direct transcription initiation approximately 50 nucleotides upstream from the 5′ border of this region. The implication that the sequences controlling specific initiation by RNA polymerase III are located in the center of the gene was strengthened when—in a plasmid containing only nucleotides 41–87 of the 5 S gene—transcription started approximately 40 nucleotides upstream from the inserted DNA fragment within vector sequences. Furthermore, two maxigenes containing insertions of either 6 or 20 nucleotides between positions +40 and +41 of the gene (i.e., outside the crucial sequence), direct transcription of an RNA species similar in size to 5 S rRNA. The usual start site present in the cloned fragment is ignored, and initiation takes place about 50 nucleotides upstream from the 5′ border of the internal element. From these results, it appears that transcription begins at a set distance measured from the 5′ end of the essential sequence within the gene. Slightly different results were obtained with a maxigene containing an insertion within this region between nucleotides +71 and +74 (Ciliberto *et al.*, 1983). This mutant gene yields a RNA transcript 15 nucleotides longer than that of the wild-type transcript. This finding emphasizes that the initiation site is measured from the 5′ end of the internal sequence; moreover, it suggests that the continuity of the internal element is not a prerequisite for accurate initiation of transcription.

The sequence around the usual start site is less important for initiation, although it influences the exact start site by an apparent preference of the RNA polymerase to begin transcription with a purine residue (Korn *et al.*, 1979; Sakonju *et al.*, 1980). Sequences 5′ to the initiation site, which are conserved among several 5 S genes (Korn and Brown, 1978), are dispensable for correct transcription initiation but exert effects on the transcription efficiency in the *in vitro* test systems (Sakonju *et al.*, 1980; Bogenhagen *et al.*, 1980; Wormington *et al.*, 1981). Thus, the position of sequences essential for transcription initiation of 5 S genes is strikingly different from the promoter elements known in prokaryotes. A prokaryotic promoter is defined as an element containing a recognition site and a stable binding site for RNA polymerase, as well as a site for transcription initiation (Pribnow, 1975). Because there was no evidence that RNA polymerase III would actually recognize and bind to the sequence within the gene, the term "control region" was chosen for the internal element (Sakonju *et al.*, 1980).

#### 4.2.2. Termination Signals

Sequence elements that cause transcription of 5 S genes to terminate were deduced by comparing either *in vitro* transcripts of cloned natural genes, which exhibit variations around the termination site (Korn and Brown, 1978), or mutant genes obtained by *in vitro* genetics (Bogenhagen *et al.*, 1980; Bogenhagen and Brown, 1981). In contrast to the sequences essential for transcription initiation, prokaryotes and eukaryotes share several features concerning elements necessary for termination. Prokaryotic termination sites are preceded by a GC-rich region with an inherent dyad symmetry, which is thought to slow down the elongation rate by RNA polymerase (Maizels, 1973; Gilbert, 1976). Termination occurs in an AT-rich region that contains a string of T residues on the noncoding strand of the gene (Rosenberg and Court, 1979). In eukaryotes, 5 S gene transcription stops within clusters of four or more T residues, also present on the noncoding strand (Denis and Wegnez, 1973; Ford and Brown, 1976; Korn and Brown, 1978; Bogenhagen and Brown, 1981). The presence of three, rather than four, consecutive T residues in a mutant gene enhances readthrough past the termination site. In addition, in order for termination to be efficient, either the presence of two GC residues or the absence of an A cluster adjacent to the stretch of T residues is required (Bogenhagen and Brown, 1981). A-rich regions following T clusters weaken termination and cause readthrough to subsequent stop signals.

Sequences preceding the termination site are of less importance in eukaryotic 5 S genes. Bogenhagen and Brown (1981) studied termination *in vitro* of mutant genes containing the internal control region and only a few nucleotides of sequences preceding the termination site in a somatic gene of *X. borealis*. Transcription of this gene, which lacked any dyad symmetry upstream from the termination site, terminated efficiently within the first T cluster following the gene. Thus, an element of dyad symmetry is not necessary to promote termination by RNA polymerase III. Another conclusion reached from this result is that a specific distance of the termination signal from the initiation site is not required.

Although RNA polymerase III by itself is unable to initiate selective transcription of 5 S genes, random transcripts generated from cloned 5 S genes by the purified enzyme (in the absence of other components) terminate precisely at the end of the gene. Thus, the enzyme itself has the capability to recognize termination signals encoded at the end of the gene (Cozzarelli *et al.*, 1983).

### 4.3. Proteins Involved in the Specific Transcription of 5 S rRNA Genes

To gain insight into the apparent multiplicity of components involved in 5 S gene transcription and into the mechanism of action of the presumed transcription factors, soluble extracts were used as the starting material for the fractionation and possibly the purification of the essential components. Initial chromatography of an ovarian extract showed that at least two components are necessary in addition to RNA polymerase III for 5 S RNA gene transcription *in vitro* (Engelke *et al.*, 1980). Chromatographic fractionation of an extract derived from cultured human cells yielded three fractions that are required, apart from the purified enzyme, to direct specific synthesis of 5 S rRNA *in vitro* (Segall *et al.*, 1980). Two of these fractions plus RNA polymerase III suffice for the transcription of tRNA genes. Similar results were recently reported for the *Xenopus* system. Shastry *et al.*, (1982) fractionated extracts derived from ovaries, unfertilized eggs, embryos, and cultured kidney cells. Interestingly, in the assay system used, the various components from different *Xenopus* tissues or developmental stages seem to be functionally interchangeable. However, since two of the components (factors IIIB and IIIC) have not been purified to homogeneity, these factors may be qualitatively and quantitatively different in various tissues.

The only transcription factor that has been obtained in pure form is the protein specific for the transcription of 5 S genes (Engelke *et al.*, 1980). The initial identification and purification of this factor (TFIIIA) was facilitated by the availability of an extract from unfertilized eggs, which directs the synthesis of tRNA—but not of 5 S rRNA—to a significant extent. When such an extract is complemented with certain fractions obtained after chromatography of an ovarian extract, transcription of 5 S genes is specifically enhanced. The purified protein has a molecular weight of about 40,000. Factors that are functionally equivalent are present in *Xenopus* embryos and kidney cells, although in much lower concentrations, which has hindered their purification thus far (see Pelham *et al.*, 1981b; Roeder *et al.*, 1981; Shastry *et al.*, 1982).

## 5. The Role of a Positive Transcription Factor (TFIIIA) in the Transcription of 5 S rRNA Genes

### 5.1. Interaction of TFIIIA with the Template

Using the footprinting technique devised by Galas and Schmitz (1978), Engelke *et al.* (1980) demonstrated the interaction of the 5 S gene specific transcription factor with the intragenic control region of a somatic and an oocyte-type 5 S gene from *X. borealis*. Gene residues 47–96 and 45–92, respectively, are protected from DNAase digestion by the factor. Proteins present in an egg extract and RNA polymerase III from *Xenopus* do not bind to any part of the 5 S coding region in this assay and have no influence on the protection pattern obtained with purified TFIIIA. From these results, a positive role of TFIIIA in transcription initiation was proposed. For the initial recognition event of the crucial DNA sequence, apparently only this factor is required. Once it is bound to the template, RNA polymerase III, by interaction with a stable complex (that contains additional transcription factors), is directed to the initiation site to start transcription at a set distance from the control region.

Sequences required for the interaction of TFIIIA with the template are limited to the internal control region, but the various portions of this region are not recognized equally by the factor. 3′ deletion mutants of the 5 S gene, which do not support accurate initiation of transcription (Bogenhagen *et al.*, 1980), also cannot bind the transcription factor (Sakonju *et al.*, 1981). Such a simple correlation does not exist, however, for genes lacking the 5′ part of the control region. Although transcription initiation is abolished in these deletion mutants (Sakonju *et al.*, 1980), TFIIIA can still interact with the remainder of the control region. These results are consistent with experiments designed to test the ability of deletion mutants to compete with the wild-type 5 S gene for transcription components *in vitro* (Wormington *et al.*, 1981). Whereas templates lacking part of the 5′ end of the internal control region reduce the efficiency of wild type gene transcription, templates in which the 3′ end has been deleted do not compete better than vector sequences. Taken together, these results suggest that the control sequence is composed of two functionally distinct regions. Initial binding of the factor probably occurs in the 3′ portion. However, an intact 5′ part of the control region is required in order for accurate initiation to take place.

Using chemically modified templates containing methylated purine bases and ethylated phosphates, Sakonju and Brown (1982) demonstrated that the major contact points of TFIIIA reside in the 3′ portion of the gene. Only one G residue (position 91) on the coding strand interacted with the protein. On the noncoding strand, close interactions were apparent with almost every nucleotide between residues 81 and 91, comprising an entire turn of the helix and, in addition, with residues 70 and 71. Although the factor protects up to residue 45 from DNAase digestion, no further close contact points were found 5′ from residue 70. However, two base changes between the oocyte type and the somatic gene in the 5′ portion of the internal control region (residues 53 and 55) influence the binding affinity of TFIIIA to the 5 S gene (Sakonju and Brown, 1982). These base changes account for a fourfold increase in competition strength of the somatic gene as compared with the oocyte-type 5 S gene but have no effect on the transcription efficiency of these genes (Wormington *et al.*, 1981).

The question of how many TFIIIA molecules bind to the intragenic control region has been addressed in several recent reports. Initial titration experiments led to the assumption that one or two factor molecules are sufficient to interact with the 5 S gene (Sakonju and Brown, 1982). Employing the same technique, Hanas *et al.* (1983) found that two TFIIIA molecules bind to the 5 S gene in a cooperative fashion. However, experiments recently reported by Bieker and Roeder (1984) and by Smith *et al.* (1984)

provide strong evidence that only one factor molecule covers the entire control region. First, employing gel exclusion chromatography of TFIIIA preincubated with a 5 S gene present in a circular plasmid, Bieker and Roeder (1984) demonstrated that one 5 S gene is saturated by binding of one factor molecule. Second, using the footprinting technique to titrate the 5 S gene with increasing amounts of the factor or purified proteolytic fragments thereof (that still bind to the gene), Smith *et al.* (1984) arrived at the same ratio of factor to DNA. Third, by titrating a fractionated extract that lacks detectable TFIIIA activity with either the intact protein or a 30,000-$M_r$ papain fragment of the factor (that retains about 20% of the transcriptional activity), maximal transcription was reached at a ratio of one factor molecule per 5 S gene under conditions in which the template DNA is limiting. Fourth, if two or more factor molecules were to bind to the internal control region, one would then expect heterodimers to occur when intact TFIIIA and proteolytic fragments are incubated with a 5 S gene in the same reaction. Such heterodimers were not observed.

Several other findings lend support to the results summarized above. The factor is a monomer in its native form. Determination of the physical properties indicates that it has a highly asymmetrical shape with an estimated size of 135 × 18 Å. Assuming a B-form DNA, about 150 Å are protected by the factor in the footprinting reaction (Bieker and Roeder, 1984). The asymmetrical shape of TFIIIA agrees well with a model of the functional domains of TFIIIA derived by examining the DNA-binding and transcriptional properties of proteolytic fragments of the factor (Smith *et al.*, 1984). The protein can be divided into three domains. A 20,000-$M_r$ domain, present at one end of the molecule (represented by a 20,000-$M_r$ trypsin fragment) contains a DNA binding site that interacts with the crucial nucleotides in the 3′ part of the internal control region (Sakonju and Brown, 1982). This fragment, however, does not support transcription of the 5 S gene. A fragment of 30,000 $M_r$ (derived by papain treatment of the factor) contains an adjacent 10,000 $M_r$ in addition to the 20,000-$M_r$ domain. This fragment covers the control region to the same extent as the intact factor but is markedly reduced in its transcriptional activity. A third domain of 10,000 $M_r$ (present in the intact factor molecule) that does not directly bind to the DNA, confers full transcription-promoting activity to the factor. This domain is a possible candidate for binding of other factors required to form a transcription complex (see below) or for the interaction with RNA polymerase III. Smith *et al.* (1984) conclude that TFIIIA must be intertwined along the major groove of the DNA helix for at least one entire turn, making close contact with nucleotides 81–91 of the noncoding strand of the gene (Sakonju and Brown, 1982). Upstream from residue 79, the factor could lie along one surface of the DNA or it could be wrapped around the helix, but in a less tight fashion than in the 3′ part, since no close contact points are observed in this part of the control region.

The intertwining of TFIIIA with the 5 S gene could explain the slight alteration in DNA configuration observed after binding of the factor (Reynolds and Gottesfeld, 1983; Hanas *et al.*, 1984; B. S. Shastry and R. G. Roeder, personal communication). Relaxation of a plasmid containing 5 S genes in the presence of the transcription factor results in an unwinding of the helix of 1–4 base pairs per gene. Although a local unwinding with concomitant strand separation of the template could facilitate 5 S gene transcription as discussed by Reynolds and Gottesfeld (1983), it is more likely that the alteration detected is caused by a slight reduction of the pitch of the DNA helix as a result of protein binding (Hanas *et al.*, 1984).

### 5.2. Interaction of TFIIIA with 5 S rRNA

In addition to its role as an initiation factor, TFIIIA serves as a storage protein for 5 S rRNA in the oocyte. An interaction of TFIIIA with RNA was first suggested when 5 S rRNA was observed to inhibit the *in vitro* transcription of 5 S genes but not of other genes transcribed by RNA polymerase III, such as tRNA genes or adenovirus VA RNA genes (Pelham and Brown, 1980). Since 5 S rRNA in the oocyte is stored in the form of ribonucleoprotein particles (7 S RNP and 42 S RNP particles), it seemed possible that one of the proteins present in the particles was the transcription factor. Honda and Roeder (1980) investigated this possibility by reaction of the particle-derived proteins with an antibody raised against the purified transcription factor in protein blots. Although both particles contain proteins roughly comigrating with TFIIIA in denaturing gels, only the 7 S RNP particle reacts with the antibody. Comparison of the cleavage patterns of TFIIIA and the 7 S particle protein by either cyanogen bromide (Honda and Roeder, 1980) or chymotrypsin (Pelham and Brown, 1980) established that the two proteins are identical. In contrast, chymotryptic digestion of the 42 S particle-derived proteins generated different peptides.

The functional equivalence of the protein present in the 7 S particle with the transcription factor was demonstrated by the specific transcription of 5 S genes in an egg extract mediated by this protein (Honda and Roeder, 1980) and its binding to the internal control region of the 5 S gene (Pelham and Brown, 1980). Interestingly, transcription can also be achieved with the complete 7 S RNP particle, whereas RNase treatment of the particle is necessary for the binding of the protein to the gene in the footprinting assay. This result can be explained by a dissociation of the complex in the cell-free system, in which the relative concentrations of 5 S RNA and 5 S DNA might favor the association of the factor with DNA.

The observation that 5 S rRNA synthesis is inhibited by addition of an excess of 5 S rRNA to an *in vitro* transcription reaction and that only the free protein can bind to the internal control region of the gene implies that the factor cannot bind to 5 S rRNA and to the gene simultaneously. Pelham and Brown (1980) observed that once 5 S gene transcription is initiated, it is relatively resistant to inhibition by 5 S rRNA. Also, even when large amounts of 5 S rRNA are added to an *in vitro* reaction, transcription persists for more than one round. From these results, Pelham and Brown (1980) concluded that probably not all the transcripts immediately associate with the factor after transcription is terminated. A study of the time course of the assembly of newly synthesized 5 S rRNA and TFIIIA *in vitro* supports this conclusion (Krämer and Roeder, 1983). As soon as 15 min after the start of 5 S rRNA synthesis *in vitro*, a fraction of the nascent transcripts are associated with the factor and can thus be precipitated by an anti-TFIIIA antibody. The maximal percentage of *in vitro* transcripts packaged into a RNP particle does not exceed 20% of the RNA, an observation also made by Pelham *et al.* (1981a). It remains unclear as to whether the association of TFIIIA with 5 S rRNA occurs during transcription or after transcription termination. However, a function of TFIIIA in transcript release can be ruled out, since certain 5 S genes are efficiently transcribed in germinal vesicle extracts, but the resulting transcripts are not bound by the factor (Brown *et al.*, 1981; Pelham *et al.*, 1981a).

### 5.3. Levels of TFIIIA During *Xenopus* Development

TFIIIA functions both as a 5 S rRNA transcription factor and as a component of a 5 S rRNA storage particle in the immature oocyte. Because of this dual function of TFIIIA, transcription of the 5 S gene could be regulated by the relative ratio of 5 S rRNA to transcription factor during oogenesis. About 50 ng of TFIIIA is present in the young oocyte, providing a sufficient amount of the factor for the activation of the multiple 5 S genes and for the storage of the transcripts (Honda and Roeder, 1980; Roeder *et al.*, 1981; Wormington *et al.*, 1983; Shastry *et al.*, 1984). The TFIIIA level remains relatively constant through stages 1–4, when large amounts of 5 S rRNA accumulate. Pulse-labeling experiments indicate that TFIIIA is mainly synthesized in previtellogenic oocytes, and no newly made factor can be detected at late stages (A. Krämer and R. G. Roeder, unpublished observations). Once ribosome assembly begins, 5 S rRNA is transferred from the particles to the ribosomes. The level of TFIIIA decreases by 20–30-fold, reaching a concentration of 2–4 ng in the mature oocyte. This possibly leads to a decrease in the rate of transcription of the 5 S genes, although they are still active (Mairy and Denis, 1971) (see above). A further drop in the level of TFIIIA (20–50-fold) is observed during the maturation of the oocyte into an egg. Accordingly, in eggs and early embryos the 5 S genes seem to be inactive (Brown and Littna, 1966; Brown *et al.*, 1981). It is unclear, however, whether the decreased level of TFIIIA is responsible for the apparent lack of 5 S rRNA synthesis during these stages, since the concentration of the factor does not increase when the 5 S genes are reactivated at later developmental stages.

In somatic cells, a protein that is immunologically and structurally related to TFIIIA, but larger by about 2000 daltons, was detected in addition to the factor (Pelham *et al.*, 1981b; Roeder *et al.*, 1981). Although a function of the TFIIIA-like protein in the transcription of the somatic-type 5 S genes in somatic cells seemed attractive, chromatographic separation of the two related proteins indicated that only the TFIIIA-size protein was active in effecting transcription of 5 S genes (Roeder *et al.*, 1981). The function of the larger TFIIIA-like protein remains unclear. Owing to the low concentration of TFIIIA and the related protein in somatic cells ($1 \times 10^3$ and $1 \times 10^4$ molecules per cell, respectively) (Wormington *et al.*, 1983; Shastry *et al.*, 1984) their purification has not yet been achieved.

### 5.4. Formation of Stable Complexes

Incubation of 5 S genes in extracts from germinal vesicles or in cytoplasmic extracts derived from *Xenopus* oocytes or human cells results in the formation of stable DNA-protein complexes (Bogenhagen *et al.*, 1982; Lassar *et al.*, 1983). When two different 5 S genes (distinguishable by the size of the transcripts generated) are successively added to an extract in saturating amounts, only the gene added first is transcribed. In contrast, simultaneous addition of two genes results in an equal amount of transcription from each template (Wormington *et al.*, 1981; Bogenhagen *et al.*, 1982; Lassar *et al.*, 1983). Once formed, the complexes are resistant to dilution of the extract, whereas addition of the template to a diluted extract results in a decrease of transcription. These observations suggested that transcription factors remain associated with the same DNA molecule for many rounds of transcription.

The role of TFIIIA in complex formation was first investigated with an extract that had been depleted of the factor by a specific antibody (Bogenhagen *et al.*, 1982): Transcription of 5 S genes (but not of tRNA genes) is largely reduced. Selective transcription of one gene over a second one added later could only be achieved if TFIIIA was present in the extract before the addition of the template. Thus, TFIIIA is required for the formation of a stable complex on the 5 S gene. However, preincubation of a 5 S gene with TFIIIA alone is not sufficient to exclude the transcription of a second 5 S gene.

In reconstitution experiments with fractionated extracts from human cells (Lassar *et al.*, 1983) or from *Xenopus* somatic cells (Setzer, 1983), it was shown that another component (TFIIIC) is necessary to promote the formation of a stable complex on the 5 S gene. (Components TFIIIC and TFIIIB represent activities present in partially purified fractions. For reasons of simplicity they are referred to as factors, even though they could consist of more than one protein). TFIIIC is itself bound in the complex and does not merely serve a catalytic function in complex formation. The site of interaction of TFIIIC with the TFIIIA–5 S gene complex is not known. Since an independent interaction of TFIIIC with the 5 S gene has not been observed, it is conceivable that by binding to the internal control region TFIIIA could create a recognition site for this factor (Lassar *et al.*, 1983). The third component required for 5 S gene transcription (TFIIIB) is sequestered into a stable complex *in vitro* only after binding of factors A and C to the gene. Since a stable binding of TFIIIB to the complex is only observed after an extended incubation period, it appears that this factor is responsible for the lag period observed in the *in vitro* transcription reaction (Birkenmeier *et al.*, 1978; Lassar *et al.*, 1983; Setzer, 1983). The transcribing enzyme, RNA polymerase III, is not an integral part of these stable complexes. However, it specifically recognizes these active complexes. After transcription termination the enzyme is probably released and rebinds preferentially to another gene organized into an active complex to initiate another cycle of transcription (Bogenhagen *et al.*, 1982; Wormington *et al.*, 1983).

The finding that stable complexes can be formed *in vitro* is consistent with the observation made by Parker and Roeder (1977) that the only component required for transcription of 5 S genes in oocyte chromatin is RNA polymerase III. The stable complexes formed *in vitro* thus resemble the state of active 5 S genes in chromatin where all the transcription factors with the exception of RNA polymerase III are stably associated with the chromatin.

The existence of stable complexes on the 5 S gene is relevant to a possible control of 5 S gene transcription. Exogeneous histones, added to a TFIIIA-depleted germinal vesical extract before the addition of a 5 S gene and TFIIIA, markedly depress transcription of this gene (Bogenhagen *et al.*, 1982). However, if TFIIIA and a 5 S gene are added to the extract concomitant with the histones, 5 S RNA synthesis is not significantly reduced. Thus, histones could block the access of the transcription factor to the gene and thereby prevent its binding to the internal control region, resulting in the formation of a stable inactive (repressed) complex. Similar conclusions were reached by transcription of *in vitro* reconstituted chromatin (Gottesfeld and Bloomer, 1982). An active template that is transcribed as efficiently as naked DNA can only be obtained when TFIIIA is present at the time of nucleosome assembly. Recent experiments suggest that histones in the form of nucleosomes play a role in the repression of the oocyte-type 5 S genes in somatic cells. The oocyte-type 5 S genes in isolated somatic cell chromatin can be activated by addition of TFIIIA after removal or degradation of histone H1 (Schlissel and Brown, 1984).

These studies suggest a simple model for the maintainance of active genes in a cell (Bogenhagen *et al.*, 1982; Gottesfeld and Bloomer, 1982; Brown, 1984). Stable complexes are formed between proteins and 5 S genes, which are maintained by strong protein–protein and protein–DNA interactions. The presence of TFIIIA is essential for the formation of active complexes. When TFIIIA is missing, inactive complexes are formed among the gene, transcription factors, and histones. This model further implies that TFIIIA *in vivo* has to be present at the time of 5 S DNA replication before nucleosomes are formed to yield active 5 S gene chromatin. When the amount of this factor is limiting, as is the case in somatic cells, then at least part of the 5 S genes would be assembled into an inactive (repressed) conformation. Provided that the replication of the somatic genes in somatic cells precedes the replication of the oocyte-type 5 S genes (which is unknown) a mechanism as outlined above could account for the inactivation of the oocyte-type 5 S genes in somatic cell chromatin. It should be emphasized that this model does not require the presence of a specific repressor for the oocyte-type 5 S genes, but rather general repressors as represented by the histones (Brown, 1984).

## 6. Concluding Remarks

Recent experiments have demonstrated the importance of reliable *in vitro* assay systems for transcription. Accurate and selective transcription of purified or cloned 5 S genes occurs in injected oocytes as well as in cell-free transcription systems. In conjunction with *in vitro* mutagenesis, these systems have been used to define sequences, encoded within the DNA template, that are essential for initiation and termination of transcription. Furthermore, proteins involved in transcription have been identified and, in one case, purified. The tissue-specific regulation of the 5 S genes, which exists *in vivo*, however, cannot be reproduced in these systems. Therefore, all the factors identified thus far can be viewed as general transcription factors that do not exhibit specificity for either somatic or oocyte-type 5 S genes. By employing reconstituted systems, which effect the faithful transcription of 5 S genes, it has become clear that the regulation of 5 S rRNA synthesis during *Xenopus* oogenesis and embryogenesis as well as in somatic cells occurs at several levels. The affinity of the 5 S gene-specific transcription factor, TFIIIA, for 5 S rRNA as well as for the 5 S gene provides a mechanism by which the transcription product can feedback-inhibit its synthesis at times when the factor becomes rate limiting. This may in fact be the case at late stages of oogenesis. Additional events may lead to the differential expression of oocyte and somatic-type 5 S genes during *Xenopus* development. First, the DNA template contains sequences that can affect the affinity of transcription factors for the respective gene. Second, regulation can occur at the level of chromatin structure, into which the template is packaged in the cell. Here, additional features, such as the time of replication of a certain class of genes and/or the availability of transcription factors at the time of chromatin assembly, might play important roles.

ACKNOWLEDGMENTS. I thank Dr. R. G.Roeder and collaborators for providing unpublished data and Dr. A. Lassar, Dr. J. Cohlberg, Dr. N. Heintz, and Dr. W. Keller for comments on the manuscript.

## References

Bieker, J. J., and Roeder, R. A., 1984, Physical properties and DNA-binding stoichiometry of a 5 S gene-specific transcription factor, *J. Biol. Chem.* **259**:6158–6164.

Birkenmeier, E. H., Brown, D. D., and Jordan, E., 1978, A nuclear extract of *Xenopus laevis* oocytes that accurately transcribes 5 S RNA genes, *Cell* **15**:1077–1086.

Bogenhagen, D. F., and Brown, D. D., 1981, Nucleotide sequence in *Xenopus* 5 S DNA required for transcription termination, *Cell* **24**:261–270.

Bogenhagen, D. F., Sakonju, S., and Brown, D. D., 1980, A control region in the center of the 5 S RNA gene directs specific initiation of transcription: II. The 3′ border of the region, *Cell* **19**:27–35.

Bogenhagen, D. F., Wormington, W. M., and Brown, D. D., 1982, Stable transcription complexes of *Xenopus* 5 S RNA genes: A means to maintain the differentiated state, *Cell* **28**:413–421.

Brown, D. D., 1984, The role of stable complexes that repress and activate eucaryotic genes, *Cell* **37**:359–365.

Brown, D. D., and Dawid, I. B., 1968, Specific gene amplification in oocytes, *Science* **160**:272–280.

Brown, D. D., and Gurdon, J. B., 1977, High-fidelity transcription of 5 S DNA injected into *Xenopus* oocytes, *Proc. Natl. Acad. Sci. USA* **74**:2064–2068.

Brown, D. D., and Gurdon, J. G., 1978, Cloned single repeating units of 5 S DNA direct accurate transcription of 5 S RNA when injected into *Xenopus* oocytes, *Proc. Natl. Acad. Sci. USA* **75**:2849–2853.

Brown, D. D., and Littna, E., 1966, Synthesis and accumulation of low-molecular-weight RNA during embryogenesis of *Xenopus laevis*, *J. Mol. Biol.* **20**:95–112.

Brown, D. D., and Weber, C. S., 1968a, Gene linkage by RNA-DNA hybridization: I. Unique DNA sequences homologous to 4 S RNA, 5 S RNA and ribosomal RNA, *J. Mol. Biol.* **34**:661–680.

Brown, D. D., and Weber, C. S., 1968b, Gene linkage by RNA–DNA hybridization: II. Arrangement of the redundant gene sequences for 28 S and 18 S ribosomal RNA, *J. Mol. Biol.* **34**:681–697.

Brown, D. D., Wensink, P. C., and Jordan, E., 1971, Purification and characteristics of 5 S DNA from *Xenopus laevis*, *Proc. Natl. Acad. Sci. USA* **68**:3175–3179.

Brown, D. D., Carroll, D., and Brown, R. D., 1977, The isolation and characterization of a second oocyte 5 S DNA from *Xenopus laevis*, *Cell* **2**:1045–1056.

Brown, D. D., Bogenhagen, D. F., Jackson, I. J., Jordan, E., Pelham, H., Sakonju, S., and Wormington, W. M., 1981, The dual 5 S gene system in *Xenopus*, *Carnegie Inst. Wash. Yearb.* **80**:205–217.

Brownlee, G. G., Cartwright, E., McShane, T., and Williamson, R., 1972, The nucleotide sequence of somatic 5 S RNA from *Xenopus laevis*, *FEBS Lett.* **25**:8–12.

Carroll, D., and Brown, D. D., 1976a, Repeating units of *Xenopus laevis* oocyte-type 5 S DNA are heterogeneous in length, *Cell* **7**:467–475.

Carroll, D., and Brown, D. D., 1976b, Adjacent repeating units of *Xenopus laevis* 5 S DNA can be heterogeneous in length, *Cell* **7**:477–486.

Ciliberto, G., Raugei, G., Constanzo, F., Dente, L., and Cortese R., 1983, Common and interchangeable elements in the promoters of genes transcribed by RNA polymerase III, *Cell* **32**:725–733.

Cozzarelli, N. R., Gerrard, S. P., Schlissel, M., Brown, D. D., and Bogenhagen, D. F., 1983, Purified RNA polymerase III accurately and efficiently terminates transcription of 5 S RNA genes, *Cell* **34**:829–835.

Davidson, E. E., 1976, *Gene Activity in Early Development*, 2nd ed., Academic Press, New York.

Delaunay, J., Wegnez, M., and Denis, H., 1975, Biochemical research on oogenesis: Comparison between the proteins of the ribosomes and the proteins of the 42 S particles from small oocytes of *Xenopus laevis*, *Dev. Biol.* **42**:379–387.

Denis, H., and Mairy, M., 1972, Recherches biochimiques sur l'oogenèse: Distribution intracellulaire du RNA dans les petits oocytes de *Xenopus laevis*, *Eur. J. Biochem.* **25**:524–534.

Denis, H., and Wegnez, M., 1973, Recherches biochimiques sur l'oogenèse; Synthèse et maturation du RNA 5 S dans les petits oocytes de *Xenopus laevis*, *Biochimie* **55**:1137–1151.

Dixon, L. K., and Ford, P. J., 1982, Persistence of nonribosome bound 5 S RNA in full-grown oocytes of *Xenopus laevis*, *Dev. Biol.* **91**:474–477.

Engelke, D. R., Ng, S. Y., Shastry, B. D., and Roeder, R. G., 1980, Specific interaction of a purified transcription factor with an internal control region of 5 S RNA genes, *Cell* **19**:717–727.

Fedoroff, N. V., 1979, Deletion mutants of *Xenopus laevis* 5 S ribosomal DNA, *Cell* **16**:551–563.

Fedoroff, N. V., and Brown, D. D., 1978, The nucleotide sequence of oocyte 5S DNA in *Xenopus laevis*. I. The AT-rich spacer, *Cell* **13**:701–716.

Ford, P. J., 1971, Non-coordinated accumulation and synthesis of 5 S ribonucleic acid by ovaries of *Xenopus laevis*, *Nature* **233**:561–564.

Ford, P. J., and Brown, R. D., 1976, Sequences of 5 S ribosomal RNA from *Xenopus mulleri* and the evolution of 5 S gene coding sequences, *Cell* **8**:485–493.

Ford, P. J., and Southern, E. M., 1973, Different sequences for 5 S RNA in kidney cells and ovaries of *Xenopus laevis*, *Nature New Biol.* **241**:7–12.

Galas, D., and Schmitz, A., 1978, DNase footprinting: A simple method for the detection of protein–DNA binding specificity, *Nucleic Acids Res.* **5**:3157–3170.

Gall, J. G., 1968, Differential synthesis of the genes for ribosomal RNA during amphibian oogenesis, *Proc. Natl. Acad. Sci. USA* **60**:553–560.

Gilbert, W., 1976, Starting and stopping sequences for the RNA polymerase, in: *RNA Polymerase* (R. Losick and M. Chamberlin, eds.), pp. 193–205, Cold Spring Harbor Laboratory Press, Cold Spring Harbor, New York.

Gottesfeld, J., and Bloomer, L. S., 1982, Assembly of transcriptionally active 5 S RNA gene chromatin *in vitro*, *Cell* **28**:781–791.

Gurdon, J. B., and Brown, D. D., 1978, The transcription of 5 S DNA injected into *Xenopus* oocytes, *Dev. Biol.* **67**:346–356.

Hanas, J. S., Bogenhagen, D. F., and Wu, C.-W., 1983, Cooperative model for the binding of *Xenopus* transcription factor A to the 5 S RNA gene, *Proc. Natl. Acad. Sci. USA* **80**:2142–2145.

Hanas, J. S., Bogenhagen, D. F., and Wu, C.-W., 1984, DNA unwinding ability of Xenopus transcription factor A, *Nucleic Acids Res.* **12**:1265–1276.

Harper, M. E., Price, J., and Korn, L. J., 1983, Chromosomal mapping of *Xenopus* 5 S genes: Somatic-type versus oocyte-type, *Nucleic Acids Res.* **11**:2313–2323.

Honda, B. M., and Roeder, R. G., 1980, Association of a 5 S gene transcription factor with 5 S RNA and altered levels of the factor during cell differentiation, *Cell* **22**:119–126.

Jacq, C., Miller, J. R., and Brownlee, G. G., 1977, A pseudogene in 5 S DNA of *Xenopus laevis*, *Cell* **12**:109–120.

Korn, L. J., and Bogenhagen, D. F., 1982, Organization and transcription of *Xenopus* 5 S ribosomal RNA genes, in: *The Cell Nucleus*, Vol. 12 (H. Busch and L. Rothblum, eds.), pp. 1–29, Academic Press, New York.

Korn, L. J., and Brown, D. D., 1978, Nucleotide sequence of *Xenopus borealis* oocyte 5 S DNA: Comparison of sequences that flank several related eukaryotic genes, *Cell* **15**:1145–1156.

Korn, L. J., Birkenmeier, E. H., and Brown, D. D., 1979, Transcription initiation of *Xenopus* 5 S ribosomal RNA genes *in vitro*, *Nucleic Acids Res.* **7**:947–958.

Krämer, A., and Roeder, R. G., 1983, The use of monoclonal antibodies for the characterization of a 5 S gene-specific transcription factor (TFIIIA) from *Xenopus laevis*, *J. Biol. Chem.258*:11915–11923.

Lassar, A. B., Martin, P. L., and Roeder, R. G., 1983, Transcription of class III genes: Formation of preinitiation complexes, *Science* **222:**740–748.

Mairy, M., and Denis, H., 1971, Recherches biochimiques sur l'oogenèse: 1. Synthèse et accumulation du RNA pendant l'oogenèse du crapaud sud-africain *Xenopus laevis*, *Dev. Biol.* **24:**143–165.

Mairy, M., and Denis, H., 1972, Recherches biochimiques sur l'oogenèse: 2. Assemblage des ribosomes pendant le grand accroissement des oocytes de *Xenopus laevis*, *Eur. J. Biochem.* **25:**535–543.

Maizels, N., 1973, The nucleotide sequence of the lactose messenger ribonucleic acid transcribed from the UV 5 promoter mutant of *E. coli*, *Proc. Natl. Acad. Sci. USA* **70:**3585–3589.

Miller, J. R., and Melton, A., 1981, A transcriptionally active pseudogene in *Xenopus laevis* oocyte 5 S DNA, *Cell* **24:**829–835.

Miller, J. R., Cartwright, E. M., Brownlee, G. G., Fedoroff, N. V., and Brown, D. D., 1978, The nucleotide sequence of oocyte 5 S DNA in *Xenopus laevis*. II. The GC-rich region, *Cell* **13:**717–725.

Ng, S. Y., Parker, C. S., and Roeder, R. G., 1979, Transcription of cloned *Xenopus* 5 S RNA genes by *X. laevis* RNA polymerase III in reconstituted systems, *Proc. Natl. Acad. Sci. USA* **76:**136–140.

Pardue, M. L., Brown, D. D., and Birnstiel, M. L., 1973, Location of the genes for 5 S ribosomal RNA in *Xenopus laevis*, *Chromosoma* **42:**191–203.

Parker, C. S., and Roeder, R. G., 1977, Selective and accurate transcription of the *Xenopus laevis* 5 S RNA genes in isolated chromatin by purified RNA polymerase III, *Proc. Natl. Acad. Sci. USA* **74:**44–48.

Pelham, H. R., and Brown, D. D., 1980, A specific transcription factor that can bind either the 5 S RNA gene or 5 S RNA, *Proc. Natl. Acad. Sci. USA* **77:**4170–4174.

Pelham, H. R., Bogenhagen, D. F., Sakonju, S., Wormington, W. M., and Brown, D. D., 1981a, Studies on the developmental control of 5 S RNA gene expression, in: *Developmental Biology Using Purified Genes*, Vol. 23 (D. D. Brown, ed.), pp. 447–461, Academic Press, New York.

Pelham, H. R. B., Wormington, W. M., and Brown, D. D., 1981b, Related 5 S DNA transcription factors in *Xenopus* oocytes and somatic cells, *Proc. Natl. Acad. Sci. USA* **78:**1760–1764.

Peterson, R. D., Doering, J. L., and Brown, D. D., 1980, Characterization of two *Xenopus* somatic 5 S DNAs and one minor oocyte-specific 5 S DNA, *Cell* **20:**131–141.

Picard, B., and Wegnez, M., 1979, Isolation of a 7 S particle from *Xenopus laevis* oocytes: A 5 S RNA–protein complex, *Proc. Natl. Acad. Sci. USA* **76:**241–245.

Picard, B., Mairy, M., Wegnez, M., and Denis, H., 1980, Biochemical research on oogenesis: 17. Composition of the 42 S storage particle of *Xenopus laevis* oocytes, *Eur. J. Biochem.* **109:**359–368.

Pribnow, D., 1975, Bacteriophage T7 early promotors: Nucleotide sequences of two RNA polymerase binding sites, *J. Mol. Biol.* **99:**419–443.

Reynolds, W. F., and Gottesfeld, J. M., 1983, 5 S rRNA gene transcription factor IIIA alters the helical configuration of DNA, *Proc. Natl. Acad. Sci. USA.* **80:**1862–1866.

Roeder, R. G., 1976, Eukaryotic nuclear RNA polymerases, in: *RNA Polymerase* (R. Losick and M. Chamberlin, eds.), pp. 285–329, Cold Spring Harbor Laboratory Press, Cold Spring Harbor, New York.

Roeder, R. G., Lee, D. C., Honda, B. M., and Shastry, B. S., 1981, Analysis of eukaryotic gene transcription *in vitro*, in: *Developmental Biology Using Purified Genes*, Vol. 23 (D. D. Brown, ed.), pp. 429–446, Academic Press, New York.

Rosenberg, M., and Court, D., 1979, Regulatory sequences involved in the promotion and termination of RNA transcription, *Annu. Rev. Genet.* **14:**319–353.

Sakonju, S., and Brown, D. D., 1982, Contact points between a positive transcription factor and the *Xenopus* 5 S RNA gene, *Cell* **31**:395–405.

Sakonju, S., Bogenhagen, S. F., and Brown, D. D., 1980, A control region in the center of the 5 S RNA gene directs specific initiation of transcription: I. The 5′ border of the region, *Cell* **19**:13–25.

Sakonju, S., Brown, D. D., Engelke, D., Ng, S. Y., Shastry, B. S., and Roeder, R. G., 1981, The binding of a transcription factor to deletion mutants of a 5 S ribosomal RNA gene, *Cell* **23**:665–669.

Schlissel, M. S., and Brown, D. D., 1984, The transcriptional regulation of *Xenopus* 5 S RNA genes in chromatin: The role of active stable trasncription complexes and histone H1, *Cell* **37**:903–913.

Segall, J., Matsui, T., and Roeder, R. G., 1980, Multiple factors are required for the accurate transcription of purified genes by RNA polymerase III, *J. Biol. Chem.* **255**:11986–11991.

Setzer, D., 1983, Analysis of the 5 S rRNA transcription complex, *Carnegie Inst. Wash. Yearb.* **82**:216–217.

Shastry, B. S., Honda, B. M., and Roeder, R. G., 1984, Altered levels of a 5 S gene-specific transcription factor (TFIIIA) during oogenesis and embryonic development of *Xenopus laevis*, *J. Biol. Chem.* **259**:11,373–11,382.

Shastry, B. S., Ng, S. Y., and Roeder, R. G., 1982, Multiple factors involved in the transcription of class III genes in *Xenopus laevis*, *J. Biol. Chem.* **257**:12979–12986.

Smith, D. R., Jackson, J. I., and Brown, D. D., 1984, Domains of the positive transcription factor specific for the *Xenopus* 5 S RNA gene, *Cell* **37**:645–652.

Wegnez, M., and Denis, H., 1973, Recherches biochimiques sur l'oogenèse: 6. Propriétes du RNA 5 S présent dans les différents compartiments cellulaires des oocytes de *Xenopus laevis*, *Biochimie* **55**:1129–1135.

Wegnez, M., Monier, R., and Denis, H., 1972, Sequence heterogeneity of 5 S RNA in *Xenopus laevis*, *FEBS Lett.* **25**:13–20.

Weil, P. A., Segall, J., Harris, B., Ng, S. Y., and Roeder, R. G., 1979, Faithful transcription of eukaryotic genes by RNA polymerase III in systems reconstituted with purified DNA templates, *J. Biol. Chem.* **254**:6163–6173.

Wormington, W. M., Bogenhagen, D. F., Jordan, E., and Brown, D. D., 1981, A quantitative assay for *Xenopus* 5 S RNA gene transcription *in vitro*, *Cell* **24**:809–817.

Wormington, W. M., Schlissel, M., and Brown, D. D., 1983, Developmental regulation of the *Xenopus* 5 S RNA genes, *Cold Spring Harbor Symp. Quant. Biol.* **47**:879–884.

Chapter 11

# Gene Expression During Oogenesis and Oocyte Development in Mammals

ROSEMARY BACHVAROVA

## 1. Introduction

The ultimate goal in the study of gene expression in oocytes is to understand the special characteristics of the genetic program that confer totipotency on the egg. Totipotency includes both the ability to generate differentiated cells of every description among its progeny and the ability to develop as an organized whole organism. What switches and circuits among genes and gene products maintain or renew such potential at each generation? What special cytoplasmic organization underlies the pattern appearing in the early embryo? Obviously, we are a long way from answering this kind of question.

This chapter focuses on the period of mammalian oocyte growth when it is likely that products necessary for meiotic maturation and early embryonic development are synthesized and accumulated in relatively large amounts. Several generalizations about gene expression in oocytes of lower organisms have emerged (reviewed in Davidson, 1976):

1. Unusually stable mRNA is accumulated and not translated in the growing oocyte, but is stored for later use in the embryo.
2. This mRNA has a large coding capacity and includes messages for housekeeping components, but may also code for proteins involved in early embryonic differentiation.
3. Where looked for, repetitive sequence transcripts have been found in the cytoplasm in unusually large amounts.
4. For organisms producing large eggs, the rate of RNA synthesis is very rapid in growing oocytes; rRNA is synthesized on amplified

ROSEMARY BACHVAROVA • Department of Cell Biology and Anatomy, Cornell University Medical College, New York, New York 10021.

rDNA sequences and heterogeneous RNA is synthesized on thousands of lampbrush chromosome loops packed with RNA polymerase.

5. The axes of embryonic development appear by fertilization or soon after; either the axes themselves or the ability to form them must to some extent be preprogrammed in the egg.

This chapter explores to what extent these generalizations apply to the mammalian oocyte.

Gene expression in early meiotic prophase oocytes will also be considered, but in this case the most that can be provided is a plausible time course for the rate of RNA synthesis and its relationship to overall chromosome condensation. Some concept of the continuity of chromosomal structure through the whole period of oocyte development is emerging but much remains to be learned in this area.

Most of the studies described in this chapter have been carried out on mice. Data from work on other species are discussed where possible.

## 2. Description of Oogenesis and Oocyte Development in Mammals

This description covers both the process of generation of oocytes and their subsequent development into mature eggs ready to be fertilized. For our purposes, it is useful to divide this long and complex series of events into two major stages, corresponding to prenatal and postnatal life in most mammals. Important biological processes occur during both phases. In the first phase, primordial germ cells appear, and oocytes are generated and traverse the early stages of meiotic prophase in which chromosomal synapsis takes place. In the second phase, small resting oocytes enter the major growth phase, the bulk of the materials upon which meiotic maturation, fertilization, and early embryonic development depend are accumulated, and meiotic maturation is completed.

Reviews of various aspects of this subject may be found in Brambell (1956), Baker (1972), Zuckerman and Weir (1977), and Jones (1978).

### 2.1. Female Germ Cells During Embryonic and Fetal Life

The germ cells (of males and females) can first be identified in presomite or early somite embryos in the caudal region of the primitive streak, in the base of the allantois, and in the proximal yolk sac (Eddy *et al.*, 1981). At least some cells of the primitive epiblast of the pregastrular embryo, when reintroduced into the blastocyst, are able to produce cells of all types, including germ cells (Gardner, 1978). Germ cells apparently are present in or arise from cells in this layer and pass through the primitive streak to appear in the proximal yolk sac and allantois. The mechanism of their determination is not clear, but they appear to be formed simultaneously with the embryonic layers rather than from any one embryonic layer.

As development proceeds, the primordial germ cells migrate via the wall of the hindgut and dorsal mesentery to the dorsal body wall and then slightly laterally to the genital ridges. Upon arrival in the gonad they intermingle with somatic cells, and in females continue to proliferate as oogonia. This phase continues for several days in mice and for at least a couple of months in humans. Subsequent events are diagrammatically represented in Fig. 1. At about 13 days of development in mice and during the third month of development in humans, oogonia start to transform into oocytes in a somewhat synchronous wave. This transition is probably initiated in the $G_1$ phase of the cell cycle; the subsequent S phase is called the premeiotic S phase, after which the oocytes then enter a long meiotic prophase. The successive stages of meiotic prophase (leptotene, zygotene, and pachytene) are traversed rapidly in mice and over several months in humans. Homologous chromosomes synapse, cross over, and form chiasmata in a manner that is essentially universal among organisms. The oocyte grows slowly during this period. By the time of birth (at 19 days' development in mice), oocytes have reached the diplotene configuration in mice and in humans.

The timing of progression of oocytes through meiosis in relation to birth varies in different mammals (Baker, 1972); in a few cases meiotic prophase is delayed entirely to the postnatal period (e.g., rabbits and cats). The degree of synchrony of development also varies, the distribution of stages seen at any one point being quite narrow in mice and rats and broad in humans.

One of the more dramatic features of oogenesis in mammals is the rapid multiplication of oogonia in the very early fetus, a process overlapping with a period of extensive spontaneous degeneration (atresia) of oogonia and pachytene and diplotene oocytes. Thus, the time course of the total number of germ cells rises to a peak, reaching roughly a million cells in most mammals, and falls thereafter (Baker, 1972). The degenerative process may be an evolutionary adaptation to the relatively very small number of offspring produced by mammals as compared with lower vertebrates.

## 2.2. Female Germ Cells in Postnatal Life: Primordial Oocytes to Ovulation

At about the time of birth, the spherical oocytes become surrounded by a few squamous somatic cells (**follicular** or **granulosa** cells) to form follicles. These are called **primordial follicles,** and the diplotene oocytes within them are called **primordial oocytes.** For most oocytes and follicles,

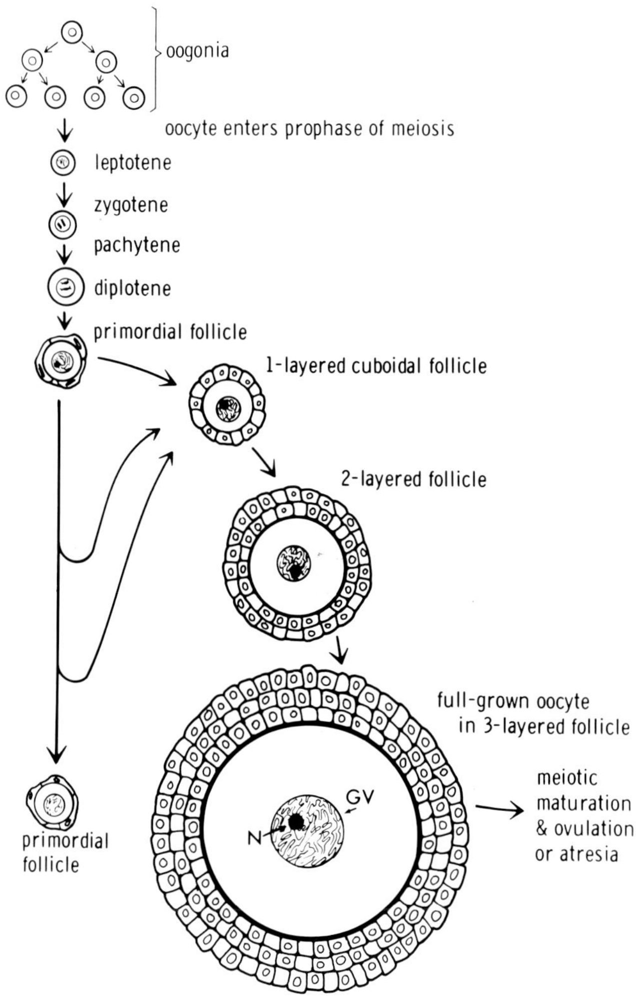


**Figure 1.** Diagram of the development of mammalian oocytes from oogonia to full-grown oocytes. Oocytes progress through meiotic prophase to the diplotene stage in a relatively synchronous wave, ending at about the time of birth. From this point, oocytes remain in primordial follicles or can move into the growth phase at any point during the prepubertal or reproductive life of the individual. Full-grown oocytes range from 70 to 120 $\mu$m in different mammals. GV; germinal vesicle (nucleus); N, nucleolus.

development is arrested at this stage, with little change over a long period. These primordial follicles and their small oocytes represent a large stockpile from which a few are selected within any given time period for development toward preovulatory follicles with fully developed oocytes. The gradual depletion of primordial follicles through spontaneous degeneration and continuous initiation of growth continues until menopause, when few remain. Those follicles that enter the growth phase increase in size by growth and replication of follicular cells and by growth of the enclosed oocyte. The increase in volume of the oocyte and associated cytoplasmic changes are the major morphological events occurring in the germ cells after birth. The primordial oocyte with a diameter of 15–20 μm grows to a full-grown oocyte of 70–150 μm in diameter in different mammals (Austin, 1961). This is an increase in volume of approximately 100 fold, resulting in the formation of one of the largest cells in the body.

The following brief description of follicle and oocyte growth applies to most mammals and is illustrated in Figs. 1–3. The oocyte and its surround-

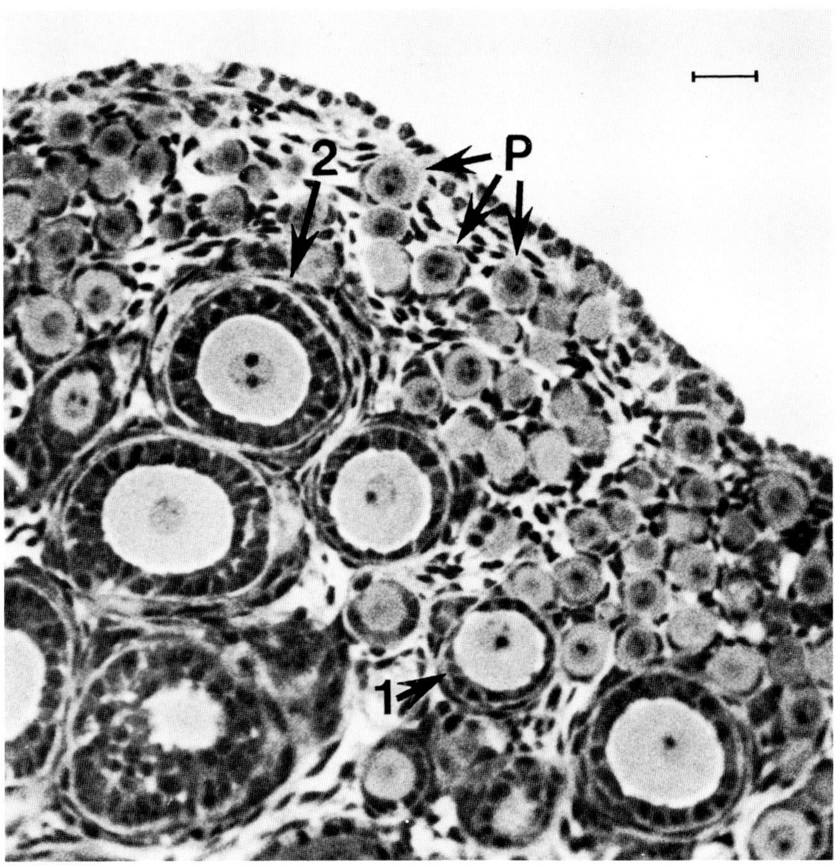


**Figure 2.** Photomicrograph of a 1-μm section of an ovary from an 8-day-old mouse fixed in glutaraldehyde, embedded in JB4 (glycol methacrylate), and stained with toluidine blue and acid fuchsin. P, primordial, 1, one-layered cuboidal, and 2, two-layered follicles. Note the dark-staining nucleoli in the germinal vesicles of growing oocytes. Phase contrast. Scale bar: 25 μm.

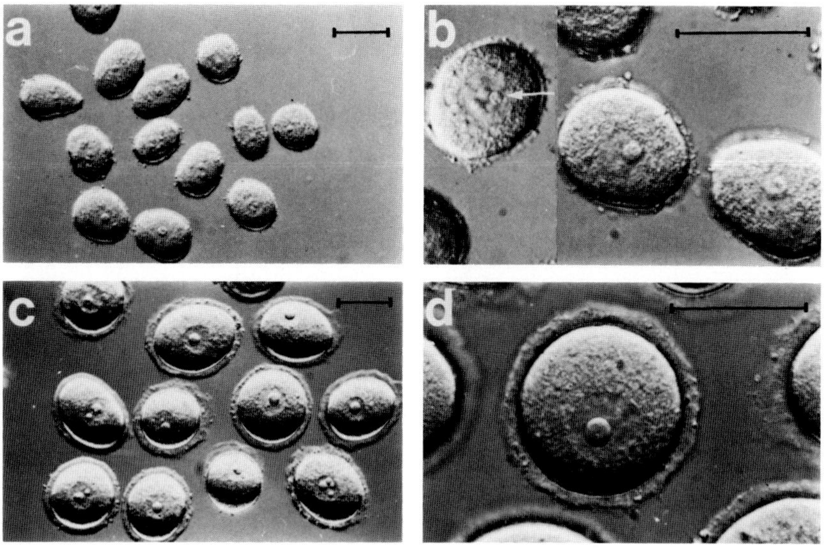


**Figure 3.** Photomicrographs of mouse oocytes at two stages of growth. (a,b) Prepared from 7-day-old mice. (c,d) From 13-day-old mice. Some variation in the diameter of oocytes from a given age is evident. The nucleolus is obvious in each oocyte as a spherical dense body in the clear nucleoplasm. The white arrow in (b) indicates the germinal vesicle with nucleoli. The zona pellucida surrounding the oocyte becomes more prominent during growth. Nomarski optics. Scale bar: 50 $\mu$m. (Reproduced from Mangia and Epstein, 1975, with permission from Dr. C. J. Epstein and Academic Press.)

ing follicular cells initiate growth approximately simultaneously. The follicular cells proliferate and undergo hypertrophy, changing from squamous cells applied to the surface of the oocyte to cuboidal cells with very basophilic cytoplasm. The growing oocyte displays a large, spherical nucleus (germinal vesicle) with a prominent nucleolus, and the nucleus increases in volume in proportion to growth of the oocyte. Approximately midway through the oocyte growth phase, as follicular cells continue to proliferate, the single cuboidal follicular cell layer transforms into two layers around the oocyte, and when the oocyte is near or at maximum size it transforms into three layers. At this point, follicle growth increases dramatically, while oocyte growth slows. After several layers of follicular cells have formed, a cavity or antrum appears asymmetrically within the follicle and expands rapidly along with continued proliferation of follicular cells to form the preovulatory follicle. Thus, follicle development is characterized by two phases: one in which follicle and oocyte grow in parallel and one in which follicle growth far outstrips oocyte growth. The stages of follicle development during the oocyte growth period are usefully described by indicating the number of follicular cell layers or the diameter of the oocyte. A classification scheme based on the number of follicular cells in the largest cross section has been devised to describe the whole period of follicle development (see Pedersen, 1970).

The oocyte chromosomes remain in a more or less diffuse diplotene configuration through the variable length of the primordial oocyte stage and through oocyte growth (see Section 4). The diplotene nucleus resembles an active interphase nucleus and supports the unique phase of oocyte growth. The end of the diplotene stage comes in response to the ovulatory surge of LH (see Chapter 1). This stimulus acting on a fully developed preovulatory follicle results in both resumption of meiosis by the oocyte and ovulation. The chromosomes move rapidly through diakinesis, metaphase I, a very short anaphase I and telophase I, and on to metaphase II of meiosis. At this point, chromosomal movements are again arrested, and the ovulated egg awaits fertilization. The period from breakdown of the germinal vesicle to metaphase II is called meiotic maturation. Details of meiotic maturation and ovulation in mammals were discussed in Chapter 7. Figure 4 illustrates full-grown mouse oocytes released from antral follicles, oocytes following breakdown of the germinal vesicle, and ovulated eggs collected from the oviduct.

Many oocytes initiate growth around the time of birth in mice (Brambell, 1927; Schultz and Wassarman, 1977b). This results in a wave of developing follicles, which may be sampled to prepare follicles or oocytes of a given stage of development by sacrificing mice of a selected age (see below). The rate of initiation of growth then falls to a level that remains steady during most of the reproductive period (Pedersen, 1970). Some primordial follicles are lost by spontaneous atresia particularly during prepubertal life. Most one- and two-layered follicles remain healthy but degenerate at multilayered or antral stages (Byskov, 1978). The first wave of developing follicles in mice reaches large size by about 20 days, but all degenerate; ovulation first occurs at about one month of age. In humans, the process of growth of small follicles and their degeneration as multilayered follicles continues throughout childhood (Peters *et al.*, 1976). At puberty, the reproductive hormonal cycle matures, and some of those multilayered follicles that happen to reach the proper stage of development at the time of the rise in follicle-stimulating hormone (FSH) continue to develop. After the luteinizing hormone (LH) surge, one or more follicles reach ovulation, and the remaining multilayered follicles degenerate. The result of atresia and ovulation continuing through the reproductive life of the female is to greatly deplete the stockpile of follicles by the time of menopause.

The control of the initiation of follicle growth is not clear, but in general the follicles near the medulla of the ovary start to grow first. Thus, the follicles may be arranged in a developmental sequence that is perhaps based on the time of arrival of germ cells in the gonad, the production line hypothesis of Edwards (1970).

The time required to traverse the stages of oocyte growth has been

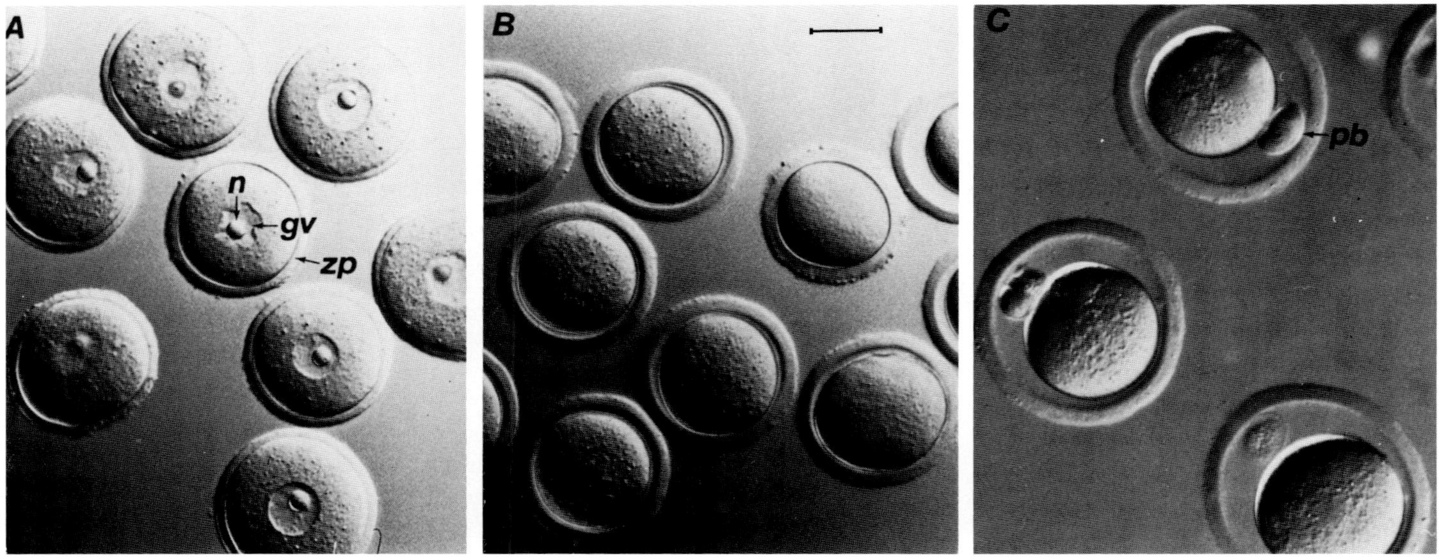


**Figure 4.** Photomicrographs of mouse oocytes. (A) Full-grown oocytes collected from antral follicles. gv, germinal vesicle; n, nucleolus; zp, zona pellucida. (B) Oocytes following GVBD. (C) Unfertilized eggs at metaphase II of meiosis collected from the oviduct. pb, polar body within the zona pellucida. Nomarski optics. Scale bar: 50 $\mu$m. (Reproduced in part from Schultz and Wassarman, 1977b, with permission from Dr. P. M. Wassarman and The Company of Biologists, Ltd.)

analyzed in mice. Because a relatively large group of follicles initiates growth quite synchronously near the time of birth, it is possible to trace the rate of development by noting the most advanced stages present on successive days after birth. Oocytes of approximately full size are seen at 14–16 days of age (Brambell, 1927; Schultz and Wassarman, 1977b), and ova can be ovulated in response to exogenous gonadotropins by 21 days of age (Gates, 1971; Kaplan *et al.*, 1982). In another approach, the rate of follicular cell division was measured using pulses of [$^3$H]thymidine and autoradiography. Then, knowing the number of follicular cells per follicle at given stages, it was calculated that about 14 days are required to progress from a one-layered cuboidal follicle to a three-layered follicle with full-grown oocyte, and only 5 more days to reach ovulation in adult mice (Pedersen, 1970). The values obtained by both methods are in good agreement. The time required for oocyte growth in humans is not known but may cover several menstrual cycles.

An approximate time course for the growth of the mouse oocyte is presented in Fig. 5. As can be seen, over the first half of the developmental period, diameter increases linearly with time and in the second half volume closely approximates a linear increase with time. Oocytes do not suddenly cease to grow at a diameter of about 70 μm, when they are usually said to have reached full size, but it is not clear how rapidly they are growing in multilayered and antral follicles. Measured as isolated fresh oocytes, they continue to increase in diameter to 21 days of age (Schultz and Wassarman, 1977b), and oocytes from antral follicles of adult mice are frequently greater than 80 μm in diameter. By contrast, as seen in paraffin sections of juvenile or adult ovaries, oocyte diameter reaches about 70 μm in three-layered follicles and increases very little thereafter (Brambell, 1928; Pedersen, personal communication, cited in Bachvarova, 1974; Bachvarova *et al.*, 1980). These different results are illustrated in Fig. 5. It is possible that expansion during this period involves primarily uptake of fluid without parallel increase in mass. Oocytes that have reached 65–70 μm are referred to as full-grown throughout this chapter, as is customary in the literature. As shown in Figs. 4 and 5, oocytes decrease in volume somewhat during meiotic maturation.

Important metabolic (see Sections 3.2 and 6.3) and developmental differences distinguish the growing from the full-grown oocyte. Full-grown oocytes, released from small or large antral follicles by pricking ovaries with sharp needles, are able to resume meiosis spontaneously *in vitro*, as evidenced by breakdown of the germinal vesicle and condensation of chromosomes (see Chapter 7). In the mouse, this property is acquired when oocytes reach a diameter of about 65 μm (Schultz and Wassarman, 1977b); in hamsters, it is acquired at 80 μm (Iwamatsu and Yanagimachi, 1975). The ability to cleave spontaneously, a common event in large oocytes undergoing atresia, as well as the related phenomenon of pseudocleavage, is attained just slightly earlier (Kingery, 1914; Wassarman *et al.*, 1977).

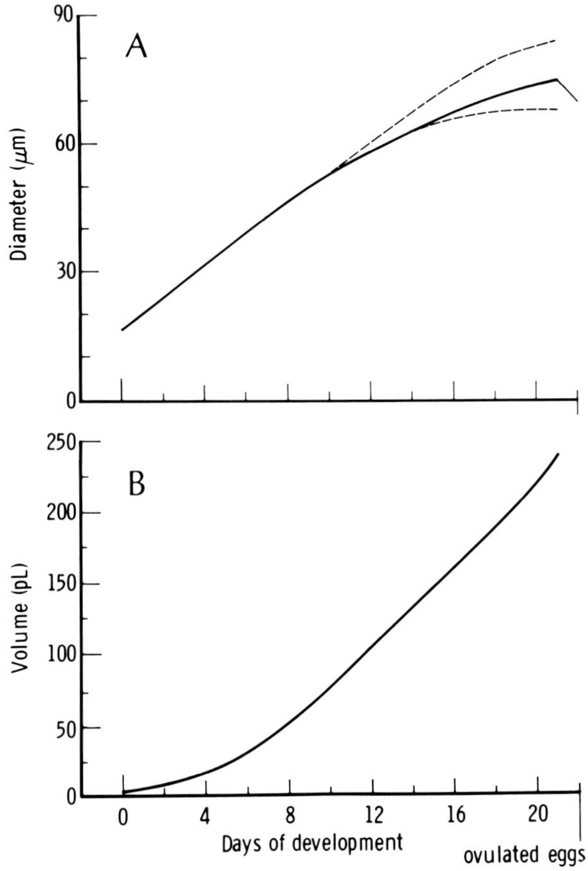


**Figure 5.** Increase in diameter and volume of living mouse oocytes with time. (A) Solid line represents the average diameter of freshly isolated oocytes prepared from mice of increasing age (equivalent to days of growth for oocytes from older animals). (Data were compiled from Eppig, 1976, Schultz and Wassarman, 1977b, and Kaplan *et al.*, 1982.) The population of growing oocytes from a given age shows some variability, e.g., a range of $\pm 8$ μm diameter is seen for oocytes from 8-day-old mice (Eppig, 1977). The dashed line above suggests the time course of growth to oocytes $>80$ μm in diameter usually found in adult mice. The dashed line below indicates the relative values found in paraffin sections (see text). Oocytes are considered full grown for about 5 days before ovulation, although diameter continues to increase slowly. Diameter decreases somewhat during ovulation. (B) Increase in volume with time, calculated from the solid line of (A).

## 3. RNA Synthesis

This section attempts to develop a picture of the time course of the rate of synthesis of RNA in developing oocytes. Our knowledge is rudimentary for meiotic prophase oocytes up to the diplotene stage. For growing oocytes, much more information is available; the parameters of synthesis and turnover of heterogeneous nuclear RNA (hnRNA) are described, and the foundation is laid for understanding the accumulation of RNA as discussed in Section 5.

### 3.1. Meiotic Prophase to the Diplotene Stage

In studying meiotic prophase, the chromosomal condensation and movements that permit pairing of homologs are usually emphasized. Nevertheless, it should not be overlooked that meiocytes are active in RNA and protein synthesis throughout most of this period in both males and females. These activities support cell growth and metabolism and in spermatocytes also help prepare for development of spermatids after meiosis. There is only one brief period coinciding with completion of chromosomal pairing when RNA synthesis may be severely depressed.

The methods used to demonstrate that RNA synthesis is occurring in early prophase oocytes include autoradiographic studies on the incorporation of labeled precursors ([$^3$H]uridine) and ultrastructural analysis of nuclear components. In mice, the desired stage can be obtained by sacrificing females on the day of pregnancy when this stage is predominant. The methods used are not capable of measuring accurately the relative or absolute rates of RNA synthesis in different stages but nevertheless give a general qualitative picture of the time course of RNA synthesis. Biochemical (rather than cytological) techniques have not yet been applied to this problem due to the difficulties of preparing sufficient quantities of purified oocyte populations from the small fetal gonads, although it may be possible to develop such methods (De Felici and McLaren, 1982). The male system is much more suitable for biochemical studies of the RNA synthesized in meiotic prophase, since rather pure populations of cells can be prepared in bulk from adult testes or from juvenile testes during the first wave of meiotic prophase (see Gold *et al.*, 1983).

An early autoradiographic study demonstrated incorporation by pachytene and early diplotene oocytes in neonatal rats (Baker *et al.*, 1969). More extensive autoradiographic analyses of [$^3$H]uridine incorporation by oogonia and leptotene through early diplotene oocytes have been reported for both the mouse and human, with comparable results in the two species (Bakken and McClanahan, 1978; Hartung and Stahl, 1978). The results for the human oocytes are reproduced in Fig. 6. Incorporation falls from a relatively high value in oogonia, through leptotene and zygotene, to reach an undetectable level and possibly true cessation at early pachytene. Incorporation rises again through mid-pachytene to reach a moderately high value similar to that in oogonia by the early diplotene stage. Nucleolar activity follows chromosome activity in general but reaches a longer minimum extending from leptotene to early pachytene in mouse oocytes and

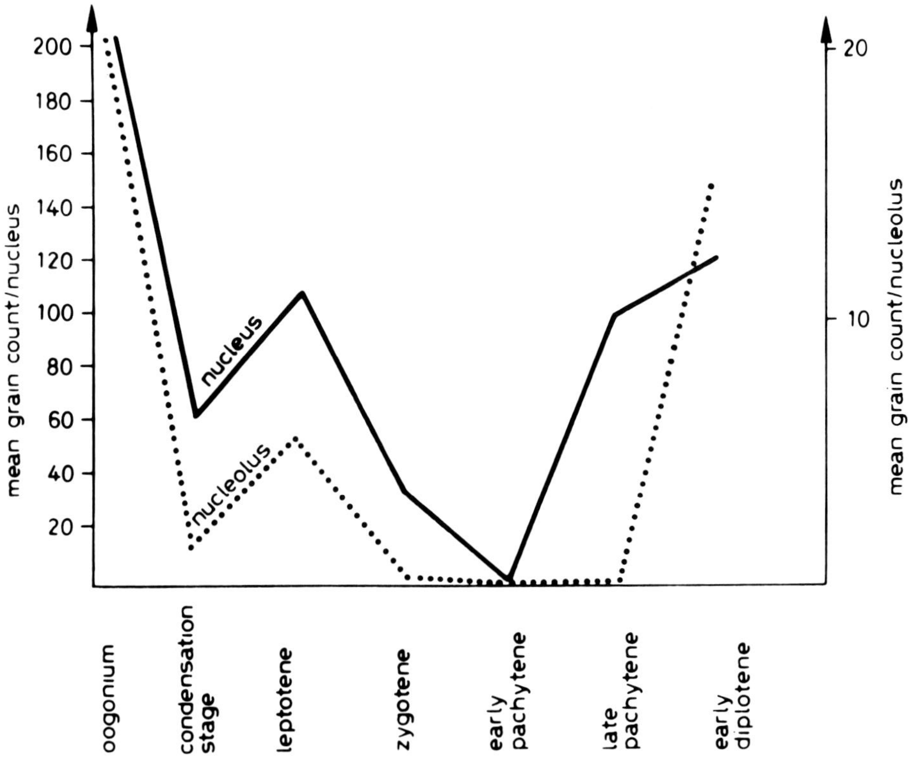


**Figure 6.** [$^3$H]uridine incorporation into human oogonia and oocytes at different stages of meiotic prophase, expressed as the mean number of grains over the nucleus and nucleolus. Small pieces of 13–19-week human fetal ovaries were incubated in complex medium containing [$^3$H]uridine for 40 min, dissociated, and fixed and air-dried spreads were prepared. (Reproduced from Hartung and Stahl, 1978, with permission from Dr. A. Stahl, and S. Karger AG, Basel.)

from zygotene to late pachytene in human oocytes. Incorporation by nucleoli of pachytene mouse oocytes has been confirmed using electron microscopic autoradiography (Mirre and Stahl, 1978).

The amount of incorporation observed in experiments such as those of Fig. 6 partly depends on the rate of penetration of precursor into the oocyte and the size of the endogenous precursor pool. Variations in these factors could produce variation in the amount of incorporation without an actual change in the absolute amount of RNA synthesized during the labeling period. In spite of these and other caveats about interpretation of the incorporation curve, its qualitative features are consistent with the structural studies described below.

RNA synthesis in human spermatocytes has been examined by similar techniques with some differences in the results (Tres, 1975). Incorporation rises from a moderate level in leptotene through zygotene to a broad peak extending through pachytene and falling during diplotene and diakinesis.

Nucleolar activity somewhat precedes chromosomal activity, rising to a peak at zygotene and falling gradually thereafter; the results for mouse spermatocytes are at least qualitatively similar (Kierszenbaum and Tres, 1974a). This pattern is almost the reverse of that in oocytes. In mice, pachytene lasts for about 1 day in females, as compared with approximately 7 days in males. Thus, it may be difficult to detect a brief period of low activity at the beginning of pachytene in males. Clearly, oocytes and spermatocytes diverge during pachytene, and the oocytes enter the very long diplotene stage, which has little analogy with any stage in spermatogenesis. Some of the RNA synthesized by pachytene spermatocytes may be rRNA and messages to be used after meiosis to support spermatid development (see Gold *et al.*, 1983). In this sense, it may be analogous to the stable RNA synthesized much later by diplotene oocytes during their growth phase that contributes to development of the embryo after meiosis.

The incorporation studies on chromosomal transcription are reinforced by analysis of ultrastructural components of the nucleus during oocyte development. Perichromatin granules revealed after EDTA bleaching of chromatin are probably hnRNP particles; such granules are present in appreciable amounts in both leptotene and pachytene mouse oocytes (Palombi and Viron, 1977). In addition, an interchromatin net of RNP material of unknown significance is present during pachytene.

The ultrastructural features of the nucleolus traced in mouse (Mirre and Stahl, 1978) and human oocytes (Hartung *et al.*, 1979) are consistent with the results of the incorporation studies. The electron-dense fibrillar component, which is believed to represent the site of transcription, is absent in leptotene to mid-pachytene in mouse oocytes, and in mid-pachytene in human oocytes. Reactivation of the nucleolus during pachytene presents a useful system for detailed analysis of the sequential appearance of its components (Mirre and Stahl, 1981).

## 3.2. Primordial, Growing, and Full-Grown Oocytes

The egg inherits a stockpile of maternal mRNA that is synthesized and stored in the growing oocyte. Translation of this mRNA is delayed until meiotic maturation or early embryogenesis. The mechanisms of its storage and activation remain an intriguing problem. In this section, studies of RNA synthesis and labeling over relatively short time periods are considered. Such studies provide a rough time course of the rate of transcription in developing oocytes. In addition, they may reveal unusual properties of overall RNA synthesis that could be related to storage of mRNA or other oocyte-specific processes. Accumulation of RNA over longer periods is discussed in Section 5.

Both cytological and, more recently, biochemical techniques have been applied to this problem. The results of a variety of studies have indicated that incorporation of [$^3$H]uridine into RNA is significant in primordial oocytes, increases in growing oocytes, is somewhat higher in late versus early growing oocytes (three-layered versus one-layered follicles), decreases sharply after full size is attained, and essentially ceases at breakdown of the germinal vesicle. This is qualitatively the expected time course of RNA synthesis if the oocyte is responsible for the synthetic activity necessary for its own growth.

Early autoradiographic studies of rat, mouse, and monkey ovaries labeled briefly *in vivo* demonstrated little incorporation by full-grown oocytes but active incorporation by primordial and growing oocytes (Oakberg, 1968; Baker *et al.*, 1969). Label is first seen over both nucleoli and nucleoplasm and within hours appears in the cytoplasm. With these studies it became clear that whereas the primordial oocyte is in a resting stage, it is in an active steady state in which the rate of synthesis presumably balances the rate of decay of ribosomes and mRNAs. Its distinct RNA-containing nucleolus also demonstrates such activity. Growing oocytes are apparently more active, having a very prominent nucleolus and abundant RNA in the nucleus (Zybina, 1968).

Incorporation of [$^3$H]uridine during the transition from early diplotene to primordial oocytes was followed by Bakken and McClanahan (1978), who found that as the oocytes become surrounded by follicular cells, incorporation by the oocyte nucleus decreases. Since at later stages the follicular cells are a major route of entry of nucleotides into oocytes, it is clearly possible that this decrease reflects changes in precursor supply and pools. However, a decreased rate of synthesis would also be reasonable, since the oocyte ceases to grow at this transition.

The time course of [$^3$H]uridine incorporation during oocyte growth has been assessed by counting the number of autoradiographic grains per nuclear cross section in ovaries fixed 1–2 hr after intraperitoneal injection of precursor (Moore *et al.*, 1974; Fourcroy, 1982). Both studies found a two- to threefold increase in growing oocytes as compared with primordial oocytes, falling back to a low level in full-grown oocytes. Moore and associates also found a steadily increasing incorporation during growth to a value in three-layered follicles that is two to three times higher than that of oocytes in one-layered cuboidal follicles. The increase per nucleus would be even larger considering the increase in nuclear volume. From these experiments, it is not clear whether the steep decline in incorporation by full-grown oocytes might be accounted for by slow penetration of precursor to the oocyte.

In general, the time course of activity revealed by incorporation studies has been supported by structural and other studies described immediately below. In addition, two approaches have been taken to measuring the rate of synthesis of RNA in mouse oocytes. One, based on the rate of accumulation of stable rRNA, is described at the end of this section. The other completely circumvents the problem of penetration of precursor into the oocytes and its dilution within the oocyte. This method, developed by

Moore (1978), involves assay of ongoing RNA polymerase activity in tissue sections. Frozen sections of ovaries were fixed in ethanol-acetone and bathed in a reaction mixture containing the four ribonucleotide triphosphates (one labeled) and suitable salts for 30 min. The precursor mixture was washed off, the tissue fixed again in ethanol-acetic acid and processed for autoradiography. The grains observed reflect elongation of RNA chains already initiated at the time of fixation. Controls showed that the reaction depended on all four ribonucleotide triphosphates and $Mg^{2+}$, that the product was digested by RNase, and that nucleoplasmic incorporation was inhibited by a low level of $\alpha$-amanitin or nucleolar activity by a low level of actinomycin D. Since the rate of elongation of RNA chains is believed to be quite constant in different cellular states, this measurement reflects the number of active polymerases or the overall rate of RNA synthesis.

Application of this method to mouse oocytes produced the results shown in Fig. 7 (Moore and Lintern-Moore, 1978). The time course of RNA synthesis is in fact very similar to that seen in the *in vivo* incorporation studies; there is a distinct rise in growing as compared with primordial oocytes, which continues during the growth stage as well; oocytes in three-layered follicles are at least two times more active than those in one-layered follicles. Again, to estimate the activity per nucleus, the grains per cross section of the nucleus should be multiplied by a factor proportional to the radius, increasing the activity of large growing oocytes as compared with small ones by another factor of about two. Finally, in full-grown oocytes RNA synthesis falls dramatically.

In general, gonadotropins do not affect oocyte growth rate or metabolism, but large doses of pregnant mare serum gonadotropin (FSH activity) stimulate nucleolar RNA polymerase activity in oocytes just reaching full size (Moore and Lintern-Moore, 1979).

Structural observations also point to a rapid and increasing rate of heterogeneous RNA synthesis in growing oocytes. The ultrastructural study of RNP components of the nucleus in meiotic prophase mouse oocytes referred to above was continued through the primordial and oocyte growth stages (Palombi and Viron, 1977). The concentration of perichromatin granules remains unchanged from pachytene to the primordial oocyte, the amount increasing somewhat due to the increase in nuclear volume. A striking increase is seen during the transition from a nongrowing to a growing oocyte. Assuming that these granules represent hnRNP and that, as is reasonable to suppose, the half-life of hnRNA does not change, such results demonstrate a greatly increased rate of hnRNA synthesis in growing over nongrowing oocytes. The concentration of granules remains similar throughout growth, suggesting an approximately eightfold increase due to the increase in nuclear volume. As oocytes reach full size, the chromatin condenses, perichromatin granules decrease greatly, but large RNA-containing bodies appear. In late full-grown oocytes, the typical perichromatin granules are still very sparse, but coarse grains and an abundance of very

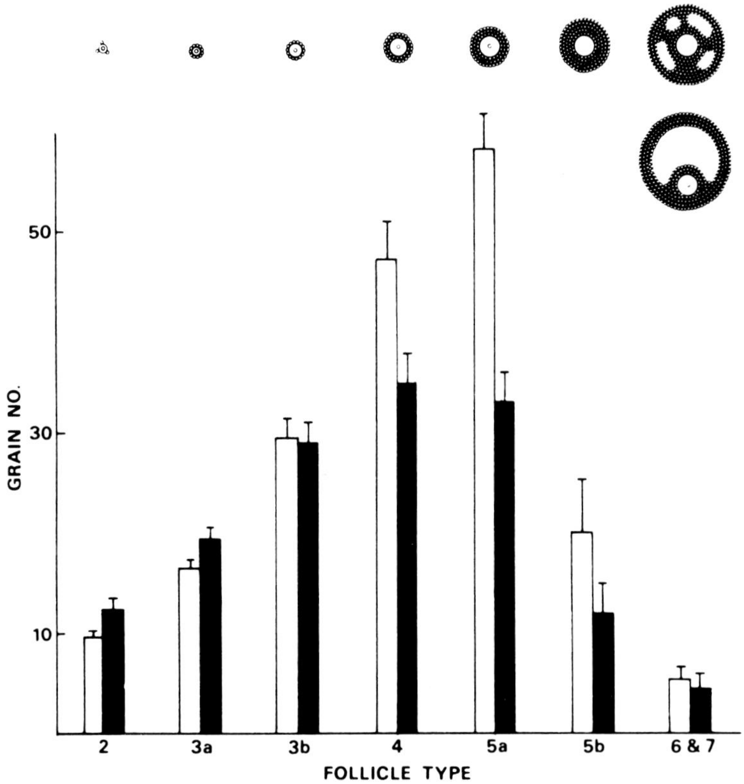


**Figure 7.** Incorporation of [$^3$H]-UMP by mouse oocytes during growth. Open bars, number of silver grains over the nucleoplasm ±SEM; closed bars, number of silver grains over the nucleolus ±SEM. The follicle types are indicated above; stages 3a–5a contain growing oocytes, types 5b–7 contain full-grown oocytes. (Reproduced from Moore and Lintern-Moore, 1978, with permission from Dr. G. P. M. Moore, The Society for the Study of Reproduction, and Journal of Reproduction and Fertility.)

fine RNP grains appear; this latter finding raised the possibility of a burst of RNA synthesis just before meiotic maturation.

Incorporation studies suggest there is no such burst of RNA synthesis in the hours before germinal vesicle breakdown, but it cannot be ruled out. That there is low but significant incorporation into nuclear and cytoplasmic RNA of full-grown oocytes has been shown in autoradiographic studies of *in vivo* labeled ovaries (see above, and Rodman and Bachvarova, 1976). However, incorporation of precursor into oocyte RNA or assays of active RNA polymerase have not been examined at closely spaced time points during the approach to breakdown of the germinal vesicle. Similarly, *in vitro* studies in which full-grown oocytes can be exposed to a high concentration of the precursor have clearly demonstrated such incorporation (Bloom and Mukherjee, 1972; Wassarman and Letourneau, 1976), but it was not shown that the incorporation was higher than that in oocytes at earlier stages further from maturation. Thus, the structural data remain the only evidence for a change in nuclear RNA metabolism in large preovulatory oocytes, but its nature remains obscure.

The above studies are consistent in showing that incorporation is no longer detectable after breakdown of the germinal vesicle. Thus, as for mitotic chromosomes, RNA synthesis is apparently arrested or severely limited in meiotic chromosomes.

The structure of the nucleolus also suggests much higher activity in growing than in nongrowing oocytes. The nucleolus of the growing oocyte is a striking feature of the cell; it is a large, spherical, densely stained vacuolated structure, as seen under the light microscope (Chouinard, 1971). Its expansion and changes at the ultrastructural level during the transition from the primordial to the growing oocyte in mice suggest unraveling and increased transcription of rDNA (Mirre and Stahl, 1981). During the later phases of growth, material accumulates in the vacuoles until, as oocytes reach full size, the nucleoli consist of a homogeneous mass of densely packed fibrils, suggesting little transcriptional activity (Chouinard, 1971, 1975; Zybina, 1975).

Quite recently, it has become possible to study RNA synthesis in growing mouse oocytes maintained *in vitro*. Such studies have permitted more detailed analysis of RNA metabolism, since labeling can be controlled, and highly labeled samples can be obtained even with the relatively small number of oocytes that can be collected. In order to draw meaningful conclusions from such studies, it is necessary to show that growth and development are maintained under *in vitro* conditions. As assayed by growth and acquisition of competence to resume meiosis, mouse oocytes continue to develop when cultured in complex medium, e.g., minimal essential medium (MEM) containing fetal calf serum in the presence of follicular cells. These conditions can be met either by culture of whole follicles dissociated from the ovary or by culture of naked oocytes over a monolayer containing follicular cells.

Ovaries of juvenile mice are used for the *in vitro* experiments in order to obtain a somewhat homogeneous population of growing follicles of known size and to avoid interference from atretic follicles and corpora lutea of older mice. To obtain whole follicles, ovaries are digested with collagenase (Eppig, 1977; Heller *et al.*, 1981) or pricked with fine needles, yielding approximately 50 follicles per ovary. Oocytes maintained in whole follicles (Fig. 8) grow at a rate equal to that *in vivo*, and 55% attain meiotic competence (Eppig, 1977, 1979). Naked oocytes are released from pronase-treated follicles during overnight culture (Bachvarova *et al.*, 1980), or may be prepared by direct digestion of ovaries with collagenase in $Ca^{2+}$, $Mg^{2+}$-free medium (Eppig, 1976), or by the use of other enzymes (Mangia and Epstein, 1975) (see Fig. 3). Oocytes free of attached follicular cells placed on a monolayer containing follicular cells continue to grow but at a

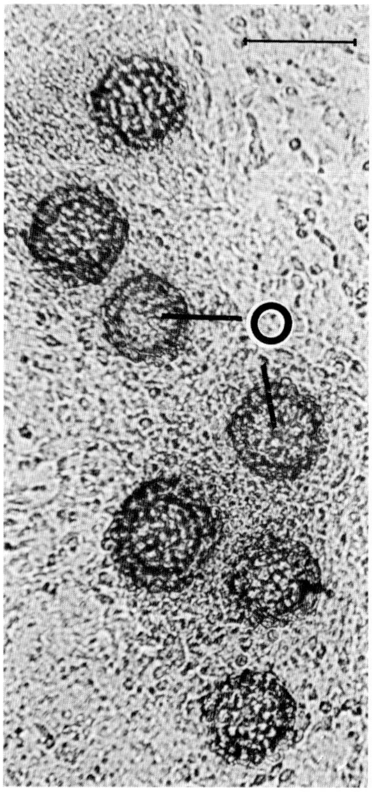


**Figure 8.** Light micrograph of mouse ovarian follicles after 6 days culture *in vitro*. Many follicular cells have migrated away from the oocytes (O) and are growing in a monolayer. However, the oocytes remain encased in a layer of follicular cells. Scale bar: 100 μm. (Reproduced from Eppig, 1977, with permission from Dr. J. J. Eppig and Academic Press.)

reduced rate, and attain meiotic competence at reasonable frequency (Bachvarova *et al.*, 1980). In either culture system, oocytes may be labeled with [$^3$H]uridine for various times, released from follicles cells if necessary, rinsed, and collected with a mouth-operated micropipette. No difference has been found in the metabolism and stability of RNA of oocytes labeled in whole follicles or labeled as free oocytes cultured in the presence of follicular cells. The incorporation per oocyte is higher when labeled within the follicle, since precursor uptake is facilitated by follicular cells (see Chapter 7). When labeled overnight in follicles, a highly labeled sample of a few hundred oocytes can be obtained quite readily.

The naked oocyte culture system has been used to characterize the kinetics of incorporation of precursor into heterogeneous RNA and rRNA during continuous labeling (Bachvarova, 1981). The aim of these experiments was to define for mid-growth–phase oocytes the approximate rate of synthesis and half-life of total heterogeneous RNA and the rate of synthesis and stability of messenger RNA (mRNA). Oocytes from 8-day-old mice labeled with [$^3$H]uridine for periods of 1–48 hr were dispersed in detergent, and the molecular-weight profile of the labeled RNA displayed on polyacrylamide gels. Incorporation into the mitochondrial fraction was not a significant fraction of the total. Incorporation into rRNA and its precursor peaks was calculated, as well as incorporation into RNA with a molecular-weight range similar to that of mRNA (less than 36 S) and into high-molecular-weight heterogeneous RNA. The latter fraction represents nuclear RNA (hnRNA), since label appears rapidly in this fraction at times when most incorporation is shown to be nuclear (Brower *et al.*, 1981). With a knowledge of the time course of total incorporation per oocyte, the course of incorporation into each RNA class can be computed. The results reproduced in Fig. 9 show that after several hours for equilibration of the nucleotide pool, 28 S and 18 S rRNA accumulate linearly, consistent with their complete stability (see Section 5.1). The absolute rate of rRNA synthesis in growing oocytes can be computed from the amount of rRNA found in the full-grown oocyte or egg and the time available for its synthesis. This value permits us to convert the slopes of the curves shown in Fig. 9 into rates of synthesis. Assuming that the nucleotide pools for nucleolar and nucleoplasmic synthesis are the same or equilibrate within a few hours,

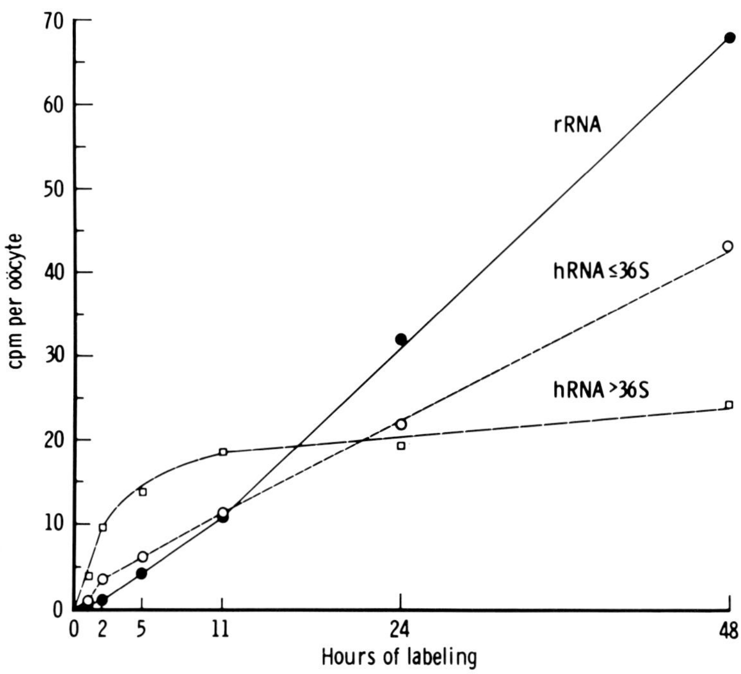


**Figure 9.** Time course of incorporation of [$^3$H]uridine into various RNA classes by growing mouse oocytes. Values for each time point were obtained by multiplying the total incorporation into RNA per oocyte by the percentage incorporation into each class determined from electrophoretic profiles. (Reproduced from Bachvarova, 1981, with permission from Academic Press.)

the rate of synthesis of hnRNA is computed from the rate of appearance of label in this class relative to rRNA. The half-life of hnRNA is estimated from the kinetics of the approach to equilibrium, using the time course of incorporation into rRNA to monitor the specific activity of the pool. From the data of Fig. 9 and our current estimate of 0.3 ng rRNA per full-grown oocyte (Kaplan *et al.*, 1982), the rates, steady-state amounts, and half-lives presented in Table I have been calculated.

The *in vitro* labeling results permit some interesting comparisons and conclusions. HnRNA has a half-life of approximately 20 min and turns over as rapidly as in somatic cells; thus, delayed processing of bulk hnRNA cannot be correlated with storage of untranslated mRNA in oocytes. The rate of synthesis of total heterogeneous RNA in oocytes is almost 10-fold higher than in mouse L cells (Brandhorst and McConkey, 1974) but still about 50-fold lower than on the lampbrush chromosomes in oocytes of the amphibian *Xenopus laevis*. The relationship between lampbrush chromosomes and the rate of synthesis of mRNA are discussed further below. In mouse oocytes, mRNA-like RNA is stable over the 2-day labeling period and accumulates at about half the rate of rRNA, an unusually high proportion of total RNA synthesis.

Recent experiments have been designed to analyze the distribution of newly synthesized cytoplasmic RNA between polysomes and the nonpolysomal fraction (De Leon *et al.*, 1983). The measurements were made on oocytes labeled in follicles and chased for 2 days to reduce reincorporation and nuclear label to an insignificant amount. The data represented in Fig. 9 show that after the initial rapid turnover, no further degradation of either ribosomes or the mRNA-like fraction occurs during the 2-day chase. Labeled oocyte lysates were centrifuged on sucrose gradients, and total as well as polyadenylated RNA in the polysomal and subpolysomal region were quantitated. Controls showed that most of the polysome material was EDTA releasable. It was calculated that about one-quarter of the newly synthesized mRNA and one-third of the ribosomes appear in polysomes, the remainder sedimenting in smaller particles. Thus, the bulk of the ribosomes and mRNA appear to be excess material accumulated for later use.

**Table I.** Synthesis and Turnover of RNA Classes in Growing Mouse Oocytes[a]

| | Steady-state amount (pg) | Half-life (min) | Rate of synthesis (pg/min) |
|---|---|---|---|
| Stable rRNA | — | — | 0.015 |
| Stable hRNA[b] less than 36 S | — | — | 0.008 |
| Unstable hRNA less than 36 S | 1.5 | 20 | 0.06 |
| hRNA greater than 36 S | 12 | 20 | 0.4 |
| Total hRNA | 13 | 20 | 0.5 |

[a]Data modified from Table 2 in Bachvarova (1981). Reproduced with permission from Academic Press.
[b]hRNA, heterogeneous RNA.

As a final note in this section, it is interesting that growing oocytes can be used for studies of transcription of injected DNA (Brinster *et al.*, 1981).

## 4. Chromosomal Structure

Chromosomal structure is affected by the level of ongoing transcription and the special requirements of meiotic prophase. This relationship is examined in this section. The condensation of mammalian oocyte chromosomes at different stages is described, and their overall size and structure are compared with that of the typical lampbrush chromosomes in growing oocytes of *Xenopus laevis*.

### 4.1. Meiotic Prophase to the Diplotene Stage

The basic features of chromosomal morphology during meiotic prophase are essentially universal (see Wilson, 1925). Von Winiwarter accurately described the sequence of events as seen in conventional histological sections of rabbit, human, and cat ovaries (von Winiwarter, 1901; von Winiwarter and Sainmont, 1909). For descriptions of chromosomal morphology in squashes or spreads, see Ohno *et al.* (1961, 1962), Luciani *et al.* (1974), Bakken and McClanahan (1978), and Speed (1982). It should be kept in mind that chromosomal morphology is best preserved in glutaraldehyde-fixed plastic-embedded ovaries. The squash technique, surface spreading, or air drying of fixed cells dropped on slides subjects the chromosomes to mechanical forces that cause partial unfolding, and the fixatives used may cause loss of some chromosomal material. These techniques are most useful for visualization of whole chromosomes and the process of synapsis.

The brief description provided here is illustrated in Figs. 10–12. In leptotene, dispersed chromatin condenses progressively into irregular threads, that can be seen by electron microscopy to contain a proteinaceous axial core (Franchi and Mandl, 1962; Tsuda, 1965; Comings and Okada, 1972; Baker and Franchi, 1967a, 1972; Speed, 1982). At zygotene, the synaptonemal complex forms with the appearance of its central element between closely apposed axial cores, in effect zippering them up; the axial cores become the lateral elements of the synaptonemal complex. When pairing is complete in early pachytene, the chromosomes have reached their shortest and thickest state, and chromomeres are seen as irregular beads along the axis (Fig. 10c). This point of maximum compaction coincides with the minimum incorporation of [$^3$H]uridine (see Fig. 6). In air-dried spreads of pachytene–early diplotene oocytes, the axis itself can become partially unwound, leaving a dense linear axial element still visible (Bachvarova *et al.*, 1982) and revealing two levels of packing of the

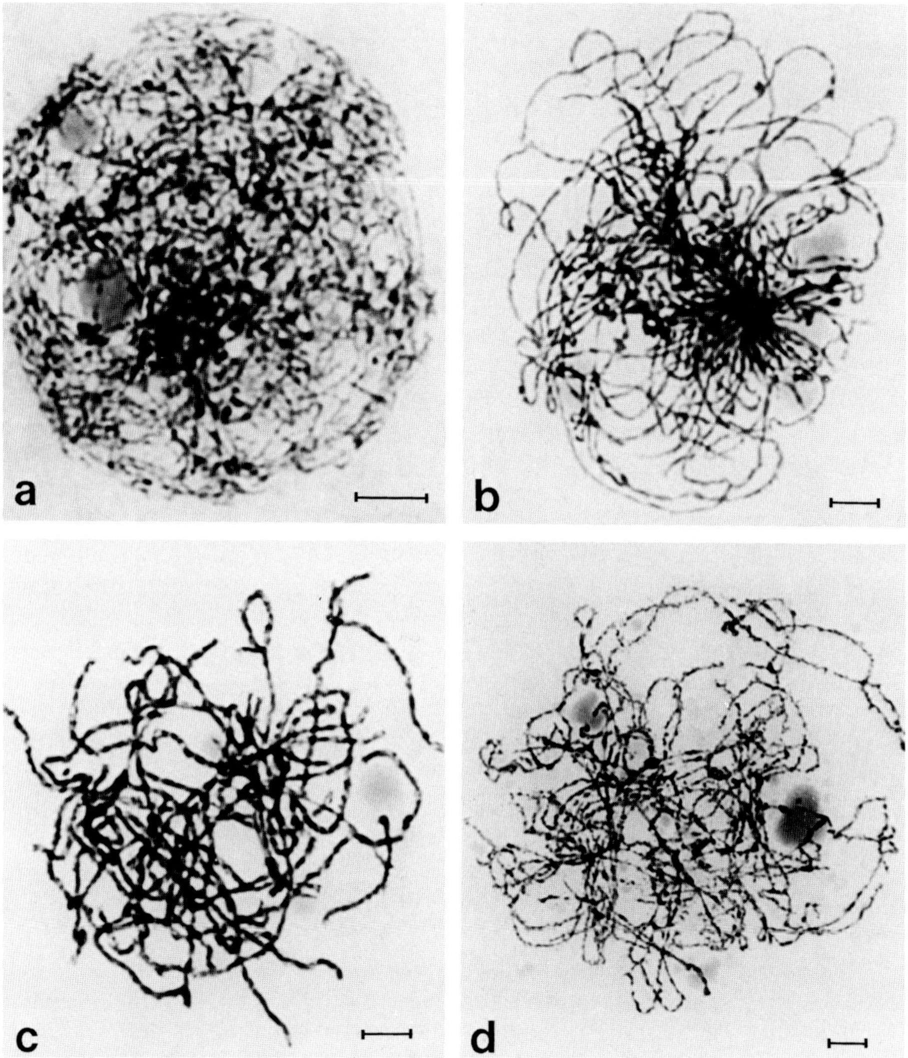


**Figure 10.** Air-dried spreads of human oocyte chromosomes. Ovarian fragments were fixed and dispersed and free oocytes were dropped on slides. The lateral spines are not visualized in this type of preparation. (a) Leptotene, (b) Leptotene-zygotene. (c) Pachytene. (d) Diplotene. Giemsa stain. Scale bar: 5 $\mu$m. (Reproduced from Luciani *et al.*, 1974 with permission from Dr. A. Stahl and from Journal of Reproduction and Fertility.)

chromatin fiber. The chromatin fibers of pachytene chromosomes observed by electron microscopic examination can be seen to form loops apparently attached to the synaptonemal complex (Baker and Franchi, 1967a; Comings and Okada, 1972; Kierszenbaum and Tres, 1974b).

As pachytene proceeds, spines begin to project laterally from the chromosome axis as visualized in sections (Bachvarova *et al.*, 1982) or squashes (Fig. 12a,d) (Ohno *et al.*, 1961, 1962; Bakken and McClanahan, 1978), giving the chromosome a fuzzy appearance and suggesting partial unwinding of chromomeres. This feature becomes more pronounced in early diplotene chromosomes, which von Winiwarter characterized as lacy or spiny. During diplotene, homologs separate, the chromosome axes elongate, and chiasmata become apparent (Fig. 10d). At the ultrastructural level, the synaptonemal complex dissociates, and the chromatin remains somewhat con-

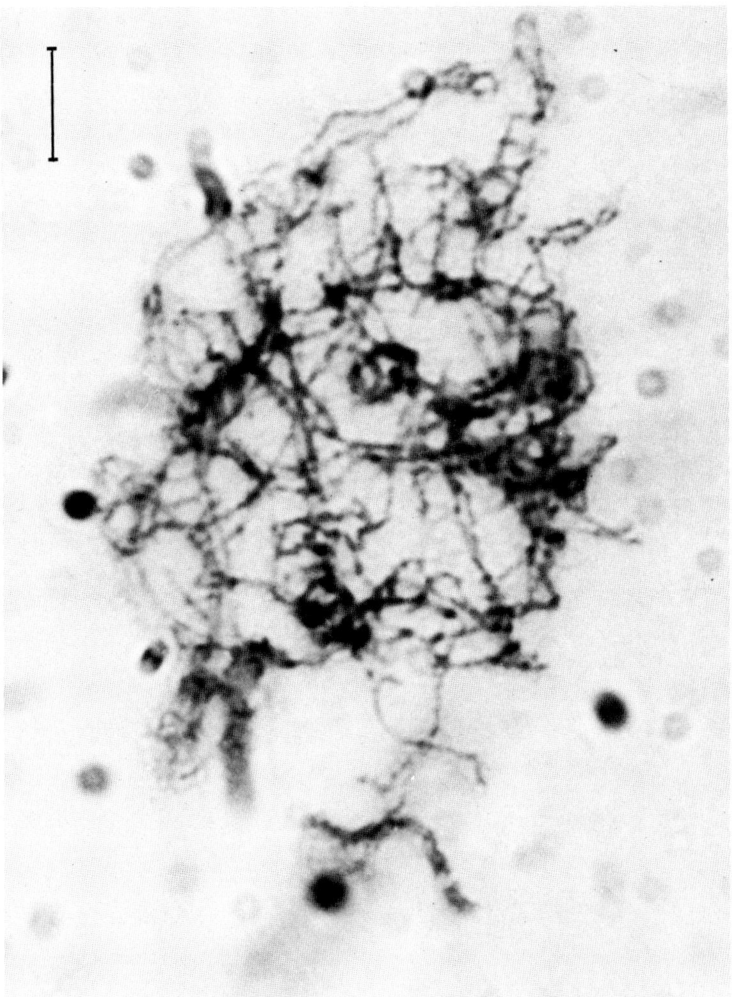

**Figure 11.** A conventional squash preparation of chromosomes from a diplotene human oocyte of a primordial follicle. Feulgen stain. Scale bar: 5 μm. (Reproduced from Baker and Franchi, 1967b with permission from Dr. T. G. Baker and Springer-Verlag.).

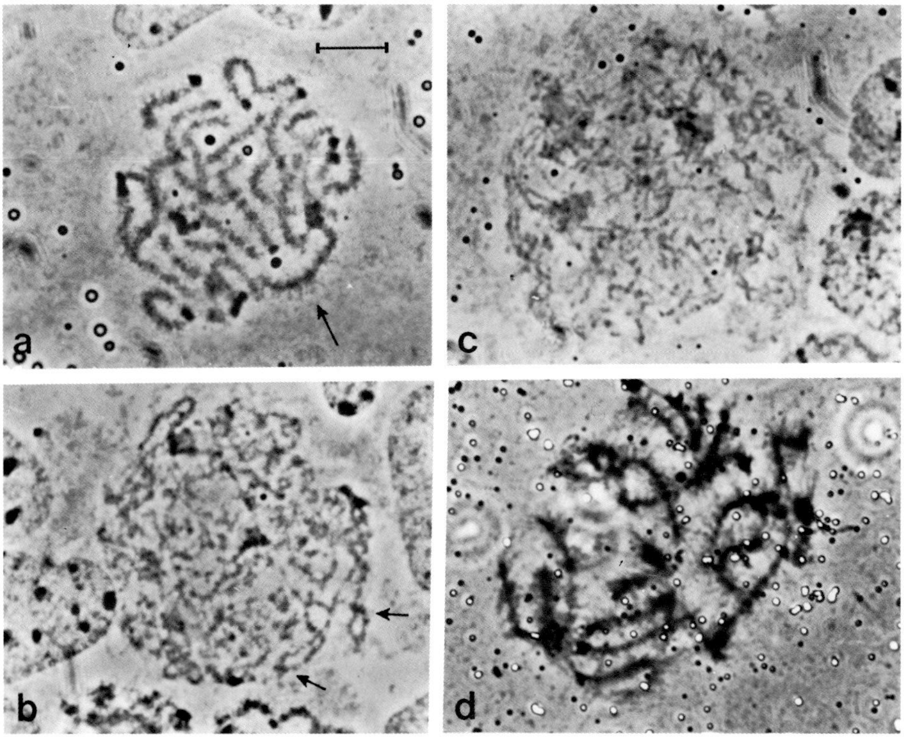

**Figure 12.** Phase-contrast micrographs of squash preparations of fetal or newborn mouse oocyte chromosomes. (a) Late pachytene, chromosomes are fuzzy, note distinct lateral projections at arrow. (b) Diplotene from a fetal mouse just before birth. Chromosomes are in the process of extending; small arrows indicate chiasmata between homologs. (c) Dictyate from a newborn mouse. Chromosomes are more extended and very fine. (d) Diplotene oocyte swelled in hypotonic solution: Lateral projections have become more obvious. Scale bar: 6 $\mu$m for (a–c), 10 $\mu$m for (d). (Reproduced from Bakken and McClanahan, 1978 with permission from Dr. A. Bakken and Springer-Verlag.)

densed around single axial cores. The loosening of at least some portions of the chromosome is accompanied by an increase in nuclear volume and by a rise in the incorporation of [$^3$H]uridine (Bakken and McClanahan, 1978). In spermatocytes, the chromosomes of mid–late pachytene also extend (Wilson, 1925), and in spread preparations they display lateral projections (Klasterska *et al.*, 1976; Oud *et al.*, 1979). This decompaction accompanies the transcriptional activity in spermatocytes (see Section 3.1).

Transcription complexes have been visualized in meiotic chromosomes of mouse oocytes spread by the Miller technique (Bakken and Hamkalo, 1978). For this technique, cells or nuclei are lysed in a solution of very low ionic strength and mild detergent, and the released chromosomes are centrifuged briefly onto a grid for electron microscopic examination. Transcription units are seen as chromatin strands carrying variable numbers of nascent RNP chains. The average rate of transcription on a transcription unit can be estimated by noting the number or density of transcribing RNA polymerases within observed units. In general, transcription units encountered in mouse oocytes from zygotene to early diplotene display only one or a few RNA polymerase molecules. Nascent RNP chains are also sparsely distributed in transcription units of pachytene mouse spermatocytes (Kierszenbaum and Tres, 1974b). In somatic cells, most of the thousands of transcribed sites are also transcribed at a low rate (Bakken and Hamkalo, 1978).

Mammalian meiotic chromosomes, which are actively engaged in transcription and are seen in the light microscope to have fibers projecting laterally from a condensed axis, are somewhat reminiscent of typical lampbrush chromosomes. Lampbrush chromosomes are found in growing oocytes of animals producing large eggs, including nonmammalian vertebrates, elasmobranches, cephalopods, some echinoderms, and mollusks (Wilson, 1925; Callan, 1957; Davidson, 1976). They are characterized by very rapid transcription on thousands of laterally projecting chromatin loops (see Chapter 9). The overall structure of amphibian lampbrush chromosomes and mouse oocyte chromosomes is compared in Fig. 13. Note that the amphibian chosen for this comparison, *Xenopus laevis*, has the same genome size as mammals. The term "lampbrush" has been used to describe both male and female pachytene–early diplotene chromosomes in mammals as seen by light microscope (e.g., Bakken and McClanahan, 1978; Kierszbaum and Tres, 1974b; Klasterska *et al.*, 1976). However, these chromosomes differ in significant ways from the typical or classically described lampbrush chromosomes. First, as shown in Fig. 13, the mammalian chromosomes remain much more condensed, the overall length of a chromosome at pachytene being severalfold shorter than the lampbrush chromosome of *Xenopus laevis*. Second, the lateral projections of *Xenopus* oocyte chromosomes are distinct loops up to a few micrometers in length (Muller, 1974) and are visible in the light microscope because each loop represents one or more very active transcription units with a very dense packing of nascent RNP chains on the chromatin fiber. The lateral projections of mammalian pachytene–early diplotene chromosomes are bundles of chromatin fibers, and complete loops are difficult to distinguish. The lateral projections are presumably nontranscribed regions. The chromatin that is transcribed must be intermingled with, and perhaps extends further laterally than, the visible projections but would clearly be too fine to resolve in the light microscope.

Such meiotic prophase chromosomes display a combination of condensation and transcription that is unusual, and they may also have a specific loop structure; perhaps they deserve a special descriptive term. In any case, the use of the term "lampbrush chromosome," which is closely associated with the very rapidly transcribing diplotene chromosomes of large growing

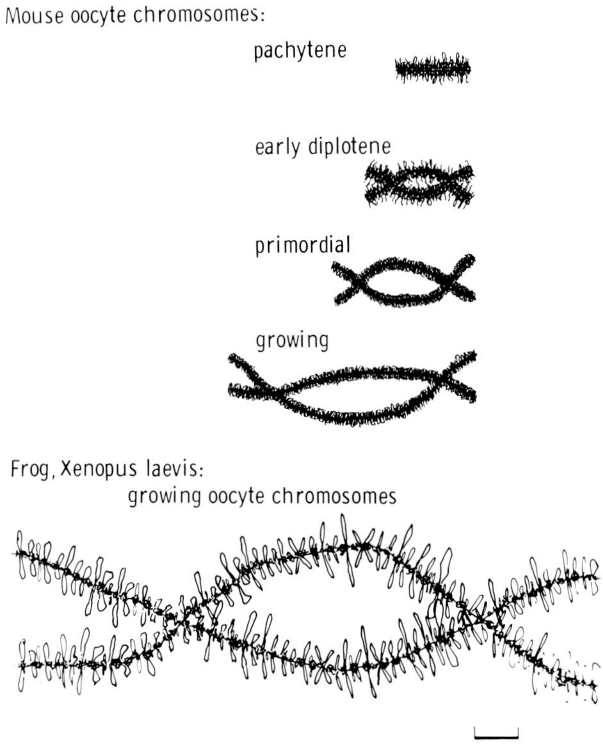


**Figure 13.** Diagram of approximate sizes and forms of mouse oocyte chromosomes at different stages, and of lampbrush chromosomes of *Xenopus laevis* as seen in the light microscope. Dimensions of mouse pachytene chromosomes are taken from Bakken and McClanahan (1978), and those of early diplotene are estimated from the same work. Dimensions of primordial and growing oocyte chromosomes can be estimated only roughly from work presented or discussed in Bachvarova *et al.* (1982). Dimensions and morphology of lampbrush chromosomes of mid-growth phase *Xenopus* oocytes are based on the work of Muller (1974) and Jamrich *et al.* (1983); lampbrush chromosomes of smaller oocytes are probably smaller. Scale bar: 5 $\mu$m.

oocytes, is confusing when applied to mammalian meiotic prophase chromosomes. It tends to imply that transcription is also very active in these cells and that they produce relatively stable mRNA (see Section 5). It is important to dissociate these properties from mammalian meiotic prophase oocytes, since there is no direct biochemical evidence available on these points. Callan (1957) has suggested that the difference between lampbrush chromosomes of growth phase amphibian oocytes and chromosomes of small oocytes and primary spermatocytes may be simply one of degree of unfolding of chromomeres; thus all these chromosomes may be termed lampbrush. However, MacGregor (1980) has pointed out that the lateral projections at pachytene may or may not represent the same regions unfolded into the loops of lampbrush chromosomes. In view of our lack of understanding of the mode of packing of the chromatin thread in either case and the fact that extremely rapid transcription is not universal in growing oocytes, Bachvarova *et al.* (1982) have proposed that the term "lampbrush chromosome" be reserved for chromosomes with looped transcription units in which a very high rate of transcription has been demonstrated by biochemical and/or electron microscopic techniques.

## 4.2. Primordial, Growing, and Full-Grown Oocytes

In many mammalian species, the chromosomes show little change in morphology as the oocytes are surrounded by follicular cells and enter the resting primordial oocyte stage. The diplotene state of the chromosomes can still be recognized in light microscopic sections; the axes of pairs of homologs are often seen to run in parallel and to join at chiasmata. This applies to a variety of mammals including the cat, cow, primates, and rabbit (Baker and Franchi, 1967b; Zybina, 1975). In squash preparations, the diplotene state is also evident (Fig. 11) (Ohno *et al.*, 1962; Baker and Franchi, 1967b); long, narrow chromosomal axes are seen delicately intertwined to form circles and "figure eights." Individual chromosomes with distinct fuzzy projections float free in lightly squashed fresh preparations (Baker and Franchi, 1967b).

By contrast, the chromosomes of rodents (rats, mice, hamsters) become more diffuse in primordial oocytes, and the pathways of individual axes cannot be followed with certainty. In plastic sections the chromatin appears quite homogeneously dispersed, much more so than in the preceding diplotene stage (Fig. 14). The texture suggests that elongate fuzzy threads are lying close together occupying most of the nuclear volume. The term **dictyate** (netlike) is now used to describe this state, although the term was applied to the chromosomes of rabbit, cat, and human oocytes by von Winiwarter. The term dictyate may be considered equivalent to diffuse diplotene, since as far as is known the homologs retain the same configuration. The transformation from diplotene to dictyate chromosomes in mice is illustrated in Fig. 12b, c and in the squash preparations of Ohno *et al.*, (1961) and the air-dried spreads of Speed (1982). The dictyate chromosomes appear as an irregular net of much finer threads. In contrast, spreads from human primordial oocytes (Fig. 11; refer also to Ohno *et al.*, 1962) are very similar to those seen at an earlier diplotene stage in the fetus (Fig. 10d), and in plastic sections the chromosomes are clearly not as homogeneously dispersed as in rodents (Fig. 14).

The difference between diplotene and dictyate primordial oocytes can also be observed at the ultrastructural level. In primordial oocytes of humans, monkeys, cows, and rabbits, sections of chromosomes exhibit clumps of chromatin fibers closely packed around an axial core (Baker and Franchi, 1967b,c, 1972; Baker *et al.*, 1969; Zybina, 1975), whereas in dic-

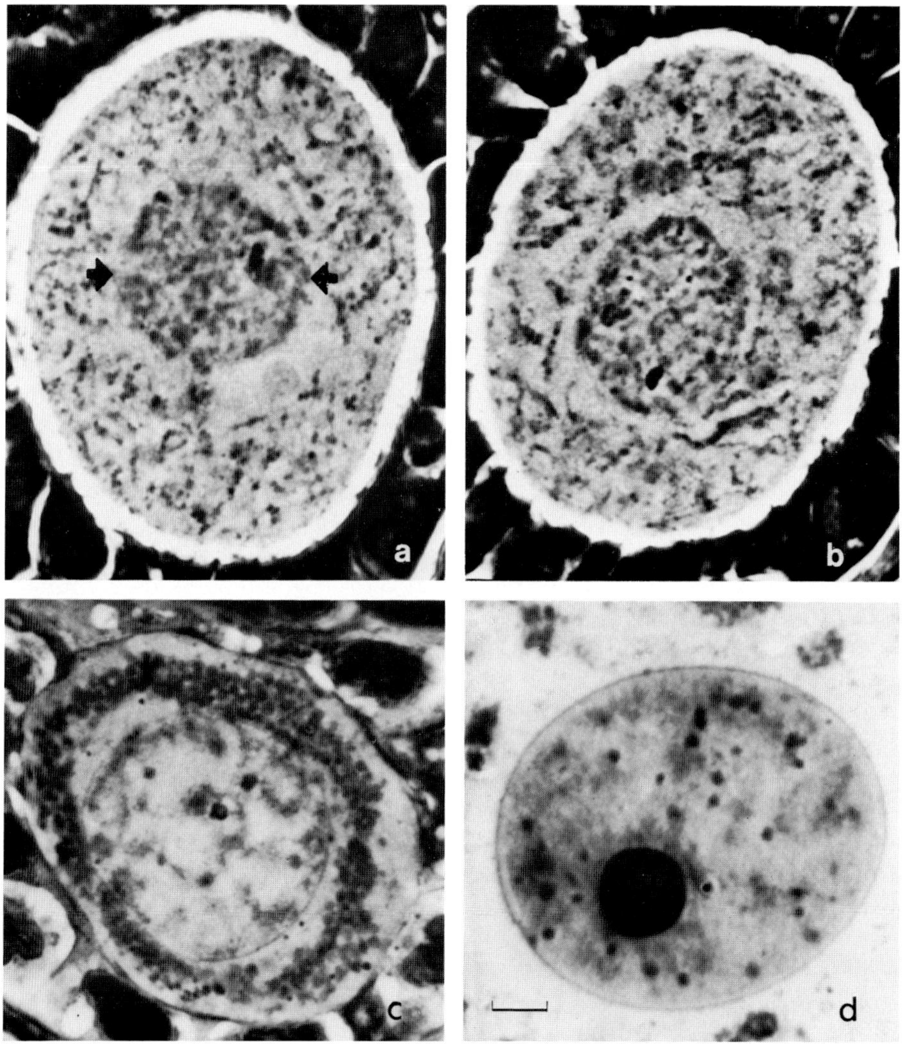


**Figure 14.** Plastic sections of growing oocytes stained with toluidine blue. (a,b) Mouse oocytes in one- and two-layered follicles. In (a) the nucleus is indicated by arrows. The nuclei of primordial oocytes resemble those of growing oocytes, except that they are smaller and the chromosomes appear to be more closely packed. (c) Monkey oocyte in a one-layered cuboidal follicle. (d) Monkey oocyte in a 4-layered follicle. The nucleus with a prominent nucleolus fills most of the field. Phase contrast. Scale bar: 5 μm. (Reproduced from Bachvarova *et al.*, 1982, with permission of Springer-Verlag.)

tyate rodent oocytes, the chromatin is more homogeneously dispersed with no axial cores evident (Franchi and Mandl, 1962; Tsuda, 1965; Baker *et al.*, 1969; Habibi and Franchi, 1978). The axial core is presumably a modified form of the same structure that appears during leptotene and becomes a lateral element of the synaptonemal complex, a structure known to be composed primarily of protein. This core can be seen very distinctly in silver-stained air-dried spreads observed in the light microscope. Using this method, the diplotene mouse oocyte has an obvious core that becomes very fine in dictyate oocytes (Speed, 1982).

Chromosomes at the primordial oocyte stage also resemble lampbrush chromosomes in some ways but at quite a different level than the resemblances at pachytene–early diplotene. The chromatin fiber as seen in electron microscopic sections is packed in quite a dense sheath around the central electron-dense axis, appearing in some cases to loop out from it. The sheath is in turn associated with granular elements. However, the axial core is probably protein rather than a chromatin fiber as in lampbrush chromosomes, and the pathway of the chromatin fiber in relationship to the axis is not clear. Moreover, no evidence has been presented for a dense RNP matrix on loops of mammalian primordial oocyte chromosomes, nor is there biochemical evidence of a very high rate of transcription. Again, it does not appear to be useful to employ a term that has such connotations.

As oocytes enter the growth phase, the germinal vesicle expands, and the chromosomes generally become more diffuse as growth proceeds (Ohno *et al.*, 1962; Baker and Franchi, 1967b.) A diagrammatic and approximate view of the changes in size and structure of the chromosomes is presented in Fig. 13. In plastic sections of ovaries fixed in glutaraldehyde, chromosomes in both monkey and mouse growing oocytes appear as thick, fuzzy threads winding through the nucleus (Fig. 14) (Bachvarova *et al.*, 1982). In the monkey, the chromosomes of small growing oocytes are still distinct threads, while in larger growing oocytes the threads extend further and interchromosomal material accumulates so that the nuclear texture becomes fairly homogeneous. In mice, a quite homogeneous texture is maintained throughout. In both species, chromosomes appear to radiate out from the heterochromatin associated with the nucleolus. In silver-stained air-dried spreads of growing mouse oocytes, fuzzy threads are more clearly visible (Fig. 15) (Bachvarova *et al.*, 1982). The fine lateral projections of earlier prophase are reduced or absent and may have dispersed until they are too fine to resolve. The threads display a denser core at their axis and are seen to cross at presumed chiasmata. Thus, the condensation of the axis present through previous stages of meiotic prophase is partially maintained even during the growth phase when chromosomes are in their most distended and active state. Figure 13 suggests a continuity of basic structure, which deserves further study.

Little change can be detected in the ultrastructure of the chromosomes in growing (as compared to primordial) oocytes of rodents. Consistent with the somewhat diffuse threads seen in the light microscope, electron micrographs show fibrillar–granular material filling the nucleus quite homogeneously (Franchi and Mandl, 1962; Tsuda, 1965; Weakley, 1966; Baker *et al.*, 1969; Chouinard, 1971). Unusual dense fibrillogranular bodies about 2 $\mu$m in diameter are quite abundant (Chouinard, 1973) and probably contain RNA (Zybina, 1968; Palombi and Viron, 1977). The ultrastructure of nuclei

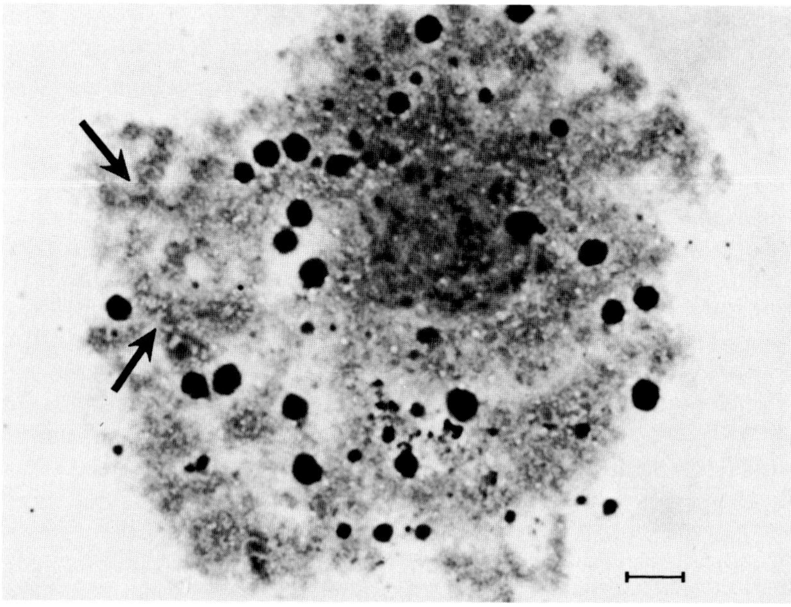

**Figure 15.** Air-dried spread of chromosomes from a growing oocyte of an 8-day-old mouse, stained with silver. Arrows indicate the linear core at the center of the chromosomes at possible chiasmata. The nucleolus is visible as a large, dark spherical region toward the upper right. About 25 very dark extranucleolar bodies of unknown function are also evident. Phase contrast. Scale bar: 5 $\mu$m. (Reproduced from Bachvarova *et al.*, 1982, with permission from Springer-Verlag.)

in growing oocytes of nonrodents has not been studied extensively. Zybina (1975) reported that in rabbit oocytes the chromatin decondenses upon growth, and the core is difficult to detect.

From studies of sensitivity to irradiation, it has been suggested that chromosome structure is similar in growing oocytes of various mammals but is different in primordial oocytes (Baker, 1971). The dictyate primordial oocytes of mice and rats are much more sensitive than the diplotene oocytes of primates. During growth, the rodent oocytes become more resistant and more similar to those of primates. Thus, in mice and rats irradiation causes selective loss of primordial oocytes and fertility declines after the growing oocytes are depleted.

At the end of oocyte growth, the chromosomes condense, and fuzzy axes become visible in the light microscope (Zybina, 1968; Baker, 1971; Palombi and Viron, 1977). The change is progressive, first becoming apparent in mouse oocytes between 14 and 16 days of age and continuing to breakdown of the germinal vesicle (Chouinard, 1975). This structural change reinforces the biochemical data provided in the previous section demonstrating that RNA synthetic activity, as well as incorporation, declines in full-grown oocytes. At breakdown of the germinal vesicle, chromosomes continue to condense and move relatively quickly through meiotic division (Donahue, 1968). Porcine chromosomes at various stages of meiotic maturation are shown in Fig. 16.

Chromatin from mid- and late-growth–phase mouse oocytes spread by the Miller technique has been examined in the electron microscope (Bachvarova *et al.*, 1982). Of the 109 transcription units found, 77 had only one nascent RNP chain and only five had four or more. Studies of incorporation and ongoing RNA polymerase activity suggest that the rate of RNA synthesis is maximal at this stage of oocyte development (see above). The increased RNA synthesis and the relatively high degree of dispersal have suggested to some (Zybina, 1968; Chouinard, 1973; Palombi and Viron, 1977) that the chromosomes are in a lampbrush state. Indeed, the rate of synthesis was shown to be 10-fold higher in growing mouse oocytes than in L cells (see above). Nevertheless, both the total rate of synthesis and the density of packing of polymerase on transcription units are roughly 50-fold lower than in the lampbrush chromosomes of growing *Xenopus laevis* oocytes (Hill, 1979). Also, the available data (Fig. 13) suggest that true lampbrush chromosomes are more extended than are chromosomes of growing mouse oocytes.

We conclude that true lampbrush chromosomes do not exist at any point in development of mouse oocytes. It has been suggested that lampbrush chromosomes with very rapid transcription over an extended period could be universal in animal oocytes, and that they may be responsible for accumulation of a product with as yet poorly understood function (Davidson, 1976). However, such chromosomes are absent in mouse oocytes, and the great variations in the rate of transcription among oocytes of different species seem to be related to the great differences in the amount of maternal mRNA accumulated in the egg. (see Section 5.4).

## 4.3. Gene Amplification

Gene amplification is one natural solution to the problem of producing a large amount of a gene product within a given period of time. Amplification of the genes for ribosomal RNA (rDNA) in growing oocytes is well known and widespread among various organisms (Tobler, 1975). The production problem is particularly acute when an RNA transcript is involved, rather than a protein which is naturally amplified during the two-step process of transcription and translation. In the amphibian, rapid transcription of the 1000-fold amplified rDNA sequences over several months produces the approximately $10^{12}$ ribosomes found in the egg.

In mammalian eggs, the need for rRNA is much less acute considering that the egg is roughly 1000 times smaller and thus the same concentration of ribosomes in the egg (and presumably an adequate level of protein syn-

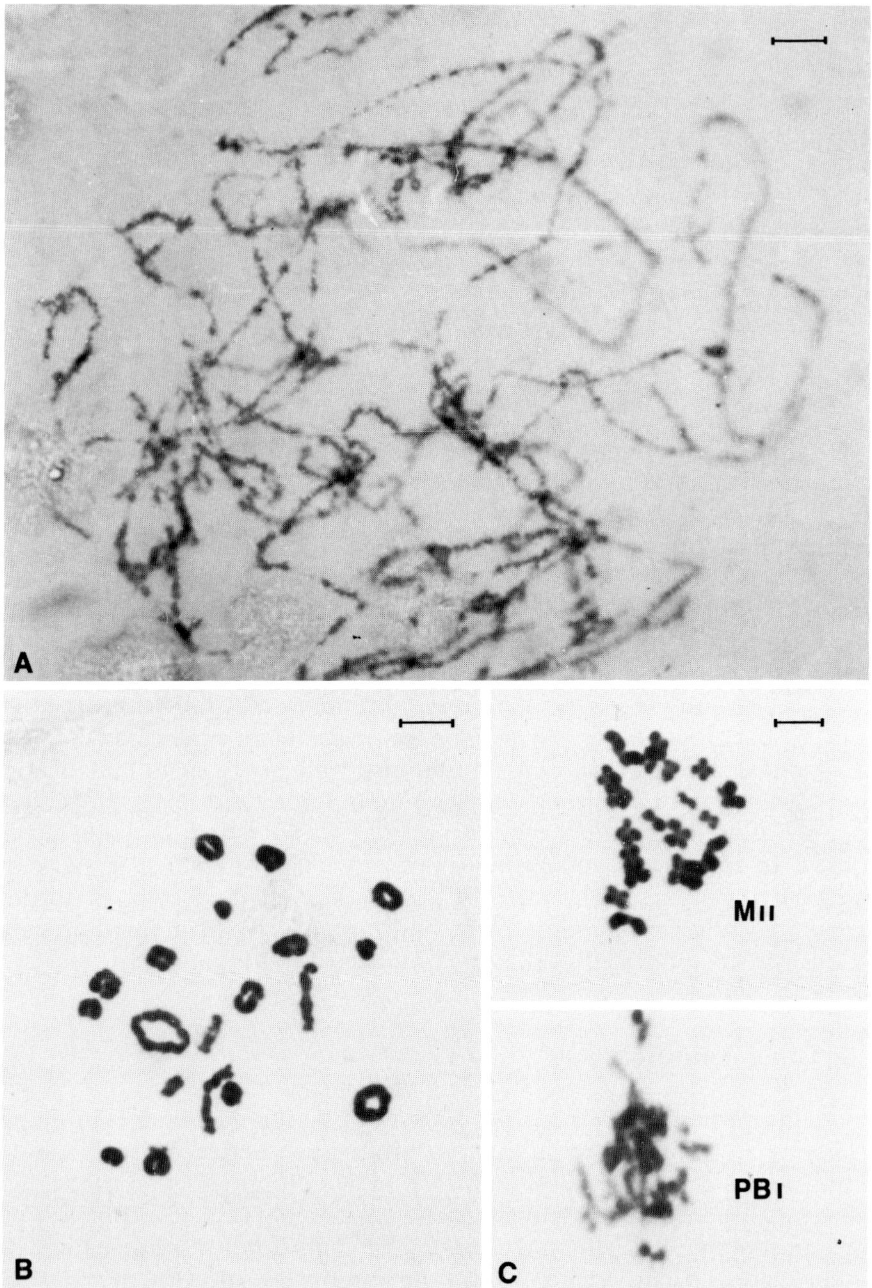


**Figure 16.** Air-dried spreads of chromosomes from porcine oocytes. (A) Freshly collected full-grown ocoyte. (B) The 19 bivalents of a maturing oocyte at diakinesis–metaphase I. (C) The two groups of chromosomes resulting from the first meiotic division of a maturing oocyte. (Upper) Metaphase II chromosomes of the oocyte. (Lower) Chromosomes of the polar body. Scale bar: 5 $\mu$m (Reproduced from McGaughey and Van Blerkom, 1977, with permission from Dr. R. W. McGaughey and Academic Press.)

thesis) could be achieved with 1000 times fewer ribosomes (see Section 5.4). Nevertheless, evidence for a low level of rDNA amplification in pachytene–diplotene oocytes in some primates has been obtained. In spreads of diplotene human fetal oocytes, micronucleoli are seen as spherical RNA-containing bodies, almost 2 $\mu$m in diameter, often associated with centromeric heterochromatin (Stahl *et al.*, 1975). Two to 40 of these are found per oocyte. Some of the micronucleoli contain silver-positive granules, known to be correlated with rDNA transcription (Hartung *et al.*, 1979). Wolgemuth *et al.* (1977, 1979, 1980) used *in situ* hybridization with purified iodinated rRNA to locate rDNA sequences in diplotene oocytes of human and baboon fetal ovaries. By counting grains over the spreads, they found a two- to threefold excess over the expected 4C amount of rDNA, and some micronucleoli were distinctly labeled. Somatic cells and spermatocytes provided a reference level of 2 or 4C, respectively. Studies on primordial or growing primate oocytes in juvenile or adult ovaries have not been reported. Mouse oocytes have been examined but have neither micronucleoli nor extra rDNA in early meiotic prophase.

The need for rDNA amplification in prophase oocytes of fetal primates seems remote, since cell volume and presumably the number of ribosomes increase slowly during these stages. If such micronucleoli continue to function during the growth phase, they might indeed play a role in mammals with relatively large eggs and a high concentration of ribosomes such as the rabbit (Table III) and possibly primates. Otherwise, it might be regarded as a vestige of evolution remaining from ancestors requiring large amounts of preformed cytoplasmic components in the egg.

In rare cases, a protein coding sequence may be amplified as part of a differentiation pathway; the genes for chorion proteins of the *Drosophila* egg are amplified in somatic cells surrounding the oocyte (Spradling and Mahowald, 1980). It is unlikely that specific genes are amplified in mouse oocytes, since it can be calculated that even abundant proteins, such as lactic dehydrogenase and zona pellucida proteins, can be produced with a moderate rate of transcription of a single gene.

## 4.4. Activity of the X Chromosomes in Female Germ Cells

In females, one of the two X chromosomes is inactivated in somatic cells, while both X chromosomes function in oocytes. This unique feature of the oocyte genetic program may coincide with other aspects of gene expression specific to the oocyte.

The cycle of X chromosome activity through one generation of the female germ cell line is now quite well understood. After fertilization, both X chromosomes of female embryos are active by the 8–16-cell stage. For example, Monk and Harper (1978) analyzed the ratio of HPRT activity (X-coded) to APRT activity (autosome-coded) in individual 8–16 cell embryos and found a bimodal distribution, indicating differential X chromosome activity in male and female embryos. Activity of X chromosomes in earlier embryos could not be measured because of the large amount of maternal mRNA coding for this enzyme (see Section 7.2). The second X is inactivated in the trophectoderm cells by the early blastocyst stage and in the primary endoderm by the late blastocyst stage; in both cases, the paternal X is selectively inactivated (see Martin, 1982). In the developing epiblast, which produces the embryo, the second X is inactivated (randomly with respect to maternal or paternal X) in greater than 85% of all cells by 5.5 to 6 days (Rastan, 1982). This was demonstrated using the Kanda technique to identify the heterochromatic X chromosome in metaphase spreads. Thus, inactivation precedes gastrulation at 6½ days and may not coincide with determination and/or differentiation as closely as previously indicated (Monk and Harper, 1979).

The X chromosome activity of the germ cells can first be tested in the oogonia of the gonad in the 11-day mouse fetus. Several methods have been used to show that the second X chromosome is inactive at this time (and thus must have been inactivated between about 6 and 10 days), but is reactivated at about 13 days, either close to or coinciding with the last S phase before meiotic prophase (Gartler *et al.*, 1980; Kratzer and Chapman, 1981; Monk and McLaren, 1981). Reactivation may be necessary but not sufficient for entry into prophase (McLaren and Monk, 1981). Activity of both X chromosomes increases during prophase. It is interesting that whereas the second X is reactivated by early prophase of female germ cells, the X and Y become heterochromatic at this time in male germ cells (Tres, 1975).

Both X chromosomes presumably remain active through early diplotene and the primordial oocyte stages. X chromosome-encoded enzymes accumulate during growth of oocytes in XX mice to twice the level found in oocytes of XO mice, indicating activity of both chromosomes (Epstein, 1972). A stock of mRNA for HPRT is also present in the egg and supports a 20-fold increase in its activity by the 8-cell stage (Monk and Harper, 1978).

Deficiencies found in XO females of mice and humans help clarify the role of the second X in these species. In mice, XO females are normal and fertile, although their overall fertility is reduced. XO human females (Turner's syndrome) are infertile with streak gonads. Oogonia are present in normal number in 4-month human fetuses (Singh and Carr, 1966), but oocytes are almost absent in newborns (Carr *et al.*, 1968), suggesting that they are eliminated in early prophase of meiosis. Careful analysis of XO mice has demonstrated that the oocyte population is only one-half normal size in 12-day-old mice, and the reduction may occur during fetal life (Burgoyne and Baker, 1981). This result suggests that loss of the second X chromosome has a quantitatively different effect on oocyte development in the two species, reducing the number more drastically in humans. The increased X chromosome activity during early meiotic prophase may provide products required for oocyte survival; human oocytes may be more sensitive to reduced amounts of such products partly because they spend a much longer time in this phase than do mouse oocytes.

Studies of embryos derived from XO mice suggest that all the eggs produced are somewhat impaired regardless of their embryonic genotypes (Burgoyne and Biggers, 1976). Thus, there are at least two periods (early meiotic prophase and oocyte growth) when activity of two X chromosomes promotes oocyte and embryo viability, one X producing amounts that may fall below a critical level. Evidence that XO embryos are preferentially lost during cleavage (Luthardt, 1976) suggests that monosomy for the sex chromosomes is deleterious at this stage. A high proportion of XO human conceptuses are lost before birth.

In view of the importance of the activity of two X chromosomes throughout most of the life cycle of female germ cells, the inactivation of the second X chromosome for a short period between about 6 and 11 days of development in mice is striking and somewhat unexpected. These observations led Monk (1981) to propose that germ cells proceed partway down the path of determination and restriction in potency in parallel with somatic cells, and then reverse these steps. Whether X inactivation reflects a step in determination is questionable in light of the early inactivation in somatic cells of the embryonic epiblast. Perhaps it is simply that both X chromosomes are active in oocytes when high levels of X chromosome-encoded products are required, whereas activity of one X chromosome is sufficient in primordial germ cells, when germ cell activities are sex independent. Nevertheless, it is striking that germ cells undergo a basic event that affects gene expression roughly in concert with somatic cells, and this event is later reversed.

## 5. Accumulation of Stable RNA During Oocyte Growth; RNA Content of the Full-Grown Oocyte and Ovulated Egg

Most or all of the components found in the full-grown oocyte are synthesized and accumulated in the growing oocyte. During its major growth phase, oocyte volume increases about 100-fold, RNA content 20-fold, and protein content 100-fold. Macromolecules of the primordial oocyte are degraded or greatly diluted during growth.

Considerable progress has been made in recent years in developing an overall understanding of the rates of synthesis, stability, and accumulation of ribosomes and mRNA in growing mouse oocytes. The outline that has emerged is relatively simple, defining this period as one of rapid gene transcription and steady accumulation of gene transcripts. These studies have progressed further in the mouse system than in other animal systems for several reasons. The period of growth from the primordial to the full-grown oocyte is well defined, continuous, and invariant in rate. The period of growth is very brief (2 weeks in mice, as compared with months in amphibians), and *in vitro* culture systems are available in which a considerable portion of the growth phase can be traversed and labeling studies carried out. Finally, juvenile mice provide a convenient source of oocytes of defined size.

Relatively stable mRNA accumulates in growing oocytes to form the pool of mRNA inherited by the embryo from the maternal genome. The time course of accumulation of mRNA in mouse oocytes will be emphasized, partly because of its apparent simiplicity and partly because this system provides a useful comparison with the best studied lower vertebrate, the frog *Xenopus*. Possible roles of the accumulated products, particularly mRNA, are considered in Section 7.

### 5.1. Total and Ribosomal RNA

During mouse oocyte growth, total RNA per oocyte increases, but the RNA concentration declines, as suggested in early cytochemical studies (Alfert, 1950; Flax, 1951). The cytoplasm of full-grown oocytes is unusually pale staining with little basophilia, whereas primordial oocytes and growing oocytes are moderately basophilic.

In order to measure the RNA content of oocytes of different sizes, microtechniques are almost essential. For this type of study, mouse oocytes of different sizes are collected via a mouth-controlled drawn glass pipette from enzymatically dispersed ovaries of juvenile mice of different ages (Mangia and Epstein, 1975; Eppig, 1976) (see Fig. 3). A yield of up to 70 oocytes per mouse can be obtained. Although oocytes are relatively large cells, each contains a very small amount of RNA. Thus, samples that can be collected in a reasonable period of time still contain too small an amount to measure by standard techniques such as phenol extraction and measurement of RNA by UV absorption.

Biochemical analysis of full-grown oocytes and eggs is simpler than for growing oocytes, partly because the cells are larger and fewer are needed. Full-grown oocytes of mice can be collected after pricking pregnant mare serum gonadotropin stimulated ovaries with fine needles. Bulk isolation procedures are also available. Mouse ovaries are pressed through a wire screen and passed through a sucrose step gradient (Sotelo and Garcia, 1977); porcine ovaries are minced with razor blades and passed through wire screens of different mesh size (Dunbar *et al.*, 1980). The latter procedure yields 50,000 oocytes in one preparation. Ovulated eggs are collected from the oviducts; cumulus cells surrounding them are removed by enzyme treatment.

Sternlicht and Schultz (1981) measured total RNA in mouse oocytes from 40 to 80 $\mu$m diameter using two microtechniques. In the first, RNA was extracted with phenol-chloroform and hydrolyzed with RNase. The ribonucleotides were phosphorylated using T4 polynucleotide kinase and

[$\gamma P^{32}$]-ATP, and the radioactivity was related to mass via a standard curve. The second assay was based on fluorescence of ethidium bromide bound to nucleic acids and gave similar values. The results replotted against time are shown in Fig. 17. Kaplan *et al.* (1982) measured the accumulation of rRNA (which comprises about 70% of total RNA) from primordial to full-grown oocytes by hybridization of extracted oocyte RNA to excess labeled cDNA complementary to rRNA. These results are also plotted in Fig. 17. Although the estimates for the absolute amount of total RNA differ somewhat (0.6 versus 0.45 ng per full-grown oocyte), the studies agree in demonstrating an increasing rate of accumulation up to 14–15 days and a diameter of 60–65 $\mu$m, and little accumulation thereafter. The concentration of RNA declines two- to threefold during the accumulation phase, and more

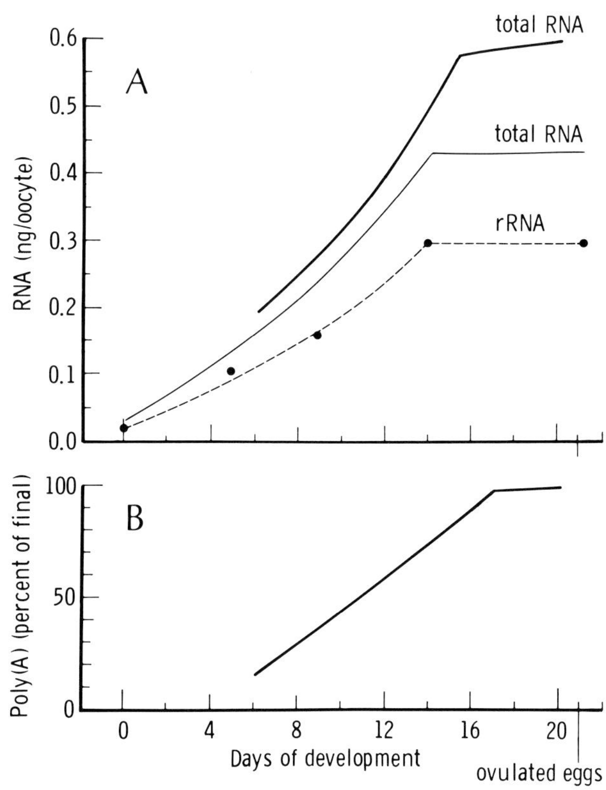


**Figure 17.** Accumulation of RNA during growth of the mouse oocyte. (A) Increases in total RNA and rRNA. Upper solid line: increase in total RNA using values reported by Sternlicht and Schultz (1981) for oocytes of different volumes. In order to plot the increase in RNA over time, oocyte volumes were converted to days of development using the data of Fig. 5. Dashed line: rRNA contents taken from Kaplan *et al.* (1982). Fine solid line: estimates for total RNA from the data of Kaplan *et al.*, given that rRNA is two-thirds of total RNA. (B) Relative amount of poly(A) adapted from data in Sternlicht and Schultz (1981) and the data of Fig. 5.

drastically during the full-grown stage as the oocyte approaches 80 μm diameter. These data are consistent with the labeling studies presented earlier in which increased nucleolar and nuclear activity was seen during growth and very little in full-grown oocytes. The cessation of accumulation appears to precede slightly what is usually considered the end of the major growth phase.

The relative amount of RNA in growing oocytes, full-grown oocytes, and early embryos has also been measured by cytophotometric quantitation of azure B-stained samples (Moore *et al.*, 1981). The data are generally consistent with the biochemical analyses, except that they suggest a peak of RNA content in early full-grown oocytes.

The various measurements of the RNA content of full-grown-mouse oocytes and ovulated eggs are listed in Table II, ranging from 0.35 to 0.6 ng per oocyte or egg. Olds *et al.* (1973) measured total acid precipitable alkali-sensitive UV-absorbing material. Piko and Clegg (1982) extracted nucleic acid and then measured the UV absorption of alkali-sensitive material, obtaining 0.35 ng per egg. Considering the difficulties of accurate measurement of small amounts of material, the values are in reasonable agreement and suggest a small drop in RNA content during maturation. Whereas Kaplan *et al.* (1982) found no significant decrease in the amount of rRNA between the 60-μm oocyte and the ovulated egg, our preliminary evidence suggests that at least 10% of total RNA is degraded during meiotic maturation (see Section 7.2).

In order to assess the rate of RNA synthesis necessary for such accumulation and the contribution of each period of oocyte growth to the RNA actually present in the egg, the stability of the RNA must be known. In an early study in which the types of RNA synthesized in growing oocytes

**Table II.** RNA Content of Full-Grown Mouse Oocytes and Eggs

| Full-grown oocytes | Ovulated eggs | Reference |
|---|---|---|
| | Total RNA | |
| | 0.55 ng | Olds *et al.* (1973) |
| 0.6 ng | | Sternlicht and Schultz (1981) |
| 0.45 ng[a] | 0.45 ng[a] | Kaplan *et al.* (1982) |
| | 0.35 ng | Piko and Clegg (1982) |
| | Poly(A) | |
| | 1.9 pg | Levey *et al.* (1978) |
| 0.9 ng | 0.7 pg | Piko and Clegg (1982) |
| 1.5 pg | | Sternlicht and Schultz (1981) |
| | Polyadenylated RNA | |
| | 21 pg[b] | Piko and Clegg (1982) |
| 90 pg | 45 pg | De Leon *et al.* (1983) |

[a]rRNA × 1.5.

[b]Using an average mRNA length of 2200 nucleotides (L. Piko, personal communication), increases to 29 pg in one-cell fertilized embryo.

were analyzed biochemically (Bachvarova, 1974), several points were established. Ovarian oocytes were labeled *in vivo* by injecting [$^3$H]uridine into the bursal sac surrounding the ovary, thereby bathing the ovary in a highly radioactive medium for about 2 hr. The amount of radioactivity present in RNA of ovulated eggs collected one to 25 days later was measured. Unfortunately, the level of incorporation obtained (about 1 cpm per egg) made biochemical analysis tedious, because large samples had to be collected. The results reproduced in Fig. 18 demonstrate that throughout the entire growth phase, 19–5 days before ovulation, significant amounts of RNA are synthesized and conserved to ovulation, i.e., each stage of oocyte growth contributes to the final mass of RNA carried in the egg. When the labeled RNA was displayed on polyacrylamide gels, it was found that 65–70% is rRNA, 18% is tRNA, and about 15% is heterogeneous RNA (hRNA), with no difference in the distribution for RNA synthesized at different stages of oocyte growth (see also Jahn *et al.*, 1976). Thus, the relative amounts of stable rRNA, hRNA, and tRNA synthesized and conserved to ovulation do not change during growth, and these amounts reflect the mass proportions of these RNA species in the egg.

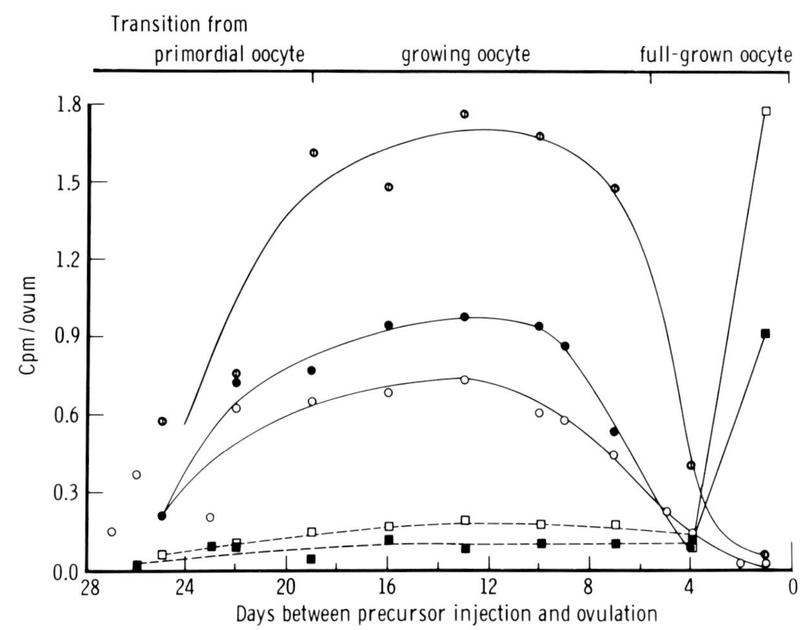


**Figure 18.** Time course of incorporation of [$^3$H]uridine into stable RNA of mouse oocytes *in vivo*. [$^3$H]uridine was injected into the ovarian bursa of adult mice and superovulated ova collected on subsequent days as indicated. Circles: incorporation into RNA with various doses of [$^3$H]uridine from 15–40 $\mu$Ci/ovary; Squares: radioactivity found in the acid soluble fraction, using 25 or 40 $\mu$Ci/ovary. (Modified from Jahn *et al.*, 1976 and reproduced with permission of Alan R. Liss.)

These data also suggest that the RNA synthesized early in growth has unusual stability, with roughly equal amounts of incorporated label being retained regardless of the stage of oocyte growth labeled. During the transition between the primordial and growing oocyte 26–19 days before ovulation, a gradual rise in label incorporated and conserved to ovulation is observed, suggesting a gradual increase in rate of synthesis and/or stability of RNA.

The stability of RNA synthesized in growing mouse oocytes has been measured directly. Ovaries were labeled *in vivo*, growing oocytes were collected 1–2 days later from the ovary, and the amount of label in RNA compared with that in ovulated eggs collected 10 days later (Jahn *et al.*, 1976). It was found that, allowing for possible reincorporation, at least 80% of the labeled RNA present 1 day after labeling is still present 10 days later in ovulated eggs. A more accurate analysis of this problem has been carried out using oocytes labeled in whole follicles and chased *in vitro* (Brower *et al.*, 1981). These studies show that total RNA label declines only 15% over 7 days, a finding confirmed by De Leon *et al.* (1983). As shown in Fig. 19, essentially 100% of the nonpolyadenylated RNA remains stable for 7 days (after allowing for initial turnover), while polyadenylated RNA turns over slowly with a half-life of 8–12 days. Thus rRNA and tRNA are both completely stable during the growth phase. This result is not unexpected, as rRNA is stable in replicating cells (Brandhorst and McConkey, 1974), while in steady-state tissues such as liver, rRNA turns over with a half-life of about 4 days (Quincey and Wilson, 1969).

Knowing that rRNA is stable, the slope of the accumulation curve in Fig. 17 is a measure of its rate of synthesis, which proceeds at a continuously high and somewhat increasing rate averaging 0.015 pg/min during oocyte growth. These data are consistent with the autoradiographic assess-

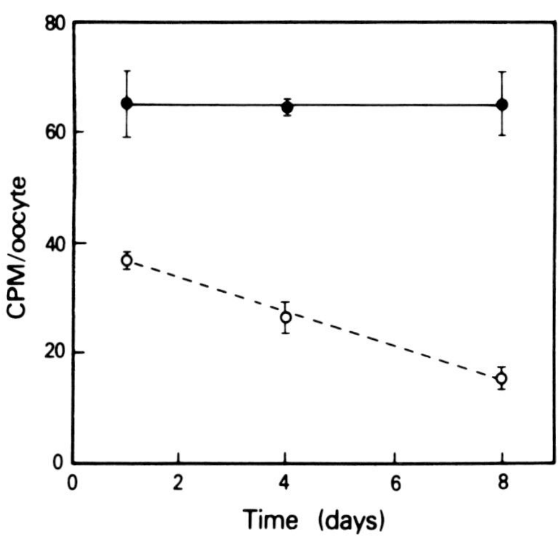


**Figure 19.** Relative stability of labeled nonpolyadenylated RNA and polyadenylated RNA during oocyte growth *in vitro*. Follicles from 10-day-old mice were incubated in medium containing [$^3$H]uridine for 5 hr and chased for various times up to 8 days. Groups of oocytes liberated from follicles were assayed for polyadenylated RNA by poly(U) Sepharose chromatography. (●) CPM/oocyte in nonpolyadenylated RNA. (○) CPM/oocyte in polyadenylated RNA. (Reproduced from Brower *et al.*, 1981 with permission from Dr. R. M. Schultz and from Academic Press.)

ment of nucleolar RNA polymerase activity described above, showing an approximately twofold increase for larger over smaller growing oocytes. Once the amount of RNA in full-grown oocytes and eggs, the number of rRNA genes in mice (300 per haploid genome), and the time available for its synthesis are known, the rate of transcription of the ribosomal RNA genes can be estimated at four transcripts per cistron per minute. This value is well within known rates (Kafatos and Gelinas, 1974) and is compatible with the lack of rDNA amplification in the mouse oocyte. However, it is possible that the low level of rDNA amplification observed in primate oocytes (see Section 4.3) could have a useful role. Primate eggs are more than 100 μm in diameter and may contain up to 20 times more rRNA than the mouse egg (see value for rabbit eggs in Table III), but the time for oocyte growth is probably longer. Thus, it is not clear whether rDNA amplification would be required in these species.

Given the rRNA content of the mouse egg, the volume of the egg, and that rRNA is found in typical ribosomes (Bachvarova and De Leon, 1977), the concentration of ribosomes can be calculated and compared to that in eggs of different species (Table III). The mouse egg has the lowest concentration of those listed, 2–3 fg/$\mu m^3$, compared with 6 fg/$\mu m^3$ in the rabbit, 4–10 fg/$\mu m^3$ in various species of sea urchin, and 6 fg/$\mu m^3$ in the frog. Thus, even among mammals, the degree of reliance on stored ribosomes varies, and the mouse appears to be particularly dependent on early activation of rRNA transcription in the embryo. For comparison, the concentration in liver is similar to that in rabbit eggs (Altman and Dittmer, 1972).

From the data presented here and in Section 3.2, the outline of rRNA accumulation in mouse oocytes is clear. During the transition from a resting primordial oocyte to a growing oocyte, the rRNA synthesized is stabilized, and its rate of synthesis probably rises. A high and somewhat increasing rate of synthesis is maintained throughout most of the growth period as the nucleolus increases in volume and complexity. Near the end of the major growth phase, rRNA synthesis declines greatly, and the

**Table III.** Total RNA Content in the Eggs of Different Animal Species[a]

| Species | Egg diameter (μm) | Egg volume (pL or $\mu m^3 \times 10^{-3}$) | RNA content per egg (ng) | RNA concn. (fg/$\mu m^3$) |
|---|---|---|---|---|
| Mouse | 70 | 180 | 0.4–0.6[b] | 2.0–3.0 |
| Rabbit | 125 | 1000 | 6.0 | 6.0 |
| Sea urchins | | | | |
| *Strongylocentrotus purpuratus* | 80 | 270 | 2.8 | 10.0 |
| *Paracentrotus lividus* | 90 | 380 | 1.4 | 3.7 |
| Frog, *Xenopus laevis* | 1100 | 700,000 | 4,000 | 5.7 |

[a]Modified from Table 6 in Piko and Clegg (1982). Reproduced with permission from Dr. L. Piko and Academic Press.
[b]See Table II.

nucleolus becomes dense and inactive. The cessation of accumulation of rRNA appears to be one of the first important metabolic changes marking the end of the major growth phase and a transition to a phase of qualitative changes leading to full competence for meiotic maturation.

As mentioned above, tRNA represents approximately 20% of the egg RNA, equivalent to 0.1 ng per egg, and it accumulates in proportion to rRNA. Boreen *et al.* (1983) have quantitated the level of labeling of 5 S RNA in mid- and late-growth-phase oocytes as about 10% of that in tRNA. The ratio of synthesis of 5 S and rRNA is then 0.02, close to the expected value of one 5 S RNA molecule per ribosome.

As described above, about one-third of the newly synthesized ribosomes appear in polysomes in mid- and late-growth-phase oocytes. This proportion remains constant when labeled ribosomes are chased over an additional 5-day period (De Leon *et al.*, 1983), suggesting that one-third of the mass of ribosomes accumulates as polysomes throughout growth. It is not clear whether the polysomes continue to increase during the full-growth oocyte stage to support the increasing protein synthesis (see Section 6.1). In any case, the fraction of ribosomes in polysomes of ovulated ova decreases to about 25% (Bachvarova and De Leon, 1977), consistent with the fall in the rate of protein synthesis during ovulation (see Section 6.3).

There is some evidence to suggest that ribosomes not engaged in protein synthesis are stored, perhaps in modified form, in the unusual fibrillar structures prevalent in rodent eggs. These structures vary in different rodents, consisting of spiral lamellae, closely spaced parallel bilaminar sheets, or (in mouse eggs) crosslinked fibrils (lattices) (see Bachvarova *et al.*, 1981 and references cited therein); such structures are absent in eggs of nonrodent mammals. They are not membranous and have a periodic structure resembling balls on a string. They accumulate during oocyte growth (Wassarman and Josefowicz, 1978) and occupy 25–30% of the volume of full-grown oocytes or eggs (Dvorak *et al.*, 1977; Garcia *et al.*, 1979). During cleavage and blastocyst formation they decrease steadily (Dvorak *et al.*, 1977), disappearing around the time of implantation (Enders, 1971; McReynolds and Hadek, 1972). This sequence clearly suggests they are stored for some function during embryogenesis. One possibility proposed by Szollosi (1972) is that they are the equivalent of yolk bodies found in eggs of some species (Brambell, 1956; Austin, 1961).

Another possibility, supported by several observations, is that the lattices of mouse eggs contain ribosomes. Their periodic repeat is large enough to include a particle of ribosome size. Particles the size of ribosomes are released from lattices spread on a liquid–air interface (Burkholder *et al.*, 1971), and RNase treatment of such spreads removes the periodic staining material (Garcia *et al.*, 1979). Ribosomes appear to be incorporated into them during their formation in oocytes (Zamboni, 1970). The density of free ribosomes or polysomes counted in electron microscopic sections decreases as the fibrillar material increases (Garcia *et al.*,

1979). Also, the buoyant density of fixed ribosomes of mouse eggs analyzed on CsCl gradients indicates that they carry extra protein under conditions in which ribosomes of somatic cells do not (Bachvarova *et al.*, 1981).

To obtain further data relevant to this question, the number of ribosomes in the egg was determined by electron microscopic morphometry and compared with the amount of RNA measured biochemically (Piko and Clegg, 1982). According to these investigators, the number of free ribosomes accounts for two-thirds of the 0.35 ng of RNA found in the egg, indicating to them that both stored and active ribosomes are visible as typical ribosomes and are not present in lattices. To reconcile these data with storage of ribosomes in lattices, the ribosome counts would have to have been too high by a factor of two and the value for total RNA too low by a factor of two. Whereas the reported errors in the data are approximately 10%, the range of values obtained by other workers do vary up to twofold (Garcia *et al.*, 1979) (see Table II). However, Piko and Clegg also found that particles in the lattices do not stain in parallel with typical ribosomes and do not show the same sensitivity to alkali treatment. Moreover, they calculated that if each potential particle of the lattices is a ribosome, the number of stored ribosomes would be about 15 times higher than the total rRNA in the egg. Thus, if lattices contain ribosomes, only a small fraction of the repeating sites would actually be occupied by ribosomes.

The evidence does not appear to be completely conclusive in either direction. Further approaches might include electron microscopic autoradiography, to determine whether RNA is contained in lattice regions, or subcellular fractionation to prepare purified lattices—a difficult task in this system.

Perhaps the most interesting functional question is whether excess ribosomes are indeed stored in the sense that they are removed from the active ribosome pool perhaps by complexing with additional protein. The small size of the polysomes in oocytes and eggs suggests that initiation of translation is quite inefficient (Bachvarova and De Leon, 1977; De Leon *et al.*, 1983). Other evidence suggests that the nontranslating ribosomes cannot associate with poly(U) to form a stable 80 S particle as can dissociated polysomal egg ribosomes or liver ribosomes (Bachvarova and De Leon, 1977). Such a functional impairment could mean that changes in protein synthesis programmed on maternal components would require activation of ribosomes (see Sections 6.3 and 7.2).

## 5.2. Messenger RNA

Information is also available for the synthesis, accumulation, half-life, and amount of polyadenylated RNA in mouse oocytes. In general, the messenger RNA class is remarkably stable and accumulates as a roughly constant proportion of total RNA, with slow turnover only of the translated fraction. The metabolism of mRNA is very complex, since presumably thousands of different genes are transcribed to produce polyadenylated RNA, possibly in shifting patterns, and a large portion of hnRNA and polyadenylated RNA turns over rapidly in the nucleus.

The most direct evidence on the time course of accumulation of polyadenylated RNA comes from studies quantitating poly(A) in oocytes of different sizes. Caution must be exercised in interpreting these data, since poly(A) content reflects polyadenylated RNA accurately only if it can be determined that the length of the poly(A) chain remains constant. The concentration of poly(A) in oocytes has been measured by two research groups using similar techniques and obtaining similar results. Sections of juvenile or adult ovaries were hybridized to [$^3$H]poly(U) under controlled conditions and the poly(A) concentration in oocytes assessed by grain density in autoradiographs. The density remains approximately constant for oocytes of 40–70$\mu$m diameter (Sternlicht and Schultz, 1981) or decreases only slightly in oocytes from one-layered cuboidal to three to four-layered follicles (Piko and Clegg, 1982). As shown in Fig. 17, when density is converted to total amount (concentration $\times$ volume) and plotted against time, its accumulation roughly parallels that for total RNA, and the rate of accumulation is faster during later oocyte growth.

The two studies are consistent in finding a decline in the concentration of poly(A) in full-grown oocytes during the 5 days of development of antral and preovulatory follicles, but Sternlicht and Schultz indicated a 25% increase in volume corresponding to a 9-$\mu$m increase in diameter of the oocyte during this period, wheras Piko and Clegg assumed oocyte volume remained constant. Since it is likely that volume does increase slowly during the full-grown stage (see Fig. 5), it is likely that poly(A) content per oocyte remains constant during this period. During meiotic maturation the density of poly(A) falls by 12%; combined with the 20% decline in volume this suggests a decrease in poly(A) amount of about one third during this period. Poly(A) content of preovulatory oocytes and ovulated eggs as assayed by solution hybridization of extracted RNA to excess [$^3$H]poly(U) shows a drop in mass of 26% (Piko and Clegg, 1982). No data are available on the length of the poly(A) chains in preovulatory oocytes, so these data do not necessarily reflect the loss in mass of mRNA.

The amount of polyadenylated RNA present in ovulated eggs can be assessed in various ways (Table II). Piko and Clegg (1982) and Clegg and Piko (1983a) measured the total amount of poly(A) and the length distribution of the poly(A) chains. Dividing the mass of 0.7 pg of poly(A) by the average length of the poly(A) chain (75 nucleotides), these workers calculated the number of polyadenylated molecules as $1.7 \times 10^7$. This value rises to $2.4 \times 10^7$ after fertilization due to cytoplasmic polyadenylation. According to their data, the average mRNA molecule is 2200 nucleotides in length (L. Piko, personal communication), giving a mass of about 21 pg of polyadenylated RNA (in the egg) and 29 pg in the one-cell embryo. Levey *et al.* (1978) found about twice as much poly(A) in the egg, leading to a higher estimate for polyadenylated RNA.

In a different approach, the mass of polyadenylated RNA in the ovulated egg was assessed using *in vivo* labeling of growing oocytes and assays of the percent of the labeled RNA of ovulated eggs in the polyadenylated fraction (Bachvarova and De Leon, 1980). It was found that 9–12% of the RNA synthesized at each stage of growth and conserved to ovulation is polyadenylated RNA; thus, its mass can be estimated to be about 10% of the total mass, or 45 pg. Given the errors inherent in these experiments, the values obtained from the two approaches are in quite good agreement.

Using the same *in vivo* labeling technique, it was estimated that about 20% of the total RNA of full-grown oocytes is polyadenylated (De Leon *et al.*, 1983). This value agrees with the fraction of polyadenylated RNA in stable labeled RNA of large oocytes maintained in follicles *in vitro* (Brower *et al.*, 1981). Thus, it appears that there is about twice as much mRNA in the full-grown oocytes as in the egg, or almost 100 pg. These results are quite consistent with the drop in poly(A) content during meiotic maturation described above and suggest total degradation or deadenylation of close to one-half the accumulated polyadenylated RNA. During ovulation about 10% of the radioactivity of *in vivo*-labeled oocytes is released as acid soluble material, suggesting breakdown of 10% of the RNA (De Leon *et al.*, 1983). Further experiments are under way to determine whether up to one-half the polyadenylated RNA is degraded during meiotic maturation. Recent studies have demonstrated, using Northern blotting and a $\beta$-actin specific probe, that mRNA for actin is present in full-grown oocytes and eggs, but that it is largely deadenylated during meiotic maturation (Bachvarova, *et al.*, 1985); this change accompanies a large decrease in $\beta$-actin synthesis (see below).

Knowing the time course of accumulation of poly(A) and presumably polyadenylated RNA, the rate of production and contribution of each stage to maternal mRNA can be estimated if its half-life is known. The *in vivo* pulse-labeling experiments described above could be explained if cytoplasmic polyadenylated RNA is synthesized as a constant proportion of, and is almost as stable as, rRNA during oocyte growth. Direct confirmation of this possibility was obtained by Schultz and co-workers. In growing oocytes of 8–12 day-old mice labeled for 5 hr in whole follicles *in vitro* and chased for up to 9 days, labeled polyadenylated RNA declines slowly, consistent with a half-life of 8–12 days (Fig. 19). The amount of reincorporation of label was judged to be small from measurements of nuclear incorporation and of label in the nucleotide triphosphate pool.

As described above, the distribution of *in vitro*-labeled ribosomes between polysomes and smaller nonpolysomal particles has been measured after short and long chase periods. Data from the same experiments were used to measure the half-lives of stored and polysomal mRNA (De Leon *et al.*, 1983). From the decline in polyadenylated RNA relative to rRNA, it was determined that polyadenylated RNA in the subpolysomal region is completely stable, whereas translated polyadenylated RNA turns over slowly with a half-life of about 6 days. Thus, stable stored mRNA is synthesized throughout growth and accumulates in proportion to its rate of synthesis. Data cited above show that the percentage polyadenylated RNA synthesized and conserved to ovulation rises only slightly during oocyte growth, and the rate of synthesis rises two-to fourfold during oocyte growth, perhaps slightly more than the rate for rRNA synthesis.

Knowing the rate of synthesis of stable heterogeneous RNA at mid-growth phase (see Table I), it is now possible to outline a plausible time course of the accumulation of mRNA in oocyte growth. Stored maternal mRNA accumulates with total RNA, at least twice as fast in late as in early growth. It represents three-quarters of the newly synthesized mRNA and is synthesized at 0.004 pg/min in mid-growth-phase oocytes, assuming that two-thirds of the stable hRNA is polyadenylated (see immediately below). Given a growth period of 15 days, the final value in full-grown oocytes is close to 80 pg per oocyte. Translated mRNA is synthesized at 0.001 g/min in mid-growth phase and accumulates slightly more slowly to a final value of about 20 pg. These values agree well with the actual measured amount of total polyadenylated RNA (Table II). The high proportion of mRNA in the egg is achieved both by a high relative rate of synthesis and by unusual stability.

Changes in mRNA content during the full-grown oocyte stage are possible but poorly defined. Protein synthesis continues, but RNA production is slow, and the percent of polyadenylated RNA in newly synthesized RNA falls (Bachvarova and De Leon, 1980; Brower *et al.*, 1981). The concentration of poly(A) in the nucleus remains high (Piko and Clegg, 1982), suggesting some stabilization of polyadenylated nuclear RNA.

The amount of additional nonpolyadenylated mRNA needs to be addressed. Comparison of the percent hRNA or polyadenylated RNA after 24 hr of *in vitro* labeling (Bachvarova, 1981; Brower *et al.*, 1981) suggests that about three-quarters of the mRNA is polyadenylated. The *in vivo* labeling method indicates that 10% of egg RNA is polyadenylated and that 15% is heterogeneous RNA. Moreover, Clegg and Piko (1983a) found a 40% increase in the number of polyadenylated molecules during the one-cell embryo stage, due to polyadenylation of stored mRNA. Thus, the total mass of maternal mRNA in the egg is up to 1.4 times the mass of polyadenylated RNA, or 40–65 pg.

Data on other mammalian species are sparse. Poly(A) comprises 0.25% of total RNA in rabbit eggs (Schultz, 1975)—a value similar to that in mouse eggs. Assuming that the poly(A) chain length is similar, the mass of polyadenylated RNA would be about 13 times higher (see Table III), or about 500 pg.

The proportion of mRNA in the mouse egg is unusually high; values for the sea urchin and amphibian range around 1–2% (Davidson, 1976). The large number of mRNA molecules results in an mRNA to ribosome ratio as high as 0.3–0.5. The reason for such a relatively large amount of mRNA is not clear, particularly in light of its rapid usage in early development (see Section 7.2).

The complexity of mRNA in animal eggs is high and remarkably uniform in a variety of species (Davidson, 1976), having a coding capacity of about 15,000 proteins. Cytoplasmic RNAs of *Xenopus* tadpoles and cultured kidney cells have a similar high complexity, which largely overlaps with egg mRNA (Perlman and Rosbash, 1978), indicating that special functions of the egg are not directly related to an unusually high information content or a large set of genes expressed specifically in eggs. The complexity of the mRNA in mammalian eggs has not been measured directly, since it is difficult to collect a large enough sample to carry out a set of measurements by liquid hybridization. A potential method to calculate the number of transcription units in oocytes is as follows. First, the average number of transcribing RNA polymerases on transcription units would be determined by means of electron microscopic analysis of spread chromatin; then, using a reasonable value for the number of nucleotides added to the growing chain per minute, the number of nucleotides polymerized per transcription unit could be calculated. Finally, the total rate of hnRNA synthesis divided by the rate per transcription unit gives the number of transcription units. The rough figures available suggest that the number of transcription units is relatively high.

## 5.3. Other RNAs

Besides RNAs of the translational machinery, cells also contain small RNAs and noncoding repetitive sequence transcripts. The most abundant stable small nuclear RNA, U1, is believed to play a role in processing of hnRNA (Busch *et al.*, 1982). Recently, it was found that the concentration of U1 RNA in nuclei of *Xenopus laevis* oocytes declines dramatically during vitellogenesis (Zeller *et al.*, 1983), although chromosomal RNA synthesis continues at a very high rate. These results raise the possibility that mRNA precursors are not properly processed in later growth. In mouse oocytes, synthesis of an RNA of molecular weight corresponding to that of U1 RNA has been detected in mid- and late-growth phases and, as tested on Northern blots using a U1 probe, the amount remains approximately constant in growing and full-grown oocytes, and in eggs (Kaplan *et al.*, 1985).

The short interspersed repetitive sequences of mammals are dominated by the Alu or Alu-equivalent family (Schmid and Jelinek, 1982). These sequences are 150–300 nucleotides long and are present in 100,000–500,000 copies per genome. Each member is at least 80% homologous to one consensus sequence. They are transcribed as parts of large RNA polymerase II transcription units but are usually lost during processing so that mRNA contains relatively little of these sequences. Certain members of this family are also transcribed by RNA polymerase III to yield discrete low-molecular-weight RNA species, such as the 7 S RNA of mammals and the 4.5 S RNA of rodents.

*Xenopus laevis* and sea urchin eggs contain unusually high concentrations of repetitive sequence transcripts (Davidson, 1976; Costantini *et al.*, 1978). Recent studies have demonstrated that these transcripts are distributed within the polyadenylated RNA molecules stored in the egg (Costantini *et al.*, 1980; Anderson *et al.*, 1982), suggesting incomplete processing of nuclear transcripts. The possibility that the removal of repetitive sequence transcripts via cytoplasmic processing or that interaction of a small RNA homologous to the repetitive sequence element could be associated with activation of stored mRNAs for translation is intriguing. Whereas this question remains almost completely unexplored in mammals, recent evidence from our research group suggests that RNA molecules in the mouse egg are not enriched in Alu-equivalent or other repetitive sequence transcripts nor in 4.5 S RNA, in comparison with liver or brain cytoplasmic RNA. (Kaplan, G., Jelinek, W. R., and Bachvarova, R. Dev. Biol. in press.)

### 5.4. Comparison with Oocytes of the Frog *Xenopus laevis*

Gene expression and accumulation of transcripts during oocyte growth has been most thoroughly studied in the frog *Xenopus laevis*. The pattern of accumulation of RNA classes has several distinct differences in this system from the pattern outlined above for the mouse. These differences are related to the different requirements for stored components to be used in early development. In the frog system, such components must be able to meet all nutritional needs of the embryo until it is able to consume food. Thus, large numbers of yolk platelets accumulate in the oocyte during its vitellogenic phase. A vast stock of ribosomes is also present in the egg, about 10,000 times more than in mouse eggs and sufficient to support development to the early tadpole stage. To produce this amount of rRNA within a reasonable period of time, the rDNA in oocytes is amplified about 1000-fold, and the extrachromosomal nucleoli are rapidly transcribed during vitellogenesis. In amphibians, this vitellogenic phase is preceded by a phase in which the multiple 5 S and tRNA genomic sequences are rapidly transcribed and accumulated in large amounts (see Davidson, 1976).

In spite of the vast amount of rRNA synthesis, it has been possible to quantify the rate of synthesis of heterogeneous RNA in frog oocytes. In fact, hnRNA is synthesized very rapidly in vitellogeneic oocytes, about 1000 times faster than in neurula cells of the same species (Anderson and Smith, 1978). Synthesis of hnRNA and accumulation of stable cytoplasmic heterogeneous RNA is at least 50 times faster than in growing mouse oocytes.

This rate is consistent with the structural evidence indicating that many thousands of lampbrush loops are closely packed with transcribing RNA polymerase. The much higher rate of production of mRNA in *Xenopus* is apparently necessary primarily to accumulate and maintain the 1000 times larger amount of mRNA found in large oocytes and eggs.

Although quite stable mRNA is produced rapidly, accumulation of rRNA and tRNA still predominates, and mRNA represents only 1–2% of egg RNA, significantly lower than the proportion of 10% found in mouse eggs. In terms of concentration, the mouse egg is able to function with a reduced concentration of ribosomes, but the concentration of mRNA is up to four times higher than in *Xenopus* eggs.

The fraction of mRNA and ribosomes used in ongoing translation is much larger in growing mouse oocytes (25% and 33%, respectively) than in *Xenopus* oocytes, in which polysomes can hardly be detected (Dolecki and Smith, 1979). This is consistent with the much-reduced storage of ribosomes in the mouse egg.

Some similarities between the two systems may also be noted. In both the mouse and the frog, hnRNA and mRNA synthesis continue at a fairly constant rate throughout oocyte growth (Anderson and Smith, 1977, 1978; Anderson *et al.*, 1982; Hill and MacGregor, 1980). In *Xenopus*, however, polyadenylated RNA accumulates during the previtellogenic phase, and the total amount found in the egg is already present before vitellogenesis (Davidson, 1976). It appears that subsequent rapid synthesis during vitellogenesis is required to maintain this amount in the face of slow turnover of the mRNA (Anderson *et al.*, 1982). In mouse oocytes, the growth period is short relative to the half-life of the message, and the mRNA level does not reach steady state. In several respects, the oocyte growth period in the mouse most closely resembles the previtellogenic phase in amphibians, i.e., synthesis of tRNA and 5 S RNA, relatively low level of rRNA synthesis, and synthesis and accumulation of polyadenylated RNA.

In both systems, a large decrease in the amount of polyadenylated RNA and changes in the degree of polyadenylation of individual mRNAs probably occur during meiotic maturation (Sagata *et al.*, 1980; Colot and Rosbash, 1982), presumably related to the changing patterns of protein synthesis during this period.

## 6. Protein Synthesis

Several questions may be clarified by studies of rates and patterns of protein synthesis in mouse oocytes. Essentially no data are available for meiotic prophase through the primordial oocyte stage, so questions will be restricted to growing and full-grown oocytes. Does the oocyte synthesize all its accumulated proteins from its own mRNAs, or are proteins imported from an external source? What oocyte-specific proteins can be detected, and what role might they play in oocytes or embryos? When do important changes in the pattern of gene expression—as judged by two-dimensional gels—occur, and can these be related to developmental changes in the oocytes?

The work of Wassarman and his group has been recently reviewed (Wassarman *et al.*, 1981).

### 6.1. Accumulation of Total Protein and Time Course of the Rate of Protein Synthesis in Growing Oocytes

The amount of total protein in oocytes of increasing size has been measured using a sensitive method based on $^{3}$H-dansylation of the protein sample (Schultz and Wassarman, 1977b). A linear increase in protein content with volume was observed, which implies an increasing rate of accumulation with time during early and mid growth phase. This is the expected result if stable polysomes are accumulating with time (see Sections 5.1 and and 5.2), and these polysomes are the source of the stable protein. The increase continues through the full-grown oocyte stage to oocytes of 85-$\mu$m diameter, suggesting that active protein synthesis is maintained throughout. The value of 30 ng protein per oocyte of maximum size, of which 4.5 ng is in the zona pellucida, correlates well with the value of 28 ng found by Brinster (1967) for ovulated eggs plus zona pellucida using the Lowry method.

Absolute rates of protein synthesis in growing mouse oocytes have been measured directly by analyses of amino acid incorporation and specific activity of amino acid pools. Schultz *et al.* (1979) incubated naked oocytes obtained by enzyme treatment of immature ovaries in a simple salt solution, energy source, and [$^{35}$S]methionine. They measured the rate of incorporation of [$^{35}$S]methionine over 5 hr and the specific activity of the cellular pool using the [$^{3}$H]fluorodinitrobenzene technique. The rate of appearance of methionine in protein and the total rate of protein synthesis can then be computed. Alternatively, these workers measured the rate of protein synthesis by observing the uptake and incorporation at two different methionine concentrations, and solving two simultaneous equations for two unknowns—the size of the endogenous pool and the rate of protein synthesis. Both methods gave similar results, and the latter was used most extensively. As shown in Fig. 20, the rate of protein synthesis increases linearly with time during oocyte growth, and somewhat more slowly during the full-grown stage, reaching 42 pg/hr in oocytes of 85 $\mu$m diameter.

Another approach was used by Mangia and co-workers (Canipari *et al.*, 1979). Oocytes were incubated in medium containing increasing concentrations of labeled leucine and—at the concentration at which no further increase in incorporation was observed—it was assumed that the internal amino acid pool had reached the same specific activity as the external

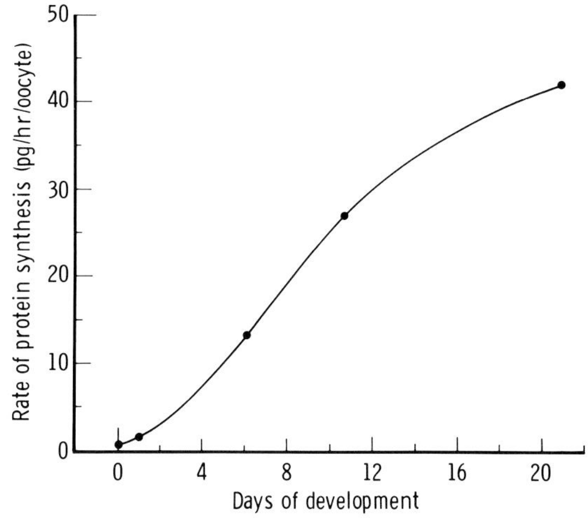


**Figure 20.** Rate of protein synthesis during growth of the mouse oocyte. Data for rate of protein synthesis in oocytes of different diameter were taken from Schultz *et al.* (1979) and converted to rate of protein synthesis on different days of development using the data of Fig. 5.

medium. As these workers point out, this method tends to underestimate the absolute rate of synthesis. The value obtained for full-grown oocytes was 48 pg/hr, similar to that of Schultz and associates, but for mid-growth phase oocytes 45 μm in diameter, their value was about twice as high.

It is difficult to know how much weight to place on the absolute values obtained, since growing oocytes maintained in simple salt solutions die within a couple of days and those maintained in complex media but without follicular cells do not grow (Bachvarova *et al.*, 1980; Brower *et al.*, 1981). Whereas oocytes may conserve relatively large pools over several hours, rapid depletion of a few crucial metabolites or a decline in general metabolic reactions in response to the very abnormal environment might result in a decreased rate of protein synthesis. It is important to realize that only the preimplantation stages of embryonic development can be traversed normally in simple salt solution plus energy source (i.e., those stages that normally develop in the oviduct and uterus). Moreover, growing oocytes require close coupling to follicular cells (see Chapter 7). Thus, Schultz and Wassarman's calculation that even if all proteins are stable, the oocyte could synthesize only 13 ng, or about one-half the protein of the egg, is not necessarily reliable.

To measure turnover of newly synthesized proteins over a longer period, mid-growth phase oocytes were labeled with [$^{35}$S]methionine for 5 hr in complex medium in the presence of follicular cells and chased for times up to 2 days (Kaplan *et al.*, 1982). The results showed that at least 40% of the labeled protein decays with a half-life of 11 hr, whereas 60% is apparently stable and contributes to accumulation of mass during growth.

From the number of ribosomes in polysomes, it can be calculated that the growing oocyte could synthesize all its protein even allowing for 40% turnover, but the range of values for the rate of peptide chain elongation

(Kafatos and Gelinas, 1974) is so wide that they cannot be used to settle the question.

Uptake of serum proteins into the oocyte has been observed (Glass, 1971), and electron microscopic observations suggest active pinocytosis at the oocyte surface (Zamboni, 1974). Whether proteins entering the oocyte are degraded to amino acids and used for resynthesis is not clear.

## 6.2. Synthesis and Accumulation of Specific Proteins in Growing Oocytes

Synthesis of the following proteins has been measured in growing mouse oocytes: lactate dehydrogenase and glucose 6-phosphate dehydrogenase (Mangia and Epstein, 1975; Mangia *et al.*, 1976; Cascio and Wassarman, 1982), creatine kinase (Iyengar *et al.*, 1983), tubulin (Schultz *et al.*, 1979), ribosomal proteins (LaMarca and Wassarman, 1979), histone H4 (Wassarman and Mrozak, 1981), zona pellucida proteins (Bleil and Wassarman, 1980a,b; Greve *et al.*, 1982), and mitochondrial proteins (Mangia and Canipari, 1977; Cascio and Wassarman, 1981). For several of these proteins, the rate of synthesis or accumulation is known to parallel that for total protein, whereas lactate dehydrogenase accumulates faster than total protein.

The concentration of some proteins is related to the unusual energy metabolism of the oocyte and early embryo (see Iyengar *et al.*, 1983 and references cited therein). Mitochondria are abundant throughout oocyte growth, presumably because they are replicating rapidly. The mouse egg contains 2–3 pg of mitochondrial DNA, which is equivalent to 100,000 mitochondria (Piko and Matsumoto, 1976), and the bovine egg contains about twice as much (Michaels *et al.*, 1982). However, at this stage, the mitochondria are metabolically quite inactive and have an unusual structure (Piko and Chase, 1973). Oxidative phosphorylation is activated during embryonic cleavage. Preceding this stage, ATP levels may be maintained partly via creatine kinase, explaining the unusual abundance of this enzyme in eggs (1.4% of total protein) (Iyengar *et al.*, 1983). The high level of lactate dehydrogenase (2–5% of total protein) may also be related to energy metabolism.

Histone H4 is synthesized in oocytes at a rate sufficient to accumulate enough histone to support two to three cell divisions. As with synthesis of rRNA, the contrast with *Xenopus* is enormous. In *Xenopus* eggs, both histone and histone mRNAs are stored in large amounts to support the very rapid cell division in the first hours of development (Davidson, 1976). The mammalian embryo has a relatively normal cell cycle of 12–20 hr during cleavage and therefore should not require a stockpile for more than perhaps one division when mRNA synthesis is low. Nevertheless, the ability to continue histone synthesis very far removed in time from bulk DNA replication is maintained in mammalian oocytes.

Synthesis of proteins coinciding on two dimensional gels with β-actin, tubulin, calmodulin, and possibly intermediate filaments has been observed, and they represent abundant spots on silver stained two dimensional gels of egg proteins (Kaplan *et al.*, 1982). All three major cytoskeletal proteins are present in eggs as tested by immunofluorescence techniques (Lehtonen and Badley, 1980).

Some abundant proteins are expected to be specific to the oocyte, such as the proteins of the cortical granules and of the fibrillar lattices in rodent eggs, but these have not yet been characterized.

The origin of the components of the zona pellucida, the acellular glycoprotein coat surrounding the egg and formed during oocyte growth, has long been controversial. Wassarman's group has characterized three major glycoproteins in isolated zonae pellucidae; they are synthesized in the growing oocyte, probably not in follicular cells, and their synthesis ceases in full-grown oocytes (Bleil and Wassarman, 1980a,b). Using immunofluorescence techniques, zona pellucida antigens are found within human and hamster growing oocytes but not in follicular cells (Bousquet *et al.*, 1981). These three proteins should be among the most rapidly synthesized in the growing oocyte, since together they account for about 15% of the total protein of the egg.

Although a large majority of individual proteins as seen on two dimensional gels of labeled cells are common to different cell types, nevertheless comparison of such patterns can define times in which major developmental changes are likely to be taking place. Comparison of the pattern of newly synthesized proteins of oocytes from 6-day, 13-day, and full-grown oocytes of mice showed a few differences among the many spots detectable (Schultz *et al.*, 1979). In another study, several new spots were identified in full-grown as opposed to growing oocytes (Kaplan *et al.*, 1982). The two-dimensional pattern in full-grown porcine oocytes from progressively larger antral follicles showed significant changes (McGaughey *et al.*, 1979). The appearance of new proteins during the full-grown oocyte stage could be dependent on activation of stored messages; the new proteins may account for the acquisition of new developmental properties as the oocytes approach meiotic maturation.

### 6.3. Protein Synthesis During Meiotic Maturation

To study metabolism during meiotic maturation, full-grown oocytes are released from antral follicles and cultured *in vitro*. Spontaneous breakdown of the germinal vesicle occurs within a few hours, and over approximately the next 12 hr oocytes progress through meiotic maturation to metaphase II at high frequency. In some cases, the *in vitro*-matured eggs can be fertilized and develop normally (Van Blerkom and McGaughey, 1978a; Eppig, 1981). The absolute rate of protein synthesis during meiotic maturation of mouse oocytes drops about 25% from 43 to 33 pg/hr (Schultz *et al.*, 1978a), which is consistent with the drop in the percent of ribosomes in polysomes.

Some of the first studies using one dimensional gels suggested that the changes in the pattern of protein synthesis were greater during meiotic maturation than during growth (Golbus and Stein, 1976; Schultz and Wassarman, 1977b; Warnes *et al.*, 1977). Two-dimensional gels of proteins newly synthesized before, during, and after meiotic maturation of oocytes from various mammals demonstrated qualitative or large quantitative differences in several to many spots (mice: Schultz and Wassarman, 1977a; Richter and McGaughey, 1981; pig: McGaughey and Van Blerkom, 1977; rabbit: van Blerkom and McGaughey, 1978a). Most of the individual proteins mentioned above continue to be synthesized as a constant fraction of total protein synthesis (Van Blerkom, 1981a), although synthesis of lactate dehydrogenase (Mangia *et al.*, 1976; Cascio and Wassarman, 1982) and actin declines (Kaplan *et al.*, 1982; Cullen *et al.*, 1980). The shift in pattern seems to be concentrated in the middle third of the maturation period, roughly coinciding with metaphase I.

Some inhibitor studies have been conducted to elucidate the immediacy of gene control over the events of meiotic maturation. Actinomycin D at 0.1–1 μg/ml arrests meiotic maturation of mouse oocytes (Bloom and Mukerjee, 1972; Golbus and Stein, 1976), but this effect is believed to be mediated by a mechanism other than inhibition of RNA synthesis (Golbus and Stein, 1976). Nevertheless, RNA synthesis in the oocyte or cumulus cells may be required for a brief period before breakdown of the germinal vesicle, since α-amanitin applied at the beginning of culture decreases the frequency of maturation in cumulus-enclosed but not denuded oocytes (Osborn and Moor, 1983). Whereas synthesis of RNA on condensed meiotic chromosomes is not absolutely excluded, the rate of incorporation falls sharply around the time of breakdown of the germinal vesicle (Rodman and Bachvarova, 1976), and other mechanisms of control of protein synthesis are more likely. Most striking is the evidence that anucleate fragments of mouse oocytes, when cultured *in vitro* along with nucleate or whole oocytes progressing through maturation, undergo many of the same changes in the two dimensional pattern of protein synthesis (Schultz *et al.*, 1978b). Thus, these changes cannot depend on liberation of translation control factors or mRNAs from the germinal vesicle, but represent a cytoplasmic system of control.

Since RNA synthesis during maturation is probably insignificant, the changes in protein synthesis suggest extensive and specific activation of stored mRNAs. The growing oocyte appears to follow one basic pattern of gene expression designed primarily to create a stored program, and changes during this growth period probably depend on changes in transcription. The utilization of the stored program may begin in the full-grown oocyte, but it is definitely set in motion during meiotic maturation. In light of the apparent changes in amount or polyadenylation of mRNA discussed above, this period appears to be particularly suitable for further studies to elucidate the mechanism of activation of the maternal program.

Several functional properties are acquired as oocytes reach full size and progress through the full-grown oocyte stage (see Section 2.2). These include the ability to undergo spontaneous or pseudo-cleavage, to undergo germinal vesicle breakdown and progression to metaphase I, and to progress from metaphase I to metaphasae II (Sorensen and Wassarman, 1976; McGaughey *et al.*, 1979). Ability to respond to sperm contact in several ways is usually acquired during meiotic maturation (Berrios and Bedford, 1979). Each of these properties presumably depends on the presence of specific proteins, the amounts and activities of which may be controlled at many levels. Some suggestive results that begin to approach these questions are available.

Inhibition of protein synthesis by puromycin (10 μg/ml) or cycloheximide 1 μg/ml) at the beginning of *in vitro* culture of full-grown oocytes arrests meiotic maturation at the circular bivalent or metaphase I stage (Golbus and Stein, 1976; Schultz and Wassarman, 1977b), suggesting that proteins required for germinal vesicle breakdown and chromosome condensation are already present. As noted above, this is one of the first capabilities acquired as oocytes reach full size. These proteins may be synthesized continuously and degraded rapidly, since inhibition of protein synthesis for a few hours before permitting resumption of meiosis does block germinal vesicle breakdown (Ekholm and Magnusson, 1979). The effects of inhibition of protein synthesis also show that those proteins required for progression from metaphase I to metaphase II are synthesized around the time of germinal vesicle breakdown. These could also be proteins with rapid turnover, or they could be synthesized from an activated preformed mRNA. The block at metaphase II in ovulated ova is maintained by active protein synthesis (Siracusa *et al.*, 1978).

The localization of newly synthesized proteins may be related to these phenomena. Some proteins synthesized up to germinal vesicle breakdown concentrate in the nucleus or on condensing chromosomes (Wassarman and Letourneau, 1976; Wassarman *et al.*, 1979; Rodman and Barth, 1979), whereas proteins synthesized after germinal vesicle breakdown are somewhat concentrated in pronuclei of the one-cell embryo (Motlik *et al.*, 1980).

## 7. Role of Accumulated Products in Embryogenesis

### 7.1. General Comparison of the Role of Maternal Components in the Mammal and Amphibian

Amphibian embryos develop in a hostile nonnutritive environment and proceed very rapidly to a stage at which they can function as an organism, that is; the stage at which they can seek and consume food. Within three days the *Xenopus* egg becomes a swimming tadpole with little change in mass. Such a transformation requires a large store of housekeeping elements, particularly to support the very rapid cleavage that occurs before gastrulation. These elements are the most abundant and best understood of the materials stored in the egg. They include components of the translation machinery, enzymes for DNA and RNA synthesis, structural components such as tubulin, and in some cases mRNAs for such proteins (see Davidson, 1976).

The egg must also be prepared to initiate processes leading to gastrulation and accompanying cell differentiation that commence relatively soon after fertilization. Transcription is required for gastrulation but does not begin until shortly before it. Thus, many of the steps in organization of the cytoplasm take place without new gene expression. The sequence of events leading from interaction of components present in the egg to the onset of differentiation is complex and not well understood (see Gerhart, 1980, for review). The point at which differential transcription becomes critical for this process is not clear. The dorsoventral axis is of particular interest in that it is established after fertilization and determines the axis of gastrulation. It first appears in the vegetal half of the embryo and is initiated by a contraction or movement of cortex relative to vegetal cytoplasm along an axis defined by a slight initial asymmetry, usually the site of fertilization. The cells arising in the vegetal half in turn influence cells in the equatorial region, inducing mesoderm, determining the site of the dorsal lip of the blastopore. Just how proteins stored in the egg or translated from maternal mRNAs could play a role in these processes remains conjectural.

The early development of mammals presents several dramatically different features, based primarily on their development within the nutritive environment of the uterus. The mammalian egg is small, particularly relative to its vertebrate relatives, having discarded most of the yolk and cytoplasmic machinery.

Development of mammals is slow, remarkably so relative to lower vertebrates. Development must be coordinated with preparation of the uterus for implantation, and the first developmental event is production of tissues that can mediate uptake of nutrients. In many cases, the trophoblast invades the uterus relatively early and establishes an organ for exchange of molecules with the maternal blood. Gastrulation occurs after a long delay at 6.5 days after fertilization in mice and at 15 days in humans, compared with 10 hr in *Xenopus* and 28 hr in the chicken.

Because the environment is nutritive, growth of the mammalian embryo can occur during embryogenesis and is a prominent and continuing feature starting even before implantation. RNA synthesis begins during early cleavage, directs the bulk of protein synthesis, and is essential for continuing development. To some extent, these early synthetic processes might be thought of as compensating for the small size of the egg, and transcription of housekeeping components is prominent in early development of mammals.

In spite of these differences, gastrulation in mammals follows the same basic plan as that of all vertebrates, very similar to that in birds and reptiles and clearly related to that in amphibians. The basic vertebrate body plan is created from a disc or sphere via extensive movements of cells and tissue layers, most prominently at the blastopore or primitive streak. In the mammal, as in other vertebrates, a mechanism must exist to create the dorso ventral and craniocaudal axes or patterns, perhaps using environmental cues to trigger selection of the actual axis. Then the pattern of components in the cytoplasm and on the cell surface must affect early transcriptional activity.

Considering the slow rate of cleavage and development in mammals, it is possible that, in addition to meeting housekeeping needs, embryonic gene expression during preimplantation development leads to the ability to form the embryonic axes, a process that clearly takes place during oocyte growth in the amphibian. There is enough time and the rate of RNA and protein synthesis is adequate for this. Inherently, it appears to be a large step in evolution to transfer such an important function from the maternal to the embryonic genome. The possibility that the early embryonic program in lower animals is a continuation and reinforcement of the maternal program (Hough-Evans *et al.*, 1977; Bravo and Knowland, 1979) makes the evolutionary transition appear more feasible.

The essentially new mechanism evolved for formation of the trophoblast could also be accommodated within the first few days of development. The tendency to form trophoblast by outer cells of the morula appears to arise from a differentiation of the cytoplasm along an outer–inner axis of polarity within each cell of the eight-cell stage (Johnson, 1981), a differentiation that could depend on preceding and accompanying RNA and protein synthesis.

## 7.2. Fate of Maternal Components

The available information on the changes in total and polyadenylated maternal RNA during meiotic maturation and early development of the mouse embryo is presented in Fig. 21. The events of meiotic maturation are associated with specific activation of maternal products, probably including maternal mRNAs. Approximately one-half the stored polyadenylated RNA is degraded or deadenylated during this period, but the amount of mRNA actually used for translation and/or totally degraded is not yet clear. A significant but not yet accurately quantified decline in total RNA occurs during this period.

Looking forward from fertilization, the surprising result is that 30–40% of the bulk RNA of the mouse egg is degraded by the late two-cell stage or about 36 hr of development (Olds *et al.*, 1973; Bachvarova and De Leon, 1980; Piko and Clegg, 1982). Volume of the embryo declines 28% by the four- to eight-cell stage (Lewis and Wright, 1935; Abramczuk and Sawicki,

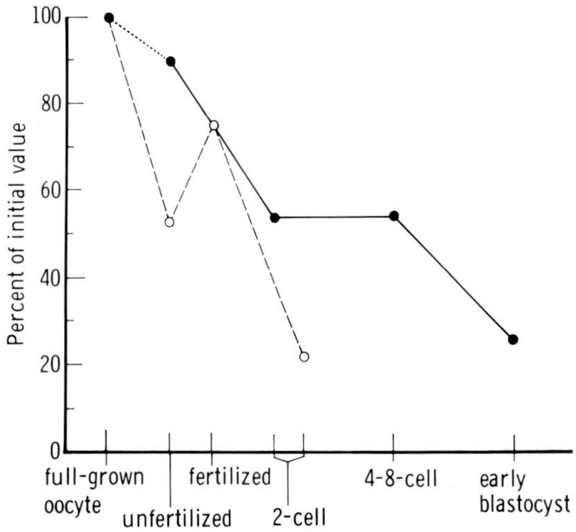


**Figure 21.** Available information on changes in maternal RNA content of oocytes during meiotic maturation and in early mouse embryos. Stages along the abscissa are spaced to reflect approximate times from HCG (stimulus for meiotic maturation). Solid line: total maternal RNA redrawn from Bachvarova and De Leon (1980). The decline during meiotic maturation has not yet been quantified accurately. Dashed line: polyadenylated RNA content taken from data in De Leon *et al.* (1983) for the decline during meiotic maturation, and from Clegg and Piko (1983a), who estimated total polyadenylated RNA in early embryos possibly including some newly synthesized polyadenylated RNA by the two-cell stage. (Data used with permission from Dr. L. Piko.)

1974) and protein content 26% by the morula (Brinster, 1967). Thus, a significant portion of the macromolecules and cytoplasmic volume accumulated during growth appear to have served their function by the time of early cleavage.

Some of the degraded molecules may provide precursors for later synthesis (yolklike function), but the decrease in volume and dissociation in time between degradation and net synthesis suggests this is not the primary role of these molecules. Rather, they appear to be associated with oocyte, fertilization, and one-cell functions and are no longer needed. If such a set of components are present in frog eggs, they may be too small a fraction to detect their loss.

Looking more closely at the one- and two-cell stages, we find that the number of polyadenylated mRNAs rises by 40% during the one-cell stage due to adenylation of preexisting mRNAs, and then falls from this value by 70% to the late two-cell stage (Fig. 21) (Clegg and Piko, 1983a; see also Levey *et al.*, 1978). Production of new mRNA proceeds at a very low level during the one-cell stage (Clegg and Piko, 1982) and becomes significant during the two-cell stage (Clegg and Piko, 1983b). Ribosomal RNA synthesis starts at the two-cell stage and produces a net increase by the eight-cell stage (Piko and Clegg, 1982).

The rate of protein synthesis remains approximately constant to the eight-cell stage. The pattern of protein synthesis shows relatively few changes at fertilization and up to the early two-cell stage (Chen *et al.*, 1980; Cullen *et al.*, 1980; Flach *et al.*, 1982). During the two-cell stage, the pattern changes dramatically (Flach *et al.*, 1982). The differences between the one-cell stage and the morula are such that "it appears virtually none of the major proteins seen in one stage are observable in the other" (Cullen *et al.*, 1980). The most dramatic changes in the pattern of protein synthesis in rabbit embryos occur during the corresponding stage (Van Blerkom and McGaughey, 1978b). It might be worth examining how many of the newly appearing proteins coincide with those synthesized in oocytes before maturation as in the frog (Bravo and Knowland, 1979). For example, the synthesis of actin declines during meiotic maturation (see above) and increases during the two-cell stage (Cullen *et al.*, 1980).

To determine the time at which the transcription of the embryonic genome becomes crucial for development, the effects of $\alpha$-amanitin, a drug that preferentially inhibits RNA synthesis by RNA polymerase II, have been studied. Embryos treated with $\alpha$-amanitin continue to develop to the two-cell stage, and during this period the amount or pattern of protein synthesis is hardly affected (Johnson, 1981). One of the most obvious changes during the one-cell stage is the appearance and increased prominence of a set of 35K proteins. Not only is this increase $\alpha$-amanitin insensitive, but *in vitro* translation shows that the mRNA is present but not translated in unfertilized eggs (Braude *et al.*, 1979) and accumulates during oocyte growth (Cascio and Wassarman, 1982). It is interesting to note that the translational activation occurs independent of fertilization, appearing with time in ovulated unfertilized eggs of mice (Van Blerkom, 1981b) and rabbits (Van Blerkom, 1979). Note also that increased translation of HPRT from maternal mRNA occurs during this stage (Monk and Harper, 1978).

In contrast to its lack of effect during the one-cell stage, $\alpha$-amanitin treatment during the two-cell stage inhibits protein synthesis by 75% and blocks the shift to a new pattern of protein synthesis almost completely (Flach *et al.*, 1982). Moreover, expression of a paternal gene has been detected at the two-cell stage (Sawicki *et al.*, 1981).

Taken together, these results are very consistent in providing an outline of maternal and embryonic gene control in early development. Translation of maternal stored mRNA, including activation of specific messages, supports protein synthesis during the one-cell stage and provides proteins necessary for the first cleavage. Note that the mammalian and amphibian embryo both support protein synthesis on maternal mRNAs for ½–1 day. In the mouse, maternal mRNAs are apparently degraded rather suddenly during the two-cell-stage to less than 30% of their maximum level, and it is not clear whether the large amount of maternal mRNA is actually used in translation before degradation. During the two-cell stage, maternal mRNAs are largely replaced on polysomes by newly synthesized messages, and a very different pattern of protein synthesis ensues. Thus, a new embryonic program of gene expression is set in motion in time to support developmental events at the eight-cell stage and later.

When bulk maternal RNA labeled during oocyte growth was followed to later embryonic stages, it was found that after the period of degradation during the two-cell stage, there is a second phase of breakdown during formation of the blastocyst, leaving 30% of the original egg RNA (Bachvarova and De Leon, 1980). Autoradiographic studies (Bachvarova and Moy, 1985) indicate that the loss of bulk maternal RNA proceeds at the same rate in the inner cell mass and the differentiating trophoblast of the blastocyst. Some maternal polyadenylated RNA remains in the blastocyst, but considering the amount of new mRNA and rRNA, both maternal components must be greatly diluted. A small amount of maternal mRNA for key regulatory proteins could still have an important function.

From the above results, it appears that important new transcriptional patterns are initiated at the two-cell stage, perhaps specifically programming subsequent differentiation. It should be mentioned briefly that the early embryonic program of transcription may be unusually simple and/or delayed in its impact: The early embryonic genome is more methylated than in differentiated tissues (Manes and Menzel, 1981; but see Singer *et al.*, 1979); DNA tumor viruses and retroviruses are not expressed in early embryonic or undifferentiated teratocarcinoma cells as they are in later differentiated cells (Linnenbach *et al.*, 1980; Jahner *et al.*, 1982); early embryos with monosomies (Magnuson and Epstein, 1981) or subjected to radiation (Spindle *et al.*, 1982) or BUdR treatment (Pollard *et al.*, 1976) can form blastocysts, but further development, particularly that of the inner cell mass, tends to be impaired. Much remains to be learned about the process of transcription and its control in early embryos.

## 7.3. Role of Oocyte Growth and Accumulation of Products in Mammals

The mammalian embryo loses a significant fraction of its total mass within 2–3 days of fertilization and is capable of supporting its own growth thereafter. Thus, the contribution of the period of oocyte growth to embryonic development appears to be minimized. Even specific products needed for differentiation could be amplified or newly created after fertilization. In fact, it is possible that the primordial oocyte before growth has essentially the same potentials as the full-grown oocyte and growth involves only amplification of the program.

One possible route to test the differentiative potential of oocytes before and after growth is to compare teratomas arising at different stages of germ cell development. Spontaneous teratomas of female mice and humans arise from full-grown oocytes in the ovary. In male mice, they can arise from early germ cells in the 12-day fetus (Stevens, 1975) and in human males, probably from spermatogonia or spermatocytes. The range of tissues produced by all these teratomas is very wide and not consistently different for tumors of different origin, suggesting that the program for expression of totipotency is available at times in the germ cell lineage other than the full-grown oocytes. Unfortunately, the potential for organized development (normal gastrulation) has not been compared among these different teratomas under optimal conditions using normal embryos as controls.

Minimum functions that must be carried out during oocyte growth and development to maturation are synthesis of products needed for meiotic maturation and fertilization. Additional possible roles are the creation of a cell of a volume large enought to fit the physical and mechanical requirements of ovulation, fertilization, and movement down the oviduct. Also, we may suppose that the program that maintains the totipotent character of the cell continues and is perhaps amplified during oocyte growth. Future work will shed light on the nature of this program, its specific expression during oocyte growth, and its relationship to early embryonic development.

## 8. Summary

Mouse oocytes progress through early meiotic prophase during fetal life and reach the diplotene stage by birth. During prepubertal and reproductive life, oocytes are continuously selected to grow from the pool of small primordial oocytes. Growing oocytes reach full size in 2 weeks, and full-grown oocytes are present in rapidly enlarging follicles for about 5 days before meiotic maturation and ovulation.

RNA synthesis during early meiotic prophase, as estimated from [$^3$H]uridine incorporation followed by autoradiography and from electron microscopic analysis of nuclear components, proceeds at a moderate rate throughout except for a brief period in early pachytene when synthesis is low or absent.

RNA synthesis continues in primordial oocytes at a moderate rate. Incorporation studies, electron microscopic analyses, and particularly measurements of ongoing RNA polymerase activity (completion of initiated chains as analysed in tissue sections) indicate a distinctly increased rate of synthesis during oocyte growth over that of primordial oocytes, followed by a decline in full-grown oocytes. During growth, this rate increases severalfold. The absolute rate of synthesis of heterogeneous nuclear RNA (using rRNA as a standard) during mid-growth is very rapid, but nevertheless still much lower than that in typical lampbrush chromosomes. Most of the hnRNA turns over with a half-life of about 20 min, as is typical in somatic cells. Newly synthesized mRNA-like RNA enters the cytoplasm at about one-half the rate of rRNA, and about one-third of the ribosomes and one-fourth of the mRNA appear in polysomes. In full-grown oocytes, the rate of synthesis falls distinctly, but a significant level of synthesis continues until it essentially ceases at breakdown of the germinal vesicle.

During meiotic prophase, chromosomes are most compact at pachytene and unfold lateral projections as RNA synthesis increases in late pachytene–early diplotene. In primordial oocytes, the diplotene state of chromosomes is obvious in most mammals, but in rodents the chromosomes are more evenly dispersed and are said to be in a dictyate state, although they are still presumably in a diplotene configuration. The chromosome core, which is present in leptotene through early diplotene stages, apparently disappears in the dictyate stage. During oocyte growth, the chromosomes extend, but sections and chromosome spreads still display fuzzy threads, suggesting that the chromsomal axis remains partially condensed throughout meiotic prophase. The structure of mammalian oocyte chromosomes as seen in the light microscope does not have those features dependent on the very rapid transcription of typical lampbrush chromosomes but could have the same basic organization—particularly in growing oocytes.

The ribosomal RNA genes of primate oocytes are amplified severalfold. It is not known whether the extra genes play a role during oocyte growth. Ribosomal DNA is not amplified detectably in mouse oocytes, and such amplification is not necessary to account for the rate of synthesis of rRNA observed.

Both X chromosomes are active in early mouse embryos, and the second X is inactivated in most cells shortly before gastrulation. In female germ cells, the second X is inactivated between 6 and 10 days of development and is reactivated at about 13 days when the oocytes enter prophase of meiosis. Both X chromosomes are active throughout further oocyte development. The absence of a second X chromosome has a much greater deleterious effect on oocyte viability in humans than in mice.

Total RNA and rRNA accumulate steadily and at a somewhat increasing rate during growth of mouse oocytes; rRNA and tRNA are stable during growth. Poly(A) also accumulates steadily, and polyadenylated RNA is unusually stable with only some slow turnover of mRNA in the translated fraction. The amount of mRNA accumulated, as calculated from its rate of synthesis and stability, is similar to the amount found in the full-grown oocyte, to date estimated by only one method at about 90 pg, or 20% of total RNA.

Compared with frog oocytes, rRNA is synthesized at an extremely low rate, and heterogeneous RNA and mRNA at a low rate. In both cases, the mRNA is unusually stable. The concentration of ribosomes in the mouse egg is three times lower and of mRNA four times higher than in the frog egg, whereas the amount of rRNA accumulated is 10,000 times less and the amount of mRNA is 1000 times less.

The rate of protein synthesis increases steadily throughout oocyte growth, but it is not known whether it is sufficient to account for all the protein accumulated by the oocyte. Studies of relative rates of synthesis of a few individual proteins have been carried out. A few changes in the two-dimensional pattern of newly synthesized protein occur during oocyte growth and during the full-grown oocyte stage when various developmental capacities are acquired. During meiotic maturation several to many changes in the two dimensional pattern appear, which are presumably programmed on maternal mRNA.

During meiotic maturation, a quite small but not accurately quantified amount of total RNA is degraded. Approximately one-half the polyadenylated RNA disappears, due to either deadenylation or degradation. The amount of polyadenylated RNA in the mature egg is variously estimated at 30–45 pg per egg. After fertilization, 30–40% of the total RNA is degraded during development to the two-cell stage. Polyadenylated maternal RNA increases somewhat to the late one-cell stage and then falls to 40% or less of its original value by the late two-cell stage. During the two-cell stage, the spectrum of proteins synthesized changes greatly, accompanying a shift from a program directed by maternal mRNA to a program directed mainly by newly synthesized embryonic RNA. It is suggested that whereas early differentiation may be a function of the distribution of maternal macromolecules in the frog embryo, molecules with similar function may be synthesized after fertilization in the mammal.

## References


Abramczuk, J., and Sawicki, W., 1974, Variation in dry mass and volume of nonfertilized oocytes and blastomeres of 1-, 2- and 4-celled mouse embryos, *J. Exp. Zool.* **188:**25–34.

Alfert, M., 1950, A cytochemical study of oogenesis and cleavage in the mouse, *J. Cell. Comp. Physiol.* **36:**381–409.

Altman, P. L., and Dittmer, D. S. (eds.), 1972, *Biology Data Book*, 2nd ed., Vol. 1, p. 378, Federation of American Societies for Experimental Biology, Bethesda, Maryland.

Anderson, D. M., and Smith, L. D., 1977, Synthesis of heterogeneous nuclear RNA in full-grown oocytes of *Xenopus laevis* (Daudin), *Cell* **11:**663–671.

Anderson, D. M., and Smith, L. D., 1978, Patterns of synthesis and accumulation of heterogeneous RNA in lampbrush stage oocytes of *Xenopus laevis* (Daudin), *Dev. Biol.* **67:**274–285.

Anderson, D. M., Richter, J. D., Chamberlin, M. E., Price, D. H., Britten, R. J., Smith, L. D., and Davidson, E. H., 1982, Sequence organization of the poly(A) RNA synthesized and accumulated in lampbrush chromosome stage *Xenopus laevis* oocytes, *J. Mol. Biol.* **155:**281–309.

Austin, C. R., 1961, *The Mammalian Egg*, Blackwell Science Publications, Oxford.

Bachvarova, R., 1974, Incorporation of tritiated adenosine into mouse ovum RNA, *Dev. Biol.* **40:**52–58.

Bachvarova, R., 1981, Synthesis, turnover, and stability of heterogeneous RNA in growing mouse oocytes, *Dev. Biol.* **86:**384–392.

Bachvarova, R., and De Leon, V., 1977, Stored and polysomal ribosomes of mouse ova, *Dev. Biol.* **58:**248–254.

Bachvarova, R., and De Leon, V., 1980, Polyadenylated RNA of mouse ova and loss of maternal RNA in early development, *Dev. Biol.* **74:**1–8.

Bachvarova, R., and Moy, K., 1985, Autoradiographic studies on the distribution of labeled maternal RNA in early mouse embryos, *J. Exp. Zool.* (in press).

Bachvarova, R., Baran, M. M., and Tejblum, A., 1980, Development of naked growing mouse oocytes *in vitro*, *J. Exp. Zool.* **211:**159–169.

Bachvarova, R., De Leon, V., Johnson, A., Kaplan, G., and Paynton, B. V., 1985, Changes in total RNA, polyadenylated RNA, and actin mRNA during meiotic maturation of mouse oocytes, *Dev. Biol.* (in press).

Bachvarova, R., De Leon, V., and Spiegelman, I., 1981, Mouse egg ribosomes: Evidence for storage in lattices, *J. Embryol. Exp. Morphol.* **62:**153–164.

Bachvarova, R., Burns, J. P., Spiegelman, I., Choy, J., and Chaganti, R. S. K., 1982, Morphology and transcriptional activity of mouse oocyte chromosomes, *Chromosoma* **86**:181–196.

Baker, T. G., 1971, Comparative aspects of the effects of radiation during oogenesis, *Mutat. Res.* **11**:9–22.

Baker, T. G., 1972, Oogenesis and ovarian development, in: *Reproductive Biology* (H. Balin and S. R. Glasser, eds.), pp. 398–437, Excerpta Medica, Amsterdam.

Baker, T. G., and Franchi, L. L., 1967a, The fine structure of oogonia and oocytes in human ovaries, *J. Cell Sci.* **2**:213–224.

Baker, T. G., and Franchi, L. L., 1967b, The structure of the chromosomes in human primordial oocytes, *Chromosoma* **22**:358–377.

Baker, T. G., and Franchi, L. L., 1967c, The fine structure of chromosomes in bovine primordial oocytes, *J. Reprod. Fertil.* **14**:511–513.

Baker, T. G., and Franchi, L. L., 1972, The fine structure of oogonia and oocytes in the Rhesus monkey (*Macaca mulatta*), *Z. Zellforsch. Mikrosk. Anat.* **126**:53–74.

Baker, T. G., Beaumont, H. M., and Franchi, L. L., 1969, The uptake of tritiated uridine and phenylalanine by the ovaries of rats and monkeys, *J. Cell Sci.* **4**:655–675.

Bakken, A. H., and Hamkalo, B. A., 1978, Techniques for visualization of genetic material, in: *Principles and Techniques of Electron Microscopy*, Vol. 9 (M. A. Hayat, ed.), pp. 84–106, Reinhold, New York.

Bakken, A. H., and McClanahan, M., 1978, Patterns of RNA synthesis in early meiotic prophase oocytes from fetal mouse ovaries, *Chromosoma* **67**:21–40.

Berrios, M., and Bedford, J. M., 1979, Oocyte maturation: Aberrant post-fusion responses of the rabbit primary oocyte to penetrating spermatozoa, *J. Cell Sci.* **39**:1–12.

Bleil, J. D., and Wassarman, P. M., 1980a, Synthesis of zona pellucida proteins by denuded and follicle-enclosed mouse oocytes during culture *in vitro*, *Proc. Natl. Acad. Sci. USA* **77**:1029–1033.

Bleil, J. D., and Wassarman, P. M., 1980b, Structure and function of the zona pellucida: Identification and characterization of the proteins of the mouse oocyte's zona pellucida, *Dev. Biol.* **76**:185–202.

Bloom, A. M., and Mukherjee, B. B., 1972, RNA synthesis in maturing mouse oocytes, *Exp. Cell Res.* **74**:577–582.

Boreen, S. M., Gizang, E., and Schultz, R. M., 1983, Biochemical studies of mammalian oogenesis: Synthesis of 5S and 4S RNA during growth of the mouse oocyte, *Gamete Res.* **8**:379–383.

Bousquet, D., Leveille, M. C., Roberts, K. D., Chapdelaine, A., and Bleau, G., 1981, The cellular origin of the zona pellucida antigen in the human and hamster, *J. Exp. Zool.* **215**:215–218.

Brambell, F. W. R., 1927, The development and morphology of the gonads of the mouse. Part I. The morphogenesis of the indifferent gonad and of the ovary, *Proc. R. Soc. Lond., Ser. B* **101**:391–409.

Brambell, F. W. R., 1928, The development and morphology of the gonads of the mouse. Part III. The growth of the follicles, *Proc. R. Soc. Lond., Ser. B* **103**:258–272.

Brambell, F. W. R., 1956, Ovarian changes, in: *Marshall's Physiology of Reproduction* (A. S. Parkes, ed.), pp. 397–542, Longmans, Green and Co., New York.

Brandhorst, B. P., and McConkey, E. H., 1974, Stability of nuclear RNA in mammalian cells, *J. Mol. Biol.* **85**:451–463.

Braude, P., Pelham, H., Flach, G., and Lobatto, R., 1979, Post-transcriptional control in the early mouse embryo, *Nature* **282**:102–105.

Bravo, R., and Knowland, J., 1979, Classes of proteins synthesized in oocytes, eggs, embryos and differentiated tissues of *Xenopus laevis*, *Differentiation (Berl.)* **13**:101–108.

Brinster, R. L., 1967, Protein content of the mouse embryo during the first five days of development, *J. Reprod. Fertil.* **13**:413–420.

Brinster, R. L., Chen, H. Y., and Trumbauer, M. E., 1981, Mouse oocytes transcribe injected Xenopus 5S RNA gene, *Science (Wash., D.C.)* **211**:396–398.

Brower, P. T., Gizang, E., Boreen, S., and Schultz, R. M., 1981, Biochemical studies of mammalian oogenesis: Synthesis and stability of various classes of RNA during growth of the mouse oocyte *in vitro*, *Dev. Biol.* **86**:373–383.

Burgoyne, P. S., and Baker, T. G., 1981, Oocyte depletion in XO mice and their XX siblings from 12 to 200 days postpartum, *J. Reprod. Fertil.* **61**:207–212.

Burgoyne, P. S., and Biggers, J. D., 1976, The consequences of X-dosage deficiency in the germ line: Impaired development *in vitro* of preimplantation embryos from XO mice, *Dev. Biol.* **51**:109–117.

Burkholder, G. D., Comings, D. E., and Okada, T. A., 1971, A storage form of ribosomes in mouse oocytes, *Exp. Cell Res.* **69**:361–371.

Busch, H., Reddy, R., Rothblum, L., and Choi, Y. C., 1982, SnRNAs, SnRNPs, and RNA processing, *Annu. Rev. Biochem.* **51**:617–654.

Byskov, A. G., 1978, Follicular atresia, in: *The Vertebrate Ovary* (R. E. Jones, ed.), pp. 533–562, Plenum Press, New York.

Callan, H. G., 1957, The lampbrush chromosomes of *Sepia officinalis L.*, *Anilocra physodes L.* and *Scyllium catulus Cuv.* and their structural relationship to the lampbrush chromosomes of amphibia, *Pubbl. Staz. Zool. Napoli* **29**:328–345.

Canipari, R., Pietrolucci, A., and Mangia, F., 1979, Increase of total protein synthesis during mouse oocyte growth, *J. Reprod. Fertil.* **57**:405–413.

Carr, D. H., Haggar, R. A., and Hart, A. G., 1968, Germ cells in the ovaries of XO female infants, *Am. J. Clin. Pathol.* **49**:521–526.

Cascio, S. M., and Wassarman, P. M., 1981, Program of early development in the mammal: Synthesis of mitochondrial proteins during oogenesis and early embryogenesis in the mouse, *Dev. Biol.* **83**:166–172.

Cascio, S. M., and Wassarman, P. M., 1982, Program of early development in the mammal: Post-transcriptional control of a class of proteins synthesized by mouse oocytes and early embryos, *Dev. Biol.* **89**:397–408.

Chen, H. Y., Brinster, R. L., and Merz, E. A., 1980, Changes in protein synthesis following fertilization of the mouse ovum, *J. Exp. Zool.* **212**:355–360.

Chouinard, L. A., 1971, A light- and electron-microscope study of the nucleolus during growth of the oocyte in the prepubertal mouse, *J. Cell Sci.* **9**:637–663.

Chouinard, L. A., 1973, An electron-microscope study of the extranucleolar bodies during growth of the oocyte in the prepubertal mouse, *J. Cell. Sci.* **12**:55–69.

Chouinard, L. A., 1975, A light- and electron-microscope study of the oocyte nucleus during development of the antral follicle in the prepubertal mouse, *J. Cell Sci.* **17**:589–615.

Clegg, K. B., and Piko, L., 1982, RNA synthesis and cytoplasmic polyadenylation in the one-cell mouse embryo, *Nature* **295**:342–345.

Clegg, K. B., and Piko, L., 1983a, Poly(A) length, cytoplasmic adenylation and synthesis of poly$(A)^+$ RNA in early mouse embryos, *Dev. Biol.* **95**:331–341.

Clegg, K. B., and Piko, L., 1983b, Quantitative aspects of RNA synthesis and polyadenylation in 1-cell and 2-cell mouse embryos, *J. Embryol. Exp. Morphol.* **74**:169–182.

Colot, H. V., and Rosbash, M., 1982, Behavior of individual maternal $pA^+$ RNAs during embryogenesis of *Xenopus laevis*, *Dev. Biol.* **94**:79–86.

Comings, D. E., and Okada, T. A., 1972, Architecture of meiotic cells and mechanisms of chromosome pairing, *Adv. Cell Mol. Biol.* **2**:310–384.

Costantini, F. D., Scheller, R. H., Britten, R. J., and Davidson, E. H., 1978, Repetitive sequence transcripts in the mature sea urchin oocyte, *Cell* **15**:173–187.

Costantini, F. D., Britten, R. J., and Davidson, E. H., 1980, Message sequences and short repetitive sequences are interspersed in sea urchin egg poly$(A)^+$ RNAs, *Nature* **287**:111–117.

Cullen, B. R., Emigholz, K., and Monahan, J. J., 1980, Protein patterns of early mouse embryos during development, *Differentiation (Berl.)* **17**:151–160.

Davidson, E. H., 1976, *Gene Activity in Early Development*, Academic Press, New York.

De Felici, M., and McLaren, A., 1982, Isolation of mouse primordial germ cells, *Exp. Cell Res.* **142**:476–482.

De Leon, V., Johnson, A., and Bachvarova, R., 1983, Half-lives and relative amounts of stored and polysomal ribosomes and poly(A)+ RNA in mouse oocytes, *Dev. Biol.* **98**:400–408.

Dolecki, G. J., and Smith, L. D., 1979, Poly(A)+ RNA metabolism during oogenesis in *Xenopus laevis*, *Dev. Biol.* **69**:217–236.

Donahue, R. P., 1968, Maturation of the mouse oocyte *in vitro*. I. Sequence and timing of nuclear progression, *J. Exp. Zool.* **169**:237–250.

Dunbar, B. S., Wardrip, N. J., and Hedrick, J. L., 1980, Isolation, physicochemical properties, and macromolecular composition of zona pellucida from porcine oocytes, *Biochemistry* **19**:356–365.

Dvorak, M., Travnik, P., and Stankova, J., 1977, A quantitative analysis of the incidence of certain cytoplasmic structures in the ovum of the rat during cleavage, *Cell Tissue Res.* **179**:429–443.

Eddy, E. M., Clark, J. M., Gong, D., and Fenderson, B. A., 1981, Origin and migration of primordial germ cells in mammals, *Gamete Res.* **4**:333–362.

Edwards, R. G., 1970, Are oocytes formed and used sequentially in the mammalian ovary? *Philos. Trans. R. Soc. Lond. Ser. B* **259**:103–105.

Ekholm, C., and Magnusson, C., 1979, Rat oocyte maturation: Effects of protein synthesis inhibitors, *Biol. Reprod.* **21**:1287–1293.

Enders, A. C., 1971, The fine structure of the blastocyst, in: *The Biology of the Blastocyst* (R. J. Blandau, ed.), pp. 71–94, University of Chicago Press, Chicago.

Eppig, J. J., 1976, Analysis of mouse oogenesis *in vitro*. Oocyte isolation and the utilization of exogenous energy sources by growing oocytes, *J. Exp. Zool.* **198**:375–382.

Eppig, J. J., 1977, Mouse oocyte development *in vitro* with various culture systems, *Dev. Biol.* **60**:371–388.

Eppig, J. J., 1979, A comparison between oocyte growth in co-culture with granulosa cells and oocytes with granulosa cell-oocyte junctional contact maintained *in vitro*, *J. Exp. Zool.* **209**:345–353.

Eppig, J. J., 1981, Preimplantation embryonic development of spontaneous mouse parthenotes after oocyte meiotic maturation *in vitro*, *Gamete Res.* **4**:3–13.

Epstein, C. J., 1972, Expression of the mammalian X chromosome before and after fertilization, *Science (Wash. D.C.)* **175**:1467–1468.

Flach, G., Johnson, M. H., Braude, P. R., Taylor, R. A. S., and Bolton, V. N., 1982, The transition from maternal to embryonic control in the 2-cell mouse embryo, *EMBO J.* **1**:681–686.

Flax, M. H., 1951, Pentose nucleic acids and proteins during oogenesis and early development of the mouse, *Anat. Rec.* **111**:465.

Fourcroy, J. L., 1982, RNA synthesis in immature mouse oocyte development, *J. Exp. Zool.* **219**:257–266.

Franchi, L. L., and Mandl, A. M., 1962, The ultrastructure of oogonia and oocytes in the foetal and neonatal rat, *Proc. R. Soc. Lond., Ser. B* **157**:99–114.

Garcia, R. B., Pereyra-Alfonso, S., and Sotelo, J. R., 1979, Protein-synthesizing machinery in the growing oocyte of the cyclic mouse, *Differentiation (Berl.)* **14**:101–106.

Gardner, R. L., 1978, Developmental potency of normal and neoplastic cells of the early mouse embryo, in: *Birth Defects* (J. W. Littlefield and J. de Grouchy, eds., F. J. G. Ebling, co-ed.), Excerpta Medica International Congress Series 432, pp. 154–166, Excerpta Medica, Amsterdam.

Gartler, S. M., Rivest, M., and Cole, R. E., 1980, Cytological evidence for an inactive X chromosome in murine oogonia, *Cytogenet. Cell Genet.* **28**:203–207.

Gates, A. H., 1971, Maximizing yield and developmental uniformity of eggs, in: *Methods in Mammalian Embryology* (J. C. Daniel, ed.), pp. 64–75, W. H. Freeman, San Francisco.

Gerhart, J. C., 1980, Mechanisms regulating pattern formation in the amphibian egg and early embryo, in: *Biological Regulation and Development*, vol. 2 (R. F. Goldberger, ed.), pp. 133–316, Plenum Press, New York.

Glass, L. E., 1971, Transmission of maternal proteins into oocytes, *Adv. Biosci.* **6**:29–58.

Golbus, M. S., and Stein, M. P., 1976, Qualitative patterns of protein synthesis in the mouse oocyte, *J. Exp. Zool.* **198**:337–342.

Gold, B., Stern, L., Bradley, F. M., and Hecht, N. B., 1983, Gene expression during mammalian spermatogenesis: II. Evidence for stage-specific differences in mRNA populations, *J. Exp. Zool.* **225**:123–134.

Greve, J. M., Salzmann, G. S., Roller, R. J., and Wassarman, P. M., 1982, Biosynthesis of the major zona pellucida glycoprotein secreted by oocytes during mammalian oogenesis, *Cell* **31**:749–759.

Habibi, B., and Franchi, L. L., 1978, Fine-structural changes in the nucleus of primordial oocytes in immature hamsters, *J. Cell Sci.* **34**:209–223.

Hartung, M., and Stahl, A., 1978, Autoradiographic study of RNA synthesis during meiotic prophase in the human oocyte, *Cytogenet. Cell Genet.* **20**:51–58.

Hartung, M., Mirre, C., and Stahl, A., 1979, Nucleolar organizers in human oocytes at meiotic prophase I, studied by the silver-NOR method and electron microscopy, *Hum. Genet.* **52**:295–308.

Heller, D. T., Cahill, D. M., and Schultz, R. M., 1981, Biochemical studies of mammalian oogenesis: Metabolic cooperativity between granulosa cells and growing mouse oocytes, *Dev. Biol.* **84**:455–464.

Hill, R. S., 1979, A quantitative electron-microscope analysis of chromatin from *Xenopus laevis* lampbrush chromosomes, *J. Cell Sci.* **40**:145–169.

Hill, R. S., and MacGregor, H. C., 1980, The development of lampbrush chromosome-type transcription in the early diplotene oocytes of *Xenopus laevis*: An electron-microscope analysis, *J. Cell Sci.* **44**:87–101.

Hough-Evans, B. R., Wold, B. J., Ernst, S. G., Britten, R. J., and Davidson, E. H., 1977, Appearance and persistence of maternal RNA sequences in sea urchin development, *Dev. Biol.* **60**:258–277.

Iwamatsu, T., and Yanagimachi, R., 1975, Maturation *in vitro* of ovarian oocytes of prepubertal and adult hamsters, *J. Reprod. Fertil.* **45**:83–90.

Iyengar, M. R., Iyengar, C. W. L., Chen, H. Y., Brinster, R. L., Bornslaeger, E., and Schultz, R. M., 1983, Expression of creatine kinase isoenzyme during oogenesis and embryogenesis in the mouse, *Dev. Biol.* **96**:263–268.

Jahn, C. L., Baran, M. M., and Bachvarova, R., 1976, Stability of RNA synthesized by the mouse oocyte during its major growth phase, *J. Exp. Zool.* **197**:161–172.

Jahner, D., Stuhlmann, H., Stewart, C. L., Harbers, K., Lohler, J., Simon, I., and Jaenisch, R., 1982, De novo methylation and expression of retroviral genomes during mouse embryogenesis, *Nature* **298**:623–628.

Jamrich, M., Warrior, R., Steele, R., and Gall, J. R., 1983, Transcription of repetitive sequences on *Xenopus* lampbrush chromosomes, *Proc. Natl. Acad. Sci. USA* **80**:3364–3367.

Johnson, M. H., 1981, The molecular and cellular basis of preimplantation mouse development, *Biol. Rev. Cambridge Philos. Soc.* **56**:463–498.

Jones, R. E. (ed.), 1978, *The Vertebrate Ovary*, Plenum Press, New York.

Kafatos, F. C., and Gelinas, R., 1974, mRNA stability and the control of specific protein synthesis in highly differentiated cells, in: *Biochemistry of Cell Differentiation* (J. Paul, ed.), MTP International Review of Science, Vol. 9, pp. 223–266, University Park Press, Baltimore.

Kaplan, G., Abreu, S. L., and Bachvarova, R., 1982, rRNA accumulation and protein synthetic patterns in growing mouse oocytes, *J. Exp. Zool.* **220**:361–370.

Kaplan, G., Jelinek, W. R., and Bachvarova, R., 1985, Repetitive sequence transcripts and Ul RNA in mouse oocytes and eggs, *Dev. Biol.* (in press).

Kierszenbaum, A. L., and Tres, L. L., 1974a, Nucleolar and perichromosomal RNA synthesis during meiotic prophase in the mouse testis, *J. Cell Biol.* **60**:39–53.

Kierszenbaum, A. L., and Tres, L. L., 1974b, Transcription sites in spread meiotic prophase chromsomes from mouse spermatocytes, *J. Cell Biol.* **63**:923–935.

Kingery, H. M., 1914, So-called parthenogenesis in the white mouse, *Biol. Bull.* **27**:240–251.

Klasterska, I., Natarajan, A. T., and Ramel, C., 1976, New observations on mammalian male meiosis, I. Laboratory mouse (*Mus musculus*) and Rhesus monkey (*Macaca mulatta*), *Hereditas* **83**:203–214.

Kratzer, P. G., and Chapman, V. M., 1981, X chromosome reactivation in oocytes of *Mus caroli*, *Proc. Natl. Acad. Sci. USA* **78**:3093–3097.

LaMarca, M. J., and Wassarman, P. M., 1979, Program of early development in the mammal: Changes in absolute rates of synthesis of ribosomal proteins during oogenesis and early embryogenesis in the mouse, *Dev. Biol.* **73**:103–119.

Lehtonen, E., and Badley, R. A., 1980, Localization of cytoskeletal proteins in preimplantation mouse embryos, *J. Embryol. Exp. Morphol.* **55**:211–225.

Levey, I. L., Stull, G. B., and Brinster, R. L., 1978, Poly(A) and synthesis of polyadenylated RNA in the primplantation mouse embryo, *Dev. Biol.* **64**:140–148.

Lewis, W. H., and Wright, E. S., 1935, On the early development of the mouse egg, *Contrib. Embryol. Carnegie Inst. Wash.* **25**:113–144.

Linnenbach, A., Heubner, K., and Croce, C. M., 1980, DNA-transformed murine teratocarcinoma cells: Regulation of expression of simian virus 40 tumor antigen in stem versus differentiated cells, *Proc. Natl. Acad. Sci. USA* **77**:4875–4879.

Luciani, J. M., Devictor-Vuillet, M., Gagne, R., and Stahl, A., 1974, An air drying method for first meiotic prophase preparations from mammalian ovaries, *J. Reprod. Fertil.* **36**:409–411.

Luthardt, F. W., 1976, Cytogenetic analysis of oocytes and early preimplantation embryos from XO mice, *Dev. Biol.* **54**:73–81.

MacGregor, H. C., 1980, Recent developments in the study of lampbrush chromosomes, *Heredity* **44**:3–35.

Magnuson, T., and Epstein, C. J., 1981, Genetic control of very early mammalian development, *Biol. Rev. Cambr. Philos. Soc.* **56**:369–408.

Manes, C., and Menzel, P., 1981, Demethylation of CpG sites in DNA of early rabbit trophoblast, *Nature* **293**:589–590.

Mangia, F., and Canipari, R., 1977, Biochemistry of growth and maturation in mammalian oocytes, *Dev. Mamm.* **2**:1–29.

Mangia, F., and Epstein, C. J., 1975, Biochemical studies of growing mouse oocytes: Preparation of oocytes and analysis of glucose-6-phosphate dehydrogenase and lactic dehydrogenase activities, *Dev. Biol.* 45:211–220.

Mangia, F., Erickson, R. P., and Epstein, C. J., 1976, Synthesis of LDH-1 during mammalian oogenesis and early development, *Dev. Biol.* **54**:146–150.

Martin, G. R., 1982, X-chromosome inactivation in mammals, *Cell* **29**:721–724.

McGaughey, R. W., and Van Blerkom, J., 1977, Patterns of polypeptide synthesis of porcine oocytes during maturation *in vitro*, *Dev. Biol.* **56**:241–254.

McGaughey, R. W., Montgomery, D. H., and Richter, J. D., 1979, Germinal vesicle configurations and patterns of polypeptide synthesis of porcine oocytes from antral follicles of different size, as related to their competency for spontaneous maturation, *J. Exp. Zool.* **209**:239–254.

McLaren, A., and Monk, M., 1981, X-chromosome activity in the germ cells of sex-reversed mouse embryos, *J. Reprod. Fertil.* **63**:533–537.

McReynolds, H. D., and Hadek, R., 1972, A comparison of the fine structure of late mouse blastocysts developed *in vivo* and *in vitro*, *J. Exp. Zool.* **182**:95–118.

Michaels, G. S., Hauswirth, W. W., and Laipis, P. J., 1982, Mitochondrial DNA copy number in bovine oocytes and somatic cells, *Dev. Biol.* **94**:246–251.

Mirre, C., and Stahl, A., 1978, Ultrastructure and activity of the nucleolar organizer in the mouse oocyte during meiotic prophase, *J. Cell Sci.* **31**:79–100.

Mirre, C., and Stahl, A., 1981, Ultrastructural organization, sites of transcription and distribution of fibrillar centers in the nucleolus of the mouse oocyte, *J. Cell Sci.* **48**:105–126.

Monk, M., 1981, A stem-line model for cellular and chromosomal differentiation in early mouse development, *Differentiation (Berl.)* **19**:71–76.

Monk, M., and Harper, M., 1978, X-chromosome activity in preimplantation mouse embryos from XX and XO mothers, *J. Embryol. Exp. Morphol.* **46**:53–64.

Monk, M., and Harper, M. I., 1979, Sequential X chromosome inactivation coupled with cellular differentiation in early mouse embryos, *Nature* **281**:311–313.

Monk, M., and McLaren, A., 1981, X-chromosome activity in foetal germ cells of the mouse, *J. Embryol. Exp. Morphol.* **63**:75–84.

Moore, G. P. M., 1978, RNA synthesis in fixed cells by endogenous RNA polymerases, *Exp. Cell Res.* **111**:317–326.

Moore, G. P. M., and Lintern-Moore, S., 1978, Transcription of the mouse oocyte genome, *Biol. Reprod.* **18**:865–870.

Moore, G. P. M., and Lintern-Moore, S., 1979, Stimulation of endogenous RNA polymerase I activity in the mouse oocyte after PMSG treatment, *Biol. Reprod.* **21**:373–377.

Moore, G. P. M., Lintern-Moore, S., Peters, H., and Faber, M., 1974, RNA synthesis in the mouse oocyte, *J. Cell Biol.* **60**:416–422.

Moore, G. P. M., Lintern-Moore, S., and Scott-Murphy, L., 1981, Relative changes in the RNA contents of oocytes and preimplantation embryos of the mouse, *Reprod. Nutr. Dev.* **21**:505–512.

Motlik, J., Kopecny, V., Pivko, J., and Fulka, J., 1980, Distribution of proteins labelled during meiotic maturation in rabbit and pig eggs at fertilization, *J. Reprod. Fertil.* **58**:415–419.

Muller, W. P., 1974, The lampbrush chromosomes of *Xenopus laevis* (Daudin). *Chromosoma* **47**:283–296.

Oakberg, E. F., 1968, Relationship between stage of follicular development and RNA synthesis in the mouse oocyte, *Mutat. Res.* **6**:155–165.

Ohno, S., Kaplan, W. D., and Kinosita, R., 1961, X-chromosome behavior in germ and somatic cells of *Rattus norvegicus*. *Exp. Cell Res.* **22**:535–544.

Ohno, S., Klinger, H. P., and Atkin, N. B., 1962, Human oogenesis, *Cytogenetics* **1**:42–51.

Olds, P. J., Stern, S., and Biggers, J. D., 1973, Chemical estimates of the RNA and DNA contents of the early mouse embryo, *J. Exp. Zool.* **186**:39–46.

Osborn, J. C., and Moor, R. M., 1983, Time-dependent effects of $\alpha$-amanitin on nuclear maturation and protein synthesis in mammalian oocytes, *J. Embryol. Exp. Morphol.* **73**:317–338.

Oud, J. L., de Jong, J. H., and de Rooij, D. G., 1979, A sequential analysis of meiosis in the male mouse using a restricted spermatocyte population obtained by a hydroxy/triaziquone treatment, *Chromosoma* **71**:237–248.

Palombi, F., and Viron, A., 1977, Nuclear cytochemistry of mouse oogenesis. 1. Changes in extranucleolar ribonucleoprotein components through meiotic prophase, *J. Ultrastruct. Res.* **61**:10–20.

Pedersen T., 1970, Follicle kinetics in the ovary of the cyclic mouse, *Acta Endocrinol. (Copenh.)* **64**:304–323.

Perlman, S., and Rosbash, M., 1978, Analysis of *Xenopus laevis* ovary and somatic cell polyadenylated RNA by molecular hybridization, *Dev. Biol.* **63**:197–212.

Peters, H., Himelstein-Braw, R., and Faber, M., 1976, The normal development of the ovary in childhood, *Acta Endocrinol. (Copenh.)* **82**:617–630.

Piko, L., and Chase, D. G., 1973, Role of the mitochondrial genome during early development in mice, *J. Cell Biol.* **58**:357–378.

Piko, L., and Clegg, K. B., 1982, Quantitative changes in total RNA, total poly(A) and ribosomes in early mouse embryos, *Dev. Biol.* **89**:362–378.

Piko, L., and Matsumoto, L., 1976, Number of mitochondria and some properties of mitochondrial DNA in the mouse egg, *Dev. Biol.* **49**:1–10.

Pollard, D. R., Baran, M. M., and Bachvarova, R., 1976, The effect of 5-bromodeoxyuridine on cell division and differentiation of preimplantation mouse embryos, *J. Embryol. Exp. Morphol.* **35**:169–178.

Quincey, R. V., and Wilson, S. H., 1969, The utilization of genes for ribosomal RNA, 5S RNA, and transfer RNA in liver cells of adult rats, *Proc. Natl. Acad. Sci. USA* **64**:981–988.

Rastan, S., 1982, Timing of X-chromosome inactivation in postimplantation mouse embryos, *J. Embryol. Exp. Morphol.* **71**:11–24.

Richter, J. D., and McGaughey, R. W., 1981, Patterns of polypeptide synthesis in mouse oocytes during germinal vesicle breakdown and during maintenance of the germinal vesicle stage by dibutyryl cAMP, *Dev. Biol.* **83**:188–192.

Rodman, T. C., and Bachvarova, R., 1976, RNA synthesis in preovulatory mouse oocytes, *J. Cell Biol.* **70**:251–257.

Rodman, T. C., and Barth, A. H., 1979, Chromosomes of mouse oocytes in maturation: Differential trypsin sensitivity and amino acid incorporation, *Dev. Biol.* **68**:82–95.

Sagata, N., Shiokawa, K., and Yamana, K., 1980, A study on the steady-state population of poly(A)$^+$ RNA during early development of *Xenopus laevis*, *Dev. Biol.* **77**:431–448.

Sawicki, J. A., Magnuson, T., and Epstein, C. J., 1981, Evidence for expression of the paternal genome in the two-cell mouse embryo, *Nature* **294**:450–451.

Schmid, C. W., and Jelinek, W. R., 1982, The *Alu* family of dispersed repetitive sequences, *Science* **216**:1065–1070.

Schultz, G. A., 1975, Polyadenylic acid-containing RNA in unfertilized and fertilized eggs of the rabbit, *Dev. Biol.* **44**:270–277.

Schultz, R. M., and Wassarman, P. M., 1977a, Specific changes in the pattern of protein synthesis during meiotic maturation of mammalian oocytes *in vitro*, *Proc. Natl. Acad. Sci. USA* **74**:538–541.

Schultz, R. M., and Wassarman, P. M., 1977b, Biochemical studies of mammalian oogenesis: Protein synthesis during oocyte growth and meiotic maturation in the mouse, *J. Cell Sci.* **24**:167–194.

Schultz, R. M., LaMarca, M. J., and Wassarman, P. M., 1978a, Absolute rates of protein synthesis during meiotic maturation of mammalian oocytes *in vitro*, *Proc. Natl. Acad. Sci. USA* **75**:4160–4164.

Schultz, R. M., Letourneau, G. E., and Wassarman, P. M., 1978b, Meiotic maturation of mouse oocytes *in vitro:* Protein synthesis in nucleate and anucleate oocyte fragments, *J. Cell Sci.* **30**:251–264.

Schultz, R. M., Letourneau, G. E., and Wassarman, P. M., 1979, Program of early development in the mammal: Changes in the patterns and absolute rates of tubulin and total protein synthesis during oocyte growth in the mouse, *Dev. Biol.* **73**:120–133.

Singer, J., Roberts-Ems, J., Luthardt, F. W., and Riggs, A. D., 1979, Methylation of DNA in mouse early embryos, teratocarcinoma cells and adult tissues of mouse and rabbit, *Nucleic Acids Res.* **7**:2369–2385.

Singh, R. P., and Carr, D. H., 1966, The anatomy and histology of XO human embryos and fetuses, *Anat. Rec.* **155**:369–384.

Siracusa, G., Whittingham, D. G., Molinaro, M., and Vivarelli, E., 1978, Parthenogenetic activation of mouse oocytes induced by inhibitors of protein synthesis, *J. Embryol. Exp. Morphol.* **43**:157–166.

Sorensen, R. A., and Wassarman, P. M., 1976, Relationship between growth and meiotic maturation of the mouse oocyte, *Dev. Biol.* **50**:531–536.

Sotelo, J. R., and Garcia, R. B., 1977, Removal and isolation of oocytes and their zona pellucida from ovaries of mature mice, *J. Cell Biol.* **75**:177a.

Speed, R. M., 1982, Meiosis in the foetal mouse ovary. I. An analysis at the light microscope level using surface-spreading, *Chromosoma* **85**:427–437.

Spindle, A., Wu, K., and Pedersen, R. A., 1982, Sensitivity of early mouse embryos to [$^3$H]-thymidine, *Exp. Cell Res.* **142**:397–405.

Spradling, A. C., and Mahowald, A. P., 1980, Amplification of genes for chorion proteins during oogenesis in *Drosophila melanogaster*, *Proc. Natl. Acad. Sci. USA* **77**:1096–1100.

Stahl, A., Luciani, J. M., Devictor, M., Capodano, A. M., and Gagne, R., 1975, Constitutive heterochromatin and micronucleoli in the human oocyte at the diplotene stage, *Humangenetik* **26**:315–327.

Sternlicht, A. L., and Schultz, R. M., 1981, Biochemical studies of mammalian oogenesis: Kinetics of accumulation of total and poly(A)-containing RNA during growth of the mouse oocyte, *J. Exp. Zool.* **215**:191–200.

Stevens, L. C., 1975, Comparative development of normal and parthenogenetic mouse embryos, early testicular and ovarian teratomas, and embryoid bodies, in: *Teratomas and Differentiation* (M. I. Sherman and D. Solter, eds), pp. 17–32, Academic Press, New York.

Szollosi, D., 1972, Changes of some cell organelles during oogenesis in mammals, in *Oogenesis* (J. D. Biggers, and A. W. Schuetz, eds.), pp. 47–64, University Press, Baltimore.

Tobler, H., 1975, Occurrence and developmental significance of gene amplification, in: *Biochemistry of Animal Development*, vol. 3 (R. Weber, ed.), pp. 91–123, Academic Press, New York.

Tres, L. L., 1975, Nucleolar RNA synthesis of meiotic prophase spermatocytes in the human testis, *Chromosoma* **53:**141–151.

Tsuda, H., 1965, An electron microscope study on the oogenesis in the mouse, with special reference to the behaviours of oogonia and oocytes at meiotic prophase, *Arch. Histol. Jpn.* **25:**533–555.

Van Blerkom, J., 1979, Molecular differentiation of the rabbit ovum. III. Fertilization-autonomous polypeptide synthesis, *Dev. Biol.* **72:**188–194.

Van Blerkom, J., 1981a, Intrinsic and extrinsic patterns of molecular differentiation during oogenesis, embryogenesis, and organogenesis in mammals, in: *Cellular and Molecular Aspects of Implantation* (S. R. Glasser and D. W. Bullock, eds.), pp. 155–176, Plenum Press, New York.

Van Blerkom, J., 1981b, Structural relationship and posttranslational modification of stage-specific proteins synthesized during early preimplantation development in the mouse, *Proc. Natl. Acad. Sci. USA* **78:**7629–7633.

Van Blerkom, J., and McGaughey, R. W., 1978a, Molecular differentiation of the rabbit ovum. I. During oocyte maturation *in vivo* and *in vitro*, *Dev. Biol.* **63:**139–150.

Van Blerkom, J., and McGaughey, R. W., 1978b, Molecular differentiation of the rabbit ovum. II. During the preimplantation development of *in vivo* and *in vitro* matured oocytes. *Dev. Biol.* **63:**151–164.

von Winiwarter, H., 1901, Recherches sur l'ovogenèse et l'organogenèse de l'ovaire des mammifères (lapin et homme), *Arch. Biol.* **17:**33–199.

von Winiwarter, H., and Sainmont, G., 1909, Nouvelle recherches sur l'ovogenèse et l'organogenèse de l'ovaire de mammifères (chat), *Arch. Biol.* **24:**1–142.

Warnes, G. M., Moor, R. M., and Johnson, M. H., 1977, Changes in protein synthesis during maturation of sheep oocytes *in vivo* and *in vitro*, *J. Reprod. Fertil.* **49:**331–335.

Wassarman, P. M., and Josefowicz, W. J., 1978, Oocyte development in the mouse: An ultrastructural comparison of oocytes isolated at various stages of growth and meiotic competence, *J. Morphol.* **156:**209–235.

Wassarman, P. M., and Letourneau, G. E., 1976, RNA synthesis in fully-grown mouse oocytes, *Nature* **261:**73–74.

Wassarman, P. M., and Mrozak, S. C., 1981, Program of early development in the mammal: Synthesis and intracellular migration of histone H4 during oogenesis in the mouse, *Dev. Biol.* **84:**364–371.

Wassarman, P. M., Ukena, T. E., Josefowicz, W. J., Letourneau, G. E., and Karnovsky, M. J., 1977, Cytochalasin B-induced pseudocleavage of mouse oocytes *in vitro*. II. Studies of the mechanism and morphological consequences of pseudocleavage, *J. Cell Sci.* **26:**323–337.

Wassarman, P. M., Schultz, R. M., and Letourneau, G. E., 1979, Protein synthesis during meiotic maturation of mouse oocytes *in vitro*. Synthesis and phosphorylation of a protein localized in the germinal vesicle, *Dev. Biol.* **69:**94–107.

Wassarman, P. M., Bleil, J. D., Cascio, S. M., LaMarca, M. J., Letourneau, G. E., Mrozak, S. C., and Schultz, R. M., 1981, Programming of gene expression during mammalian oogenesis, in: *Bioregulators of Reproduction* (G. Jagiello and H. J. Vogel, eds.), pp. 119–150, Academic Press, New York.

Weakley, B. S., 1966, Electron microscopy of the oocyte and granulosa cells in the developing ovarian follicles of the golden hamster (*Mesocricetus auratus*), *J. Anat.* **100:**503–534.

Wilson, E. B., 1925, *The Cell in Development and Heredity*, Macmillan, New York.

Wolgemuth-Jarashow, D. J., Jagiello, G. M., and Henderson, A. S., 1977, The localization of rDNA in small nucleolus-like structures in human diplotene oocyte nuclei, *Hum Genet.* **36:**63–68.

Wolgemuth, D. J., Jagiello, G. M., and Henderson, A. S., 1979, Quantitation of ribosomal RNA genes in fetal human oocyte nuclei using rRNA:DNA hybridization in situ. Evidence for increased multiplicity, *Exp. Cell Res.* **118:**181–190.

Wolgemuth, D. J., Jagiello, G. M., and Henderson, A. S., 1980, Baboon late diplotene oocytes contain micronucleoli and a low level of extra rDNA templates, *Dev. Biol.* **78:**598–604.

Zamboni, L., 1970, Ultrastructure of mammalian oocytes and ova, *Biol. Reprod.* Suppl. 2, 44–63.

Zamboni, L., 1974, Fine morphology of the follicle wall and follicle cell-oocyte association, *Biol. Reprod.* **10:**125–149.

Zeller, R., Nyffenegger, T., and De Robertis, E. M., 1983, Nucleocytoplasmic distribution of snRNPs and stockpiled snRNA-binding proteins during oogenesis and early development in *Xenopus laevis, Cell* **32:**425–434.

Zuckerman, S., and Weir, B. J., 1977, *The Ovary,* 2nd ed., Academic Press, New York.

Zybina, E. V., 1968, The structure of nucleus and nucleolus during ovogenesis of mice, *Tsitologiya* **10:**36–42.

Zybina, E. V., 1975, An electron microscope study of the lampbrush chromsomes and their activity products during the rabbit's oogenesis, *Tsitologiya* **17:**875–880.

Chapter 12

# Informational Content of the Echinoderm Egg

BRUCE P. BRANDHORST

## 1. Introduction

By the early years of this century, experimental embryologists had demonstrated that the cytoplasm of the egg contains informational materials that are not always evenly distributed and whose segregation during cleavage can result in the divergence and limitation of the developmental potentials of the blastomeres that come to contain them. Moreover, it was found that many processes of early embryonic development (i.e., cleavage, changes in cell shape, morphogenesis, and cellular specializations such as growth of cilia) may occur in the absence of nuclear genomic activity. The definition of the role and structure of DNA led to the demonstration of the role of messenger RNA (mRNA) in protein synthesis by the early 1960s. Shortly thereafter, experiments on the sea urchin led to the development of the hypothesis that the egg contains a store of maternal mRNA, which supports protein synthesis during early embryonic development. Compelling evidence for the extensive utilization of stored maternal mRNA was developed, and the cytoplasmic informational molecules defined by the earlier experimental embryologists are frequently thought to include stored mRNAs. A considerable mass of other materials, particularly proteins, as well as an elaborate structural organization in the egg also serve roles in the maternally established or influenced processes of embryogenesis.

In the sea urchin egg, most of the maternal mRNA is not actively engaged in protein synthesis. After fertilization, there is a rapid, extensive increase in the rate of protein synthesis, which is largely due to the recruitment of maternal mRNA into polysomes. For several decades, sea urchin eggs and embryos have been among the most completely described cells at

BRUCE P. BRANDHORST • Department of Biology, McGill University, Montreal, Quebec H3A 1B1, Canada.

both the physiological and molecular levels (Davidson *et al.*, 1982). Thus, sea urchin eggs have been the classical system for investigations of the nature and role of maternal mRNA. This chapter summarizes the current state of knowledge of the properties of maternal RNA in the sea urchin, its accumulation during oogenesis, and its utilization during embryogenesis. In addition, information is summarized about the metabolism of egg proteins during embryogenesis. The echinoderm embryo does not grow during embryogenesis, and there is no significant change in mass of protein and RNA until the pluteus larva begins to feed. During embryonic development, materials stored in the egg–including informational macromolecules–are used, degraded, and replaced with either similar or different molecules. Investigations of maternal macromolecules of echinoderms other than sea urchins have been far less extensive. Studies of the properties and utilization of maternal mRNA in asteroid oocytes are referred to, particularly when they provide an interesting contrast to echinoid eggs.

## 2. Oogenesis in Sea Urchins

### 2.1. Temporal and Morphological Considerations

Mature sea urchin eggs range in size from about 70–180 $\mu$m in diameter (Giudice, 1973; Hinegardner, 1975). Starfish oocytes are included in this range, but some are considerably larger. Both meiotic maturation divisions occur within the ovary of sea urchins, and mature haploid eggs can be retained for weeks or months before spawning (Bryce, 1902). By contrast, starfish oocytes undergo meiotic maturation during hormone-induced spawning (see Chap. 1). In *Strongylocentrotus purpuratus* urchins, $1–10 \times 10^6$ or more eggs are produced per spawning (average $3–6 \times 10^6$), and spawning can take place several times during a season of 4–6 months (Leahy *et al.*, 1978; Hough-Evans *et al.*, 1979). A large gravid female contains mature eggs (80 $\mu$m in diameter), as well as $0.5–2 \times 10^6$ vitellogenic oocytes (50–80 $\mu$m) and about $10^8$ previtellogenic oocytes (5–50 $\mu$m) and oogonia (5–7 $\mu$m) (Leahy *et al.*, 1978, 1981; Hough-Evans *et al.*, 1979). The large excess of mature eggs in gravid females as well as frequent spawning cycles indicate that vitellogenesis requires 5–15 days. The period of previtellogenic oogenesis is less well defined, but it is apparently no more than a few weeks or months long during the spawning season.

The sea urchin ovary consists of three layers: an outer ciliated peritoneal epithelium; a middle layer consisting of layers of connective tissue, smooth muscle, nerve elements, and highly vacuolated cells; and an inner germinal epithelium consisting of oogonia, which give rise to oocytes, as well as accessory cells, which are sometimes termed "nutritive phagocytes" (Wilson, 1940; Tennent and Ito, 1941; Harvey, 1956; Holland and Giese, 1965; Pearse and Giese, 1966; Verhey and Moyer, 1967; E. Anderson,

1968; Chatlynne, 1969). The ultrastructural morphology of oogenesis in sea urchins has been described in several of the previous references (reviewed in Giudice, 1973; Piatigorsky, 1975). The morphology of the sea urchin oocyte is typical of most oocytes containing moderate amounts of yolk. It is worth noting that the cytoplasm of sea urchin eggs contains annulate lamellae and "heavy bodies" (Harris, 1967), which are reported to contain RNA (Conway, 1971). It is sometimes speculated that macromolecular information utilized during embryogenesis resides in these structures (e.g., Afzelius, 1957; Sanchez, 1966, 1968; see Chapter 4 for a discussion of the possible functional significance of these structures).

### 2.2. Lampbrush Chromosomes

Lampbrush chromosomes appear during oogenesis in a wide variety of organisms (see Chapters 9 and 11). Despite some confusion about their presence in sea urchin oocytes, I am not aware of any confirmed evidence for their existence. Careful attempts to identify lampbrush chromosomes by electron microscopic evaluation of chromatin spreads have failed to identify them in oocytes of *S. purpuratus* and several starfish under conditions in which active ribosomal genes are easily identified (O. Miller, B. Hough-Evans, and E. Davidson, unpublished observations). By contrast, lampbrush chromosomes have been observed in the oocytes of the starfish *Echinaster sepositus* (Delobel, 1971). These oocytes are very large (400 μm in diameter), suggesting that lampbrush chromosomes may be a feature of only large oocytes (see Chapter 9).

## 3. Stored Maternal Messenger RNA in Sea Urchin Eggs

### 3.1. Entry of Maternal mRNA into Polysomes after Fertilization

The rate of protein synthesis increases dramatically upon fertilization of sea urchin eggs. When eggs preloaded with radioactive amino acids are fertilized, the rate of incorporation of the label into protein is rapidly increased (Nakano and Monroy, 1958; Epel, 1967). The rate of protein synthesis begins to increase almost immediately upon fertilization (Epel, 1967; Raff *et al.*, 1981) and is 15–30-fold higher within 2 hr (Humphreys, 1969; Regier and Kafatos, 1977; Goustin and Wilt, 1981). Protein synthesis continues to increase gradually throughout most of embryogenesis (Goustin and Wilt, 1981). This extensive, rapid increase in the rate of protein synthesis is unusually large for eggs and made possible the first experiments indicating the existence of stored maternal mRNA in eggs. The rate of protein synthesis increases normally upon fertilization even in eggs incubated with the inhibitor of RNA synthesis, actinomycin D, at a concentration sufficient to inhibit most synthesis of large, heterogeneous RNA (Gross and Cousineau, 1963; Gross *et al.*, 1964; Greenhouse *et al.*, 1971). Protein synthesis also increases in response to parthenogenetic activation of anucleate merogones (Brachet *et al.*, 1963; Denny and Tyler, 1964). In the absence of nuclear transcriptional activity, any mRNA translated in the zygote must be of maternal and/or cytoplasmic origin (see Chamberlain, 1970). Thus, these observations led to the development and widespread acceptance of the hypothesis that sea urchins use maternal mRNA for proteins synthesized during early embryonic development.

Convincing evidence that stored maternal mRNA is rapidly recruited into polysomes after fertilization came from two types of investigations: quantitative studies of RNA metabolism and cell-free translation of sea urchin RNA. Humphreys (1969) demonstrated that the efficiency of translation of mRNA, which is defined as the number of protein molecules produced per mRNA molecule per unit time, changes little, if at all, upon fertilization of *Lytechinus pictus* eggs. Thus, the very rapid 10–15-fold increase in the rate of protein synthesis requires comparable increases in the number of ribosomes engaged in protein synthesis as well as the number of mRNA molecules being translated. Recently, Brandis and Raff (1978) and Hille and Albers (1979) concluded that a two- to threefold increase in translational efficiency occurs upon fertilization, which is due to increases in the rates of polypetide chain initiation and elongation. These measurements of ribosomal transit times are inconsistent with some calculations based on measurements of absolute rates of protein synthesis as well as other measurements of transit times based on other methods (Humphreys, 1969; Goustin and Wilt, 1981, 1982). Whatever the resolution of this continuing controversy, it is clear that the amount of RNA being translated increases rapidly and extensively after fertilization.

The source of this additional mRNA translated in zygotes was shown to be a pool of maternal mRNA stored in the egg by Humphreys (1971), who determined that the fraction of ribosomes in polysomes increases 30-fold to about 20% during the first 2 hr after fertilization of *L. pictus*. Similar data have recently been obtained by hybridization of cloned ribosomal DNA to fractions of RNA extracted from a polysomal gradient and blotted onto nitrocellulose paper (Wells *et al.*, 1981). Since about 2% of the RNA in polysomes is mRNA in a variety of organisms, it can be calculated that about 6 pg, or 0.4% of the total RNA of the egg, must be recruited into polysomes within 2 hr after fertilization (Humphreys, 1971). The mass of newly synthesized RNA labeled with [$^3$H]adenosine and associated with polysomes during this period was measured, taking into account changes in the specific activity of the ATP precursor pool. Newly synthesized RNA can account for only a small fraction of the mass of RNA recruited into polysomes (Humphreys, 1971). In fact, most of the RNA synthesized by the zygote is of mitochondrial origin (Wilt, 1970; Craig, 1970; Chamberlain and Metz, 1972; Devlin, 1976; Ruderman and Schmidt, 1981).

A second type of investigation led to the direct demonstration of stored, translationally inactive RNA in eggs. Cell-free translation in heterologous protein-synthesizing systems of RNA extracted from eggs established that they contain a substantial fraction (3–4% by mass) of template-active RNA (Slater and Spiegelman, 1966; Jenkins *et al.*, 1973). Gross *et al.* (1973) demonstrated that RNA extracted from 20–40 S particles is translated into histones. Moreover, competition hybridization experiments confirmed that a substantial mass of histone mRNA sequences reside in the postribosomal supernatant of a lysate of sea urchin eggs (Farquhar and McCarthy, 1973; Skoultchi and Gross, 1973; Lifton and Kedes, 1976). After fertilization, histone mRNA sequences move into polysomes, although not immediately (see Section 7.3). Taken together, these investigations demonstrate that a large mass of mRNA that is inactive in sea urchin eggs enters polysomes after fertilization to account for most, if not all, of the increase in protein synthesis occurring during the first 2–3 hr after fertilization.

Since these pioneering investigations on sea urchins, a wide variety of other eggs have been shown to contain a store of maternal mRNA utilized during embryonic development. In most cases, the increase in protein synthesis upon fertilization (or meiotic maturation), if any, is less dramatic (and, consequently, less amenable to analysis) than in sea urchins, but stored maternal mRNA may be a universal feature of eggs.

### 3.2. Utilization of Maternal mRNA during Early Embryonic Development

A variety of investigations suggest that early embryonic development in echinoids depends almost exclusively upon utilization of maternal mRNA. Interspecies hybrid embryos of echinoderms (as well as androgenetic hybrids) show an almost exclusively maternal character in developmental timing, morphogenesis, and enzyme activity before gastrulation, even when the paternal genome is retained and is transcriptionally active (reviewed by Chen and Baltzer, 1975, and Davidson, 1976) (see also Section 8.2). These observations suggest that most developmental events of early embryogenesis are dependent on maternal gene products—most likely including maternal mRNA—present in the egg. Sea urchin embryos reared in actinomycin D undergo normal, though somewhat retarded, cleavage and morphogenesis until arresting as blastulae at about the time of hatching (Gross *et al.*, 1964; Giudice *et al.*, 1968). Moreover, the pattern of protein synthesis is not detectably altered in early embryos by actinomycin D treatment (Terman and Gross, 1965; Terman, 1970). These observations have led to the interpretation that most protein synthesis during the first several hours of embryonic development is dependent on the translation of maternal mRNA stored in the egg and recruited into polysomes after fertilization, but that RNA transcribed from the embryonic genome replaces much, if not all, of the maternal mRNA in polysomes at around the time of hatching or during the formation of primary mesenchyme. The persistence of maternal RNA is discussed further in Section 8.

## 4. Characteristics of Maternal RNA Stored in Eggs

### 4.1. Organization of the Sea Urchin Genome

#### 4.1.1. Interspersion of Repetitive and Single-Copy Sequences

Consideration of the size, complexity, and organization of nuclear DNA is required for an understanding of the properties of maternal RNA. The haploid genome of *S. purpuratus* contains 0.81 pg DNA; about 75% of the sequences are present in aproximately one copy per haploid genome (Graham *et al.*, 1974). Thus, the sequence complexity of this single-copy DNA is about $6.1 \times 10^8$ nucleotide pairs (Graham *et al.*, 1974). The remainder of the DNA consists of families of similar or identical sequences that are repeated from a few to several thousand times (Graham *et al.*, 1974). About 75% of the repetitive sequence mass consists of units of about 300 nucleotide pairs interspersed with single-copy sequences averaging about 1000 nucleotide pairs (Graham *et al.*, 1974); some repetitive sequence tracts are much longer. Most of the single-copy DNA is interspersed with these short repetitive sequences. This type of "short-period" interspersed sequence organization is found widely among metazoan animals (Davidson *et al.*, 1975a) and probably applies to all echinoderms (e.g., Smith and Boal, 1978; Smith *et al.*, 1980). In gastrula-stage embryos, most of the mass of mRNA translated in polysomes is transcribed from single-copy DNA (Goldberg *et al.*, 1973). Nuclear RNA synthesized in embryos consists of long tracts of interspersed repetitive and single-copy sequences (Smith *et al.*, 1974), and the single-copy sequences represented in polysomes of these embryos are mostly derived from DNA having this typical short-period interspersion pattern (Davidson *et al.*, 1975b). Many of these repetitive sequences are undoubtedly included in the intervening sequences (introns), which are transcribed as part of the primary transcription unit, but which are spliced out during maturation of the mRNA; primary transcripts in oocytes must have a similar structure.

#### 4.1.2. Characteristics of Repetitive Sequences

Sea urchin genomic DNA includes about 5000 different families of repetitive sequences. Because of sequence divergence within families, it is impossible to define precisely the number of families and their repetition frequency based on analyses of DNA renaturation kinetics. The definition of members of widely divergent repeat families become arbitrarily dependent on renaturation conditions. In order to overcome these uncertainties, a variety of individual sequence families have been analyzed in the genome of *S. purpuratus* using repetitive DNA clones (Klein *et al.*, 1978). For 26 cloned repeats, reiteration frequences in copies per genome range broadly from a few to $>10^5$, having an average of about 500. Analyses of the kinetics of reannealing of repetitive DNA tracers with genomic DNA indicate that the frequency distribution of these cloned repeat families is representative of the entire genome (Klein *et al.*, 1978). Estimates of sequence divergence of these cloned repeat families were made by comparing the thermal stabilities of heteroduplexes formed with genomic DNA with those of renatured cloned fragments. Most of the families lack any highly divergent relatives and show a nucleotide sequence divergence of 8% or less (Klein *et al.*, 1978). Some cloned repeat families, however, are considerably more divergent and include some very distantly related family members (Klein *et al.*, 1978). In some cases, this divergence is because the family contains subfamilies consisting of small repeat-sequence subelements, which vary in number and arrangement between members of the family (Scheller *et al.*, 1981a; Posakony *et al.*, 1981). The cloned repeat families are found also in closely related species but usually in considerably different frequencies (Moore *et al.*, 1978, 1981). Between more distantly related species (e.g., *S. purpuratus* and *L. pictus*, which diverged approximately $1.75 \times 10^8$ years ago), most large repeat families of one species are represented by only a few copies, if at all, in the other species (Moore *et al.*, 1978). These observations imply that many repetitive sequence families have arisen by recent DNA multiplication. Deletion, insertion, and local duplication, as well as point mutations, are probably involved in the generation of these partially related repeat sequences (Scheller *et al.*, 1981a). Both the long- and particularly the short-repetitive sequence families examined are widely distributed in the genome. They are not confined to a few chromosomal domains (Anderson *et al.*, 1981). Thus, there must be a mechanism for relatively rapid dispersion to distant chromosomal sites during the evolution of these newly enlarged families, possibly by transposition of excised circular DNA elements (Posakony *et al.*, 1981). The evolutionarily dynamic nature of these repetitive sequence families should be borne in mind when considering their possible functions, if any.

## 4.2. Sequence Complexity of Egg RNA

### 4.2.1. Saturation Hybridization of Single-Copy DNA Tracers

The single-copy sequence complexity of RNA can be most accurately estimated by saturation hybridization of a prepared radioactive single-copy DNA tracer to RNA in excess (Galau *et al.*, 1974). The fraction of DNA complementary to egg RNA—about 6% (assuming asymmetric transcription)—

indicates that the complexity of this RNA is about $3.7 \times 10^7$ nucleotides for *S. purpuratus* (Galau *et al.*, 1976a; Hough-Evans *et al.*, 1977). Similar complexities have been measured for RNA of other echinoid eggs as well as starfish oocytes; these measurements are summarized in Table I. Moreover, very similar complexities have been determined for mature oocytes of a wide variety of other animals (reviewed by Thomas *et al.*, 1981; Kosvedi and Smith, 1982).

The average size of maternal RNA molecules can be used to estimate the number of different RNA molecules from the complexity. Because of technical difficulties in preparing undegraded maternal RNA free of ribosomal and tRNA contamination, the size distribution of polyadenylated egg RNA is usually considered representative of maternal mRNA. This is a reasonable assumption, since essentially all the sequence complexity of maternal RNA is included in the poly(A)$^+$ RNA fraction (Costantini *et al.*, 1980; Duncan and Humphreys, 1981a), which can be conveniently prepared by affinity chromatography on oligo(dT)-cellulose or poly(U)-Sepharose. Poly(A)$^+$ RNA of eggs is heterogeneous in size (Wilt, 1977; Duncan and Humphreys, 1981b). The best present estimate of the number average size distribution of maternal RNA, which is analyzed under denaturing conditions to prevent aggregation, is 3000 nucleotides (Duncan and Humphreys, 1981b; Davidson *et al.*, 1982). Thus, the number of different average-size poly(A)$^+$ RNA molecules in the sea urchin egg is approximately 12,000. Average RNA sizes of 1200–2000 nucleotides have been used in some previous calculations (e.g., Galau *et al.*, 1974; Davidson, 1976; Wilt, 1977; Shepherd and Nemer, 1980), resulting in larger calculated numbers of different RNA molecules. Newly synthesized poly(A)$^+$ RNA translated in polysomes of later embryos has a similar heterogenous size distribution and averages 2000–3000 nucleotides in length as well (Kung, 1974; Brandhorst *et al.*, 1979). It is possible that much of the complexity of egg RNA resides in very large molecules (see Section 4.7), suggesting that $10^4$ different egg RNA molecules may be an overestimate.

### 4.2.2. Kinetics of cDNA Hybridization: Abundant RNA Sequences

RNA complexities can also be estimated by analyses of the kinetics of hybridization of a cDNA probe, which is transcribed from poly(A)$^+$ RNA by reverse transcriptase (RNA-dependent DNA polymerase), to an excess of template RNA. This methodology does not provide an accurate measure of total sequence complexity, but it does provide an estimate of the relative abundance of sequences within the population (assuming that RNA sequences are transcribed in proportion to their prevalence in the population). On the other hand, saturation hybridization of single-copy DNA tracer to RNA is primarily sensitive to the most diverse sequences, which are relatively rare in the RNA population (Galau *et al.*, 1974); such investigations do provide an estimate of the relative abundance of the complex

**Table I.** Characteristics of Echinoderm Egg RNAs

| Species | Egg diameter[a] (μm) | Genome size[b] (bp × $10^{-8}$) | RNA mass (ng/egg) | RNA complexity (nucleotides × $10^{-7}$) | Complex RNA[c] (% of total RNA) | Average prevalence[d] (copies/egg × $10^{-3}$) | Rare transcript concentration[e] (pM) |
|---|---|---|---|---|---|---|---|
| *Strongylocentrotus purpuratus*[f] | 80 | 8.1 | 3.3[k] | 3.7 | 0.5–1.5 | 1.6 | 9.9 |
| *Arbacia punctulata*[g] | 79 | 7.2 | 2.3[k] | 3.0 | 1–2 | 1.4 | 9.0 |
| *Tripneustes gratilla*[h] | 90 | 13 | 1.8 | 3.4 | 2 | 1.9 | 8.1 |
| *Lytechinus pictus*[i] | 111 | 8.8 | 3.9[l] | 3.1 | 2 | 4.1 | 10.0 |
| *Pisaster ochraceus*[j] | 163 | 6 | 15 | 4.3 | 2–3 | 17 | 12.4 |

[a]From Hinegardner (1975); there is substantial variation.
[b]From Hinegardner (1974) and Thomas *et al.* (1981); expressed as base pairs of DNA per haploid genome.
[c]The fraction of RNA driving the hybridization reaction: the complex class of rare sequences. Eggs also contain a population of more abundant sequences.
[d]Calculated from the average mass of complex class RNA and the complexity of that RNA.
[e]Calculated by Kosvedi and Smith (1982).
[f]Data from Galau *et al.* (1976) and Hough-Evans *et al.* (1977).
[g]Data from Anderson *et al.* (1976).
[h]Data from Duncan and Humphreys (1981a).
[i]Data from Wilt (1977). Calculations are based on the kinetics of hybridization of cDNA to its poly(A) RNA template.
[j]Data from Kosvedi and Smith (1982).
[k]Calculated from data of Whiteley (1949).
[l]Data from Brandhorst (1980). A value of 1.8 ng is commonly used in the literature.

class of RNA (see Section 6 for an explanation of the values that appear in Table I).

The kinetics of hybridization of cDNA to sea urchin poly(A)$^+$ RNA indicate that various sequences are present in a broad range of abundance (Wilt, 1977; Duncan and Humphreys, 1981a). The application of this type of analysis to *L. pictus* egg poly(A)$^+$ RNA led to a measured complexity of $3.6 \times 10^7$ nucleotides, very similar to determinations of RNA complexity in other echinoid eggs using single-copy DNA hybridization (Wilt, 1977). A similar investigation by Duncan and Humphreys (1981a) on *Tripneustes gratilla* led to a measured RNA complexity of only $1.4 \times 10^7$ nucleotides. That this estimate is less than half the value determined by single-copy DNA hybridization (Duncan and Humphreys, 1981a) (see Table I) can be explained by uncertainties inherent in the technique and the observation that a large fraction of the rare sequences are underrepresented in the poly(A)$^+$ RNA fraction (Duncan and Humphreys, 1981a).

The kinetics of cDNA hybridization indicate that some RNA sequences are considerably more abundant than the rare sequences, which make up most of the complexity (Wilt, 1977; Duncan and Humphreys, 1981a). The hybridization curves can be resolved into at least three components. About 50 average-size poly(A) molecules are present in the egg in an abundance 10–100 times higher than the rare sequences, which are each present in a few thousand copies per egg on average (see Section 6 and Table I). A few (2–3) are present in 1000-fold higher abundance. These extremely prevalent transcripts are probably all of mitochondrial origin (O'Brochta *et al.*, 1981; Wells *et al.*, 1982; Brandhorst, unpublished observations). Together, these more abundant sequences constitute about one-half the mass of maternal RNA (Wilt, 1977; Duncan and Humphreys, 1981a).

Another indication of the complexity of maternal mRNA is the observation that in the first few hours after fertilization (when the mRNA translated is predominantly of maternal origin) about 1000 newly synthesized polypeptides can be detected by two-dimensional electrophoresis (Brandhorst, 1976; Tufaro and Brandhorst, 1979). These translation products apparently correspond to the more prevalent maternal mRNAs and perhaps to some of the more efficiently translated rare mRNAs as well (Brandhorst, 1976; Lasky *et al.*, 1980;Bédard and Brandhorst, 1983).

## 4.3. Repetitive Sequence Representation in Egg RNA

Costantini *et al.* (1978) prepared DNA tracers representing the short interspersed repetitive sequences of *S. purpuratus*. When reacted with excess total egg RNA, about 80% of the tracer hybridized. This indicates that most of the short repetitive sequence families are represented in egg RNA and that transcripts of both strands of many of these repeats are present (otherwise a maximum of 50% of the DNA tracer could react). Analysis of the kinetics of the reaction indicates that some repeat sequence families (about 15%) are highly prevalent (more than $10^5$ copies per egg, on average), some (about 40%) are moderately prevalent ($3 \times 10^4$ copies per egg), whereas others are rare (about $10^3$ copies per oocyte). Thus, many short repetitive-sequence families are represented in egg RNA in abundance similar to that of most single-copy sequences, but a much larger fraction of short repetitive-sequence families than single-copy sequences are abundantly represented. Most of these prevalent repeat transcripts are derived from 500–1000 repeat families, most having 20–200 copies per haploid genome (Costantini *et al.*, 1978). The abundance of repetitive sequence transcripts in egg RNA is not related to their genomic repetition frequencies (Costantini *et al.*, 1978).

Individual cloned repetitive sequences were also analyzed by Costantini *et al.* (1978). Both strands of all nine repeat clones analyzed were represented in egg RNA, with a prevalence of 3000–$10^5$ copies per oocyte. These cloned repeats, which correspond to short interspersed families a few hundred nucleotides in length, are predominantly represented in very large egg RNA molecules several thousand nucleotides in length (Costantini *et al.*, 1978; Posakony *et al.*, 1983). The structural features of these transcripts are considered further in Sections 4.7 and 5.

About 35% of the long (noninterspersed) repetitive sequence families are represented in egg RNA (Costantini *et al.*, 1978). Some of these are derived from large families of structural genes, such as the histone genes, which are repeated hundreds of times as tandemly arranged clusters (Kedes and Birnstiel, 1971; Hentschel and Birnstiel, 1981). Other smaller families of structural genes are represented in egg RNA, such as those coding for actins (Crain *et al.*, 1981; Scheller *et al.*, 1981b; Infante and Heilman, 1981; Merlino *et al.*, 1981) and tubulins (Raff *et al.*, 1972; Alexandraki and Ruderman, 1981). Whether all transcripts of long repetitive sequences represent structural genes is unknown. Most of the transcripts of these sequences are rare (Costantini *et al.*, 1978). About one-sixth of the reactive cDNA transcribed from poly(A)$^+$ RNA is complementary to repetitive DNA sequences (Duncan and Humphreys, 1981a). However, only the 3′ ends of the RNA are represented in the cDNA.

## 4.4. Polyadenylation

Messenger RNA molecules of eukaryotic cells are often polyadenylated at their 3′ termini, which is the result of a nuclear post-transcriptional processing event. Upon fertilization, the poly(A) content of the egg increases until by 90 min after fertilization about 0.09% (approximately 1.5 pg for *L. pictus*) of the cellular RNA is poly(A), at least a doubling of the amount in eggs (Slater *et al.*, 1972, 1973; Wilt, 1973; Mescher and Humphreys, 1974). Since all poly(A) tracts are attached to RNA molecules, the polyadenylation of RNA increases after fertilization. This polyadenylation must occur on maternal RNA in the cytoplasm, since it occurs even in parthenogenetically activated anucleate merogones (Wilt, 1973).

Affinity chromatography indicates that about 2.4% of the egg RNA has poly(A) tracts sufficiently long to bind to poly(U)-Sepharose (Wilt, 1977). Only about 1.5% of the RNA bound to oligo(dT)-cellulose even at low temperature (Wilt, 1977); oligo(dT) chromatography does not bind oligo(A) tracts efficiently and probably represents a more accurate measure of the polyadenylated fraction of egg RNA (see Section 4.6).

The size distribution of poly(A) following fertilization changes little, increasing in number average length by one-third or less to 60 nucleotides for *L. pictus* (Wilt, 1977) and 85 nucleotides for *T. gratilla* (Dolecki *et al.*, 1977). Thus, the increase in poly(A) content should be largely due to an increase in the number of RNA molecules bearing poly(A) tracts. The measured increase after fertilization in the fraction of RNA binding to oligo(dT) or poly(U) is only 20–30% (Wilt, 1977). The reason for the apparent discrepancy in poly(A) content and polyadenylated RNA content is uncertain, but it might be due to an increase in the lability of large poly(A)$^+$ RNA molecules after fertilization. By 2 or 3 hr after fertilization, all the poly(A) is newly synthesized (Dolecki *et al.*, 1977). Thus, the maternal complement of poly(A) turns over completely after fertilization, due to a combination of tract shortening and elongation (Dolecki *et al.*, 1977; Wilt, 1977). The poly(A)$^+$ RNA molecules of 2-hr zygotes include the same sequences in the same relative amounts as the poly(A)$^+$ RNA molecules of eggs (Wilt, 1977; Duncan and Humphreys, 1981a). Thus, the active metabolism of poly(A) after fertilization does not result in the extensive recruitment or loss of poly(A)$^+$ RNA sequences. By 2 hr after fertilization, at least 80% of the poly(A) is associated with polysomes (Slater *et al.*, 1973; Wilt, 1973; Dolecki *et al.*, 1977).

The role of this very dynamic metabolism of poly(A) is unknown. Inhibition of the increase in poly(A) content by treatment of eggs with cordycepin does not alter the increase in protein synthesis following fertilization (Mescher and Humphreys, 1974); thus, increased polyadenylation of RNA does not appear to be involved in the translational activation of maternal mRNA. Poly(A)$^+$ RNA does appear to be preferentially translated after fertilization (Duncan and Humphreys, 1981a). Translation of mRNA is not required for the increase in poly(A) content after fertilization, since it is not inhibited by pactamycin (Dolecki *et al.*, 1977).

Sea urchins contain a substantial mass of mRNA lacking poly(A). Cell-free translation of egg RNA fractionated on oligo(dT)-cellulose indicates that there are approximately equal amounts of translatable mRNA having and lacking poly(A) (Ruderman and Pardue, 1977; Bédard and Brandhorst, unpublished observations). With the exception of histone mRNAs, which lack poly(A), the two fractions of mRNA code for a similar set of proteins (Ruderman and Pardue, 1977; Brandhorst *et al.*, 1979). Poly$(A)^+$ RNA of eggs selected on oligo(dT) includes the same sequences as poly(A)-deficient RNA, although many rare sequences are underrepresented in the poly$(A)^+$ RNA fraction; 50–80% of the mass of heterogeneous RNA sequences are in the poly(A)-deficient fraction (Costantini *et al.*, 1980; Duncan and Humphreys, 1981a). About one-half the newly synthesized RNA entering polysomes of later stage embryos also lacks detectible poly(A) tracts (Nemer, 1975; Fromson and Duchastel, 1975; Brandhorst *et al.*, 1979). The fraction of the mass of RNA in polysomes of zygotes that is polyadenylated is unknown.

When meiotic maturation is induced by 1-methyladenine in starfish oocytes (see Chapter 1), GVBD is accompanied by an increase in the rate of protein synthesis (Houk and Epel, 1974) and a large increase in poly(A) content (Jeffery, 1977). This increase in poly(A) is inhibited by actinomycin D (Jeffery, 1977). Whereas this observation led to the conclusion that the increase in poly(A) content is due to accumulation of newly synthesized poly$(A)^+$ RNA, it is unlikely that RNA synthesis is sufficiently active to account for the very rapid and extensive increase. Thus, recruitment of maternal RNA (Rosenthal *et al.*, 1982) in asteroid oocytes is accompanied by increases in poly(A) content, but these events occur during maturation, rather than after fertilization, as in echinoids. Protein synthesis and polyadenylation of RNA in starfish oocytes are considered in Section 7.4.

## 4.5. Capped 5′ Termini

Most eukaryotic mRNA molecules have a methylated cap ($m^7G^{5'}$ pppXp-) at their 5′ ends, which is added and methylated posttranscriptionally. The methylated cap is necessary for the efficient translation of most cellular mRNAs in cell-free translation systems (Both *et al.*, 1975; Muthukrishnan *et al.*, 1975). The translational inactivity of most mRNA in sea urchin eggs led to the hypothesis that stored mRNA might be uncapped or unmethylated. This is apparently not so. The RNA of unfertilized eggs is translated in a wheat germ lysate even if methylation is inhibited (Hickey *et al.*, 1976; Pirrone *et al.*, 1977). Moreover, its translation is inhibited by a methylated cap analogue (Hickey *et al.*, 1976) and by $\beta$-elimination of the cap structure (Pirrone *et al.*, 1977). These results indicate that at least a portion of the stored maternal RNA of the egg is capped and methylated. It is possible that some maternal mRNAs acquire a methylated cap only after fertilization or that the cap structure changes after fertilization. Sea urchin embryos actively cap and methylate preexisting RNA molecules (Sconzo *et al.*, 1977).

### 4.6. Oligo(A) and Oligo(U) Tracts in Egg RNA

Maternal RNA of sea urchin eggs includes oligo(A) tracts about 10 nucleotides in length (Duncan and Humphreys, 1981b). They are not detectable in RNA having 3′ tracts of poly(A) 50–120 nucleotides in length and they are not detected in polysomal RNA after fertilization. These oligo(A)-containing molecules were selected free of poly(A)$^+$ RNA by affinity chromatography on poly(U)-Sepharose. They include essentially all the sequences represented in poly(A)$^+$ RNA (or polysomal RNA of zygotes) in a similar abundance distribution (Duncan and Humphreys, 1981b). There may be as many as 60 oligo(A) tracts per average (3000 nucleotide) oligo(A)-containing RNA molecule. The mass of oligo(A) tracts is conserved throughout cleavage stages without new synthesis of oligo(A), indicating that their associated maternal transcripts also persist at least until blastula stage. The possibility that oligo(A)-containing RNA is an incompletely processed class of mRNA precursor is discussed in Section 5.

Maternal molecules of sea urchin eggs also contain oligo(U) tracts, though they are considerably less abundant than oligo(A) tracts (Duncan and Humphreys, 1983). About one-half the oligo(U) tracts are associated with large, heterogeneous RNA molecules of the egg, and these tracts disappear rapidly and extensively after fertilization. Oligo(U) tracts are commonly represented in nuclear RNA of sea urchins (Dubroff, 1977) and mammalian cells (Molloy *et al.*, 1972) but have also been reported to be in cytoplasmic RNA (Dubroff, 1980; Korwek *et al.*, 1976). It has been speculated that the interaction of oligo(U) tracts with poly(A) of RNA might be involved in control of the translational activity of mRNA (Heywood and Kennedy, 1976; Duncan and Humphreys, 1983).

### 4.7. Interspersion of Repetitive and Single-Copy Sequences in Maternal RNA

Most transcripts of repetitive sequences detectible in the sea urchin egg are concentrated in the poly(A)$^+$ RNA fraction (Costantini *et al.*, 1980). On average, about one-half the transcripts of each prevalent repeat family are polyadenylated. These polyadenylated repetitive sequences are usually interspersed and colinear with single-copy sequences, as demonstrated by the following investigations of Costantini *et al.* (1980). Poly(A)$^+$ RNA of eggs was purified, denatured, and then allowed to reanneal to a low Cot (concentration $\times$ time) of 600 msec. When spread for electron microscopic study, complex multimolecular structures were observed, as shown in Fig. 1. These structures are networks of molecules held together by complementary short repetitive sequence duplexes. Almost no such structures were observed when RNA had not been first reannealed. Under

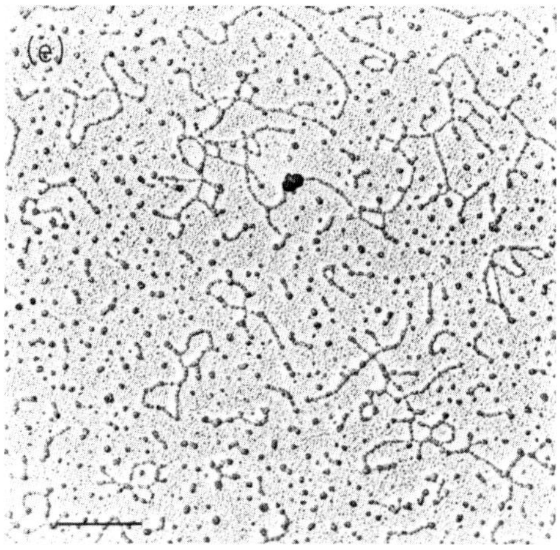


**Figure 1.** Electron microscopy of renatured egg RNA. Poly(A)$^+$ RNA extracted from *S. purpuratus* eggs was renatured (Cot 600) according to the method of Costantini *et al.* (1980). After renaturation, double-stranded RNA was selected by ethanol/cellulose chromatography according to the method of Anderson *et al.* 1982 and spread for electron miscroscopy. The scale bar represents a single-stranded RNA of 1000 nucleotides. (Reproduced with permission from Anderson *et al.*, 1982).

the annealing conditions used, duplexes can form only between repetitive sequences, most of which are represented by both strands in egg RNA. Most single-copy transcripts, on the other hand, are asymmetrically represented (Hough *et al.*, 1975; Lev *et al.*, 1980) and consequently do not form duplexes even at high Cot. The networks shown in Fig. 1 are very similar to structures formed when denatured DNA having a short-period repeat-sequence interspersion pattern is renatured at low Cot (Chamberlin *et al.*, 1975). Most of the complexes observed have six or more ends, indicating that they consist of three or more RNA molecules, most of which must bear more than one repeat element—many of them internally located on the transcript. That the RNA of these structures that is not in duplex form is transcribed from single-copy DNA was demonstrated by isolation and characterization of repeat-containing poly(A)$^+$ RNA. This RNA consists of 85–90% (on average) single-copy-sequence transcripts, but includes short, interspersed repetitive-sequence transcripts.

About 65% of the mass of RNA prepared as in Fig. 1 consists of structures including at least two annealed molecules (Costantini *et al.*, 1980). At least 35% (perhaps twice as much) of the mass of egg poly(A) isolated by virtue of its inclusion of repetitive sequences has this interspersed anneal-able character. About 70% of the poly(A)$^+$ RNA of eggs behaves as RNA containing double stranded tracts when analyzed by ethanol-cellulose chromatography after annealing at low Cot (Anderson *et al.*, 1982). If the annealed RNA were partially degraded, these values might even be underestimates of the fraction of poly(A)$^+$ RNA molecules that contain interspersed repeats. Thus, it appears that at least two-thirds of the mass of poly(A)$^+$ RNA of *S. purpuratus* eggs includes interspersed repetitive sequences. Essentially all the different single-copy sequences of egg RNA reside on molecules that include colinear repeats; 70–80% of these sequences are concentrated in this RNA fraction. These large interspersed transcripts are a major form by mass for most of the single-copy sequences (Costantini *et al.*, 1980).

As discussed in Section 4.1.2, about 90% of the mass of short repetitive-sequence transcripts is derived from about 500–1000 different repetitive-sequence families (Costantini *et al.*, 1978). On the other hand, at least half the mass of single copy sequence transcripts and nearly all their complexity are included in the rare complex class of maternal RNA. It follows that many of the prevalent repeat transcripts are associated with rare single-copy-sequence transcripts. Thus, most repetitive-sequence families must be represented on many different rare maternal transcripts (probably between 10 and 100). Consequently, the prevalent transcripts of each repeat family must be derived from many distinct transcription units, each including different single-copy sequences (Costantini *et al.*, 1980). Each repetitive-sequence family defines a set of transcripts from distant regions of the genome (Anderson *et al.*, 1981) that share this sequence. Much of the maternal RNA, then, consists of overlapping sets of repeat-containing transcripts. The functional significance of these sets or "gene batteries" is unknown, but it has been speculated that they might be recognized by other molecules, thereby mediating the regulation of gene expression (Britten and Davidson, 1969; Davidson and Britten, 1979).

The RNA used in Fig. 1 had a mass average size of about 5000 nucleotides (Costantini *et al.*, 1980). Posakony *et al.*, (1983) hybridized several cloned short, interspersed repeat DNAs to blots of egg poly(A)$^+$ RNA that had been denatured and separated by size. Examples of these RNA gel blots are shown in Fig. 2. The repeat-containing transcripts are nearly all quite large, ranging in size from 3 to 15 kb and include a variety of transcripts of different sizes. When these RNA gel blots were probed with separated strands of the cloned repeat DNAs, both strands were extensively represented, but the patterns were strand specific, as shown in Fig. 2. This confirms that both strands of individual repeat sequence families are transcribed, but that transcription is probably asymmetrical. That is, any particular repetitive sequence element is probably part of a single transcription unit having only one coding strand. As shown in Fig. 2, members of the spectrum of these transcripts containing a particular repetitive sequence change during development, and in some cases the total mass of the set of repeat-containing transcripts is augmented or depleted during development (see also Davidson and Posakony, 1982). Scheller *et al.* (1978) have shown that each of nine cloned repeats has embryonic stage-specific and adult tissue-specific patterns of representation. In older embryos and adult tissues, these prevalent repeat-containing transcripts are largely confined to the nucleus, and few can be detected on polysomes (Goldberg *et al.*, 1973; Scheller *et al.*, 1978). It is likely, then, that the repeat-containing transcripts observed in total poly(A)$^+$ RNA of later stage embryos (Fig. 2) reside in nuclei, but this has not been directly demonstrated. The cyto-

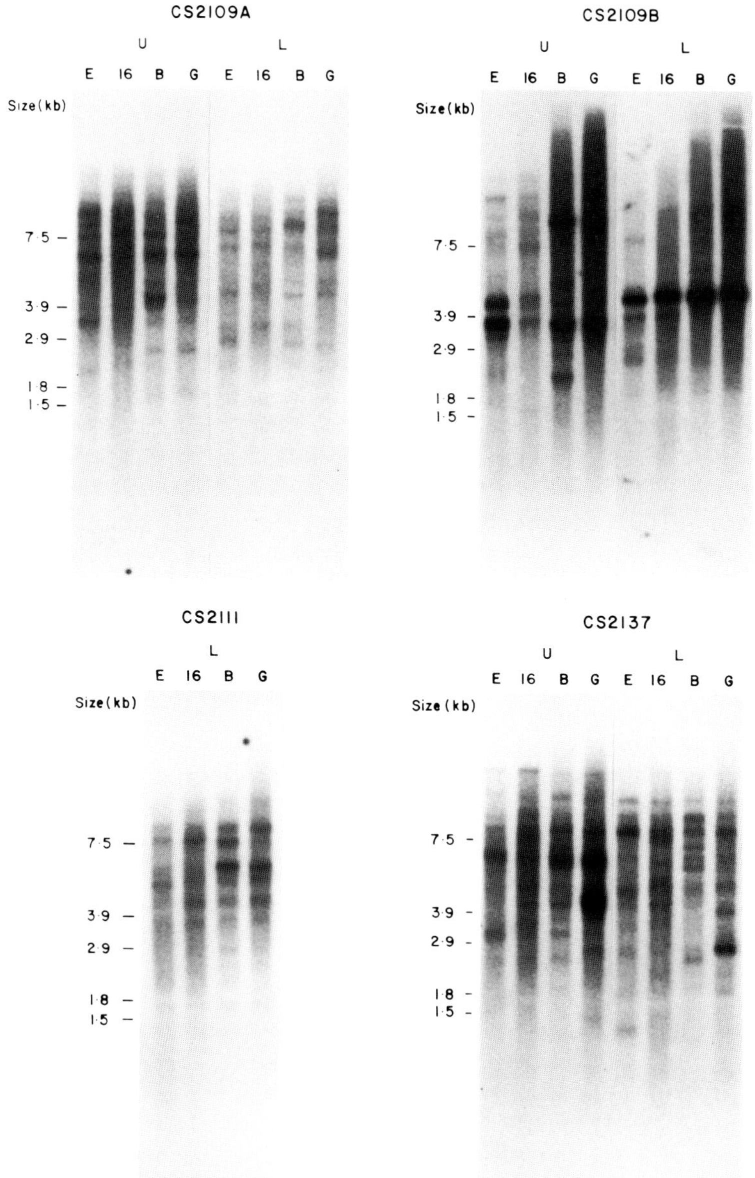


**Figure 2.** Gel blot hybridizations of strand-separated cloned repeat sequence DNA probes with poly(A)$^+$ RNA of developing embryos. Total poly(A)$^+$ RNA was prepared from *S. purpuratus* eggs (E), 16-cell stage embryos (16), blastulae (B), and gastrulae (G). Each lane of the gel was loaded with 2 $\mu$g poly(A)$^+$ RNA; electrophoresis and blotting were carried out on a formaldehyde-agarose gel according to Posakony *et al.* (1983). Cloned repetitive DNA sequences probes were strand separated electrophoretically, labeled with $^{32}$P, and hybridized with the blotted RNA. Approximate sizes of the RNA are indicated. U, upper strand; L, lower strand. (Reproduced with permission from Posakony *et al.* (1983). Copyright: Academic Press Inc., London, Ltd.)

plasmic localization of these large, interspersed repeat-containing transcripts in the egg might thus be a developmental accident, the result of the rupture of the germinal vesicle during meiotic maturation. Contrary to this possibility is the observation that the sequence complexity of nuclear RNA of oocytes and embryos is about five times that of the maternal RNA of the egg cytoplasm (Hough *et al.*, 1975; Hough-Evans *et al.*, 1977, 1979; Kleene and Humphreys, 1977). In *Xenopus* oocytes, a similar class of cytoplasmic poly(A)$^+$ RNA molecules includes interspersed repeat sequence transcripts even after manual removal of the intact nucleus (Anderson *et al.*, 1982). Indeed, these repeat-containing molecules are actively synthesized and accumulate in the cytoplasm during oogenesis of *Xenopus* (Anderson *et al.*, 1982).

Posakony *et al.* (1983) selected cloned cDNAs randomly transcribed from the entire set of egg poly(A)$^+$ RNA sequences (rather than only 3′ ends) that contained each of several cloned short, repetitive sequences. Single-copy DNA fragments of several of these cDNA clones were prepared, labeled with $^{32}$P, and hybridized to blots of poly(A)$^+$ RNA similar to those depicted in Fig. 2. As shown in Fig. 3, only one or a few discrete transcripts were observed, in contrast to the complex pattern observed for the adjacent repetitive sequences shown in Fig. 2. These single-copy DNA probes are represented asymmetrically in the maternal RNA, as shown in Fig. 3. Thus, these observations confirm the interpretation that different members of the interspersed repeat families are transcribed asymmetrically as part of different transcription units, which include adjacent single-copy sequences and members of other repeat families. A single repeat-containing transcription unit might give rise, though, to a more complex pattern of transcripts than that observed in Fig. 3. If a primary transcript is partially processed into a series of intermediates, a variety of smaller transcripts would correspond to the same transcription unit. Only those that retain the sequences used as probes would appear in Fig. 3. The DNA sequences of the single-copy probes 10HB and 18AR1 used in Fig. 3 contain multiple stop codons in all reading frames, indicating that they are not mRNA coding sequences (Posakony *et al.*, 1983). The appearance of new transcripts sharing the same single-copy sequence during development is most easily explained by new transcription and alternative processing, although in the case of 18AR1 the appearance of a smaller transcript could be indicative of processing of the preexisting larger transcript.

The very large sizes of many egg RNA molecules containing interspersed repeats (as shown in Fig. 2) suggest that calculations of the number of different RNA molecules based on an average size of 3000 nucleotides (see Section 4.2) may be an overestimate. The number of different transcription units represented in maternal RNA might be only about 4000–5000 (Posakony *et al.*, 1983). The egg poly(A)$^+$ RNA that predominates by mass (those of $1–5 \times 10^3$ nucleotides) may include only a fraction of the sequence complexity.

The arrangement of repetitive sequences in these large, interspersed

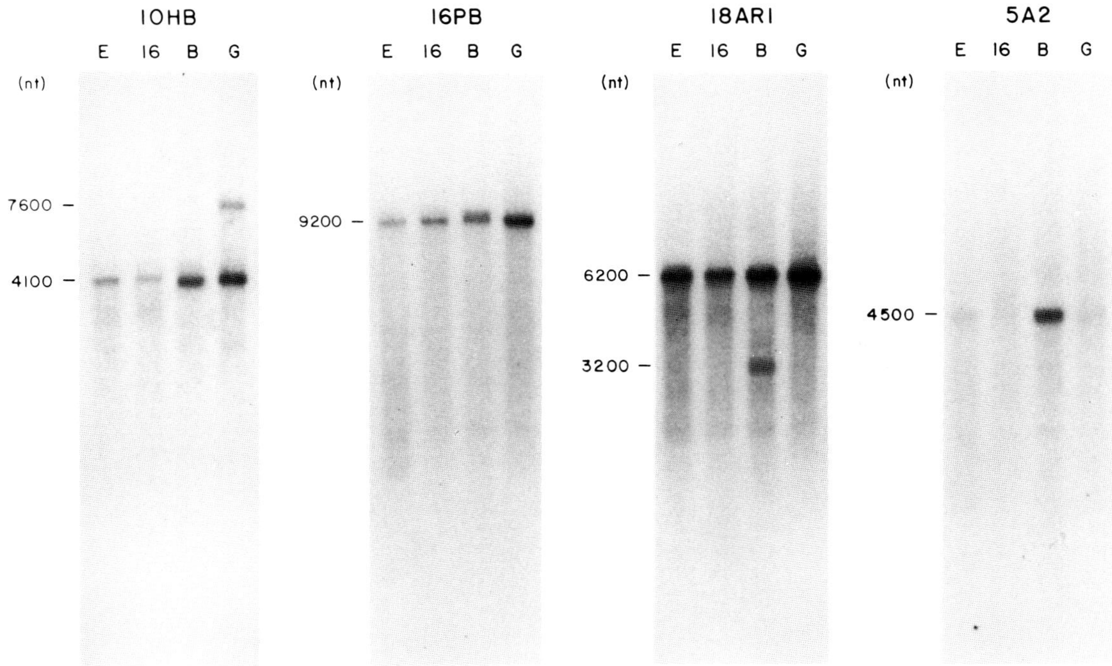


**Figure 3.** Gel blot hybridization of poly(A)$^+$ RNA with single copy DNA probes from repeat containing egg transcripts. Total poly(A)$^+$ RNA was prepared from eggs and embryos of *S. purpuratus* as in Fig. 2. On each lane 2$\mu$g of RNA was electrophoretically separated and blotted. Several cloned DNAs from a cDNA libary prepared from egg poly(A)$^+$ RNA were selected by hybridization to cloned repetitive DNA sequence probes. These cDNA clones were restriction mapped and single copy sequence fragments adjacent to the cloned repetitive sequences were prepared and labeled with $^{32}$P. These probes were hybridized separately to the RNA gel blots as shown. Probes 10HB, 16PB, and 18ARI were each derived from different egg transcripts including the 2109A repeat; 5A2 was derived from a clone containing the p2137 repeat. Approximate sizes of the transcripts detected are indicated. (Reproduced with permission from Posakony *et al.*,1983).

maternal RNA molecules is not known in detail. They could be predominantly near the ends, or they could be scattered throughout, much like the interspersion of intervening sequences (introns) between coding sequences (exons) of structural genes (Davidson *et al.*, 1982).

## 5. Functional Significance of Maternal RNA Molecules Resembling Nuclear Precursors of Messenger RNA

The structural features of the maternal RNAs, described in Sections 4.6 and 4.7, indicate that many bear a striking resemblance to nuclear RNA precursors of mRNA. Some are very large, like the nuclear RNA of sea urchin embryos (Kung, 1974); some are polyadenylated, whereas others are not. Some include tracts of oligo(A) and oligo(U), like the nuclear RNA of sea urchin embryos (Dubroff and Nemer, 1975; Dubroff, 1977) and other organisms (Nakazato *et al.*, 1974). Many have an interspersed repeat organization similar to that of sea urchin nuclear RNA (Smith *et al.*, 1974); both strands of most of these repetitive sequences are represented in the RNA as in sea urchin nuclear RNAs (Costantini *et al.*, 1978; Scheller *et al.*, 1978), so they can form complexes by annealing (Federoff *et al.*, 1977).

An investigation of the maternal transcripts of a particular gene of *S. purpuratus*, gene 88, reinforces this interpretation that maternal RNA and nuclear RNA are structurally similar. Gene 88 is predominantly represented by a rare 9.5-kb transcript in the egg cytoplasm (Lev *et al.*, 1980; Thomas *et al.*, 1982). Most of the gene 88 transcript is transcribed from DNA sequences repeated to various extents, but the 3′ end is transcribed from about 2 kb of single-copy sequence (Thomas *et al.*, 1982). By the 16-cell stage, nearly all the mass of the single-copy sequence at the 3′ end sediments with polysomes (Lev *et al.*, 1980; Thomas *et al.*, 1982). These are mostly a set of transcripts smaller than the egg transcripts, suggesting that processing, which conserves the 3′ end, has occurred, although nonrandom degradation during polysome preparation has not been excluded. By mid-blastula stage, gene 88 transcripts are no longer detectable in the cytoplasm, but they are detectable as nuclear transcripts throughout development (Lev *et al.*, 1980). RNA gel blots indicate that blastula and gastrula nuclear RNA preparations contain gene 88 transcripts predominantly 9.5 kb in length, which have the same 3′ end (and are transcribed from the same strand) as the 9.5-kb gene 88 transcript of the egg (Lev *et al.*, 1980; Thomas *et al.*, 1982). Thus, it is quite likely that these nuclear RNA molecules are initiated at the same site as the gene 88 egg transcripts and have the same structure. More detailed mapping of the nuclear and maternal gene 88 transcripts is required to rule out the possibility that they have different structures (due to differences in processing) that are fortuitously of the same size and share 3′ ends. Moreover, more information about the structure of the polysomal gene 88 transcripts and about the kinetics of their generation is required before it can be concluded that they are derived by processing from the large egg transcript.

These considerations have led to the interpretation that much of the maternal RNA of sea urchins is similar to unprocessed or partially processed precursors of mRNA (Costantini *et al.*, 1980; Duncan and Humphreys, 1981b; Davidson *et al.*, 1982; Davidson and Posakony, 1982; Thomas *et al.*, 1982). If so, they might be processed after fertilization to become actively translated mRNAs. It is not yet clear whether these large repeat-containing transcripts can be translated without further processing. The cloned interspersed repeats included in this maternal RNA that have been sequenced all have multiple stop codons in all reading frames (Posakony *et al.*, 1981, 1983), but some of the single-copy sequences might give rise to intact polypeptides. The annealable fraction of poly(A)$^+$ RNA of *Xenopus* oocytes is hardly translated in wheat germ or rabbit reticulocyte cell-free systems, even after denaturation (Richter *et al.*, 1983). Moreoever, this RNA fraction does not compete with globin mRNA nor does it enter polysomes as extensively as the nonannealable (single-stranded) poly(A)$^+$ RNA fraction when injected into *Xenopus* oocytes (Richter *et al.*, 1983). These observations suggest that this annealable RNA is not translated in *Xenopus* oocytes and probably not in embryos either. By contrast, the annealable poly(A)$^+$ RNA fraction of sea urchin eggs inhibits protein synthesis when added to the reticulocyte lysate as a double-stranded complex but is efficiently translated when denatured at high temperature (R. Moon, personal communication). To conclude that the annealable fraction of maternal RNA is not translated in early sea urchin embryos will require the demonstration that RNA having these structural features is absent from polysomes of the embryos; such a demonstration will be complicated by the anticipated lability of such large RNA molecules during polysome preparation.

Some maternal mRNAs (e.g., those coding for histones and actin) are of the same size as mature mRNA in polysomes of older embryos (Lifton and Kedes, 1976; Crain *et al.*, 1981; Scheller *et al.*, 1981b). By contrast, a family of mRNAs coding for ectodermal proteins synthesized in embryos include colinear repetitive sequence transcripts after maturation via splicing (Bruskin *et al.*, 1981, 1982; Carpenter *et al.*, 1983). Thus, some mRNAs including interspersed repeats can be actively translated in embryos.

If these maternal RNA molecules really are incompletely processed precursors of mRNA, do they become processed during embryogenesis? Since about 70% of the single-copy sequences of egg RNA are found in polysomes of early embryos (Hough-Evans *et al.*, 1977), it is likely that most of the transcription units giving rise to these molecules are included in the functional mRNA of embryos. Could the existence of large, untranslatable transcripts in the egg account for the low level of protein synthetic activity in eggs? This is unlikely, because the egg contains a substantial fraction of mRNA that is efficiently translated in exogenous and endogenous cell-free translation systems into proteins characteristic of the embryo (Jenkins *et al.*, 1973; Ruderman and Pardue, 1977; Infante and Heilman, 1981; Winkler and Steinhardt, 1981; Bédard and Brandhorst, 1984). Presumably these translatable molecules account for most or all of the very rapid increase in protein synthesis following fertilization. On the other hand, the large interspersed maternal transcripts might be processed during later embryonic development to give rise to functional mRNAs; the behavior of the gene 88 egg transcript is consistent with this possibility. If such processing does occur, some intriguing possibilities arise. These transcripts could be selectively processed as a mechanism for regulating the pattern of protein synthesis during later development. Moreover, they might be localized in particular regions of egg cytoplasm and segregated during cleavage, perhaps accounting for embryonic determination phenomena. The basis of such recognition and/or regulation might reside in the repetitive sequences.

The similarity of the cytoplasmic maternal transcripts to nuclear RNA suggests that they might be RNA leaked from nuclei. The limited complexity of the maternal RNA indicates that it is not a random representation of nuclear RNA transcribed by oocytes (Hough-Evans *et al.*, 1979). It is possible, though, that the processing of nuclear RNA transcripts destined to be stored mRNA is very inefficient and is not completed in the nucleus until after fertilization, when the efficiency might increase (see Section 10). Wickens and Gurdon (1983) have shown that SV40 viral precursor RNA is very inefficiently processed when injected into the germinal vesicle of *Xenopus* oocytes and that it enters the cytoplasm without being processed. Also consistent with inefficient processing is the observation that newly synthesized RNA enters the cytoplasm much more slowly in unfertilized eggs than in embryos, although this RNA cannot account for the mass of maternal RNA (Brandhorst, 1980).

Other possible functions of the large interspersed maternal RNA have been proposed and discussed (Costantini *et al.*, 1980; Thomas *et al.*, 1981; Davidson *et al.*, 1982; Davidson and Posakony, 1982). These transcripts may return to the nuclei of embryonic blastomeres, perhaps to serve a gene regulatory role. They might be involved in processing of newly synthesized RNA. They may have a role in monitoring the quantity or quality of cytoplasmic transcripts. They could represent functional mRNAs for which chain termination signals usually operating have been ignored, producing long 3′ extensions. This last possibility is consistent with the observation that during transcription of histone genes of urodele lampbrush chromosomes, readthrough of the usual chain termination signals occurs giving rise to transcripts having very long 3′ tails (Diaz *et al.*, 1981). However, these histone gene transcripts do not accumulate in the cytoplasm. Such tails might be without function when attached to properly spliced, functional mRNA. Finally, it is possible that many of these repeat-containing maternal RNA molecules have no function at all. They might be idle orphans indiscriminately spawned by selfish DNA.

The cloned single-copy sequences analyzed by Posakony *et al.*, (1983) show extensive intraspecific sequence polymorphism. Moreover, the abundance and sizes of repeat-containing transcripts sharing a given single-copy sequence are quite different in two closely related species, *S. purpuratus* and *S. franciscanus* (Posakony *et al.*, 1983). These observations, and the noncoding nature of the single copy sequences analyzed by Posakony *et al.* (1983), indicate that a significant portion of each interspersed maternal RNA consists of nontranslatable sequences, at least part of which can diverge rapidly. The sequences of two cloned 2109A repeat-family members terminate with inverted repeat elements, and at least one is surrounded by direct repeats (Posakony *et al.*, 1983). These structural features suggest that the 2109A repeat elements are transposable (e.g., Will *et al.*, 1981; Flavell and Ish-Horowicz, 1981; Potter, 1982). The rapidity of changes (on an evolutionary time scale) in family sizes and organization of individual repeat families (Moore *et al.*, 1978; Scheller *et al.*, 1981a; Posakony *et al.*, 1981) (see Section 4.1) suggest that many repeat elements do transpose; this could account for their dispersal throughout the genome during or after sequence multiplication. These transposing repeat elements should sometimes become harmlessly embedded in active transcription units as part of intervening sequences or noncoding tails. If the transcripts of such transcription units are not fully processed or do not require further processing, they could be expected to appear as part of the maternal RNA population. If, on the other hand, insertion of the repetitive sequences prevents processing and/or translation of the transcript, it might still accumulate during oogenesis, when all cytoplasmic RNA might be quite stable, and serve no purpose whatsoever. This could be allowed if the structural genes were part of a small gene family, only one of which might need to remain functional. If repeat sequences are represented in RNA simply because they happen to have been transposed into the transcription unit of an active gene, their transcriptional expression would have the same developmental pattern as that of the structural gene. Thus, changes in repeat sequence representation observed for nuclear RNA would be the consequence, rather than mediator, of developmental regulation of gene expression. These considerations do not exclude the possibility that the sets of transcription units defined by common repetitive sequence families have some functional significance, at least in some cases.

An understanding of the relationship of these large interspersed maternal RNA molecules to mRNA molecules and nuclear precursors of mRNA requires a more complete knowledge of the structures of some cloned polysomal mRNAs of embryos and the large maternal RNA molecules with which they share sequence homologies. An understanding of the functional significance, if any, of the repetitive sequence transcripts included in maternal RNA is likely to provide a key to understanding the mysterious presence of interspersed repetitive DNA sequences in animal cells.

## 6. Mass of Maternal RNA Stored in the Egg

Estimates of the mass of stored maternal mRNA are dependent on the operational definition of mRNA. The amount of template-active mRNA in sea urchin eggs has been estimated by cell-free translation in systems derived from bacteria and eukaryotes to be 4–5% of the total RNA (Slater and Speigelman, 1966; Jenkins *et al.*, 1973; Moon *et al.*, 1982). Wilt (1977) showed that about 1.5% by mass of the RNA of eggs of *L. pictus* has tracts of poly(A) sufficiently long to bind oligo(dT)-cellulose. Other estimates of the fraction of RNA that is polyadenylated range from 0.75% to 2% (Nemer *et al.*, 1974; Duncan and Humphreys, 1981a). A larger fraction of egg RNA binds to poly(U)-Sepharose (Wilt, 1977), but the excess binding is probably due to selection of oligo(A) tracts (Wilt, 1977; Duncan and Humphreys, 1981b). Since RNA containing oligo(A) tracts is underrepresented in polysomes of embryos (Duncan and Humphreys, 1981b), the oligo(A)-containing RNA apparently does not function as mRNA, unless it is processed during later development. Within 2–3 hr after fertilization, nearly all the poly(A)$^+$ RNA is in polysomes, and this RNA includes molecules that are poly(A)-deficient in eggs (Wilt, 1977; Dolecki *et al.*, 1977). Thus, most of the mass of poly(A)$^+$ RNA in the egg can be considered to be mRNA. In addition, maternal mRNA includes a similar mass of poly(A)-deficient RNA that is translatable in cell-free systems (Ruderman and Pardue, 1977). The fraction of mRNA molecules containing poly(A) in polysomes of zygotes is unknown but is about one-half in later embryos (Brandhorst *et al.*, 1979). These considerations suggest that 2–4% of the egg RNA is translated during the first few hours of embryogenesis. There could also be a large class of maternal mRNA that is not actively translated until later in development, possibly after processing. RNA in the postribosomal supernatant of early embryos is translatable in cell-free systems (Infante and Heilman, 1981), albeit less efficiently than polysomal RNA (Rudensey and Infante, 1979).

The mass of maternal mRNA can also be estimated from the rate of hybridization of a single-copy tracer to excess egg RNA (Galau *et al.*, 1974). Thus, as shown in Table I, about 1–2% of the egg RNA drives the hybridization reaction. The uncertainty in these estimates is due to ignorance of the length of hybridizing RNA molecules (Galau *et al.*, 1976b) and to inaccuracies in estimating the rate constant of the pseudo-first-order reaction. In the sea star *Pisaster ochraceus*, 2–3% of the oocyte RNA drives the reaction (Kosvedi and Smith, 1982). Nearly all the sequence complexity of egg RNA is in the rare or complex class of RNA, which drives the reaction (see Section 4.2). These rare sequences are each present on average in 1500–4000 copies per egg in sea urchins. Rare sequences are present in similar concentrations in the various sea urchin eggs, in the larger oocytes of *P. ochraceus*, and in a variety of other vertebrate and invertebrate eggs (reviewed in Thomas *et al.*, 1981; Kosvedi and Smith, 1982) (see also Table I).

A prevalent class of RNA also exists in the egg, representing about half of the mass of heterogeneous RNA (RNA that is large but not rRNA) but less than 10% of the complexity (Wilt, 1977; Duncan and Humphreys, 1981a; Lasky *et al.*, 1980). Thus, the total mass of heterogeneous RNA in the sea urchin egg estimated by nucleic acid hybridization kinetics is 2–4% of the RNA or 66–132 pg in *S. purpuratus*. This estimate is in close agreement with estimates of the amount of template-active RNA, suggesting that most of the heterogenous RNA of the egg can function as mRNA. It should be noted, however, that some single-copy RNA sequences in the egg are not represented in polysomes of early embryos (Hough-Evans *et al.*, 1977) (see Section 8.1).

Thus, estimates based on a variety of definitions of mRNA suggest that approximately 2–4% of the RNA of the sea urchin egg is actually translated after fertilization. From about 2–6 hr after fertilization about 20% of the ribosomes of the egg are polysomes (Humphreys, 1971; Wells *et al.*, 1981; Goustin and Wilt, 1981). If 2–4% of the egg RNA is actively translated mRNA during this period, then 10–20% of the mass of RNA in polysomes is mRNA, far higher than the 2% typical of a variety of other cells (reviewed in Humphreys, 1971). By gastrula stage, 3–4% of the RNA in polysomes is newly synthesized mRNA (Brandhorst and Humphreys, 1972; Galau *et al.*, 1976a), and an unknown amount of maternal mRNA may still be translated. It follows that either the mass of RNA entering polysomes after fertilization is greatly overestimated (unlikely) or that polysomes in sea urchin embryos are underloaded with ribosmes relative to other animal cells. Underloading of ribosomes on sea urchin mRNA could occur for either or both of two reasons. Peptide initiation could be slow relative to elongation, or the coding sequence to which ribosomes bind may constitute an unusually small fraction of the average sea urchin mRNA molecule. The latter possibility is consistent with the large size and peculiar structure of much of the mass of maternal RNA, particularly if these molecules have long untranslated tails (see Sections 4.7 and 5). Hille *et al.* (1981) have reported that the treatment of early embryos with emetine, an inhibitor of polypeptide chain elongation, does not lead to an increase in the size or number of polyribosomes. This indicates that initiation is not rate limiting, consistent with there being a short, fully loaded coding sequence on a large underloaded maternal mRNA molecule. Thus, much of the heterogeneous egg RNA is either not translated during early embryonic development or is inefficiently translated on a mass basis probably due to its peculiar structure. Electron microscopic analyses of polysomes prepared from zygotes provide no evidence for polysomes having long tails free of ribosomes (Martin and Miller, 1983).

## 7. Activation of Translation of Maternal Messenger RNA

### 7.1. Masking and Other Hypotheses

The mechanism by which maternal RNA of sea urchins is stored in a translationally inactive state in the egg and then recruited into polysomes after fertilization is not understood. The most favored hypothesis, first proposed by Spirin (1966), is that the maternal mRNA is "masked" by some factor—probably protein—before fertilization, and an early postfertilization event leads to the release of this mask (reviewed by Raff, 1980; Raff and Showman, 1983). The masked mRNA is thus presumed to be in the form of translationally inactive RNP particles. Strong evidence for the masked mRNA hypothesis came from investigations in which RNPs derived from eggs were found to be inefficiently translated in a heterologous cell-free system as compared to deproteinized RNA extracted from these particles or mRNA extracted from polysomes (Jenkins *et al.*, 1978; Ilan and Ilan, 1978). The translational activity of the egg RNPs depended on how they were prepared and on the ionic conditions used for the *in vitro* translation (Jenkins *et al.*, 1978; Ilan and Ilan, 1978). This conclusion has recently been challenged by Moon *et al.*, (1982) in a thorough investigation. They concluded that, under a wide variety of conditions, RNA in egg RNPs is not significantly less efficiently translated than RNA purified from these RNPs. From these reports it is evident that the translational activity of RNPs depends on the method used for their preparation and the conditions used for their translation. Since there are no criteria available to define what the appropriate conditions are, cell-free translation analyses are unlikely to provide a suitable test of the masked mRNA hypothesis.

Several investigations of the physical properties and protein composition of mRNPs of eggs and RNPs released from polysomes of embryos have failed to provide consistent and compelling evidence for major differences in their structure (Kaumeyer *et al.*, 1978; Young and Raff, 1979; Moon *et al.*, 1980; Moon, 1983). No putative masking protein has yet been identified.

A different type of experiment suggests that maternal mRNA is not masked in *Xenopus* oocytes. When various amounts of rabbit globin mRNA were injected into ooctytes, the overall rate of protein synthesis was unaltered. The globin mRNA competitively decreased the incorporation of amino acids into endogenous proteins (Laskey *et al.*, 1977). Injection of reticulocyte polysomes, on the other hand, led to an overall increase in protein synthesis and did not reduce the rate of synthesis of endogenous proteins. These experiments indicate that there is a limited translational capacity saturated in the oocyte. Thus, the availability of translatable mRNA is not rate limiting, and the limiting component is present on polysomes. These observations are difficult to reconcile with the masked mRNA hypothesis. Similar conclusions have recently been drawn for sea urchin eggs injected with globin mRNA (A. Colin and M. Hille, personal communication).

There are a variety of other possible mechanisms of translational restriction in the sea urchin egg. Maternal mRNA may be physically sequestered, perhaps bound to (or within) the membranous vesicles that abound in sea urchin eggs. The sea urchin $\alpha$-type histone mRNAs are sequestered in the female pronucleus (see Section 7.3). Whereas other maternal mRNAs are not sequestered in the pronucleus (Venezky *et al.*, 1981), they might be sequestered in other structures such as the "heavy bodies" or annulate lamellae (Harris, 1967; Afzelius, 1957) or other membranous structures (Mano and Nagano, 1970). Another interesting possible form of sequestration involves the detergent-insoluble cytoskeleton of the egg. Most embryonic polysomes—unlike most translatable maternal mRNAs of the egg—are associated with the cytoskeleton, at least under certain ionic conditions (Moon *et al.*, 1983). Fertilization might promote an interaction of maternal mRNA with elements of the cytoskeleton, possibly facilitating its translation.

Some maternal mRNA might be in a structurally untranslatable form. Cell-free translation in heterologous systems of egg RNA (reviewed in Section 3) indicates that much of it is translatable. Moreover, careful analysis of the structural features of histone mRNAs of eggs and embryos revealed no differences (Lifton and Kedes, 1976). Minor alterations in the mRNA, such as loss of some secondary structure, might result in translational activation. Translational activation might require the enhancement of the activity of a translational effector, particularly one that remains associated with the mRNA to which it is bound (thereby maintaining the size distribution of polysomes). Such a mechanism is consistent with the experimental observations of Laskey *et al.*, (1977), and the activation could be dependent on intracellular ionic or pH conditions (Grainger *et al.*, 1979).

## 7.2. RNA Synthesis and Translation in Eggs

Unfertilized sea urchin eggs synthesize proteins at a low rate. A comparison of the proteins labeled with [$^{35}$S]methionine by two-dimensional electrophoresis indicates that virtually all the 1000 polypeptides detected are synthesized in the same relative amounts both before and after fertilization, in spite of massive recruitment of mRNA into polysomes (Brandhorst, 1976; Tufaro and Brandhorst, 1979; Bédard and Brandhorst, 1983). Whereas some proteins were not included in these analyses (particularly basic proteins such as histones), most of the mass of newly synthesized protein appeared on the gels, although the fraction of this mass giving rise to autoradiographically detectible spots is unknown (Brandhorst, 1976). Clearly, the increased rate of protein synthesis upon fertilization is not due to recruitment of a qualitatively distinct set of sequences into polysomes. On the contrary, larger numbers of the same sequences are translated after fertilization. Since the efficiency of translation of mRNA does not change much upon fertilization (Humphreys, 1969) (see also Section 3), the same sequences must exist in the egg in two states: fully translatable or translationally inactive.

A clue to the distinction between active and inactive mRNA in the egg comes from investigations of RNA metabolism in sea urchin eggs, which actively synthesize RNA (Brandhorst, 1980). Whereas a substantial fraction of the RNA synthesized in eggs is of mitochondrial origin (Ruderman and Schmidt, 1981), the rate of RNA synthesis in the nucleus is similar to the highest rates observed per haploid genome during embryonic development (Brandhorst, 1980). Much of this newly synthesized egg RNA is large, heterogeneous in size, and polyadenylated; some slowly accumulates in the cytoplasm, and some becomes associated with polysomes. At steady state, about 3.7% of the mass of RNA in polysomes is newly synthesized heterogeneous RNA, indicating that much—possibly all—of the RNA translated in eggs is newly synthesized. The labeled RNA of eggs that is not in polysomes cannot be chased into polysomes after fertilization, i.e., it does not behave as stored maternal mRNA. These observations indicate that the egg contains two populations of mRNA sharing most sequences: an unstable, translatable population actively synthesized by the egg, and a stable, translationally inactive population accumulated during oogenesis. These two populations can thus be distinguished by their times of synthesis. It might be that the oocyte uses this temporal distinction to select and store maternal mRNA, perhaps by providing a "masking" and stabilizing factor only during an appropriate period of oogenesis (Brandhorst, 1980).

### 7.3. Pronuclear Sequestration of Maternal Histone Messenger RNA

The rate of protein synthesis in sea urchin eggs begins to increase within 5 min of fertilization (Epel, 1967; Raff *et al.*, 1981), and the initial mobilization of maternal mRNA into polysomes is completed by 2–3 hr (see Section 3). The maternal mRNA coding for the major histone variants synthesized during early embryonic development (the $\alpha$-histone subtypes) do not follow these kinetics of mRNA recruitment. They do not begin to enter polysomes until 1.5–2 hr after fertilization; they begin to appear in polysomes about the time of first cleavage (Woods and Fitschen, 1978; Wells *et al.*, 1981). These histone mRNAs, which are detected on RNA blots by hybridization with specific cloned histone DNA probes, were always found only in nucleated fragments of eggs produced centrifugally or surgically (Showman *et al.*, 1982, 1983). *In situ* hybridization indicates that histone mRNAs are highly concentrated in the egg pronucleus, whereas poly(A) and specific polyadenylated mRNAs are distributed evenly throughout the cytoplasm (Venezky *et al.*, 1981). Upon mitosis, these histone mRNAs are released from the nucleus and begin to appear on polysomes. This release from nuclei does not appear to require nuclear envelope breakdown, since it is not inhibited by 6-dimethylaminopurine or aphidicolin, which prevent this breakdown (Showman *et al.*, 1982, 1983). These histone mRNAs, however, are not localized in unfixed nuclei prepared from lysates of eggs (Showman *et al.*, 1983). Consequently, the observation that histone mRNAs are localized in small RNP particles in the egg cytoplasm (K. W. Gross *et al.*, 1973; Skoultchi and Gross, 1973) is an artifact of nuclear leakage during cellular fractionation.

The $\alpha$-histone mRNAs are not detectable in the germinal vesicles or cytoplasm of sea urchin oocytes isolated from ovaries, indicating that they accumulate in the egg pronucleus after the completion of meiosis (Angerer *et al.*, 1984). Thus, accumulation of stored materials occurs even after the completion of meiotic maturation, which is normally regarded as the completion of the period of oogenesis. The egg contains a store of mRNA coding for another set of histone variants as well, the CS-histone subtypes, which are synthesized only during oogenesis and the first few cleavages (Newrock *et al.*, 1977; Childs *et al.*, 1979). The CS-mRNAs do not appear to be sequestered in the egg pronucleus (Showman *et al.*, 1982). Unfertilized sea urchin eggs actively synthesize histone mRNAs having electrophoretic mobilities similar to those of the $\alpha$-subtypes (Brandhorst, 1980; Ruderman and Schmidt, 1981). The synthesis of only the CS-type histones can be detected in eggs (Herlands *et al.*, 1982). Thus, there is a distinction made by the egg in the synthesis, localization, and utilization of different types of histone mRNAs.

While $\alpha$-histone genes in tandemly repeated arrays have been detected in all species of the several classes of echinoderms investigated, the storage of $\alpha$-histone mRNA in the pronucleus has been observed only for the more recently established echinoid orders (Raff *et al.*, 1984). In asteroids and ophiuroids, $\alpha$-histone mRNAs accumulate during cleavage but are not detectably stored in oocytes. Eggs of the primitive cidaroid sea urchins also lack a store of $\alpha$-histone mRNAs. On the other hand, all echinoderm oocytes and eggs apparently contain CS-histone mRNA. Thus the pronuclear sequestration of $\alpha$-histone mRNA for use during early cleavage is a relatively recent event, since the modern species in which it occurs all arose during the late Triassic (i.e., about 200 million years ago). Even in these modern echinoids, most of the mass of $\alpha$-histone mRNA is synthesized during cleavage and the necessity for the pronuclear store of maternal $\alpha$-histone mRNA is not established: Anucleate fragments of eggs can develop relatively normally when fertilized by sperm, which do not contain a store of histone mRNA. These observations suggest that the pattern of synthesis and accumulation of histone mRNA (and, perhaps many other mRNAs) during early embryogenesis is established upon meiotic maturation rather than fertilization. The pronuclear sequestration of $\alpha$-histone mRNAs may be required to ensure that only CS-histones are present during the first cleavage cycle when remodeling of sperm chromatin occurs.

### 7.4. Translational Changes During Meiotic Maturation of Starfish Oocytes

After induction by 1-methyladenine of maturation of oocytes of the starfish *Asterias forbesi*, there is a change in the pattern of protein synthesis as analyzed by electrophoresis (Rosenthal *et al.*, 1982). These changes occur rapidly upon GVBD and are translationally mediated. The same population of mRNAs is present before and after germinal vesicle breakdown (GVBD), but the distribution of sequences on polysomes changes (Rosenthal *et al.*, 1982). The changes in protein synthesis are not due to release of mRNAs (or selective factors required for their translation) sequestered in the GV, since 1-methyladenine stimulates similar changes in anucleate fragments of oocytes (Martindale and Brandhorst, 1983). These results suggest that some mRNA sequences are selectively translated in oocytes or eggs. Alternatively, the changes may be due to the nonselective activation of translation of a different, but overlapping, population of maternal mRNA sequences stored in the oocyte. Such an explanation is consistent with the increase in the rate of protein synthesis, which also begins upon GVBD, and the continued detection of synthesis of oocyte-enriched polypeptides following GVBD, albeit at reduced relative rates.

A change in the state of polyadenylation of some mRNA sequences upon GVBD has been observed (Rosenthal and Brandhorst, unpublished observations). Messenger RNAs that are more actively translated before maturation tend to lose poly(A) tracts, whereas those that are more actively translated after GVBD become more completely polyadenylated. Similar changes in protein synthesis and state of polyadenylation have been observed upon GVBD in oocytes of the surf clam, *Spisula* (Rosenthal *et al.*, 1983). Upon fertilization of *Asterias* eggs, no similar changes in patterns of protein synthesis occur for several hours, as in sea urchins. It is likely that the pattern of protein synthesis (and possibly transcription as well) characteristic of early echinoderm embryos is established during meitoic maturation, although quantitative adjustments can be made post-transcriptionally, depending on the metabolic state of the egg. These patterns of protein synthesis in the egg and early embryos are dependent on maternal information.

## 8. Stability of Maternal Messenger RNA

### 8.1. Persistence of Maternal RNA Sequences During Embryogenesis

A variety of investigations indicate that many sequences included in the stored maternal RNA of eggs remain throughout embryonic development of sea urchins. In many cases the maternal mRNA sequences are replaced by the same sequences transcribed from the embryonic genome; in some cases, however, the maternal mRNA may be quite stable (see Section 8.2). Hough-Evans *et al.* (1977) prepared a single-copy DNA fraction enriched for sequences complementary to RNA extracted from eggs. This DNA was hybridized with RNA in excess prepared from eggs and either total cytoplasm or polysomes of embryos of various stages. Figure 4 summarizes the results of these reactions, which are sensitive only to changes in the complex class of rare mRNAs. All the rare sequences present in the egg are maintained in embryos for several hours, but there is a gradual decline in the fraction of egg sequences detectable in the cytoplasm during embryonic development; by pluteus stage, only about 40% of the egg sequences remain detectable. In early embryos, 73% of the egg sequences are detected on polysomes, and these decline during development as well. This indicates that not all RNA sequences in the egg are functionally utilized as mRNA at any particular stage, suggesting a class of translationally inactive RNAs in embryos (see Section 8.3) or that translationally inactive sequences are eliminated from some maternal mRNAs by processing before they enter polysomes. By the pluteus stage, the complexity of polysomal and total cytoplasmic RNA coincide (see also Galau *et al.*, 1976a).

Virtually all the rare sequences in polysomes at gastrula and pluteus stages are present in the egg as well (Galau *et al.*, 1976a). On the other hand, the sequence complexity of polysomal RNA of blastulae is greater than the complexity of maternal sequences in blastulae (Galau *et al.*, 1976a; Hough-Evans *et al.*, 1977). This implies that at blastula stage a unique set of

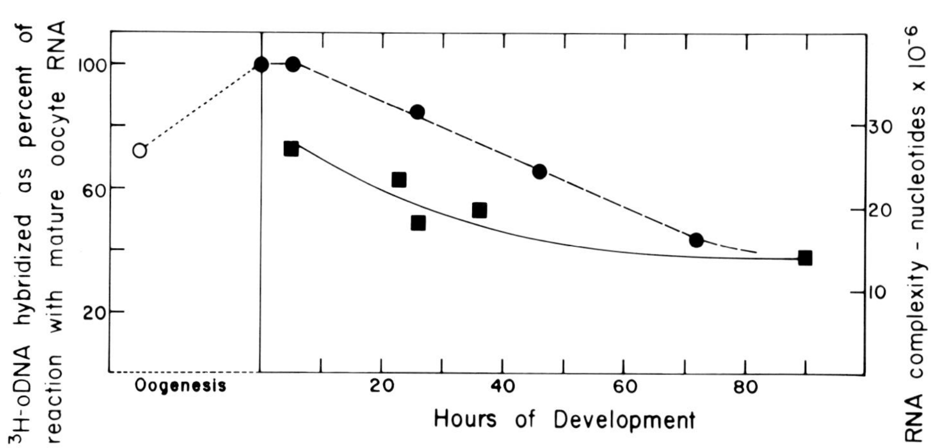


**Figure 4.** Representation of egg RNA sequences during embryonic development. A single-copy DNA tracer enriched for egg RNA sequences was hybridized to excess RNA prepared from *S. purpuratus* eggs, ovaries, and embryos of various stages. Saturation values are plotted as percentages of the level of hybridization of the tracer with egg RNA. (○) ovary RNA; (●) eggs and cytoplasmic RNAs; (■) polysomal RNAs. (Reproduced with permission from Hough-Evans *et al.*, 1977.)

sequences derived by transcription of the embryonic genome is being translated, but this set is not retained at later stages (Galau *et al.*, 1976a; Hough-Evans *et al.*, 1977; Shepherd and Nemer, 1980).

The developmental representation of more prevalent RNA sequences has been analyzed by the use of recombinant cDNA libraries (Laskey *et al.*, 1980; Flytzanis *et al.*, 1982). More than 90% of several hundred cloned abundant poly(A) RNA sequences of the egg (i.e., those present in the range of $5 \times 10^4$–$10^6$ copies/egg) are present in similar abundance in the cytoplasm of plutei; there is usually an apparent decline in abundance at the gastrula stage, suggesting that many molecules present at the pluteus stage have been synthesized during embryonic development to replace maternal RNAs that have decayed (Flytzanis *et al.*, 1982). These observations indicate that the abundant maternal RNA sequences are even more preserved than are rare sequences during embryonic development.

The observed persistence of RNA sequences during embryogenesis suggests that many of these shared sequences might code for ubiquitous proteins, those having a "housekeeping" role. Contrary to this possibility are observations that less than 20% of the rare embryonic sequences of the gastrula are detectable in several adult tissues (Galau *et al.*, 1976a) and that only 15% of the moderately prevalent cloned mRNA sequences of the pluteus are detectable in adult coelomocytes (Xin *et al.*, 1982). These results indicate that substantial regulation of the mRNA population does occur during sea urchin development.

Another analytical approach for comparing mRNA sequences is two dimensional electrophoretic analyses of newly synthesized proteins. As shown in Fig. 5, such investigations indicate that most proteins that are synthesized in eggs and translated from stored maternal mRNAs after fertilization are synthesized in similar relative amounts throughout embryonic development until at least early pluteus stage (Brandhorst, 1976; Bédard and Brandhorst, 1983; Brandhorst *et al.*, 1983). About 20% of approximately 800 polypeptides analyzed in *S. purpuratus* or *L. pictus* undergo detectable changes in relative rates of synthesis; only 10% undergo changes of at least 10-fold, and only about 1% undergo changes of more than 100-fold (Bédard and Brandhorst, 1983; Brandhorst *et al.*, 1983). Very few changes occur during the first 10–12 hr of development, but many changes occur between the time of hatching and the beginning of gastrulation. In particular, nearly all the polypeptides of zygotes whose synthesis becomes undetectable during development disappear before gastrulation (Brandhorst *et al.*, 1983), suggesting that their respective pools of maternal mRNA have been exhausted by then and are not replaced. Overall, there is a slight increase during embryogenesis in the number of polypeptides detectably synthesized.

Several polypeptides have been identified whose synthesis is greatly enriched in ectoderm, endoderm, or mesenchyme of urchin embryos (Bru-

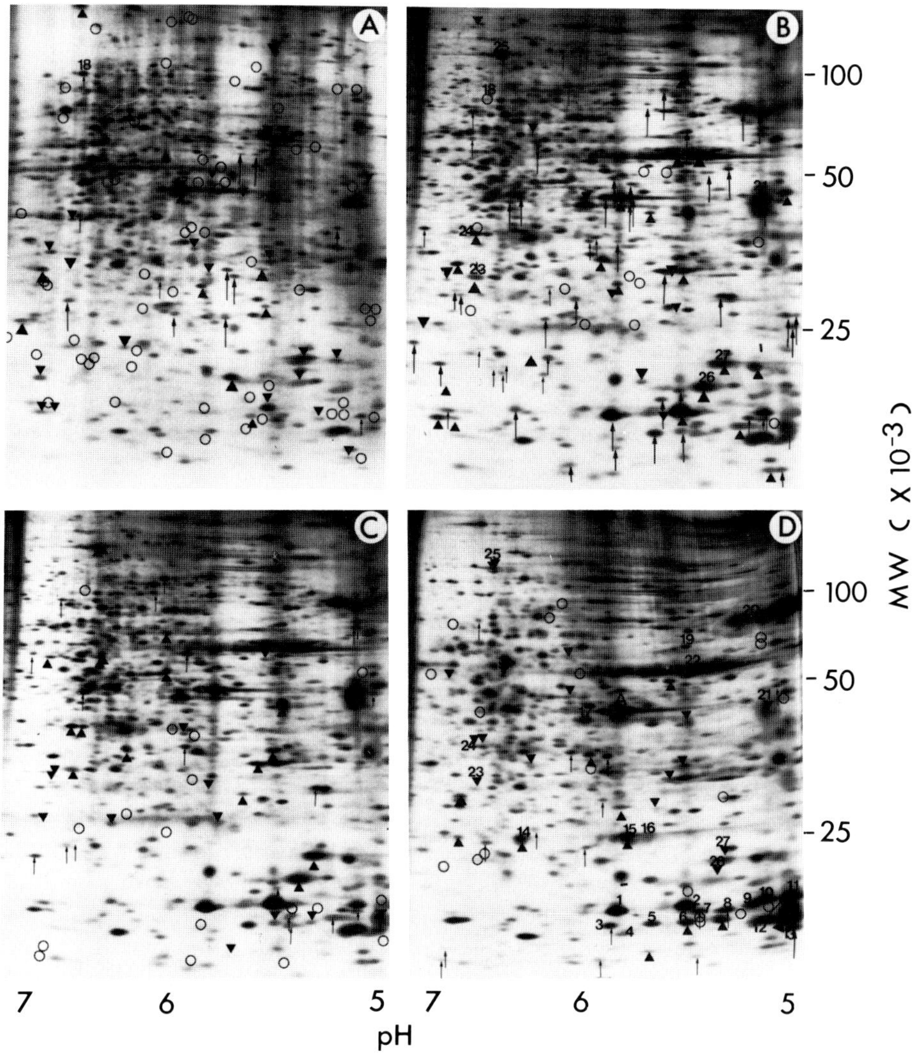


**Figure 5.** Comparison of proteins synthesized by zygotes and advanced embryos. Proteins were extracted from S. *purpuratus* embryos pulse-labeled with [$^{35}$S]methionine and separated by two-dimensional gel electrophoresis. Autoradiography exposures were adjusted for equal emissions from total radioactive protein. Two-cell stage embryos shown in (A) are compared with late mesenchyme blastulae in (B), and the late mesenchyme blastulae (C) are compared with plutei in (D). Arrows point to spots (corresponding to newly synthesized polypeptides) detected at one stage but not at the other stage compared; circles indicate the corresponding area of the missing spot on the gel compared. Triangles pointing up indicate that the relative intensity of that spot is greater at that stage, while triangles pointing down indicate that the relative intensity of that spot is less at that stage. The relative rates of synthesis of about 80% of the polypeptides did not change between fertilization and the pluteus stage. Numbered spots are discussed in Bédard and Brandhorst (1983).

skin *et al.*, 1982; Bédard and Brandhorst, 1983; Brandhorst *et al.*, 1983; Harkey and Whiteley, 1982). Synthesis of most of these polypeptides is barely detectable or absent in eggs but increases greatly during embryogenesis. It is thus likely that these tissue-specific proteins are specified by mRNAs that are absent or rare in the maternal mRNA population, but which accumulate during development; in some cases this has been directly demonstrated (Bruskin *et al.*, 1982; Fregien *et al.*, 1983). Adult tissues display patterns of protein synthesis very different from the rather conserved pattern of eggs and embryos (Bédard and Brandhorst, unpublished observations).

## 8.2. Persistence and Decay of Maternal RNA Molecules

The lack of a convenient method for distinguishing maternal transcripts from those synthesized in embryos has made it difficult to analyze the metabolism of maternal RNA in embryos. Quantitative investigations of the accumulation of newly synthesized RNA in embryos suggest that by gastrula stage most or all of the mRNA being translated into polysomes is newly synthesized, implying that maternal mRNA has disappeared (Brandhorst and Humphreys, 1972; Galau *et al.*, 1977). In particular, 3–4% of the mass of RNA in gastrula polysomes is newly synthesized mRNA; there is, however, no direct measurement of the total mass of mRNA in gastrula polysomes, which may be underloaded with ribosomes (see Section 6). Thus, a substantial mass of maternal mRNA might persist in polysomes of advanced embryos.

The kinetics of accumulation of several newly synthesized cloned RNA sequences in the cytoplasm of embryos during development have been analyzed recently making appropriate corrections for precursor pool specific radioactivities (Cabrera *et al.*, 1983). These studies demonstrate that various newly synthesized rare sequences turn over rapidly. By the time of gastrulation, their entire mass can be accounted for by newly synthesized transcripts, indicating that the corresponding maternal transcripts have decayed and been replaced. The accumulation of several RNA sequences, which are abundant in the egg and throughout embryonic development, behave quite differently; they accumulate slowly as very stable transcripts. Even by prism stage, accumulation of newly synthesized RNA cannot account for the total mass of these sequences, indicating that the maternal RNA molecules have persisted. These investigations did not directly analyze the metabolism of the maternal mRNA molecules. The recent identification of cloned cDNA probes for maternal RNA sequences that disappear during early embryonic development should make it possible to determine directly the stability of some maternal mRNAs (Lev *et al.*, 1980; Brandhorst, unpublished observations).

A fundamental unanswered question is whether maternal and newly synthesized embryonic forms of a mRNA have the same stability in the embryo or are metabolically distinct. The amount of histone H3 mRNA is constant during the first four cleavages (Wells *et al.*, 1981), during which time the contribution of new synthesis is negligible (Maxson and Wilt, 1981). Thus, the half-life of maternal H3 mRNA must be at least several hours during the first few hours of development. Accumulation of newly synthesized histone mRNA between the 16- and 200-cell stages indicate that it has a half-life of at least 3 hr (Maxson and Wilt, 1982). Later, the accumulated newly synthesized histone mRNA disappears rapidly with a half-life of 1.5–2 hr, indicating that the stability of histone mRNA is regulated during cleavage (Maxson and Wilt, 1982). Since the maternal histone mRNA content is quickly overwhelmed by newly synthesized mRNA, its metabolism cannot be analyzed even indirectly.

Another experimental approach to determining the persistence of maternal RNA molecules during development is based on the use of interspecies hybrid embryos. In three different sea urchin hybrid crosses, the synthesis of most paternally distinct proteins is never detected even in plutei (Tufaro and Brandhorst, 1982). RNA sequence analyses indicate that most prevalent paternal cytoplasmic RNA sequences of gastrula stage embryos are missing, or reduced in prevalence, in hybrid embryos in a cross of *S. purpuratus* eggs with *L. pictus* sperm (Tufaro and Brandhorst, 1982). Many rare paternal RNA sequences and some specific paternal proteins are found in the hybrid embryos by gastrula stage. One possible interpretation of these observations is that many abundant maternal mRNAs persist throughout embryonic development without being significantly replaced by newly synthesized transcripts, in which case these paternal sequences should not be abundant in the hybrid embryos. Contrary to this interpretation is the observation that some specific paternal mRNA sequences (detected on RNA blots with cloned DNA probes), accumulate greatly in mass in embryos of the maternal species but not in hybrid embryos (Tufaro and Brandhorst, unpublished observations). Thus, the expression of at least some paternal genes is perturbed in hybrid embryos. This observation indicates that interpretation of paternal gene expression in hybrid embryos should be cautious (e.g., Showman and Whiteley, 1980). It remains possible that some paternal mRNAs do not appear in hybrid embryos because they correspond to persistent maternal transcripts.

## 8.3. Selective Translation of Maternal Messenger RNA in Embryos?

The possibility that some maternal transcripts persist through a long period of embryonic development raises the possibility that they might function as mRNA only at later embryonic stages. Bédard and Brandhorst (1984) have addressed this possibility by comparing the *in vitro* translation products of RNA extracted from polysomes and postribosomal RNP particles of embryos of various stages to proteins synthesized *in vivo*. Approximately half the products of translation of these mRNAs in a rabbit reticulocyte lysate analyzed by two-dimensional electrophoresis correspond to polypeptides synthesized *in vivo*. Among these, nearly all show a close correlation between their relative prevalence in the cytoplasm and their relative rates of synthesis *in vivo*, i.e., there are few examples of selective translational regulation of maternal mRNAs (see also Infante and Heilman, 1981; Lau and Lennarz, 1983). These analyses would not, however, detect the presence of maternal transcripts stored in a translationally inactive form, which might become activated by structural modification. The presence of a large amount of maternal RNA resembling incompletely processed precursors of mRNA indicates that this possibility must be seriously considered, although there is no direct evidence for it. Embryos of other species apparently do use selective translation of maternal mRNAs to alter patterns of protein synthesis substantially during embryonic development (e.g., Braude *et al.*, 1979; Brandhorst and Newrock, 1981; reviewed by Raff and Showman, 1983).

## 9. Distribution of Maternal RNA Sequences in Eggs and Embryos

The experimental interpretation that eggs or early embryos contain localized morphogenetic determinants that are segregated during embryonic development has led to speculation that these factors include stored maternal RNAs. The four micromeres of the sea urchins arising from the fourth cleavage will differentiate into primary mesenchyme cells when cultured in isolation (Okazaki, 1975). The micromeres are thus determined and must be different in some way from the remainder of the embryo, although such differences could be very subtle. At the 16-cell stage, when most protein synthesis is due to translation of maternal mRNA and the pattern of protein synthesis is almost identical to that of the egg, micromeres synthesize the same detectable patterns of polypeptides as the other blastomeres (Tufaro and Brandhorst, 1979). As the micromeres differentiate, their pattern of protein synthesis becomes quite distinct (Harkey and Whiteley, 1982). Sequence comparisons using saturation hybridization of single-copy DNA probes indicate that some rare maternal RNA sequences are lacking in micromeres (Rodgers and Gross, 1978), but these sequences are not among those found in polysomes at the 16-cell stage (Ernst *et al.*, 1980). These RNA sequences might then correspond to the elusive morphogenetic determinants; it is not known whether they later appear on polysomes. The major proteins (detected by silver staining of two-dimensional electrophoretic gels) of micromeres are not distinguishable in type or quantity from those of macromeres and mesomeres (Brandhorst and Tufaro, unpublished observations). Proteins synthesized in all three types of blastomeres at the 16-cell stage include the Spec proteins, which in later embryos are synthesized only in dorsal ectoderm cells derived from mesomeres (Bruskin *et al.*, 1982); there is extensive accumulation of Spec mRNAs in these cells after hatching and during gastrulation to a much higher level than in eggs and early embryos.

The unfertilized egg has a defined animal–vegetal polarity at the time of fertilization (Schroeder, 1980). The basis of this polarity is not known, but it is likely to be associated with the structural arrangement of the cytoskeleton of the egg. Either the cytoskeleton or localization of specializations of the plasma membrane may provide orientation for the cytoplasmic rearrangements that occur during early cleavage (e.g., Schroeder, 1980). As such, the structural organization of the egg can be considered an informational legacy of oogenesis that serves to organize the morphogenetic events of early embryonic development.

## 10. Synthesis and Accumulation of RNA During Oogenesis

### 10.1. Synthesis and Accumulation of Messenger RNA

Little is known about the metabolism of mRNA during oogenesis in echinoderms, largely because of technical difficulties associated with labeling oocytes *in vivo* and in preparing staged oocytes free of contaminating cells for short-term culture. Vitellogenic oocytes are enriched in the ovary for a few weeks after induced shedding of mature eggs during the productive season. During the rest of the year, the ovary contains oogonia and previtellogenic oocytes (see Section 2.1). Giudice and colleagues have developed methods for preparing previtellogenic and vitellogenic oocytes fractionated by size (Giudice *et al.*, 1972; Cognetti *et al.*, 1979). These oocytes, prepared from ovaries of *Paracentrotus lividus*, actively incorporate radioactive nucleoside precursors into RNA, including ribosomal RNA and nonnucleolar heterodisperse RNA (Sconzo *et al.*, 1972). Since follicle cells of the ovary are not very active in incorporating nucleosides, there is little contamination of the newly synthesized oocyte RNA even when the cultures are not pure (Sconzo *et al.*, 1972). More detailed analyses indicate that vitellogenic oocytes of *L. pictus* accumulate large heterodisperse RNA (presumably mRNA or precursors thereof), histone mRNA, 4 S RNA, and mitochondrial RNA in the cytoplasm during a pulse of [$^{3}$H]uridine (Ruderman and Schmidt, 1981). Failure to detect newly synthesized rRNA in the cytoplasm after 2 hr of labeling, in spite of active synthesis of rRNA precursors (Sconzo *et al.*, 1972; Griffith *et al.*, 1981), was probably due to the relatively slow processing of rRNA by oocytes (see Section 10.2); mRNA precursors may also be slowly processed in these oocytes, suggesting that long labeling times are required to gain an impression of the steady-state distribution of newly synthesized RNA in the cytoplasm.

Some of the RNA synthesized by isolated previtellogenic oocytes, including histone mRNAs, is associated with polysomes (Ruderman and Schmidt, 1981). The store of CS-type histones in mature eggs is probably the result of translation of histone mRNAs during oogenesis (Poccia *et al.*, 1981; Herlands *et al.*, 1982). Piatigorsky (1975) reviewed investigations of RNA metabolism of oocytes *in situ* by injection of labeled precursors into the adult urchin.

A single-copy DNA tracer enriched for egg RNA sequences was hybridized to excess RNA prepared from previtellogenic oocytes of *S. purpuratus* (Hough-Evans *et al.*, 1979); less than one-half the egg RNA sequences were detected in the oocytes. Members of the complex class of RNA sequences in these oocytes are present in about 600 copies each, their concentration relative to total RNA being similar to that in the mature egg. RNA prepared from ovaries of reproductive females after shedding has a sequence complexity between that of previtellogenic oocytes and eggs (Hough-Evans *et al.*, 1977), suggesting a sequential appearance of new RNA sequences during the vitellogenic phase of oogenesis, although contamination by nongerminal cells could be responsible for the increased complexity. Previtellogenic oocytes contain no detectable histone mRNA, whereas the egg contains about $10^6$ copies of H2B and H3 mRNAs but only about one tenth of that amount of H1 mRNA (Mauron *et al.*, 1982). Thus, it is clear that many maternal RNA sequences accumulate during vitellogenesis. It remains unclear as to how and to what extent the oocyte distinguishes between mRNA utilized for protein synthesis as it is growing and mRNA stored for use during embryogenesis. The complexity of nuclear RNA of previtellogenic oocytes of *S. purpuratus* is about $1.6 \times 10^8$ and includes virtually all the sequences found in the nuclear RNA of gastrulae (Hough-Evans *et al.*, 1979).

## 10.2. Synthesis and Accumulation of Ribosomal RNA

Sea urchin eggs contain about 2 ng of rRNA, corresponding to about $4 \times 10^8$ ribosomes. Sea urchins have 50–200 ribosomal genes per haploid genome (Wilson *et al.*, 1976; Griffith and Humphreys, 1979), and no amplification of rRNA genes has been detected in echinoid oocytes (Vincent *et al.*, 1969; Griffith *et al.*, 1981). The rate of rRNA synthesis in isolated oocytes of *T. gratilla* is about 110,000 ribosome equivalents per hour (Griffith *et al.*, 1981). Each ribosomal gene must produce an rRNA precursor molecule every 6 min. This does not require an unusual rate of initiation, elongation, or polymerase packing but implies that most ribosomal genes are actively transcribed in oocytes. At the observed rate, it would take 4–5 months for the oocyte to accumulate its complement of ribosomes.

The rate of rRNA synthesis in oocytes is about 40 times higher per gene than in nonfeeding embryos, which do not accumulate significant amounts of rRNA (Emerson and Humphreys, 1970; Galau *et al.*, 1977; Surrey *et al.*, 1979; Griffith and Humphreys, 1979). The synthesis of rRNA cannot be detected in eggs (Brandhorst, 1980; Griffith *et al.*, 1981). On the other hand, large amounts of ribosomal precursors are found in both eggs and oocytes, approximately 100–300 times more than in embryos and larvae (Griffith *et al.*, 1981). This suggests that rRNA precursor synthesized during oogenesis remains in an unprocessed form in the egg or that the rates of synthesis and processing are coordinately reduced to a great extent in eggs. This observation is intriguing in light of the content of cytoplasmic maternal RNA having characteristics of incompletely processed precursors of mRNA.

## 11. Metabolism of Egg Proteins During Embryonic Development

A large mass of protein accumulates during oogenesis and can be considered part of the stored information content of the egg. Several hundred polypeptides are detected when extracts of eggs are resolved by two-dimensional electrophoresis and stained with silver (Bédard and Brandhorst, 1983). The patterns of stained proteins of eggs and plutei of *S. purpuratus* are compared in Fig. 6. These patterns are even more similar than the patterns of newly synthesized proteins, but some important developmental changes are obvious. Several very prevalent egg proteins decline gradually in mass and are greatly reduced or undetectible by pluteus stage. None of these proteins is detectably synthesized during embryonic development; they are synthesized and stored during oogenesis and are then processed or degraded during embryonic development. These proteins presumably include those in yolk granules. Yolk proteins have been partially characterized in several species of sea urchin (Ii *et al.*, 1978; Ozaki, 1980, 1982; Harrington and Easton, 1982). It has been proposed that the major yolk glycoproteins accumulate in the coelomic fluid (Harrington and Easton, 1982) and are taken up by oocytes, possibly after passing through the accessory cells (Ozaki *et al.*, 1982). Yolk proteins have not yet been identified on the two-dimensional gels shown in Fig. 6.

Most polypeptides detectable by staining do not change significantly in mass during development (Bédard and Brandhorst, 1983). Some are never detectably labeled with [$^{35}$S]methionine during embryonic development, suggesting that they are quite stable. Many proteins ultimately found in chromatin of embryos are never detectably synthesized during embryogenesis (Kuhn and Wilt, 1981). The tubulins of embryonic cilia are mainly derived from a pool existing in the egg (Bibring and Baxandall, 1981), although tubulins are also actively synthesized during development. The average rate of protein synthesis during development (Goustin and Wilt, 1981) indicates that about 1% of the mass of protein turns over each hour

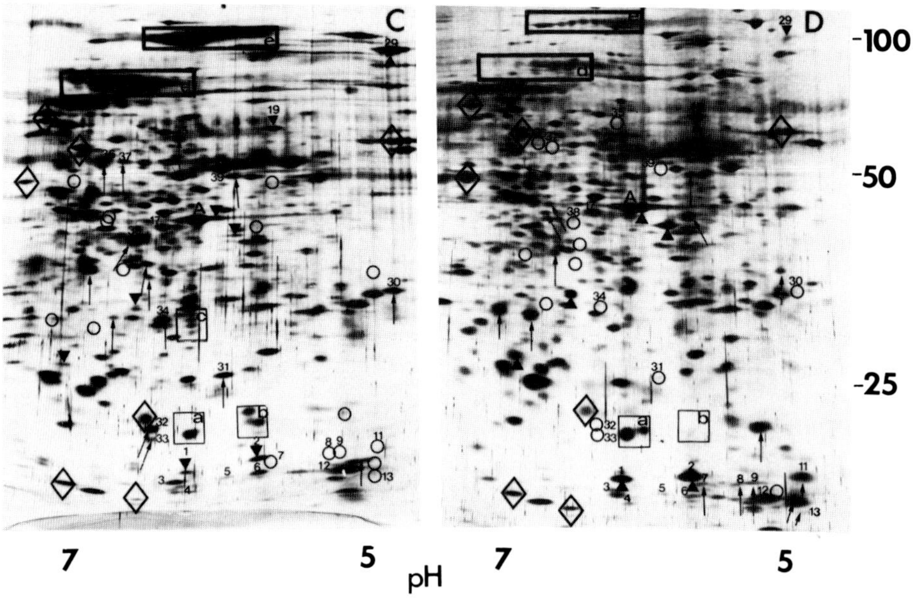

**Figure 6.** Comparison of egg and pluteus proteins. Proteins of S. *purpuratus* eggs (left) and plutei (right) were separated by two-dimensional electrophoresis and stained with silver. The symbols used are the same as in Fig. 5. The clusters of spots in boxes d and e, each of which probably represents post-translational modifications of the same polypeptide, declined markedly during development and were not labeled with [$^{35}$S]methionine in eggs or embryos. Other proteins never detectably labeled, but that did not change detectibly in mass, are enclosed in "diamonds." (Reproduced from Bédard and Brandhorst, 1983.)

in embryos. Other major polypeptides are actively synthesized during embryonic development, but the patterns of labeling with amino acids indicate that there is a diversity of patterns of protein metabolism during development (Bédard and Brandhorst, 1983; Brandhorst *et al.*, 1983). The extent to which the unfertilized egg is prepared for embryonic development is demonstrated by the observation that only a few polypeptides increase substantially in mass, and some are never synthesized by embryos (Bédard and Brandhorst, 1983). Among those that accumulate are several enriched in particular tissues (Bruskin *et al.*, 1982; Bédard and Brandhorst, 1983). The differences in the patterns of stored and newly-synthesized proteins of early embryos indicate that at some period of oogenesis the population of translated mRNA in the oocyte is substantially different from that stored in the egg (unless uptake of proteins from outside the oocyte is extensively utilized).

In the case of many proteins, including the histones, actins, and tubulins, the embryo uses mRNA translated during oogenesis, mRNA stored in eggs and translated during embryogenesis, and mRNA synthesized during embryogenesis. It is not clear whether the use of these multiple strategies is required to meet some period of peak demand, whether it provides failsafe insurance that required proteins will be available, or whether products of the various classes of mRNA are utilized in different ways as yet too subtle to recognize.

## 12. Speculations on the Roles of Stored Maternal RNA

Eggs contain a store of maternal mRNA in order to provide for the protein synthesis carried out during early embryonic development. If the sea urchin zygote were able to produce stable mRNA at the maximum rate per nucleus observed for the accumulation of mRNA in the cytoplasm of embryos (Brandhorst and Humphreys, 1972; Galau *et al.*, 1977), it would take a few days to accumulate the amount of mRNA estimated to enter polysomes in the first 2–3 hr after fertilization. If the elevated rate of protein synthesis of the zygote is required (or provides selective advantage by minimizing a period of development during which the embryo is most vulnerable to environmental shock and predation), the use of stored maternal RNA becomes a necessity until cleavage produces enough nuclei to produce sufficient newly synthesized mRNA for translation. The translational inactivity of the maternal mRNA in the sea urchin egg (or starfish oocyte) is consistent with the state of metabolic quiescence and conserves energy.

It is likely that other functional roles for stored maternal mRNA have evolved, at least in some organisms. The possibility that maternal mRNAs are nonuniformly distributed in the egg and segregated during cleavage remains an attractive hypothesis. When patterns of protein synthesis must change qualitatively during early embryonic development, the regulation of those changes probably cannot be based on transcription of single-copy genes. Consequently, the utilization of maternal mRNA has sometimes become regulated post-transcriptionally, as during maturation of oocytes (Rosenthal *et al.*, 1980, 1982) or during early development of mammalian or *Ilyanassa* embryos (Braude *et al.*, 1979; Brandhorst and Newrock, 1981). Since few changes in patterns of protein synthesis occur during early embryonic development of sea urchin embryos (Bédard and Brandhorst, 1983), translational regulation of maternal mRNA is apparently not extensive (see Section 8.3), although regulation of the stability of specific mRNAs may be critical (e.g., Maxson and Wilt, 1982). During the period of sea urchin development when patterns of protein synthesis are rapidly changing, there are sufficient nuclei to produce rapidly the requisite amounts of new mRNAs. The uncertain role of maternal RNA containing interspersed repetitive sequences and oligo(A) tracts leaves open the possibility that maternal RNA produced during oogenesis has a crucial role during development aside from the utilization of maternal mRNA for protein synthesis during the period of rapid cleavage (see Section 5).

## 13. Summary

The sea urchin egg contains a store of mRNA synthesized during oogenesis but translated only after fertilization, which accounts for a large, rapid increase in the rate of synthesis of largely the same set of proteins synthesized by eggs. Starfish oocytes contain a population of stored maternal mRNA that becomes actively translated upon GVBD and codes for a set of proteins distinct from that synthesized by oocytes. The sequence complexity of RNA in echinoderm eggs is about $3.5 \times 10^8$ nucleotides, enough to code for about 12,000 different mRNAs averaging 3 kb in length. About 2–4% of the egg RNA functions as mRNA during early embryonic development; most of the sequences are rare, represented in a few thousand copies per egg, but some are considerably more abundant. Many of the stored RNA sequences accumulate during the period of vitellogenesis, which lasts a few weeks. The mechanisms of storage and translational activation of maternal mRNA are not well understood. Histone mRNAs are sequested in the egg pronucleus until first cleavage, but other mRNAs are widely distributed in the cytoplasm.

The population of maternal RNA includes many very large molecules having interspersed repetitive sequence transcripts colinear with single-copy sequences. The structural features of much of the cytoplasmic maternal RNA is thus reminiscent of incompletely processed nuclear precursors of mRNA. The functional role of these strange molecules is not understood, but many interesting possibilities have been considered. For instance, they may be segregated into different cell lineages during cleavage and/or they may become translationally activated by selective processing during development. Maternal mRNA appears to be underloaded with ribosomes when translated, possibly because the coding sequences are short relative to the size of the mRNA.

Most abundant and many rare mRNA sequences persist during embryonic development. The rare sequence molecules are replaced by newly synthesized RNA, but some abundant maternal transcripts appear to persist throughout embryonic development. Most of the proteins present in the egg do not change significantly in mass during development, but a few decline or accumulate substantially. Together, these observations indicate that much of the information for embryogenesis is stored in the egg, although substantial changes in gene expression occur during development.

ACKNOWLEDGMENTS. I thank Dr. E. H. Davidson, Dr. R. A. Raff, Dr. W. H. Klein, and Dr. J. Ruderman for providing preprints of their publications. I thank E. H. Davidson and Academic Press for permission to use material reproduced in the figures and J. Rigg for providing original photographs. I thank B. R. Hough-Evans for providing useful information. Many other colleagues, particularly P.-A. Bédard, F. Tufaro, and J. Posakony, contributed information and stimulated ideas appearing in this manuscript. M. Bannet provided critical help with the manuscript, and S. Gauthier typed it expertly. My research has been supported in part by grant A9595 from the National Science and Engineering Research Council of Canada.

## References

Afzelius, B. A., 1957, Electron microscopy of the basophilic structures of the sea urchin egg, *Z. Zellforsch.* **45**:660–675.

Alexandraki, D., and Ruderman, J. V., 1981, Sequence heterogeneity, multiplicity and genomic organization of α- and β-tubulin genes in sea urchins, *Mol. Cell Biol.* **1**:1125–1137.

Anderson, D. M., Galau, G. A., Britten, R. J., and Davidson, E. H., 1976, Sequence complexity of the RNA accumulated in oocytes of *Arbacia punctulata*, *Dev. Biol.* **51**:138–145.

Anderson, D. M., Scheller, R. H., Posakony, J. W., McAllister, L. B., Trabert, S. G., Beall, C., Britten, R. J., and Davidson, E. H., 1981, Repetitive sequences of the sea urchin genome. Distribution of members of specific repetitive families, *J. Mol. Biol.* **145**:5–28.

Anderson, D. M., Richter, J. D., Chamberlin, M. E., Price, D. H., Britten, E. J., Smith, L. D., and Davidson, E. H., 1982, Sequence organization of the poly(A) RNA synthesized in lampbrush chromosome stage *Xenopus laevis* oocytes, *J. Mol. Biol.* **155**:281–309.

Anderson, E., 1968, Oocyte differentiation in the sea urchin *Arbacia punctulata*, with particular reference to the cortical granules and their participation in the cortical reaction, *J. Cell Biol.* **37**:514–531.

Angerer, L. M., DeLeon, D. B., Angerer, R. C., Showman, R. M., Wells, D. E., and Raff, R. A., 1984, Delayed accumulation of maternal histone mRNA during sea urchin oogenesis, *Dev. Biol.* **101**:477–484.

Bédard, P.-A., and Brandhorst, B. P., 1983, Patterns of protein synthesis and metabolism during sea urchin embryogenesis, *Dev. Biol.* **96**:74–83.

Bédard, P.-A., and Brandhorst, B. P., 1984, Populations of translatable messenger RNA in sea urchin embryos (submitted).

Bibring, T., and Baxandall, J., 1981, Tubulin synthesis in sea urchin embryos II. Ciliary A tubulin derives from unfertilized egg, *Dev. Biol.* **83**:122–126.

Both, G. W., Banerjee, A. K., and Shatkin, A. J., 1975, Methylation dependent translation of viral messenger RNAs *in vitro*, *Proc. Natl. Acad. Sci. USA* **72**:1189–1193.

Brachet, J., Decroly, M., Ficq, A., and Quertier, J., 1963, Ribonucleic acid metabolism in unfertilized sea urchin eggs, *Biochim. Biophys. Acta* **72**:660–662.

Brandhorst, B. P., 1976, Two-dimensional gel patterns of protein synthesis before and after fertilization of sea urchin eggs, *Dev. Biol.* **52**:310–317.

Brandhorst, B. P., 1980, Simultaneous synthesis, translation, and storage of mRNA including histone mRNA in sea urchin eggs, *Dev. Biol.* **79**:139–148.

Brandhorst, B. P., and Humphreys, T., 1972, Stabilities of nuclear and messenger RNA molecules in sea urchin embryos, *J. Cell Biol.* **53**:474–482.

Brandhorst, B. P., and Newrock, K. M., 1981, Post-transcriptional regulation of protein synthesis in *Ilyanassa* embryos and isolated polar lobes, *Dev. Biol.* **83**:250–254.

Brandhorst, B. P., Verma, D. P. S., and Fromson, D., 1979, Polyadenylated and nonpolyadenylated messenger RNA fractions from sea urchin embryos code for the same abundant proteins, *Dev. Biol.* **71**:128–141.

Brandhorst, B. P., Tufaro, F., and Bédard, P.-A., 1983, Expression of maternal and embryonic genes during sea urchin development, in: *Molecular Aspects of Early Development* (G. M. Malacinski and W. H. Klein, eds.), pp. 39–60, Plenum Press, New York.

Brandis, J. W., and Raff, R. A., 1978, Translation of oogenetic mRNA in sea urchin eggs and early embryos. Demonstration of a change in translational efficiency following fertilization, *Dev. Biol.* **67**:99–113.

Braude, P., Pelham, H. Flach, T., and Lobatto, R., 1979, Post-transcriptional control in the early mouse embryo, *Nature* **282**:102–105.

Britten, R. J., and Davidson, E. H., 1969, Gene regulation for higher cells: A theory, *Science* **165**:349–357.

Bruskin, A. M., Tyner, A. L., Wells, D. E., Showman, R. M., and Klein, W. H., 1981, Accumulation in embryogenesis of five mRNAs enriched in the ectoderm of the sea urchin pluteus, *Dev. Biol.* **87**:308–318.

Bruskin, A. M., Bédard, P.-A., Tyner, A. L., Showman, R. M., Brandhorst, B. P., and Klein, W. H., 1982, A family of proteins accumulating in ectoderm of sea urchin embryos specified by two related cDNA clones, *Dev. Biol.* **91**:317–324.

Bryce, T. H., 1902, Maturation of the ovum in *Echinus esculentus*, *Q. J. Microsc. Sci.* **46**:177–224.

Cabrera, C. V., Lee, J. J., Ellison, J. W., Britten, R. J., and Davidson, E. H., 1984, Regulation of cytoplasmic mRNA prevalence in sea urchin embryos: Rates of appearance and turnover for specific sequences, *J. Mol. Biol.* **174**:85–111.

Carpenter, C. D., Bruskin, A. M., Spain, L. M., Eldon, E. D., and Klein, W. H., 1983, The 3′ untranslated regions of two related mRNAs contain an element highly repeated in the sea urchin genome, *Nucleic Acid Res.* **10**:7829–7842.

Chamberlain, J. P., 1970, RNA synthesis in anucleate egg fragments and normal embryos of the sea urchin, *Arbacia punctulata*, *Biochim. Biophys. Acta* **213**:183–193.

Chamberlain, J. P., and Metz, C. B., 1972, Mitochondrial RNA synthesis in sea urchin embryos, *J. Mol. Biol.* **64**:593–607.

Chamberlin, M. E., Britten, R. J., and Davidson, E. H., 1975, Sequence organization in *Xenopus* DNA studied by electron microscope, *J. Mol. Biol.* **96**:317–328.

Chatlynne, L. G., 1969, A histochemical study of oogenesis in the sea urchin *Strongylocentrotus purpuratus*, *Biol. Bull.* **136**:167–184.

Chen, P. S., and Baltzer, F., 1979, Morphology and biochemistry of diploid and androgenetic haploid (merogonic) hybrids, in: *The Sea Urchin Embryo Biochemistry and Morphogenesis* (G. Czihak, ed.), pp. 424–472, Springer-Verlag, Berlin.

Childs, G., Maxson, R., and Kedes, L. H., 1979, Histone gene expression during sea urchin embryogenesis: Isolation and characterization of early and late messenger RNAs of *Strongylocentrotus purpuratus*, *Dev. Biol.* **73**:153–173.

Cognetti, G., Platz, R. D., Meistrich, M. L., and DiLiegro, I., 1979, Studies on protein synthesis during sea urchin oogenesis, *Cell Diff.* **5**:283–291.

Conway, C. M., 1971, Evidence for RNA in the heavy bodies of sea urchin eggs, *J. Cell Biol.* **51**:889–893.

Costantini, F. D., Scheller, R. H., Britten, R. J., and Davidson, E. H., 1978, Repetitive sequence transcripts in the mature sea urchin oocyte, *Cell* **15**:173–187.

Costantini F., Britten, R. J., and Davidson, E. H., 1980, Message sequences and short repetitive sequences are interspersed in sea urchin egg poly(A)$^+$ RNAs, *Nature* **287**:111–117.

Craig, S. P., 1970, Synthesis of RNA in non-nucleate fragments of sea urchin eggs, *J. Mol. Biol.* **47**:615–618.

Crain, W. R., Durica, D. S., and Van Doren, K., 1981, Actin gene expression in developing sea urchin embryos, *Mol. Cell Biol.* **1**:711–720.

Davidson, E. H., 1976, *Gene Activity in Early Development*, 2nd ed., Academic Press, New York.

Davidson, E. H., and Britten, R. J., 1979, Regulation of gene expression: Possible role of repetitive sequences, *Science* **204**:1052–1059.

Davidson, E. H., and Posakony, J. W., 1982, Repetitive sequence transcripts in development, *Nature* **297**:633–635.

Davidson, E. H., Galau, G. A., Angerer, R. C., and Britten, R. J., 1975a, Comparative aspects of DNA organization in metazoa, *Chromosoma* **51**:253–259.

Davidson, E. H., Hough-Evans, B. R., Klein, W. H., and Britten, R. J., 1975b, Structural genes adjacent to interspersed repetitive DNA sequences, *Cell* **4**:217–225.

Davidson, E. H., Hough-Evans, B. R., and Britten, R. J., 1982, Molecular biology of the sea urchin embryo, *Science* **217**:17–26.

Delobel, N., 1971, Étude déscriptive des chromosomes en écouvillon chez *Echinaster sepositus*, *Ann. Embryol. Morphol.* **4**:383–396.

Denny, P. C., and Tyler, A., 1964, Activation of protein biosynthesis in nonnucleate fragments of sea urchin eggs, *Biochem. Biophys. Res. Commun.* **14**:245–249.

Devlin, R., 1976, Mitochondrial poly(A) RNA synthesis during early sea urchin development, *Dev. Biol.* **50**:443–456.

Diaz, M. O., Barsacchi-Pilone, G., Mahon, K. A., and Gall, J. G., 1981, Transcripts from both strands of a satellite DNA occur on lampbrush chromosome loops of the newt *Notopthalmas*, *Cell* **24**:649–659.

Dolecki, G. J., Duncan, R. F., and Humphreys, T., 1977, Complete turnover of poly(A) on maternal mRNA of sea urchin embryos, *Cell* **11**:339–344.

Dubroff, L. M., 1977, Oligouridylate stretches in heterogeneous nuclear RNA, *Proc. Natl. Acad. Sci. USA* **74**:2217–2221.

Dubroff, L. M., 1980, Oligomeric sequences in the cytoplasmic RNA of sea urchin embryos, *Biochim. Biophys. Acta* **607**:115–121.

Dubroff, L. M., and Nemer, M., 1975, Molecular classes of heterogeneous nuclear RNA in sea urchin embryos, *J. Mol. Biol.* **95**:455–476.

Duncan, R., and Humphreys, T., 1981a, Most sea urchin maternal mRNA sequences in every abundance class appear in both polyadenylated and nonpolyadenylated molecules, *Dev. Biol.* **88**:201–210.

Duncan, R., and Humphreys, T., 1981b, Multiple oligo(A) tracts associated with inactive sea urchin maternal mRNA sequences, *Dev. Biol.* **88**:211–219.

Duncan, R., and Humphreys, T., 1983, Oligo(U) sequences present in sea urchin maternal RNA decrease following fertilization, *Dev. Biol.* **96**:258–262.

Emerson, C., and Humphreys, T., 1970, Regulation of DNA-like RNA and the apparent activation of ribosomal RNA synthesis in sea urchin embryos: Quantitative measurements of newly synthesized RNA, *Dev. Biol.* **23**:86–112.

Epel, D., 1967, Protein synthesis in sea urchin eggs: A "late" response to fertilization, *Proc. Natl. Acad. Sci. USA* **57**:899–906.

Ernst, S. G., Hough-Evans, B. R., Britten, R. J., and Davidson, E. H., 1980, Limited complexity of the RNA in micromeres of 16-cell sea urchin embryos, *Dev. Biol.* **79**:119–127.

Farquhar, M. N., and McCarthy, B. J., 1973, Histone mRNA in eggs and embryos of *Strongylocentrotus purpuratus*, *Biochem. Biophys. Res. Commun.* **53**:515–522.

Federoff, N., Wellauer, P. K., and Wall, R., 1977, Intermolecular duplexes in heterogeneous nuclear RNA from HeLa Cells, *Cell* **10**:597–610.

Flavell, A. J., and Ish-Horowicz, D., 1981, Extra-chromosomal copies of the eukaryotic transposable element *copia* in cultured *Drosophila* cells, *Nature* **292**:591–595.

Flytzanis, C. N., Brandhorst, B. P., Britten, R. J., and Davidson, E. H., 1982, Developmental patterns of cytoplasmic transcript prevalence in sea urchin embryos, *Dev. Biol.* **91**:27–35.

Fregien, N., Dolecki, G. J., Mandel, M., and Humphreys, T., 1983, Molecular cloning of five individual stage and tissue-specific messenger sequences from sea urchin pluteus embryos, *Mol. Cell Biol.* **3**:1021–1031.

Fromson, D., and Duchastel, A., 1975, Poly(A)-containing polyribosomal RNA in sea urchin embryos: Changes in proportion during development, *Biochim. Biophys. Acta* **378**:394–404.

Galau, G. A., Britten, R. J., and Davidson, E. H., 1974, A measurement of the sequence complexity of polysomal messenger RNA in sea urchin embryos, *Cell* **2**:9–20.

Galau, G. A., Klein, W. H., Davis, M. M., Wold, B. J., Britten, R. J., and Davidson, E. H., 1976a, Structural gene sets active in embryos and adult tissues of the sea urchin, *Cell* **7**:487–505.

Galau, G. A., Britten, R. J., and Davidson, E. H., 1976b, Studies on nucleic acid reassociation kinetics. III. Rate of excess RNA-DNA hybridization compared to rate of DNA renaturation, *Proc. Natl. Acad. Sci. USA* **74**:1020–1023.

Galau, G. A., Lipson, E. D., Britten, R. J., and Davidson, E. H., 1977, Synthesis and turnover of polysomal mRNAs in sea urchin embryos, *Cell* **10**:415–432.

Giudice, G., 1973, *Developmental Biology of the Sea Urchin Embryo*, Academic Press, New York.

Giudice, G., Mutolo, V., and Donatuti, G., 1968, Gene expression in sea urchin development, *Wilhelm Roux Arch. Entuicklungmech. Org.* **161**:118–139.

Giudice, G., Sconzo, G., Bono, A., and Albanese, I., 1972, Studies on sea urchin oocytes I. Purifications and cell fractionation, *Exp. Cell Res.* **72**:90–94.

Goldberg, R. B., Galau, G. A., Britten, R. J., and Davidson, E. H., 1973, Nonrepetitive DNA sequence representation in sea urchin embryo messenger RNA, *Proc. Natl. Acad. Sci. USA* **70**:3516–3520.

Goustin, A. S., and Wilt, F. H., 1981, Protein synthesis, polyribosomes, and peptide elongation in early development of *Strongylocentrotus purpuratus*, *Dev. Biol.* **82**:32–40.

Goustin, A. S., and Wilt, F. H., 1982, Direct measurement of histone peptide elongation rate in cleaving sea urchin embryos, *Biochim. Biophys. Acta* **699**:22–27.

Graham, D. E., Neufeld, B. R., Davidson, E. H., and Britten, R. J., 1974, Interspersion of repetitive and non-repetitive DNA sequences in the sea urchin genome, *Cell* **1**:127–137.

Grainger, T. L., Winkler, M. M., Shen, S. S., and Steinhardt, R. A., 1979, Intracellular pH controls protein synthesis rate in the sea urchin egg and early embryo, *Dev. Biol.* **68**:396–406.

Greenhouse, G. A., Hynes, R. P., and Gross, P. R., 1971, Sea urchin embryos are permeable to actinomycin, *Science* **171**:686–689.

Griffith, J. K., and Humphreys, T. D., 1979, Ribosomal ribonucleic acid synthesis and processing in embryos of the Hawaiian sea urchin, *Tripneustes gratilla*, *Biochemistry* **18**:2178–2185.

Griffith, J. K., Griffith, B. B., and Humphreys, T., 1981, Regulation of ribosomal RNA synthesis in sea urchin embryos and oocytes, *Dev. Biol.* **87**:220–228.

Gross, K. W., Jacobs-Lorena, M., Baglioni, C., and Gross, P. R., 1973, Cell-free translation of maternal messenger RNA from sea urchin eggs, *Proc. Natl. Acad. Sci. USA* **70**:2614–2618.

Gross, P. R., and Cousineau, G. H., 1963, Effects of actinomycin D on macromolecular synthesis and early development of sea urchin eggs, *Biochem. Biophys. Res. Commun.* **10**:321–326.

Gross, P. R., Malkin, J. L., and Moyer, W. A., 1964, Templates for the first proteins of embryonic development, *Proc. Natl. Acad. Sci. USA* **51**:407–414.

Harkey, M. A., and Whiteley, A. H., 1982, Cell-specific regulation of protein synthesis in the sea urchin gastrula: A two dimensional electrophoretic study, *Dev. Biol.* **93**:453–462.

Harrington, F. E., and Easton, D. P., 1982, A putative precursor to the major yolk protein of the sea urchin, *Dev. Biol.* **94**:505–508.

Harris, P., 1967, Structural changes following fertilization in the sea urchin egg, *Exp. Cell Res.* **48**:569–581.

Harvey, E. B., 1956, *The American Arbacia and Other Sea Urchins*, Princeton University Press, Princeton, N.J.

Hentschel, C. C., and Birnstiel, M. L., 1981, The organization and expression of histone gene families, *Cell* **25**:301–313.

Herlands, L., Allfrey, U. G., and Poccia, D., 1982, Translational regulation of histone synthesis in the sea urchin *Strongylocentrotus purpuratus*, *J. Cell Biol.* **94**:219–223.

Heywood, S. M., and Kennedy, D. S., 1976, Purification of myosin translational control RNA and its interaction with myosin messenger RNA, *Biochemistry* **15**:3314–3319.

Hickey, E. D., Weber, L. A., and Baglioni, C., 1976, Translation of RNA from unfertilized sea urchin eggs does not require methylation and is inhibited by 7-methylguanosine-5′-monophosphate, *Nature* **261**:71–73.

Hille, M. B., and Albers, A. A., 1979, Efficiency of protein synthesis after fertilization of sea urchin eggs, *Nature* **278**:469–471.

Hille, M. B., Hall, D. C., Yablonka-Reuveni, Z., Danilchik, M. V., and Moon, R. T., 1981, Translational control in sea urchin eggs and embryos: Initiation is rate limiting in blastula stage embryos, *Dev. Biol.* **86**:241–249.

Hinegardner, R., 1974, Cellular DNA content of the Echinodermata, *Comp. Biochem. Physiol.* **49B**:219–226.

Hinegardner, R., 1975, Care and handling of sea urchin eggs, embryos and adults, in *The Sea Urchin Embryo. Biochemistry and Morphogenesis* (G. Czihak, ed.) pp. 10–25, Springer-Verlag, Berlin.

Holland, N. D., and Giese, A. C., 1965, An autoradiographic investigation of the gonads of the purple sea urchin (*Strongylocentrotus purpuratus*), *Biol. Bull.* **128**:241–258.

Hough, B. R., Smith, M. J., Britten, R. J., and Davidson, E. H., 1975, Sequence complexity of heterogenous nuclear RNA in sea urchin embryos, *Cell* **5**:291–299.

Hough-Evans, B. R., Wold, B. J., Ernst, S. G., Britten, R. J., and Davidson, E. H., 1977, Appearance and persistence of maternal RNA sequences in sea urchin development, *Dev. Biol.* **60**:258–277.

Hough-Evans, B. R., Ernst, S. G., Britten, R. J., and Davidson, E. H., 1979, RNA complexity in developing sea urchin oocytes, *Dev. Biol.* **69**:258–269.

Houk, M. S., and Epel, D., 1974, Protein synthesis during hormonally induced meiotic maturation and fertilization in starfish oocytes, *Dev. Biol.* **40**:298–310.

Humphreys, T., 1969, Efficiency of translation of mRNA before and after fertilization in sea urchins, *Dev. Biol.* **20**:435–458.

Humphreys, T., 1971, Measurements of messenger RNA entering polysomes upon fertilization of sea urchin eggs, *Dev. Biol.* **26**:210–218.

Ii, I., Deguchi, K., Lawashima, S., Endo, S., and Veta, N., 1978, Water-soluble lipoproteins from yolk granules in sea urchin eggs. I. Isolation and several properties, *J. Biochem.* **84**:737–749.

Ilan, J., and Ilan, J., 1978, Translation of maternal message ribonucleoprotein particles from sea urchin eggs in a cell-free system and product analysis, *Dev. Biol.* **66**:375–385.

Infante, A. A., and Heilmann, L. J., 1981, Distribution of messenger ribonucleic acid in polysomes and nonpolysomal particles of sea urchin embryos: Translational control of actin synthesis, *Biochemistry* **20**:1–8.

Jeffery, W. R., 1977, Polyadenylation of maternal and newly synthesized RNA during starfish oocyte maturation, *Dev. Biol.* **57**:98–108.

Jenkins, N. A., Taylor, M. W., and Raff, R. A., 1973, *In vitro* translation of oogenetic messenger RNA of sea urchins and picornavirus with a cell-free system from sarcoma-180, *Proc. Natl. Acad. Sci. USA* **70**:3287–3291.

Jenkins, N. A., Kaumeyer, J. F., Young, E. M., and Raff, R. A., 1978, A test for masked message: The template activity of messenger ribonucleoprotein particles isolated from sea urchin eggs, *Dev. Biol.* **63**:279–298.

Kaumeyer, J. F., Jenkins, N. A., and Raff, R. A., 1978, Messenger ribonucleoprotein particles in unfertilized sea urchin eggs, *Dev. Biol.* **63**:266–278.

Kedes, L. H., and Birnstiel, M. L., 1971, Reiteration and clustering of DNA sequences complementary to histone mRNA, *Nature New Biol.* **230**:165–169.

Kleene, K. C., and Humphreys, T., 1977, Similarity of hnRNA sequences in blastula and pluteus stage sea urchin embryos, *Cell* **12**:143–155.

Klein, W. H., Thomas, T. L., Lai, C., Scheller, R. H., Britten, R. J., and Davidson, E. H., 1978, Characteristics of individual repetitive sequence families in the sea urchin genome studied with cloned repeats, *Cell* **14**:889–900.

Korwek, E. L., Nakazato, H., Venkatesan, S., and Edmonds, M., 1976, Poly(uridylic acid) sequences in messenger ribonucleic acid of HeLa cells, *Biochemistry* **15**:4643–4649.

Kosvedi, I., and Smith, M. J., 1982, Sequence complexity in the maternal RNA of the starfish *Pisaster ochraceus* (Brandt), *Dev. Biol.* **89**:56–63.

Kuhn, O., and Wilt, F. H., 1981, Chromatin proteins of sea urchin embryos, *Dev. Biol.* **85**:416–423.

Kung, C. S., 1974, On the size relationship between nuclear and cytoplasmic RNA in sea urchin embryos, *Dev. Biol.* **36**:343–356.

Laskey, R. A., Mills, A. D., Gurdon, J. B., and Partington, G. A., 1977, Protein synthesis in oocytes of *Xenopus laevis* is not regulated by supply of messenger RNA, *Cell* **11**:345–351.

Lasky, L. A., Lev., Z., Xin, J. H., Britten, R. J., and Davidson, E. H., 1980, Messenger RNA prevalence in sea urchin embryos measured with cloned cDNAs, *Proc. Natl. Acad. Sci. USA* **77**:5317–5321.

Lau, J. T. Y., and Lennarz, W. J., 1983, Regulation of sea urchin glycoprotein mRNAs during embryonic development. *Proc. Natl. Acad. Sci. USA* **80**:1028–1032.

Leahy, P. S., Tutshulte, T. C., Britten, R. J., and Davidson, E. H., 1978, A large-scale maintenace system for gravid purple sea urchins *(Strongylocentrotus purpuratus), J. Exp. Zool.* **204**:369–380.

Leahy, P. S., Hough-Evans, B. R., Britten, R. J., and Davidson, E. H., 1981, Synchrony of oogenesis in laboratory-maintained and wild populations of the purple sea urchin *Strongylocentrotus purpuratus, J. Exp. Zool.* **215**:7–22.

Lev, Z., Thomas, T. L., Lee, A. S., Angerer, R. C., Britten, R. J., and Davidson, E. H., 1980, Developmental expression of two cloned sequences coding for rare sea urchin embryo messages, *Dev. Biol.* **76**:322–340.

Lifton, R. P., and Kedes, L. H., 1976, Size and sequence homology of masked maternal and embryonic histone messenger RNAs, *Devl. Biol.* **48**:47–55.

Mano, Y., and Nagano, H., 1970, Mechanism of release of maternal messenger RNA induced by fertilization of sea urchin eggs, *J. Biochem.* **67**:611–628.

Martin, K. A., and Miller, O. L., 1983, Polysome structure in sea urchin eggs and embryos: An electron microscopic analysis, *Dev. Biol.* **98**:338–348.

Martindale, M. Q., and Brandhorst, B. P., 1983, Germinal vesicle breakdown and changes in protein synthesis in starfish oocytes, *Dev. Biol.* **101**:512–515.

Mauron, A., Kedes, L., Hough-Evans, B. R., and Davidson, E. H., 1982, Accumulation of individual histone mRNAs during embryogenesis of the sea urchin *Strongylocentrotus purpuratus, Dev. Biol.* **94**:425–434.

Maxson, R. E., and Wilt, F. H., 1981, The rate of synthesis of histone mRNA during the development of sea urchin embryos *(Strongylocentrotus purpuratus), Dev. Biol.* **83**:380–386.

Maxson, R. E., and Wilt, F. H., 1982, Accumulation of the early histone messenger RNAs during the development of *Strongylocentrotus purpuratus, Dev. Biol.* **94**:435–440.

Merlino, G. T., Water, R. D., Moore, G. P., and Kleinsmith, L. J., 1981, Change in expression of the actin gene family during early sea urchin development, *Dev. Biol.* **85**:505–508.

Mescher, A., and Humphreys, T., 1974, Activation of maternal mRNA in the absence of poly(A) formation in fertilized sea urchin eggs, *Nature* **249**:138–139.

Molloy, G. R., Thomas, W. L., and Darnell, J. E., 1972, Occurrence of uridylate-rich oligonucleotide regions in heterogeneous nuclear RNA of HeLa cells, *Proc. Natl. Acad. Sci. USA* **69**:3684–3688.

Moon, R. T., 1983, Poly(A)-containing messenger ribonucleoprotein complexes from sea urchin eggs and embryos: Polypeptides associated with native and UV-crosslinked mRNPs, *Differentiation* **24**:13–23.

Moon, R. T., Moe, K. D., and Hille, M. B., 1980, Polypeptides of nonpolysomal messenger ribonucleoprotein complexes of sea urchin eggs, *Biochemistry* **19**:2723–2730.

Moon, R. T., Danilchik, M. V., and Hille, M. B., 1982, An assessment of the masked message hypothesis. Sea urchin egg messenger ribonucleoprotein complexes are efficient templates for *in vitro* protein synthesis, *Dev. Biol.* **93**:389–403.

Moon, R. T., Nicosia, R. F., Olsen, C., Hille, M. B., and Jeffery, W. R., 1983, The cytoskeletal framework of sea urchin eggs and embryos: Developmental changes in the association of messenger RNA, *Dev. Biol.* **95**:447–458.

Moore, G. P., Scheller, R. H., Davidson, E. H., and Britten, R. J., 1978, Evolutionary change in repetition frequency of sea urchin DNA sequences, *Cell* **15**:649–660.

Moore, G. P., Pearson, W. R., Davidson, E. H., and Britten, R. J., 1981, Long and short repeats of sea urchin DNA and their evolution, *Chromosoma* **84**:19–32.

Muthukrishnan, S., Both, G. W., Furuichi, Y., and Shatkin, A. J., 1975, 5′-Terminal 7′-methylguanosine in eukaryotic mRNA is required for translation, *Nature* **255**:33–37.

Nakano, E., and Monroy, A., 1958, Incorporation of $^{35}$S-methionine in the cell fractions of sea urchin eggs and embryos, *Exp. Cell Res.* **14**:236–244.

Nakazato, H., Edmonds, M., and Kopp, D. W., 1974, Differential metabolism of large and small poly(A) sequences in the heterogeneous nuclear RNA of HeLa cells, *Proc. Natl. Acad. Sci. USA* **71**:200–204.

Nemer, M., 1975, Developmental changes in the synthesis of sea urchin embryo messenger RNA containing and lacking polyadenylic acid. *Cell* **6**:559–570.

Nemer, M., Graham, M., and Dubroff, L. M., 1974, Co-existence of non-histone messenger RNA species lacking and containing polyadenylic acid in sea urchin embryos, *J. Mol. Biol.* **89**:435–454.

Newrock, K. M., Alfageme, C. R., Nardi, R. V., and Cohen, L. H., 1977, Histone changes during chromatin remodeling in embryogenesis, *Cold Spring Harbor Symp. Quant. Biol.* **42**:421–431.

O'Brochta, D. A., Spieth, J., Raff, R. A., Brandhorst, B. P., Klein, W. H., and Wells, D., 1981, Mitochondrial origin of an extremely prevalent set of transcripts in sea urchin embryos, *Biol. Bull.* **161**:321.

Okazaki, K., 1975, Spicule formation by isolated micromeres of the sea urchin embryo, *Am. Zool.* **15**:567–581.

Ozaki, H., 1980, Yolk proteins of the sand dollar, *Dendraster excentricus, Dev. Growth Diff.* **22**:365–372.

Ozaki, H., 1982, Vitellogenesis in the sand dollar *Dendraster excentricus, Cell Diff.* **11**:315–317.

Pearse, J. S., and Giese, A. C., 1966, Food, reproduction and organic constitution of the common Antarctic echinoid *Sterechinus neumayeri* (Meissner) *Biol. Bull.* **130**:387–401.

Piatigorsky, J., 1975, Gametogenesis, in: *The Sea Urchin Embryo, Biochemistry and Morphogenesis* (G. Czihak, ed.), pp. 42–98, Springer-Verlag, Berlin.

Pirrone A. M., Spinelli, G., Acierro, P., Errera, M., and Giudice, G., 1977, The RNA of unfertilized sea urchin eggs is capped, *Cell Diff.* **5**:335–342.

Poccia, P., Salik, J., and Krystal, G., 1981, Transitions in histone variants following fertilization and evidence for a maternal store of cleavage stage histones in the sea urchin egg, *Dev. Biol.* **82**:287–296.

Posakony, J. W., Scheller, R. H., Anderson, D. M., Britten, R. J., and Davidson, E. H., 1981, Repetitive sequences of the sea urchin genome. Nucleotide sequences of cloned repeat elements, *J. Mol. Biol.* **149**:41–67.

Posakony, J. W., Flytzanis, C. N., Britten, R. J., and Davidson, E. H., 1983, Interspersed sequence organization and developmental representation of cloned poly(A) RNAs from sea urchin eggs, *J. Mol. Biol.* **167**:361–389.

Potter, S. S., 1982, DNA sequence of a foldback transposable element in *Drosophilia, Nature* **297**:201–204.

Raff, R. A., 1980, Masked messenger RNA and the regulation of protein synthesis in eggs and embryos, in: *Cell Biology: A Comprehensive Treatise*, Vol. 4 (D. M. Prescott and L. Goldstein, eds.), pp. 107–136, Academic Press, New York.

Raff, R. A., and Showman, R. M., 1983, Maternal messenger RNA; Quantitative, qualitative, and spatial control of its expression in embryos, in: *The Biology of Fertilization* (C. B. Metz and A. Monroy, eds.), Academic Press, New York.

Raff, R. A., Colot, H. V., Selvig, S. E., and Gross, P. R., 1972, Oogenetic origin of messenger RNA for embryonic synthesis of microtubule proteins, *Nature* **235**:211–214.

Raff, R. A., Brandis, J. W., Huffman, C. J., Koch, A. L., and Leister, D. E., 1981, Protein synthesis as an early response to fertilization of the sea urchin egg: A model, *Dev. Biol.* **86**:265–271.

Raff, R. A., Anstrom, C. T., Huffman, C. J., Leaf, D. S., Loo, A. H., Showman, R. M., and Wells, D. E., 1984, Origin of a gene regulatory mechanism in the evolution of echinoderms. *Nature* **310**:312–314.

Regier, J. C., and Kafatos, F. C., 1977, Absolute rate of protein synthesis in sea urchins with specific activity measurements of radioactive leucine and leucyl-tRNA, *Dev. Biol.* **57**:270–283.

Richter, J. D., Smith, L. D., Anderson, D. M., and Davidson, E. H., 1983, Interspersed poly(A) RNAs of amphibian oocytes are not translatable *J. Mol. Biol.* **173**:227–241.

Rodgers, W. H., and Gross, P. R., 1978, Inhomogeneous distribution of egg RNA sequences in the early embryo, *Cell* **14**:279–288.

Rosenthal, E. T., Hunt, T., and Ruderman, J. V., 1980, Selective translation of mRNA controls the pattern of protein synthesis during early development of the surf clam, *Spisula solidissima*, *Cell* **20**:487–494.

Rosenthal, E. T., Brandhorst, B. P., and Ruderman, J. V., 1982, Translationally mediated changes in patterns of protein synthesis during maturation of starfish oocytes, *Dev. Biol.* **91**:215–220.

Rosenthal, E. T., Tansey, T. R., and Ruderman, J. V., 1983, Sequence-specific adenylations and deadenylations accompany changes in the translation of maternal RNA after fertilization of *Spisula* oocytes, *J. Mol. Biol.* **166**:309–327.

Rudensey, L. M., and Infante, A. A., 1979, Translational efficiency of cytoplasmic nonpolysomal messenger ribonucleic acid from sea urchin embryos, *Biochemistry* **18**:3056–3063.

Ruderman, J. V., and Pardue, M. L., 1977, Cell-free translation analysis of messenger RNA in echinoderm and amphibian development, *Dev. Biol.* **60**:48–68.

Ruderman, J. V., and Schmidt, M. R., 1981, RNA transcription and translation in sea urchin oocytes and eggs, *Dev. Biol.* **81**:220–228.

Sanchez, S., 1966, Inhibition du développement de l'oeuf d'oursin après action de l'actinomycin D sur l'ovocyte, *Compt. Rend. Soc. Biol.* **160**:143–145.

Sanchez, S., 1969, Effets de l'actinomycin D sur les constituents cellulaires de l'ovocyte d'oursin, *Exp. Cell Res.* **50**:19–31.

Scheller, R. H., Costantini, F. D., Kozlowski, M. R., Britten, R. J., and Davidson, E. H., 1978, Specific representation of cloned repetititve DNA sequences in sea urchin RNAs, *Cell* **15**:189–203.

Scheller, R. H., Anderson, D. M., Posakony, J. W., McAllister, L. B., Britten, R. J., and Davidson, E. H., 1981a, Repetitive sequences of the sea urchin genome. Subfamily structure and evolutionary conservation, *J. Mol. Biol.* **149**:15–39.

Scheller, R. H., McAllister, L. B., Crain, W. R., Durica, D. S., Posakony, J. W., Britten, R. J., and Davidson, E. H., 1981b, Organization and expression of multiple actin genes in sea urchin, *Mol. Cell Biol.* **1**:609–628.

Schroeder, T. E., 1980, Expressions of the postfertilization polar axis in sea urchin eggs, *Dev. Biol.* **79**:428–443.

Sconzo, G., Albanese, I., and Giudice, G., 1972, Studies on sea urchin oocytes, II. Synthesis of RNA during oogenesis, *Exp. Cell Res.* **72**:95–100.

Sconzo, G., Roccheri, M. C., DiLeberto, M., and Giudice, G., 1977, Studies on the structure and possible function of the RNA cap in developing sea urchins, *Cell Diff.* **5**:323–334.

Shepherd, G. W., and Nemer, M., 1980, Developmental shifts in frequency distribution of polysomal mRNA and their posttranscriptional regulation in the sea urchin embryo, *Proc. Natl. Acad. Sci. USA* **77**:4653–4656.

Showman, R. M., and Whiteley, A. H., 1980, The origin of echinoid hatching enzyme messenger RNA, *Dev. Growth Diff.* **22**:305–314.

Showman, R. M., Wells, D. E., Anstrom, J., Hursh, D. A., and Raff, R. A., 1982, Message-specific sequestration of maternal histone mRNA in the sea urchin egg, *Proc. Natl. Acad. Sci. USA* **79**:5944–5947.

Showman, R. M., Wells, D. E., Anstrom, J. A., Hursh, D. A., Leaf, A. S., and Raff, R. A., 1983, Subcellular localization of maternal histone mRNAs and the control of histone synthesis in the sea urchin embryo, in: *Molecular Aspects of Early Development* (G. M. Malacinski and W. H. Klein, eds.), pp. 109–130. Plenum Press, New York.

Skoultchi, A., and Gross, P. R., 1973, Maternal histone messenger RNA: Detection by molecular hybridization, *Proc. Natl. Acad. Sci. USA* **70**:2840–2844.

Slater, D. W., and Spiegelman, S., 1966, An estimation of the genetic messages in the unfertilized echinoid egg, *Proc. Natl. Acad. Sci, USA* **56**:164–170.

Slater, D. W., Slater, I., and Gillespie, D., 1972, Post-fertilization synthesis of polyadenylic acid in sea urchin embryos, *Nature* **240**:333–337.

Slater, I., Gillespie, D., and Slater, D. W., 1973, Cytoplasmic adenylation and processing of maternal RNA, *Proc. Natl. Acad. Sci. USA* **70**:406–411.

Smith, M. J., and Boal, R., 1978, DNA sequence organization in the common Pacific starfish *Pisaster ochraceus*, *Canad. J. Biochem.* **56**:1048–1054.

Smith, M. J., Hough, B. R., Chamberlin, M. E., and Davidson, E. H., 1974, Repetitive and non-repetitive sequences in sea urchin heterogeneous nuclear RNA, *J. Mol. Biol.* **85**:103–126.

Smith, M. J., Lui, A., Gibson, K. K., and Etzkorn, J. K., 1980, DNA sequence organization in the starfish *Dermasterias imbricata*, *Can. J. Biochem.* **58**:352–360.

Spirin, A. S., 1966, On "masked" forms of messenger RNA in early embryogenesis and in other differentiating systems, *Curr. Top. Dev. Biol.* **1**:1–38.

Surrey, S., Ginzburg, I., and Nemer, M., 1979, Ribosomal RNA synthesis in pre and post-gastrula stage sea urchin embryos, *Dev. Biol.* **71**:83–99.

Tennent, D. H., and Ito, T., 1941, A study of the oogenesis of *Mespilla globulus* (Linne), *J. Morphol.* **69**:347–404.

Terman, S. A., 1970, Relative effect of transcription-level and translation-level control of protein synthesis during early development, *Proc. Natl. Acad. Sci. USA* **65**:985–992.

Terman, S. A., and Gross, P. R., 1965, Translation level control of protein synthesis during early development, *Biochem. Biophys. Res. Commun.* **21**:595–600.

Thomas, T. L., Posakony, J. W., Anderson, D. M., Britten, R. J., and Davidson, E. H., 1981, Molecular structure of maternal RNA, *Chromosoma* **84**:319–335.

Thomas, T. L., Britten, R. J., and Davidson, E. H., 1982, An interspersed region of the sea urchin genome represented in both maternal poly(A) RNA and embryo nuclear RNA, *Dev. Biol.* **94**:230–239.

Tufaro, F., and Brandhorst, B., 1979, Similarity of proteins synthesized in isolated blastomeres of early sea urchin embryos, *Dev. Biol.* **72**:390–397.

Tufaro, F., and Brandhorst, B. P., 1982, Restricted expression of paternal genes in sea urchin interspecies hybrids, *Dev. Biol.* **92**:209–220.

Venezky, D. L., Angerer, L. M., and Angerer, R. C., 1981, Accumulation of histone repeat transcripts in sea urchin egg pronucleus, *Cell* **24**:385–391.

Verhey, C. A., and Moyer, F. H., 1967, Fine structural changes during sea urchin oogenesis, *J. Exp. Zool.* **164**:195–226.

Vincent, W., Halvorsen, H., Chen, H., and Shin, D., 1969, A comparison of the ribosomal gene amplification in uni- and multinucleate oocytes, *Exp. Cell Res.* **57**:240–250.

Wells, D. E., Showman, R. M., Klein, W. H., and Raff, R. A., 1981, Delayed recruitment of maternal mRNA in sea urchin embryos, *Nature* **292**:477–478.

Wells, D. E., Bruskin, A. M., O'Brochta, D. A., and Raff, R. A., 1982, Prevalent RNA sequences of mitochondrial origin in sea urchin embryos, *Dev. Biol.* **92**:557–562.

Whiteley, A. H., 1949, The phosphorous compounds of sea urchin eggs and the uptake of radio-phosphate upon fertilization, *Am. Nat.* **83**:249–267.

Wickens, M. P., and Gurdon, J. B., 1983, Posttranscriptional processing of simian virus 40 late transcripts in injected frog oocytes, *J. Mol. Biol.* **163**:1–26.

Will, B. M., Bayev, A. A., and Finnegan, D. J., 1981, Nucleotide sequence of terminal repeats of 412 transposable elements of *Drosophila melanogaster*, *J. Mol. Biol.* **153**:897–915.

Wilson, F., Blin, N., and Stafford, D., 1976, A denaturation map of sea urchin ribosomal DNA, *Chromosoma* **58**:247–253.

Wilson, L. P., 1940, Histology of the gonad wall of *Arbacia punctulata*, *J. Morphol.* **66**:463–476.

Wilt, F. H., 1970, The acceleration of RNA synthesis in cleaving sea urchin embryos, *Dev. Biol.* **23**:444–455.

Wilt, F. H., 1973, Polyadenylation of maternal RNA of sea urchin eggs after fertilization, *Proc. Natl. Acad. Sci. USA* **70**:2345–2349.

Wilt, F. H., 1977, The dynamics of maternal poly(A)-containing mRNA in fertilized sea urchin eggs, *Cell* **11**:673–681.

Winkler, M. M., and Steinhardt, R. A., 1981, Activation of protein synthesis in a sea urchin cell-free system, *Dev. Biol.* **84**:432–439.

Woods, D. E., and Fitschen, W., 1978, The mobilization of maternal histone messenger RNA after fertilization of the sea urchin egg, *Cell Diff.* **7**:103–114.
Young, E. M., and Raff, R. A., 1979, Messenger ribonucleoprotein particles in developing sea urchin embryos, *Dev. Biol.* **72**:24–40.
Xin, J.-H., Brandhorst, B. P., Britten, R. J., and Davidson, E. H., 1982, Cloned embryos mRNAs not detectably expressed in adult sea urchin coelomocytes, *Devl. Biol.* **89**:527–531.

Chapter 13

# Genetic Analysis of Oogenesis and the Role of Maternal Gene Expression in Early Development

KENNETH D. KONRAD, LEE ENGSTROM, NORBERT PERRIMON, and ANTHONY P. MAHOWALD

## 1. Introduction

Oogenesis is a complex process of cytodifferentiation, involving a variety of cell types, that results in the production of the female gamete. In higher organisms, this gamete contributes nearly all the cytoplasm composing the developing zygote. The developmental significance of this cytoplasmic contribution has been revealed by a large body of experimental evidence indicating that the maternal genome plays a primary role in the embryogenesis of most organisms—at least until gastrulation, the stage at which overt cellular differentiation generally begins (reviewed by Davidson, 1976).

Molecular analyses of the gene expression associated with oogenesis have also indicated a prominent role of maternal gene products in embryogenesis. Eggs from a variety of organisms display a high degree of RNA complexity (Thomas *et al.*, 1981; Davidson, 1976) and—in the case of sea urchins and amphibians—most egg transcripts have been shown to overlap with those present during embryogenesis and later development (Flytzanis *et al.*, 1982; Perlman and Rosbash, 1978). Similarly, two-dimensional gel analyses have demonstrated that the most prominent proteins synthesized during sea urchin development are initially encoded by maternal transcripts (Brandhorst, 1976) (see Chapter 12). Examination of mutations that affect oogenesis and early development can indicate the function of individual members of this complex set of maternal transcripts. Genetic anal-

KENNETH D. KONRAD, LEE ENGSTROM, NORBERT PERRIMON, and ANTHONY P. MAHOWALD • Developmental Biology Center, Department of Developmental Genetics and Anatomy, Case Western Reserve University, Cleveland, Ohio 44106 ysis may therefore represent a fruitful approach to identifying maternal gene products that provide developmental information to the embryo.

Mutations in genes required for ovarian functions can produce a variety of phenotypes. Some do not significantly affect the viability of adults carrying the mutation, but rather produce sterility or a maternal effect in their progeny as a primary phenotype. Such mutations appear to disrupt gene functions that are vitally required exclusively or primarily during oogenesis or embryogenesis and may be of special interest to developmental geneticists. Systematic efforts to investigate this gene expression by mutagenesis have been undertaken in *Drosophila melanogaster* (Gill, 1963; Kaplan *et al.*, 1970; Bakken, 1973; Rice, 1973; Gans *et al.*, 1975; Mohler, 1977; Anderson and Nüsslein-Volhard, 1983; T. Schüpbach *et al.*, personal communication; J. D. Mohler *et al.*, in preparation; Komitopoulou *et al.*, 1983), and similar gene functions have been detected in screens for mutations disrupting embryogenesis in *Caenorhabditis elegans* (Wood *et al.*, 1980; Miwa *et al.*, 1980; Cassada *et al.*, 1981; Isnenghi *et al.*, 1983). In the case of *Drosophila*, it appears that nearly all loci that can be mutated to produce female sterility as a primary phenotype have been identified; thus, it will soon be possible to consider the full range of effects produced by such mutations in oogenesis and early development.

Most genes required for oogenesis and embryogenesis are not uniquely required for these processes. Many mutations produce sterility as one of a variety of pleiotropic effects (Lindsley and Grell, 1968) and may represent "leaky" mutations in constitutively vital genes. Most mutations in such genes, however, produce lethality before any effects on ovarian function can become manifest. The maternal expression of genes identified by zygotic lethal mutations can be examined in *Drosophila* using embryonic germ cell transplants or by the production of germ line clones homozygous for lethal mutations in heterozygous females and—in both *Drosophila* and *Caenorhabditis*—by the isolation of temperature sensitive zygotic lethal mutations. These approaches permit identification of many more developmental processes that are dependent on materials stored in the oocyte.

This chapter considers the maternal gene expression required for oogenesis and early development as revealed by female sterile mutations, maternal effect mutations, and zygotic lethal mutations, which affect these processes in *Drosophila* and other organisms. It concentrates on the types of developmental defects associated with a representative selection of such mutations and discusses their contribution to our understanding of the significance of maternal gene products in early embryogenesis.

## 2. Female Sterile Mutations

Most of the available information on the genetic analysis of oogenesis has been obtained from mutations isolated in saturation screens for female sterility in *Drosophila melanogaster*, and this section will deal exclusively with this organism. Mutations isolated in these screens disrupt gene functions that are not absolutely required for adult viability and, therefore, may be utilized primarily or exclusively for oogenesis. It must be emphasized, however, that although genetic evidence suggests that some of these mutations may be null, no female sterile mutation has, as yet, been conclusively demonstrated to eliminate gene activity. Without this information, classification of mutations may be misleading. Some of the mutations that appear to disrupt gene functions specific to oogenesis may be hypomorphic (reduced gene activity), and more extreme alleles could show the gene function involved to be required constitutively or more than once during development. In fact, a sizeable proportion of female sterile mutations exhibit pleiotropic effects on adult morphology or viability.

The germ cells of *Drosophila* are precociously segregated from the soma at the posterior tip of the early embryo and, during gastrulation, are carried internally within the posterior midgut invagination. About 50% of these cells migrate to the lateral mesoderm, where they are subsequently surrounded by mesodermal cells (Underwood *et al.*, 1980a). Differentiation of the somatic cells comprising the ovary begins during the third larval instar.

Germ cell differentation in *Drosophila* (Fig. 1) begins after pupariation with the establishment of an anterior stem cell population. Clonal (Wieschaus and Szabad, 1979) and morphological (King, 1970; Carpenter, 1979) analyses indicate that an average of about three stem cells are present per ovariole. During later pupal stages and throughout adult life, each stem cell within the ovarioles continues to divide, forming another stem cell and a cystoblast. The cystoblast undergoes four mitotic divisions with incomplete cytokinesis (Brown and King, 1964), and the resulting syncytial 16-cell clusters become surrounded by the proliferating, mesodermally derived follicle cells. Although at least two of the 16 germ-line cells undergo synaptonemal pairing, only one completes meiotic recombination and becomes the oocyte (Carpenter, 1979; Grell and Generoso, 1982). At the posterior end of the germarium, stage 1 egg chambers containing the oocyte and 15 nurse cells bud off and begin egg formation. The nurse cells will grow and simultaneously transfer cytoplasm to the growing oocyte through interconnecting ring canals, while the surrounding follicle cells will form the vitelline and chorion layers of the egg shell. King (1970) subdivided this dynamic process into 14 morphologically identifiable stages (Fig. 1).

The morphology of ovarian differentiation illustrates the complex cellular interactions that must be coordinated at the level of the egg chamber and the organism to produce a viable egg in *Drosophila*. Mutations that disrupt ovarian function may therefore affect a variety of gene expression, some of which is specific to either germ line or somatic cells. Two methods are available for determining the dependence of a mutation on either cell type, both of which result in the formation of a genetically mosaic ovary.

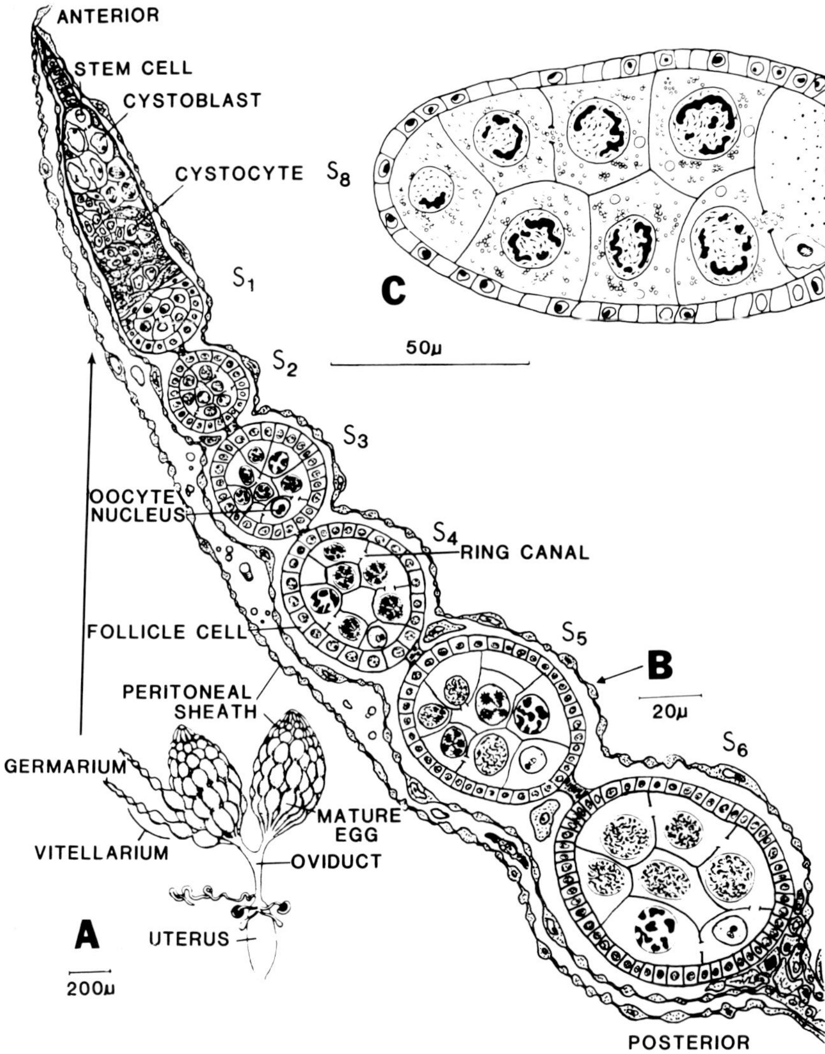


**Figure 1.** Oogenesis in *Drosophila*. (A) Dorsal view of the internal reproductive system of the *Drosophila* female. Each ovary consists of 15–20 ovarioles, two of which are separated from the peritoneal sheath of the left ovary. (B) Anterior portion of an ovariole and its epithelial vestments showing stages 1–6 (S1–S6) previtellogenic egg chambers. (C) Drawing of a stage 8 (S8) egg chamber with the oocyte to the right. Yolk accumulation begins at this stage. (D) Drawing of a stage 10B egg chamber with the oocyte (right) containing yolk granules (dark circles). The surrounding follicle cells are now columnar over the oocyte and squamous over the nurse cells, and a few have migrated between the nurse cells to become border cells at the nurse cell–oocyte junction. (E) Drawing of a stage 12 (S12) egg chamber. The nurse cells have emptied much of their contents into the oocyte and now form a cap on the anterior end (left) of the oocyte. The vitelline envelope has formed, and chorion formation has begun

The first method is **transplantation of germ cells between mutant and nonmutant embryos** (Van Deusen, 1977). If the defect is germ-line dependent, mutant germ cells in a nonmutant host will produce the same phenotype as a mutant female. If, however, the defect is somatic-cell dependent, the mutant germ cells will produce normal eggs in a nonmutant host. Reciprocal transplantations should produce reciprocal results. The second method is the **dominant female sterile technique** (Wieschaus, 1980, Perrimon and Gans, 1983), which uses germ-line dependent dominant female

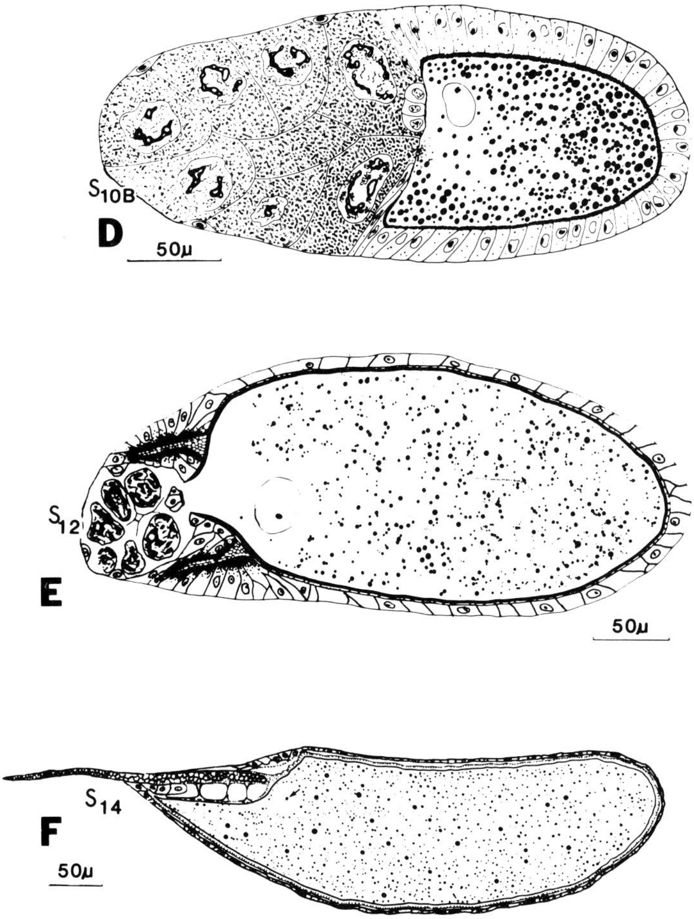


around the oocyte at this stage. (F) Drawing of a mature oocyte (S14) with filaments and chorion completed. The oocyte nuclear envelope has disappeared, leaving the karyosphere (dense chromatin, small arrow) within an island of clear cytoplasm. (Modified from King, 1970, 1975.)

sterile (Dfs) mutations that block oogenesis at early stages. When mitotic recombination is induced in larvae heterozygous for both a recessive mutation that disrupts ovarian function and a Dfs mutation on the homologous chromosome arm, some of the resulting flies will contain germ line clones that lack the Dfs mutation and are homozygous for the recessive mutation (Fig. 2). Mutations that are germ-line dependent will produce the same phenotype both in germ-line clones and in homozygous flies, whereas germ-line clones of somatic cell-dependent mutations will result in normal eggs.

Section 2.1 examines mutations that affect somatic cells required for ovarian function, most of which result in female sterility. Section 2.2 deals with mutations affecting germ-line differentiation, egg chamber organization, and egg polarity, all of which produce female sterility. Section 2.3 discusses zygotic lethal mutations, which have been shown to produce female sterile phenotypes when expressed in the germ line.

## 2.1. Female Sterile Mutations Affecting Somatic Cell Functions

Some female sterile mutations affecting somatic tissues indirectly disrupt ovarian function. Examples include mutations that disrupt mating behavior (Hall *et al.*, 1980) and genital tract functions (e.g., egg retention) (King and Mohler, 1975). Other mutations interfere with the differentiation of the somatic cells that make up the ovary. Mutations that interfere with normal sex determination of these cells can result in sterile females possessing morphologically male or intersexual gonads (Lauge, 1980). Such

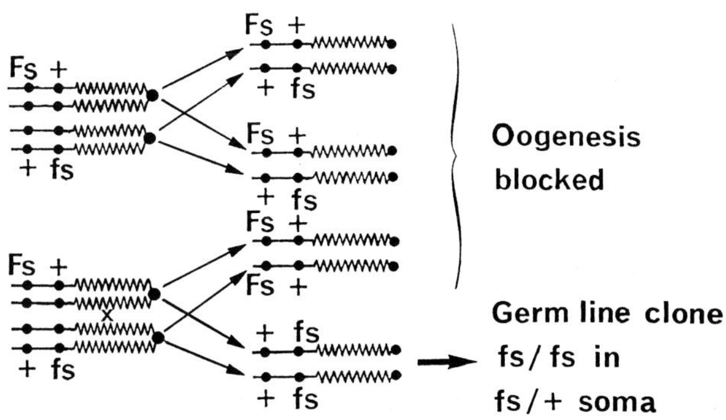


**Figure 2.** Diagram of mitotic recombination using a dominant female sterile mutation (Fs) to generate germ-line clones homozygous for recessive female sterile mutations. The upper portion illustrates normal mitotic chromosome segregation producing cells containing the Fs gene. The lower portion illustrates mitotic recombination (X) and the resulting segregation of a homozygous (fs/fs) female sterile germ-line cell. Nearly 70% of recombination events occur within the heterochromatic portion of the chromosome. (See Perrimon and Gans, 1983, for details).

mutations do not appear to affect differentiation of the female germ line (Marsh and Wieschaus, 1978; Schüpbach, 1982), although contradictory evidence has been obtained (Cline, 1983a), and one such mutatation, *transformer-2*, does affect differentiation of the male germ line (Belote and Baker, 1983).

Vitellogenesis is a major function of both the fat body and follicle cells (Brennan *et al.*, 1982), and a number of mutations have been observed to alter the amount of vitellogenin in the hemolymph or its accumulation in the oocyte. A mutation of the latter type, *apterous*, appears to reduce vitellogenin uptake indirectly through a defect in juvenile hormone production (Postlethwait and Weiser, 1973) and has pleiotropic effects on adult morphology. A number of mutations that affect vitellogenin accumulation, however, are not pleiotropic (cf. Bakken, 1973). Although all such mutations isolated to date have resulted in female sterility, it is interesting to note that vitellogenin-deficient silkworm eggs grown in male hosts were able to develop normally after parthenogenetic activation (Yamashita and Irie, 1980). This result indicates that the organization of developmental information within those eggs does not depend on normal yolk accumulation. The topic of vitellogenesis is considered in detail in Chapter 2.

Another major function of the follicle cells is the formation of the egg envelopes. Using the Dfs technique, Perrimon and Gans (1983) examined 25 mutations that result in the production of collapsed eggs; about one-half these mutations were observed to be dependent on expression in somatic cells. Several of these mutations are known to reduce synthesis of chorion proteins and—whereas some are not pleiotropic—two appear to affect amplification of chorion protein genes and display pleiotropic effects on adult viability (Komitopoulou *et al.*, 1983; Orr *et al.*, 1984).

The germ-line-dependent mutations affecting chorion structure may affect assembly of chorion proteins rather than their synthesis. In one mutation of this type the chorion defect appears to be secondary to a failure in yolk uptake (Waring *et al.*, 1983). This mutation has been shown to be germ-line dependent both by pole cell transplants (Waring *et al*, 1983) and by the Dfs technique (N. Perrimon and P. DiMario, unpublished observations). The latter investigators have characterized more than 20 alleles of this mutation, none of which has pleiotropic effects.

The migration of follicle cells around the growing oocyte is affected by two mutations with pleiotropic effects on adult morphology and by a temperative sensitive (TS) zygotic lethal mutation (King, 1970; King and Mohler, 1975). Although initially considered to affect primarily the follicle cells, both *tiny* and *diminutive* have been shown to be germ-line dependent: *tiny* by both pole cell transplants (DiMario and Hennan, 1982) and the Dfs technique (Perrimon and Gans, 1983) and *diminutive* by the Dfs technique (N. Perrimon, unpublished observation). These mutations, as well as the germ-line-dependent mutations affecting chorion formation, illustrate that some follicle cell functions require an interaction with the oocyte.

Furthermore, they demonstrate the difficulty in predicting the germ-line or somatic cell dependence of a mutation by its phenotype.

Not all mutations that produce changes in egg morphology result in female sterility. Two examples of this type of mutation are known, both of which are somatic cell dependent and have no pleiotropic effects. The recently isolated mutation *unicorn* results in the fusion of the normally paired chorionic filaments into a single filament (Perrimon and Mahowald, 1985). These otherwise normally-shaped eggs are fully fertile. The mutation *short egg* results in eggs that are more rotund than normal. It has been suggested that this mutation may disrupt arrays of microtubules in the follicle cells, which are responsible for establishing and maintaining the normally elongated shape of the egg (Wieschaus *et al.*, 1981). Many of these eggs hatch and complete development, indicating that the exact shape of the egg is not a critical parameter in establishing developmental information during oogenesis. It is interesting to note that physically removing the follicle cells from eggs of another Dipteran insect, *Heteropeza pygmaea*, causes the eggs to become round (Went and Junquera, 1981). A recently isolated female sterile mutation in *Drosophila*, *kugelei*, produces completely round eggs that never leave the ovary (C. Nüsslein-Volhard, personal communication). This mutation has not, as yet, been tested for somatic cell or germ-line dependence.

The nonpleiotropic mutations affecting somatic ovarian cells indicate that a number of the functions of these cells require specialized gene expression. Analysis of such mutations should yield valuable insights into the mechanisms of egg formation. This complex process also requires a wide variety of gene expression used by other cell types. Some of these gene functions may be discovered by analysis of pleiotropic mutations that disrupt oogenesis.

The mutations affecting somatic ovarian cells also suggest that their role is primarily to support the germ line (i.e., providing nutritive yolk and protective egg shells). No mutations producing specific defects in the developmental potential of the embryo have been observed to be somatic cell dependent. In contrast, a number of germ-line-dependent mutations produce such defects.

### 2.2. Female Sterile Mutations Affecting Differentiation of the Germ Line

No mutations have been isolated that specifically affect the migration of the germ cells. However, mutations in the *agametic (ag)* locus result in the selective death of germ cells shortly after their migration to the lateral mesoderm (see Section 3.2 for *grandchildless* phenotype). This mutation is not pleiotropic and has been shown to be germ-line dependent by pole cell transplants (Engstrom *et al.*, 1982).

Other mutations appear to affect the division and differentiation of cystocytes. Mutations of this type generally produce what has been termed an "ovarian tumor" phenotype with abnormal numbers of cystocytes or nurse cells. Pleiotropic mutations that are recessive (e.g., *fused*) and dominant (e.g., *Fs(2)D*), as well as nonpleiotropic mutations (e.g., *female sterile (2)B* [*fes*]; *ovarian tumor* [*otu*]), can produce this phenotype (King, 1970; King and Riley, 1982). Other mutations, most of which appear to be nonpleiotropic, affect recombination or dysjunction of oocyte chromosomes (see King and Mohler, 1975).

Mutations have also been isolated that affect the selection of the oocyte from among the 16 cystocytes. A recently isolated mutation of this type causes a failure of oocyte formation, and the resulting egg chambers contain 16 nurse cells (T. Schüpbach, personal communication). Mutations in several loci result in the formation of two oocytes, one at each end of the follicle (Perrimon and DiMario, unpublished observations). However, these mutations are not fully penetrant for this phenotype.

A number of mutations, most of which are pleiotropic, produce cytologically detectable abnormalities in nurse cell chromosomes (King and Mohler, 1975). In one of these, *singed*, the nurse cell nuclei contain less than normal amounts of DNA (King, 1970), suggesting that the level of polyploidization in these cells may be affected. This mutation, which results in the production of small but normal-shaped eggs, has been shown to be germ-line dependent by the Dfs technique (Perrimon and Gans, 1983).

Other mutations that disrupt egg morphology can appear to affect egg polarity. A germ-line-dependent mutation, *fs(1)K10*, results in eggs displaying ventrally fused, enlarged chorionic filaments (Wieschaus *et al.*, 1978). Examination of egg chamber morphology in mutant females reveals that tall, columnar follicle cells, normally found only dorsally at the anterior end of the oocyte, are, instead, symmetrically arranged (Fig. 3). This phenotype has been interpreted as resulting from a dorsalization of egg chamber polarity. Interestingly, a small percentage of *fs(1)K10* eggs are fertilized, and the resulting embryos also display a dorsalized cuticular pattern (see Section 3.3). The same egg phenotype is produced by females homozygous for the mutation or hemizygous for the mutation over a deficiency, suggesting that the mutation may be null for gene activity. No effects on adult morphology or viability have been observed, and five additional alleles of *fs(1)K10* have been isolated, none of which produces a stronger phenotype. Since multiple lethal alleles of flanking loci have also been isolated (Perrimon, *et al.*, 1984b), it appears unlikely that mutations in this locus will be pleiotropic, making *fs(1)K10* a good candidate for a gene required exclusively during oogenesis and early embryogenesis.

Another mutation, *dicephalic*, affects the anteroposterior polarity of

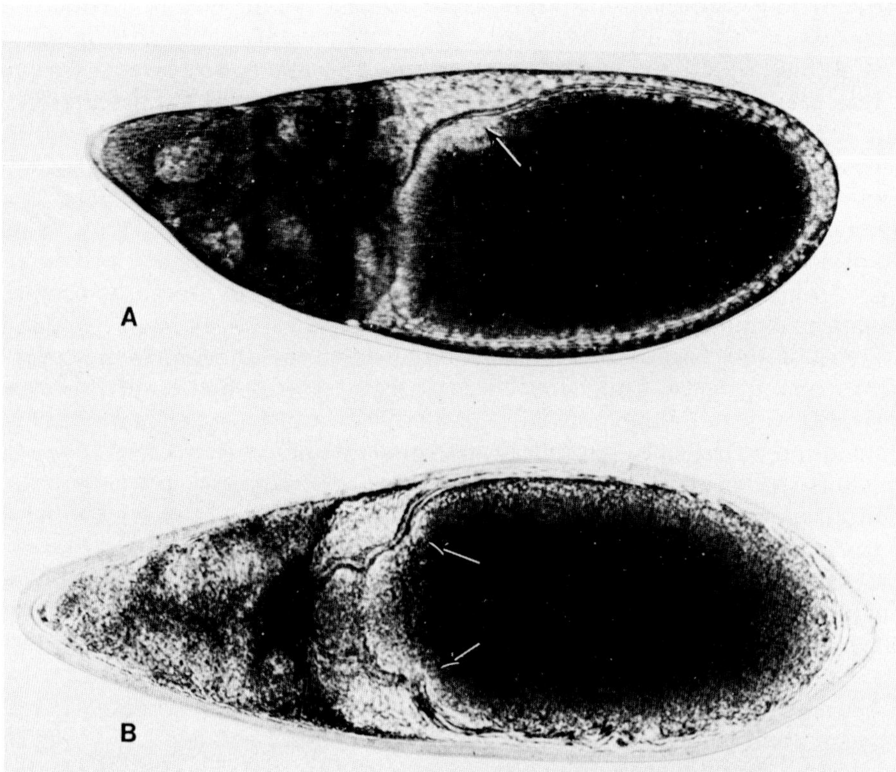


**Figure 3.** Whole mounts of stage 11 wild-type (A) and *fs(1)K10/fs(1)K10* (B) egg chambers. In the normal egg chamber, the follicle cells are much more columnar on the dorsal-anterior oocyte surface (arrow) than on the ventral surface. In the *fs(1)K10* chamber, however, the epithelium is columnar in both the dorsal and ventral anterior aspects (arrows).

the egg (Lohs-Schardin, 1982). The ovaries of females homozygous for this mutation, which is not fully penetrant, often contain egg chambers having a centrally located oocyte and nurse cells at both ends (Fig. 4). Many of the eggs produced by these females display a micropyle at each end and, when fertilized, produce embryos with two anterior ends (see Section 3.3). The selective nature of the defects associated with these latter two mutations indicates that the asymmetric construction of the egg may require the expression of genes specific for that process.

The recently isolated mutation *spindle* results in the production of some eggs that lack chorionic filaments as well as the concave curvature characteristic of the dorsal side of wild-type eggs (Anderson and Nüsslein-Volhard, 1984). It has been suggested that this phenotype results from a ventralization of egg chamber polarity; however, the ovarian morphology of mutant females displays a wide range of defects, including ovarian tumors and egg chambers with the *dicephalic* phenotype. This phenotype is similar to that produced by the TS dominant female sterile mutation *Fs(2)D*. At 25°C, females carrying this mutation produce ovarian tumors. However, at 18°C, some eggs with the *spindle* phenotype are also produced (N. Perrimon, unpublished observations). Thus, it appears that mutations in genes having more general functions can also produce specific changes in egg polarity. Whether these mutations have an effect on embryonic polarity cannot be determined, since eggs with the *spindle* phenotype do not develop.

All the female sterile mutations mentioned in Sections 2.1 and 2.2 disrupt gene functions that are not absolutely required for adult viability. Some of these mutations selectively disrupt specialized cell functions, such as egg shell synthesis, vitellogenin uptake and meiosis, that may require gene functions specific to oogenesis. Interestingly, the establishment of egg chamber polarity also appears to use gene functions uniquely required for that process. The characterization of these latter gene functions is expected to help elucidate the mechanisms of pattern formation and represents a topic of intense interest for developmental geneticists. Pleiotropic mutations also disrupt these processes and in many instances produce specific phenotypes. This illustrates that developmental processes

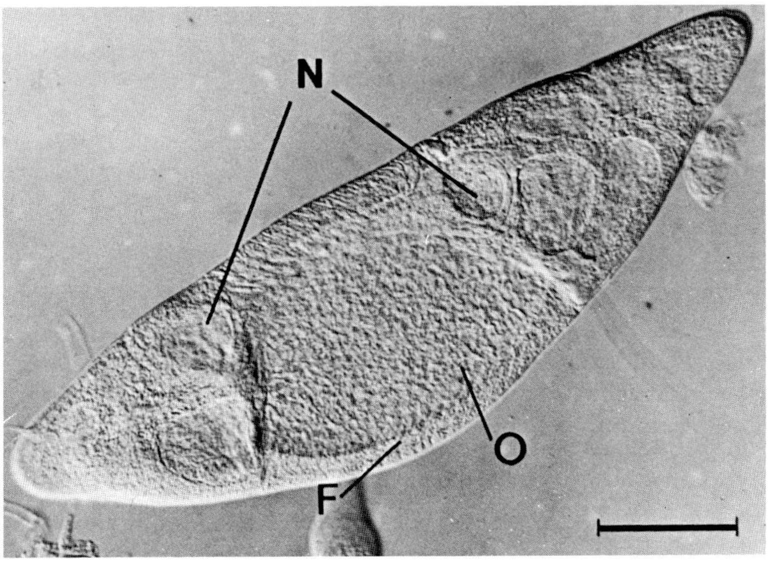


**Figure 4.** Interference contrast image of a stage 10B *dicephalic* egg chamber (compare with wild-type, Fig. 1D). The oocyte (O) is located in the center of the chamber with clusters of nurse cells (N) both anteriorly and posteriorly. Follicle cells are indicated (F). The phenotype first becomes evident when the oocyte is seen to occupy a lateral rather than a normal posterior position (see Fig 1B,C). Scale bar: 100 $\mu$m. Photomicrograph from Lohs-Schardin, 1982.)

involving specialized gene functions require the coordinated expression of many genes also required in other contexts. Most mutations in such genes result in lethality; in the next section we shall examine the effects of some zygotic lethal mutations when expressed during oogenesis.

### 2.3. Female Sterile Mutations That Are Also Zygotic Lethal Mutations

The *Drosophila* genome is thought to contain approximately 5000 genes (cf. Garcia-Bellido and Ripoll, 1978). Extensive mutagenesis studies have indicated that approximately 85% of these genes mutate to a lethal state (Shannon *et al.*, 1972; Hochman, 1973; see also R. A. Raff and Kaufman, 1983). Recent studies have indicated that expression of most genes identified by zygotic lethal mutations is required for normal oogenesis.

Garcia-Bellido and Moscoso Del Pardo (1979) postulated that the reduced viabilility of offspring from deficiency-heterozygous females is due to the hypodosage of genes uncovered by the deficiency. This possibility was further pursued by Robbins (1980, 1983), who examined the effects of combined reductions in maternal and zygotic expression of vital genes. Maternal gene expression was reduced through heterozygosity for sex-linked deficiencies or lethal point mutations, whereas zygotic expression was varied in male offspring hemizygous for the same genetic defect and carrying a duplication for that chromosome region as well. The duplication was located on the fourth chromosome and displayed position effect variegation. Using the variegation to manipulate experimentally the zygotic expression of genes within the duplication, Robbins was able to show that all the deficiencies and 10 of 13 lethal mutations displayed maternal–zygotic interactions. Viability was affected when both maternal and zygotic expression of these genes was reduced, but not when either maternal or zygotic gene activity was normal.

Two additional methods have been used to examine the maternal expression of zygotic lethal mutations. One method involves TS mutations, with ovarian effects being examined after shifting homozygous females to the restrictive temperature. The other method uses the production of homozygous lethal germ-line clones by the Dfs technique.

King and Mohler (1975) refer to a study by H. A. Schneiderman *et al.* (personal communication) of the effects of TS lethal mutations on female fertility. Females homozygous for about one-half the mutations examined remained fertile at the restrictive temperature; however, 22 mutations at 16 loci produced female sterility. Mutations at 15 of these loci produced a variety of ovarian defects including abnormal follicle cell migration and stage-specific degeneration of follicle cells or nurse cells. The remaining mutation resulted in a maternal effect lethality (see Section 3.5).

Germ-line clones of zygotic lethal mutations have been examined by

Garcia-Bellido and Robbins (1984a), who found that mutations in 13 of 15 vital loci in the *zeste-white* region of the X chromosome produce an effect on germ-line function. Perrimon *et al.* (1984a) have similarly examined 48 loci distributed over a large region of the X chromosome. Maternal expression was found to be required for 35 of these genes. Germ-line clones of many of these mutations produced germ-line lethality before vitellogenesis. Ovarian tumors and collapsed eggs were also observed; two of the mutations produced a maternal effect lethality (see Section 3.5). Many of the germ-line clones initially produced a more moderate phenotype followed by a progressive deterioration to the more extreme phenotypes mentioned earlier. This phenomenon may be the result of a perdurability of wild-type gene expression present before the formation of the clone. The continuous decline in phenotype produced by these mutations suggests that the expression of these genes is required continuously throughout oogenesis and early development. The zygotic lethal phase of 24 of the maternally expressed genes is embryonic or larval, similarly suggesting that the maternal contribution is normally sufficient to reach that stage of development. This interpretation has also been used to explain the survival until hatching of the majority of cell-autonomous, cell-lethal mutations in *Drosophila* (Ripoll, 1977; Ripoll and Garcia-Bellido, 1979).

On the basis of these results, it appears that oogenesis in *Drosophila* requires the expression of at least 75% of the genes that can mutate to lethality. The possibility that many zygotic lethal mutations represent defects in general metabolism or "housekeeping" functions is not unlikely, considering the complexity of metabolic pathways. The viability of both somatic and germ line cells would be expected to require the expression of such genes. It is interesting to note that many germ-line clones of zygotic lethal mutations disrupt specific stages of oogenesis and resemble the phenotypes displayed by female sterile mutations. This observation supports the possibility that some female sterile mutations represent hypomorphic alleles of zygotic lethal mutations (Gans *et al.*, 1975), which produce female sterility because of the high metabolic activity required for oogenesis (Bischoff and Lucchesi, 1971). Further analysis at the molecular level will be required to elucidate a solution to this problem.

## 3. Maternal Effect Mutations

Many mutations that affect gene expression during oogenesis are not revealed as defects in ovarian function but rather as a maternal influence on the development of progeny from females carrying such mutations. Most maternal effect mutations that have been described either display no other phenotype or allow the survival of adults carrying the mutation. Mutations of this type are discussed in Sections 3.1–3.4 according to the developmental stage at which their associated defects first become apparent. Zygotic lethal mutations for which maternal effects have been detected will be described in the final part of this section.

Maternal effect mutations can be further subdivided into strict maternal effect mutations that disrupt maternal gene expression required for normal embryogenesis, and rescuable maternal effect mutations that disrupt maternal gene expression sufficent for normal embryogenesis. In the latter class of mutations the maternal deficiency can be "rescued" by zygotic gene expression if the wild-type gene is supplied by the sperm.

Whereas such classifications are convenient for purposes of discussion, it must be emphasized that, as is the case with female sterile mutations, no maternal effect mutations have, as yet, been conclusively demonstrated to be null. In several cases that we shall consider multiple alleles of a single locus can be placed into different categories, illustrating the ambiguities that can result from the classification of mutations for which the null phenotype is unknown.

### 3.1. Maternal Effect Mutations Affecting Embryogenesis before Cellular Differentiation

One of the best-known examples of a maternal effect is a mutation affecting the orientation of the cleavage spindle in spirally cleaving zygotes of the land snail *Lymnaea* and, as a result, the direction of coiling of the adult snail shell (Boycott *et al.*, 1930). The coiling is usually dextral, but occasional individuals display sinistral coiling. Genetic analysis has demonstrated that the direction of coiling is determined by a single locus with a wild-type dextral (+) allele and a recessive sinistral (s) allele. This is observed as a strict maternal effect mutation, because the coiling is dependent not on the genotype of the developing snail but on the genotype of its mother. If a +/+ female is crossed to a s/s male the +/s F1 progeny will all display dextral coiling. If a s/s female is crossed to a +/+ male the +/s F1 progeny will all display sinistral coiling. When dextral +/s F1 siblings or sinistral +/s F1 siblings are mated, the F2 progeny from either cross all display dextral coiling, even though 25% of the F2 progeny from each cross are genetically s/s. These F2 s/s females will then produce all sinistral progeny regardless of the genotype of the sperm. This mutation does not appear to have any other effect on the organism. Since the maternally inherited direction of shell coiling is determined by the orientation of the cleavage spindle, it is apparent that in this organism the orientation of the initial cleavages is determined during oogenesis. The sinistral cleavage and shell coiling in mutant embryos has been reported to be corrected to the dextral pattern after injection of cytoplasm from early dextral embryos. The injections were successful only if they were made before the formation of the second polar body (Freeman and Lundelius, 1982).

Another maternal effect mutation disrupting early development has been described in a different species of *Lymnaea*. Embryos from females homozygous for this mutation undergo cleavage and form blastulae. Gastrulation, however, fails to occur, and the embryos remain permanent blastulae (Arnolds, 1982). This mutation is strictly maternal and produces no detectable phenotype, other than sterility, in homozygous females.

The embryonic phenotype produced by the latter mutation in *Lymnaea* is similar to that produced by the *o* mutation in axolotl (Brothers, 1979). Females homozygous for *o* produce eggs that are fertilized and undergo cleavage and blastula formation. In mid- to late blastula, however, the rate of cell division begins to become abnormally slow, and the embryos invariably cease development at some point during gastrulation. This phenotype is independent of the genotype of the sperm. The cytoplasmic deficiency of these mutant eggs has also been shown to be completely corrected by the injection, before the first cleavage, of cytoplasm or nucleoplasm from wild-type eggs. This mutation may disrupt a gene function required constitutively because it produces pleiotropic effects including male sterility as well as a reduction in adult vitality and their capacity to undergo regeneration (Humphrey, 1966). Several other maternal effect mutations disrupting development before gastrulation have also been isolated in axolotl (Malacinski and Brothers, 1974; E. C. Raff *et al.*, 1976).

Since these latter mutations result in a maternally inherited embryonic lethality, they can be classified as maternal effect lethal (mel) mutations. In *Drosophila*, mel mutations producing eggs of normal appearance that fail to hatch comprise a large subclass of female sterile mutations (Wright, 1970; Bakken, 1973; Zalokar *et al.*, 1975; Rice, 1973). The defects produced by most mel mutations characterized to date become apparent before completion of the cellular blastoderm.

In normal *Drosophila* development, the zygote nucleus undergoes rapid divisions in the absence of cytokinesis, forming a syncytium. By the end of the ninth nuclear division, most of the nuclei have migrated to the periphery of the egg, forming a syncytial blastoderm. After four more divisions membranous furrows form and separate the nuclei, creating a cellular blastoderm that, upon completion, is immediately followed by gastrulation.

Some of the mel mutations produce eggs that are not fertilized or that show no signs of nuclear division. In other mel mutations, fertilization takes place, but no syngamy occurs, and development generally arrests after very few nuclear divisions. Interestingly, however, several mel mutations that may cause a failure of syngamy produce eggs that can undergo extensive development as haploids (Zalokar *et al.*, 1975). Most mel mutations display mitotic abnormalities that may become evident either very early or as late as the final mitosis in the syncytial blastoderm. Several of these mutations appear to be associated with defects in the mitotic spindles (Bakken, 1973). One mutation of this type, *paralog*, has been extensively analyzed (Thierry-Mieg, 1982) and found to be pleiotropic. At 29°C, females homozygous for this mutation are sterile, producing eggs that display mitotic defects. At lower temperatures many of the eggs form a cellular blastoderm but fail to form pole cells. Such embryos hatch and complete development but result in sterile adults (see *grandchildless* phenotype, Section 3.2), which also display a variety of cuticular defects.

Other mel mutations affect the migration of the nuclei to the periphery or the formation of the cellular blastoderm. Three of the latter mutations described by Rice and Garen (1975) indicate that the process of cellularization may have specific regional requirements. Embryos from females homozygous for one of these (*mel-1*) form pole cells but fail to form a cellular blastoderm. Nevertheless, the nuclei of the syncytial blastoderm begin to undergo recognizable morphogenetic movements. This mutation has been shown to be germ-line dependent by pole cell transplants (Regenass and Bernhard, 1978). In *mel-6* embryos, the cellular blastoderm forms only at the anterior and posterior ends (Fig. 5A), whereas *mel-3* embryos are missing cells in a sizeable posterior, dorsal portion of the blastoderm as well as small patches of cells in a surrounding portion of the remaining blastoderm (Fig. 5B). This phenotype of *mel-3* is TS. Temperature-shift experiments have shown that the temperature-sensitive period (TSP) for the maternally-inherited embryonic lethality occurs during the last 10–12 hr of oogenesis (Rice, 1973). At the temperature permissive for female fertility, however, a variety of physical defects in the offspring of homozygous females were observed. In addition to a reduced rate of adult eclosion, many of the adults displayed an abnormality or absence of a variety of

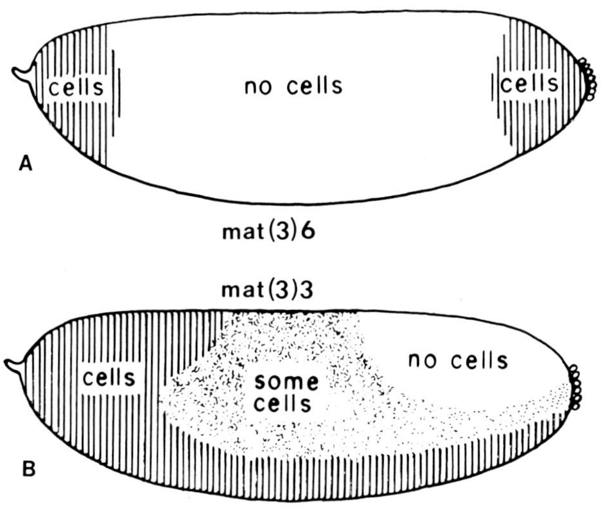


**Figure 5.** Diagrammatic representation of blastoderm defects in embryos from *mat(3)6* (A) and *mat(3)3* (B) females. The dorsal surface is at the top, and the posterior end with pole cells is to the right. The distribution of blastoderm cells on the surface would be uniform in wild-type embryos. (Adapted from Rice and Garen, 1975.)

abdominal or thoracic structures. This result may be explained by the observation that many of the embryos produced by homozygous females at the permissive temperature appeared to be missing patches of cells from the cellular blastoderm. It is not known whether these mutations affecting cellularization reflect localized requirements for different gene products or a differential sensitivity to concentrations of similar gene products.

A systematic genetic analysis of the maternal contribution of gene products required for embryogenesis is also being performed in *Caenorhabditis* (Wood *et al.*, 1980; Miwa *et al.*, 1980; Cassada *et al.*, 1981; Isnenghi *et al.*, 1983). The contribution of maternal and zygotic expression of 55 temperature-sensitive embryonic lethal mutations has been evaluated using the following genetic crosses. Hermaphrodites heterozygous for a mutation are raised at the restrictive temperature and permitted to self-fertilize. The gametes in these worms all develop in the background of the wild-type allele. The resulting eggs are kept at the restrictive temperature, eliminating zygotic expression of the gene in homozygous offspring. If these offspring survive, a sufficient parental contribution of the wild-type gene product can be inferred; if they die, a requirement for zygotic expression is indicated. Hermaphrodites homozygous for the genes, raised at the permissive temperature and shifted to the restrictive temperature just before the developmental stage at which oogenesis begins, are also mated to wild-type males. If no outcross offspring survive, maternal expression of the wild-type gene is required.

Out of 55 genes examined, 53 have been shown to be expressed maternally. Mutations in 32 of these genes can be classified as strict maternal effect lethals (maternal gene expression necessary and sufficient for embryogenesis) and in 17 as rescuable maternal effect lethals (either maternal or zygotic gene expression sufficient for embryogenesis). Four mutations disrupt the expression of gene activity required both maternally and zygotically for normal embryogenesis to occur (Isnenghi *et al.*, 1983). Most of these mutations are pleiotropic and also display gonadogenesis defects or postembryonic lethal phases. The maternal expression of the zygotic lethal mutations is considered in Section 3.5.

Strict mel mutations in 10 genes (*B2, B244, emb-1, emb-3, emb-6, emb-7, emb-8, emb-20, emb-27, emb-30*) and one of the rescuable mel mutations (*emb-2*), however, do not display defects other than those observed in embryogenesis. In the case of the *emb-6* gene, this classification is questionable because one allele, *hc65*, has been reported as being strictly maternal and required only for embryogenesis (Miwa *et al.*, 1980), whereas another allele, *g-36*, is male rescuable and displays a larval lethality (Isnenghi *et al.*, 1983). Temperature-shift experiments have been reported for mutations in nine of these genes and indicate that for eight of the strict mel mutations, the TSP occurs before the two-cell stage, but for the rescuable mel mutation, the TSP extends to the eight-cell stage. The embryonic development of strict mel mutations in seven of the genes has been closely examined (Schierenberg *et al.*, 1980; Wood *et al.*, 1980) and an early terminal phenotype noted for mutations in two others (Isnenghi *et al.*, 1983). In *emb-1* and *emb-27*, the eggs fail to divide after fertilization. In *emb-8* and *emb-30*, the egg shell appears to be osmotically sensitive, and the embryos undergo few or no divisions. Two others display defects in the cleavage pattern. In *B2* the first cleavage is symmetric rather than asymmetric, and in *B244* the first cleavage is variably skewed. Although the cleavage pattern of mutants in the four remaining loci is essentially normal, all of them display early abnormalities in timing or sequence of cell divisions, nuclear morphology, or blastocoel formation. None of these mutant embryos undergoes normal morphogenesis.

Fourteen of the TS mel mutations—two rescuable and the others strict—allow growth to the adult stage of homozygous hermaphrodites shifted to the restrictive temperature after the embryonic TSP. These hermaphrodites, however, are sterile and display a wide range of defects in gonadogenesis (cf. Hirsch and Vanderslice, 1976). It appears that these mutations disrupt gene activity that is required both in gonadogenesis and embryogenesis but not for adult viability. The phenotypes and TSPs for these mel mutations are not remarkably different from those described above. In general, the strict mel mutations display early cleavage defects and have the earliest TSPs, whereas the rescuable mel mutations may display either normal or abnormal early cleavages and have later TSPs. All these embryos also fail to undergo normal morphogenesis; the most common terminal phenotype is at a premorphogenic stage—after cellular proliferation. Whereas these results indicate that the wild-type products of these genes are not necessary for cellular proliferation to proceed, they do not imply that their primary role is in morphogenesis. All such mutations that have been closely examined display defects in the timing and sequence of otherwise normal early cleavages. Subtle variations in these developmental steps may have a profound effect on the ability of these embryos to proceed through morphogenesis.

The maternal effect mutations mentioned so far produce very early developmental defects and appear to disrupt processes involved in cellular proliferation rather than cellular differentiation. Since many of these mutations appear to disrupt gene functions required uniquely or primarily in early embryogenesis, the early cleavages may require genetic information distinct from cell divisions that occur later in development. Alternatively, they may require higher concentrations of similar gene products. The maternal genome plays a pre-eminent role in supplying the materials required for these early developmental processes. In *Drosophila*, all known mutations that disrupt development before the formation of the cellular blastoderm are maternally inherited; these represent most of the maternal effect mutations that have been isolated to date. Sections 3.2 and 3.3 discuss examples of a minority of maternal effect mutations, which, in contrast, appear to interfere with cellular determination and differentiation.

## 3.2. Maternal Effect Mutations Affecting Germ-Line Determination

The production of germ cells in many invertebrates has long been known to be dependent on ooplasmic localizations (reviewed in Nieuwkoop and Sutasurya, 1981; Mahowald and Boswell, 1983). In *Drosophila*, a specific basophilic cytoplasm containing distinctive organelles termed polar granules appears at the posterior tip during the terminal stages of oogenesis. The critical demonstration that this region is a functional germ plasm preformed during oogenesis comes from the transplantation experiments of Illmensee *et al.* (1976). Small quantities of ooplasm containing the distinctive germinal granules were transplanted to the anterior tip of two to four nuclei-stage embryos. Subsequently, the host embryos were shown to have ectopic pole cells by serial section analysis. In order to substantiate these ultrastructural observations, the ectopic pole cells were transplanted to the posterior tip of a genetically distinct host embryo, and the presence of germ cells derived from the induced germ cells was verified genetically.

Since these experiments indicated that the germ plasm could function autonomously, even in abnormal regions of the embryo, it appears evident that maternal effect mutations should be obtained that would affect specifically the ability of the embryo to form germ cells. These mutations have been termed *grandchildless (gs)* because the first generation would be normal except for the absence of germ cells and their consequent sterility. Although there have been many examples of mutations that produce a gs phenotype, only recently have fully penetrant mutations been discovered, and even in these instances the mutations have not affected the germ plasm exclusively.

The various gs mutations that have been analyzed are listed in Table I (see Mahowald and Boswell, 1983, for a further discussion of these various mutations). In the case of the original gs (Spurway, 1948), polar granules appear in mutant oocytes but are then selectively lost. After fertilization, nuclei fail to move into the posterior polar plasm, and no pole cells form. Although the cellular blastoderm forms, it appears defective, because it is extremely fragile (Mahowald *et al.*, 1979).

Most of the other grandchildlesslike mutations produce severe developmental abnormalities in somatic portions of the embryo. Moreover, they show no detectable structural defect in the germ plasm of the embryo. Thus, it is possible that the effect on the germ plasm is actually secondary to a more fundamental defect.

The recently discovered mutation *tudor* (Wieschaus and Nüsslein-Volhard, unpublished results), however, appears more specific. Six alleles have been found with a range of effects on the size of polar granules (Boswell *et al.*, in preparation). The most extreme alleles, including the original mutation, have no detectable polar granules at any time and never form pole cells. One allele forms very small granules but stills fails to form pole cells. Finally, one allele has polar granules of $<50\%$ of the size and

**Table I.** Grandchildless Mutations in Drosophila

| Name | Penetrance (% agametic) | Embryonic development |
|---|---|---|
| *gs* (*subobscura*) | 90–100% | Fragile blastoderm; very little lethality |
| *gs87* | TS<br>17% at 15°C<br>60% at 28.5°C | TS<br>38% lethal at 15°C<br>83% lethal at 28.5°C |
| *par* | Sterile at 29°C<br>40% at 23°C | 60% embryonic lethality at 23°C |
| *gs(2)M* | TS<br>40% agametic at 28.5°C | 40–60% late embryo lethals at 28.5°C |
| *gs(1)N26* | TS<br>25°C: no pole cells<br>18°C: 50% no pole cells | Localized defects in blastoderm, especially posterior tip in 40% |
| *gs(1)N441* | TS<br>25°C: 70% no pole cells<br>18°C: 20% no pole cells | No visible embryonic defect |
| *ag* | Semidominant;<br>TS<br>50% agametic at 25°C | No embryonic defects except on pole cells |
| *tudor* | 100% | 30–40% embryonic lethals; knirpslike phenotype |

quantity of wild type and occasionally forms pole cells and functional gametes. These effects on polar granule formation are fully penetrant.

In spite of the apparent specificity of these alleles on germ plasm formation, all six mutations show a concomitant recessive maternal effect lethal phenotype that is incompletely penetrant (Boswell *et al.*, in preparation). Approximately 30–40% of the embryos from females homozygous for any of the alleles die before hatching. Interestingly, the severity of the embryonic deficiency produced is correlated with the severity of the effect of the mutation on the germ plasm. Those mutations lacking polar granules produce lethal embryos with an extreme abdominal segment deficiency, usually resulting in one to four segments. At the other extreme, the incompletely penetrant allele produces the same frequency of lethal embryos, but they lack any detectable morphological defects.

Since the severity of the two effects produced by each allele is related, presumably some underyling interrelationship must exist between the establishment of the germ plasm and the formation of the segmental pattern of the embryo. This result is surprising, because the transplantation data mentioned earlier had clearly indicated that germ plasm function should be autonomous. Possibly the extensive mutagenesis experiments now in progress will show the presence of other loci specific for germ plasm function.

## 3.3. Maternal Effect Mutations Affecting Somatic Cell Determination

### 3.3.1. Mutations Affecting Embryonic Polarity

A minority of the mel mutations in *Drosophila melanogaster* characterized to date do not cause visible defects in embryonic development until after formation of the cellular blastoderm. Both strict and rescuable mel mutations appear in this class, and many of these mutants appear to affect maternal gene products required for normal cellular differentiation.

Strict mel mutations that interfere with somatic cell determination often appear to affect the embryo in a global manner. For example, two types of mutations that disrupt normal anteroposterior polarity have been identified. One of these, *dicephalic (dic)*, has already been mentioned in Section 2.2, as it also produces some egg chambers and egg shells with two anterior ends (Lohs-Schardin, 1982) (Fig.4). A small percentage of *dic* eggs develop, and most of the resulting embryos display a mirror-image duplication of anterior structures at each end. It appears, therefore, that the anteroposterior polarity of the embryo is at least partially dependent on the asymmetrical construction of the egg.

The recessive mutation *bicaudal* (Nüsslein-Volhard, 1977) also affects anteroposterior polarity but has an effect opposite to that of *dicephalic*. Whereas the morphological features of oogenesis are normal, embryos produced from *bicaudal* eggs display mirror-image duplications of the posterior end (Fig. 6). The amount of anterior or posterior development displayed by different *dic* or *bicaudal* embryos is variable. The production of symmetrical ends within individual embryos therefore suggests that both ends have been affected coordinately. Recently two additional loci, one dominant and the other semidominant, which produce a bicaudal-like phenotype, were found (Nüsslein-Volhard *et al.*, 1982).

A relatively large number of strict mel mutations that disrupt dorsoventral polarity have also been identified. Two of these, *dorsal (dl)* (Nüsslein-Volhard, 1979; Anderson and Nüsslein-Volhard, 1984a) and *gastrulation defective (fs(1)gd)* (Konrad *et al.*, in preparation), have been extensively characterized. Embryos from eggs produced by females homozygous for these mutations form a normal cellular blastoderm but fail to undergo gastrulation. The ventral furrow is absent and—instead of undergoing germ band elongation—the blastoderm cell layer forms a series of transverse lateral folds that continue to deepen, eventually forming a twisted string of interconnected vesicles (Fig. 7A). A cuticle forms that lacks normal lateral and ventral structures and that appears to be dorsal-

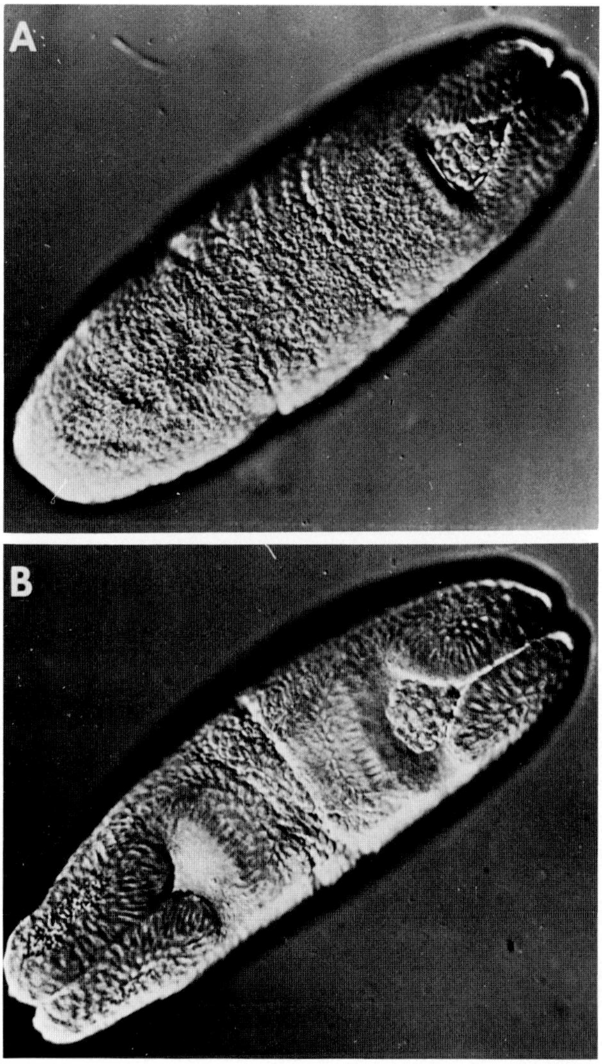


**Figure 6.** Interference contrast images of wild-type (A) and *bicaudal* (B) embryos at gastrulation. Both embryos are seen in a dorsal view with their posterior ends oriented toward the top. Note the triangular posterior mid-gut invagination containing pole cells in both embryos. A symmetrical anterior invagination without pole cells is also seen in the *bicaudal* embryo. (Photomicrographs from Nüsslein-Volhard, 1979.)

ized. The formation of rudimentary anterior and posterior structures indicates that a normal anteroposterior polarity has been maintained.

The ventral blastoderm cells in these embryos are not lost but appear to undergo a change in developmental fate. This has been demonstrated for blastoderm cells of embryos displaying a *dl dominant* phenotype (Nüsslein-Volhard *et al.*, 1980). Embryos from females heterozygous for *dl* at 29°C develop nearly normal ventral cuticular structures but still fail to form a ventral furrow and lack mesoderm-derived tissue. UV-microbeam ablation of mid-ventral cells in normal embryos generally produces no cuticular defects, because those cells represent mesodermal precursors that become internalized during ventral furrow formation. In contrast, UV ablation of mid-ventral cells in *dl-dominant* embryos produces defects in the ventral cuticle.

Alleles at both loci also produce embryos with some ventral structures, suggesting that the dorsalization in these mutations is not as severe. If we consider the variations in phenotype as representing levels of gene activity, we can hypothesize that as the activity of these loci decreases, the fate of the blastoderm becomes increasingly dorsalized.

Two other mutations have been found to affect dorsoventral polarity in the embryo. However, they produce different phenotypes. The mutant *fs(1)K10* described earlier results in eggs that appear dorsalized (Fig. 3). A

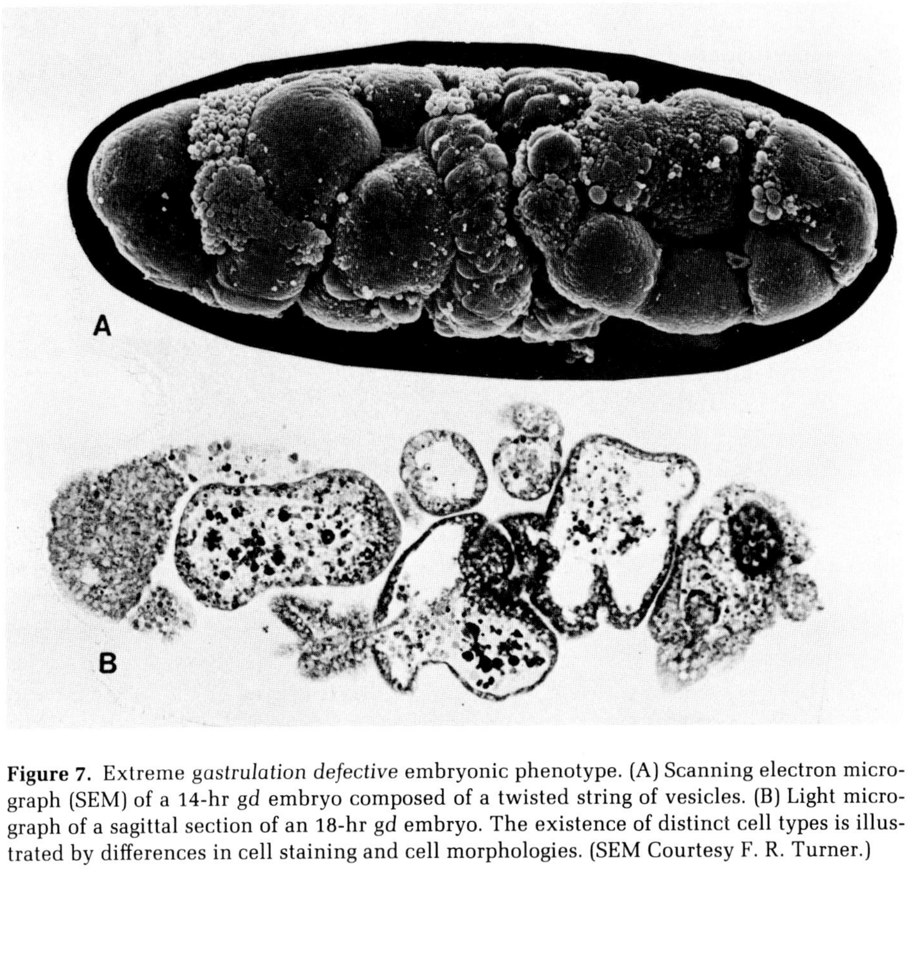

**Figure 7.** Extreme *gastrulation defective* embryonic phenotype. (A) Scanning electron micrograph (SEM) of a 14-hr *gd* embryo composed of a twisted string of vesicles. (B) Light micrograph of a sagittal section of an 18-hr *gd* embryo. The existence of distinct cell types is illustrated by differences in cell staining and cell morphologies. (SEM Courtesy F. R. Turner.)

few percent of these eggs are fertilized; the resulting embryos display a dorsal-like phenotype as well (Wieschaus *et al.*, 1978; Wieschaus, 1980). Thus, dorsoventral polarity of the embryo may also depend on the asymmetrical construction of the egg. Females heterozygous for the dominant mutation *Toll* produce embryos in which ventral cuticular elements expand into the dorsal region, so that the embryos appear ventralized (Anderson and Nüsslein-Volhard, 1984a).

The global manner in which these mel mutations change embryonic cell determination convincingly fits models hypothesizing the existence of a graded system of positional information rather than localized determinants (Nüsslein-Volhard, 1979; Nüsslein-Volhard *et al.*, 1982). The mutations are thought to act early in development to disrupt the gradients to varying degrees in either the anteroposterior axis or the dorsoventral axis. The concept of positional information is, however, rather abstract, and the mechanism by which these mutations affect cellular determination is unknown. For example, considerable evidence exists indicating that midventral cells, which normally form mesoderm, are not determined at the time of cellularization and require specific cell interactions to achieve a determined state (Konrad and Mahowald, 1983). It is therefore possible that, rather than disrupting a global system of positional information, mutations such as *dl* and *fs(1)gd* could disrupt the ability of cells to interpret or respond to those determinative cues. Observations made in our laboratory in a developmental analysis of these mutations support the latter possibility. Although *dorsal* and *fs(1)gd* embryos fail to form the invaginations that normally produce internal tissues, we have observed differentiation of internal cells by 8–10 hr after fertilization. It is not clear how these cells become internalized, but it is possible that they are either shed from the blastoderm surface or pinched off from the deep transverse folds. By 16–24 hr of development differentiated structures corresponding to endodermal and neural derivatives can be identified throughout the embryos (Fig. 7B). The presence of nervous tissue can also be detected by histochemical analysis by staining for acetylcholinesterase (ACh) activity. Contractile fibers have also been observed in the vesicles at low frequency.

A developmental fate map for the blastoderm cells of wild-type *Drosophila* embryos has been established by a number of techniques (Poulson, 1950; Lohs-Schardin *et al.*, 1979; Underwood *et al.*, 1980b). Although neural precursors that contribute to the brain have been mapped to the anterodorsal region, precursors for the ventral nerve cord, and all mesodermal and endodermal structures are derived from cells on the ventral surface or posterior tip. Thus, the presence of internal structures throughout *dl* and *fs(1)gd* embryos does not support a dorsalized determination of all cells at blastoderm. It appears possible that the ventral blastoderm cells in these embryos are incapable of undergoing gastrulation. If, however, they become fortuitously placed in an environment in which they are permitted to differentiate in a relatively normal fashion, they can form internal structures. This differentiation or their dorsalized cuticular fate may be a response to a variety of cellular interactions.

We have been able to define a TSP for one allele of *fs(1)gd*, which begins late in oogenesis and extends to 90 min postfertilization—the time at which nuclei migrate to the surface to form the syncytial blastoderm. The TSPs of a second, weaker allele of *fs(1)gd* and the *dl dominant* phenotype also begin late in oogenesis, but they extend to 120 and 150 min postfertilization, respectively. The early stage at which activity of these loci is required may provide a reasonable explanation for their strict maternal nature.

A major difference between these loci is observed in experiments testing the effect of injections of wild-type cytoplasm or RNA into the mutant embryos before cellular blastoderm formation. This perturbation partially rescues the embryonic phenotype of *dl* eggs as well as those from females homozygous for a number of other dorsalizing maternal effect mutations, but has no effect on *fs(1)gd* embryos (Anderson and Nüsslein-Volhard, 1984a, 1984b). This result suggests, at least for *dl*, that the maternal gene products required to establish embryonic dorsoventral polarity are not irreversibly distributed during oogenesis.

The phenotype displayed in embryos from females homozygous for the most extreme alleles of both *dl* and *fs(1)gd* appears to represent null gene activity, as the same phenotype is produced when the females carry either allele over a deficiency for that region of the genome. In addition, although a sizable number of alleles have been isolated for both loci, no zygotic lethal alleles or alleles with phenotypes other than female sterility have been observed. Multiple lethal alleles of loci flanking both of them have been obtained. These two sets of mel mutations are, to date, the best characterized of such mutations, and the loci they disrupt are among the most convincing candidates for maternally expressed genes required uniquely in embryogenesis.

Mutations in at least eight more loci, including *easter*, have been found with phenotypes similar to *dl* and *fs(1)gd* (Anderson and Nüsslein-Volhard, 1984a). The large number of genes producing this phenotype suggests that they are not "selector genes" involved in determinative binary switching decisions (Kauffman, 1977; see E. C. Raff and Kaufman, 1983) but rather that they encode a complex series of gene products that may interact, directly or indirectly, to produce the cellular organization required for normal cellular movements and dorsoventral determination.

### 3.3.2. Mutations Affecting Other Forms of Cell Determination

Mel mutations also affect global cellular determination events that do not involve embryonic polarity. A strict mel mutation isolated in the amphibian *Pleurodeles waltlii* produces an early embryonic phenotype strikingly similar to that produced by the dorsalizing mutants in *Drosoph-*

*ila* (Beetschen and Fernandez, 1979; Bluemink and Beetschen, 1981). Females homozygous for *ac* produce embryos that undergo normal cleavage but fail to gastrulate. The ectoderm in the animal hemisphere of these mutant embryos becomes furrowed with irregular grooves. In many embryos these furrows deepen and become continuous producing a brain-like appearance. Such embryos die shortly after this stage.

Gene products of the *daughterless (da)* locus of *Drosophila* appear to be required for establishing dosage compensation and sex determination in all the somatic cells of the embryo (Cline, 1983b). In embryos from *da* homozygous females, the female-specific *Sex-lethal* locus appears not to be activated, causing all female offspring to die sometime after the cellular blastoderm stage. This is a strict maternal effect mutuation. Although *da* produces no morphological defects in homozygous adults, the homozygous *da* female progeny from crosses of heterozygous flies survive at only one-half the rate of their heterozygous siblings (Sandler, 1975).

An opposite but perhaps analogous effect is produced by the strict maternal effect mutation *sonless (snl)*. Homozygous *snl* females are phenotypically normal. However, they produce less than 1% male offspring. The interaction of *snl* with sex-altering zygotic mutations shows this effect to be dependent on the sex of the embryonic cells. The mutation *double sex* causes both homozygous females and males to develop into intersex adults. When such flies are produced from homozygous *snl* females the feminized male offspring become more viable, whereas the masculinized females become less viable. The same effect on female offspring is seen with *transformer*, which causes female homozygotes to develop into sterile phenotypic males (Colaianne and Bell, 1972).

Another locus, *extra sex combs (esc)* is essential for the establishment of embryonic segmental identity (Struhl, 1981) as well as other functions. This locus was originally defined by a recessive mutation that caused sterility and a partial transformation of the distal portion of the second and third legs into first legs in homozygous adult males. Subsequently, it was discovered that mutations in this locus also produced a rescuable maternal effect. In the absence of maternal and zygotic expression of this gene product, the correct number of larval segments form, but they all resemble abdominal segment 8, suggesting that the *esc*$^+$ product may act as a repressor of the *bithorax* gene complex (see Lawrence, 1981). Eggs from females homozygous for *esc*, when fertilized by sperm carrying one copy of *esc*$^+$, show a range of more moderate segmental transformations and—when fertilized by a sperm carrying two copies of *esc*$^+$—can develop into adult flies. Most of these paternally "rescued" adults, however, still display patchy posterior transformations of adult segments. This developmentally late manifestation of the defect in maternal gene expression suggests that determinative restrictions on both adult and embryonic cell precursors may occur through similar mechanisms. Temperature-shift experiments with a TS allele have shown that *esc* affects establishment of segmental identity during a brief period of several hours between the end of blastoderm formation and germ-band shortening (Struhl and Brower, 1982).

Both *almondex (amx)* (Shannon, 1972; Jimenez and Campos-Ortega, 1982) and *fs(1)R1* (Romans *et al.*, 1976; S. G. LaBonne *et al.*, unpublished observations) show a maternal effect phenotype resembling the nervous system hypertrophy characteristic of *Notch* (Poulson, 1940; Wright, 1970) and other neurogenic mutants (Lehmann *et al.*, 1981). In embryos from females homozygous for *amx* or *fs(1)R1*, the whole ventral ectoderm forms neural tissue rather than only a portion becoming neural. Both mutations produce a maternal effect that can be defined as rescuable. In the case of *fs(1)R1*, hatching occurs after fertilization by wild-type sperm in some eggs produced by females homozygous for one allele but not a second allele (Romans *et al.*, 1976). In mutations affecting both loci, however, the phenotype of eggs that fail to hatch becomes less extreme when a wild-type allele is introduced by the sperm. Recent results from our laboratory (S. G. LaBonne and A. P. Mahowald, unpublished observations) indicate that both mutations are also partially rescued by microinjection of wild-type cytoplasm into preblastoderm embryos. These mutations both produce a pleiotropic effect in homozygous and hemizygous flies that is seen as a narrowing in the width of the eye.

The sex-linked mutation *fused (fu)* also displays a rescuable maternal effect on cellular determination among a number of pleiotropic effects. Young females homozygous for *fu* produce a small number of eggs that can hatch if fertilized by wild-type sperm. Many of the "rescued" heterozygous daughters that result from this cross still display abnormal abdominal segmentation. In eggs fertilized by mutant sperm (Counce, 1956; A. P. Mahowald, unpublished observations) the ectodermal cells of the posterior compartment of each abdominal segment differentiate as a mirror image of the anterior compartment, thereby producing a gooseberry pattern for the ventral larval cuticle (cf. Nüsslein-Volhard and Wieschaus, 1980). As the *fu* homozygous females age, they produce fewer eggs and develop tumorous ovaries (King, 1970). This phenotype, which is not a maternal effect, is also produced by a number of female sterile mutations described in Section 2.2. Thus, it appears that the *fu* gene product is required both for ovarian organization and embryonic segmentation.

A few maternal effect genes that appear to disrupt cellular determination affect limited portions of the blastoderm fate map, in contrast to the more global defects produced by the mutations listed above. A detailed description of *fs(1)1502* has been made by Komorowska (1980), and a new allele has been analyzed by F. R. Turner and E. C. Stephenson in our laboratory (unpublished observations). Embryos from females homozygous for either of these alleles form a normal blastoderm; however, they fail to form the clypeolabrum or anterior-most head segment. As a result the embryos form defective mouthparts and rarely hatch, and those that do hatch are unable to feed. The phenotype of both alleles is strictly maternal.

Mutations in a sizable group of loci that also selectively disrupt the determination of a portion of the cellular blastoderm include *fs(1)E1* and *fs(1)E2*. Embryos from females homozygous for these mutations fail to develop structures from the posterior end of the blastoderm and display anterior defects as well. The first visible defect in these embryos, which we have termed "pole hole," actually occurs before completion of the cellular blastoderm as a very small cytoplasmic deficiency at the posterior tip below the pole cells (Fig. 8A). Subsequently, however, the differentiation of posterior cells representing about 20% of the cellular blastoderm, becomes abnormal. These embryos do not form a posterior midgut and hindgut invagination and consequently lack a major portion of the gut. Posterior cuticular structures are also missing (Fig. 8B). This cuticular phenotype is similar to that produced by *fs(2)torso* briefly described by Nüsslein-Volhard *et al.*, (1982). At least six additional loci have been observed to produce the *torso* phenotype as a maternal effect (C. Nüsslein-Volhard, personal communication; T. Schüpbach, personal communication). One of these is also a zygotic lethal and is discussed in Section 3.5. It is not yet known if *torso* and the other torsolike mutations also produce the pole hole phenotype. It appears, however, that the formation of the posterior midgut invagination and the differentiation of cells in the posterior region of the embryo require a complex set of gene products. This situation may be analogous to that previously discussed concerning formation of the ventral furrow and establishment of dorsoventral polarity.

Finally, we have discovered *fs(1)E3* (L. Engstrom *et al.*, unpublished observations), in which embryos from homozygous females display a limited transformation of a small region of the posterior blastoderm into ill-defined new structures that protrude from the posteroventral surface of the embryo during postgastrulation stages of development. This mutation produces a rescuable maternal effect that, in contrast to the previously mentioned loci, is completely rescued by a wild-type allele introduced with the sperm. Six alleles of this mutation have been analyzed, all of which produce a similar phenotype.

It is not known whether the selective effects of the last few mutations are due to a disruption of localized determinants or to a differential sensitivity to changes in concentration of common gene products. Most maternal effect mutations that appear to disrupt cellular determination or differentiation do so in a global or general manner. This observation, in conjunction with the ability to "rescue" many of the mutations by cytoplasmic injection or by the wild-type allele supplied by the sperm, suggests that determination of somatic cells in *Drosophila* is not controlled directly by maternal substances prelocalized in the egg. It rather appears to be an epigenetic process resulting from progressive interaction of a number of gene products.

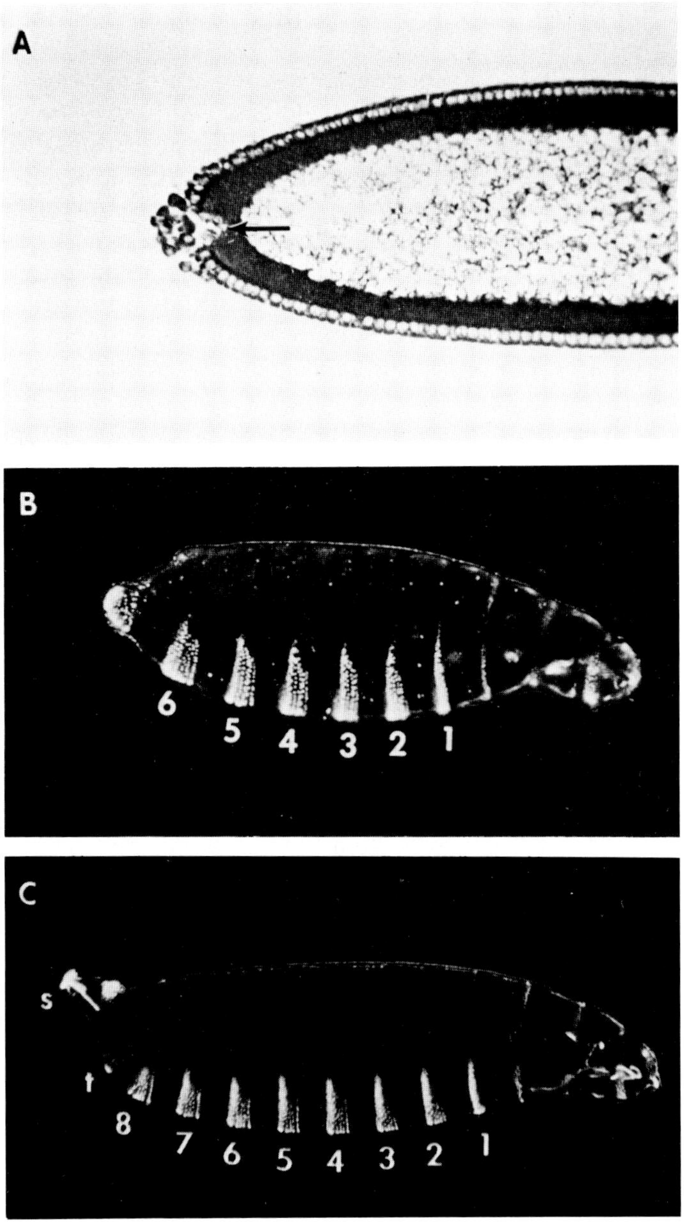


**Figure 8.** Defects associated with *fs(1)E1* and *fs(1)E2*. (A) Photomicrograph of a mid-sagittal section of a blastoderm embryo showing the pole-hole phenotype. Note the absence of blastoderm cells (arrow) under the pole cells, which have formed normally. (B) Cuticular pattern of a 24-hr (nonhatching) pole-hole embryo illustrating the presence of normal segmentation from anterior (right) through the sixth abdominal segment. (C) Note that most posterior structures including the spiracles and anal tuft (s and t, respectively) in the normal embryo are missing. Variable head defects are also found in pole-hole embryos.

### 3.4. Maternal Effect Mutations Affecting Embryonic Metabolism

Although metabolic defects cannot be ruled out as the basis for any of the mutations mentioned above, such defects are clearly involved in a number of rescuable maternal effect mutations. Such mutations produce relatively late developmental defects and do not cause specific defects in cell differentiation but rather disrupt the general metabolism of the embryo. The *cinnamon (cin)* and *maroonlike (mal)* loci both modify eye color and xanthine dehydrogenase (XDH) activity (Sayles *et al.*, 1973; Baker, 1973; Browder and Williamson, 1976). Either maternal or zygotic expression of the wild-type alleles of these loci produces normal eye color. Whereas the *mal* mutation has no effect on embryonic viability, zygotes from *cin* homozygous females die before hatching when fertilized by mutant sperm and hatch normally when fertilized by wild-type sperm. The mechanism by which these mutations affect XDH activity, as well as that of a number of different metabolic enzymes, is unknown. It is possible, however, that the maternal effects of these loci (i.e., enzyme activity in homozygous mutant progeny of mothers having the wild-type allele) represent a very long perdurability of a maternal gene product. No maternal effects of mutations at the *rosy* locus, the XDH structural cistron, have been observed.

Similar phenotypes are produced by some of the mutations in the X-linked *deep orange (dor)* locus. Females homozygous for the original *dor* mutation display a dark eye color and a rescuable maternal effect lethality. Embryos from these females die around the time of gastrulation when fertilized by *dor* sperm and about one-half the heterozygous female offspring survive when fertilized by a wild-type sperm. Injection of cytoplasm from wild-type eggs into *dor* eggs at the syncytial blastoderm stage permits about one-third of the eggs to continue development past gastrulation to a more advanced stage of embryogenesis (Garen and Gehring, 1972). Other alleles of *dor* produce the eye color phenotype in fertile females, and some produce zygotic lethality in larval or pupal stages (Bischoff and Lucchesi, 1971), suggesting that the moderate phenotypes are produced by hypomorphic mutations. This mutation has been shown to be germ-line dependent by pole-cell transplants (Marsh *et al.*, 1977).

The *rudimentary (r)* locus encodes a polypeptide that is enzymatically active in three steps of pyrimidine biosynthesis. Adults homozygous or hemizygous for mutations in this X-linked locus display a wing defect. Most alleles also display a rescuable maternal effect lethality. Homozygous females are sterile when mated to *r* males, but heterozygous female offspring survive after fertilization with wild-type sperm. The mutant flies have a nutritional requirement for pyrimidines. Homozygous *r* females mated to *r* males can produce offspring of both sexes when their medium is enriched with cytidine (Norby, 1970). Furthermore, injection of wild-type cytoplasm or pyrimidines into *r* eggs from homozygous females can rescue the embryonic lethality (Okada *et al.*, 1974). Another auxotrophic mutation displaying a rescuable maternal effect lethality is a yeast-requirement mutation isolated by Falk and Nash (1974).

The lethal phase of the rescuable maternal effect mutation *abnormal oocyte (abo)* has not been determined. When females homozygous for this mutation are mated to males carrying an attached XY chromosome and no free Y, very few of the resulting X/O males survive. This maternal effect is rescued by the Y chromosome or by increasing the maternal dose of the heterochromatic region of the X chromosome (Sandler, 1970; Parry and Sandler, 1974). It has been postulated that this mutation affects ribosome accumulation in the oocyte by influencing the expression of rDNA, which is located in heterochromatin of the X and Y chromosomes. Molecular studies have supported this hypothesis (Yedvobnick *et al.*, 1980; Graziani *et al.*, 1981).

The mutations mentioned in this section further illustrate the role of the maternal genome in providing material required for cellular metabolism in early embryogenesis. Some of these mutations are observed as maternal effect lethals rather than as zygotic lethals, as the metabolic defects can be overcome by dietary intake or alternate metabolic pathways, which are either unavailable or insufficient in the early embryo. Some zygotic lethal mutations have also been observed to produce a maternal effect when expressed during oogenesis; these are examined in the next section.

### 3.5. Maternal Effect Mutations That Are Also Zygotic Lethal Mutations

The evidence that the majority of zygotic lethal mutations in *Drosophila* also affect maternal gene expression required during oogenesis and early development was reviewed in Section 2.3. Most TS mutations arresting embryonic development of *Caenorhabditis elegans* display postembryonic lethal phases as well (Wood *et al.*, 1980; Cassada *et al.*, 1981; Isnenghi *et al.*, 1983). Both strict and rescuable maternal effects are produced by these mutations. Most of the phenotypes and TS periods are similar to those described earlier for mutations exclusively affecting embryogenesis (see Section 3.1). However, several interesting exceptions have been observed. A number of the mutations producing rescuable maternal effects display TS periods well before gastrulation, implying that the zygotic expression of these genes begins very early in development. Four mutations disrupt gene activity that is required both maternally and zygotically for embryogenesis to proceed. It has been suggested that expression of such genes may be required continuously throughout embryogenesis for essential metabolic functions (Hirsch, 1979).

Ten TS lethal mutations that cause arrest during larval stages when continuously raised at the restrictive temperature were also examined for parental effects on survival (Wood *et al.*, 1980). For these tests, eggs were collected from heterozygous hermaphrodites at the permissive temperature and shifted to the restrictive temperature before or just after hatching. For four of these larval lethal mutations, homozygous progeny survived beyond the normal lethal phase, implying that the combined maternal gene products of both the wild-type allele and the mutant allele at the permissive temperature influenced development as late as the second larval moult.

Maternal effects have also been observed for several TS mutations in *Drosophila*. One of these—*l(1)ts6225*—is a TS lethal (King and Mohler, 1975) (see Section 2.2). At the permissive temperature, females homozygous for this mutation are fertile. When transferred to 29°C, they become sterile after 1 or 2 days, producing eggs that are normal in appearance but blocked in either fertilization or syngamy. When these females are returned to the permissive temperature, fertility returns—but only after several days (T. Cline, personal communication, cited in King and Mohler, 1975). Three other TS lethal mutations with maternal effects have been observed. Fausto-Sterling *et al.* (1977) reported that the mutation *l(1)tsl* disrupted mid-gut formation and dorsal closure in embryos from females shifted to the restrictive temperature. Another locus, *l(3)1902* has been characterized as a small disc mutation by Shearn *et al.* (1978), who found two alleles that produced a maternal effect disrupting embryonic development before or during gastrulation. Two separate TSPs, the first during oogenesis and early embryogenesis and the second around the pupal stage, have also been described for *fs(1)h* (Forquignon, 1981).

Other zygotic lethal mutations in *Drosophila* have also been observed to display maternal effects on early development. Two of these, *mus-105* and *mus-109* were originally identified by viable alleles producing mutagen-sensitive chromosomal aberrations. Lethal alleles of these mutations have subsequently been isolated (Baker *et al.*, 1982). Females homozygous for the three viable alleles of *mus-109* display a strict maternal effect lethality, producing embryos that die early in embryogenesis, whereas one viable allele of *mus-105* displays a similar maternal effect, which is rescuable. The lethal alleles of these mutations result in larvae that have defective imaginal discs and die at the larval–pupal boundary. These workers interpret these results as indicating that (1) these loci specify essential mitotic functions, and (2) their maternal expression is sufficient for the embryonic divisions. However, the division of the imaginal disc cells, which are expendable for larval survival, becomes abnormal. Baker and associates further report that as many as 50% of the late lethal mutations they have examined may specify essential mitotic functions. Most or all of these genes would also be expected to be expressed during oogenesis.

Using the Dfs technique, Garcia-Bellido and Robbins (1983) observed that 3 of 13 zygotic lethal mutations that also disrupt maternal gene expression in germ-line clones (see Section 2.3) exhibit a maternal effect. Two of these were described as disrupting early embryonic development, whereas the third produced eggs that either were not fertilized or did not develop. Using a similar approach, Perrimon *et al.* (1984a) observed a zygotic mutation with a pupal lethal phase, *EA75*, in which female embryos derived from germ line clones (*EA75*/Y) exhibit the *torso* phenotype described in Section 3.3 (Fig. 8). Male embryos (*EA75*/Y) die at an earlier stage, indicating a partial "rescue" by zygotic expression of the wild-type allele. Three alleles of this mutation have now been observed to be fully penetrant for this germ-line phenotype (N. Perrimon *et al.*, unpublished observations). Thus, gene activity identified by zygotic lethal mutations can also be required for what appear to be determinative decisions in embryogenesis. Two other lethal mutations examined in this study have also been found to be fully penetrant for maternal effects disrupting early embryogenesis. Further analysis will be required to determine whether such gene products are multifunctional or whether homologous functions are required more than once during development.

Three zygotic lethal mutations that are fully penetrant for specific embryonic lethal phases have also been shown to have maternal components. Denell (1982) found evidence of a maternal effect of the dominant mutation *Polycomb* (*Pc*), which produces an embryonic phenotype similar to that described for *esc* in Section 3.3.2. The phenotype of *Pc1*/*Pc3* embryos is more extreme when the female is heterozygous for the *Pc3* allele than when heterozygous for *Pc1*. *Notch* and *mastermind* produce a nervous system hypertrophy similar to that of *amx* described in Section 3.3.2, but the degree of neuralization produced by these two mutations is different. Using the Dfs technique, Jimenez and Campos-Ortega (1982) were able to demonstrate that germ-line clones of these mutants produce identical phenotypes and that the difference in zygotic phenotypes is due to the maternal expression of these genes. In addition, whereas the maternal effect of *mastermind* was found to be completely rescuable by zygotic expression of the wild-type gene, both maternal and zygotic expression of the *Notch* locus is required for normal embryogenesis. However, female embryos, containing a wild-type allele, derived from *Notch* germ-line clones displayed a more normal phenotype—indicating a partial "rescue" similar to that seen with *EA75*.

Other zygotic lethal mutations with specific embryonic phenotypes have been similarly examined and, in contrast, display no maternal effect. These include *krüppel* (Wieschaus, 1980), *bithorax* (Kerridge and Dura, 1982), and four others (Perrimon, unpublished observations), indicating that many determinative events in *Drosophila* embryogenesis require exclusively zygotic gene expression.

## 4. Conclusions

The mutations reviewed here indicate that maternal expression of the majority of the genome in both *Drosophila* and *Caenorhabditis* is required for normal oogenesis and early embryogenesis. These gene functions include those that appear to be specific for these processes as well as those required constitutively or more than once during development. As would be expected, maternal effect mutations have revealed that a major function of maternal gene expression is to provide materials necessary for the rapid cleavages of early development and the cellular metabolism of the early embryo before the initiation of zygotic gene transcription. Many of these maternal gene products, however, perdure for a considerable time and can influence development after zygotic gene transcription begins. Maternal gene activity also provides developmental information to the early embryo and appears to play a critical role in the specification of such parameters as embryonic polarity and sex determination as well as the initiation of major morphogenetic movements and segmental identity.

The mechanisms through which the maternal genome contributes to these determinative events are unknown. Very few maternal gene products appear to affect specifically localized cell determination. One of the best-characterized examples of a localized, maternal cell determinant is that of germ plasm in *Drosophila* and even in this case, nearly all mutations isolated to date that affect germ cell determination are pleiotropic. Furthermore, the majority of maternal effect mutations disrupting early determinative decisions do so in a global manner, and the phenotypes they produce are most easily explained by the presence of gradients of determinant information in the egg. Virtually nothing is known about how such gradients might be established. It is interesting to note, however, that similar phenotypes are often produced by maternal effect mutations in different loci. For example there are now ten *dorsal*-like, seven *torso*-like, and three *bicaudal*-like loci. In addition, at least nine loci acting both maternally and zygotically produce a *Notch*-like phenotype (Lehmann *et al.* 1983; S. G. LaBonne *et al.*, unpublished observations). These observations illustrate the complex integrative nature of determinative events in embryogenesis. Furthermore, they suggest that the determinative information provided by the maternal genome may, in fact, arise epigenetically through the interactions of numerous gene products.

In this context, it is not surprising that many genes that are required constitutively or more than once during development are also required during oogenesis and early embryogenesis. It is somewhat surprising that some pleiotropic zygotic lethal mutations are fully penetrant for specific embryonic defects when expressed in the germ line. Since only a small proportion of zygotic lethal mutations have, to date, been examined for maternal effects, it is probable that many more loci exist that make a major contribution to embryonic determination. Many zygotic lethal mutations are likely to represent defects in essential metabolic functions; further analysis of such mutations should answer the question of whether altered metabolism can result in defects in spatial organization and cellular determination (cf. Wood *et al.*, 1980).

Zygotic lethal mutations also affect segment number and polarity

(Nüsslein-Volhard and Wieschaus, 1980) as well as identity (Lewis, 1978; Wakimoto and Kaufman, 1981). In contrast to maternal effect mutations, the phenotypes of some zygotic mutations affecting cell determination suggest the existence of genetic programs in which regulatory gene products are directly responsible for determinative events (see R. A. Raff and Kaufman, 1983). Since the maternal expression of at least a few such zygotically expressed genes is not required and the number of segments appears to be established progressively between fertilization and blastoderm formation (Schubiger *et al.*, 1977), interactions between the maternal and zygotic genomes must be required in these processes. It is possible that the maternal genome provides general polarity cues in the early embryo that are interpreted by the zygotic genome to produce more specific determinative events. Models suggesting how this might occur have been proposed (Capdevilla and Garcia-Bellido, 1981; Kaufman and Wakimoto, 1982). An important task that lies ahead for developmental geneticists is to investigate the nature of the interactions of gene products that produce the determinative decisions of early development.

## References


Anderson, K. V., and Nüsslein-Volhard, C., 1984a, Genetic analysis of dorsal-ventral embryonic pattern in *Drosophila*, in: *Primers in Developmental Biology* (G. Malacinski and S. Bryant, eds.), Macmillan, New York.

Anderson, K. V., and Nüsslein-Volhard, C., 1984b, Information for the dorsal–ventral pattern of the *Drosophila* embryo is stored as maternal mRNA, *Nature*, **311**:223–227.

Arnolds, W. J. A., 1982, A mutation with a maternal effect in *Lymnaea stagnelis*, L., affecting development of bilateral symmetry and dorsoventrality and preventing gastrulation, *Proc. Kon. Ned. Akad. Wetensch. C* **85**:1–20.

Baker, B. S., 1973, The maternal and zygotic control of development by *cinnamon*, a new mutant in *Drosophila melanogaster*, *Dev. Biol.* **33**:429–440.

Baker, B. S., Smith, D. A., and Gatti, M., 1982, Region-specific effects on chromosome integrity of mutations of essential loci in *Drosophila melanogaster*, *Proc. Natl. Acad. Sci. USA* **79**:1205–1209.

Bakken, A. H. 1973, A cytological and genetic study of oogenesis in *Drosophila melanogaster*, *Dev. Biol.* **33**:100–122.

Beetschen, J. C., and Fernandez, M., 1979, Studies on the maternal effect of the semi-lethal factor *ac* in the salamander *Pleurodeles waltlii*, in: *Maternal Effects in Development* (D. R. Newth and M. Balls, eds.), pp. 269–286, Cambridge University Press, New York.

Belote, J. M., and Baker, B. S., 1983, The dual functions of a sex determination gene in *Drosophila melanogaster*, *Dev. Biol.* **95**:512–517.

Bischoff, W. L., and Lucchesi, J. C., 1971, Genetic organization of *Drosophila melanogaster*: Complementation and fine structure analysis at the deep orange locus, *Genetics* **69**:453–466.

Bluemink, J. C., and Beetschen, J.-C., 1981, An ultrastructural study of the maternal-effect embryos of the *ac/ac* mutant of *Pleurodeles waltii* showing a gastrulation defect, *J. Embryol. Exp. Morphol.* **63**:67–74.

Boycott, A. E., Diver, C., Garstang, S. L., and Turner, F. M., 1930, The inheritance of sinistrality in *Limnaea peregra* (Mollusca, Pulmonata), *Philos. Trans. R. Soc. Ser. B* **219**:51–131.

Brandhorst, B. P., 1976, Two-dimensional gel patterns of protein synthesis before and after fertilization of sea urchin eggs, *Dev. Biol.* **52**:310–317.

Brennan, M. D., Weiner, A. J., Goralski, T. J., and Mahowald, A. P., 1982, The follicle cells are a major site of vitellogenin synthesis in *Drosophila melanogaster*, *Dev. Biol.* **89**:225–236.

Brothers, A. J., 1979, A specific case of genetic control of early development: The *o* maternal effect mutation of the Mexican Axolotl, in: *Determinants of Spatial Organization* (S. Subtelny and I. R. Konigsberg, eds.), pp. 167–183, Academic Press, New York.

Browder, L. W., and Williamson, J. H., 1976, The effects of *cinnamon* on xanthine dehydrogenase, aldehyde oxidase, and pyridoxal oxidase activity during development in *Drosophila melanogaster*, *Dev. Biol.* **53**:241–249.

Brown, E. H., and King, R. C., 1964, Studies on the events resulting in the formation of an egg chamber in *Drosophila melanogaster*, *Growth* **28**:41–81.

Capdevilla, M. P., and Garcia-Bellido, A., 1981, Genes involved in the activation of the bithorax complex in *Drosophila*, *Wilhelm Roux Arch.* **90**:339–350.

Carpenter, A. T. C., 1979, Synaptonemal complex and recombination nodules in wild-type *Drosophila melanogaster* females, *Genetics* **92**:511–541.

Cassada, R., Isnenghi, E., Culotti, M., and von Ehrenstein, G., 1981, Genetic analysis of temperature-sensitive embryogenesis mutants in *Caenorhabditis elegans*, *Dev. Biol.* **84**:193–205.

Cline, T., 1983a, The interaction between daughterless and sex-lethal in triploids: A lethal sex-transforming maternal effect linking sex determination and dosage compensation in *Drosophila melanogaster*, *Dev. Biol.* **95**:260–274.

Cline, T., 1983b, Functioning of the genes daughterless *(da)* and sex-lethal *(Sxl)* in *Drosophila* germ cells, *Genetics* **104**:s16.

Colaianne, J. J., and Bell, A. E., 1972, The relative influence of sex of progeny on the lethal expression of the sonless gene in *Drosophila melanogaster*, *Genetics* **72**:293–296.

Counce, S. J., 1956, Studies on female-sterility genes in *Drosophila melanogaster*, II. The effects of the gene *fused* on embryonic development, *Z. Ind. Abstam.-Vererbungslehre* **87**:462–481.

Davidson, E., 1976, *Gene activity in Early Development*, Academic Press, New York.

Denell, R. E., 1982, Homeosis in *Drosophila*: Evidence for a maternal effect of the Polycomb locus, *Dev. Genet.* **3**:103–113.

DiMario, P., and Hennen, S., 1982, Analysis of the *Drosophila* female sterile mutant, *tiny*, by means of pole cell transplantation, *J. Exp. Zool.* **221**:219–224.

Engstrom, L., Caulton, J. H., Underwood, E. M., and Mahowald, A. P., 1982, Developmental lesions in the *agametic* mutant of *Drosophila melanogaster*, *Dev. Biol.* **91**:163–170.

Falk, D. R., and Nash, D., 1974, Sex-linked auxotrophic and putative auxotrophic mutants of *Drosophila melanogaster*, *Genetics* **76**:755–766.

Fausto-Sterling, A., Weiner, A. J., and Digan, M. E., 1977, Analysis of a newly isolated temperature sensitive, maternal effect mutation of *Drosophila melanogaster*, *J. Exp. Zool.* **200**:199–210.

Flytzanis, C. N., Brandhorst, B. P., Britten, R. J., and Davidson, E. H., 1982, Developmental patterns of cytoplasmic transcript prevalence in sea urchin embryos, *Dev. Biol.* **91**:27–35.

Forquignon, F., 1981, A maternal effect mutation leading to deficiences of organs and homeotic transformations in the adults of Drosophila, *W. Roux Arch.* **190**:132–138.

Freeman, G., and Lundelius, J. W., 1982, The developmental genetics of dextrality and sinistrality in the gastropod *Lymnaea peregra*, *W. Roux Arch.* **191**:69–83.

Gans, M., Audit, C., and Masson, M., 1975, Isolation and characterization of sex-linked female sterile mutants in *Drosophila melanogaster*, *Genetics* **81**:683–704.

Garcia-Bellido, A., and Moscoso Del Pardo, J., 1979, Genetic analysis of maternal information in *Drosophila*, *Nature* **278**:346–348.

Garcia-Bellido, A., and Ripoll, P., 1978, The number of genes in *Drosophila melanogaster*, *Nature* **273**:399.

Garcia-Bellido, A., and Robbins, L. G., 1983, Viability of female germ line cells homozygous for zygotic lethals in *Drosophila melanogaster*, *Genetics* **103**:235–247.

Garen, A., and Gehring, W., 1972, Repair of the lethal developmental defect in *deep orange*

embryos of *Drosophila* by injection of normal egg cytoplasm, *Proc. Natl. Acad. Sci. USA* **69**:2982–2985.
Gill, K. S., 1963, Studies on oogenesis in *Drosophila melanogaster, J. Exp. Zool.* **152**:251–278.
Graziani, F., Vicani, L., Boncinelli, E., Malva, C., Manzi, A., and Mariani, C., 1981, Selective replication of ribosomal DNA after loss of the abnormal oocyte phenotype in *Drosophila melanogaster, Proc. Natl. Acad. Sci. USA* **78**:7662–7664.
Grell, R. F., and Generoso, E. E., 1982, A temporal study of the ultrastructural level of the developing pro-oocyte of *Drosophila melanogaster, Chromosoma* **87**:49–75.
Hall, J. C., Tompkins, L., Kyriacou, C. P., Siegel, R. W., von Schilcher, F., and Greenspan, R., 1980, Higher behavior in *Drosophila* with mutants that disrupt the structure and function of the nervous system, in: *Development and Neurobiology of Drosophila* (O. Siddiqi, P. Babu, L. M. Hall, and J. C. Hall, eds.), pp. 425–455, Plenum Press, New York.
Hirsch, D., 1979, Temperature sensitive maternal effect mutants of early development in *Caenorhabditis elegans*, in: *Determinants of Spatial Organization* (S. Subtelny and I. R. Konigsberg, eds.), pp. 149–165, Academic Press, New York.
Hirsch, D., and Vanderslice, R., 1976, Temperature-sensitive developmental mutants of *Caenorhabditis elegans, Dev. Biol.* **49**:220–235.
Hochman, B., 1973, Analysis of a whole chromosome in *Drosophila, Cold Spring Harb. Symp. Quant. Biol.* **38**:581–589.
Humphrey, R. R., 1966, A recessive factor (o, for ova deficient) determining a complex of abnormalities in the Mexican axolotl, *Ambystoma mexicanum, Dev. Biol.* **13**:57–76.
Illmensee, K., Mahowald, A. P., and Loomis, M. R., 1976, The ontogeny of germ plasm during oogenesis in *Drosophila, Dev. Biol.* **49**:40–65.
Isnenghi, E., Cassada, R., Smith, K., Denich, K., Radnia, K., and von Ehrenstein, G., 1983, Maternal effect and temperature sensitive period of mutations affecting embryogenesis in *Caenorhabditis elegans, Dev. Biol.* **98**:465–480.
Jimenez, F., and Campos-Ortega, J. A., 1982, Maternal effects of zygotic mutants affecting early neurogenesis in *Drosophila, Wilhelm Roux Arch.* **191**:191–201.
Kaplan, W. D., Seecof, R. L., Trout, W. E. III, and Pasternack, M. E., 1970, Production and relative frequency of maternally influenced lethals in *Drosophila melanogaster, Am. Nat.* **104**:261–271.
Kauffman, S., 1977, Characteristic waves, compartments and binary decisions in *Drosophila* development, *Am. Zool.* **17**:631–648.
Kaufman, T. C., and Wakimoto, B. T., 1982, Genes that control high level developmental switches, in: *Evolution and Development* (J. T. Bonner, ed.), pp. 189–205, Dahlem Konferenzen, Springer-Verlag, New York.
Kerridge, S., and Dura, J. M., 1982, Lethal *bithorax* complex mutations of *Drosophila melanogaster* show no germ line maternal effects, *Genet.* **3**:207–214.
King, R. C., 1970, *Ovarian Development in Drosophila melanogaster*, Academic Press, New York.
King, R. C., 1975, The cell cycle and cell differentiation in the Drosophila ovary, in: *Results and Problems in Cell Differentiation*, Vol. 7 (J. Reinert and H. Holtzer, eds.), pp. 85–109, Springer-Verlag, New York.
King, R. C., and Mohler, J. D., 1975, The genetic analysis of oogenesis in *Drosophila melanogaster*, in: *Handbook of Genetics*, Vol. 3 (R. C. King, ed.), pp. 757–791, Plenum Press, New York.
King, R. C., and Riley, S. F., 1982, Ovarian pathologies generated by various alleles of the *otu* locus in *Drosophila melanogaster, Dev. Genet.* **3**:69–89.
Komitopoulou, K., Gans, M., Margaritis, L. H., and Kafatos, F. C., 1983, Isolation and characterization of sex-linked female-sterile mutants in *Drosophila melanogaster* with special attention to eggshell mutants. *Genetics* **105**:897–920.
Komorowska, B., 1980, L'effet d'une mutation de sterilité femelle sur la cephalogenèse de *Drosophila melanogaster* pendant la vie embryonnaire, Thesis, University of P. and M. Curie, Paris.

Konrad, K. D., and Mahowald, A. P., 1984, Genetic and developmental approaches to understanding determination in early development, in *Molecular Aspects of Early Development* (G. Malacinski and W. Klein, eds.), pp. 167–188, Plenum Press, New York.

Lauge, G., 1980, Sex determination, in: *The Genetics and Biology of Drosophila*, vol. 2 (M. Ashburner and T. R. F. Wright, eds.), pp. 34–106, Academic Press, New York.

Lawrence, P. A., 1981, The cellular basis of segmentation in insects, *Cell* **26**:3–10.

Lehmann, R., Dietrich, U., Jimenez, F., and Campos-Ortega, J. A., 1981, Mutations of early neurogenesis in *Drosophila, Wilhelm Roux Arch.* **190**:226–229.

Lehmann, R., Jimenez, F., Dietrich, U., and Campos-Ortega, J. A., 1983, On the phenotype and development of mutants of early neurogenesis in *Drosophila melanogaster, Wilhelm Roux Arch.* **192**:62–74.

Lewis, E. B., 1978, A gene complex controlling segmentation in *Drosophila, Nature* **276**:565–570.

Lindsley, D. L., and Grell, E. H., 1968, Genetic variations in *Drosophila melanogaster, Carnegie Inst. Wash.* Publ. No. 627.

Lohs-Schardin, M., 1982, *Dicephalic*—a *Drosophila* mutant affecting polarity in follicle organization and embryonic patterning, *Wilhelm Roux Arch.* **191**:28–36.

Lohs-Schardin, M., Cremer, C., and Nüsslein-Volhard, C., 1979, A fate map for the larval epidermis of *Drosophila melanogaster:* Localized cuticle defects following irradiation of the blastoderm with an ultraviolet lazer microbeam, *Dev. Biol.* **73**:239–255.

Mahowald, A. P., and Boswell, R. E., 1983, Germ plasm and germ cell development in invertebrates, in: *Current Problems in Germ Cell Differentiation* (A. McLaren and Wylie, eds.), Cambridge University Press, New York.

Mahowald, A. P., Caulton, J. H., and Gehring, W. J., 1979, Ultrastructural studies of oocytes and embryos derived from female flies carrying the *grandchildless* mutation in *Drosophila subobscura, Dev. Biol.* **69**:118–132.

Malacinski, G. M., and Brothers, A. J., 1974, Mutant genes in the Mexican Axolotl, *Science* **184**:1142–1147.

Marsh, J. L., Van Deusen, E. B., Wieschaus, E., and Gehring, W. J., 1977, Germ line dependence of the *deep orange* maternal effect in *Drosophila, Dev. Biol.* **56**:195–199.

Marsh, L., and Wieschaus, E., 1978, Is sex determination in germ line and soma controlled by separate genetic mechanisms?, *Nature* **272**:249–251.

Miwa, J., Schierenberg, E., Miwa, S., and von Ehrenstein, G., 1980, Genetics and mode of expression of temperature-sensitive mutations arresting embryonic development in *Caenorhabditis elegans, Dev. Biol.* **76**:160–174.

Mohler, J. D., 1977, Developmental genetics of the Drosophila egg. I. Identification of 50 sex-linked cistrons with maternal effects on embryonic development, *Genetics* **85**:259–272.

Nieuwkoop, P. D., and Sutasurya, L., 1981, *Primordial Germ Cells in the Invertebrates*, Cambridge University Press, Cambridge.

Norby, S., 1970, A specific nutritional requirement for pyrimidines in rudimentary mutants of *Drosophila melanogaster, Hereditas* **66**:205–214.

Nüsslein-Volhard, C., 1977, Genetic analysis of pattern formation in the embryo of *Drosophila melanogaster.* Characterization of the maternal-effect mutant *bicaudal, Wilhelm Roux Arch.* **183**:249–268.

Nüsslein-Volhard, C., 1979, Maternal effect mutations that alter the spatial coordinates of the embryos of *Drosophila melanogaster,* in: *Determinants of Spatial Organization* (S. Subtelny and I. R. Konigsberg, eds.), pp. 185–211, Academic Press, New York.

Nüsslein-Volhard, C., Lohs-Schardin, M., Sander, K., and Cremer, C., 1980, A dorsal-ventral shift of embryonic primordia in a new maternal-effect mutant of *Drosophila, Nature* **283**:474–475.

Nüsslein-Volhard, C., and Wieschaus, E., 1980, Mutations affecting segment number and polarity in *Drosophila, Nature* **287**:795–801.

Nüsslein-Volhard, C., Wieschaus, E., and Jurgens, G., 1982, Segmentierung. *Drosophila* Eine genetische analyse, in: *Verhandlungen der deutschen Zoologischen Gesellschaft*, pp. 91–104, Gustav Fischer Verlag (in press).

Okada, M., Kleinman, I. A., and Schneiderman, H. A., 1974, Repair of a genetically caused defect in oogenesis in *Drosophila melanogaster* by transplantation of cytoplasm from wild-type eggs and injection of pyrimidine nucleosides, *Dev. Biol.* **37**:55–62.

Orr, W., Komitopoulou, K., and Kafatos, F. C., 1984, Mutants supressing in *trans* chorion gene amplification in *Drosophila*, *Proc. Natl. Acad. Sci. USA* **81**:3773–3777.

Parry, D. M., and Sandler, L., 1974, The genetic identification of a heterochromatic segment on the X chromosome of *Drosophila melanogaster*, *Genetics* **77**:535–539.

Perlman, S., and Rosbach, M. N., 1978, Analysis of *Xenopus laevis* ovary and somatic cell polyadenylated RNA by molecular hybridization, *Dev. Biol.* **63**:197–212.

Perrimon, N., and Gans, M., 1984, Clonal analysis of the tissue specificity of recessive female sterile mutations of *Drosophila melanogaster* using a dominant female sterile mutation, *Fs(1)K1237*, *Dev. Biol.* **100**:365–373.

Perrimon, N., and Mahowald, A. P., 1985, A study of follicular cells in *Drosophila*, using the technique of clonal analysis (submitted).

Perrimon, N., Engstrom, L., and Mahowald, A. P., 1984a, Analysis of the effects of zygotic lethal mutations on the germ line functions in *Drosophila Dev. Biol.* **105**:404–414.

Perrimon, N., Engstrom, L., and Mahowald, A. P., 1984b, Developmental genetics of the 2E–F region of the *Drosophila* X chromosome: A region rich in "developmentally important" genes, *Genetics* **108**:559–572.

Postlethwait, J. H., and Weiser, K., 1973, Vitellogenesis induced by juvenile hormone in the female sterile mutant *apterous-four* in *Drosophila melanogaster*, *Nature New Biol.* **244**:284–285.

Poulson, D. F., 1940, The effect of certain chromosomal deficiencies on the embryonic development of *Drosophila melanogaster*, *J. Exp. Zool.* **83**:271–325.

Poulson, D. F., 1950, Histogenesis, organogenesis and differentiation in the embryo of *Drosophila melanogaster* Meigen, in: *Biology of Drosophila* (M. Demerec, ed.), pp. 168–274, Wiley, New York.

Raff, E. C., Brothers, A. J., and Raff, R. A., 1976, Microtubule assembly mutant, *Nature* **260**:615–617.

Raff, R. A., and Kaufman, T. C., 1983, *Embryos, Genes and Evolution*, Macmillan, New York.

Regenass, V., and Bernhard, P., 1978, Analysis of the *Drosophila* maternal effect mutant *mat(3)1* by pole cell transplantation experiments, *Mol. Gen. Genet.* **164**:85–91.

Rice, T. B., 1973, Isolation and characterization of maternal effect mutants: An approach to the study of early development in *Drosophila melanogaster*, Ph.D. thesis, Yale University Press, New Haven, Conn.

Rice, T. B., and Garen, A., 1975, Localized defects of blastoderm formation in maternal effect mutants of *Drosophila*, *Dev. Biol.* **43**:277–286.

Ripoll, P., 1977, Behavior of somatic cells homozygous for zygotic lethals in *Drosophila melanogaster*, *Genetics* **86**:357–376.

Ripoll, P., and Garcia-Bellido, A., 1979, Viability of homozygous deficiencies in somatic cells of *Drosophila melanogaster*, *Genetics* **91**:443–453.

Robbins, L. G., 1980, Maternal-zygotic lethal interactions in *Drosophila melanogaster*: The effects of deficiencies in the zeste-white region of the X-chromosome, *Genetics* **96**:187–200.

Robbins, L. G., 1983, Maternal zygotic lethal interactions in *Drosophila melanogaster*: Zeste-white region single-cistron mutations, *Genetics* **103**:633–648.

Romans, P., Hodgetts, R. B., and Nash, D., 1976, Maternally influenced embryonic lethality: allele specific genetic rescue at a female fertility locus in *Drosophila melanogaster*, *Can J. Genet. Cytol.* **18**:773–781.

Sandler, L., 1970, The regulation of sex chromosome heterochromatic activity by an autosomal gene in *Drosophila melanogaster*, *Genetics* **64**:481–493.

Sandler, L., 1975, Studies on the genetic control of heterochromatin in *Drosophila melanogaster*, *Israel J. Med. Sci.* **11**:1124–1134.

Sayles, C. D., Browder, L. W., and Williamson, J. H., 1973, Expression of xanthine dehydrogenase activity during embryonic development of *Drosophila melanogaster*, *Dev. Biol.* *33*:213–217.

Schierenberg, E., Miwa, J., and von Ehrenstein, G., 1980, Cell lineages and developmental defects of temperature-sensitive embryonic arrest mutants in *Caenorhabditis elegans*, *Dev. Biol.* **76**:141–159.

Schubiger, G., Moseley, R. C., and Wood, W. J., 1977, Interaction of different egg parts in determination of various body regions in *Drosophila melanogaster*, *Proc. Natl. Acad. Sci. USA* **74**:2050–2053.

Schüpbach, T., 1982, Autosomal mutations that interfere with sex determination in somatic cells of *Drosophila* have no direct effect on the germline, *Dev. Biol.* **89**:117–127.

Shannon, M. P., 1972, Characterization of the female-sterile mutant *almondex* of *Drosophila melanogaster*, *Genetica* **43**:244–256.

Shannon, M. P., Kaufman, T. C., Shen, W. M., and Judd, B. H., 1972, Lethality patterns and morphology of selected lethal and semi-lethal mutations in the zeste-white region of *Drosophila melanogaster*, *Genetics* **72**:615–638.

Shearn, A., Hersperger, G., and Hersperger, E., 1978, Genetic analysis of two allelic temperature sensitive mutants of *Drosophila melanogaster* both of which are zygotic and maternal effect lethals, *Genetics* **89**:341–353.

Spurway, H., 1948, Genetics and cytology of *Drosophila subobscura:* IV. An extreme delay in gene action causing sterility, *J. Genet.* **49**:126–140.

Struhl, G., 1981, A gene product required for correct initiation of segmental determination in *Drosophila*, *Nature* **293**:36–41.

Struhl, G., and Browder, D. L., 1982, Early role of the $esc^+$ gene product in the determination of segments in *Drosophila*, *Cell* **31**:285–292.

Thierry-Mieg, D., 1982, Paralog, a control mutant in *Drosophila melanogaster*, *Genetics* **100**:209–237.

Thomas, T. L., Posakony, W. J., Anderson, D. M., Britten, R. J., and Davidson, E. H., 1981, Molecular structure of maternal RNA, *Chromosoma* **84**:319–335.

Underwood, E. M., Caulton, J. H., Allis, C. D., and Mahowald, A. P., 1980a, Developmental fate of pole cells in *Drosophila melanogaster*, *Dev. Biol.* **77**:303–314.

Underwood, E. M., Turner, F. R., and Mahowald, A. P., 1980b, Analysis of cell movements and fate mapping during early embryogenesis in *Drosophila melanogaster*, *Dev. Biol.* **74**:286–301.

Van Deusen, E. B., 1977, Sex determination in germ line chimeras of *Drosophila melanogaster*, *J. Embryol. Exp. Morphol.* **37**:173–185.

Wakimoto, B. T., and Kaufman, T. C., 1981, Analysis of larval segmentation in lethal genotypes associated with the Antennapedia gene complex in *Drosophila melanogaster*, *Dev. Biol.* **81**:51–64.

Waring, G., DiOrio, J. P., and Hennen, S., 1983, Isolation of a germ line dependent female sterile mutation that affects yolk specific sequestration and chorion formation in *Drosophila*, *Dev. Biol.* **100**:452–463.

Went, D. F., and Junquera, P., 1981, Embryonic development of insect eggs formed without follicular epithelium, *Dev. Biol.* **86**:100–110.

Wieschaus, E., 1980, A combined genetic and mosaic approach to the study of oogenesis in *Drosophila*, in: *Development and Neurobiology of Drosophila* (O. Siddiqi, P. Babu, L. M. Hall, and J. C. Hall, eds.), pp. 85–94, Plenum Press, New York.

Wieschaus, E., and Szabad, J., 1979, The development and function of the female germ line in *Drosophila melanogaster:* A cell lineage study, *Dev. Biol.* **68**:29–46.

Wieschaus, E., Marsh, J. L., and Gehring, W. J., 1978, *fs(1)K10*, a germ-line dependent female sterile mutation causing abnormal chorion morphology in *Drosophila melanogaster*, *Wilhelm Roux Arch.* **184**:75–82.

Wieschaus, E., Audit, C., and Masson, M., 1981, A clonal analysis of the roles of somatic cells and germ line during oogenesis in *Drosophila*, *Dev. Biol.* **88**:92–103.

Wood, W. E., Hecht, R., Carr, S., Vanderslice, R., Wolf, N., and Hirsch, D., 1980, Parental effects

and phenotypic characterization of mutations that affect early development in *Caenorhabditis elegans*, *Dev. Biol.* **74**:446–469.
Wright, T. R. F., 1970, The genetics of embryogenesis in *Drosophila*, *Adv. Genet.* **15**:262–395.
Yamashita, O., and Irie, K., 1980, Larval hatching from vitellogenin-deficient eggs developed in male hosts of the silkworm, *Nature* **283**:385–386.
Yedvolnick, B., Kriden, H. M., and Levine, B. I., 1980, Analysis of the autosomal mutation *abo* and its interaction with the ribosomal DNA of *Drosophila melanogaster*: The role of X-chromosome heterochromatin, *Genetics* **95**:661–672.
Zalokar, M., Audit, C., and Erk, I., 1975, Developmental defects of female-sterile mutants of *Drosophila melanogaster*, *Dev. Biol.* **47**:419–432.

# Index